T0190309

CAMBRIDGE LIBRARY COLLECTION

Books of enduring scholarly value

Technology

The focus of this series is engineering, broadly construed. It covers technological innovation from a range of periods and cultures, but centres on the technological achievements of the industrial era in the West, particularly in the nineteenth century, as understood by their contemporaries. Infrastructure is one major focus, covering the building of railways and canals, bridges and tunnels, land drainage, the laying of submarine cables, and the construction of docks and lighthouses. Other key topics include developments in industrial and manufacturing fields such as mining technology, the production of iron and steel, the use of steam power, and chemical processes such as photography and textile dyes.

Electric Illumination

Two years after Thomas Edison patented his electric light bulb, the 1881 International Exposition of Electricity in Paris, featuring many spectacular lighting displays, showcased the potential of this technology for commercial and domestic use. The accompanying International Congress of Electricians also agreed on international standards for units of electrical resistance, potential and current. In its wake, James Dredge (1840–1906), editor of the British periodical *Engineering*, compiled this illustrated overview of electrical technology and its application to lighting. First published in two volumes between 1882 and 1885, and using material that had previously appeared in *Engineering*, as well as new articles by various contributors, this substantial work reflects the complexities and possibilities of a propitious technological development. Among other topics, Volume 1 covers electrical units, methods of generation, conductors, and various kinds of lamp. The appendices give abstracts of British electrical patents from 1837 to 1872.

Cambridge University Press has long been a pioneer in the reissuing of out-of-print titles from its own backlist, producing digital reprints of books that are still sought after by scholars and students but could not be reprinted economically using traditional technology. The Cambridge Library Collection extends this activity to a wider range of books which are still of importance to researchers and professionals, either for the source material they contain, or as landmarks in the history of their academic discipline.

Drawing from the world-renowned collections in the Cambridge University Library and other partner libraries, and guided by the advice of experts in each subject area, Cambridge University Press is using state-of-the-art scanning machines in its own Printing House to capture the content of each book selected for inclusion. The files are processed to give a consistently clear, crisp image, and the books finished to the high quality standard for which the Press is recognised around the world. The latest print-on-demand technology ensures that the books will remain available indefinitely, and that orders for single or multiple copies can quickly be supplied.

The Cambridge Library Collection brings back to life books of enduring scholarly value (including out-of-copyright works originally issued by other publishers) across a wide range of disciplines in the humanities and social sciences and in science and technology.

Electric Illumination

VOLUME 1

EDITED BY JAMES DREDGE

CAMBRIDGE
UNIVERSITY PRESS

University Printing House, Cambridge, CB2 8BS, United Kingdom

Cambridge University Press is part of the University of Cambridge.

It furthers the University's mission by disseminating knowledge in the pursuit of
education, learning and research at the highest international levels of excellence.

www.cambridge.org
Information on this title: www.cambridge.org/9781108070638

© in this compilation Cambridge University Press 2015

This edition first published 1882
This digitally printed version 2015

ISBN 978-1-108-07063-8 Paperback

This book reproduces the text of the original edition. The content and language reflect
the beliefs, practices and terminology of their time, and have not been updated.

Cambridge University Press wishes to make clear that the book, unless originally published
by Cambridge, is not being republished by, in association or collaboration with,
or with the endorsement or approval of, the original publisher or its successors in title.

["ENGINEERING" SERIES.]

ELECTRIC ILLUMINATION

BY

CONRAD COOKE. M. F. O'REILLY.

JAMES DREDGE. S. P. THOMPSON.

AND

H. VIVAREZ.

EDITED BY

JAMES DREDGE.

(Chiefly compiled from "ENGINEERING.")

With Abstracts of Specifications having reference to Electric Lighting.

PREPARED BY

W. LLOYD WISE

Member of Council of the Institute of Patent Agents.

VOL. I.

LONDON:

OFFICES OF "ENGINEERING," 35 & 36, BEDFORD STREET, STRAND, W.C.

NEW YORK: JOHN WILEY AND SONS.

PREFACE.

- •• -

THERE is no industrial question which engrosses so much public attention at the present time as that of Electric Lighting. The extensive employment of this mode of illumination in streets and other places of public resort, has familiarised the world at large with its leading characteristics, while the splendid international Exhibition of Electricity held last year in Paris, and, to a less extent, the more recent show at the Crystal Palace, have given a fair idea of the almost endless variety of mechanical combinations by which—with more or less success—the Electric Light can be obtained. The most striking, though probably the least satisfactory, evidence of the lively interest taken by the public in this development of practical science, is to be found in the fact that during the past few months an enormous amount of money has been applied for and obtained, for the formation of public companies, to acquire from different inventors the patent rights of apparatus for the production and utilisation of the electric current. That some of these companies have a useful and profitable career before them is not open to question, but it is doubtful whether few, perhaps if any, of the patents thus sold, are worth the money demanded and paid for them; in some cases because the invention has but small intrinsic value, in others because its originality, and therefore the validity of the patents protecting it, is open to question, and in all, or nearly all cases, because, with the rapid march of invention in this direction, towards which so many able minds are turned, progress is so constant and so quick, that what appears almost boundless in its ingenuity and usefulness to-day, is rendered obsolete by an invention of greater ingenuity and wider usefulness to-morrow.

It is to be regretted, therefore, that the public have rushed after electric lighting speculations with more enthusiasm than good judgment, for this movement must be attended with a certain amount of disappointment and loss, and therefore will be followed by an unreasonable reaction, that may delay for a time the legitimate development of this great source of light and energy, which, however, appears certain in the early future, to have applications almost as broad as its most enthusiastic admirers claim for it.

It is of interest to recall briefly what has been done in this direction during the past few years. In this country the modern revival of electric illumination dates from the time when, in 1873, Mr. Conrad Cooke, employing a Gramme generator, installed the signal-light on the summit of the Clock Tower of the Houses of Parliament. The same year in Paris a factory was first lighted permanently, and also in the same year Messrs. Siemens Bros., of Berlin, maintained a signal on the top of the great central dome of the Vienna International Exhibition. Between that date and 1877, when M. Hyppolite Fontaine published his excellent and still standard work, *Eclairage à Electricité*, considerable progress had been made in France. Thus the works of MM. Sautter, Lemonnier and Co., of Paris; of MM. Menier and Co., Noisiel; several cotton factories; the harbour works at Havre, and many other places, had been lighted with success. Of course, long before this time the electric light had been used for lighthouse illumination, but such installations are apart from ordinary industrial applications. The great popular movement, however, dates from 1878, when, during the French Universal Exhibition, Gramme and Jablochkoff rendered some of the most important parts of Paris more brilliant and more beautiful than had been conceived possible, the centre of attraction and admiration having been the Avenue de l'Opera, including the façade of the Grand Opera at one end, and the Place du Theatre Français at the other. At the same time, also, the success of Edison was prematurely trumpeted forth from across the Atlantic, with the disastrous result of temporarily depreciating the value of gas shares to an absurd but alarming extent.

From this time, the applications of electric lighting to industrial purposes progressed rapidly, the number of successful methods increasing also, while the employment of lamps on the incandescence-arc system,— producing light of moderate intensity, though relatively more costly to maintain than that given by the more powerful arc lamps,— indicated a promising means for applications to domestic lighting. It was not, however, till the practically simultaneous announcements of the Swan and Edison incandescence systems, that the problem of applying the electric current to domestic illumination, appeared to be approaching solution. Though little more than two years have passed since that time, the forms first put forward by both these inventors, now seem crude, clumsy and imperfect, compared to the beautiful productions exhibited last year at the Palais de l'Industrie, and more recently at the Crystal Palace. Both Mr. Swan and Mr. Edison have abundantly proved the entire suitability of their systems,—under proper management,—to domestic lighting, but for popular use, much remains to be done before this result can be obtained, and possibly it will be found that no mode of electric lighting can be thoroughly adapted for domestic purposes, until an efficient system of storage batteries has been devised. Success in this direction appears to be very close at hand, and the question, apart from Planté's investigations, has been studied with highly promising results by Faure, Swan, Sellon, Volckmar, Kabath and others.

It may be assumed, therefore, that the mechanical production of electric currents, and their utilisation as a source of light, has reached that stage of practical development from which it must necessarily advance, with a constantly widening circle of usefulness and efficiency.

The interest excited by the present active progress in invention relating to electric illumination, naturally causes us to overlook that early and very striking period, nearly 40 years ago, when the attention of many inventors was turned towards the same direction as at present. The names of Staite, Wright, King, and a few others are well known, though books hitherto published have done but scant justice to the great ingenuity of these men. There are, besides, a number of other inventors whose ideas are recorded in the Patent Office, which may be studied with advantage now,—especially by inventors,—who will find that there exist from an early date, many patented proposals which clash dangerously with some modern inventions. It would be invidious to give here any examples of early anticipation; the reader may form his own conclusions in studying the following pages.

The reason why inventors of the early period, to which reference has just been made, were baffled, despite their ingenuity, was because they had no cheap and practical mode of generating the current necessary for the production of light, and were obliged to obtain their power from batteries. Yet even in this direction something had been effected: a great deal of very suggestive work was done in the view of utilising battery currents to generate motive power, whereas, the very machines constructed for this purpose would have converted mechanical energy into electricity. Thus, Elias, in 1844, did not dream of generating currents, but only of utilising them, although the motor he devised might have been made available for the former far more useful purpose. Paciuotti agaiu devised his celebrated machine rather as a motor, than as a generator of electricity. Reference to this volume will show, however, that the idea of converting mechanical into electrical energy occupied the minds of inventors at quite an early date: thus, King and Poole, in 1846; Dujardin, in 1847; Henley, in 1849, and many others, realised the possibility of generating useful currents by making coils revolve in close proximity to permanent magnets. Most of these inventions were imitations, more or less close of Pixii and his followers Clarke, Saxton, Page and Stöhrer, and it was not till 1854 that we find any record of an entirely new departure. In that year Sören Hjorth patented a generator in which the principle of augmenting the power of electric currents by the reaction of electro-magnets upon each other, is described so clearly that it is evident the inventor appreciated the value of the dynamic principle, and thus anticipated by about 13 years, the practically simultaneous discoveries of Varley, Wheatstone and Siemens. With this brilliant exception, and that of another inventor who, in 1858, also proposed " to employ the electro-magnets in obtaining induced electricity which supplies *wholly* or partially the electricity necessary for polarising the electric magnets, which electricity would otherwise be required to be obtained from batteries and other known sources," inventors confined themselves for another 10 years to the perfection of magneto-electric machines. Among these inventors stand out prominently the names of Holmes and Wilde. A description of the magneto-electric machine of M. Worms de Romilly, which was patented in France, in March, 1866, will be found on page 672. The specification under which the invention was secured, is a very remarkable one, as it contains the first intimation of much that has been done by subsequent inventors. The patent claims three classes of armatures, which are illustrated and described in considerable detail—the ring armature, the drum armature, and a second form of ring armature, in which the wire is coiled upon a deep and narrow ring, so as to present a broad surface of coils to the fixed magnets, at right angles to the axis of rotation. Further, in the principal arrangement described, the field magnets—permanent, not electro-magnets, however—are arranged in a parallel series, and running longitudinally around the armature, and two such series of magnets are employed, one placed without and the other within the ring, so as to utilise both sides of the coils. The armatures described, suggest at once those of Gramme, of Alteneck and of Schuckert, while the arrangement of the field magnets resembles not a little that of De Meritens and of Jablochkoff; again, the mode of utilising the inner as well as the outer sides of the ring, recalls the recent proposals of Jürgensen and others. Unfortunately, it was impossible to place the notice of this very comprehensive invention in its chronological order, and it has therefore been appended to the end of the book. To place before the reader, a record, as unbroken as possible, of all that has been done by inventors for generating the electric current, and in apparatus for utilising it, as a luminous power, has been one of the main objects of the present volume. To do this with sufficient completeness to satisfy the general reader, would have been comparatively an easy task, but it has been desired to render the work, above all, useful to inventors, and to those interested in invention; for this purpose it was necessary to analyse every patent specification bearing on the subject that has been filed in this country. It would have been impossible within the limits of this volume to have presented abstracts of all patents relating to the subject in question up to the present date; it has been found therefore necessary to divide them, reserving the specifications of the last 10 years for a succeeding volume.

Another object of the present work is to explain in the simplest possible language, and yet without any sacrifice of scientific accuracy, the principles which govern the conversion of mechanical into electrical energy; in other words, the mechanical production of electric currents. This task, as well as that of analysing the subject in a somewhat more advanced method, has been undertaken, and it is hoped with success, by Professor Silvanus Thompson. The chapter on the Voltaic Arc is by Mr. M. F. O'Reilly, and contains a large amount of information, part of which has never before been published, while the remainder lies scattered through many books. These three chapters—the Voltaic Arc, the Mechanical Production of Currents, and the Theory of Dynamo-electric Generators, together with the preceding ones on Electrical Units, and the Measurement of Currents—will, it is hoped, be found to contain much useful matter, expressed in such a way as to be clearly comprehensible to the non-technical reader, but without any sacrifice of practical value. The remaining contents of the volume naturally divide themselves into GENERATORS of Electricity; CONDUCTORS, for the distribution of the current; CARBONS, and the various apparatus for utilising the latter, in connection with the current, either in the form of ARC, INCANDESCENCE-ARC, or INCANDESCENCE Lamps.

It is too much to hope that all systems hitherto proposed or carried into practice, have been included in these pages, but it is believed that the omissions, inadvertently made, are few and unimportant, and that a large number are now presented to the reader for the first time, many of them being of special interest.

With reference to the Appendix containing the ABSTRACTS of PATENT SPECIFICATIONS on subjects bearing on electric lighting, it may be permitted to speak with more confidence than of the body of this work. Of the usefulness of a faithful abstract of this nature there is no doubt, and that it has been both faithfully and ably executed, the professional reputation of the gentleman—Mr. W. Lloyd Wise—who undertook the heavy work, is an ample guarantee. As it was found impossible to publish, as has been said above, abstracts of all patents up to date, in the present volume, an ample index of those bearing on the subject from 1872 to 1882 has been appended, together with a list of all United States' patents connected with electric lighting.

The present writer finds it an agreeable task to acknowledge in detail the assistance he has received in the preparation of this work, other than that already referred to. He has to thank Mr. Conrad W. Cooke for contributions on some of the more important systems,—notably those of Brush, Edison, Weston, De Meritens and Maxim,—and for the interesting description of the Pacinotti machine, as well as for many valuable suggestions. To M. Vivarez he is indebted especially for the chapter on the Jablochkoff system, the account of the manufacture of the Jablochkoff candle, for a part of the chapters on Electrical Units and on Conductors, as well as for the descriptions of some generators and lamps. In addition to the chapter on the Voltaic Arc, Mr. M. F. O'Reilly has rendered aid in collecting information on the early generators, and in the description of the successive inventions of Gramme. The kind assistance of Mr. J. Munro and Mr. B. A. Raworth is also acknowledged. Recognition should be given to various printed sources whence information has been drawn; these are Fontaine's *Eclairage á Electricité;* the Comte du Moncel's *Applications de l'Electricité; La Lumière Electrique;* and Dr. H. Schellen's *Die magnet-und dynamo-elektrichen Maschinen;* the publisher of the latter book has kindly furnished several engravings.

A large proportion of the present work is reproduced from what has already appeared, scattered through the pages of ENGINEERING; it has, however, been subjected to careful revision, and in many cases to considerable amplifications. About one-third, however, besides the Abstracts of Specifications, appears in these pages for the first time.

It is intended that this volume shall be followed by a second, dealing largely with Applications of the Electric Light, as well as with a number of cognate subjects, it has been found impossible to touch upon here: such as Cost of Production and Maintenance, Photometry, Secondary Batteries, Accessories, Motive Power, &c.

JAMES DREDGE.

OFFICES OF " ENGINEERING,"
    August. 1882.

CONTENTS.

—•—•—

SECTION I.

————— ——— ———

SECTION II.

SECTION III.

APPENDIX.

Abstracts of Patents relating to Electric Lighting, granted in the United Kingdom, from 1837 to 1872;
List of Patents granted in the United Kingdom, 1872 to 1882; List of Patents relating to
Electric Lighting in the United States; Index.

LIST OF ILLUSTRATIONS.

xiv

xvi

xviii

LIST OF ILLUSTRATIONS IN APPENDIX.

SECTION I.

———•••———

B

ELECTRIC ILLUMINATION.

I.

ELECTRICAL UNITS.

THE principle of the conservation of energy, announced for the first time by Helmholtz, controls all problems in the measurement of force, and plays an important part in the application of the system of so-called absolute units, to which the British Association has lent its name, and to the consideration of which the Congress, at the recent Electrical Exhibition of Paris, devoted much attention, with the object of establishing it on an international basis.

The word "energy" is applicable to all physical manifestations —mechanical work, production of heat, light, &c. Conservation results from the important fact that energy expended is always to be found integrally in some other expressions of work—calories, chemical action, and so forth. It is an embodiment of the old aphorism, "Rien ne se perd, rien ne se crée." It is evident that the measure of the various forms of energy ought to be obtained by a system of units, intimately dependent one on another, but the establishment of such a system is the work of time; several centuries, for instance, were required, and a revolution, resulting in the sweeping away of prejudice, before a first series of geometrical units, for the measure of space and weight could be arranged in logical classification.

Nearly a century ago, a national convention in France marked a great era in progress by the creation of the metric system, which takes as a base for measures of length, area, volume, weight, and mass, the metre—equal to the forty-millionth part of the length of the terrestrial meridian.

Whether the metric units do, or do not, remain exact physical representations of a certain quantity in nature, is of no real importance, since, if they change, they enter into the system of universally accepted though arbitrary units. And this wide acceptation, apart from its practical convenience, constitutes the real value of the metric system, which has been tested by the experience of more than a century, and has been adopted by so many countries. The appreciation of this fact was demonstrated at the Paris Congress by an almost unanimous decision, to base all new international standard electric measurements on the metric system. A similar view had been taken a number of years before by a committee of the British Association, which co-ordinated the previous ideas of Weber, Gauss, and some other German *savants*, in electrical measurements, for, although the metric system is not generally adopted in this country and the United States, it has long served here as the basis of scientific calculation, and has thus become popularised to a certain extent. The British Association based its system of measurements on the three units of length, mass, and time, for which it adopted the centimetre, the gramme, and the second, whence the symbol, C. G. S., by which this system of units is commonly designated.

It was in 1873 that the C. G. S. unit was called into existence, and it was the result of the labour of the Committee that had been appointed to consider the question of absolute units. This Committee was composed of Sir W. Thomson, Professor G. C. Foster, Professor J. C. Maxwell, Mr. G. F. Stoney, Professor Fleeming Jenkin, Dr. Siemens, Mr. F. J. Bramwell, and Professor Everett, the latter acting as secretary. The principal work of this Committee was the selection and nomenclature of units of force and energy, and especially of electrical and magnetic units. In their report, to which it may be appropriate to refer in some detail, the selection of the fundamental units was definitely proposed, but their nomenclature was only provisionally suggested. The Committee pointed out that up to that date, when it was desired to specify a magnitude in absolute measure, three fundamental units were necessary to indicate length, mass, and time, and they considered that it would be highly desirable that one definite fundamental unit should be devised to express the three dimensions by a single term. "We think that in the selection of each ·kind of derived unit, all arbitrary multiplications and divisions by powers of 10, or other factors, must be rigorously avoided, and the whole system of fundamental units of force, work, electrostatic and electromagnetic elements must be fixed at one common level, that level, namely, which is determined by direct derivation from the three fundamental units once for all selected." Those recommended and adopted were, as stated above, the gramme, centimetre, and second. The peculiar fitness of the fundamental units chosen, will be recognised when it is remembered that an intimate relation exists between the centimetre and gramme, the latter being the weight of a cubic centimetre of distilled water measured at the temperature of its maximum density, viz., 4 deg. C. It was this connexion that led to the selection—despite their small values—of the C. G. S. units, by the British Association Committee, there being no such inter-dependent connection between the metre and the kilogramme.

In their report the committee pointed out, that the Ohm as represented by the original standard cell is approximately 10^9 C. G. S. units of resistance ; the Volt is approximately 10^8 C. G. S. units of electromotive force ; and the Farad is approximately 10^{-9} C. G. S. units of capacity. For defining quantities multiplied or divided by one million, the prefixes *mega* and *micro* were suggested, so that a Megohm denoted one million Ohms, and a Microfarad, one-millionth part of a Farad. Thus the sign 10^6 would represent the prefix " mega," and the sign 10^{-6} the prefix " micro ;" or 1,000,000 and ·000001 respectively. The C. G. S. unit of force is that " force which acting on a gramme of matter for a second generates a velocity of one centimetre a second." This unit is named the *Dyne*. The weight of a gramme at any part of the earth's surface is about 980 dynes, or rather less than a kilodyne. The C. G. S. unit of work, or energy, represents the work done by the dyne in passing through the distance of a centimetre. This value was named the erg. The kilogrammetre is equal to 98,000,000 ergs, and the gramme centimetre is equal to 980 ergs. The C. G. S. unit of power is, the power of doing work at the rate of one erg per second. The equivalent given above assumes the value of g (the acceleration of a body falling *in vacuo*) to be 980 C. G. S. units of acceleration, but to obtain an exact result the value of g at the station where the calculation is made must of course be accurately ascertained.

It will be seen from the foregoing, which summarises the report of the Committee, that the features introduced by the British Association were the unit of force and the unit of work. As regards the former, the force most easily measured is the attractive force of the earth on bodies upon its surface. It is readily deduced from the weight of the object. Now, as is well known, this weight varies with the geographical position of the body, being greater near the poles than at the equator. On this account it 'was somewhat objectionable as a fundamental unit. An invariable unit was desired and was found in that force which acting on unit mass (one gramme) for unit time (one second) generates unit velocity (one centimetre per second). This force, as mentioned above, was called the *dyne*, and according to the preceding statement, it will be seen that taking

p as the weight of a gramme, and g the acceleration, then $\dfrac{p}{g} = \dfrac{\text{dyne.}}{1}$

The weight of the gramme is thus equal to g dynes, and at Paris where $g = 981$ centimetres, it is equal to 981 dynes.

The idea of work follows as a necessary consequence on that of force. Work, it is needless to say, is the product of the intensity of the force by the distance through which it operates. The most generally employed expression of this energy is the horse-power. It was introduced by Watt, and is equal to 33,000 foot-pounds per minute, or 550 per second. This is equivalent to 7.46×10^9 ergs per second. The horse-power is a very arbitrary unit; its continuance is justified only by its widely extended use. In France the horse power or "cheval vapeur" is measured in kilogrammetres—the unit of work which is equal to one kilogramme raised to a height of one metre; the foot-pound is one-seventh nearly of a kilogrammetre. The English horse-power is equivalent to 76 kilogrammetres, while the French "force de cheval" is only 75 kilogrammetres. These units, the foot-pound and the kilogrammetre, do not harmonise with the C. G. S. system, in which they are replaced by the *erg*, which is measured as the product of the dyne (the unit of force) by the centimetre (unit of length). An erg is therefore equal to a *dyne-centimetre.*

The practical fault of this unit is, that it is too small, so that work, such as is usually understood by the term, would have to be expressed in numbers altogether too large to be convenient. Thus the kilogrammetre, by the C. G. S. system, is equal (in Paris) to 98,100,000 ergs, and the horse-power (75 kilogrammetres) to 7,357,500,000 ergs. This difficulty was met, as has been stated, by devising a second unit—the meg-erg, equal to one million ergs. The kilogrammetre is therefore equal to 98.1 meg-ergs. A minor unit, the "ergten," equal to 10 ergs, was also introduced. This, Mr. W. H. Preece suggested, should be called a "Watt."

The heat units are based on the same system; they are—the *degree*, which measures temperature, and the *gramme-degree*, which is the quantity of heat necessary to raise one gramme of water from zero to 1 deg. (Centigrade). In the old system the kilogramme of water was employed, the new heat unit or *Calorie* is therefore one thousand times smaller than the old.

The preceding remarks serve to introduce the consideration of the new electrical units which have been created by the decisions of the Paris Electrical Congress. In obtaining a view of the phenomena attending the propagation of an electric current, it is necessary to be familiar with certain primary ideas. Unfortunately the definitions of these are somewhat unsatisfactory, owing to our very imperfect knowledge of the nature of the electric current itself.

These ideas are (1) *Potential,* or the condition of a body with respect to electricity ; (2) *Electromotive force,* or that mysterious power which tends to produce a transfer of electricity from one point to another of a conductor ; (3) *Conductivity,* or the facility which a body offers to the flow of electricity ; (4) *Resistance,* or the degree of difficulty experienced by a current in its passage through a body ; (5) *Intensity,* or strength of a current which must be directly proportional to the electromotive force and inversely to the resistance ; (6) *Quantity,* which is plainly the product of the current strength by the time that it lasts ; (7) *Capacity,* or the quantity of electricity necessary to raise the potential of a conductor from zero to unity.

The discussion of an international system of electrical units was one of the chief occupations of the Congress at Paris. The first sitting, on the 15th of September, 1881, was devoted to the organisation of the work ; the second, on the 20th of September, was abruptly dissolved by the reception of the news of the death of President Garfield. At the third sitting, on the 21st of September, M. Mascart, secretary of the Commission charged with the investigation of the question, informed the Congress of the points that had been discussed, and the decisions that had been arrived at No time had been lost in this important work. Under the presidency of M. T. B. Dumas, the Commission held four important *seances*—on the 16th, 17th, 19th, and 21st of September. England was represented by Sir W. Thomson, Messrs. Warren de la Rue, Spottiswoode, Ayrton, Everett, and Moulton ; France by MM. Mascart, Levy, Raynaud, and Lippmann ; Germany by Dr. Siemens, MM. Helmholtz, Clausius, Widemann, and Förster ; Italy by M. Govi, and Russia by M. Stoletow. The results of the four sittings of the Commission are contained in the following resolutions, which were moved at the third meeting :—

1.   In electrical measurements, the three fundamental units shall be adopted : Centimetre, Gramme and Second (C. G. S.)

2.   The practical units, the Ohm and the Volt, will preserve their actual values : 10^9 for the Ohm and 10^8 for the Volt.

3.   The unit of resistance, the Ohm, will be represented by a column of mercury of one square millimetre section, at the temperature zero Centigrade.

4.   An international commission shall be appointed to ascertain, by new experiments, the length of the column of mercury of one square millimetre section at the temperature of zero Centigrade, which will represent the value of the Ohm.

5.   The current produced by a Volt through an Ohm shall be called an *Ampère.*

6.   A *Coulomb* shall be the quantity of electricity defined by the condition that an Ampère gives a Coulomb per second.

7. A *Farad* shall be the capacity defined by the condition that a Coulomb in a Farad gives a Volt.

These conditions were unanimously adopted by the Congress, after an explanation by Sir W. Thomson on their meaning and value.

It may be useful now to consider these units one by one, and to endeavour to make their meaning clear by a few elementary considerations.

The term *Electrical Potential* was introduced into our scientific nomenclature in 1828 by Mr. George Green, of Nottingham.   Potential, in mechanics, means the power of doing work; the electrical potential of any point in a body, or in space, is defined as the quantity of work done in bringing unit electrification from an infinite distance up to that point.   Thus the potential at A may be different from that at B. If A be higher than B, then, on connecting them by a conductor, a current will flow from A to B, and continue until the potentials are equalised.   There is an analogy in the flow of water through pipes, where there is necessarily a difference of level, that corresponds to the difference of potential; this difference of level produces a hydrostatic pressure, that is electromotive force; when the tap is turned, the water flows out, that is the current.   We then conclude that, wherever there is a difference of potential, there is electromotive force, and on completing the circuit (the analogue of opening the tap), a current will be established.

The terms, *Electromotive Force* and *Difference of Potential*, are thus not exactly synonymous, and it is useful to distinguish between them.

The unit of electromotive force was called the Volt by the British Association, in honour of the great Italian physicist Volta. It is a close approximation to the electromotive force of a Daniell's cell (about ·95 of it). In other words, a Daniell's cell is a little more than a Volt (1·106 according to Siemens, and 1·079 according to Latimer Clark). At the General Post-Office in London, a standard cell has been adopted, consist- of a Daniell's element arranged as in the annexed sketch, (Fig. 1), in

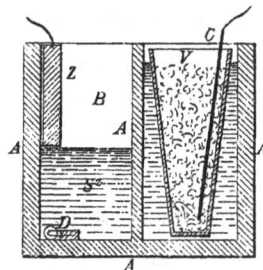

Fig. I.

which A A is a vessel with two compartments. In the first is placed to half the depth a saturated solution of sulphate of zinc, and a plate of zinc descending almost to the level of the liquid. In the second compartment is a porous jar V, filled with crystals of sulphate of copper, into which is plunged the copper plate C. A small cylinder of zinc D is placed as shown at the bottom of the compartment B of the vessel. In the position shown in the sketch, the cell is inoperative. To make it work, the jar V is lifted from the second compartment and placed in the first, raising the level of the solution and submerging the zinc, thus setting the cell in action. When the experiment has been concluded, the jar V is restored to its original position. The rod D receives the deposit of copper from the sulphate which has percolated through the porous jar V ; the solution in B is thus always kept clear. The Latimer Clark element is also sometimes employed as a standard, its electromotive force being 1·457 Volts.

The *Resistance* of a circuit to the passage of an electric current varies directly as its length and inversely as its cross section. The unit employed to measure resistance is the Ohm, which is represented approximately by the resistance of a telegraph wire of galvanised iron 100 metres in length and 4 millimetres in diameter, or by a column of mercury 1·05 metres long and one square millimetre section, or again by 48 metres of pure copper wire 1 millimetre in diameter. The first standards for resistance were made in 1867 of wire coils, composed of alloys of gold and silver, platinum and silver, and platinum and iridium, the conductivity of which is but little affected by changes of temperature. These wires were from five to eight-hundredths of a millimetre in diameter and from 1 metre to 2 metres long; they were insulated from one another by a casing of silk and paraffin. Nine years later these coils were subjected to a series of comparative tests, when it was found, with the exception of the platinum and iridium alloy which had changed, they remained absolutely unaltered from their original condition. Even admitting, however, the absolute permanence of these alloys, which was practically shown by experiment, the adoption of such standards is open to objection. The paraffin insulator prevents the metallic thread from placing itself easily in equilibrium with the temperature of the surrounding air. Besides that, the paraffin, especially when the coil is plunged in ice, may become affected by moisture and render the measurements unreliable. These defects in practical working led to the design of other standard resistances of silver platinum wire, the coils of which are separated from each other by sheets of hard rubber pierced with holes, the insulation being effected by the air. The coils thus composed are placed in a cylindrical box formed of two copper capsules stamped out of the solid and screwed one into the other; this apparatus can be submerged in water without any detriment.

In addition to the unit of resistance adopted by the British Association, there existed until the decision of the Paris Congress, the Siemens unit, measured by the resistance of a column of mercury of one square millimetre section, and one metre in length, at the temperature of zero Centigrade. This unit was equivalent to ·955 Ohm. Though this unit may have been officially abandoned, it will be observed that its principle has been substantially retained, since the Congress expressed the hope that the value of the Ohm should be represented by a column of mercury. It is true, as Professor Helmholtz remarked, that the mercury has to be contained within a glass tube, which forms a fragile measuring apparatus, and one liable to destruction; and, moreover, that it is impossible to be certain that some want of truth may not exist in the tube, which would affect the value of the standard. On the other hand, Sir W. Thomson called attention to the unavoidable imperfections in solid metal standards, and Dr. Siemens reminded the Commission that the mercury standard could be reproduced geometrically in all parts of the globe, and that in any case an approximation, much superior to that possible with wire standards, would be obtained. These considerations determined the decision of the Congress.

In the system adopted, the two units of electromotive force and of resistance, expressed by the C. G. S. system, would have too small a value for ordinary purposes. It was resolved, therefore, that the Volt shall be practically expressed in 100 millions of C. G. S. units of electromotive force, or 10^8, and the Ohm in 1,000 millions of C. G. S. units, or 10^9. As an auxiliary resistance unit, the megohm, or 10^{15}, is often employed.

The *Intensity* of a current is also readily measured by the chemical action which it is capable of producing. Faraday established the law that the intensity of a current is the same at all points of the circuit which it traverses, and that the chemical action is proportional to this intensity. There is no occasion to refer here to the demonstration of the law now definitely accepted by science, but attention may be called to an important fact which is a consequence of it. The constancy in the intensity of the chemical work, at all points of a circuit, tends to show that electricity should not be regarded as energy *per se*, but as its vehicle. According to the comparison of M. Marcel Deprez, it is no more a form of energy, than a column of water under pressure, or a driving belt. In fact it is not lost, but is always found the same at different points of the circuit, the potential only varying.

When the relations existing between resistance, intensity, quantity, and electromotive force are defined without reference to any but fundamental units, it is found that the measurement of resistance involves simply a velocity, that is, the quotient of a length by a time, thus: $\frac{\text{centimetre}}{\text{second.}}$ The Ohm, therefore, is $10^9 \frac{\text{centimetres}}{\text{second}}$, or a velocity equal to the distance from the equator to the pole traversed in one second. It is difficult to explain what the unit is without discussing the whole subject of electrical measurement, but the following definition by Mr. Fleeming Jenkin, will serve to show its dependence on the absolute units:—

"The resistance of the $\frac{\text{centimetre}}{\text{second}}$ is such, that the current generated in a circuit of that resistance by the electromotive force due to a straight bar one metre long moving across the magnetic field of unit intensity, perpendicular to the lines of force and its own direction, with a velocity of one centimetre per second, would, if doing no other work, develop in that circuit in one second of time a total amount of heat equivalent to one absolute unit of work,"

The three quantities, electromotive force, resistance, and intensity, are connected together by a simple law discovered by Ohm, and confirmed by Pouillet, Deprez, &c. This law is expressed as follows :—

$$\text{Intensity} = \frac{\text{Electromotive force}}{\text{Resistance}}.$$

The unit of intensity is therefore equal to $\dfrac{\text{Volt}}{\text{Ohm}}$ according to the definition adopted, and consequently is equal to one-tenth of the absolute unit of current. This unit has long been employed in England under the name of Weber, while in Germany a unit ten times smaller was also called by the same name, and was due to a system of measurement adopted by Weber and Gauss, in which the fundamental units were the milligramme and millimetre. The Paris Congress, in order to avoid the inconvenience of a permanent confusion arising from these two Webers, deemed it advisable to suppress the name completely and to replace it by the *Ampère*. In relation to this choice Professor Helmholtz said that the selection was amply justified by the important works of the great *savant*, to whom the world owed so clear a knowledge of electro-magnetic phenomena; it had, moreover, the advantage of combining the name of a Frenchman with those illustrious ones from Germany, England and Italy which have been adopted to designate other units. The name of Œrsted had indeed been proposed, but was rejected as being somewhat cumbrous; that of Gauss was also suggested, but was not considered appropriate, as his name is not associated with magneto-electric investigations. Probably, when a unit of magnetic intensity for the field of dynamo-electric machines has been decided upon, it will receive the name of Gauss.

The approximate measure of the Ampère is the intensity capable of precipitating 4 grammes of silver per hour, or 1·19 grammes of copper, or of decomposing ·09378 grammes of water per second. If this is compared to the Siemens unit of intensity, which is equal to a $\dfrac{\text{Daniell element}}{\text{Siemens' unit}}$, it will be found that this latter unit is equal to $1\cdot16\,\dfrac{\text{Volt}}{\text{Ohm}}$, and that it is capable of depositing 1·38 grammes of copper per hour. According to Ohm's law, it will be seen that the Ampère is equal to one-tenth C. G. S. unit of current.

Standards for intensity have not been hitherto made; but the values are closely ascertained by graduated apparatus, such as the Gaiffe and Deprez galvanometers, or by the electro-dynamometer of Weber.

The *Quantity* of electricity varies plainly as the intensity, and also as the time, hence it varies as their product. The *Coulomb* is the new unit of quantity, and may be defined as the quantity of electricity given by an Ampère in a second. The quantity will also depend upon the electromotive force, and the capacity of the body for holding or accumulating electricity, hence Quantity = Capacity × Electromotive force. The *capacity* is defined by the relation

$$\text{Capacity} = \frac{\text{Quantity}}{\text{Electromotive force}}$$

and hence

$$\text{Unit Capacity (the Farad)} = \frac{\text{Unit of quantity}}{\text{Unit of electromotive force}} = \frac{\text{Coulomb}}{\text{Volt}}$$

According to the above expressions it will be seen that the Coulomb represents 10^{-1} C. G. S. units of quantity, and that the Farad represents 10^{-9} C. G. S. units of capacity. The Farad is rather large, and consequently the microfarad is preferred for practical measurements of capacity. The microfarad is therefore 10^{-15} C. G. S. units of capacity.

Such, explained in an elementary manner, is the system of units adopted by the Paris Congress; before concluding their work, however, two propositions were made, which should be referred to; the first resolution, moved by M. Widemann, was as follows :—

"The Congress of Electricians expresses the sincere hope that the French Government will place itself in communication with other Powers, with the object of forming an executive committee to undertake the necessary investigations for permanently deciding on a series of units."

The second proposition, made by M. Förster, had reference to making this committee supplementary to the International Bureau of Weights and Measures, and ran as follows :—

"In recognising the great services that the International Bureau of Weights and Measures may offer in the researches of the International Commission of Electric Units, and in the preservation of standards, the Congress considers it desirable to leave to this Commission, which will be nominated diplomatically, the decision that should be taken on this matter."

After these two resolutions had been unanimously voted, the Congress concluded its labours of investigation of the system of electrical units, which will in the future be carried forward by an International Commission.

II.

THE MEASUREMENT OF ELECTRICAL INTENSITY.

SINCE the applications of electricity have been so largely developed for industrial purposes, the necessity has arisen for appliances by which the intensity of the currents employed, whether for electroplating, for the production of light, or the transmission of power, can be estimated continuously; and these measurements should be made with as much ease as the pressure of steam is observed on a pressure guage, or the speed of an engine by means of a tachymeter. For such a purpose, and to perform the necessary work, the apparatus for electrical measurement must fulfil several essential conditions.

It should be sensitive, and give instantaneous indications by the movement of a needle over a graduated dial, without any manipulations being necessary, such as many galvanometers adapted for scientific investigations, require. It is indeed indispensable that all the changes in the intensity of the current should be recorded instantaneously and without the intervention in any way of the observer, otherwise this would occupy a certain time, and might interfere with the observation of variations, which though of small extent are, of course, necessary to be known. Moreover, it is necessary that the apparatus should be small and compact, and not so delicate that it cannot be confided to the hands of any ordinary operator. Finally, it should not only record relative, but absolute values, so that it may be unnecessary to make any calculations or to refer to any tables.

In the galvanometer we have selected for illustration, the foregoing requirements have been combined in a practical form. It consists of a magnetised needle, subjected to the action of the current to be measured, and the displacement which results is shown by an index on a graduated dial. The oscillations of this index, which would render the reading tedious and uncertain, are reduced by the action of a powerful magnetic field. This is the industrial galvanometer of M. Marcel Deprez, the well-known French physicist, with whose numerous investigations and admirable inventions the scientific world is familiar. The apparatus was perfected in the year 1880, and it is already widely used; it has therefore been in the hands of practical men for a sufficiently long period to be thoroughly tested, and it has earned a high reputation. The instrument is known by the name of the "fish-bone" galvanometer, on account of the form of its principal organ—the needle—which, composed of a thin blade of soft iron 10 centimetres (3·94 in.) long, and 3 centimetres (1·18 in.) wide, is divided, after the manner of a double comb, by a number of slots made with a saw, which thus convert it into

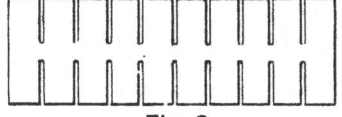

Fig. 2.

a series of ten needles connected in the middle, as shown in Fig. 2. M. Deprez considers that this system polarises better than a needle formed of a solid undivided blade. This multiple needle oscillates around a horizontal axis formed by two knife-edges resting on two small supports. One of these supports is visible at A on the base in Fig. 3. At the centre of the needle is a

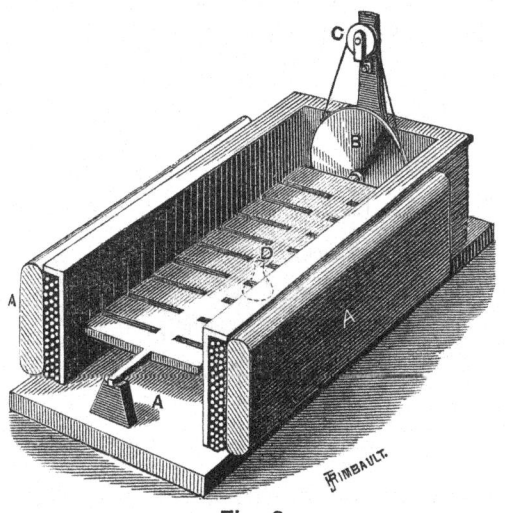

Fig. 3.

small spherical counterweight D, which tends always to pull it into a horizontal position when the current that displaces it, ceases to act. The needle is enclosed in a frame that entirely surrounds it; on this frame are wound two distinct circuits, one of which is formed of a thin sheet of insulated copper, 10 square millimetres in section. This copper-sheet only has four convolutions, and has a very low resistance. The other circuit, on the contrary, is formed of fine wire of great length, and has consequently a very high resistance. The first circuit serves for the passage of currents of high intensity, the second receives the weak currents.

Around the frame is a powerful magnet A which directs the needle. The bend of this magnet is cut away in the sketch, Fig. 3, in order to show the section of the enclosing frame and the mode of carrying the needle. At its further end the latter is made fast with a grooved pulley B, which is coupled to the pulley C, one-fifth of the diameter of B, by means of a practically inextensible wire, one end of which is fixed to the pulley B in such a way as to prevent any slipping. From this arrangement it results that the angles described by the pulley C are exactly five times greater than those described by the pulley B. Consequently, with a deviation of 10 deg. in the needle D, there will be a deviation of 50 deg. on the pulley C and of the light index fixed to it. The deviations of the needle may be considered as proportional to the intensity of the current, as also may those of the index, in such a way that, knowing the intensity of a current corresponding to a given angle for the graduation, the whole graduation of the scale may be deduced.

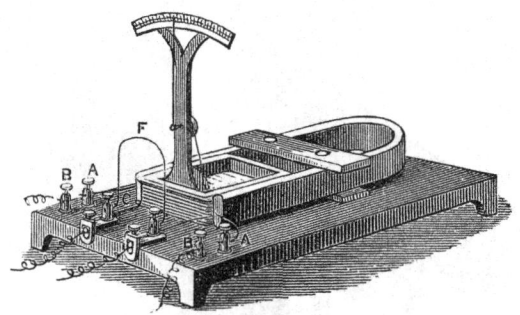

Fig. 4.

Fig. 4 shows the complete apparatus mounted on its stand, on which are six terminals. Those marked A receive the two ends of the fine wire wound on the frame ; they communicate with the terminals B, by which is introduced the current to be measured. The terminals C are in connection with the copper strip wound on the frame, and also receive the conducting wire through which the current passes ; they can, when desirable, be connected by the wire F, which allows the greater portion of the current to pass, if it becomes too powerful for the instrument. The dial is graduated in half degrees ; it indicates at will Volts or Ampères, in such a way that the galvanometer can be employed to measure electromotive force as well as intensities. The graduation is effected as follows:—Suppose, for example, it is decided that a half Volt or Ampère shall be represented by each degree. Let us consider first the graduation for electromotive force. The terminals B are first placed in communication (and consequently the long and fine wire circuit of the galvanometer) with a battery of very small internal resistance, the electromotive force of which is accurately known. According to the hypothesis controlling the graduation—that is to say, one degree for each half Volt—the needle ought to stop at the division n; this will not be the case, and the needle will mark the division n^1. If n^1 is greater than n (and the graduation should be arranged with this object) the needle should be brought back from the division n^1 to the division n, by throwing into the circuit a resistance, the value of which should be regulated gradually by trial until the needle is set against the division n. This resistance is conveniently composed of a wire rolled around a small ebonite core, in such a way that the convolutions are all insulated from each other. The difficulty in this method lies in the fact that it is not possible to assign to a given battery an unvarying electromotive force. Between two cells apparently identical, variations of from one to three hundredths are found. But this difficulty is more theoretic than practical; for every-day work an approximation within a few hundredths is all that is necessary, and the difference resulting in the graduation is certainly not appreciable to the eye. Moreover, to secure results as exact as possible, a Latimer Clark battery may be used for the graduation; this battery remains very constant when the circuit is closed, for the short time necessary for experiment.

The following is the method adopted in graduating for intensities: The current from five large surface Bunsen's cells is brought by the terminals C to the low resistance circuit on the case of the galvanometer, at the same time that a 1 Ohm resistance is thrown in. Remembering that the intensity of a current is constant at all points of the circuit, and that the law of Ohm, $I=\dfrac{E}{R}$, gives in the present case (where R=1) I=E, it will be seen that the intensity of the current will be measured exactly by the difference of potentials at the extremities of the resistance equal to one Ohm, this difference can be read off very exactly with the help of a second galvanometer, graduated for Volts as explained above, and which is connected to the terminals of the resistance coil. If it be desired that the degree indicated by the galvanometer index needle should agree with that corresponding to the base of graduation previously decided on, the needle is removed from the division n^1 at which it stopped, to the division n on which it should be fixed, by introducing between the terminals C a stout red copper wire, bent as shown in the sketch Fig. 5, and the ends of which engage in the terminals C. A derivation from the current is established by this wire, the amount of which can be varied by introducing the ends more or less deeply into the terminals. Its proper position can be thus regulated by trial, in such a way as to set the position of the needle at the point of graduation previously fixed upon. Instead, however, of employing two galvanometers for the graduation of Ampères, one only need be used. If two currents are passed successively in the two circuits on the case of the galvanometer, so that the needle stops respectively at the points n and n^1 of the divided arc, and if afterwards these two currents are passed simultaneously through the two circuits, the needle will stop at the division n^{11}, so that $n^{11} = n^1 + n$. Then, in the case of the graduation in Ampères, if we wish to know the intensity of the current that traverses the resistance coil, we place first, in the same circuit, the battery, the coil, and the galvanometer, and note the degree n, on which the needle stops ; then we place the two terminals of the resistance coil in connection with the second circuit on the galvanometer case. The needle will deviate a certain number of degrees n^1 corresponding to the difference of potentials at the ends of the coil, and will consequently indicate the intensity of the current, in the case where the resistance introduced is equal to unity.

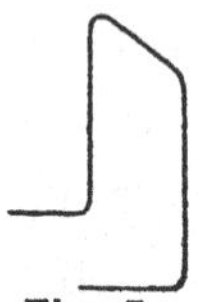

Fig. 5.

A second galvanometer, which may be taken as a representative instrument, is the dead-beat galvanometer, devised by Messrs Ayrton and Perry, and which was described by the inventors before the Society of Telegraph Engineers, substantially as follows :—

" The small instrument shown in Fig. 6 is very dead-beat, this result being attained partly by the lightness of the needle and pointer, and partly from its moving in a very strong permanent magnetic field. The needle is balanced, and consequently the deflections are about the same for any position of the instrument. By a proper arrangement of the coils, we have suceeded in making the deflections directly proportional to the current, and in the instrument 1 deg. deflection is produced by a current of two Webers, the greatest deflection, 45 deg., being produced therefore by a current of 90 Webers. The main peculiarity of the instrument, however, is the following :—The thick wire coiled round the needle, and through which the electric light current circulates, is in reality a strand or little cable composed of ten insulated wires. Each of

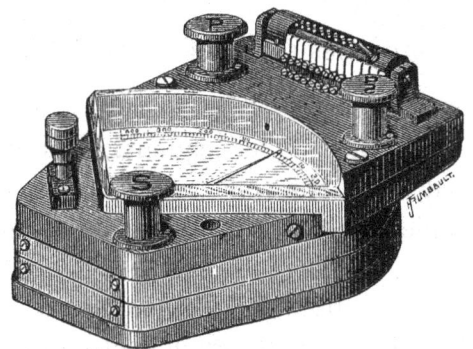

Fig. 6.

these wires having the same resistance, equal portions of the strong current flows through each of them. To produce a deflection even of 5 deg. requires a current of 10 Webers, but by means of a simple commutator these 10 wires, which have hitherto been joined in multiple arc, can, by a mere turn of the hand, be connected in series, and now a current of exactly one-tenth part, or one Weber, will produce 5 deg. deflection. To ascertain, then, the real value of any deflection, all we have to do is as follows :—Turn the commutator to series, and send a current from a single cell—a Daniell or a Grove—of which the electromotive force E, but not necessarilly the resistance, is known. A certain deflection $a°$ is produced. Now take out the plug of the 1 Ohm resistance coil attached to the instrument, and a deflection b is obtained. Then the resistance of the instrument, the wires, and the cell is

$$\frac{b°}{a°-b°} \text{ Ohms,}$$

or the deflection $a°$ in the instrument is produced by

$$\frac{\text{E } (a°-b°)}{b°} \text{ Webers}$$

when the wires are joined in series, or by

$$\frac{\text{E } (a°-b°)}{b°} \times 10 \text{ Webers}$$

when in parallel circuit.

"Thus with a Grove's cell a deflection of 7·4 deg. is obtained on the instrument when the coils are in series, and 4·1 deg. when the

1 Ohm resistance is inserted. The resistance, then, of the coils, the connecting wires, and the galvanometer

$$\frac{b^\circ}{a^\circ - b^\circ} = 1\cdot 24 \text{ Ohms}:$$

or as the electromotive force of a Grove's cell is $1\cdot 8$ Volts, a current of $\frac{1\cdot 8}{1\cdot 24}$, or $1\cdot 45$ Webers, produces a deflection of $7\cdot 4$ deg., or one Weber will produce about 5 deg. with the coils in series, and therefore $\frac{1}{2}$ deg. with the coils in multiple arc. Consequently, currents from 0 to 9 Webers can be measured when the coils are in series, and from 0 to 90 Webers when in multiple arc, without any calculation or reference to tables.

"While, then, with the ordinary instruments employed for measuring strong currents, the absolute value of any deflection can only be checked by employing a current of known strength, and as strong as the one the instrument is employed to measure, with this instrument the strong current necessary to produce any given deflection when the coils are in multiple arc, is exactly ten times as strong as the comparatively weak and easily producible current of known strength which produces the same deflection when the coils are in series.

"To render it impossible for the electric light current to be passed through the coils when in series, the binding screws, marked P P, and to which the wires from the dynamo-electric generator are attached, are only in circuit when the commutator is turned to parallel; the common screw, marked S, and the third screw, also marked S, being only in circuit when the commutator is turned to series. Neither the coils of the galvanometer nor the 1 Ohm resistance coil, then, can be damaged by the commutator being left accidentally in a wrong position."

Almost in the same category with this class of instruments, electrical counters find a place; these latter being specially designed with a view of recording the flow of an electric current, and forming a part of the whole system of canalisation and distribution from central stations to various users. Mr. Edison, whose inventive and practical genius characterises almost everything he undertakes, has devoted much time and attention to this subject, and naturally so, as he has long conceived, and is now carrying out on a large scale, the distribution of light and power in New York from central stations, and a reliable counter becomes in such a scheme, an absolute necessity for each service wire. Having studied the matter in its entirety, he aimed at devising an apparatus at once simple and convenient, enabling an accurate measure of flow to be made, analogous to the recording instruments of water and gas consumption. Edison's electrical counter is based on the law of Faraday that established the ratio between the intensity of an electric current and the chemical action which results from it. To the counter a branch circuit is taken from the main circuit, and of a low intensity (generally one hundreth); this fractional current is passed through a voltameter, where it deposits a certain amount of copper, the weight of which may be regarded as representing the quantity of electricity that has flowed through. Such is the general principle of the Edison counter, which up to the present time has been carried out in two special forms.

The first of these is shown in Fig. 7, and is adapted for intermittent measurements. The apparatus is composed, as will be seen, of two

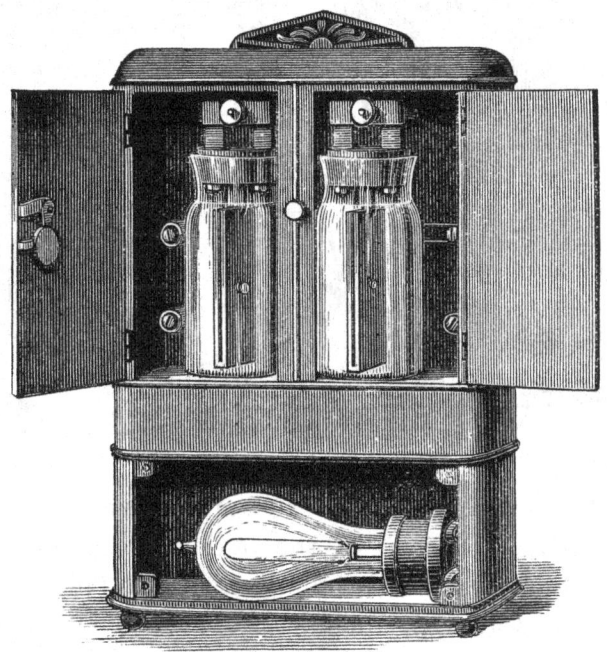

Fig. 7.

voltameters, filled with a solution of sulphate of copper, in which two electrodes that can be removed and weighed, are plunged. Underneath these voltameters is placed a little incandescence lamp, laid on its side, which is thrown into circuit automatically, when the temperature falls to 0 deg. Centigrade, or thereabouts, and thus becomes lighted, the heat produced preventing freezing and deposition of crystals in the voltameters. The second arrangement is automatic, and constantly indicates the consumption from minute to minute, the result being recorded on a counter. The instrument is shown in Fig. 8, and is a remarkable example of originality in invention, and of the desire of Edison to introduce electricity for general domestic use without making any great apparent change in the arrangements and fittings to which the public

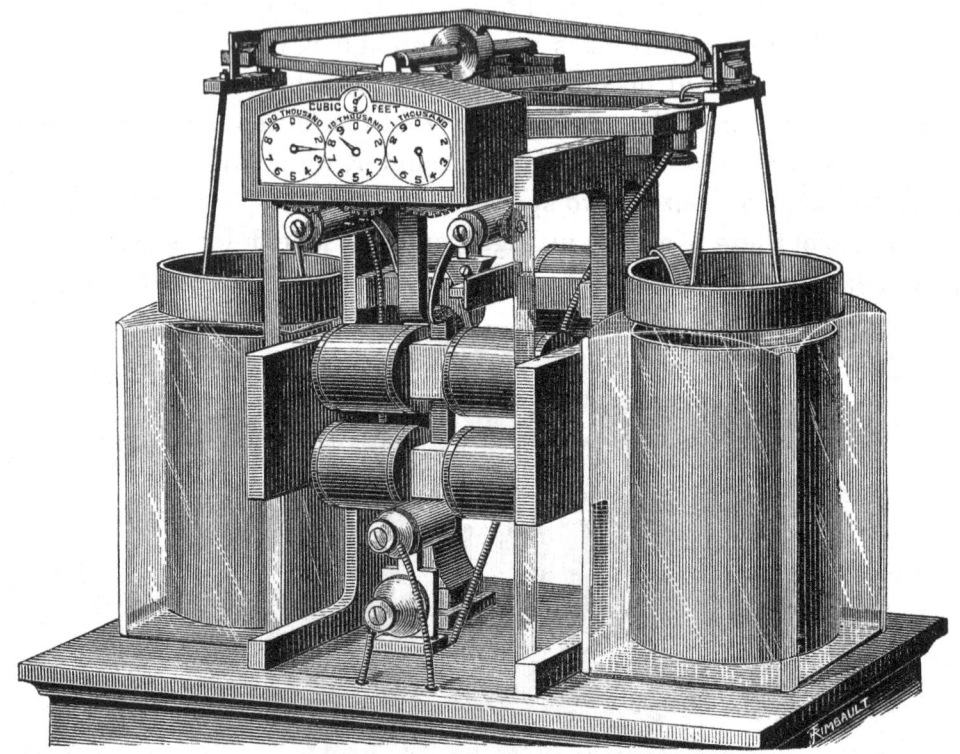

Fig. 8.

are habituated. It will be seen that the apparatus has a system of counter dials, exactly like those of the ordinary gas meter, so that to the consumer nothing is changed in appearance. This measurement of electricity as if it were gas, though strange at first sight, will be understood when it is remembered that the unit has been fixed by the Edison Company, at the quantity of electricity which gives a light equal to that produced by one cubic foot of gas. The automatic counter consists of a scale beam carrying at each end a copper cylinder, both of which act as electrodes and are plunged into two voltameters, in each of which is a fixed electrode. They are furnished with a small hook-shaped armature, which enters into a bath of mercury whatever may be the inclination of the beam. The current to be measured can be passed through each of the voltameters in such a way that on one of the copper cylinders a deposit of copper is made, while the other, on the contrary, becomes a soluble electrode. The balance is affected by these variations in weight, and the beam is depressed to one side. When the beam is tilted to a degree corresponding to a weight determined beforehand, an electro-magnet comes into action which reverses the current and changes the operations of the voltameters; the copper previously deposited is redissolved, and a second oscillation is produced in the opposite direction. In this manner a continued series of movements is obtained, which is transmitted to the counter by an ordinary train of gearing.

A further reference to both these simple and ingenious counters will be made hereafter, in describing the Edison system of electric lighting by incandescence.

THE VOLTAIC ARC.

T was in March of the year 1800 that Alexander Volta, F.R.S.,* then Professor of Natural Philosophy in the University of Pavia, announced to Sir Joseph Banks, President of the Royal Society, the construction of his first pile. That date is one of the most memorable in the annals of science, for it marks the beginning of an era of unparalleled activity, discovery and invention.

Humphry Davy, then a young man of two-and-twenty, was just beginning to attract public notice. In February, 1801, he obtained the appointment of assistant lecturer in the recently founded Royal Institution. His first thoughts were naturally for his favourite subject, and he eagerly applied the new source of power to the investigation of the properties of the elements. He had at his disposal three batteries: one of 24 plates of copper and zinc, 12 in. square; a second of 100 plates, 6 in. square; and a third of 150 plates, 4 in. square. The liquid used was a solution of alum, to which were added a few drops of nitric acid. Connecting together the last two batteries, he formed one of 250 pairs, and it was with this that he succeeded, on October 6, 1807, in obtaining from caustic potash "small globules of metallic lustre," which proved to be the new metal potassium. A few days later, he reduced, by a similar process, sodium from its hydrate soda.

It was subsequently to these discoveries that "a few zealous cultivators and patrons of science" subscribed the funds necessary for the formation of a great battery to be placed in the laboratory of the Royal Institution.† This was in the year 1808.‡

The new battery consisted of 2,000§ cells, arranged in 200 porcelain

---

* Volta was elected a foreign member of the Royal Society in 1792.

† Dove says the money was voted by Parliament; he is also in error when he says that it was with this battery the alkalies were decomposed.

‡ Tyndall—"Notes of a Course of Lectures on Voltaic Electricity."

§ Some text-books say 3,000; this is an error.

troughs, one of which may still be seen in the Royal Institution. The fluid was a mixture of sixty parts of water with one of nitric and one of sulphuric acid. The plates were zinc and copper, square in form and 32 in. in surface. Such was. the battery—now of historic interest—from which the first flashes of the electric light were obtained.

We have no difficulty in believing that the effects were "brilliant and impressive." Davy himself has left us* a brief description of his celebrated experiment, which was made before the members of the Royal Institution in 1810.† "When pieces of charcoal," says he, "about an inch long and one-sixth of an inch in diameter were brought near each other (within the thirtieth or fortieth part of an inch) a bright spark was produced and more than half the volume of the charcoal became ignited to whiteness; and, by withdrawing the points from each other, a constant discharge took place through the heated air in a space equal at least to four inches, producing a most brilliant ascending arch of light, broad and conical in form in the middle."

The carbon-points used by Davy were pencils of common charcoal. As such they must have wasted away rapidly, and, no regulating apparatus having been devised for adjusting the distance between them, the light must necessarily have been of short duration. In fact, it remained for 34 years a brilliant but sterile laboratory experiment. It was not until 1844 that that eminent physicist Léon Foucault replaced the soft friable charcoal by the hard compact carbon found in gas retorts; and, availing himself of the newly-invented and powerful battery of Professor Bunsen,‡ succeeded in producing a steady, continuous light. It was publicly exhibited in Paris by M. Deleuil, and soon its fitness was recognized for illuminating public works, squares, theatres, and lighthouses. Foucault found it serviceable even for photographic purposes. Using a battery of 46 cells, he compared the chemical effects of the light upon a sensitive plate of silver iodide with those of the lime-light, and found them to be as thirty-four to one.

Its general introduction was, however, retarded by the inconveniences attending the manipulation of powerful galvanic batteries, such as the labour of charging and discharging, the weakening of the current, and the evolution of deleterious gases; indeed, it may be questioned whether,

---

* Collected Works, vol. iv.

† Davy showed the carbon-light on a small scale as early as 1802.—*Jour. Roy. Inst.*

‡ Grove's battery was invented in 1836. In 1840, Cooper replaced the sheet of platinum by a rod of carbon; and, in 1842, Bunsen gave the battery its present convenient and practical form.

without the development of Faraday's grand discovery of magneto-induction, it would ever have become a. grand commercial success.

Before proceeding to the physical properties of the electric light, we must consider a few of the leading principles involved in its production.

When a current passes through, say, a copper wire, only part of the chemical action that goes on in the battery is transmitted as electricity, the remainder appearing as heat in the conductor. If we use a platinum wire, a greater fraction of the current will run down into energy of a lower form, and the wire may be raised to a red and even to a white heat. The more the conductor resists the passage of the current, the greater the quantity of heat generated; and, as the resistance varies directly as the length and inversely as the cross section, the heat developed will vary in the same proportion. Owing to the very high resistance of carbon, pencils of considerable thickness may be used for electrodes, and their poles still rendered red-hot by a strong current. On then introducing a small air-space, by separating them a little, we increase the total resistance in circuit, and may thus raise the poles to incandescence.

The resistance of conductors is also affected by their temperature; that of metals increases, whilst that of liquids decreases as they are heated. It is for this reason that a galvanic cell gives a stronger current when tepid than when cold. Du Moncel* has found that in general mediocre conductors of electricity, whether solid or liquid, become less resisting as their temperature rises. Carbon belongs to this class; the diminution in its resistance is, however, very small, being for coke about three ten-thousandths for each degree between 26 deg. and 270 deg. Cent., as determined by the experiments of Borgman.

The heat developed in any part of the circuit varies with the time and also with the strength of the current. This law, which was discovered by Dr. Joule, may be written

$$H = C^2 \, R \, T \times 0\cdot 24$$

where R denotes the resistance in Ohms, C the current in Ampères, T the time, and H the heat in gramme-degrees.

We have already incidentally noticed that in order to start the electric light, it is necessary to bring the carbons into actual contact. This is due, in great measure, to the nature of the battery current, which is characterised by its low electromotive force. When a path, even infinitesimally small, is open, the current will invariably force its way

---

* Mémoire sur la conductibilité des corps médiocrement conducteurs.

through; but, if there be a break of continuity, a considerable difference of potential will be required to enable it to leap across.

Thus, with 5,880 cells of his chloride of silver battery, Dr. Warren de la Rue obtained a spark of one-seventeenth of an inch, and, with the full strength of the 11,000, only one-seventh of an inch.*

When contact has been once made, the carbons may be separated by a short air-space without extinguishing the incipient light. The current is at first interrupted, but at the same instant there is developed, by the self-induction of the primary, an "extra current" which has sufficient electromotive force to enable it to pass over the small interpolar space In doing so, it carries along with it fine particles detached from the electrodes by its abrasive action, and which its heat immediately volatilises. Though this carbon vapour has a high resistance, it serves nevertheless as a medium for the re-transmission of the current. The gap may now be gradually widened, the arc will glow all the more intensely, and a considerable length of the electrodes be raised to vivid incandescence.

If interrupted for a moment—say, the fortieth or the twenty-fifth part of a second, as found by Le Roux—the arc will relight itself on closing again the circuit. This is, doubtless, on account of the conducting power of the heated vapour that still remains, low as it may be, and also of the warm air that surrounds the hot carbons. This also explains the continuance of the electric light when generators are employed which produce alternating currents. These are, absolutely speaking, discontinuous; the interruption, however, lasting only for a very small fraction of a second, does not affect the steadiness of the light.

The arc is a phenomenon of conduction, not of disruption. It is an integral part of the current, characterised by the same properties and governed by the same laws as any other part of the circuit. This important point was established by Matteucci in 1850.

The arc is also a form of energy appearing in one part of the circuit and is equivalent to the chemical work done in the battery or to the heat given out by the coal consumed in the furnace. The exact relation, however, between the current, the heat and light produced, though of great value both from a scientific and economic point of view, has not yet been determined.

Since the luminosity of flame depends, in great measure, upon the

* Phil. Trans., vol. clxxi.

presence of incandescent particles, by far the greater portion of the electric light is emitted by the white-hot carbons, although their temperature, as we shall presently see, is lower than that of the arc itself.

The light may be produced not only in air but also under the surface of water and other non-conducting liquids, in oils, and *in vacuo*, from which we infer that it is due to the incandescence and not the oxidation of the carbons.

If we take two carbon points and enclose them in a glass globe, which we then exhaust and hermetically seal, we shall find that we may make them intensely luminous, as often as we please, without effecting any material change in their aggregate volume. But the case is otherwise in the presence of air. The carbons then waste away by slow but real combustion. Some of the various appearances observed are shown in Fig. 9. They are not, however, equally oxidised, the positive being consumed twice as rapidly as the negative. This necessitates an automatic

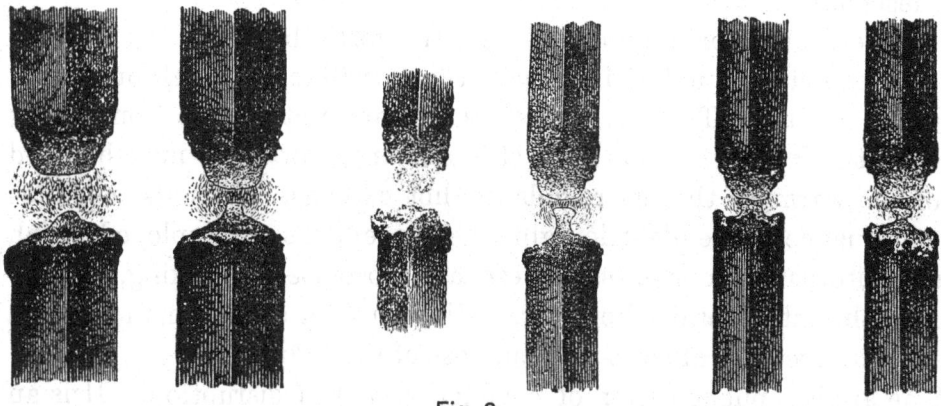

Fig. 9.

regulator to prevent the displacement of the luminous focus. There is also a marked difference between the carbons especially *in vacuo*. The negative, in consequence of " stalagmitic " additions received from the positive, elongates, becoming conoidal at its apex, Fig. 10, whilst the positive shortens and becomes slightly concave. Liquid spherules are frequently noticed forming on the glowing surface of the carbons. They become white hot, move slowly towards the poles and then suddenly glide across the arc. This transference takes place in both directions, but it is more abundant from the positive, owing, no doubt, to the disintegrating action of the current. These globules consist of mineral substances, chiefly alkaline silicates, which exist as impurities in the carbon. When the latter is prepared chemically, no such fitful movements are observed, and there is greater steadiness in the light. The transfer of these

Fig. 10.

particles represents a definite amount of work, which is taken from the whole energy that might otherwise have been converted into heat.

These foreign substances are much less fixed than the carbon; they are more readily converted into vapour by the arc, and round it they form the comparatively dull flickering flame so frequently noticed.

This deformation of the positive carbon is a matter of some importance. The crater-like cavity which is gradually formed is precisely the seat of the most fervid heat; it glows with an intense brilliancy, and yet in the ordinary arrangement of the carbons its radiation is directed to the ground, and therefore lost for illuminating purposes. A committee of English electricians, including Professor Tyndall, Mr. Sabine, and Mr. Douglas, some time since examined this point and recommended as the best disposition for the electrodes that shown in Fig. 11, in which the axis of the negative lies in the same vertical plane as the edge of the positive carbon. The white-hot dome, acting

Fig. 11.

then as a reflector, considerably increases the light projected in any desired azimuth.

The appearance of the arc-discharge varies with the form of the terminals between which it occurs. Thus between a point and a disc it is conical, whilst between two points it is somewhat globular. In some "candles," the Jablochkoff for instance, the thin carbon-rods are vertical and parallel to each other, so that their upper extremities are in the same horizontal plane. In this case the discharge is curved, forming a brilliant arch between the two poles. The current tends to take the path of least resistance from the tip of one rod to that of the other, but is bent upwards by the ascending currents of heated air and by the electro-magnetic action of the ascending current in one rod on the descending current in the other, until its elasticity just equilibrates their disturbing action.

The carbons, however, are generally fixed in the same vertical plane. When the lower is made positive, the rectilineal path of the discharge is disturbed as before by the heated air and gases. They combine to lengthen it, bend it upwards and outwards, and consequently also to diminish the stability of its equilibrium by diminishing the resistance it opposes to agencies tending to break it up.

It is more usual to make the upper the positive carbon. The current being then directed downwards, repels the rising gases, and the arc accordingly becomes shorter and narrower, more compressed and less expanded, and, therefore, also more concentrated and intense. It also offers less resistance to the passage of the current, but a greater resistance to rupture. The same is true of the Jamin "candle," in which the carbons are parallel, and the arc at their lower extremity.*

Of all artificial sources of heat, the electric arc is the most intense. The least fusible substances yield to its energy. "Platinum," wrote Davy in the account which he has left of his famous experiment, "was melted as readily as wax in the flame of a common candle; quartz, the sapphire, magnesia, lime, all entered into fusion." The diamond—a very refractory body—when placed in the arc, becomes white hot, swells out, fuses, and gradually transforms into a black crumbling mass, and carbon itself has been softened, so that it could be easily bent and welded.

Dr. William Siemens has recently availed himself of this great

* See *Comptes Rendus*, April 28, 1879.

energy to construct a model electrical furnace in which masses of iron and steel in considerable quantities are melted in a few minutes.*

It is not an easy problem to determine the temperature of the voltaic arc. Indeed, when we go beyond a certain range, the numerous constants introduced into the empirical formulæ become elements of uncertainty. It was shown by the committee appointed by the British Association to inquire into the methods for determining high temperatures, that the indications of the thermo-electric current, when the same substance radiates at different temperatures, form a parabolic curve when plotted in terms of the temperature. From this it is inferred that the total radiation is nearly proportional to the square of the temperature. Assuming the accuracy of this law, Professor Dewar compared the radiation from the positive pole of a Siemens' arc with that emitted by the same surface when heated with a large oxyhydrogen blowpipe, and found it to be as ten to one. The temperature of the carbon point would, therefore, be three times greater in the former than in the latter case. Assuming the temperature of the oxyhydrogen flame to be 2,000 deg. Cent., Professor Dewar gives 6,000 deg. Cent. as that of the incandescent carbons.†

It has been said that the temperature of the arc approaches that of the sun. How far the statement is correct may be inferred from the following. Ericsson affirms that the solar temperature must be between four and five million Fahrenheit degrees, Secchi reduces it to 250,000 deg. Fahr.; Violle,‡ to whom the French Academy awarded its prize in 1877, maintains that it cannot exceed 2,500 deg. Cent., whilst Rosetti, a distinguished Italian physicist, has lately set it down at 10,000 deg. Cent., or at 20,000 if we allow for the absorption of the solar atmosphere.§

The heat condition of the carbons has been investigated by Professor Rosetti.‖ He measured their temperature at the moment of ignition and found that it was affected by their thickness, the number of cells in the battery, and the manner in which they are connected. Thus in one series of experiments, the mean temperature of the positive

* Paper read before the Society of Telegraph Engineers, January, 1880.

† Proc. Roy. Soc., January, 1880.

‡ *Comptes Rendus*, 1877. Although the prize was awarded to M. Violle, and "encouragements" to several other competitors, still the Academy declared that the problem was not solved.

§ *Philosophical Magazine*, 1879.

‖ *Journal de Physique*, August, 1879.

carbon was 2,980 deg. Cent. ; by using a thinner electrode, it rose to 3,065 deg. Cent. ; and, by changing this for a third of still smaller dimensions, it reached 3,136 deg. Cent. The effect of the number of the cells is shown in the following table :—

With 50 cells, the temp. of the + carbon was 2,190 deg. C.

,,	60	,,	,,	,,	,,	2,334	,,
,,	70	,,	,,	,,	,,	2,536	,,
,,	80	,,	,,	,,	,,	2,784	,,

Foucault had concluded from photometric measurements that a battery of 80 Bunsen cells loses one-third of its strength in less than three hours. Professor Rosetti was thus obliged to make a large number of determina-tions, including all the varying conditions mentioned above, before arriving at any definite conclusion. In every case, however, he found the temperature of the positive carbon to be higher than that of the negative ; and in summing up his experiments, he concluded that the temperature of the positive pole is not less than 3,200 deg. Cent., whilst that of the negative is 2,500·deg. Cent. This difference in the thermal conditions of the carbon-terminals is observed only when battery currents are used for producing the light; it vanishes when magneto-electric or dynamo machines are employed.

Since the arc is an integral part of the circuit, its resistance must vary according to its temperature, its length,* the size of the carbons, and the number and arrangement of the cells in the battery. Mr. Preece gives it as varying between 1 and 3 Ohms ;† Professors Ayrton and Perry found it to be 12 Ohms, when the current was produced by 60 Grove cells, and 30 Ohms when the battery contained 122 cells. Du Moncel puts it down as between 30 and 40 Ohms, in the ordinary conditions necessary for a good light.‡

The spectrum of the voltaic arc presents several points of much interest. We must distinguish the spectrum of the carbons from that of the arc proper. The former being that of light emitted by an incandescent solid, is continuous from the red to the violet. Unlike that of the sun, it is not intersected by dark lines, but is one brilliant ribbon of pure vivid colour. If we could subject to prismatic analysis

---

* The length of the arc is measured from the summit of the negative carbon to the highest point of the positive cavity.

† *Telegraphic Journal*, February, 1879.

‡ *Eclairage Electrique.*

a ray of sunlight, which had not passed through the solar atmosphere, we should likewise obtain a perfectly continuous spectrum. In passing through the photosphere, it is sifted by the glowing vapours that compose this circumsolar region, and deprived of a great number of elementary radiations. This absorptive action of the photosphere reveals itself in the spectroscope, by the comparatively dark "Fraunhofer" lines. If we send the electric beam through an artificial vapour-atmosphere, the same selective action will take place, and the spectrum will become discontinuous.

The spectrum of the electric light differs again from that of the sun in another important respect, viz., the extent of its invisible radiation. This difference is due to our atmosphere, and especially to the vapour suspended in it.

In 1859, Professor Tyndall showed, for the first time, that aqueous vapour was not perfectly transparent to radiant heat, as had been believed up to that time, but that it exercised a very marked absorption on rays of low refrangibility. His results met with many objectors, especially among German professors, and gave rise to a warm controversy led by Professor Magnus, of Berlin. The question has, however, been set at rest by Professor Tyndall's latest researches, in which he scrupulously eliminated all the possible sources of error pointed out by his opponents, and also by his recent experiments on the action of an intermittent beam upon volatile liquids and their vapours.

It thus appears that while the solar beam makes its way through the successive strata of our atmosphere, it is deprived of a considerable portion of its heat-rays. An interesting illustration has recently been afforded by the experiments of Professor Langley. In his observatory, situated on Mount Whitney (United States), at an altitude of 12,000 ft., he has studied on the grandest scale possible the distribution of energy in the solar spectrum, and has found, as he says, in a letter written last October to Professor Tyndall: " An enormous extension of the ultra-red rays beyond the point to which they have been followed below, and being made on a scale different from that of the laboratory—one indeed as grand as nature can furnish—and by means wholly independent of those usually applied to the research, must, I think, when published, put an end to any doubt as to the accuracy of the statements so long since made by you as to the absorbent power of water vapour over the greater part of the spectrum,

D

and as to its predominent importance in modifying to us the solar energy."*

These ultra-red rays vibrate in periods so slow that they are insufficient to impress the retina and produce the sensation of vision. Professor Tyndall was the first to devise a means of quickening their motion and rendering them visible. Dissolving iodine in carbon disulphide, he obtained a solution impervious to light but perfectly transparent to heat. Interposing this ray filter, as he happily termed it, in the path of the electric beam, the obscure heat alone passed through. Converging this by a lens of rock-salt, a highly diathermanous substance, on a platinum spiral placed at the dark focus, it gradually became heated up to incandescence. This is the phenomenon of calorescence, a name given to it by Professor Tyndall to distinguish it from the analogous phenomenon of rendering visible the rays beyond the violet, which Professor Stokes called fluorescence.

The distribution of heat in the spectrum of the electric light may be studied by means of a linear thermopile and a delicate calibrated galvanometer. By slowly moving the pile from the violet to the red and even into the dark space beyond, we obtain gradually increasing deflections. If we denote the length of the spectrum by a straight line and draw pependiculars proportional to the readings of the galvanometer, we shall form, by connecting all these points, a continuous curve. The space it encloses is called a heat spectrum or thermogram. That shown in Fig. 12 is plotted as the mean of twelve sets of careful

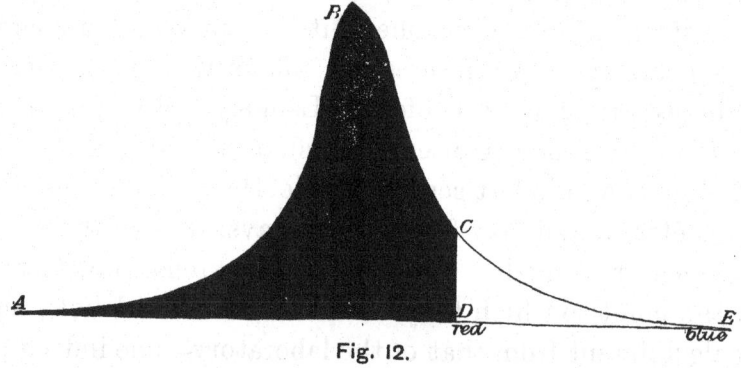

Fig. 12.

measurements made by Professor Tyndall.† It will be seen that the curve rises slowly as we recede from the violet until we reach D C; the

* Bakerian Lecture, 1881.   † Phil. Trans., 1868.

limit of the visible spectrum, then rapidly until it attains a maximum at B, whence it falls abruptly at first and afterwards fades imperceptibly away. The area of the dark space is 7·7 times that of the other, whence we infer that the energy of the invisible is nearly eight times that of the visible spectrum of the electric light.

In another experiment, Professor Tyndall endeavoured to imitate the action of aqueous vapour on the solar rays, by sending the electric beam through a layer of water of suitable thickness and then examining the spectrum of the emergent light. He found that the thermogram, obtained in this way, resembled that derived from the spectrum of the sun, the invisible calorific action being reduced by the water from eight times to twice the visible.

On the other hand, our atmosphere exercises a strong action on the rays of high refrangibility, completely intercepting, as M. Cornu has shown, the greater part of the violet and ultra-violet rays. It follows that if we could look at the sun from a point beyond our atmosphere, the totality of the solar radiation which we would then receive, would give us, as Professor Langley says, " a sensation to which we would affix the word ' blueness ' rather than ' yellowness ' or ' whiteness.' "*

From all this, we infer that the electric is much richer than the solar light which reaches us, especially in what has been called the chemical and calorific rays. In point of brilliancy, Foucault estimated it as half, while more recent investigations make it rather less than one-third that of the sun.

This richness of the electric light in rays of all wave lengths has been recently utilised by Dr. William Siemens in horticultural experiments. He has found that it is very favourable to the development of the green colouring matter of plants, that it quickens their growth and promotes the ripening of the fruit. It would also appear that, at least for short periods, plants do not require a rest during the twenty-four hours, but make rapid progress when subjected alternately to solar and electric light.† Further experiments are now being carried on to ascertain which rays of the spectrum are chiefly concerned in forming and developing the various constituents of the life-history of plants.

The spectrum of the arc is not of the same kind as that given by the carbons. It is not only discontinuous but consists only of a number of bright lines. This is precisely the spectrum peculiar to gases, whence

* The " Sun," by Professor Young—Appendix. † Proc. Roy. Soc., 1880.

we conclude that the voltaic arc proper consists of highly-heated carbon vapour. The bright lines are noticeable in the violet and ultra-violet; in the blue, they are specially numerous and brilliant. On comparison with known spectra, these " carbon lines," as they have been called, are found to coincide with those which are characteristic of acetylene and cyanogen, and accordingly Professor Dewar has shown* that these bodies are produced in the arc, the vapour of the atmosphere supplying the hydrogen and nitrogen necessary for their formation.

The electric light thus emits white light from the carbons and a bluish purple from the arc. The combination of the two gives it its characteristic tinge of bluish white.

* Proc. Roy. Soc., 1878.

THE MECHANICAL PRODUCTION OF ELECTRIC CURRENTS.

S an introduction to the description of the various magneto and dynamo-electric generators which have been devised, it will be valuable to lay down, in the simplest and most intelligible way, the principles which are concerned in the mechanical production of electric currents. Every one knows that electric lights are now produced from powerful currents of electricity generated in a machine containing magnets and coils of wire, and driven by a steam engine, or gas engine, or water-wheel. But of the thousands who have heard that a steam engine can thus provide us with electric currents, how many are there who comprehend the action of the generator or dynamo-electric machine ? How many, of engineers even, can explain where the electricity comes from, or how the mechanical power is converted into electrical energy, or what the magnetism of the iron magnets has to do with it all ? Take any one of the dynamo-electric machines of the present date—the Siemens, the Gramme, the Brush, or the Edison machine—of each of these there exist descriptions excellent in their way, and sufficient for men already versed in the technicalities of electric science. But to those who have not served an apprenticeship to the technicalities—to all but professed electricians—the action of these machines is almost an unknown mystery. As, however, an understanding of the how and the why of the dynamo-electric machine or generator is the very A B C of electrical engineering, an exposition of the fundamental principles of the mechanical production of electric currents demands an important place in the current science of the day. It will be our endeavour to expound these principles in the plainest terms, while at the same time sacrificing nothing in point of scientific accuracy or of essential detail.

The modern dynamo-electric machine or generator may be regarded as a combination of iron bars and copper wires, certain parts of the machinery being fixed, whilst other parts are driven round by the application of mechanical force. How the movement of copper wires and iron bars in this peculiar arrangement can generate electric currents is the point which we are proposing to make clear. Friction has nothing to do with the matter. The old-fashioned spark-producing " electrical machine" of our youthful days, in which a glass cylinder or disc was rotated by a handle whilst a rubber of silk pressed against it, has nothing in common with the dynamo-electric generator, except that in both something turns upon an axis as a grindstone or the barrel of a barrel-organ may do. In the modern "dynamo" we cannot help having friction at the bearings and contact pieces, it is true, but there should be no other friction. The moving coils of wire or " armatures " should rotate freely without touching the iron pole-pieces of the fixed portion of the machine. In fact friction would be fatal to the action of the " dynamo." How then does it act? We will proceed to explain without further delay. There are, however, three fundamental principles to be borne in mind if we would follow the explanation clearly from step to step, and these three principles must be laid down at the very outset.

1. The first principle is that the existence of the energy of electric currents, and also the energy of magnetic attractions, must be sought for not so much *in the wire* that carries the current, or *in the bar* of steel or iron that we call a magnet, as *in the space that surrounds* the wire or the bar.

2. The second fundamental principle is that the electric current, is, in one sense, quite as much a *magnetic* fact as an electrical fact; and that the wire which carries a current through it has magnetic properties (so long as the current flows) and can attract bits of iron to itself as a steel magnet does.

3. The third principle to be borne in mind is that to do work of any kind, whether mechanical or electrical, requires the expenditure of energy to a certain amount. The steam engine cannot work without its coal, nor the labourer without his food; nor will a flame go on burning without its fuel of some kind or other. Neither can an electric current go on flowing, nor an electric light keep on shedding forth its beams, without a constant supply of energy from some source or other.

The last of these three principles, involving the relation of electric currents to the work they can do and to the energy expended in their production, will be treated of separately and later. Meantime we resume the task of showing how such currents can be produced mechanically, and how magnetism comes in in the process.

Surrounding every magnet there is a " field " or region in which the magnetic forces act. Any small magnet, such for example as a compass needle, when brought into this field of force, exhibits a tendency to set itself in a certain direction. It turns so as to point with its north pole toward the south pole of the magnet, and with its south pole toward the north pole of the magnet; or, if it cannot do both these things at once, it takes up an intermediate position under the joint action of the separate forces and sets in along a certain line. Such lines of force run through the magnetic "field" from one pole of the magnet to the other in curves. If we define a line of force as being the line along which a free north-seeking magnetic pole would be urged, then these lines will run from the north pole of the magnet round to the south pole,

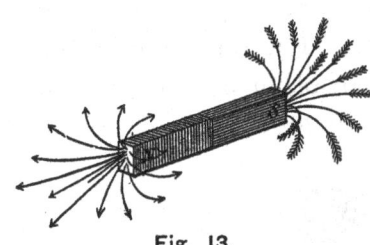

Fig. 13.

and pass through the substance of the magnet itself. In Fig. 13 a rough sketch is given of the lines of magnetic force as they emerge from the poles of a bar magnet in tufts. The arrow heads show the direction in which a free north pole would move. These lines of force are no fiction of the imagination like the lines of latitude and longitude on the globe; they exist and can be rendered visible by the simplest of expedients. When iron filings are sprinkled upon a card or a sheet of glass below which a magnet

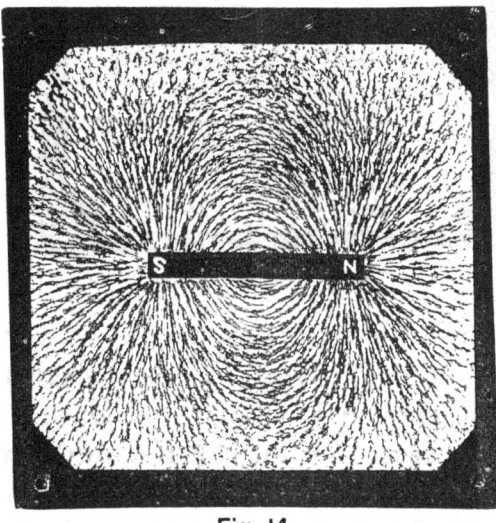

Fig. 14.

is placed, the filings set themselves—especially if aided by a gentle tap—along the lines of force. Fig. 14 is a reproduction from nature of this very experiment, and surpasses any attempt to draw the lines of force artificially. It is impossible to magnetise a magnet without also in this fashion magnetising the space surrounding the magnet; and the space thus filled with the lines of force possesses properties which ordinary unmagnetic space does not possess. These lines give us definite information about the magnetic condition of the space where they are. Their direction shows us the direction of the magnetic forces, and their density shows us the strength of the magnetic forces; for, where the force is strongest, there we have the lines of force most numerous and most strongly delineated in the scattered filings. To complete this first consideration of the magnetic field surrounding a magnet, we will take a look at Fig. 15, which reproduces the lines of filings, as they settle in the field of force opposite the end of a bar magnet. The repulsion of the north pole of the magnet upon the north poles of other magnets would be, of course, in lines diverging radially from the magnet pole.

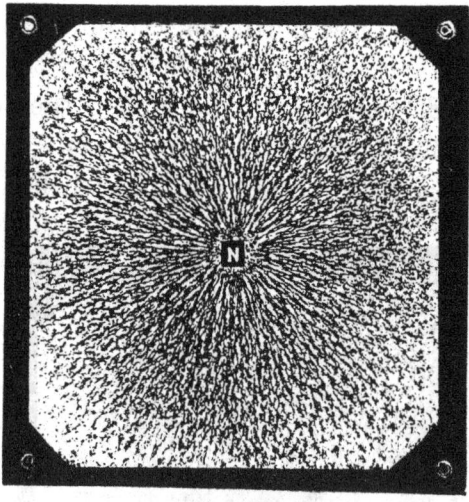

Fig. 15.

We will next consider the space surrounding a wire through which a current of electricity is flowing. This wire has magnetic properties so long as the current continues, and will, like a magnet, act on a compass needle. But the needle never tries to point toward the wire, its tendency is always to set itself broadside to the current and at right angles to it. The "field" of a current flowing up a straight wire is, in fact, not unlike the sketch shown in Fig. 16, where, instead of tufted groups, we have a sort of magnetic whirl to represent the lines of force. The lines of force of the galvanic field are, indeed, circles or curves which enclose the conducting wire, and their number is proportional to the strength of the current. In the figure, where the current is supposed to be flowing up the wire (shown by the dark arrows), the little arrows show the direction in which a free north pole would be urged round the wire;* a south pole would, of course, be urged round the wire in the contrary direction. Now, though when we look at the telegraph wires, or at any wire carrying a current of electricity, we cannot *see* these whirls of magnetic force in the surrounding space, there is no doubt that they exist there, and that a great part of the energy spent in starting an electric current is spent in producing these magnetic whirls in the surrounding space. There is, however, one way of showing the existence of these lines of force; similar, indeed, to that adopted for showing the lines of force in the field surrounding a magnet. Pass

Fig. 16.

the conducting wire up through a hole in a card or a plate of glass, as

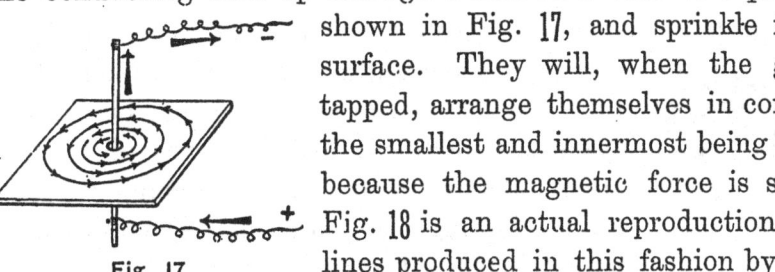

Fig. 17.

shown in Fig. 17, and sprinkle filings over the surface. They will, when the glass is gently tapped, arrange themselves in concentric circles, the smallest and innermost being the best defined because the magnetic force is strongest there. Fig. 18 is an actual reproduction of the circular lines produced in this fashion by iron filings in the field of force surrounding an electric current.

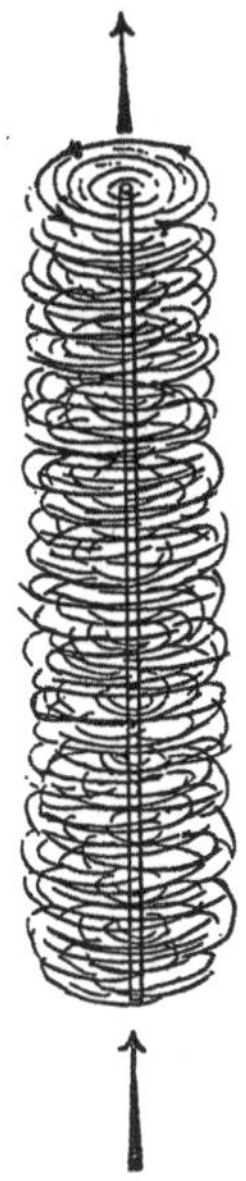

This experimental evidence must suffice to established two of the three fundamental points stated at the outset, for they prove conclusively that the electric current may be treated as a magnetic phenomenon, and that both in the case of the pole of a magnet, and in that of the wire which carries a current, a portion, at any rate, of the energy of the magnetic forces exists outside the magnet or the current, and must be sought in the surrounding space.

Having grasped these two points, the next step in our argument

---

* It will not be out of place here to recall Ampère's ingenious rule for remembering the direction in which a current urges the pole of a magnetic needle. " Suppose a man swimming in the wire with the current, and that he turns so as to face the needle, then the north pole of the needle will be deflected towards his left hand,"

is to establish the relation between the current and the magnet, and to show how one may produce the other.

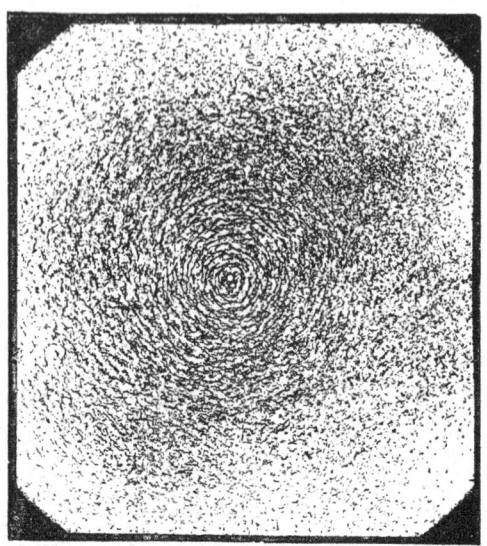

Fig. 18.

If we wind a piece of copper wire into a helix or spiral, as in Fig. 19,

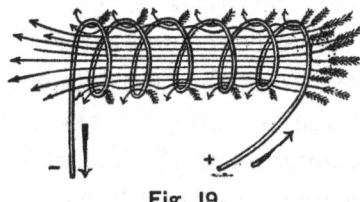

Fig. 19.

and pass a current of electricity through it, the magnetic whirls in the surrounding space are modified, and the lines of force are no longer small circles wrapping round the conducting wire. For now the lines of force of adjacent strands of the coil merge into one another, and run continuously through the helix from one end to the other. Compare this figure with Fig. 13, and the similarity in the arrangement of the lines of force is obvious. The front end of the helix acts, in fact, like the north pole of a magnet, and the further end like the south pole. If a small bar of iron be now pushed into the interior of this helix, the lines of force will run through it and magnetise it, converting it into an *electro-magnet*. The magnetic "field" of such an electro-magnet is shown in Fig. 20, which is reproduced from the actual figure made by iron filings. To magnetise the iron bar of the electro-magnet as strongly as possible the wire should be coiled many times round, and the current should be of maximum strength. This mode of making an iron rod or bar into a powerful magnet is adopted in every dynamo-electric machine. For, as will be presently explained, very powerful magnets are required, and these magnets are most effectively made by sending the electric currents through spiral coils of wire wound (as in Fig. 20) round the bars that are to be made into magnets.

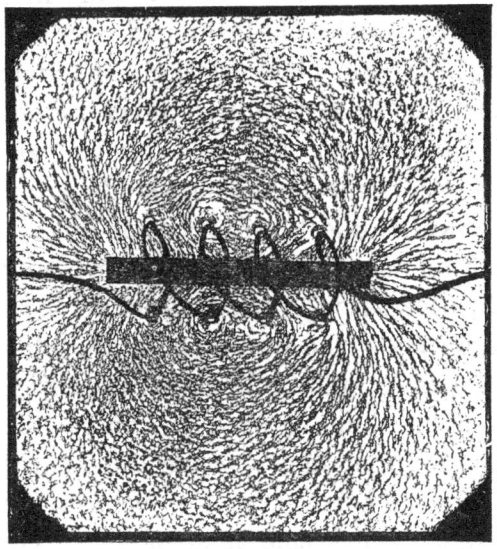

Fig. 20.

The reader will at this point probably be ready to jump to the conclusion that magnets and currents are alike surrounded by a sort of magnetic atmosphere, and such a view may help those to whom the subject is fresh, to realise how such actions as we have been describing can be communicated from one magnet to another, or from a current to a magnet. Nevertheless, such a conclusion would be both premature and inaccurate. Even in the most perfect vacuum these actions still go on, and the lines of force can still be traced. It is probably more correct to conclude that these magnetic actions are propagated through space, not by special magnetic atmospheres, but by there being movements and pressures and tensions in the *æther* which is believed to pervade all space as a very thin medium, more attenuated than the lightest gas, and which, when subjected to electro-magnetic forces, assumes a peculiar state, and gives rise to the actions which have been detailed in the preceding paragraphs.

We have now described how the electric current, or rather the wire through which it flows, possesses for the time magnetic properties, and is surrounded by lines of force ; and have further shown how magnetism can be produced from electric currents, a spiral coil of wire through which a current is sent acting like a magnet.

The next point to be studied is the magnetic property of a single loop of the wire through which an electric current flows. Fig. 21

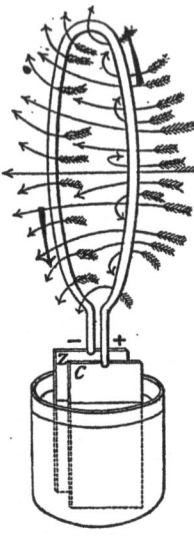

represents a single voltaic cell, containing the usual plates of zinc and copper dipping into acid to generate a current in the old-fashioned way. This current flows from the zinc plate through the liquid to the copper plate, and from thence it flows round the wire ring or circuit back to the zinc plate. Here the lines of magnetic force in the surrounding space are no longer only whirls like those drawn in Figs. 16 and 18, for they react on one another and become nearly parallel where they pass through the middle of the ring. The thick arrows show the direction of the electric current, the fine arrows are the lines of magnetic force, and show the paths along which a free north pole would be urged. All the front face, where the arrow-heads are, will

Fig. 21.

be like the north pole of a magnet. All the other face of the ring will be like the south pole of a magnet. Our ring resembles a flat magnet, one face all north pole, the other face all south pole. Such a magnet is sometimes called a "magnetic shell." *

Since the circuit through which the current is flowing has these magnetic properties, it can attract other magnets or repel them according to circumstances.

If a magnet be placed near the circuit, so that its north pole N is opposite that side of the circuit which acts as a south pole, the magnet and the circuit will attract one another. The lines of force that radiate from the end of the magnet, curve round and coalesce with some of those of the circuit. It was shown by the late Professor Clerk-Maxwell that every portion of a circuit is acted upon by a force urging it in such a direction as to make it enclose within its embrace the greatest possible number of lines of force. This proposition, which has been termed "Maxwell's Rule," is very important, because it can be so readily applied to so many cases, and will enable one so easily to think out the actual reaction in any particular case. The rule is illustrated by the sketch

---

* The rule for telling which face of the magnetic shell (or of the loop circuit) is north and which south in its magnetic properties is the following: If as you look at the circuit the current is flowing in the same apparent direction as the hands of a clock move, then the face you are looking at is a south pole. If the current flows the opposite way round to the hands of the clock then it is the north pole face that you are looking at.

shown in Fig. 22, where a bar magnet has been placed with its north pole opposite the south face of the circuit of the cell. The lines of force of the magnet are drawn into the ring and coalesce with those due to the current. According to Faraday's mode of regarding the actions in the

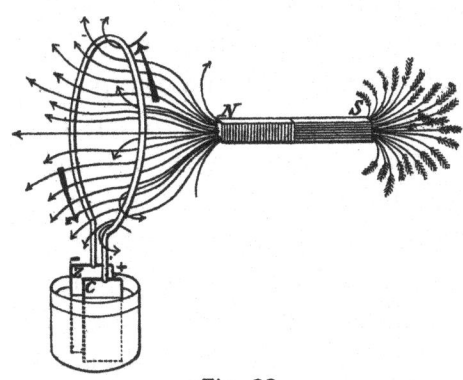

magnetic field there is a tendency for the lines of force to shorten themselves. This would occur if either the magnet were pulled into the circuit, or the circuit were moved up towards the magnet. Each attracts the other, and whichever of them is free to move will move in obedience to the attraction. And the motion will in either case be such as to increase the total number of lines of force that pass through the

Fig. 22.

circuit. Lest it should be thought that Fig. 22 is fanciful or overdrawn, we reproduce an actual magnetic " field " made in the manner described in the preceding article. Fig. 23 is a kind of sectional view of Fig. 22, the circuit being represented merely by two circular spots or holes above and below the middle line, the current flowing towards the spectator through the lower spot, and passing in front of the figure to the

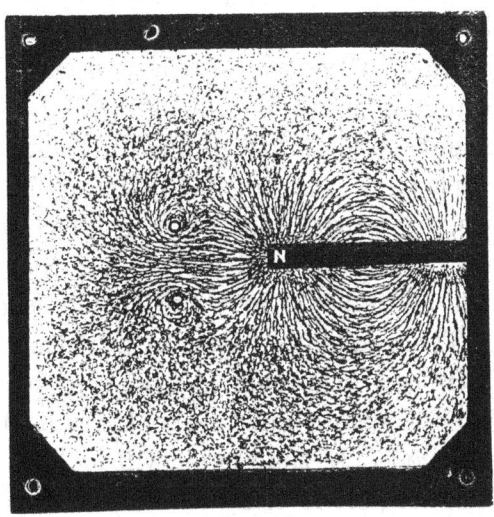

Fig. 23.

upper hole where it flows down. Into this circuit the pole N is attracted, the tendency being to draw as many lines of force as possible into the embrace of the circuit.

So far as the reasoning about these mutual actions of magnets and currents is concerned, it would therefore appear that the lines of force are the really important feature to be understood and studied. All our reasons about the attractions of magnets could be equally well thought out if there were no corporeal magnets there at all, only collections of lines of force. Bars of iron and steel may be regarded as convenient conductors of the lines of force; and the poles of magnets are simply the places where the lines of force run out of the metal into the air or *vice versâ*. Electric currents also may be reasoned about, and their magnetic actions foretold quite irrespective of the copper wire that acts as a conductor: for here there are not even any poles; the lines of force or magnetic whirls are wholly outside the metal. There is an important difference, however, to be observed between the case of the lines of force of the current, and that of the lines of force of the magnet. The lines of force of the magnet *are* the magnet so far as magnetic forces are concerned; for a piece of soft iron laid along the lines of force thereby becomes a magnet and remains a magnet as long as the lines of force pass through it. But the lines of force crossing through a circuit are not the same thing as the current of electricity that flows round the circuit. You may take a loop of wire and put the poles of magnets on each side of it so that the lines of force pass through in great numbers from one face to the other, but if you have them there even for months and years the mere presence of these lines of force will not create an electric current even of the feeblest kind. There must be *motion* to induce a current of electricity to flow in a wire circuit.

Faraday's great discovery was, in fact, that when the pole of a magnet is moved into, or moved out of, a coil of wire, the motion produces, while it lasts, currents of electricity in the coil. Such currents are known as "induced currents"; and the action is called magneto-electric "induction." The momentary current produced by plunging the magnet pole into the wire coil or circuit is found to be in the opposite direction to that in which a current must be sent if it be desired to attract the magnet pole into the coil. If the reader will look back to Fig. 22 he will see that a north magnet pole is being attracted in from behind into a circuit round which, as he views it, the current flows in an opposite sense to that in which the hands of a clock move round. Now, compare this figure with Fig. 24, which represents the generation of a momentary current induced by the act of moving the north pole N towards a wire

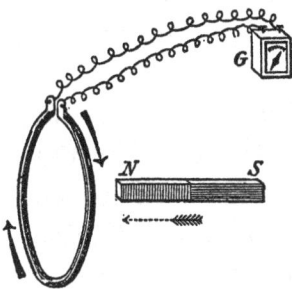

ring, which is in this case connected with a little detector galvanometer G. The momentary current flows round the circuit (as seen by the spectator from the front) in the *same* sense as the movement of the hands of a clock. The induced current which results from the motion is found, then, to be in a direction exactly opposed to that of the current that would itself produce the same movement of the magnet pole. If the north pole

Fig. 24.

instead of being moved towards or into the circuit, were moved away from the circuit, this motion will also induce a transient current to flow round the wire, but this time the current will be in the same sense as that in Fig. 22, in the opposite sense to that in Fig. 24. Pulling the magnet pole away, sets up a current in the reverse direction to that set up by pushing the pole nearer. In both cases the currents only last while the motion lasts.

Now, on a previous page it has been pointed out that the lines of force of the magnet indicate not only the direction, but the strength of the magnetic forces. The stronger the pole of the magnet is, the greater will be the *number of lines of force* that radiate from its poles. The strength of the current that flows round a circuit is also proportional to the number of lines of force, which are thereby caused to pass (as in Fig. 21) through the circuit. The stronger the current, the more numerous the lines of force that thread themselves through the circuit. When a magnet is moved, with relation to a circuit near it, it is found that any alteration in the number of lines of force that cross the circuit is accompanied by the production of a current. Referring once more to Fig. 22, we will call the direction of the current round the circuit in that figure the *positive* direction; and, to define this direction, we may remark that if we view the circuit from such a point as to look along the lines of force in their own direction, the direction of the current round the circuit will appear to be the same as that of the hands of a clock moving round a dial. If the magnet N S be now drawn away from the circuit, so that fewer of its lines of force pass through the circuit, experiment shows the result, that the current flowing in circuit will be, for the moment, increased in strength, the *increase* in strength being proportional to the rate of *decrease* in the number of lines of force. So, on the other hand, if the magnet were pushed up toward the circuit, the current in the circuit would be momentarily reduced in strength, the decrease in strength in the current being proportional to the rate of increase in the number of lines of force.

Similar considerations apply to the case of the simple circuit and the magnet shown in Fig 24. In this circuit there is no current flowing so long as the magnet is at rest; but if the magnet be moved up toward the circuit so as to *increase* the number of lines of force that pass through the circuit, there will be a momentary "inverse" current induced in the circuit, and it will flow in the *negative* direction. While if the magnet be moved away, the *decrease* in the number of lines of force would result in a transient "direct" current, or one flowing in the *positive* direction.

It would be possible to deduce these results from an abstract consideration of the matter from the point of view of the principle of conservation of energy. But we prefer to reserve this point until a general notion of the action of dynamo-electric machines has been given.

The following principles or generalised statements follow as a matter of the very simplest consequence from the foregoing considerations :—

(*a*) To induce a current in a coil of wire by means of a magnet there must be relative motion between coil and magnet.

(*b*) Approach of a magnet to a coil, or of a coil to a magnet, induces currents in the opposite direction to that induced by recession.

(*c*) The stronger the magnet the stronger will be the induced currents in the coils.

(*d*) The more rapid the motion, the stronger will be the momentary current induced in the coils (but the time it lasts will, of course, be shorter).

(*e*) The greater the number of turns in the coil, the stronger will be the total current induced in it by the movement of the magnet.

These points are of vital importance in the action of dynamo-electric generators. It remains, however, yet to be shown how these transient and momentary induction currents can be so directed and manipulated as to be made to combine into a steady and continuous supply. To bring a magnet pole up towards a coil of wire is a process which can only last a very limited time; and its recession from the coil, also cannot furnish a continuous current, since it is a process of limited duration. In the earliest machines in which the principle of magneto-electric induction was applied, the currents produced were of this momentary kind, alternating in direction. Coils of wire fixed to a rotating axis were moved past the poles of a magnet. While the coil was approaching, the lines of force were increasing, and a momentary inverse current was set up, which was immediately succeeded by a momentary direct current as the coil receded from the pole. Such machines, on a small scale, are still to be found in opticians' shops for the purpose of giving people shocks. On a large scale alternate-current machines are still employed for many purposes in electric lighting, as, for example, for use with the Jablochkoff candle. Large alternate-current machines are made also by Wilde, Gramme, Siemens, De Meritens, and others.

It will have been observed that in all the actions which we have been describing, whereby electric currents are generated in coils of wire, by expending mechanical power in moving them past the poles of magnets, or in moving the poles of magnets past the coils, it is the *relative motion* of magnet and coil that is concerned. We have insisted upon explaining this action by observing the lines of force in the surrounding space, and laid down the rule that the strength of the current so induced in the circuit, was proportional to the rate of decrease in the number of lines of force that cross through the circuit. It matters not whether the magnet move towards the coil or the coil towards the magnet, the result is the same, an increase in the lines of force threading through the coils, a decrease in the potential energy of the system, and the production of a momentary inverse current. In the actual construction, however, of almost every kind of dynamo-electric machine, the magnets are fixed while the coils move. This is merely for the sufficient reason that the coils are usually lighter than the magnets, and, therefore, it would be absurd, from a mechanical point of view, to make the moving parts heavier than the fixed parts. In the generators of Pacinotti, Gramme, Wilde, Siemens, Brush, and Edison, the magnets are fixed while the coils rotate. In Lontin's dividing machine, and in a few other generators, however, there is an exception. The magnets attached to a central axis revolve, while the numerous coils in which it is desired to generate currents are fixed upon an external frame.

The next step in explaining how the production of continuous currents in one direction is accomplished in a typical generator, will be

E

a consideration of the effect of moving a ring of wire up to a magnet and passing it over its poles.

In Fig. 25 is shown a single ring or coil of wire to serve as a circuit, and beyond it a magnet is shown with its north pole facing the ring. From what has been already laid down, it is easy to say what will be the result of moving the ring along towards the magnet. While the ring approaches the north pole, as in the first case of the figure, the lines

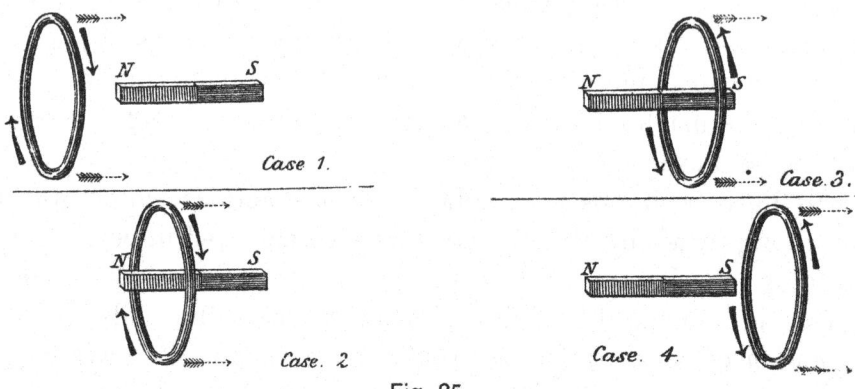

Fig. 25.

of force that thread through the ring will be increasing in number, and there will be an *inverse* current (*i. e.*, one which is in a direction inverse to that which would cause the pole N to be attracted; compare the figure of case 1 with Fig. 22, and note the direction of the arrows). Moreover, even after the ring has been moved as far as the pole N there will still be an inverse current so long as the ring by moving forward is increasing the number of lines of force that it embraces. Comparison with Fig. 16 and Fig. 23 will show that there will be an increase in the number of lines of force until the ring has got as far as the centre or neutral line of the magnet. The second case of Fig. 25 represents matters up to that point. But now if the forward motion of the ring be continued towards the pole S and beyond it, the number of lines of force that thread themselves through the ring will diminish, and therefore from the moment that it passes the middle point of the magnet, a direct current will be induced in it, as shown by the arrows in cases 3 and 4, and this will continue as long as the ring recedes from S.

Again, suppose we move our ring or coil between the two powerful magnetic poles, as indicated in Fig. 26. We will suppose that our coil is moving in a circular path between the poles S and N, which are respectively a south and a north pole. The lines of force will run

across in almost straight lines from N to S. We will consider what
occurs in the coil as it moves round the upper half of the circular path,
its first position being on the left. At first it lies horizontal with its
plane very nearly along the lines of force, so that almost none of them
pass through it. As it rises and turns round, the number of lines of force
that thread through the coil, increases until it arrives at c^1 where the
number that pass through is a maximum. All this time an inverse
current will, therefore, be generated in it. After passing c^1, and
although it is getting nearer N, the number of lines of force that
cross through the coil are decreasing, because it is gradually turning
round flat-ways again. From c^1 onwards there will, therefore, be
a direct current induced in the coil. Nor does the direct current
cease when the coil arrives at the point nearest N, for now, though
not shown in Fig. 26, as the coil goes along the lower semi-circle

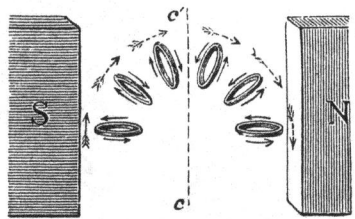

Fig. 26.

the lines of force will begin again to thread through it, but this time
they enter the other side of the coil to that which they formerly entered.
There will, therefore, be an increase in the number of lines of force
applied negatively, which is, of course, equivalent to a still further
decrease. This will go on until the coil arrives at c, where the negative
increase is a maximum. All the way round the right-hand half of
the figure, from c^1 to c, there will be a direct current induced in the
coil. From c onward and round the left-hand half of the figure up
to c^1 there will be an inverse current generated. To render such a
current available, therefore, as a continuous supply, some device is
needed to turn the inverse currents on the left-hand half so that they
shall flow into the wires—that go, say, to feed an electric light—in
the reverse way to that in which the direct currents flow into them.
For if we *invert* that which is already *inverse* we thereby bring it into
the *direct* relation. The device for accomplishing this end is called a
commutator, and the line $c\,c^1$, which marks out the boundary between

E 2

the currents that are direct and those that are inverse, is known as the
" diameter of commutation."

In the earliest and simplest generators or magneto-electric
machines, where but one rotating coil was employed, the commutator
consisted of a simple piece of brass or copper tube slit longitudinally into
two portions, and fixed upon the axis of revolution so as to revolve
with it, the two halves of the split tube being fixed upon a small cylinder
of ivory, wood, or other insulating material. One-half of the tube was
connected with one end of the wire of the coil, the other half with
the other end. Against the split tube pressed two contact springs, or
"brushes," to take off the currents. Suppose that the arrangements
of the magnets and of the rotating coil was such that, while the armature
was rotating in the direction shown in Fig. 27, the induced currents

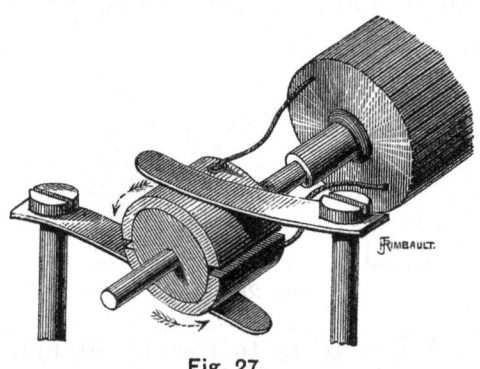

Fig. 27.

in the coils flow toward the commutator in the upper half of the
revolution, and flow from the commutator in the lower half of the
revolution, then each half of the split tube will, as it passes over the
top of the axis, deliver to the upper contact-spring or brush the current
flowing into it, while the lower contact-brush will always be feeding the
return currents back into the lower half of the revolving commutator.

In the modern generators, however, it is not enough to merely
commute the alternating currents of a single revolving coil. A number
of small currents flowing along a circuit one after the other, even though
they all flow in the same direction, do not constitute a continuous flow if,
as in the case considered above, there is a momentary interruption of the
circuit when the brushes or contact-springs pass from one half of the
split-tube commutator to the other. And even if this change of contact
were instantaneous, there would not be a uniform flow, because when the

coils approach and pass through the position of commutation, inductive action in them has almost ceased. Each of the two currents generated in the coil during one revolution begins gently, rises to a maximum and then dies away again, the commutation being effected at the period when the current has died out. To procure a continuous and steady current then, it is necessary to combine in one armature a large number of coils, so arranged that they shall be brought successively into operation; the current beginning in one coil before it has ceased in another. The most perfect result would be attained if it were possible to have some of the coils always passing through the point of maximum efficiency, and others coming up to take their places, so that the enforced idleness of the coils as they pass through the position of commutation should be compensated for by the activity of others, which at that moment are in the position where the induction of currents is at a maximum.

Such compound armatures may be classified, for the sake of convenience, under four typical forms, according to the structure upon which the many coils are wound. These four typical forms are : (A) *Ring* armatures; (B) *Drum* armatures; (C) *Pole* armatures; and (D) *Disc* armatures.

(A) *Ring armatures.* The suggestion to wind a large number of coils of wire round a ring, which by its rotation should bring first one coil and then another into the magnetic field, is due originally to Paccinotti, whose machine will be illustrated in its appropriate place in this work.

In 1870, and without any knowledge of Paccinotti's work, M. Gramme devised the well-known ring armature that bears his name. Ring armatures are also employed in the more recent generators of Brush, Bürgin, Shuckert, and others. It will be sufficient here to take one general case, and show how the principles laid down above are applied and developed.

Fig. 28 is a skeleton diagram of the working parts of a Gramme dynamo-electric generator; it is drawn from a model devised by Professor Silvanus Thompson, of Bristol. The ring armature is the principal feature. It rotates between the poles S and N of an electro-magnet of a U form with vertical limbs. The commutator consists of a number of separate bars or strips of copper fixed round the periphery of the axis, while above and below are the contact-brushes which touch the uppermost and lowermost pieces of the commutator respectively. In the model, the ring is enwrapped by a continuous helix of twelve turns; each one of these twelve turns being connected with one of the twelve copper strips of the commutator. In the drawing the armature is supposed to rotate in the same direction as the hands of a clock. The separate turns or coils in the upper half of the ring armature are therefore moving from S toward N, while those of the lower half are moving from N towards S. As in Fig. 26, so here, there will be direct currents generated in all the coils

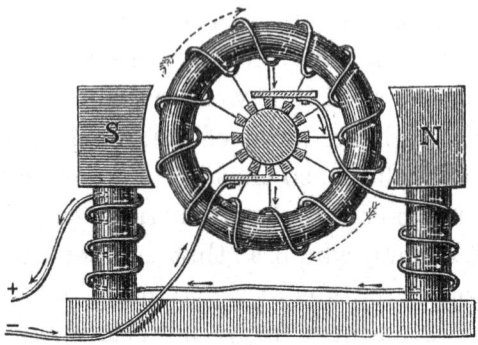

Fig. 28.

that are on the right-hand of the diameter of commutation $c\ c^1$, and inverse currents in all the coils on the left. The little arrows show the directions of these currents. By following out the various turns of the spiral it will be seen that all the separate currents induced in the individual turns of the spiral on the left of $c\ c^1$ are in the same direction along the wire, and, therefore, all help one another in producing a strong current, though the electromotive force in each may not be equal. A row of horses hitched on to one another in tandem fashion all facing one way, all help to produce one strong pull, though the individual horses may not be equally strong. In these coils, as shown in Fig. 28, the strongest induced electromotive force is, as a matter of fact, in the coils that are passing closest to the poles. In the right-hand half of the spiral the individual currents are all flowing in the other direction. Up each side of the ring armature, therefore, there is a strong current flowing, drawing electricity from the lowest point and urging it towards the topmost point. Now the bars of the commutator, which are here seen end-on in a ring surrounding the axis, are all separate and insulated from one another and from the axis. If electricity were to flow into one of them it would have to flow back again, for there is nowhere for it to flow to; the bars of the commutator lead nowhere. This is, however, not the case for the two bars of the commutator that occupy the highest and lowest points. They touch against the contact brushes which communicate with wires of the external circuit. So the electricity that flows up both right and left of the ring armature can flow into the upper contact brush and thence into the leading wire of the circuit, which therefore has a constant current urged through it. A corresponding action goes on at the lower contact brush; from this point electricity is drawn both to the right and to the left, being supplied by the return wire of the circuit. The current thus furnished is practically continuous; for, though the contact is made first against one bar of the commutator and then against another, as fast as they arrive at the line $c\ c^1$, it is clear that the currents on the two sides of the ring armature will always be flowing towards that point at which the contact is made. Neither is there any breach in the continuity of the current generated, because one bar of the commutator does not cease to touch the brushes until the next has come up to contact, and because there always are a certain number of turns of the spiral, both on the right and on the left, in which the elementary currents are being induced. In the actual dynamo-electric generators, the coils of the armatures are, of course, made up of many turns of wire. In the common Gramme ring, for example, there are hundreds or thousands of turns, and these are grouped in twenty-four sections, the wire at the end of each section being joined to the wire at the beginning of the next, and the junction connected directly with one of the twenty-four bars of the commutator.

The commutator itself presents the appearance shown in Fig. 29. Similar commutators (or " collectors ") are used in almost all the principal

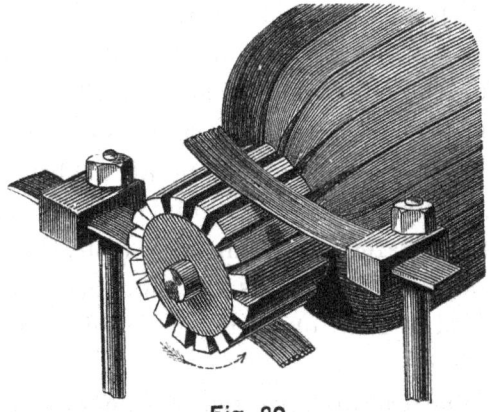

Fig. 29.

forms of generators, Gramme, Siemens, Edison, Bürgin, in fact all the machines that furnish " direct " continuous currents, excepting only the

Brush, in which the peculiar manner of connecting the coils involves a special kind of collector.

(B) *Drum armatures.* Instead of winding the coils of wire round a ring they may be wound lengthways over a drum or spindle, the wire being carried along the drum parallel to its axis, across the end, back along the drum on the side opposite, and so round to the starting point, the separate turns or groups of turns being spaced out at regular intervals all round the drum. This method of winding is illustrated by a sketch of a very simply-wound skeleton drum armature, Fig. 30. Here there are

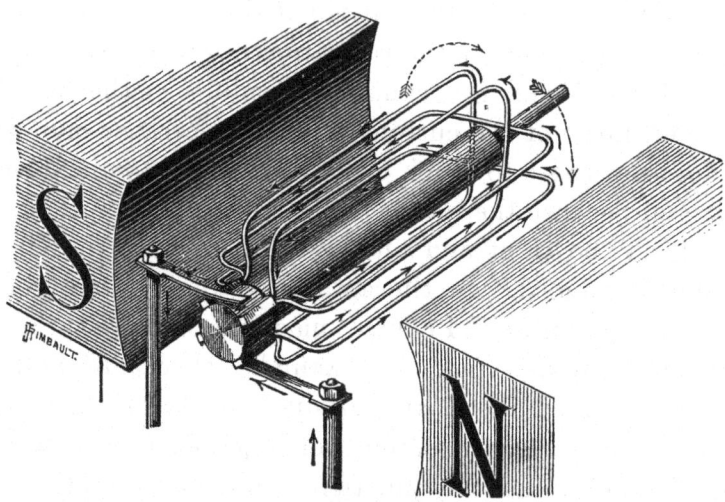

Fig. 30.

but four turns of coil. Each coil begins at one of the copper bars of the commutator, passes along the drum, across the back end, returns to the front end, and is there joined to the next adjacent bar of the commutator to that from which it started. The coils of this armature, when rotated in the magnetic field, cut the lines of force, each separate loop or coil having currents induced in it as it passes round. In each of the wires as it rises past the S pole, currents are generated which flow towards the front, whilst in the other half of their revolution in descending past the N pole the currents generated in them flow from the front towards the back The method of joining the coils to the commutator bars ensures that the currents shall follow one another and flow into the upper contact-brush.

The first drum armature was invented by Siemens of Berlin, and in its present developed form was devised, in 1872, by Herr von Hefner Alteneck, of the firm of Siemens and Halske, but the method of coupling up the coils then adopted was somewhat complicated, and is described elsewhere in detail. A sectional view of one form of this armature is given in Fig. 31.

Fig. 31.

The coils here, are wound along a wooden cylinder fixed upon steel spindles. Drum armatures have been employed also by other constructors, notably by Edison, who in his gigantic generator has replaced the wire coils by rods or bars of solid copper connected across the ends by copper discs or washers. The drum armature is more readily constructed and wound than the ring armature, but is less compact. In Maxim's and some other generators, an armature is employed, which is a kind of hybrid between a ring and a drum armature; for, though the wires are wound along a drum, the drum is itself a cylinder of iron, through the interior of which the wires return, being wound round it as in the ring armatures.

(C) *Pole armatures.* The third type of armature is that in which the rotating coils are wound upon radial bars of iron, as upon the arms of electro-magnets. Fig. 32 gives a scheme of this arrangement

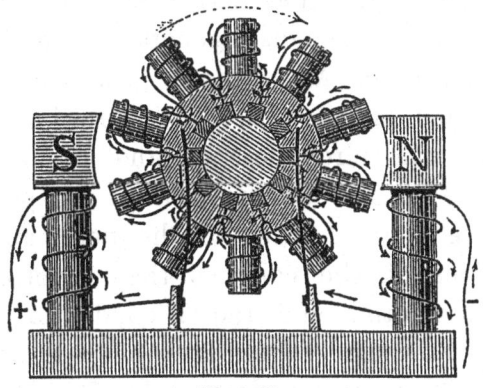

Fig. 32.

and shows a ten-pole armature with its commutator. The greatest possible number of lines of force pass through the coils when the pole on which they are wound lies in the same direction as the lines of force of the magnetic field, or horizontally. Hence in this case the "diameter of commutation" is horizontal, and the brushes must touch the commutator on the right and left sides. Consider a pole placed on the extreme left. As it rises and moves over the top, the number of lines of force that run through it and thread through the coils wound round it, diminishes, and there will be currents induced in it which will be of maximum strength as it nears the top. From this position onward, the number of lines of force will increase, but with negative sign, until the pole is on the extreme right. During the lower half of its revolution the action is reversed. The result is a flow of currents as before into one contact brush and out of the other. Pole armatures were first employed by Pixii and Saxton, but only singly or in pairs. An early representative type, in which a larger number of them had a multiple commutator, is that made by Niaudet, who, however, set the poles in a ring round a central wheel. The radial arrangement due to T. Allan, of Edinburgh, in 1852, was modified and developed later by Lontin. The Wallace-Farmer generator has a pole armature, but with the poles arranged as in Niaudet's machine. The Weston armature occupies an intermediate position between the pole armature and the drum armature, the iron drum being built up of a number of discs with teeth or projections around their peripheries, between which projections the coils are wound, each projection acting in succession as a pole, and arranged somewhat like that patented by Pulvermacher in 1849.

(D) *Disc armatures.* Of the disc armature there are but few examples, one being the Arago-disc generator, and another being an experimental armature of Mr. Edison. In each case the coils are wound in the plane of the disc, the wire being carried for a certain distance along the circumference and then turning radially toward the centre. The disc armature must be placed so that its axis is parallel to the lines of force of the field, and these are cut by the radial portions of the coils as they revolve. In Edison's armature the circumferential and radial conductors are substantial bars of metal. The particular advantages offered by this form are not yet ascertained. A generator devised by Dr. Hopkinson has an armature intermediate in character between the disc and the pole form, the flat poles being combined as sectors of a disc.

Up to this point nothing has been said specifically about the best way of making the magnet poles S and N, as powerful as possible. All our later arguments simply treated N as the north pole and S as the south pole of a strong magnet; and, in the earliest machines for producing currents thus by the mechanical movement of coils in the presence of magnets, the magnets used to produce the requisite " field " of force were ordinary magnets of hard steel. Hence the name of "magneto-electric machines," assigned to the older instruments invented by Pixii, Saxton, Holmes, Nollet, and Siemens. Hence also arose the name " magneto-electricity " for the currents produced by induction in these machines ; for it was at first thought by some, that the electric currents that were generated in these older machines, and which consisted of rapidly alternating currents, were due to some kind of electricity other than that which flowed smoothly and continuously out of a voltaic battery. But we know now that electricity, whether the source be frictional, voltaic, magnetic, or thermal, is all the same. The term magneto-electricity is therefore happily dropping out of use. A little later, machines were made in which permanent steel magnets were replaced by electromagnets. Generators on this principle were constructed by Wilde and by Ladd. A small magneto-electric machine with steel magnets furnished a current which was used to excite the field magnets of a large machine. Then followed the famous quadruple independent discoveries of Hjorth, Varley, Siemens, and Wheatstone, that there was no need of a separate exciting machine, but that the generator might serve as its own excitor. Suppose the magnet whose poles are S N, to be *very weak*, the currents induced in the coils of the rotating armature will also be very weak. But lead them round the magnet in a spiral, and they will exalt its magnetism, and if it become more powerful, it will react on the coils and induce more powerful currents. So, without any permanent steel magnet to begin with, with only masses of iron rendered (as all masses of iron on the earth are) feebly magnetic by the action of the magnetism of the earth, this " action and reaction " principle serves when the armature is set rotating, to raise the field magnets, in a few instants, to the highest degree of magnetisation which they are capable of acquiring. The model (Fig. 28) shows this, for the wire attached to the upper brush, into which the currents generated in the armature are continually flowing; is wound round the upright limb of the soft iron field magnet on the right in a spiral, and then crosses to the left and forms another spiral round the other limb, before it passes out to convey the current into the leading wire of the circuit. Such a machine differs only from a magneto-electric machine in the particular that it uses its own current to magnetise its own magnets ; and for the sake of distinction machines of this kind have been called *dynamo-electric machines*. But the fact is that all generators, whether they thus supply themselves with gnetism or not, are as truly magneto-electric in their action as if they worked with magnets of permanent steel; for, as we have seen, there must be in all of them a *field of magnetic lines of force* in which the coils move. All the generators are in this sense "magneto-electric." Moreover, in all these generators the energy of the currents is supplied by the mechanical energy expended in producing rotation of the armature. It is not the magnets that supply the constant flow of power. If of steel they are just as powerful at the end as at the beginning. If they are of iron they are no weaker at the beginning than at the end. It is harder work to push the coils across the magnetic field when there is the current flowing in them, than would be the case if there were no current, and this extra work the steam engine has to do. The work so done is not, however, lost, as is the work done in overcoming friction : it reappears as the energy of the currents of electricity thereby generated. The *mechanical power* of the steam engine is the real source of the electric power, and therefore all the generators, whether their magnets be soft iron or hard steel, are in this sense "dynamo-electric."

In concluding this section let us recapitulate the various points in our argument concerning the mechanical production of electric currents :

Firstly.—Magnetic forces must be studied as they exist in the magnetic field or space surrounding the magnet.

Secondly.—A wire that carries a current also possesses a magnetic field in the space surrounding it, and the stronger a current is, the stronger will be its field, and the more numerous the lines of force that thread through the circuit.

Thirdly.—If you alter the strength of the field of a current that flows in a wire, by moving a magnet pole in that field, you alter the strength of the current in the wire.

Fourthly.—If there be no current in the wire to begin with, and you move a magnet near the wire (thereby altering the number of lines of force in the field round the wire), the motion of the magnet will evoke a current in the wire.

Fifthly.—Increasing the number of magnetic lines of force that cross a circuit, induces in that circuit a transient inverse current ; while decreasing the number of magnetic lines of force that cross the circuit, induces in that circuit a transient direct current.

Sixthly.—When coils of wire rotate across the lines of force of a magnetic field, the currents induced in the coils during half their journey are direct; during the other half inverse.

Seventhly.—By using a suitable commutator, all the currents, direct and inverse, can be turned into the same direction in the wire that goes to supply the currents to the external circuit.

Eighthly.—The current of the generator may be utilised to magnetise its own magnets by being made to flow round them in spiral coils, thereby magnetising them far more powerfully than any permanent steel magnet can be magnetised.

Ninthly.—In every case the energy of the electric currents generated is supplied by the mechanical power employed to drive the generator.

To sum up the matter still more briefly we will add a few words. An electric current flowing in a circuit of wire may be regarded as a magnetic whirl in the space surrounding the wire. If, then, by moving the coil of wire past a magnet we set up magnetic whirls in the space surrounding the coil, we set up electric currents in the wires themselves. Dynamo-electric generators are machines for mov ng coils of wires past the poles of magnets, there being special arrangements, *firstly*, to procure the setting up of very powerful magnetic whirls around the coils of wire, and therefore of very strong electric currents in the wires themselves; and, *secondly*, to turn all these currents into on direction so as to flow in one steady stream through the circuit.

Having explained as clearly as possible "how electric currents can be produced mechanically, and how magnetism comes in the process," it remains now to explain the relation of electric currents to the work they can do, and to the energy expended in their production.

We laid down. as a fundamental principle, that to do work of any kind, whether mechanical or electrical, requires the expenditure of an equivalent amount of energy. And that just as a steam engine cannot work without using fuel, or a labourer without food, so an electric current cannot go on flowing, nor an electric light keep on shedding forth its beams, without a supply of energy from somewhere or other.

Now although, as already explained, we use magnets in order to generate currents of electricity in rotating coils of wire, a magnet is not in itself a source of power. It will do no work for us until we have done an equal amount of work on it. We must pull its keeper away from it before it can pull the keeper back and do work. It is just the same with other forces. An iron weight is not in itself a source of power. It will do no work for us—it will not even drive a clock—until we do some work on it. We must lift it (and spend our energy in the process) from the ground, working our muscles to overcome the pull of gravity before it can do the work of driving the wheels of the clock in its descent. In generating electric currents from magnets in the manner explained in the preceding pages, we, or our steam engines, have to supply the necessary energy. We spend this energy in moving something in opposition to a resisting force. This something happens to be a coil of wire, or a combination of such coils. The force happens in this case to be a magnetic force, and the result of the motion happens to be (by the particular arrangements of the coils and magnets) the setting up of magnetic whirls round the wire, or what we otherwise call an electric current in the wire. But it is we (or our steam engines) that do the work. Take a parallel case in mechanics: house bells are often arranged so that you pull the bell against a resisting spring which when you leave go, flies back and rings the bell. Now here the spring possesses a certain power called elasticity, and were it not for this it would not fly back and ring the bell. But the elasticity does not of itself do anything. The spring is not in itself a source of power. It is only when we do work in overcoming its resisting force that it reacts or does work for us; and all the time it is we who supply the energy.

Again, an electric current while it lasts represents a certain amount of active energy. A heavy railway truck shunted along a railway line by the push of a powerful engine represents, while it moves, a certain amount of active energy, for if it crashed against a wall it might knock it down. But presently friction brings it to rest. The electric current also comes rapidly to rest if left to itself, being stopped in the fraction of a second by the resistance it meets with in the metal wire through which it flows. To keep the truck speeding on mile after mile requires a continuous expenditure of power by an engine. To keep the electric current flowing steadily hour after hour requires a continuous expenditure of energy in the generator.

This brings us back to the whole question of the mechanical production of electric currents in the dynamo-electric machine, and to the point how the mechanical energy of a rotating axis (derived from burning coal in a steam engine) gets converted into the electrical energy of a current flowing invisibly along a wire, which seemingly is quite a different sort of power.

We must again point out that the energy of a current is to be conceived as existing in the space *outside* the wire, quite as much as in the wire itself; that the current itself is always accompanied by magnetic whirls in the space round the wire, and that if we can only set up and maintain magnetic whirls in the space surrounding the wire, we set up and maintain currents in the wire itself. Now the way to set up magnetic whirls is to move magnets. And as we cannot set up or maintain magnetic whirls without expending energy, it is clear that we shall have to spend energy in moving the magnets.

Let us go back a little way in our study, far enough to take up again the case of the action of an electric current on a magnet. Referring to Fig. 22, it will be remembered that this diagram illustrated the very important experimental fact that where a magnet and an electric current flowing in a circuit act on one another they tend to move so that the number of lines of force that thread through the circuit, increases to a maximum.

Now the lines of force of a magnet have a definite direction of their own, and the positive direction along them is that in which a free north pole would be urged. A free south pole would be urged, of course, in the opposite direction along the lines of force. The action of our simple voltaic circuit (see Figs. 21 and 22) upon a free north pole would be to suck it in from behind, and to drive it right through the circuit, as the arrows show, along the lines of force to an infinite distance in front. The action upon a free south pole would be just the reverse; it would be sucked in from the front and driven out behind. If we began with a single free north pole at an indefinitely great distance behind the circuit, it would be gradually drawn up, and more and more of its radiating lines of force would be drawn into the embrace of the circuit. Imagine yourself looking towards the circuit from the point of space occupied by this ideal magnet-pole; as the magnet-pole is drawn up nearer and nearer you see the circuit subtending an ever-increasing space in the horizon of your vision, just as a sixpenny piece stuck against a window pane covers up more and more of the landscape as you bring your eye nearer and nearer to it. Think of the circuit as the base of a cone having the magnet-pole at its apex. As the pole is drawn nearer the apex grows blunter; its solid angle is ever increasing. Now it can be shown mathematically that the *work* done by the circuit in drawing up the pole in this way from an indefinitely great distance, is proportional to the solid angle thus subtended by the circuit. This rule is as important in its way as is the rule that the force with which the circuit draws the pole, is proportional to the number of lines of force that are drawn into the embrace of the circuit.

We must now apply one of the first principles of dynamics to the case. *Potential energy always tends to run down to a minimum.* Suppose we have done work by winding up a spring, or by lifting a weight to the top of an inclined plane, or by pulling away from a magnet the iron keeper that was attached to it. In each case the work done is *stored up* as "potential energy;" and in each case there is a tendency for the energy so stored to run down. Our spring will uncoil itself; our heavy weight will descend; our magnet will forcibly draw its keeper back; and each of these actions will result in the work of producing motion. To pull our magnet-pole back out of the circuit to which it is attracted, will require an expenditure of work; it will pull back like a spring; but when we leave go, its potential energy will tend to run down and it will be drawn up again into the circuit.

It is clear then that by pushing the magnet-pole up towards the circuit we diminish the mutual energy of the system, while by pulling the magnet-pole back from the circuit or coil of wire, we increase the mutual energy of the system.

But now arises the question: in what part of the system is it that the energy is thus increased or diminished? Is it in the magnet-pole, or in the circuit, or in the space between them? The magnet-pole is certainly not the seat of the change, for its strength remains the same in all the operations. The change of energy must, therefore, be either in the strength of the current or in the strength of the "field" of magnetic force that surrounds it. And as you cannot change one without changing the other, we conclude that both are affected.

The application of this very simple principle leads us, therefore, inevitably to the conclusion that when we push the pole up nearer to the circuit, (and diminish the potential energy of the system) this very action diminishes—so long as it lasts—the strength of the current in the wire coil that forms the circuit; and that on the other hand when we pull back the magnet-pole out of the circuit (and increase the potential energy of the system) we thereby increase for the time the strength of the current in the circuit.

And if the original current flowing round the coil were *nil?* What then? The argument still holds good, but with this important difference. Seeing that the lines of force of the magnet are the only ones originally belonging to the system, the only way in which they can increase is by a current starting in the circuit, and the only way they can diminish is by a counter-current starting in the circuit. This is precisely the result which experiment proves to take place. For, as shown on a previous page, when a magnet-pole is pushed up into a coil, it produces a momentary inverse current, and when it is drawn away, it produces a momentary direct current in the coil.

To produce powerful currents by a dynamo-electric generator it would therefore appear that arrangements must be made for the revolving coils to cut as many lines of force as possible in the shortest possible time. This implies, (1) that the armature should rotate very rapidly; (2) that it should be placed in a position where the field of magnetic force is very strong; (3) that it should consist of many turns of wire, each enclosing as much area as possible; and (4) it should offer a very small resistance to the current. These conditions are not all equally attainable at once. We will examine them separately.

(1). *The Armature should rotate very rapidly.*—The strength of the induced current being proportional to the rate of decrease in the number of lines of force that traverse the circuit, the advantage of rapid rotation is obvious. In the case of those machines in which the armature revolves in a magnetic field of constant strength (as in the case where the field magnets are of hard steel), the strength of the current is almost exactly proportional to the number of revolutions per second. But if the armature be very heavy, or the driving machinery unequal to its work, there is a limit to the speed that can be maintained with efficiency. For example, a small Siemens generator may be driven at a speed of 2,500 revolutions per minute, and yield a current of about 21 Ampères, while in contrast to this, Edison's large dynamo-electric generator exhibited, at the Paris Exhibition, produced a current of 900 Ampères with a speed of only 350 revolutions per minute.

(2). *The Armature should be placed in a very strong field of magnetic force.*—It is for this end that the field magnets are constructed to concentrate the lines of force as much as possible across the space

F

where the armature revolves. In the case of the Gramme ring the iron in the armature itself aids in intensifying the field of force. Edison has recognised the advisability of making the field magnets of his dynamo-electric generators on a more massive scale. In the large generator mentioned above, the field magnets weigh nearly 10 tons, and consist of seven cylindrical coils, 8 ft. long, terminating in enormous blocks of soft iron as cheeks.

(3). *The Armature should consist of many turns of wires, each enclosing as much area as possible.*—Each turn of the wire adds to the inductive effect, for if each cuts across the lines of force of the field, there will be an electromotive force set up in each. If there are a hundred turns of wire in the coil that is moved, there will be a force urging a current forward a hundred times as great as if the coil consisted of but one turn. It does not follow that the whole current will be a hundred times as great, because the hundred turns will offer a hundred times more resistance to the flow of a current (supposing the wire to be of the same thickness as the " one turn "). Practice only can dictate the choice of the thickness of the wire and the number of turns. Where currents of very great electromotive force are desired a thin wire of many turns is right ; but where (as in electro-plating) it is desired to have a great quantity of current with a low electromotive force, the turns are few and a very thick wire is used. Again, if the area enclosed by the turns is large, a great many lines of force can pass through them ; but to make room for such large turns, the magnet-poles must be wide apart, and this makes the " field " weaker, as the lines of force are not so concentrated as before. It is here again a matter for experience to determine. One point may be noticed, however, namely, that if we try to double the power of a coil by doubling the diameter of the turns, instead of doubling their number, we enclose four times as great an area. In Edison's monster generator already referred to as furnishing the extraordinary current of 900 Ampères, it is not desired to raise the electromotive force much above 100 Volts, and the coils are, therefore, comparatively few. And instead of being made of stout copper wire they are constructed of solid copper bars bolted round the periphery of a drum, so as to offer an excessively small resistance to the currents.

(4). *The coils of the armature should offer a very small resistance to the current.*—The importance of this rule is obvious to any one who is acquainted with Ohm's law. The waste of energy occasioned by the heating effect of the current on every part of the circuit that offers resistance to the flow, is a sound economic reason for keeping down the useless resistance of the armature.

Lastly, of these considerations respecting the relation of the electric current to the mechanical energy spent in producing it, we would recall the all-important law of Joule, which is that the energy of a current is proportional to the square of its strength. If you can by any means double the strength of a current you will increase fourfold its power to do work. This law resembles the law in mechanical science, to the effect that the energy of a moving body (or its *vis viva*, as the old books used to say) is proportional to the square of its velocity. An example will illustrate the point. Suppose a cannon ball, weighing 1 lb , to be shot out of a gun with a velocity of 100 ft. per second. A certain quantity of energy, in the form of the explosive activities of the gunpowder, must be expended on it to give it that velocity, and by virtue of its velocity and its weight it is capable of doing a certain amount of destructive work. To shoot *two* such cannon balls, at the same rate of 100 ft. per second, will obviously require the expenditure of twice as much gunpowder, and the two balls will do twice as much damage in the end. Now, instead of sending *two* cannon balls at the same rate of 100 ft. per second, let us send *one* cannon ball at the rate of 200 ft. per second. Will the same quantity of gunpowder suffice? Will the effective work be the same? No, indeed. You will find that you want a *fourfold* charge of gunpowder to import a *double* velocity; and that by virtue of the double velocity the destructive work will also be increased *fourfold.* Treble the velocity, while the weight moved remains the same, and you increase its effective energy ninefold; but then you will require a ninefold charge of powder to produce that trebled velocity. The energy of a moving body is proportional to the *square* of the velocity with which it moves.

Now, in like manner, the energy of the electric current is proportional to the square of the strength of the flow, that is to say, is proportional to the square of the rate at which electricity is conveyed through the circuit. And, as the heating effect of a current is proportional to its energy, it follows that the heat produced by a current is proportional to the square of the strength of the current.

F 2

Again, it has been shown repeatedly by experiment that, in a generator in which the magnets are either permanent, or excited by a separate current, the strength of the current through a circuit of constant resistance, is proportional to the speed of the machine, that is to say, to the velocity with which its armature rotates. Now, by the foregoing principle it is certain that the mechanical work done in driving the armature will be proportional to the square of the velocity of the rotation. It is, therefore, intelligible that the energy of the current should be proportional to the square of the strength of the current, if the current be proportional to the speed. For, by the principle of the conservation of energy, the work which the generator will do can never be more than the work done by the engine in driving it, and should be precisely equal to it in amount, if there were none lost in the process.

V.

THE THEORY OF DYNAMO-ELECTRIC GENERATORS.

IN the following pages, now presented to the reader, there is developed a theory of the action of dynamo-electric machines of a somewhat novel character. It is based upon the experimental fact that all these machines are reversible, that is to say, that they can work either as generators of electric currents when driven by mechanical power, or as mechanical motors when supplied with electric currents. The main outlines of the treatment here adopted are derived from some important researches of M. Antoine Breguet, of Paris. Starting from first principles and simple experimental data, we shall propound as successive links in a logical chain of reasoning, the various steps by which Faraday's discovery of magneto-electric induction is carried to its fullest development in modern dynamo-electric machinery. We shall be enabled by this means to clear up a number of points hitherto regarded as comparatively obscure in the operation of such machines. We shall, for example, be enabled to explain a point hitherto supposed to be irreconcilable with theory, namely the practice of electrical engineers in setting the " brushes " or collectors of the current in the dynamo-electric machines in an oblique position against the commutator, and unsymmetrically with respect to the field-magnets of the instrument. We shall also be able to point out how this angular position of the brushes is affected by the question whether the machine is to be used as a generator or as a motor. The real part played by the iron core of the rotating armature is also elucidated, some light is thrown upon the different methods of winding up the coils of the armatures, and a means is provided of judging of their respective merits or demerits.

The first principle to be laid down is that of the essential reversibility of all dynamo-electric and magneto-electric machines. Whether they have ring armatures like the well-known generators of Gramme and Brush, or drum armatures like those of Siemens and Edison, or pole armatures like those employed in the Lontin machine and in some of the alternate-current machines of Gramme and Siemens, still they are all reversible.

Rotated by mechanical means they supply a current of electricity derived from the energy of a steam engine or other motor. But if you supply them with a current of electricity they will, conversely, rotate and turn the electricity back into mechanical energy. The same thing is true of all electo-magnetic engines or electro-motors, such as those of Elias Page, and Froment. They are intended to rotate by electricity, but if you rotate them by mechanical means they will furnish in turn a current of electricity. This principle of reversibility extends even to some unsuspected cases. Very early on in the history of magnetism, Barlow found he could cause a wheel or disc of copper to rotate between the poles of a magnet by sending a current at the same time perpendicularly through the disc from the axis to the circumference, where it passed into a pool of mercury arranged to make electric contact with as little friction

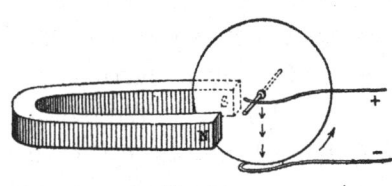

Fig. 33.

as possible (Fig. 33). In 1831, Faraday performed the converse experiment, and found that by mechanically rotating a copper disc between the poles of a magnet, he thereby generated a current in a wire, the two ends of which touched the circumference and the axis of the wheel which were

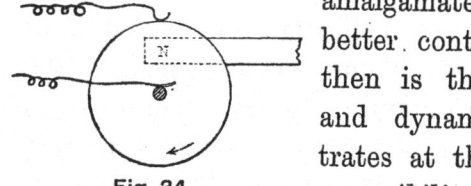

Fig. 34.

amalgamated over with mercury so as to insure better contact with the wire (Fig. 34). Here then is the simplest kind of electro-motor and dynamo-electric generator, and it illustrates at the outset the important principle of *reversibility.*

The second matter which claims notice is the mutual interaction of a magnet and a conductor which carries a current. We sometimes speak loosely of the displacement of a current caused by a magnet, when we mean that the conductor carrying the current is displaced. Although the term is convenient, it is scarcely accurate; for we must distinguish between mechanical force, or that which tends to move matter, and electromotive force (so called), which tends only to move electricity in a conductor. The mechanical reactions between magnets and current conductors which turn machines, are obviously of the former class. These reactions, therefore, deserve to be studied from a nearer point of view, by applying a principle enunciated by Faraday concerning magnetic attractions, and as we have already seen, extended by Professor S. P. Thompson to the case of the attraction between currents. Faraday, as we have already seen, first recognised the significance of the so-called lines of magnetic force, which are seen crossing in curves through every magnetic field when iron filings are sprinkled over it. Without necessarily attributing to these lines any physical existence, we may conveniently employ them as Faraday did, to investigate and to describe the actions between magnets and magnetic bodies. Faraday laid down the following properties as those possessed by these lines of force: Firstly, the lines of force tend to shorten themselves. Secondly, lines of force lying in the same direction side by side repel one another. To these M. Breguet added that a line of force when it passes through iron or other metal capable of magnetic susceptibility, must be regarded as if shorter than one of equal actual length passing through air, so that the " tendency to shorten " may exhibit itself by a tendency to run through a magnetic substance near at hand.

By means of these simple principles, Faraday was able to deduce the laws of magnetic attraction and repulsion from the figures formed by the lines of iron filings when sprinkled over the magnetic field whose properties were thus to be investigated.

One of the figures obtained by Professor Thompson (*vide* Fig. 35) enables us to study the action of Barlow's wheel; and this figure M. Breguet takes as the basis of his theory of the dynamo-electric machines. The two square spots show the poles of the magnet, and a point a little way from them represents a metallic conductor perpendicular to the plane of the figure, and traversed by a current which passes downwards through the round spot. This current produces a magnetic " field " all round it, which if the magnet were not present would consist of lines of force disposed in concentric circles. But in presence of the magnet and its radiating lines of force there is a mutual reaction, the nature of which can be learned by simply looking at the figure formed by the iron filings. The tendency of the lines to shorten would assuredly urge the conductor towards the poles of the magnet; and in Barlow's wheel, where this goes on continually in the part of the disc lying between the axle and the mercury cup, the simple attraction becomes a movement of rotation. A current passing in the opposite direction through the wire would obviously be urged in the contrary direction, or away from the magnet. Conversely, if the conductor were

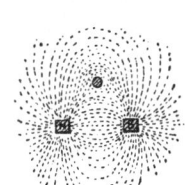

Fig. 35.

mechanically moved further away from the magnet, a current would be generated of opposite direction to that which caused the motion, the electromotive force of the current being proportional to the number of lines cut by the conductor in a second of time.

Following M. Breguet we next pass to a consideration of the first and simplest of electro-motors, as shown in Fig. 36. This

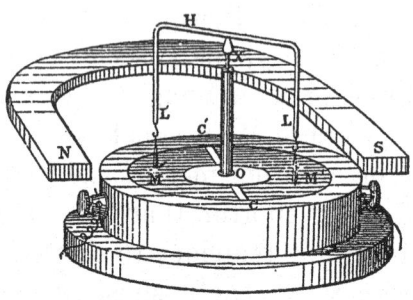

Fig. 36.

apparatus consists of a metallic conductor bent twice at right angles and balanced on a centre at a point X. The vertical branches carry little metallic jointed appendages which dip into the halves of a circular mercury cup divided across by a diametral partition into two portions connected respectively with the poles of a battery; and the whole is placed between the poles N S of a magnet so fixed that the line joining the two poles is at right angles to the diametral line. The conductor rotates upon its centre so long as the current passes. We may consider separately the action of the magnet upon the two vertical portions L L¹ and on the horizontal portion H. The action upon the two

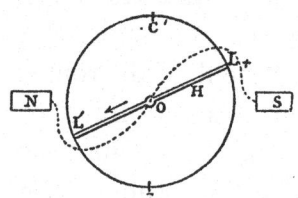

Fig. 37.

vertical portions of the conductor is best studied by taking a plan of the apparatus, as shown in Fig. 37, which gives in diagrammatic outline all the working parts. The arrow indicates the direction of the current, which therefore ascends at L and descends at L¹. Comparison between the conditions which here exist, and those which gave the magnetic figure, Fig. 35, with iron filings, will show a series of lines of force, of which the most characteristic will be the S-shaped curve shown by the dotted lines. Hence the tendency will be to displace L¹ towards the top of the figure and L towards the bottom, and their positions of equilibrium will be respectively C¹ and C   But if the inertia of movement carry the contact-breakers past these points and make them touch the opposite mercury cups, the current will be reversed, L being drawn towards C¹ and L¹ towards C. Hence there will be a continuous rotation, the direction of the current in the conductor being reversed at every half revolution, when the moving wire passes through the position of equilibrium at C C¹. A little consideration will show that the action upon the horizontal portion H is similar, and adds itself to the forces producing rotation.

Now suppose that instead of the simple wire bent twice at right angles, a conductor be taken having the vertical branches prolonged into two arcs, and carrying, as in Fig. 38, four little jointed contact pieces.

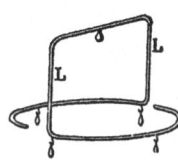

We shall now require the mercurial cups to extend over 90 deg. of arc, *i.e.*, extending from C¹ only as far as S, and from C only as far as N. Had we taken eight little contact pieces, each mercury cup need only have occupied one-eighth of the circle, or 45 deg. of arc. If the number were indefinitely increased, then the arcs subtended by the mercurial contact cup might be diminished down to mere points at C¹ C. These considerations hold equally good in the converse cases where the instrument is used as a generator of currents.

Fig. 38.

One further step remains to be considered before we pass on to the application of these matters to the Gramme machine. Instead of the single wire we may take a wire coiled upon a frame, as in Fig. 39, and having many turns. On each wire of this coil there will be a similar action, hence the total force of rotation will be proportionately greater when an equal current is used. In all the

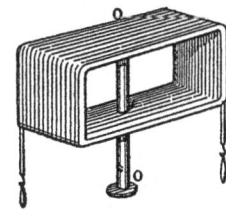

various arrangements hereafter to be described, every single wire may be considered in a similar way to represent a coil, and hence the figures may be made as simple as possible. It will be noticed that the single flat coil of Fig. 39, if wound upon an iron spindle and frame, virtually constitutes an armature of the type introduced by Siemens, and applied by him in the early magneto-electric machines, and also

Fig. 39.

recently employed by M. Marcel Déprez and others in the construction of electro-motors.

We have now to pass on to a more detailed consideration of Siemens dynamo-electric generator, the joint invention of Dr. Werner Siemens and of Herr Friederich von Hefner Alteneck. The special feature in it, the armature, with its peculiarly wound coils of which we shall presently speak, is due to the genius of the latter gentleman, and is in consequence sometimes spoken of as "Alteneck's armature" to distinguish it from the earlier and simpler longitudinal armature with cross section like a double-headed T employed in the older Siemens machines.

Not to anticipate, however, we must refer the reader to the very simple machine shown in Fig. 36, in which a single wire bent twice at right angles, is made to rotate electro-magnetically between the poles of a horseshoe magnet. A current enters this wire by a mercury cup at one side, and leaves it by a similar cup at the other; the direction of the current through the wire being automatically reversed at every half revolution as the wire swings round, thus alternately attracting up the wire towards the pole of the magnet, and repelling it as it passes away on its circular path.

Suppose next that four such wires are suspended around the same pivot, and disposed so as to make equal angles with one another, each extremity of each wire being provided as before with a small contact-piece (Fig 40). What are now the conditions of rotation? A little

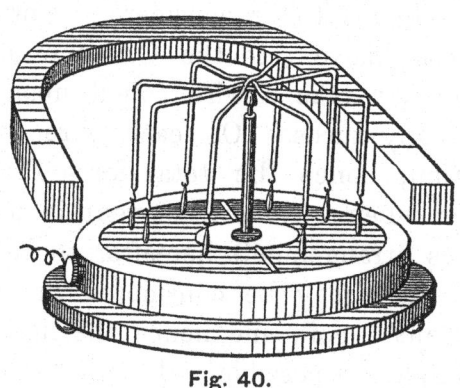

Fig. 40.

consideration will show us that if the mercury cups extend as before, and as in Fig 41, to 180 deg. on each side, the arrangement will be in every way inferior to the one formerly considered; for the current is now shared between four conductors, in each of which there will be but one-quarter of the total current, and they cannot all be situated in the most advantageous position of the field where the attraction or repulsion of the magnet is the greatest. Moreover, some of them will positively be pulled forward while others are pulled backward. So, though the current is just as strong as before, the altered distribution of the current is wholly disadvantageous, and the apparatus is both worse and heavier than before. Suppose, however, that the mercury cups are diminished till they subtend, as in Fig. 42, angles of but 90 deg., only two of the

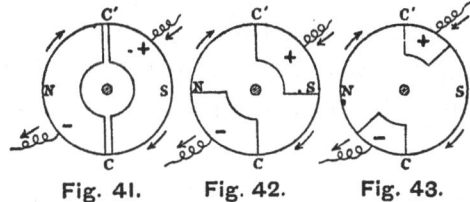

Fig. 41.   Fig. 42.   Fig. 43.

conductors can dip simultaneously into the cup, and one will enter it just as another leaves it. Hence half the current will now traverse each; and if the sectorial mercury cups are judiciously placed so that their edge of first contact lies along $C\,C^1$, which we may call the "diameter of commutation," at right angles to the line joining the magnet poles, the two conductors in contact can never be far from the point where the attractive force of the magnetic field will have the largest effective leverage upon them. Better still will the machine be if the mercurial sectors are still further diminished, as in Fig. 43, down to 45 deg. Now only one conductor can dip in at once, but it will take the whole current, and as it passes out of contact the next will pass in. The increased weight of the fourfold conductor over the single bent wire of the first arrangement is more than compensated for by the advantage of always having in the most advantageous part of the field, that conductor which is being acted upon. Instead of simply receiving an impulse once every half turn, the impulses come eight times during every revolution, and the rotation will therefore be much more uniform.

Once more let us remember what we pointed out previously, namely, that every single wire we have considered may be replaced by a coil of many strands, and we shall realise the advance now made towards an efficient electro-motor.

Here, also, we may pause to note that, if by the principle of reversibility we use mechanical power to rotate this system of conductors (or "armature" as we may call it), as it lies in the magnetic field, we shall obtain induced currents. And this new generator will possess corresponding advantages over the simpler arrangement that had only one bent wire, inasmuch as it will give both a stronger and a steadier supply of electricity; there being now eight successive currents generated during one revolution instead of two as before, and these currents will be more powerful, since the conductors in which they are generated are being moved through the most advantageous region of the field.

It is of course unnecessary to suppose the contacts to be made by cups of mercury, except that for light experimental bits of apparatus, the freedom from friction thus gained is of service. If the eight ends of these wires were brought down and soldered to a metallic collar on the axis, the collar being slit into eight separate parts, each of which successively came into contact with metallic brushes occupying the position relatively of the sectorial mercury cups, the same end would be attained, though with a little increased friction. With a larger number of conducting wires, the number of segments of the metallic collar or commutator would be increased, and their angular width proportionally diminished. This is in fact the kind of " commutator " which is used upon almost all the dynamo-electric generators in use.

We are now prepared to discuss the various systems of winding a Siemens armature, to which under the heading of drum armatures we have briefly referred on page 56. In attempting to explain its nature, M. Breguet devised a simpler system, so closely resembling it that he at first thought it identical. He took a single stout wire and bent it up in the manner represented in Fig. 44, the two ends being soldered to one

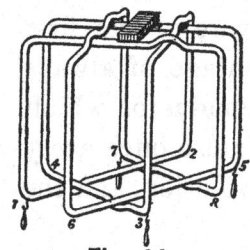

Fig. 44.

another, and the separate bends so insulated as not to touch one another where they cross. Four little contact pieces were added at the corners 1, 3, 5, and 7, to dip into mercury cups like those of the preceding apparatus, but with quadrantal sectors, as shown. Now here the whole current obviously traverses each branch of the conductor, and the rotation will take place with more than redoubled energy; for the impulses will last during a whole quarter of a revolution, and there will always be two wires attracted, and two repelled by each pole of the magnet. The same system is represented diagrammatically in Fig. 45, where, however, a metallic collar slit into eight portions is supposed to replace the little contact pieces used for dipping into mercury cups. Of course more than eight vertical wires might be employed. Any regular polygon having an even number of sides would answer; but the octagon with star-like points is

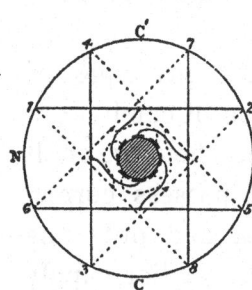

Fig. 45.

very simple and effective, and answers all purposes. As before, each single wire of the simple experimental apparatus may be replaced by a coil of many turns; and the whole may be wound upon a longitudinal cylinder or core.

The actual method of winding up the cylindrical drum armature of a Siemens generator—the method invented by Hefner Alteneck—is represented in a similar diagram in Fig. 46; from which it will be seen that the arrangement adopted has 16 vertical conductors, and that it is an unsymmetrical one.

Curiously enough, a German engineer, Herr Frölich, who, like M. Breguet, intended to describe the Siemens (or Alteneck) armature, discovered another system of winding up the wires, which is shown in diagram in Fig. 47. Here also there are 16 vertical conductors arranged in pairs at the points of a regular octagon, and crossing the

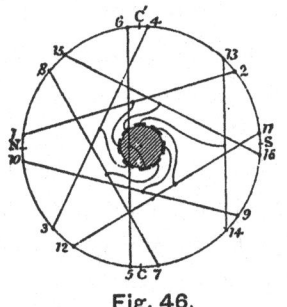

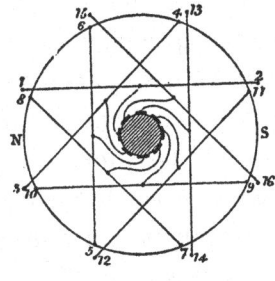

Fig. 46.   Fig. 47.

octagon by its diagonals at one end of the armature and by long chords crossing in the form of an eight-pointed star at the other.

There are, in fact, a large variety of ways of winding coils upon a longitudinal armature, of which that adopted by Siemens is one, and not the best one. M. Breguet says he has found no fewer than eight. Of these there is one which appears to be better than all the others. It is represented in Fig. 48. Its superiority consists in requiring the employment of a shorter length of wire to attain the same effects; which of course means not only a reduction of first cost, but a saving in the wasteful heating effects of internal resistance. In this system the portions of the coils which cross the ends of the armature to unite the 16 vertical wires cross the octagon along short chords. As these end portions of the coils contribute little or nothing to the effective

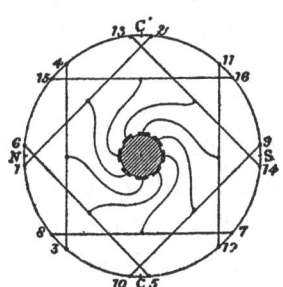

Fig. 48.

work of the motor or generator it is a clear advantage that they should be as short as possible. M. Breguet has given the following as a table of the proportional lengths of wire necessary to make up end portions of equal-sized armatures on the four systems :—

System of Herr Frölich (Fig. 47).... .... 30·8
  ,, Alteneck and Siemens (Fig. 46).... 30·5
System devised by M. Breguet (Fig. 45) .... 28·4
  ,, ,, ,, (Fig. 48) .... 26·0

From which it appears that the last described arrangement is superior to all the others. In Edison's newest generators, the drum armature resembles that of the Siemens machines in many respects. But the longitudinal coils of wire are replaced by longitudinal copper bars, arranged along the periphery of the drum parallel to its axis, and each connected at one end to one bar of the commutator, and at the other end to the longitudinal bar that lies on the opposite side of the cylinder. The method of connecting is therefore symmetrical, but for further details the reader is referred to the special notices of the Edison generators.

There only remains one point to add, namely, that the power of all these machines can be increased not only by increasing the number of turns of wire, and arranging them as advantageously as possible, but by increasing the intensity of the magnetic field in which they move by introducing iron cores into the very middle of the coils.

We have now to apply the foregoing principles in developing a theory of the Gramme machine, and we shall in conclusion give some further considerations of extreme importance upon a fact known to all practical electrical engineers, but hitherto completely unexplained by theory, namely, that it is necessary to displace the commutator brushes of dynamo-electric machines from the position of symmetry, the angular displacement varying with the velocity of rotation, and with the work which is being done by the machine.

Suppose, firstly, that a wire bent in the form shown in Fig. 49, was placed upon the apparatus described in the preceding pages and illustrated in Figs. 36 and 41. Under what conditions can it rotate ? If a current passes through this wire, it will ascend branches 1 and 3 and descend

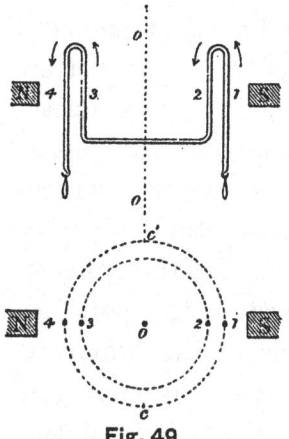

Fig. 49.

branches 2 and 4. It is clear that there will be a tendency to move 2 in an opposite direction to 1; hence the forces tending to drive the conductor round will be two couples acting in opposite senses, urging 1 and 4 in one direction, and 2 and 3 in the other direction round the axis. The couple 2 and 3 will be, however, the weaker of the two, since the length between 2 and 3 is less, and they are further removed from the most intense part of the magnetic field than are 1 and 4. The total force of rotation will clearly be the difference of the two opposing couples; and the arrangement, as it is, is evidently worse than the arrangement given in Fig 41. The rotation of conductors 1 and 4, is just *pro tanto* hindered by the opposing forces on 2 and 3. If only the action of the magnetic field on these two branches could be reduced to nothing, the effective force of rotation will obviously be increased. Now there is one way, and but one, to *screen off* the magnetic field from the branches 2 and 3, and that is by interposing a screen of iron in the form of a ring. To understand fully how the iron ring can act as a magnetic screen for the wires within it, it is needful to comprehend the nature of magnetic substances in general in respect of their behaviour as screens.

We have already spoken of magnetic lines of force, as Faraday defined them, and their properties. The most important of these properties in the present connection is that they pass by preference through a magnetic substance, which, so to speak, conducts them better. Iron is about a million times as magnetic as air, hence we find that the lines of force in a magnetic field are very greatly altered in form by the presence of a mass of iron, as they have a tendency to so arrange themselves that they may run as far as possible through iron and as little as possible through the air or the surrounding space. For example, if Fig. 50 represents the magnetic "field" between two poles N and S, in which also a hollow cylinder of iron has been placed, it is found, by the method of sprinkling iron filings over a card laid in the field, that instead of assuming the usual simple arcs passing across from one pole to the other, the lines of force are bent about in a remarkable manner. They curve round so as to meet the ring, travel along in the substance of the iron as far as possible, then emerge at

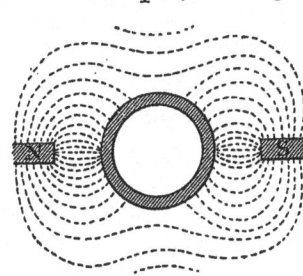

Fig. 50.

the other side to curve round sharply into the other pole. *The entire space within the ring is destitute of lines of force* and is thus screened off from magnetic influences. The interior of a hollow ball of iron is in like manner screened from all external magnetic influences; a property of which advantage was taken by Sir W. Thomson in the construction of certain galvanometers specially designed for use on board cable-laying ships. The cylinder of iron in Fig. 50 would serve equally well as a screen to the interior portions whether at rest or in rotation round its axis, for, even if rotating, the lines of force would prefer to pass through the iron rather than cross the interior air-filled space. The external field might be somewhat deformed in symmetry during the rotation, in consequence of iron requiring *time* to part with its magnetism, but this would not affect the interior space were no magnetic forces are.

It will be convenient for our purpose to consider the effects produced in the magnetic field by a cylinder of iron so short that it may practically be considered a *ring*. This is shown in section in Fig. 51. Here we may notice several points. First the peculiar grouping of the lines of force.

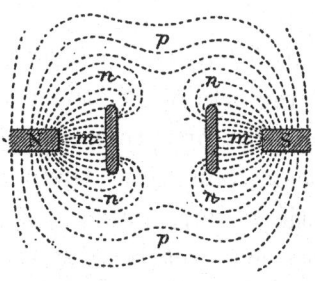

Fig. 51.

Most of them in the intensest parts of the field run straight from the pole to the ring, thence round through the iron of the ring invisibly, and emerge again at the opposite side. These groups of lines, marked *m*, indicate that the magnetic force is concentrated into that part of the field immediately opposite the poles. Nextly, there are certain groups *n* of less intensity which pass above or below the edges of the ring, to curve round into it on the inner edge, and which in like manner run round the substance of the ring and emerge at the opposite side. Lastly, there are some outlying lines of force marked *p* which pass by above and below, and which are of no importance. It is particularly to be noticed that no lines cross the interior of the ring to go from one pole to the other, so that the ring still acts as a screen.

Now suppose such a ring to be placed within the bent conductor of Fig. 49, the arrangement will assume the form shown in Fig. 52, where branches 1 and 4 of the conductor are exterior to the ring, 2 and 3 interior. Clearly, as far as 1 and 4 are concerned, they are more advantageously situated than before, for the iron ring intensifies the portion of the magnetic field in which they are situated. As to the

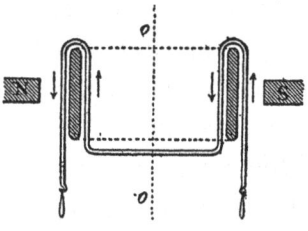

Fig. 52.

branches 2 and 3, their condition is now wholly changed. They are chiefly screened from magnetic actions, the only lines that cross them being those of the groups '*b*, but these lines, instead of being merely lines of force running across from N to S, are lines which actually curve round and cross them as if coming from S to N. Consequently the magnetic forces on 2 and 3 act in the reverse sense to what they did when there was no iron ring, and the force of rotation acting on 2 and 3 now tends to spin them round in the *same sense* as 1 and 4. The model illustrating this beautiful theoretical demonstration of the part played by the iron ring, was shown to the Société de Physique by M. Breguet, at one of its meetings during 1879, and its performance was so perfect as to leave nothing to be desired. A further modification shown in Fig. 53, having

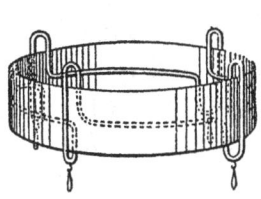

Fig. 53.

four equidistant turns, is a still nearer approach to the Gramme ring. It only requires that a coil of many turns should be substituted for each single fold of wire, and that the number of such coils upon the iron ring should be increased, to obtain the true parallel. In that case, however, the commutator pieces would be made of small arc and proportionately numerous, as described on a

preceding page. Up to this point we have treated the matter as if the Gramme armature were intended to rotate as in an electro-motor, converting an existing electric current into motion. The principle of *reversibility* already laid down by us shows, however, that all the considerations advanced apply equally well to the case in which the motion of the armature in the field is employed to generate a current of electricity. As is well known, the Gramme machine, besides being an admirable generator, is capable of being advantageously used as an electro-motor. It matters not, moreover, from a theoretical point of view, whether or not the iron ring is fixed, or rotates with the coils of wire upon it; though for practical reasons the latter is of course always the case.

We are now prepared to discuss the question whether the Gramme armature with its ring, or the Siemens armature with its longitudinally wound coils, is the more advantageous. It will be admitted that the only effective portions of the coils of the Siemens armature are the wires parallel to the axis of rotation, and that those portions which cross the ends radially, are comparatively useless, as they cut few lines of force, and add to the total resistance. Also, in the Gramme armature the effective portions of the coils are those external to the ring, the wires along the screened internal face cutting few lines of force. Hence, given two rival machines having equally intense magnetic fields, equal velocities of rotation, and therefore equal electro-motive forces, that machine will have the advantage in which the shortest wire can be coiled into the greatest number of effective turns. It must be remembered that one turn of the Siemens armature corresponds to two turns on the Gramme armature taken at opposite points on one diameter of the ring. Hence, calling the longitudinal dimension of either armature l, and its diameter d, the length of wire necessary to make one complete turn of the Siemens armature will be $2 (l + d)$, while that necessary to make the equivalent two turns of the Gramme ring will be $4 l$ (neglecting the thickness of the iron ring, which may be relatively small). Hence, the Gramme will be better if $4 l$ is less than $2 (l + d)$; the Siemens better if $4 l$ is greater. If $4 l = 2 (l + d)$, or if $l = d$, then they will be of equal power. Hence, it follows that of necessity the Siemens armature should be long in proportion to its diametral thickness, while the Gramme will be best if the width of the ring be greater than its thickness parallel to the axis. The difficulty of winding up the Gramme ring prevents full advantage being taken of this point. In the flatter rings of the Brush machine and of the Schuckert machine, the solid sectorial cheeks of iron in the armature enable the winding up to be accomplished on a ring much more nearly approaching a disc form, than the ring used in the Gramme machines.

We will, in concluding these general considerations, devote a little attention to a fact of great importance known to all practical electrical engineers, but hitherto completely unexplained by theory, namely, that it is necessary to displace the commutator brushes of dynamo-electric machines from their position of symmetry. Experience has dictated that to obtain the best possible results it is necessary to adjust the brushes which take the current from the commutator, to an oblique position, different from that dictated by theory. Moreover the amount of this angular displacement is found to vary with different speeds of the machine, and with the work done in the circuit. Neglect of this matter leads to sparkling at the commutators and consequent wear and waste. In the Brush machine, in the Edison, and some other forms there is indeed a particlar adjustment to enable the brushes to be set at will. In the generator of Mr. Hiram Maxim there is a current regulator intended to keep the current delivered by the machine at a constant value, the brushes being actuated by an automatic adjustment whose object is not to place the brushes at the most advantageous point, but to lose some of the current at the commutator, if that current be too strong. There can be little doubt of the detriment to commutator and armature involved in this method of regulation. Another point, which has hitherto been quite unexplained, is that if the machine were being employed as a generator of currents, this displacement of the brushes must be a lead, or a displacement in the direction of the rotation, while in the case in which the machine was used as a motor, the brushes must be displaced in the opposite direction to that of the rotation, or with a negative lead. In spite of these plainly incompatible conditions, it has always been customary to attribute the practice thus dictated empirically by experience, to the slowness with which the soft iron of the armatures loses its magnetism, or in other words to the retardation of demagnetisation. If we go back to the simplest of all the forms of movable rotating conductors we have considered, it will be apparent that the best moment for changing the direction of the current in the conductor is that instant when a perpendicular common to the conductor and to the effective lines of force in the field, passes through the axis of rotation. Thus in Fig. 54, if

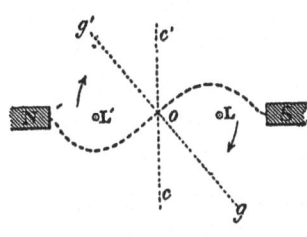

Fig. 54.

the lines of force ran straight across the field from pole N to pole S, then clearly the best points to reverse the current would be at c and c_1, or on a diameter at right angles to N S. Now, up to this point it has been *assumed* that the magnetic field is symmetrical across between the poles, and that the greatest number of effective lines of force run across straight from N to S, or at least in gentle arcs symmetrically above and below the line N S. As a matter of fact this is never the case, since the presence of the conductors carrying the vertical currents, is sufficient to introduce serious modifications in the positions of the lines of force. Around the conductors which carry the electric currents there is also a

G 2

"field" of magnetic force in which the lines of force are arranged, not radially as about the poles of magnets but in concentric circles. The presence of the current then changes the total directions of the forces at work in the field, and throws the lines of force into new places. In fact, as we have already mentioned, the principal lines running across the field from N to S will be distorted, as shown in Fig. 54, into an S-shape, illustrating unmistakeably the tendency to move in opposite directions the conductors L^1 and L, in which the current is flowing in opposite ways. From the principle laid down above, it follows that the most advantageous position to reverse the current is, therefore, not along the diameter $c\ c^1$, but along the diameter $g\ g^1$, drawn at right angles to the chief line of force at the centre. The angle between these two diameters is clearly the angle at which the contact brushes ought to be displaced, in order that the current may be reversed at the most favourable moment. We are here considering the case where a current is being used to produce motion, that is to say, where the Gramme machine is being used *as a motor.* Now the form of the S-shaped line will depend on the relative strengths of that part of magnetism in the field due to the field magnets N S, and that due to the current. If the current is relatively weaker, the S-shaped curve will be more nearly a straight line, if the current be powerful (relatively to the magnets), then the line running from N to S will be a well-formed S. Now, this clearly implies that if the current is strong, relatively to the magnet, the angular displacement of the contact brushes, $c^1\ o\ g^1$, will be great, while, if the current is relatively weak, the displacement of the brushes will be small. Again, it is well known that when an electro-motor is running with a great velocity there is a reaction current set up in it in an opposite direction to that which produces the motion and tending to produce an opposing electromotive force. If an electro-motor is allowed to run very rapidly, this opposing induced current will reduce the supplied current to a fraction of its original strength. The maximum work is done by an electro-motor when this reaction current just halves the original current. Hence if an electro-motor whose armature is running in an invariable magnetic field be doing light work, and therefore be running very quickly, the reaction current will reduce the total current in the conductors relatively to the power of the magnets, and under such circumstances, the displacement of the contact brushes will be small. If the motor be doing heavy work, and its velocity be therefore slow, the reaction current will be feeble and the total current strong in proportion to the magnets, hence the brushes must be displaced through a large angle. The rule, therefore, for dynamo-electric machines used as *motors* is that the contact brushes must be displaced in an inverse sense to that of the rotation of the armature, and with an angular displacement, which is greater as its velocity is less, or as the work done by the motor is greater.

Now consider the case where the machine is used as a generator. If the armature be driven round in the field mechanically, the current generated will be in the opposite direction to that of a current which we have supposed to produce a rotation in the same sense. Hence, if we think of the magnetic forces due to the current thus induced in the conductors, we shall see that they will produce a displacement of the lines of force in the field, but this displacement will be the converse of the former case ; the chief line of force crossing the centre of the field will be S-shaped as before, but will be reversed in position as compared with that shown in Fig. 54, and will pass from N above L^1, below L, and so to S. In this case the diameter of commutation $g^1 g$ will cross the line $c^1 c$ from right to left, and the contact brushes must be displaced in the same sense, as the rotation of the armature, or must be set so as to have a true lead. Here again the question of the relative strength of the magnetic forces due to the current and to the magnets comes into play. If the field magnets be simply permanent steel magnets, or if they are electro-magnets excited by a current generated independently from another source, then the electromotive force in the rotating armature will be, as shown by the researches of Mascart and Angot, and others, simply proportional to the velocity of rotation. In that case, if the resistance of the circuit is constant, the strength of the current will also rise proportionally to the velocity and the deformations of the field, and hence the angular displacement of the contact brushes must be greater as the velocity is greater. The rule is therefore for magneto-electric machines, and for those dynamo-electric machines in which the field magnets are separately excited, when used as generators, that the contact brushes must be displaced in the same sense as that of the rotation, and with an angular displacement which is greater as the velocity is greater.

Finally, take the more common case where the field magnets of the dynamo-electric generator are excited by the current generated, or are included in the circuit of the machine. With small velocities, before the electro-magnets have nearly attained their maximum magnetisation, the strength of the magnets will increase as the strength of the current. Hence the relative intensities of magnets and of current grow almost at the same rate, and the displacement of the contact brushes need only be small. With greater velocities, however, and stronger currents the magnets will begin to approach their condition of saturation, when any increase in the strength of the current no longer produces anything like a corresponding increase in the power of the magnets. Under such circumstances it is clear the deformation of the lines of the field will become greatly exaggerated, and the displacement of the contact brushes must be very great. M. Breguet tells us that with magnets too small in proportion to the armature, and with a velocity of 1,770 revolutions per minute, he has succeeded in obtaining conditions which necessitated that the contact brushes should have a lead of 70 deg. A simple calculation leads to the conclusion that for great velocities the tangent of the angle of the lead should be proportional to the number of revolutions per minute.

Before quitting the subject, we will just state what part is really played in this phenomenon of disymmetry of commutation by the tardiness of the iron ring in receiving or parting with its magnetism. The presence of the iron ring exercises two influences quite distinct from one another. Firstly, since it requires time to magnetise or demagnetise it, it will necessitate that the contact brushes be displaced a little in advance of their theoretical position, or with an increased lead. This displacement, always in the sense of the rotation, will, therefore, diminish the total displacement where the machine is used as a motor, but will increase it where used as a generator. In either case, however, its influence is quite small. M. Breguet found experimentally that even with the enormous speed of 1,770 revolutions per minute the displacement due to this source did not exceed 10 deg. Secondly, the presence of the iron ring increases the intensity of the magnetic field, concentrating a greater number of lines of force in it, and, therefore, tending to reduce the deformation in its symmetry. This is a most important influence, and it will be seen that in the case where the machine is used as a generator the presence of iron in the armature tends to bring back the most favourable position of the contact brushes towards the position of symmetry, that is, tends to *diminish* the angular displacement which must otherwise be observed. The action of the iron of the ring is, therefore, absolutely the contrary to that commonly attributed to it; for, so far from necessitating a displacement of the contact brushes in the direction of the rotation, it absolutely serves to diminish, and that very considerably, the angular displacement necessitated by the disymmetry of the magnetic field.

The researches of M. Breguet placed the whole theory of electric generators of the Gramme type in a new and intelligible light. They not only cleared up the discrepancies hitherto existing, but laid down the basis upon which all future dynamo-electric machines must be constructed; they showed the part played by the iron ring as a magnetic screen for the first time; and by the aid of the reasonings now put forward one begins to see the *rationale* of many details previously dictated only by empirical practice.

SECTION II.

—•••—

6. *Magneto and Dynamo-Electric Generators.*

VI.

MAGNETO AND DYNAMO-ELECTRIC GENERATORS.

E have already noticed that the manipulation of batteries is attended with difficulties which militate against their employment for the production of the powerful currents required for the electric light. It is true that the grand principle of electro-magnetic induction had been discovered by Faraday years before Grove or Bunsen constructed the batteries that bear their name; but, fertile germ as it was, it nevertheless had not yet crossed the threshold of the physicist's laboratory.

Faraday was pre-eminently a man of pure science. He felt a keen pleasure in the pursuit of natural truths, and would not allow himself to turn therefrom to follow up their applications. He, therefore, contented himself with studying the new phenomena and investigating their laws, leaving to others the care of their practical development. "I have rather been desirous," he writes, "of discovering new facts and new relations than of exalting those already obtained, being assured that the latter would find their full development hereafter."

It was in 1831, while holding the post of Director of the Laboratory of the Royal Institution, that he devoted his attention to what he called the "evolution of electricity from magnetism." Œrsted had shown, in 1820, that freely suspended magnets took up a definite position under the action of a current; Arago and Sturgeon had converted bars of soft iron into strong magnets by passing a current round them, and Ampère, by the aid of mathematical analysis and a few well-chosen typical experiments, had within the short period between the July and September of the same year, by a feat unequalled in the history of scientific discovery, reduced all magnetic phenomena to the mutual attractions and repulsions of minute electrical currents. All these were, to Faraday's penetrating mind, strongly suggestive of reciprocal action, and he accordingly sought with eagerness to obtain electrical currents from magnets. Round the two halves of a welded iron ring, Fig. 55, he wrapped two separate coils of insulated wire, one of which was connected with a battery, and the other with a galvanometer. His own terse description of

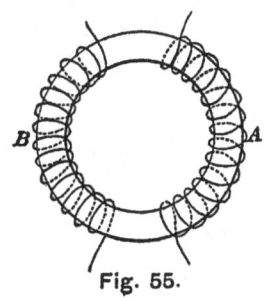

Fig. 55.

this experiment, as we find it in his laboratory note-book, runs as follows : " Charged a battery of ten pairs of plates 4 in. square. Made the coil on B side one coil, and connected its extremities by a copper wire passing to a distance, and just over a magnetic needle (3 ft. from wire ring), then connected the ends of one of the pieces on A side with battery; immediately a sensible effect on needle. It oscillated and settled at last in original position. On breaking connection of A side with battery, again a disturbance of the needle."

These effects were sudden and evanescent, and plainly caused by the act of magnetisation and demagnetisation. They were precisely what Faraday was in search of. On the second day of his experiments, he wrote to a friend : " I think I have got hold of a good thing, but can't say. It may be a weed instead of a fish that, after all my labour, I may at last pull up." On the ninth day he worked with the great magnet of the Royal Society, consisting of 450 bar-magnets, each 15 in. long, and was led to the conception of those magnetic curves which on that memorable day (November 4, 1831) he called " lines of force." On the tenth day of his research, he was fully satisfied of his success, and convinced that he had really evolved electricity from magnetism.

This was undoubtedly a grand discovery, and one which Jacobi does not hesitate to place on a level with gravitation in point of importance. In reading over Faraday's own account of these experiments, one cannot but be struck by the wonderful manner in which— by experiment alone—he completely disentangled all the complex phenomena of induction.

Faraday collected all the facts which he had observed into a paper entitled " Experimental Researches in Electricity," which was read before the Royal Society, on November 24, 1831.

It may be worth noticing that Faraday's spark—the first of its kind—was obtained, as we have seen, from an electro-magnet. Shortly afterwards, in January, 1832, Nobili obtained the spark from a bar of soft iron influenced by a steel magnet; and in the March of the same year, Professor Forbes succeeded in producing it from a soft iron magnet " made " by contact with a lodestone.

As a thorough grasp of electro-magnetic induction is necessary to understand the play of electric forces in the generators which we have to describe, we shall add a few remarks to what we have already said.

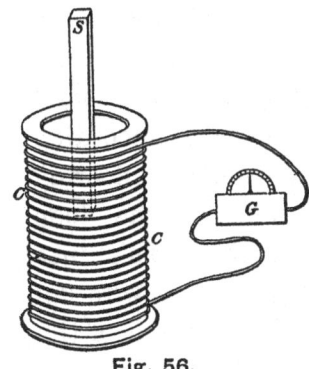

Fig. 56.

Let C, Fig. 56, be a coil of insulated wire, the two ends of which are connected with the galvanometer G, placed at such a distance as not to be affected by the magnet. On plunging the N pole of a steel magnet into the hollow of the coil the needle appears to receive a sudden jerk, say to the right; it will oscillate for a brief space of time and then come to rest at zero; withdrawing the pole, the needle swings off at once to the left. When the magnet is stationary, it has no effect on the needle; the slightest displacement, however, suffices to produce a corresponding quivering motion or deflection. Results of a similar character, but opposite in direction, would be produced by the S pole of the magnet. If then we hold the coil upright and drop the magnet through, the galvanometer will be affected by two opposite motions successively, and their resultant will, therefore, be *nil*.

The production of these currents is easily explained by Faraday's beautiful conception of "lines of force." We have already seen that when a closed circuit is urged across a magnetic field, a current is induced, the strength of which will vary as the number of lines of force cut through. Applying this principal to our magnet, we see that while plunging it into the coil the number of lines of force cut by the various convolutions of the wire, increases until the magnetic equator is just within the coil; any subsequent movement, whether forward or retrograde, would plainly diminish this number, and consequently give rise to a current opposite to the former in direction. Similar results may be obtained by moving the coil towards the magnet, supposed to be stationary.

These effects may be intensified by introducing a bar of soft iron within the coil, as shown in Fig. 57. Every approach or withdrawal of a pole alters the magnetic condition of the iron, increasing or diminish-

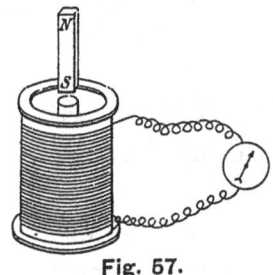

Fig. 57.

ing its lines of force, and every such change manifests itself by the production of a magneto-electric current. In the telephone, these currents are practically infiinitesimal, yet they are competent to produce the marvellous results with which we are so familiar. The bar or "core" is often advantageously replaced by a bundle of soft iron wires or thin rods.

Since we have reason to consider the earth as a magnet, we should expect that it would act inductively upon movable coils. Accordingly, if

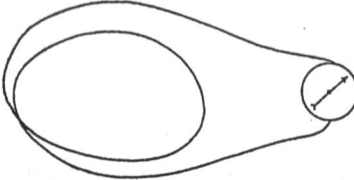

we take a loop of wire, Fig. 58, and connect its extremities with a sensitive recording instrument, say a Thomson's reflecting galvanometer, every motion of the loop which will cause it to cut the earth's lines of force, will evoke within the wire an

Fig. 58.

induced current which will reveal its presence by the displacement of the spot of light on the scale. If the displacement be to the right when the loop is turned from A to B, it will be to the left when tilted from B to A. This is the principle of Faraday's " rectangle " and Delezenne's circle.

Palmieri was the first to use a flat coil as an earth-inductor. The currents he obtained were sufficiently strong to decompose water and give violent shocks.

Again, in Ampère's theory, magnets are assimilated to currents, whence we would naturally expect voltaic currents to act inductively upon one another. Faraday called this induction " volta-electric," to distinguish it from that due to magnets which he had previously called

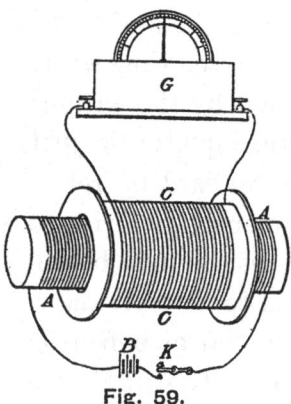

magneto-electric. Let A, Fig. 59, be a coil of wire connected through the key K with the battery B; C is a second and larger coil connected with the galvanometer G. On pressing down the key, the current circulates round the " primary " A, and at the same instant it generates an electric current in the " secondary " C, which is opposite in direction to the primary; hence the term " inverse " as applied to it. The needle soon comes to zero and remains there as long as the

Fig. 59.

currrent is steadily flowing round A. Releasing K, the primary circuit is interrupted and the needle swings away from its position of rest in the opposite direction, thus denoting a " direct " current.

The primary circuit being closed, then any increase in the strength of the battery current will give rise to an " inverse " and every decrease, to a " direct " current in the secondary wire C.

If we lift out the primary coil gradually, we practically diminish the effectual inducing current and thus obtain a "direct" induced current,

If we slowly replace it we increase the active primary and obtain an "inverse" secondary. Induction continues so long as the relative displacement of the coils lasts.

These phenomena are likewise due to the magnetic field which is produced as soon as the circuit is complete. The voltaic current, like the hypothetical currents of Ampère, is capable of surrounding itself with a system of lines of force which increases with its strength and faithfully responds to every fluctuation.

When the circuit is "made," these lines of force start into existence, and at the same time induce a current opposite in direction to that which produced them; opening the circuit, they instantly vanish, thereby giving rise to a direct current. Every change in the strength of the battery, or in the proximity of the primary to the secondary coil, is attended by variations in the intensity of the field, and these necessarily evoke corresponding induced currents.

The Russian philosopher, Lenz, discovered, in 1833, the law* which governs all these inductive inter-actions. It may be stated as follows:— The currents induced by the relative movements, either of two circuits or of a circuit and a magnet, are always in such directions as to produce mechanical forces tending to stop the motion which produces them.

Thus, for instance, when we introduce a magnetic pole into a hollow coil, the induced current is inverse of the Ampèrian current circulating round that pole. Moreover, we know that currents, opposite in direction, repel each other, hence the secondary current reacts on the primary in such a way as to oppose the motion which produced it.

We have already remarked that Faraday limited himself, on principle, to pure, in contradistinction to applied, science, being convinced that every valuable discovery would ultimately receive the development of which it was susceptible. His own investigations of 1831 afford a happy illustration of this, for shortly after the publication of the remarkable paper in which they are described, we find the magneto-spark already multiplied and exalted by ingenious appliances. Though these differ from one another in some particulars they agree in using permanent magnets as inductors. In later machines these are replaced by electro-magnets, and with advantage, as large masses of soft iron are thus

---

*This celebrated law was communicated to the Academy of Sciences of St. Petersburg, on November 29, 1833; it appeared in *Poggendorff's Annalen*, in 1834. This accounts for the different dates which we find in different text-books.

readily converted into a powerful magnetic magazine by the passage of a strong current. Subsequent discoveries showed that the rotating armature itself could supply this current and that voltaic batteries could be entirely dispensed with. All machines in which mechanical is converted into electrical energy, belong to one of these two classes; those formed with permanent magnets being called magneto-electric, and the others dynamo-electric machines. The electricity produced is undoubtedly the same in both cases, the inductor only is different. The names suggest this difference and are therefore not without their utility in our scientific nomenclature.

In the present section we shall describe, approximately in chronological order, the more important generators of both kinds, referring the reader to the Abridgments of Patent Specifications contained in this volume, for others of a less striking character.

The first to construct an apparatus embodying the principles of magneto-electric induction was Hyppolyte Pixii, of Paris. It was made in 1832, and exhibited before the Academy of Sciences at its meeting of September 23rd. This "machine," as it is usually called, is shown in Fig. 60; in it a double coil of insulated wire, 3,000 ft. in length, is

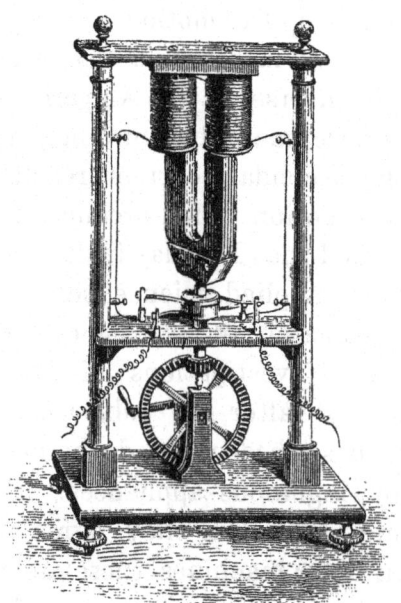

Fig. 60. (Pixii, 1832.)

wound round a massive soft iron core and fixed to a horizontal cross-piece; below is a compound permanent magnet movable about a vertical axis to which a rapid rotation may be communicated by means of a handle and pair of bevel wheels.

The laws of electro-magnetic induction, so clearly enunciated by Faraday, enable us to follow with ease the genesis and development of the currents. Let the relative position of the magnet and the coils be as shown in Fig. 61, and the direction of rotation from left to right. It must be remembered that the Ampèrian currents circulate in a north-seeking pole *with*, and in a south-seeking pole *against* the hands of a watch. The N pole acts inductively upon the overhanging core making its lower extremity a S pole. The Ampèrian currents set up round its molecules, flow with the hands of a watch, and as the magnet

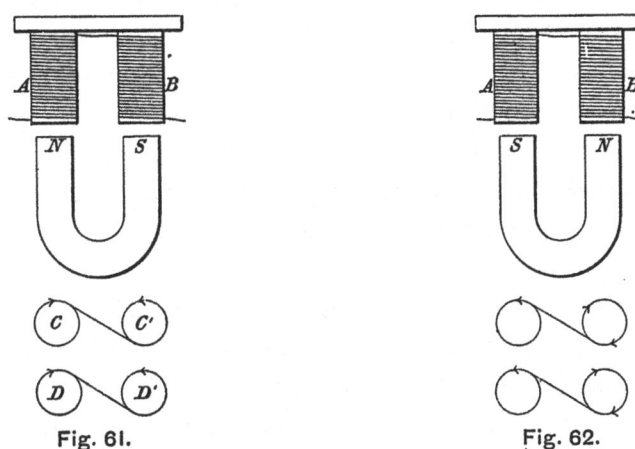

Fig. 61.   Fig. 62.

is receding the current induced in the neighbouring wire A is in the same direction, viz., that shown at C, and will continue with gradually diminishing strength until it finally vanishes when the magnet has just turned through 90 deg. The core is then equally and oppositely influenced by the two equidistant poles of the magnet, and is, therefore, in its neutral condition.

During the next quarter of a revolution N is approaching B, and as the south polarity which it induces is continually increasing, it follows that the current developed in the surrounding helix will be the inverse of that which produced it; in other words, it will be against the hands of a watch, as indicated at C'. These two currents, however, do not neutralise, but increase each other, and flow in the same direction through the double coil, owing to the manner in which the wire is wrapped round the two parts, and which is shown at C C'.

We must now return to our starting-point and consider the effects due to the S pole. Acting inductively on the core of B, it makes its lower extremity a N pole. The Ampèrian currents are then against the hands of a watch, and as the strength of the pole is falling on account of the withdrawal of S, the current developed in the wire will be in the same direction, and, therefore, as shown at D'. While moving over the second half of its semi-revolution S will be approaching A, whose north polarity will, therefore, be increasing. The current evoked in the helix will oppose that which produced it and will consequently flow with the hands of a watch, as figured at D.

We thus see that while N is receding from A, and S from B, we have the two currents C and D', which unite to form a single current-flowing, say, from right to left; while they approach B and A respectively, we have the currents C' and D, which combine to give a current in the same direction.

The intensity of these currents will depend upon the strength of the steel magnets and the closeness with which they graze the soft iron cores. The electromotive force will depend upon the rapidity with which they spin round.

Continuing the rotation of the magnet we see that the S pole now starts from A, Fig. 62, whence we may infer that the induced currents will present themselves in exactly the opposite order to that in which we met them at first, and will by their coalescence give rise to a single current contrary in direction to that produced during the first half of the revolution.

Besides the rapid changes in the magnetism of the core, there is still another cause concerned in the generation of the currents in the double coil, viz., the changes in the character of the magnetic field. An analysis, similar to the above, would show that the rotation of the lines of force due to the magnet, induces currents in the coil which, at every moment, are of the same sign as those produced by the rapid rise and extinction of magnetic force in the core, and, therefore, intensify the effects due to the latter.

Pixii's apparatus is therefore a discontinuous alternating current machine. Absolute continuity was not attainable from such an arrangement; but by means of an ingenious device, called the commutator, it was found possible to rectify one of the currents so as to make them both flow through an external circuit invariably in the same direction. This rectification is of no importance when the current is employed for medical purposes, springing mines, firing torpedoes, or the like; but it is of primary moment when used for electro-plating, electro-typing, or for effecting chemical decomposition in general.

The commutator is shown in Fig. 60. It consists of a brass cylinder arranged co-axially with the spindle with which it rotates. It is divided into two symmetrical parts, which are insulated from each other. Two contact springs, shown in the figure, are disposed on each side, and may press either against the same or alternate sections. Its action will be understood from Figs. 63 and 64. Let us suppose the

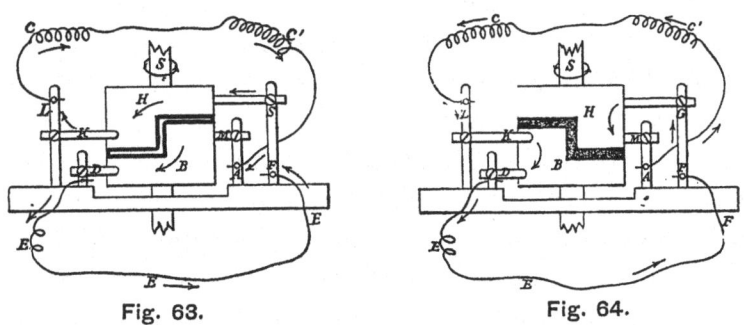

Fig. 63.   Fig. 64.

currents to circulate through the coils C C' in the direction indicated by the arrows, Fig. 63. Emerging from C' it descends to A, thence through the vertical rod and horizontal spring to B, which is insulated from the other half H, after which it pursues its course through the small spring D and the external circuit E E to F, up to S, across H, through K to L, and the coils again.

When the spindle has completed a semi-revolution, the contacts which are then automatically made are shown in, Fig. 64. It will be remembered that the current through the coils is now from C' to C. It then passes to L, K, B, thence round to D, the only path open to it, through the external circuit E E, up to F, through G, across H to M, down to A, and then back through the coils again. It will be noticed that in both cases the direction of the current through E E is the same, and the object of the commutator is attained. This machine gave sparks and severe shocks, a feeble charge was also given to a Leyden jar, and Faraday himself decomposed water with it.

Pixii's machine was the first " practical development" of Faraday's investigations, and is, to our mind, more deserving of special notice in the history of magneto-electricity than many of the modifications and

H 2

adaptations made by subsequent constructors.   The earliest of these was
Mr. Saxton, who exhibited his machine at the meeting of the British
Association in 1833.   He had the happy idea of fixing the heavier and
rotating the lighter parts of the apparatus.   The magnetic battery was
placed in a horizontal position and the double coil or armature, spun
before its poles.   He also added a variety of appendages to facilitate the
exhibition of its electrical, chemical, luminous, and physiological effects.

In November, 1834, Mr. E. M. Clarke, magnetical instrument
maker, of London, introduced a new feature into these machines.   It
occurred to him that different effects ought to be produced by using wire
of different thicknesses, and accordingly he obtained, from thick copper
bell-wire, brilliant sparks but no perceptible shocks, while with very long
thin wire the shocks were powerful but the sparks feeble.   He was thus
led to construct two sorts of armatures, the one for intensity and the
other for quantity.   In his largest armatures, the former contained 1,500
yards of insulated wire, one-ninety-sixth of an inch in diameter, and the
latter 40 yards, one-eighteenth of an inch in diameter.   With these he
was able to delight and astonish the scientists of his time ; he obtained
large bright sparks and brilliant scintillations from a steel file, gave
shocks, charged a Leyden jar, deflected the leaves of an electroscope,
ignited powder, and decomposed water.

All these effects must have suggested to reflecting minds the identity
of the voltaic, and what was called "magneto" electricity.   The
difference is not one of kind but merely of electromotive force.

Clarke returned (Fig. 65) to the vertical position for the compound
magnet, causing the armature to rotate close to its surface laterally, and
not in front of its poles.   Though the arrangement of Saxton seems the
more favourable, still it is usual to overlook his merits and to speak of all
similar apparatus for the generation of magneto-electric currents as
" Clarke's machines."

Faraday's discovery elicited attention not only in the Old but
also in the New World.   Dr. Charles G. Page—one of the pioneers
of American science—constructed, in 1835, what he called a " new
powerful " magneto-electric machine.   We have already seen that if
we place a steel magnet within a hollow coil and successively approach
and withdraw a bar of soft iron, the galvanometer will faithfully respond
to every phase of the movement.   When within the influence of the
uppermost pole, the mass of soft iron becomes a temporary magnet,

its strength, and therefore also the intensity of the "field" due to the opposite poles, increases so long as the movement of approach continues. During the recession, there is a gradual falling off and redistribution of the lines of force. These two classes of changes give rise to direct and inverse currents in the wire coil. This is the principle of Dr. Page's apparatus, in which the strength of the induced currents depends upon

Fig. 65. Clarke.

the intensity of magnetisation of the horseshoe magnet, and also on its proximity to the revolving bar. This distance is adjustable by means of a screw, and may be read off on a scale. This method of regulating the intensity of the current is important, when the machine is used for medical purposes.

It was with this apparatus that M. Verdet made one of his important investigations in magneto-electricity, viz., the effect of time in the development of the current. He found that the inversion does not occur exactly at the moment when the soft iron bar passes in front of the poles, but somewhat later. This has been accepted as showing that the generation of the currents is not absolutely instantaneous. In coils of wire the retardation is due to the inductive effects of the successive convolutions and is practically infinitesimal, but in metallic masses it is caused by the coercive force found even in the "softest" iron, and is quite appreciable.

We have already remarked that in the case of magneto-electric machines the electromotive force of the induced currents is directly proportional to the speed with which the armature is driven; but now we have to add a limitation : the rise, subsidence, and reversal in the magnetism of the soft iron cores all require a definite time, and when the speed coincides with this period, the electromotive force has its maximum value; if we continue raising the speed, we diminish the electromotive force until at last it vanishes. This curious result has been observed in dynamo-machines, and has been called the " phenomenon of demagnetisation."

We now come to a highly interesting apparatus which, though not used as an electric generator by its inventor, displays such striking originality and is so very suggestive of later dynamo-electric discoveries that it naturally finds its place among the early machines. This important apparatus is the electro-motor of Elias, of Amsterdam. It was constructed in 1842, and marks a new era in electrical science.

It may be considered as a prototype of the Gramme machine, although it was constructed for a different purpose, being, as stated above, in fact an electro-motor, while the Gramme machine is, it is almost

Fig. 66.   Elias.

unnecessary to say, an electric generator. The striking resemblance which exists between the one invention and the other makes it, however, necessary to describe the Elias machine, and to explain the difference between it and the Gramme generator. Figs. 66, 67, 68, convey an exact idea of the model that was exhibited at the recent Paris Electrical Exhibition, and which was contributed by the Ecole Polytechnique of Delft, in the Dutch Section. The perspective illustrations, Figs. 66 and 68, show the machine very clearly, and the section, Fig 67, explains the construction still further. In this apparatus there is an exterior ring made of iron, about 14 in. in diameter and 1·5 in. wide, divided into six equal sections by six small blocks which project from the inner face of the ring, and which act as so many magnetic poles. On each of

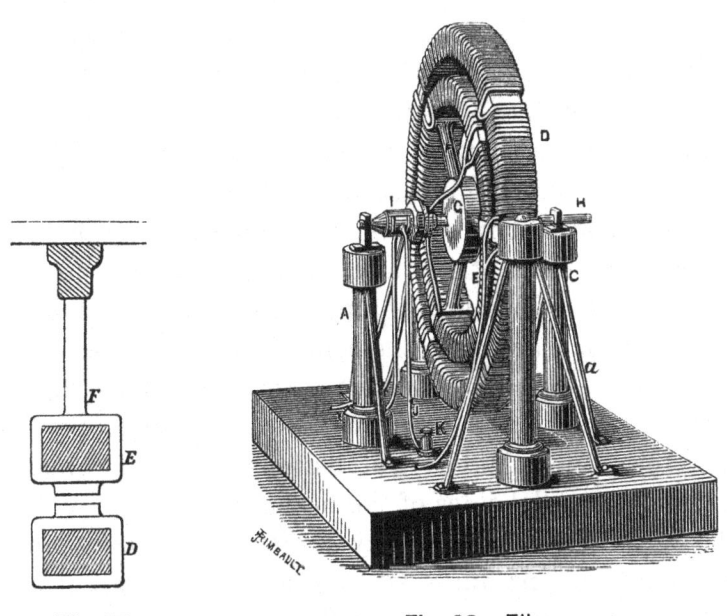

Fig. 67.   Fig. 68. Elias.

the sections between the blocks is rolled a coil, of one thickness only, of copper wire about ·04 in. in diameter, enclosed in an insulating casing of gutta-percha, giving to the conductor thus protected a total thickness of ·20 in.; this wire is coiled, as shown in the illustration. It forms 29 turns in each section, and the direction of winding, changes at each passage in front of a pole-piece. The ends of the wire coinciding with the horizontal diameter of the ring are stripped of the gutta-percha, and are connected to copper wires which are twisted together and around two copper rods, which are placed vertically, their lower ends entering two small cavities made in the base of the apparatus. The circuit is thus continuous with two ends at opposite points of the same diameter. The ring is about 1·1 in. thick, and is fixed, as shown, to two wooden columns B B by two blocks of copper *a*.

It will be seen from the mode of coiling the wire on this ring, that if a battery be connected by means of the copper rods, the current, by magnetising the outer ring, will create six consecutive poles on the various projecting blocks. The inner ring E is about 11 in. in outside diameter, and is also provided with a series of six projecting pieces which, in their rotation, pass in succession very close to the projecting pole-pieces of the outer ring.

In the intermediate spaces, the distance between the coils of the inner and outer ring is ·40 in. The latter is movable, and is supported by three wooden arms F fixed to a boss G, which is traversed by a spindle supported in bearings by the columns A and C. A coil is wound around this ring in exactly the same way as that on the outer ring, the wire being of the same size, and the insulation of the same thickness. The insulating material is also stripped from the ends of the wire at points of the diameter opposite each other, and the six sections form a continuous circuit. At the two points of junction they are connected with a hexagonal commutator placed on the central spindle; one end corresponding to the sides 1, 3, and 5, and the other to the sides 2, 4, and 6. Two copper rods, J, fixed on the base to two plates of copper furnished with binding screws, are widened and flattened at their upper ends to rest against opposite parallel sides of the hexagon.

It will be seen that if the battery is put in circuit, the current in the inner ring will produce six consecutive poles, the names of which will change as the commutator plates come into contact successively with the sides of the hexagon. Consequently, if at first the pole-pieces opposite each other are magnetised with the same polarity, a repulsion between them will be set up which will set the inner ring in motion, and the effect will be increased on account of the attraction of the next pole of the outer ring. At the moment when the pole-piece thus attracted comes into the field of the pole of opposite polarity, the action of the commutator will change its magnetisation, whilst that of the pole-piece on the fixed ring always remains the same; the same phenomenon of repulsion will be produced, and the inner ring will continue its movement in the same direction, and so on. To the attractive and repulsive action of the magnetic poles has to be added the reciprocal action of the coils around the two rings, the action of which is similar. From this brief explanation, the differences between the Elias and the Gramme machines will be understood. The Dutch physicist did not contemplate the production of a current; he utilised two distinct sources of electricity to set the inner ring in motion, and did not imagine that it was possible, by suppressing one of the inducing currents and putting the ring in rapid rotation, to obtain a continuous current. It has happened very many times, that inventors living in different countries, and strangers to one another, have been inspired with the same idea, and have followed it by similar methods, either simultaneously or at different periods, without arriving at the same results. It is not sufficient for the seed to be the same; it must have fallen in good ground, and be cultivated with care; while in one place it scarcely germinates, in another it may produce a vigorous plant and abundant fruit.

The current of thought and invention continued for another decade, undiverted from its old channels, and manifesting itself by instruments and machines all modelled on the Pixii and Saxton type.

In a patent filed in 1840 by Wheatstone and Cooke, one such is described, though it is referred to as being " of ordinary construction." The drawings show a horizontal horseshoe electro-magnet, revolving in front of a vertical compound horseshoe magnet, practically similar to the Saxton machine. The following year, Wheatstone patented for telegraphic purposes the grouping of several of these generators with the armature revolving on one spindle; the drawings show five such machines thus grouped, the rotating coil so arranged that the current in any one coil commenced before the currents in the other coils had ceased. The coils were coupled together in parallel circuit, the positive and negative commutator brushes being connected to the same leading wires respectively.

In 1842, Mr. Woolrich, of Birmingham, patented the old idea of a pair of coils revolving on a horizontal spindle before the poles of a powerful permanent compound magnet.

Schottlaender, in 1843, described an armature formed of permanent magnets, placed concentrically around, and parallel to, a horizontal axis. This armature revolved within a circle of magnets similarly arranged, and with insulated wire coiled around each branch. The action of the generator is thus described by the inventor: " When the poles of the inner circle magnets stand opposite to the contrary poles of the outer magnets no electricity passes, but on moving the inner circle of magnets and bringing similar poles in opposition, an electric current passes through the wire coils. By communicating a rapid motion to the inner circle of magnets, a rapidly intermitting flow of electricity passes off from one wire to another, its intensity being proportional to the number and size of the magnets, the quantity of covered wire, and the velocity of revolution."

In 1844, Stöhrer, of Liepzig, devised a magneto-electric generator, chiefly for purposes of demonstration. It is substantially a modification and development of the old Pixii machine, as will be seen by reference to the illustration, Fig. 69, which shows the latest form. There are

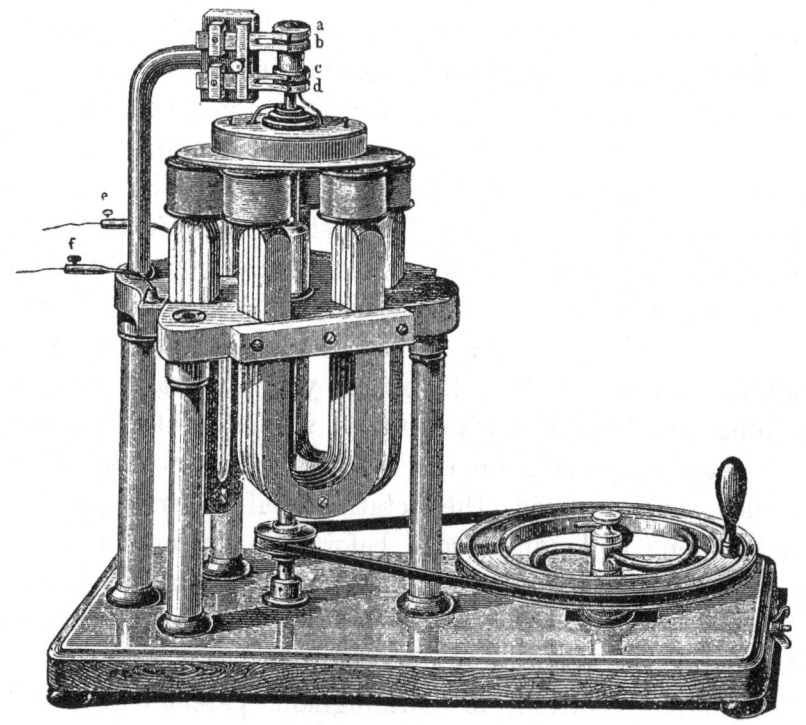

Fig. 69. Stohrer.

three permanent horseshoe magnets arranged vertically, and in such a manner that their six poles form the angles of a hexagon in a horizontal plane. The armature consists of an iron disc mounted on a vertical shaft, and carrying six induction coils, the wires of which are so wound that as the coils successively approach the magnetic poles, during the revolution of the armature, a current of the same direction is induced in all, and likewise a current of uniform direction, but opposite to the former as they recede. In this way there are twelve changes of current during each revolution of the armature. The commutator employed shows considerable originality, and has been followed in several modern generators. In order to avoid complication, a machine with only one magnet may be assumed. Immediately above the armature disc, a brass sleeve i is fixed to the shaft C, Figs. 70 and 71, and enveloping it is a second sleeve o, also of brass; these two are insulated from each other

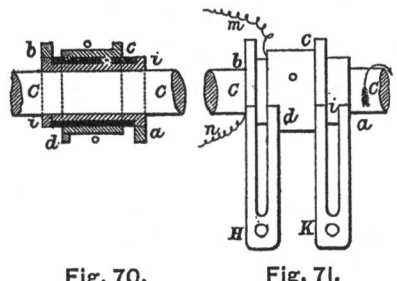

Fig. 70.   Fig. 71.

by boxwood. A steel half-ring a is attached to the upper end of the sleeve i, and a similar steel ring b is secured to the lower end of i; each of these pieces extends half round the sleeve, the two being opposite to each other. Similarly there are two half-rings c and d attached to the outer sleeve. Two springs of the form shown at H and K are kept in contact with the steel rings, alternately with the one on the inner and that on the outer sleeve, and serve to collect the current. The ends of the wires from the induction coil are connected to the sleeve, the wire m going to the outer one o, and the wire n to the inner one i, as in Fig. 71. With such a commutator arranged for one magnet only, two opposite currents would be obtained at each revolution, and then would be harmonised by the commutator springs H and K connected to the conductors. If three or more magnets were employed, the number of sleeves would have to be increased, one being necessary for each pole; the steel contact pieces would then cease to be half-rings, being shortened so that they should not overlap each other, and the number of collecting springs H and K would have also to be increased.

In 1845, Wheatstone and Cooke jointly patented an invention that was very favourably received at the time, was considerably used, and several times re-patented. This consisted in substituting for the permanent magnet previously employed in the familiar generator, an electro-magnet excited by a battery, " such battery being so used, by keeping up voltaic magnetism in such machines, a far greater effect is produced than such battery could produce if used in electric telegraphs without the intervention of such machines." This patent, by the way, is interesting in another sense, as it embodies the process of enclosing conductors, in lead tubes formed around them by hydraulic pressure.

In 1846, some further improvements by E. A. King, of London, were made, especially in the construction of the armature coils. The ordinary wire was dispensed with, and replaced by rings of rolled copper. These rings, or discs, were perforated in the centre to admit the core, and were slit from the centre to one edge. The two parts were then bent away from each other and soldered to the edges of the preceding and following discs. Connected up in this way, they swept round the core just like the thread of a screw. Fig. 72 is a front view of King's machine; *d d d* are the steel magnets fixed upon a circular frame with their poles directed to the centre. Eight coils were arranged in the

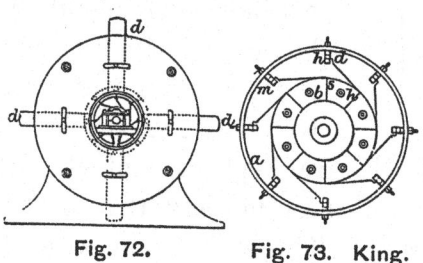

Fig. 72.   Fig. 73. King.

circumference of a brass wheel, and rotating with the axis, revolved through the magnetic field, thus giving rise to eight distinct currents. These were rectified by the commutator shown on a large scale in Fig. 73. It consisted of eight segments insulated from one another, and each provided with a spring *d* fastened to a bolt *h*. The armature coils are connected with these segments and with a brass " connection ring " placed at the opposite extremity of the axis. This ring has also its collecting spring, and in this way the eight currents developed by the armature may be collected as one, or may be separated into two or four circuits.

Mr. M. Poole, of London, was the first to patent (in 1847) in this country, a modification in the arrangemet of these small current generators, in which the armature coils, instead of being rotated, are moved to and fro, in front of the permanent magnets, the core coming into contact with the magnet at each stroke. Some rather ingenious details were included in this arrangement; the coils were composed of three concentric rings, the end of each section being led to a terminal so that a part of the wire might be cut out of the circuit by moving a key. The successive portions of each coil were of different diameter, the finest wire being nearest the centre, so that as they were further removed from the magnetic influence, the resistance set up was reduced. In 1847, Mr. W. H. Hatcher patented a similar arrangement of reciprocating coils, as did also M. Dujardin, of Lille.

We now come to a new period in the development of these machines, a period in which it was sought to utilise them for a variety of industrial purposes. The initiative in this progressive movement is due to a descendant of the celebrated Abbé Nollet, viz., M. Nollet, Professor of Physics in the Military School, Brussels, who in 1849, conceived the idea of constructing a Clarke's machine on a large scale. Some experience, both electrical and mechanical, had already been acquired, and with the aid of Van Malderen, an intelligent workman, Nollet embodied all this experience in the plans which he drew up of the new generator. But his health giving way, he did not live to carry them out. The idea was neglected for some time, but it was too fruitful to be entirely abandoned, and six years later it was revived by an Anglo-French Company, the *Compagnie de l'Alliance*, from which Nollet's machine received its usual appellation. We shall return to it a little further on.

Meanwhile the improvement of magneto-electric generators continued to receive some attention in England. In 1850 appears the first notice in the Patent Office of using hollow cores for electro-magnets. This is in a patent communicated to A. V. Newton, of London, for a magneto-electric generator, among many other things. Beyond the use of hollow cores the generator presented no novelty.

Mr. E. C. Shepard, of London, describes in his specification of 1850, what is practically the Stöhrer generator in its complete form. The armature consisted of a disc carrying eight bobbins that revolved horizontally in front of four horizontal compound permanent magnets. The currents were rectified by a commutator similar to that of Stöhrer already described.

Mr. W. Millward, of Birmingham, patented, in 1851, a magneto-generator, the arrangement of which is shown in Figs. 74 and 75. It consists of a brass ring or wheel *a*, mounted on a vertical spindle, and carrying upon the rim a number of small electro-magnets *b b*. Around the wheel and arranged radially, was a series of permanent magnets *c c*, the number of the latter being half that of the coils. A commutator was fixed to the vertical spindle of this machine, which was specially intended to generate currents for exciting a large electro-magnet used for magnetising steel bars. The following passage occurs in Mr. Millward's specification, which is of interest : "The power of such an electro-magnet depends upon the strength of the permanent magnets used in the machine, . . it will be therefore evident that by having two sets of permanent magnets and charging them in such a machine, that their supporting power may be increased by continual passes or charges from the electro-magnet thus produced."

The arrangement of generator proposed in 1852 by Mr. E. C. Shepard

Fig. 74.

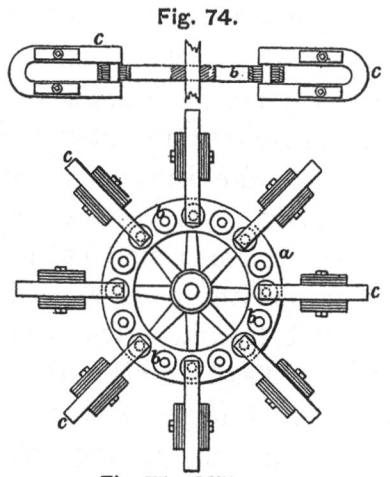

Fig. 75.   Millward.

is somewhat similar to that of Millward, but is more completely carried out. This generator is illustrated in Fig. 76. Upon a horizontal shaft B were mounted a number of wheels E E; around the periphery of each of these wheels was placed a series of small electro-magnets C C, the cores of which consisted of a soft iron tube with wooden discs on each end; these coils were held in place by insulated brass clips as shown. Outside each alternate wheel was a ring F, to which were bolted a number of horseshoe permanent magnets D D, built up each of a number of thin plates and bolted radially to the rings F, as seen in the drawing, the poles of the magnets lay between the electro-magnets on the wheels; there were half the number of horseshoe magnets as there were of the electros C. As the shaft B and the wheels on it revolved, one half of the helices or electros were traversed by a current in one direction, and the other half by a current in the opposite direction. Upon each of the wheels were four conducting-plates or rings p^1, n^1, p^2, n^2 for collecting the current from the two sets of coils, the latter being suitably connected to the rings for this purpose. Four rods or wires, of which two are shown, *f f*, served to connect these four rings of each wheel to the four rings *g g g g* in the commutator, in each of which rings there were 16 segments arranged alternately to be

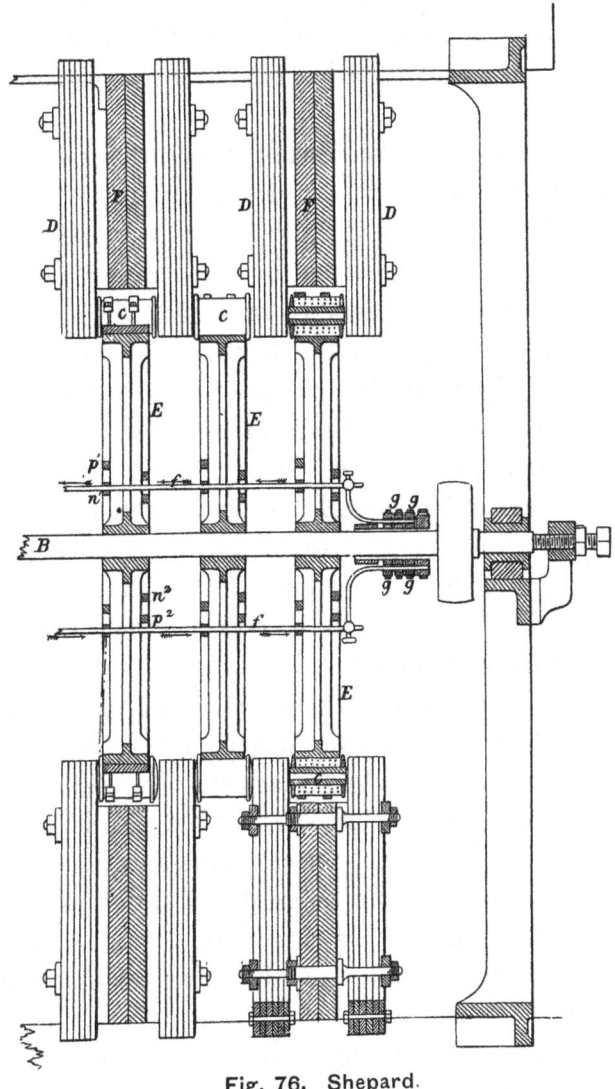

Fig. 76. Shepard.

conducting or non-conducting. The current generated was led off from the commutator by two brushes or flat springs, so arranged, that as a non-conducting segment of the revolving commutator quitted contact with one spring, a conducting segment made contact with the other. The inventor of this machine, though not displaying absolute originality, carried his idea farther, and in more practical shape than any one had done up to that date.

In 1852, E. B. Bright, of Liverpool, and C. F. Bright, of Manchester, described an arrangement for reciprocating coils with the addition of a special commutator. In the same year, Mr. Watt, of London, patented a generator with two horseshoe electro-magnets excited by a battery current, the armature having a combined revolving and reciprocating movement.

In 1852 also, W. Petrie, of Woolwich, brought forward a somewhat novel and ingenious arrangement of generator, in which several iron strips were inserted longitudinally into a cylindrical core of brass. This core was divided into sections by iron rings, between which, insulated coils were wound, preferably on tubes loose upon the brass core; the coils were suitably coupled together. Each of the collars was caused to revolve between the poles of a permanent magnet, the series of magnets being arranged with alternate poles.

Mr. D. A. Sonnenburg and P. Neehten, of Bremen, about the same time (1852) devised a small magneto-machine, consisting of two pairs of permanent magnets fixed opposite each other, and with their poles just so far apart as to allow of the insertion and rotation of a coil armature of the ordinary Clarke type between them.

The magneto-electric generator of T. Allan, of Edinburgh, possesses considerable interest, as anticipating several arrangements of a much later date. It was described for the first time in 1852, and consisted of a series of permanent horseshoe magnets disposed radially about a common centre, with their poles alternating. A horizontal spindle passed through the centre of the circle thus enclosed by the poles, and carried in the plane of the magnets, a series of radiating bars made of soft iron, which served as cores for insulated coils; two or more such sets were employed, the revolving armatures being mounted on the same spindle.

Bain, of Hammersmith, in 1852, modified the Pixii machine to adapt it for the purpose of driving clocks. He employed six electro-magnets, arranged round the same centre. In front of these, two steel magnets, set back to back, were mounted vertically and caused to revolve.

Belford's magneto-electric machine patented in 1853, is illustrated in the annexed Figs. 77 and 78, the former being partly a vertical section through the centre of the generator, and the latter a horizontal section through the revolving armature carrying the electros. It comprises two series of compound permanent magnets arranged radially around a common centre, with the north and south poles of each series placed alternately, and poles of a similar nature placed opposite each other in the two series. The opposing poles of the upper and lower series are placed far enough apart to allow of a disc being set between them; this disc is mounted on a vertical spindle revolving in a bearing in the top plate of the machine, and on a step in the bottom plate. In openings in the armature are placed two series of electro-magnets, there being as many of these coils in the inner circle as there are permanent magnets ranged round the circle, and twice as many coils in the outer circle. The

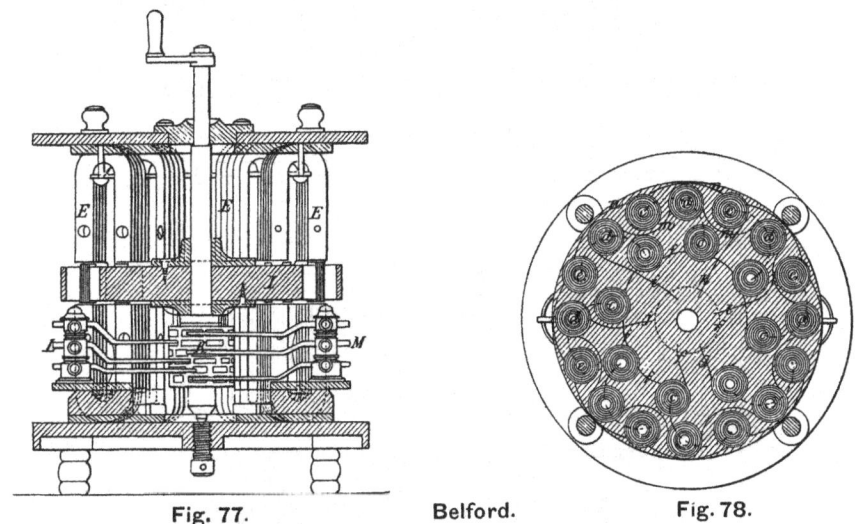

Fig. 77.   Belford.   Fig. 78.

coils in each row are at the same distance apart, and they are so distributed that the centres of no two coils, of which one is in the outer and the other in the inner circle, can be opposite the centre of the poles of the magnets at the same time. The coils are divided into three series, of which one includes the eight inner ones *c c*, and the 16 outer coils make two circuits *d* and *e*. The plan Fig. 78 shows the mode proposed for connecting the coils to obtain currents of high potential. The outer end of each coil is connected to the outside end of the next coil of the series; and the inner ends are similarly connected so as to leave only two terminal wires. For producing currents of low potential the inner end of one half of the coils in each series and the outer ends of the other half were connected to form one pole, and the remaining ends formed the

I

other pole. The commutator shown in Fig. 77 consisted of a cylinder K formed of non-conducting material, with blocks of copper set in it ; there were six rows of these inlaid segments, two for each series of coils, and the number of segments was the same as that of the magnets ranged around the circle. The collecting-brushes or springs, six in number, were attached to the standard L M, and when the centres of the coils were

Fig. 79.

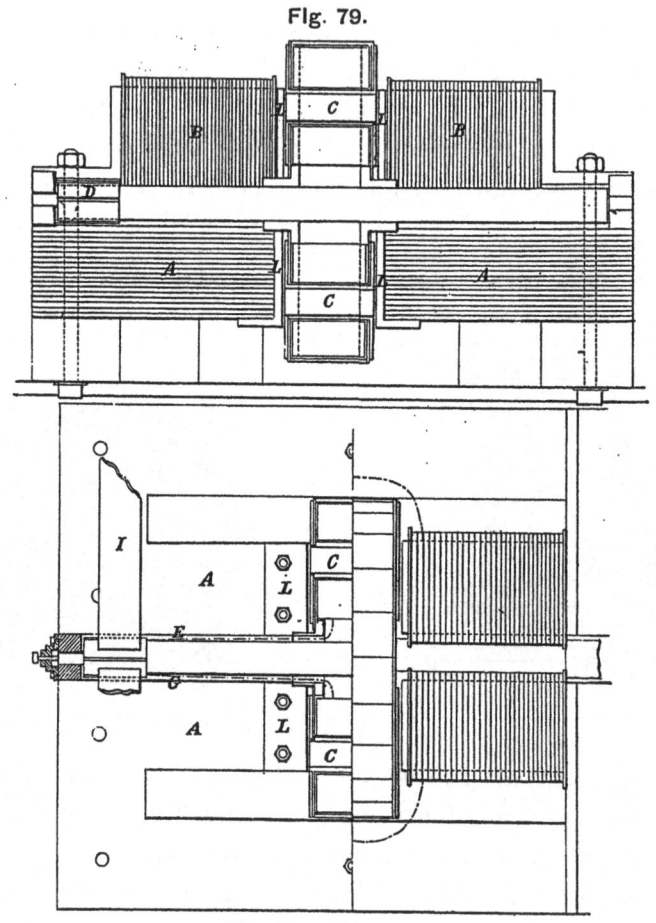

Fig. 80. Hjorth.

opposite the poles of the magnets these springs rested on the insulating material between the copper segments. These insulating distance pieces were arranged spirally and passed under the springs in the same order that the coils with which they were connected, passed the poles of the magnets ; in this way the currents always flowed from the copper segments and springs to the standards L and M.

We now come to one of the most interesting suggestions for the mechanical production of electric currents, in a historical point of view. This is the provisional specification of Sóren Hjorth, of Copenhagen, dated October, 1854. In this specification the principle of augmenting electric currents by the reaction of electro-magnets upon each other, is clearly described. We reproduce the drawings attached to this patent, and the specification almost verbatim. The latter refers to three figures that do not appear in the drawings annexed to it, but form part of Hjorth's succeeding specification, No. 2,199, to which reference is made later on. The explanation probably is that the two inventions were originally included in one provisional specification, and were divided, either as an afterthought or at the suggestion of the Patent Office officials. The three figures referred to, relate mainly to the second invention, and were consequently added to that specification.

Fig. 79 represents a sectional view of the "battery" seen endways, Fig. 80 is a half plan and half sectional plan of battery, Fig. 81, sections of commutator for battery, Fig. 82, a diagram showing the peculiar shape

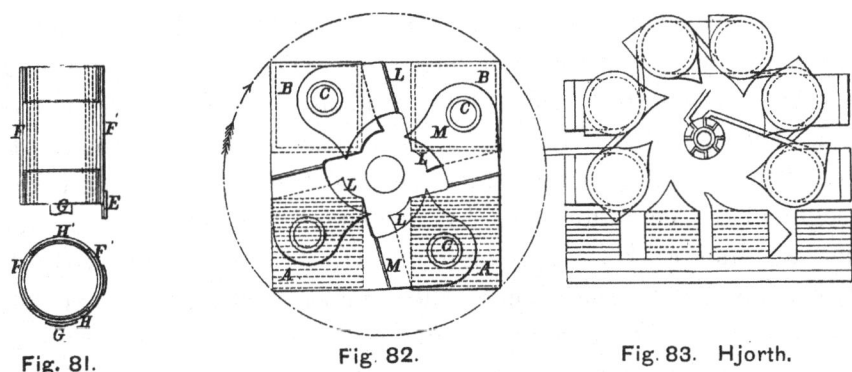

Fig. 81.   Fig. 82.   Fig. 83. Hjorth.

of the false poles of magnets and armatures; Fig. 83, is a diagram showing the mode of applying more magnets and armatures.

The main feature of the battery consists in applying one, two, or several permanent magnets A, of cast iron and shaped as shown, in connection with an equal number or more electro-magnets B, shaped as indicated in the drawing, in such a manner that the currents induced in the coils of the revolving armatures are allowed to pass round the electro-magnets; consequently the more the electro-magnets are excited in the said manner, the more will the armatures C be excited, and more electricity of course induced in the respective coilings; and while a mutual and accelerating force is thus produced in this manner between the electro magnets and the armatures, an additional or secondary current is at the same time induced in the coiling of the electro-magnets by the motion of the armatures; the said current flowing in the same direction as that of the primary current after having passed the commutator. The direction of the current induced in the coils of the armatures, will of course be reversed according to the change of the ir-respective polarities and the commutator D is therefore applied for the purpose of causing the same to flow constantly in one direction. This commutator consists, as shown in Fig. 81, of four or more segments of copper, each opposite pair of which is fixed respectively to one or two smaller rings of different breadth, applied inside the said segments in such a manner that while two opposite segments are fastened to the broad ring placed between two narrow rings, the other opposite segments are fastened to the latter rings in order to form communication between the same, while the said commutator is insulated from the shaft; each opposite pair of segments is also insulated from the other pair. The conductor E extending from the coils of the armature is fixed to the segment F, communicating by means of the said ring with F¹, and the other conductor G to the segment H communicating with H¹. The current induced in the coils of the armatures, and flowing through the conductor E, extending from the coils of the armature and fixed to the segment F, communicates by means of the said ring F¹, and the other conductor G to the segment H, communicates with H¹. The current induced in the coils of the armatures and flowing through the conductor E to the segment F, during one stage of polarity, will during the succeeding change, flow towards the segment H through the conductors G, and so on; the armatures having in the meanwhile made one-quarter of a revolution, the conductor I between the commutator and coils of the electro-magnets will now, after the next change of polarity having taken place, be in contact with the segment F, while the conductor K communicating with the " external circuit " will be in contact with H, and the current having previously flowed towards F, but now flowing towards H, will thus pass in the same direction as it did previously, and of course constantly in one direction.

Fig. 82 represents iron plates or false poles L, applied on the ends of the permanent and electro-magnets, shaped in such a manner

as that a large surface with sharp edges is exposed to the points
of attraction of the armature, while sharp points of different lengths
are exposed to the points of separation, and the edges extending
from the said sharp points are chamfered off towards the points of
attraction, the latter having the same length. Corresponding iron
plates M are applied on the ends of the armatures, and shaped in
such a manner, as shown in Fig. 82, that the points of attraction
expose a large surface with sharp edges to the false poles of the
magnets, being at the same time as distant from the centre as possible,
while the points of separation are near the centre and pointed ; and
the edges of the plates from the said points towards the points
of attraction, are chamfered off with the view to avoid reactionary
currents, and at the same time to facilitate the motion of the armature.
The permanent magnets A and electro-magnets B are shaped in
such a manner, as shown in Fig. 4, that the very ends· of the poles
face the revolving armatures, while this shape of the magnets also allows
the arrangement of the poles on the one side of the revolving armature
to be all north, and those on the other side to be all south. While
steel magnets also may be applied instead of cast-iron magnets, the
permanent magnets may be coiled like the electro-magnets, which also
will serve to make them more permanent.

In November, 1854, N. Knight, of London, described certain improvements in magnetic apparatus, among which is small generator of
which a sketch is given in Fig. 84. Instead of coiling wire over the poles

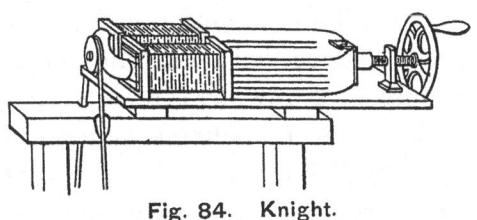

Fig. 84. Knight.

of the permanent magnets, the inventor proposed to make the coils
independent and fixed to a suitable frame, and to slide the poles of the
magnet within them, means being provided to regulate the depths to
which the poles were plunged by means of an adjusting screw. Before
the outer ends of the electro-magnets, a soft iron armature was caused to
revolve rapidly, and a current was generated in this manner.

The *Alliance* Company was formed in 1855, its object being to effect the decomposition of water on a large scale, and to use the gases thus obtained for lighting purposes.   The scheme was indeed extravagant and the hopes of all interested in it, illusory.   It would have been better economy and much more in harmony with sound practical science to have burnt the coal used for driving the machine, in suitable retorts, and employed as illuminants the duly prepared products of combustion.   Be that as it may, capitalists were found ready to subscribe to the preposterous scheme, and the necessary moneys were advanced for the construction of the machines.   The Company wisely secured the services of Joseph Van Malderen, who engaged to build up a generator after the drawings and instructions of his master, Professor Nollet.   The results, however, did not meet the expectations of the sanguine, but showed up their futility, and the Company soon failed.   Shortly afterwards, it was reorganised under the able direction of M. Berlioz for the specific purpose of utilising the magneto-current for electric lighting and electro-chemical decomposition.   The original machine was overhauled and various improvements introduced at the suggestion of Professors Holmes, Combe, Du Moncel, and M. Masson.

This machine, of which a diagram is given in Fig. 85, consisted of a framework of cast iron carrying 40 horseshoe steel magnets, arranged radially and symmetrically around a central axis in eight rows of five magnets each, their poles all being directed towards the axis.   In the spaces between the magnets, upon a horizontal axis, revolved four brass disc wheels, each carrying 16 induction bobbins, each of which consisted of a tube of soft iron wound with insulated copper wire, all being wound in the same direction.   The magnets were so arranged around the machine that opposite poles succeeded each other both in each circular series as well as in each horizontal row, so that at each revolution of the machine, the core of each of the 64 bobbins had its magnetic polarity reversed 16 times, and as it is impossible to reverse the polarity of a magnetised bar without reversing the direction of an electric current induced in a wire wound round it, it follows that at each revolution of the machine no less than 1,024 electric impulses, alternating in direction, were transmitted into the external circuit of the machine.

In 1855, we come to the second specification of Hjorth, which appears to have formed a part of that already referred to, and dated October, 1854.   In this, however, the principles controlling the action of dynamo-electric machines are even more clearly explained, and the illustration shows how closely the inventor had argued out the subject, how far he was in advance of his time, and how much he anticipated subsequent invention. With the exception of the permanent magnets embracing the armature, the machine illustrated in the patent, now nearly 30 years old, is a very close approximation of several of recent date

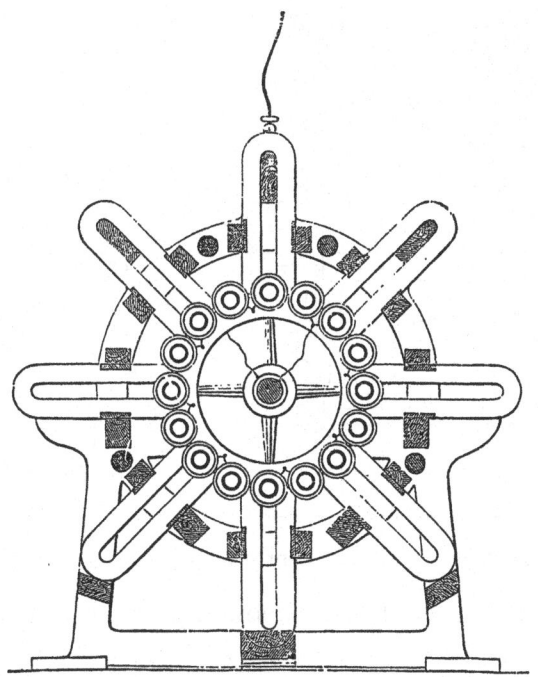

Fig. 85. Alliance.

figured in these pages. It was moreover intended to be a true auto-exciting generator, and combined in one machine, the permanent magnets which Wilde, years after, placed in a distinct machine with the Siemens armature, to excite the electro-magnets of his historical generator. The following description of the Hjorth generator is taken from his specification :—

The invention consists in certain improved arrangements for producing electricity in great quantity and intensity by means of armatures brought in succession within the action of permanent magnets. For this purpose a series of armatures, in a wheel revolving at a slow speed, are brought in succession between the poles of permanent magnets and the poles of a series of electro-magnets, surrounded with spiral rings or coils of copper wire, within cylinders also coiled with wires. The accompanying Figs. 86 and 87, show an end view and plan of my improved magneto-electric battery where *a*, is the revolving wheel; *b*, the armatures surrounded with coiled tubes having false poles, pointed and chamfered off at the points of separation, whilst the false poles of the fixed magnets are inlaid with the brass plates *j* at

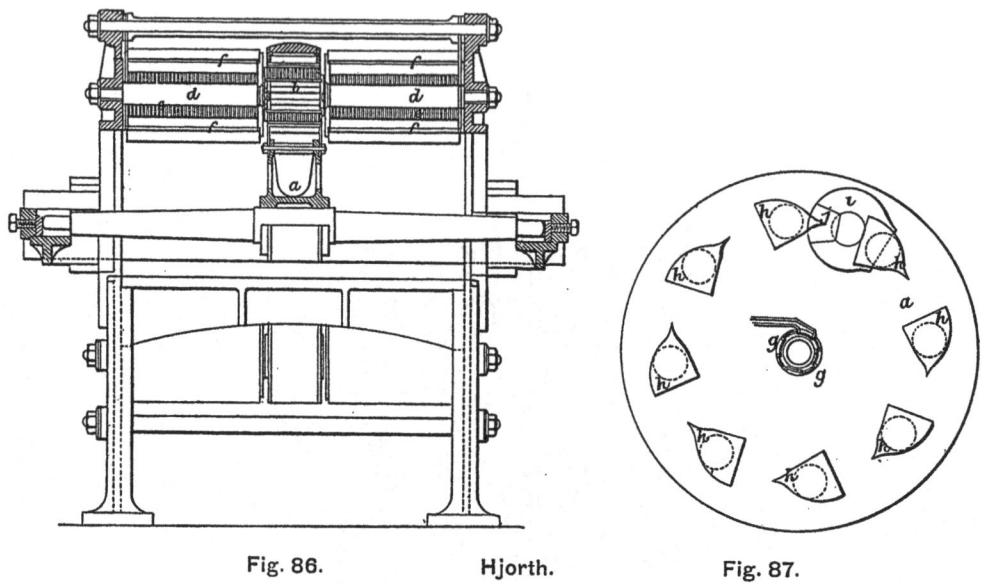

Fig. 86.   Hjorth.   Fig. 87.

the points of separation; *c c*, the permanent magnets, between the poles of which the wheel revolves; *d d*, a series of electro-magnets; *e e*, spiral coils surrounding the electro-magnets; *f f*, cylinders surrounding the coils, coiled on the outside with wires; *g g*, the commutator.

The action of the battery is as follows:—The permanent magnets acting on the armatures brought in succession between their poles, induce a current in the coils of the armatures, which current, after having been caused by the commutator to flow in one direction, passes round the electro-magnets, charging the same and acting on the armatures. By the mutual action between the electro-magnets and the armatures an accelerating force is obtained, which in the result, produces electricity greater in quantity and intensity than has heretofore been obtained by any similar means. The general appearance of this generator is shown in Fig. 88.

A valuable innovation in magneto-electric machines—the most important since the days of Pixii—was made and patented in this country, in 1856, by Dr. Werner Siemens, of Berlin. Struck by the importance of keeping the rotating coils in an intense magnetic field, he devised the arrangement since known as the Siemens armature. It

Fig. 88. Hjorth.

consisted of a long cylinder of soft iron, cut as represented in Fig. 89, with a deep, wide longitudinal groove. In the hollow thus formed, the wire is wound like thread on a shuttle, and is prevented by a number of transverse bands from being displaced by the centrifugal force due to its rotatory motion. The armature was then inserted between the poles of a series of horseshoe magnets and, when set in rapid rotation, developed a powerful current.

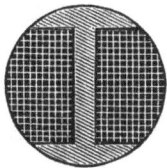

Fig. 89.

Considered as an electro-magnet, its poles are the two uncut portions of the cylinder. Its small compact form allows of its being placed in an intense field and revolved at a very high speed. Like all previous

armatures, it gives discontinuous alternating currents.   The rectification
is effected by a commutator mounted on the spindle and rotating with it.
This armature deservedly met with great favour, and is largely used at
the present day in electro-motors.

We have just seen that Professor Holmes had been consulted
in the construction of the *Alliance* machines.   This was a public
recognition of his ability as a practical electrician, of which he gave
further proof on his return to England, by designing several improved
forms of the *Alliance* generator.

His first patent was taken out in 1856, and his proposed generator
is shown in Figs. 90 to 94. Fig. 90 is a front elevation of a machine

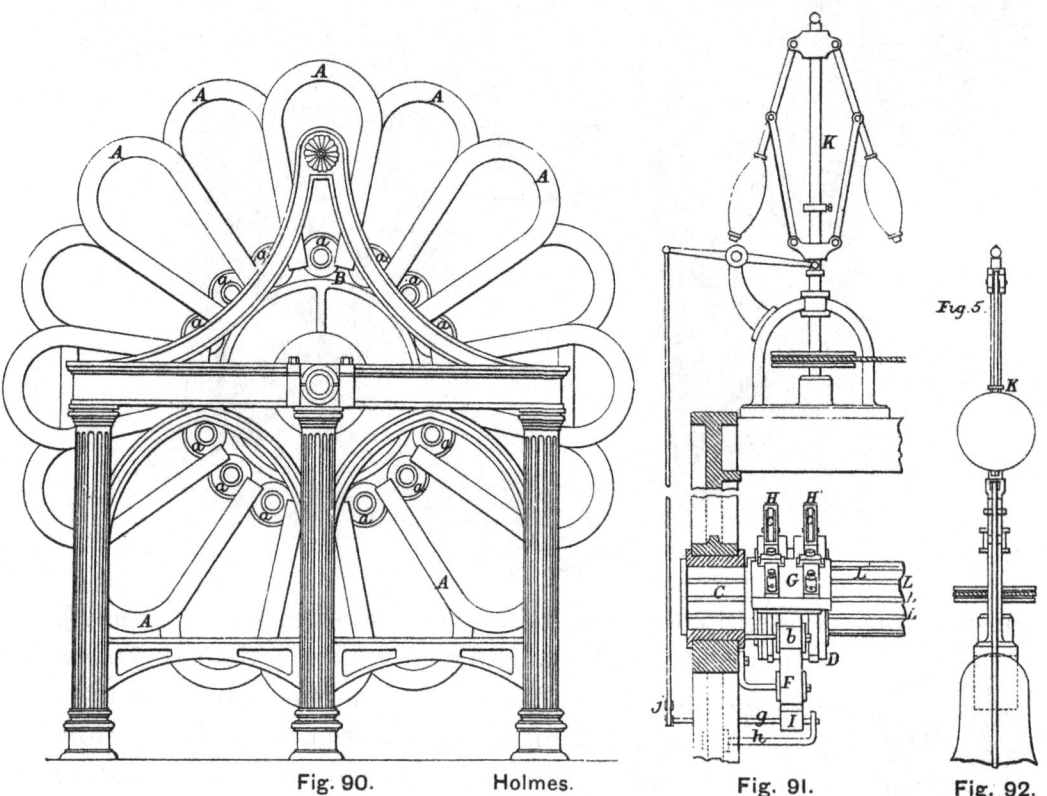

Fig. 90.   Holmes.   Fig. 91.   Fig. 92.

with six discs armed radially, with electro-magnets *a a*. The permanent
magnets A are fixed to the framing. The poles of the magnets are
spaced at equal distances apart, and the electros are so spaced around
each disc that all their centres are always in a similar position with
respect to the corresponding poles of the permanent magnets at
every part of a revolution. The method of connecting the helices

on the disc is shown in the diagram Fig. 94. The commutator is clearly figured in the drawing, and will be found described in detail among the abstracts of Patent Specifications at the end of this volume.

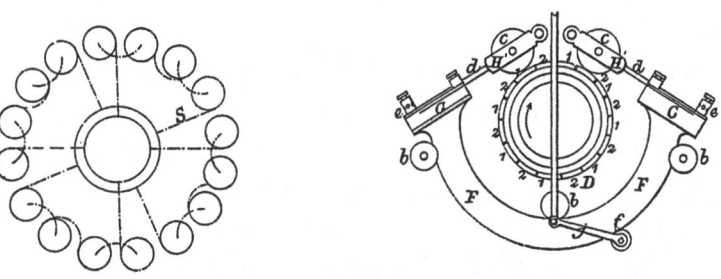

Fig. 93.   Fig. 94.

Figs. 90 to 94. Holmes.

We illustrate the machine here as it is of interest, being the first of a series of arrangements proposed by Mr. Holmes, and the germ from which have sprung generators that have done excellent service.

Professor Holmes lost no time in submitting to the Trinity House a proposal for the introduction of his machine into lighthouses, urging the superiority of the new over the old light in point of brilliancy and penetrating power. The Elder Brethren referred the case to Faraday, their scientific adviser, who, after watching the behaviour of the light with keen interest, reported it to be "so intense, so abundant, so concentrated and focal, so free from under-shadows and from flickering that one cannot but desire it should succeed."

Notwithstanding this favourable recommendation of so high an authority, nearly eighteen months elapsed before any further steps were taken in the matter. The second trial was made at the South Foreland on the 28th of March, 1859. The new and improved machine used on this occasion worked well, the light was bright and steady, and Faraday was able to write in his report "the light produced is powerful beyond any other that I have yet seen so applied, and in principle may be accumulated to any degree; its regularity in the lantern is great, its management and its care may be confided to attentive keepers of the ordinary degree of intellect and knowledge."

But the innovation of Professor Holmes was too great not to challenge another prolonged ordeal. A trial of six months was demanded by the Elder Brethren, during which the light should be entirely under their control. The test was severe, yet Professor Holmes readily acquiesced, and the result fully justified his expectations. It was finally admitted—and the honour is certainly great—that he had practically demonstrated the fitness and sufficiency of the magneto-electric current for lighthouse illumination. His machine was permanently installed at Dungeness, and the electric beam shot forth for the first time on the 6th of June, 1862.

In 1856, Mr. W. T. Henley refers in a specification to a generator in which both permanent magnets and coils were stationary, and a current was set up in the coils by a revolving plate which alternately connected each of the cores to the north and south poles of the magnet, causing reversions of polarity, and electrical currents.

Professor Holmes, in 1857, designed a second machine, intended to yield "a maximum amount of electricity at a low speed of working." Figs. 95 to 97 illustrate this machine, in which a series of permanent

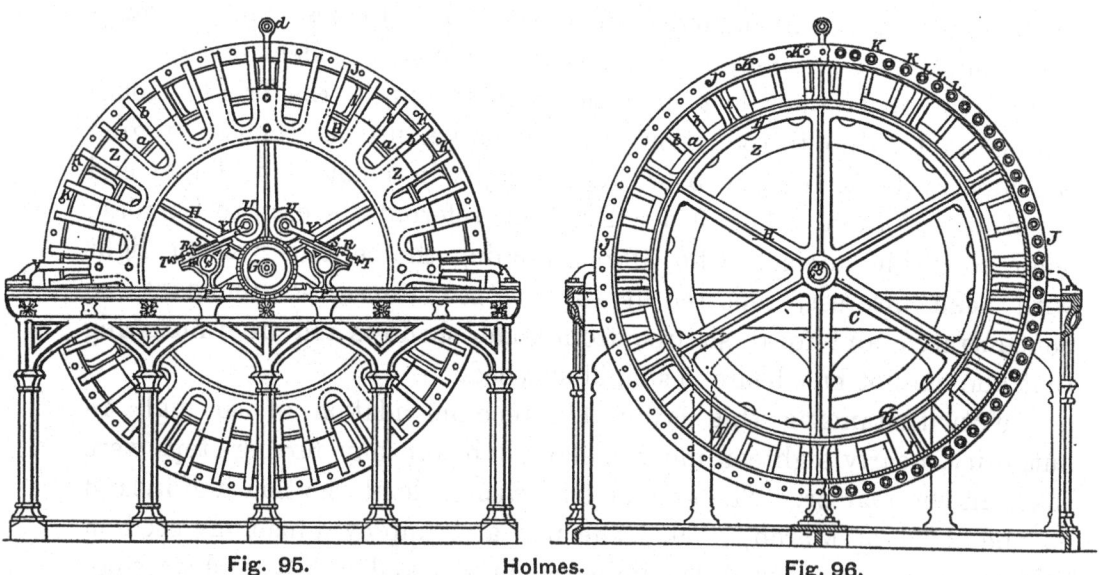

Fig. 95.   Holmes.   Fig. 96.

magnets are arranged radially in circular discs, three or more such discs being mounted rigidly on the frame of the machine, while between them, revolved other discs carrying a number of coils secured between the projecting rim of the wheels, round the peripheries of which they are placed. Fig. 95 is a front elevation of the generator, and Fig. 96 is a section showing one of the wheels around which the coils are arranged. Upon the shaft G are keyed two wheels H, to the rim of each of which is bolted a brass ring, from which projects a number of radial arms, securing the brass trough shaped ring J containing the coils. These are wound on soft iron hollow cores, each core being secured to the rim J by bolts passing through the sides of the rim and the core. The flanges at the end of the cores are made large enough to project slightly beyond the rim J of the wheel, so as to bring them into closer contact with the permanent magnets, and the edges of these flanges are rounded off. The wires forming the coils L are connected in the usual manner, and the terminal wires are carried down through one of the arms H, and through the passage M made in the shaft G. The commutator described for this machine is shown in Fig. 97. It consists of two

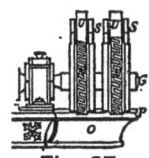

wheels N, fast on the end of the shaft G, and insulated from it. Each of these wheels is made in two parts, put together so as to form only one disc, and near the periphery of each part, diagonal projections are formed

Fig. 97. locking into each other; the spaces left between the teeth prevent contact being established between parts requiring to be insulated. Two of these commutator wheels are keyed on the shaft G, and each is connected with the terminal wires brought through the shaft G, with each series of coils " that is with that series which is shown opposite the spaces between the limbs." On each side of the commutator N is a bracket O, carrying a non-conducting piece P. To each of these pieces is fixed the standard Q having a socket R, which receives the square part of a forked rod S, carrying the roller U made of brass; as will be seen, there are two such rollers resting on each commutator. The conductors leading away the current are attached to the rod S. According to the inventor's specification, a maximum speed of 85 revolutions was sufficient to obtain a powerful and continuous current, the reduction in speed being made up by the increase in the number of magnets and coils. In the machine illustrated, there are three series of permanent magnets, with 20 in each series, and 80 coils in each of the two wheels, or twice the number of helices as there were poles in each of the magnetic discs.

In 1858, we find a very important clause in a patent taken out in November of that year by an inventor whose name is unfortunately not known, but whose agent was J. H. Johnson, of London. In the specification, the author, going a step beyond Hjorth, whose arrangement has been already described, points out clearly the possibility of replacing in magneto-electric machines, the permanent by electro-magnets, thus anticipating by eight years the grand, simultaneous and independent discoveries of Siemens and Wheatstone and Varley. There is no ambiguity in the statement; the words are clear and precise. The inventor says :—"It is also proposed to employ the electro-magnets in obtaining induced electricity which supplies *wholly* or partially the electricity necessary for polarising the electro-magnets, which electricity would otherwise be required to be obtained from batteries or other known sources." It does not appear that this remarkable proposal met with success; it was not until the re-assertion of the principle, in 1867, that its importance and fertility of application were recognised.

The generator of G. W. Beardslee, of New York, patented in 1859, calls for some mention. On a vertical shaft were mounted three sets of permanent magnets arranged with radial bars like the spokes of a wheel. These were placed a sufficient distance apart to admit of fixing between them a series of electro-magnets, so that in rotating, the permanent magnets passed between the electros and induced currents in the coils which were led to a commutator

Wheatstone appears as the patentee of another generator in 1860. In this one, electro-magnets were placed one on each pole of a series of permanent magnets that were ranged around a circle, in the centre of which was a spindle carrying a soft iron armature, the breadth of which was somewhat greater than the distance between two adjacent coils. The wires of these latter were so connected that the currents produced simultaneously in each coil, two by the approach, and two by the receding of the armature, were in the same direction.

We now come to a new era in the development of magneto-electricity, an era of discovery, fruitful invention and steady progress. Most of the machines that we have hitherto described yielded currents alternating in direction and instantaneous in duration; but Dr. Pacinotti, of Florence, showed, as early as 1860, how currents absolutely continuous and always in the same direction could be obtained.

His machine was indeed the prototype of nearly all the direct-current electric generators of the present day, and especially those, such as the Gramme, Brush and De Meritens machines, in which the rotating armature is of annular form; and when it is considered what a large number of the well-known electric generators are founded upon this discovery, it must be a matter of general gratification that the International Jury of the recent Paris Electrical Exhibition (1881) gave to Dr. Antonio Pacinotti one of their highest awards.

The original machine designed by Dr. Pacinotti, in the year 1860, and which is illustrated by Figs. 98 to 102 formed one of the most interesting exhibits in the Paris Exhibition, and conferred upon the Italian Section a very distinctive feature; it is probable that while all were interested in examining it, there must have been many who could not help being impressed with the fact just referred to, that it took something away from the originality of design in several of the generators exhibited in various parts of the building.

This very interesting machine was first illustrated and described by its inventor in the *Nuovo Cimento*, in the year 1864, under the title " A Description of a small Electro-Magnetic Machine," and from this description the information and diagrams contained in these pages are derived, though the perspective view (Fig. 98) is taken from the instrument itself.

Fig. 98. Pacinotti.

In this valuable historical communication, the author commences by describing a new form of electro-magnet consisting of an iron ring around which is wound a single helix of insulated copper wire completely covering the ring, and the two ends of the annular helix being soldered together, an annular magnet is produced, enveloped in an insulated helix forming a closed circuit, the convolutions of which are all in the same direction. If in such a system any two points of the coil situated at opposite ends of the same diameter of the ring, be connected respectively with the two poles of a voltaic battery, the electric current, having two courses open to it, will divide into two portions traversing the coil around each half of the ring from one point of contact to the other. The direction of the current in each portion will be such as to magnetise the iron core, so that its magnetic poles will be situated at the points where the current enters and leaves the helix, and a straight line joining these points may be looked upon as the magnetic axis of the system. From this construction it is clear that by varying the position of the points of contact of the battery wires and the coil, the position of the magnetic axis will be changed accordingly, and can be made to take up any diametrical position with respect to the ring, of which the two halves (separated by the diameter joining the points of contact of the battery wires with the coil) may be regarded as made up of two semi-circular horseshoe electro-magnets having their similar poles joined. To this form of instrument the name "Transversal electro-magnet" (*Elettro-calamita transversale*) was given by its inventor, to whom is undoubtedly due the merit of having been the first to construct an electro-magnet, the position of whose poles could be varied at will by means of a circular commutator.

By applying the principle to an electro-magnetic engine, Dr. Pacinotti produced the machine which is now illustrated.

The armature consists of a turned ring of iron having around its circumference 16 teeth of equal size and at equal angular distance apart, as shown in Fig. 99, forming between them as many spaces or

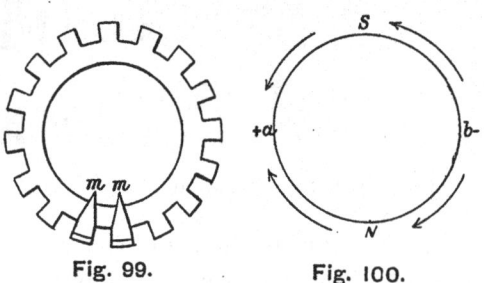

Fig. 99.   Fig. 100.

notches, which are filled up by coiling within them helices of insulated copper wire *r r r*, and between them are fixed as many wooden wedges, *m m*, by which the helices are firmly held in their place. All the coils are wound round the ring in the same direction, and the terminating end of each coil is connected to the commencing end of the next or succeeding helix; the junctions so made are attached to conducting wires which are gathered together close to the vertical shaft on which the armature ring is fixed, passing through holes at equal distances apart in a wooden collar fixed to the same shaft, and being attached at their lower extremities to the metallic contact pieces of the commutator *c* shown at the lower part of Fig. 101, which is an elevation of the machine, while Fig. 102 is a plan of the same apparatus.

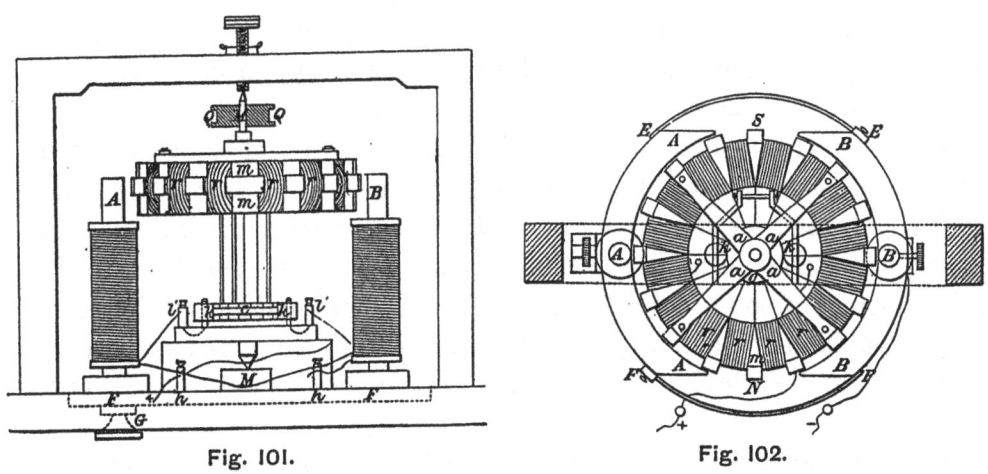

Fig. 101.   Fig. 102.

The commutator consists of a small boxwood cylinder having around its cylindrical surface two rows of eight holes, one above the other, in which are fitted 16 contact-pieces of brass that slightly project above the surface of the wood, the positions of those in the upper circle alternating or " breaking joint " with those in the lower, and each contact-piece is in metallic connection with its corresponding conducting wire, and, therefore, with the junction of the helices on the armature. Against the edge of the commutator are pressed, by means of adjustable levers, two small brass contact-rollers *k k*, which are respectively connected with the positive and negative poles of the voltaic battery (either through or independent of the coils of a fixed electro-magnet to which we shall presently refer), and the magnetic axis of the ring will lie in the same plane as the line joining the points of contact of the battery and rotating helix, this axis remaining nearly fixed notwithstanding the rotation of the iron ring in which the magnetism is induced.

In the apparatus shown in Figs. 101 and 102, the armature rotates between the two vertical limbs A B of a fixed electro-magnet furnished with extended pole-pieces A A, B B (Fig. 102), each of which embraces about six of the armature coils. The fixed electro-magnet is constructed of two vertical iron cylindrical bars A and B, united at their lower

K

extremities by a horizontal iron bar F F, the one being rigidly and permanently attached to it, while the other is fastened to it by a screw G passing through a slot so that the distance of the pole-pieces from one another, and from the armature ring, is capable of adjustment.

The connections of the machine, which are shown in Fig. 101, are made as follows : The positive current entering by the attachment screw *h*, passes by a wire to the right-hand commutator screw *l*, to the right-hand roller *k*, through the commutator to the ring, around which it traverses to the left-hand roller *k*¹ and screw *l*¹ to the magnet coil A, and thence through the coil of the magnet B to the terminal screw *h* on the right hand of the figure. This method of coupling up is of very great historical interest, for it is the first instance on record of the magnet coils and armature of a machine being included in one circuit, giving to it the principle of construction of a dynamo-electric machine, thus antedating, in publication, by two years, the interesting machines of Siemens, Wheatstone, and Varley, and preceding them in construction by a still longer period.

With this apparatus Dr. Pacinotti made the following interesting experiments with the object of determining the amount of mechanical work produced by the machine (when worked as an electro-magnetic engine) and the corresponding consumption of the elements of the battery. Attached to the spindle of the machine was a small pulley Q Q (Fig. 101) for the purpose of driving another pulley by means of a cord on a horizontal spindle, carrying a drum on which was wound a cord attached to a weight ; on the same spindle was also a brake and brake-wheel, the lever of which was loaded so as just to prevent the weight setting the whole system into motion when no current was flowing. In this condition, when the machine was set in motion by connecting the battery, the mechanical work expended in overcoming the friction of the brake was equal to that required to raise the weight ; and, in order to obtain the total work done, all that was necessary was to multiply the weight lifted by the distance through which it was raised. The consumption of the battery was estimated at the same time by interposing in the circuit a sulphate of copper voltameter, of which the copper plate was weighed before and after the experiment. The following are some of the results obtained by Dr. Pacinotti in experimenting after the manner just described : With the current from a battery of four small Bunsen elements, the machine raised a weight of 3·28 kilos. to a height of

8·66 m. (allowing for friction), so that the mechanical work was equivalent to 28·45 kilogrammetres. During the experiment the positive plate of the voltameter lost in weight ·224 gramme, the negative gaining 235 gramme, giving an average of chemical work performed in the voltameter of 229 gramme, and multiplying this figure by the ratio between the equivalent of zinc to that of copper, and by the number of the elements of the battery, the weight of zinc consumed in the battery was computed at ·951 gramme, so that to produce one kilogrammetre of mechanical work 33 milligrammes of zinc would be consumed in the battery. In another experiment made with five elements, the consumption of zinc was found to be 36 milligrammes for every kilogrammetre of mechanical work performed. In recording these experiments Dr. Pacinotti points out that although these results do not show any special advantage in his machine over those of other construction, still they are very encouraging, when it is considered that the apparatus with which the experiments were made, were full of defects of workmanship, the commutator among other things, being eccentric to the axis, causing the contacts between it and the rollers to be very imperfect and unequal.

In his communication to the *Nuovo Cimento*, Dr. Pacinotti states that the reasons which induced him to construct the apparatus on the principle which we have just described, were : (1) That according to this system the electric current is continuously traversing the coils of the armature, and the machine is kept in motion not by a series of intermittent impulses succeeding one another with greater or less rapidity, but by a constantly acting force producing a more uniform effect. (2) The annular form of the revolving armature contributes (together with the preceding method of continuous magnetisation) to give regularity to its motion and at the same time reduces the loss of motive power, through mechanical shocks and friction, to a minimum. (3) In the annular system no attempt is made suddenly to magnetise and demagnetise the iron core of the rotating armature, as such changes of magnetisation would be retarded by the setting up of extra currents, and also by the permanent residual magnetism which cannot be entirely eliminated from the iron ; and with this annular construction such changes are not required, all that is necessary being that each portion of the iron of the ring should pass, in its rotation, through the various degrees of magnetisation in succession, being subjected thereby to the influence of the electro-dynamic forces by which its motion is produced. (4) The polar extension pieces of the fixed electro-magnet, by embracing a sufficiently large number of the iron projecting pieces on the armature ring, continue to exercise an influence upon them almost up to the point at which their magnetisation ceases when passing the neutral axis. (5) By the method of construction adopted, sparks, while being increased in number, are diminished in intensity, there being no powerful extra currents produced at the breaking of the circuit, and Dr. Pacinotti points out that when the machine is in rotation a continuous current is induced in the circuit which is opposed to that of the battery; and this leads to what, looked at by the light of the present state of electric science, is by far the most interesting part of Dr. Pacinotti's paper, published as it was more than 17 years ago.

In the part referred to, Dr. Pacinotti states that it occurred to him that the value of the apparatus would be greatly increased if it could be altered from an electro-magnetic to a magneto-electric machine, so as to produce a continuous current. Thus, if the electro-magnet A B (Figs. 101 and 102) be replaced by a permanent magnet, and the annular armature were made to revolve, the apparatus would become a magneto-electric generator, which would produce a continuous induced current always in the same direction; and, in analysing the action of such a machine, Dr. Pacinotti observes that, as the position of the magnetic field is fixed and the iron armature with its coils rotates within it, the action may be regarded as the same as if the iron ring were made up of two fixed semi-circular horseshoe magnets with their similar poles joined, and the coils were loose upon it and were caused to rotate over it.

In explanation of the physical phenomena involved in the induction of the electric currents in the armature when the machine is in action as a generator, Dr. Pacinotti makes the following remarks: Let us trace the action of one of the coils in the various positions that it can assume in one complete revolution; starting from the position marked N, Fig. 100, and moving towards S, an electric current will be developed in it in one direction while moving through the portion of the circle N a, and after passing the point a, and while passing through the arc a S, the induced current will be in the opposite direction. This direction will be maintained until the point b is reached, after which the currents will be in the same direction as between N and a; and as all the coils are connected together, all the currents in a given direction will unite and will give to the combined current a direction indicated by the arrows in Fig. 100. In order to collect this current (so as to transmit it into the external circuit), the most efficient position for the collectors will be at points on the commutator at opposite ends of a diameter which is perpendicular to the magnetic axis of the magnetic field. With reference to Fig. 100, we imagine either that the two arrows to the right of the figure are incorrectly placed by the engraver, or that Dr. Pacinotti intended this diagram to express the direction of the current throughout the whole circuit, as if it started from *a*, and after traversing the external circuit, entered again at *b*, thus completing the whole cycle made up of the external and internal circuits.

Dr. Pacinotti calls attention to the fact that the direction of the current generated by the machine is reversed by a reversal of the direction of rotation, as well as by shifting of the position of the collectors from one side to the other of their neutral point, and concludes his most interesting communication by describing experiments made with it in order to convert it into a magneto-electric machine. "I brought," he says, "near to the coiled armature the opposite poles of two permanent magnets, and I also excited by the current from a battery, the fixed electro-magnets (see Figs. 101 and 102), and by mechanical means I rotated the annular armature on its axis. By both methods I obtained an induced electric current, which was continuous and always in the same direction, and which, as was indicated by a galvanometer, proved to be of considerable intensity, although it had traversed the sulphate of copper voltameter which was included in the circuit."

Dr. Pacinotti goes on to show that there would be an obvious advantage in constructing electric generating machines upon this principle, for by such a system electric currents can be produced which are continuous and in one direction, without the necessity of the inconvenient and more or less inefficient mechanical arrangements for commutating the currents and sorting them, so as to collect and combine those in one direction, separating them from those which are in the opposite. He also points out the reversibility of the apparatus, showing that as an electro-magnetic engine it is capable of converting a current of electricity into mechanical motion capable of performing work, while as a magneto-electric machine it is made to transform mechanical energy into an electric current, which in other apparatus forming part of its external circuit, is capable of performing electric, chemical, or mechanical work.

All these statements are matters of everyday familiarity at the present time, but it must be remembered that they are records of experiments made 20 years ago, and as such they entitle their author to a very distinguished place among the pioneers of electric science. It is somewhat remarkable that they did not lead him straight to a discovery of the "action and reaction" principle of dynamo-electric magnetic induction to which he approached so closely, and it is also a curious fact that so suggestive and remarkable a communication should have been written and published as far back as 1864, and that it should not have produced a revolution in electrical science.

Fig. 103. Pacinotti.

Before taking leave of this interesting record of Pacinotti's investigations and discoveries, we may refer to two other forms of machines devised by him about the same date, and which were shown also in the Paris Electrical Exhibition. They illustrate forcibly the direction in which his investigations were turned, and are interesting from more than an archæological point of view. In the first of these, Fig. 103, a revolving armature is mounted on a horizontal shaft between four field magnets; in the second, Fig. 104, a vertical armature of the Siemens type is made to revolve before the pole-pieces of two horizontal electro-magnets; the commutator is placed above the armature; the current is collected by the brushes which are mounted as shown, and the angles of which are adjustable. It is to be noticed also that the intensity of the magnetic field can be varied at will, by means of the screws in the front of the machine, and by turning which the distance between the armature and the field magnets can be increased or diminished.

Fig. 104. Pacinotti.

Mr. Wilde meanwhile continued actively to improve his machines. In his specification of 1861, an arrangement is described of combined permanent magnets and coils, between two sets of which an armature revolves, inducing a direct current. In a second patent, dated the same year, a pair of permanent magnets were bent into a semi-circular form, and placed with their poles opposite each other. On the inside of each pole was fixed an electro-magnet, pointing to the centre of the circle enclosed by the curved permanent magnets. A spindle revolved at this centre, carrying an armature. In another form, the permanent magnets were arranged like the spokes of a wheel, the electros projecting from them at right angles. In front of these a star-shaped armature revolved.

In a third patent by the same inventor, dated the same year, a generator is described in which a similar idea is worked out. Reference may here be made to Tyer's patent of 1861, in which the same leading idea is described, though the application is somewhat different. A sufficiently detailed abridgement of Tyer's specification will be found on another page.

The Wilde magneto-electric generator, patented in February, 1863, bears a striking resemblance to the much earlier Siemens machine already noticed. It is illustrated by Fig. 105, and is thus substantialy described by the inventor: The cylindrical soft iron armature k is grooved on each side in the direction of its length to receive the coil; at one end of the armature is the disc pulley and at the other the commutator l. This commutator is formed of two cylinders of hardened

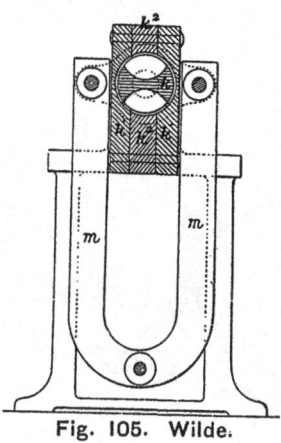

Fig. 105. Wilde.

steel, insulated from each other and cut diagonally so as to run more freely over the springs bearing on it. The armature k revolves in a cylinder formed by two iron pole-pieces k^1, and a centre-piece k^2 of some non-conducting material; outside these pole-pieces is the row of horse-shoe magnets m. It should be mentioned that this little generator is embodied in a patent for magneto-electric telegraphs, and no claim is made for the armature, which was, of course, substantially that of Siemens.

Through the repeated exertions of Professor Holmes and Mr. Wilde, magneto-electric machines were brought up to a great degree of power and perfection. A fresh impetus was now given and a new field opened to constructive talent and skill by the development of the dynamo-electric principle which we have seen so closely approximated by Elias, of Amsterdam, as far back as 1842, virtually introduced by Hjorth in his designs of 1854, and clearly enunciated by an unknown English electrician in 1858; (see page 125).

This principle consists, as we have already explained in treating on the Mechanical Production of Electric Currents, in dispensing entirely with permanent magnets, and in using the traces of magnetism which are found even in the softest iron, in order to set up a feeble current in the coils of the revolving armature. This current, infinitesimal as it may be, is transmitted round the electro-magnets, thereby slightly augmenting their magnetisation. This new increment acts on the coils, inducing a perceptibly stronger current which, in turn, intensifies the field. By a series of rapid interactions, the electro-magnets are soon charged to saturation, and a current of considerable energy is generated in the armature.

The first machine embodying this principle that was ever *constructed* was made by Mr. A. Stroh for Sir Charles Wheatstone in the

Fig. 106. Wheatstone.

summer of 1866. It is illustrated in Fig. 106, and consists of two flat plates of soft iron, fixed in an upright position with a Siemens armature between their pole-pieces. It is provided with the usual commutator for the rectification of the currents before their transmission round the electro-magnets. This machine was shown at the Royal Society, on February 14th, 1867, on which occasion Sir Charles Wheatstone read his celebrated paper "On the Augmentation of the Power of a Magnet by the reaction thereon of Currents induced by the Magnet itself."

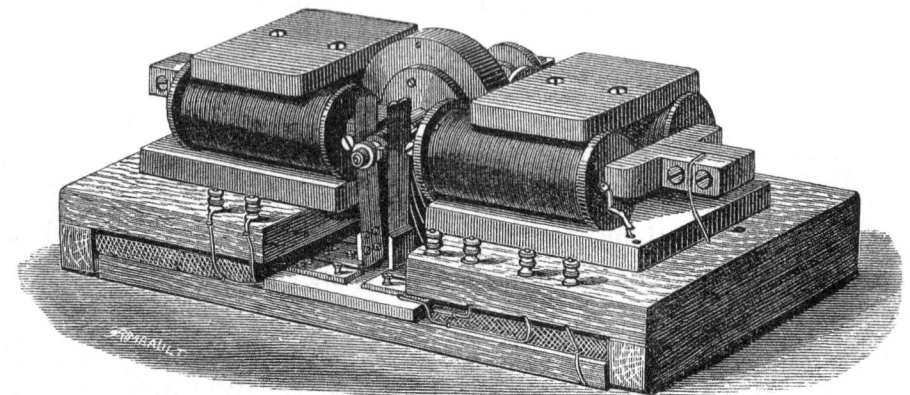

Fig. 107.

On the 24th of December, 1866, Mr. S. Alfred Varley provisionally patented an accumulative machine; his specification, however, was

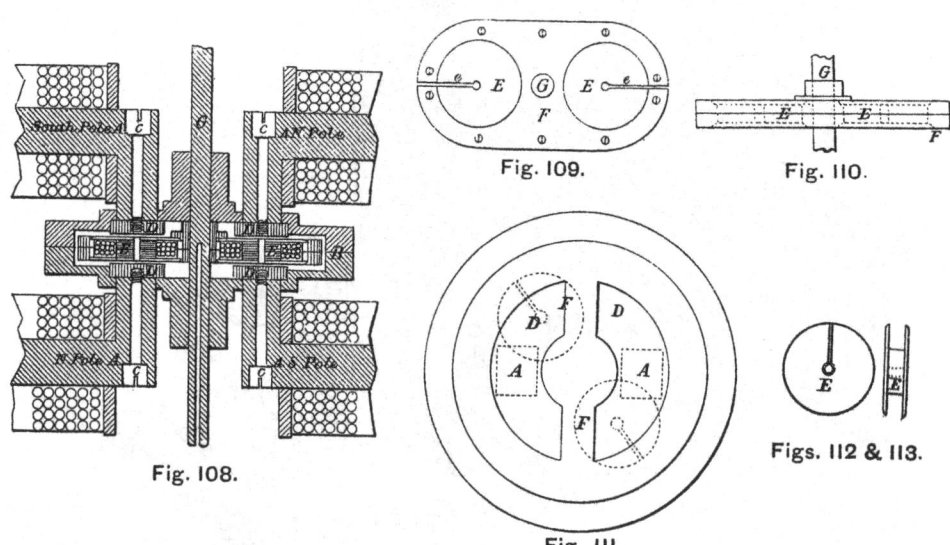

Fig. 108.　　Fig. 109.　　Fig. 110.　　Fig. 111.　　Figs. 112 & 113.

Fig. 107 to 113.　Varley.

published only in the following July. A general view of this machine is given in Fig. 107; the construction is fully illustrated by the detailed Figs. 108 to 113.

We may preface the description of this generator by the following extract from Mr. Varley's specification :—

" We construct our apparatus as follows :—We wrap soft iron bars with insulated wire in a similar way to an ordinary electro-magnet ; these bars may be made ∪-shaped, and become electro-magnets when the apparatus is in use ; we also construct iron bobbins of such a length that they will pass just freely between the poles of the electro-magnets and wrap them with insulated wire.

" In constructing the apparatus we prefer generally to use two electro-magnets and two bobbins ; the bobbins are mounted on an axle and revolve between the poles of the electro-magnets, and are so arranged on the axle that when one of the bobbins is between the poles of one of the electro-magnets the other bobbin is between the poles of the other, and *vice versâ*. On the axle carrying the bobbins there is a commutator ; the ends of the insulated wire surrounding the bobbins are connected to this commutator, and through it to the insulated wire of the electro-magnets, forming the whole into one electric circuit. Before using the apparatus we generally send an electric current through the electro-magnets ; the object of this is to secure a small amount of permanent magnetism in the direction we wish in the soft iron cores of the electro-magnets. On revolving the axle the bobbins become slightly magnetised in their passage between the poles of the permanent magnets, and as the electro-magnets are so arranged that the north pole of the one magnet is opposite the south pole of the other, the bobbins are magnetised first in one direction and then in the other, generating weak currents (corresponding to the direction in which the bobbins are magnetised) in the insulated wire surrounding them. The commutator changes the connections of the electro-magnets at the same time as the magnetism of the bobbins is reversed, and consequently the electric currents developed in the bobbins always flow in one direction through the insulated wire of the electro-magnets. The effect of the current passing through the electro-magnets is to increase their magnetism and to magnetise in a higher degree the bobbins when passing between the poles of the electro-magnets ; more powerful currents are now developed in the insulated wire surrounding the bobbins, which in their turn induce a higher degree of the magnetism in the electro-magnets ; in this way the magnetism developed on the electro-magnets and the bobbins act and react on one another, causing the circulation of increasing quantities of electricity.

" In constructing this apparatus to prolong the contact between the bobbins and the poles of the electro-magnets we generally arm the poles of the electro-magnets or the bobbins with horns of soft iron."

Of the detailed illustrations, Fig. 107 is a general view of the machine, and Figs. 108 to 113 show the more important parts drawn to a larger scale. Fig. 108 is a section of the poles of the field magnets and shows the way the armature bobbins are mounted to revolve between these poles; A A A A are the poles of the soft iron magnets which pass through the sides of the brass box B, and are bolted by means of the screws C to pieces of soft iron D attached to the inside of the brass box, and which become extensions of the magnetic poles; the brass box in two halves is fastened together by screws; Fig. 111 shows the inside of one of the halves and the soft iron polar extension D.

The construction of the iron bobbins E is shown in Figs. 112 and 113, which represent a top and side view of one of the bobbins before being wrapped with insulated wire; the total width of the bobbin from outside to outside is only half-an-inch; the object of making it so short was to reduce the resistance to magnetic polarisation. As will be seen a slot is cut from the centre to the periphery on one side, to prevent the development of electric currents in the iron.

Figs. 109 and 110 show the frame in which the bobbins are mounted; G is the axle attached to the frame by means of which the bobbins are revolved.

One end of the insulated convolutions of each of the bobbins is connected with the metal work of the frame, and the other ends are led through a hole bored in the centre of the axle, and connected to an ordinary reversing commutator mounted on the axle against which four springs press, which close the circuit through the electro-magnets.

It was not only in London that the dynamo-electric principle suggested itself to electricians. As early as January 17th, 1867, Dr. Werner Siemens read a paper before the Academy of Sciences, in Berlin, in which it was clearly expounded. On February 4th, his eminent brother, Dr. William Siemens, of London, sent to the Royal Society, a communication " On the Conversion of Dynamic into

Electrical Force without the use of Permanent Magnetism." This paper and that of Sir Charles Wheatstone, which was sent in ten days later, were read on the same evening, viz., February 14th.

It is usual to ascribe to Siemens and Wheatstone the merit of simultaneous and independent discovery, or rather re-discovery of the reaction principle, thus completely ignoring the claims of Mr. S. Alfred Varley. The question of priority and originality was discussed in a lengthy correspondence in *Engineering*, in November 1877 ; and we think it would be difficult to summarise this valuable contribution to the literature of the subject better than in the words of Mr. Robert Sabine, " Professor Wheatstone says he was the first to complete and try the reaction machine. Mr. S. A. Varley was the first to put the machine officially on record in a provisional specification, dated December 24th, 1866, which was, therefore, not published until July 1867. Dr. Werner Siemens was the first to call public attention to the machine in a paper read before the Berlin Academy on the 17th of January, 1867."

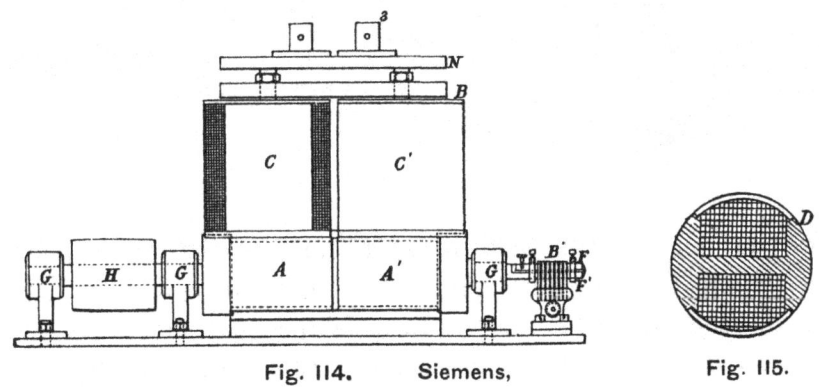

Fig. 114.   Siemens,   Fig. 115.

The machine of Dr. Werner Siemens, which was protected by a patent taken out in this country by Dr. William C. Siemens is illustrated by Figs. 114 and 115. A cylindrical armature similar to that already described (see page 121) as having been patented in 1856, was rotated in front of the poles of electro-magnets, in a direction opposite to that in which it would move if actuated by a current passing through the electros. The mechanical power employed in rotating this armature developed currents within its coil, which were led off by a commutator and were used for lighting and other purposes. It is stated that a magnetic impulse must be first imparted to the electro-magnets " either by the momentary insertion of a galvanic battery, however small, into the system, or by a touch from a permanent magnet, or by dipping the iron bar of the apparatus in a direction parallel to the earth's magnetic axis." After the excitation had been once effected, the residual magnetism was sufficient for future developments of currents. In the illustrations, two out of four field magnets are shown at C C', with pole pieces A A', between which the armature D revolves upon a shaft running in the bearings C, and moved by the pulley H. The commutator is shown at the opposite end of the shaft where it is longitudinally slotted, and an insulating material introduced between the lower portion J F'. One of the brushes is marked B'. The piece B connecting the field magnets carries an insulated plate N; upon this are placed four mercury cups O, O³,

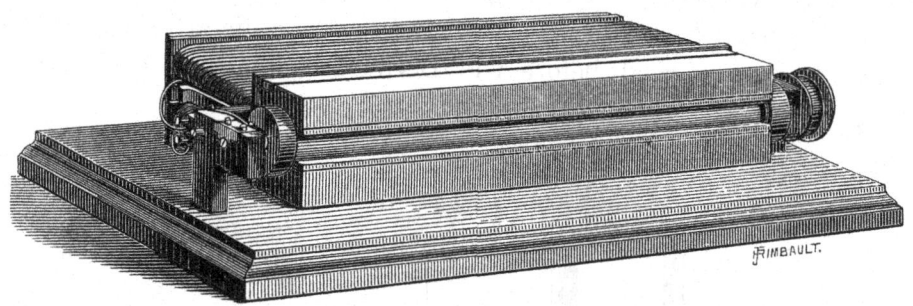

Fig. 116.   Siemens.

to which one end of the field magnet coils are connected, and by means of which the magnets can be coupled in series or parallel, according to whether it is desired to produce intensity or quantity. Fig. 116 shows one of the first generators of this kind actually made by Dr. Siemens. The electro-magnets are placed horizontally, and the armature revolves between the poles which are recessed so as to embrace the greater portion of the armature.

On March 23rd, 1867, Mr. Wilde patented a generator with which his name has been for so many years associated. As this may be regarded as an historical patent we reproduce a very close analysis of the specification. The improved electro-magnetic generator consists of an even number of cylindrical electro-magnets *a* (Fig. 117) arranged on two rings or frames *d*, their coils being joined together so that when an electric current passes through them their poles become alternately N and S, both as regards adjacent and opposite electro-magnets. The fixed electro-magnets are excited by the magneto-electric generator.

Armatures.—Between the two circles *g* of electro-magnets and concentrically with them, electro-helices with iron cores *f* are caused to revolve. The iron cores of both fixed and revolving electro-magnets are provided with overlapping pole pieces *j* (Fig. 118) to preserve the continuity of the magnetic circuit. In addition, the invention consists in a method of effecting the exciting of the electro-magnets from their

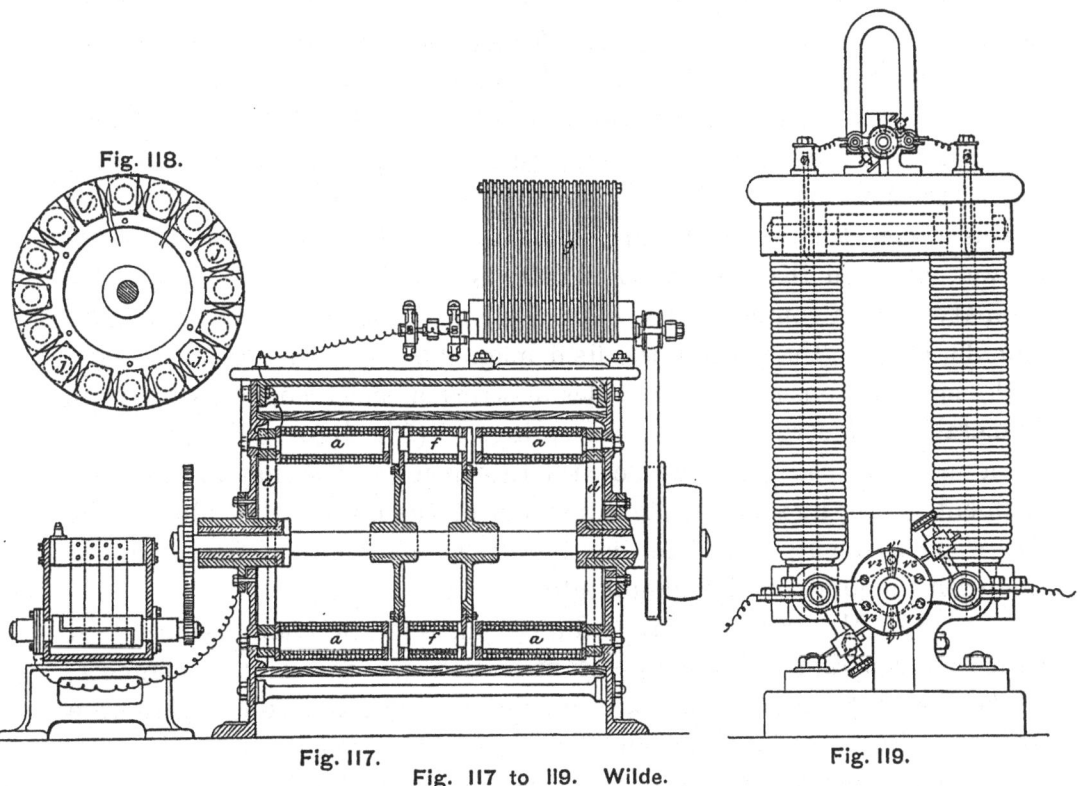

Fig. 118.

Fig. 117.

Fig. 117 to 119.   Wilde.

Fig. 119.

own residual magnetism, and for this purpose the alternating current from one or two of the revolving bobbins, after being turned in one direction by a commutator, is transmitted through the coils of the fixed electro-magnets, the current from the remaining bobbins being applied to the production of light or other dynamic effects. Another method is to put two distinct coils on each armature, one to be used to excite the electro-magnets, and the other for producing the current required outside the machine. Fig. 118 shows the first method, where the coils *j j* are used to excite the fixed electro-magnets.

Another method is to have two distinct circles of armatures, the current from one to be used for exciting the fixed electro-magnets, and that from the other to be available for outside work.

An arrangement is described and illustrated whereby the residual magnetism of the large magnet of an electro-magnetic generator is made available for inducing a current in the small armature which, in turn, excites the large magnet.

Commutator.—The commutator (Fig. 117) is cylindrical and detached from the generator, but geared thereto in such a way that it changes the direction of the current as many times as the magnetism of the armatures is reversed. In Fig. 119 another form of commutator is illustrated, the purpose of which is to effect the short circuit between the coils of the armatures and the coils of the electro-magnet on a second disc, and thus avoid the destructive action of the sparks when the short circuit is made on the commutator itself, as was done in that described in Mr. Wilde's patent of 1865. The ordinary form of commutator is employed, and is placed on the armature spindle, close to the poles of the field magnets. The currents are taken off by springs, consisting of bundles of strips of sheet copper, which are in metallic connection with similar springs V^3 (Fig. 119). The latter bear on a metallic disc fixed to the armature spindle, and having the two segments V^1 on its surface, the remainder of the circle being filled up by the wood segments V^2. The centres of the metal segments V^1 are in line with the joints in the commutator.

In 1867, Mr. Ladd constructed a dynamo-electric machine furnished with two separate armatures, one of which fed the electro-magnets, whilst the other supplied the external circuit. The electro-magnets were soft-iron plates two feet long and one foot wide, fixed horizontally with their similar poles opposite each other. Notwithstanding such small dimensions, the current produced was equal to that of 25 or 30 Bunsen cells. The machine was shown at the Paris Exhibition of 1867, where it attracted much attention and was awarded a silver medal.

Figs. 120 and 121 illustrate the arrangement devised by Mr. F. H. Holmes, in 1868 ; being a side view and a longitudinal section respectively. The rotating magnets c to c^4 may be of circular soft iron plates with steel arms welded radially to their peripheries ; that is when they are to be permanent magnets. If they are to be electro-magnets, the plates and arms are of wrought or cast iron, and the space between their hollow faces contains a coil of insulated wire D, wound on the axis parallel to the side of the plate. The coils F have iron cores, and the

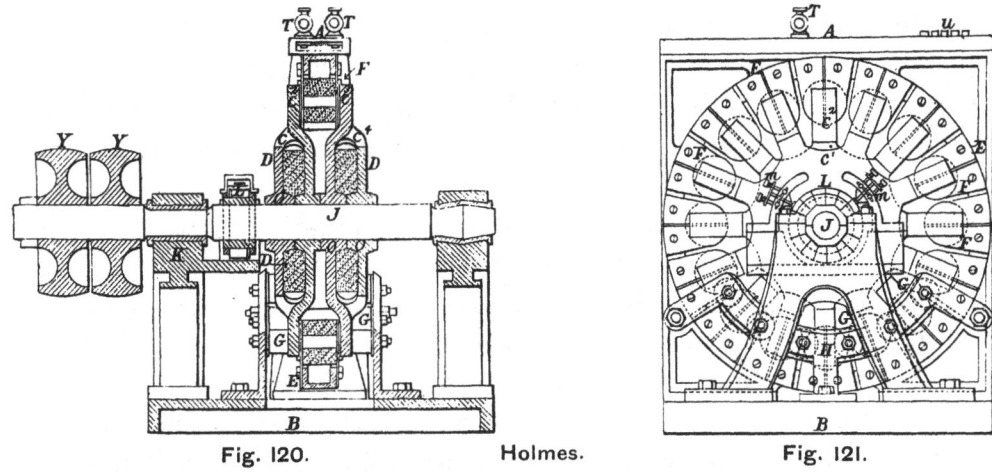

Fig. 120.   Holmes.   Fig. 121.

pole-pieces are split radially and an insulating piece is introduced. When permanent magnets are employed fine wire coils only are used. By suitable connection, any number of pairs of coils c^1 can be used to excite the rotating electro-magnets.

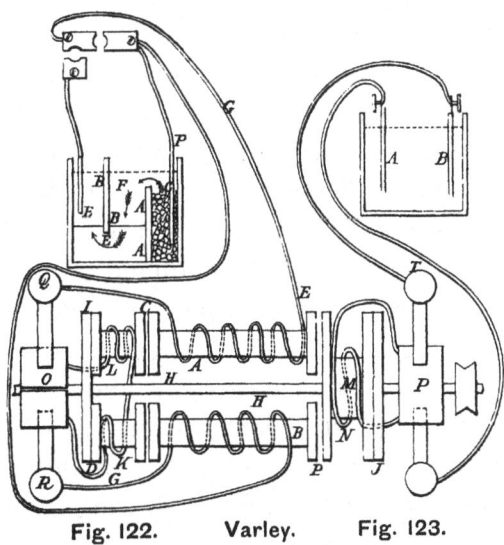

Fig. 122.   Varley.   Fig. 123.

Figs. 122 and 123 show the arrangement of an " electro-dynamic " machine, described and patented by Octavius Varley and Frederick H. Varley, in February, 1869. It consists of two electro-magnets formed on deep and narrow cores placed parallel to each other, and supported by

L

broad flanges at the ends. The shaft H, preferably of brass, lies between the two electros and parallel to them ; it projects beyond the end flanges of the magnets, and means are provided to rotate it rapidly. Near each end of this shaft are mounted the discs I J, carrying on them the electro-magnets X K, L, M, and N, the insulated wires of which " are attached to reversing commutators O and P, forming two rotary magnetic armatures." On passing a current through the electro-magnets A B, and causing the shaft H to revolve, a dynamic current is set up in the coils of wire surrounding the short bars of the rotary magnetic armature, which flows in alternate directions, but which are commutated by O and P to flow from the terminals Q, N, S, T in one direction only. The current set up in one of the armatures flows through the wire G to the electros A and B, and reacting on the rotary armature strengthens the continuous current generated during the rotation of the shaft. The current generated in the other armature flows into the " induction plates " A and B, where the charge is stored up.

We have now passed in review all the principal proposals for the mechanical production of electric currents, prior to 1870 ; many of minor interest, not noticed in the preceding pages, will be found described, in their chronological order, among the abstracts of patent specifications at the end of the volume. Most of the earlier attempts, however ingenious, proved unpractical ; some few bearing the stamp of genius are clearly anticipatory of later and successful developments ; others again dating from this earlier period, have steadily developed into practical types of to-day. But, with a few exceptions, the interest attached to the inventions and propositions prior to 1870 is mainly historical. With that date, however, the problem of producing electrical currents by mechanical means enters a new phase, and with the first patent of Gramme, this branch of industrial science commences a fresh epoch. Important systems became developed and adopted for electric lighting, and as in pursuing our investigations it will be found that these systems will require a more extended and detailed description, we shall take each of them separately, following approximately, but not strictly, their chronological order.

GRAMME.

The currents generated by the dynamo machines of Wheatstone, Varley, and Siemens, though very energetic, were, nevertheless, instantaneous and alternating. The honour of devising a practical generator yielding absolutely continuous currents belongs to M. Zenobie Théophile Gramme, of Paris. His first patent was taken out in 1870, and his first machine was submitted to the Academy of Sciences in July, 1871, when it elicted warm commendation from the members of that learned body.

Adopting the soft iron ring of Pacinotti, he wrapped round it consecutive lengths of insulated wire, thus forming a number of short distinct coils. The ends of these were brought out and formed into one circuit by joining with metallic sectors, which were themselves connected with the novel commutating arrangement of the machine. Fig. 124 represents a Gramme armature partly in section and partly

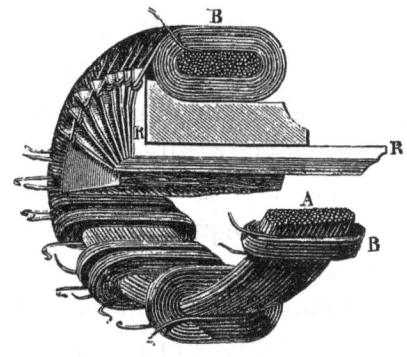

Fig. 124.

dissected, so as to show its principle of construction. The inner ring or core A is made of a continuous length of iron wire wound into a solid circular hank upon a wooden form or mould attached to a lathe, and which is removed when the ring is complete; the wire while it is being coiled, passes through a bath of a bituminous substance, which binds the whole together, as well as tends to insulate the convolutions of the wire. By this means the cross currents within the core are reduced to a minimum, magnetisation is more rapidly accomplished, and the lines of magnetic force within the ring are maintained parallel in direction to the convolutions of the iron wire and therefore with the annular axis of the ring. Around the iron wire ring so formed, are wound the lengths of insulated copper wire, one of which is shown at the lower part of the figure marked B. These induction coils are wound close together, and the contiguous ends of each adjoining pair of coils are joined to the metallic strip or conductor R R, there being as many of these conductors as there are coils wound on the ring. The opposite ends of these conductors are in metallic connection with as many insulated copper strips arranged side by side around a circle, so as to form sectors of the commutator cylinder upon which the metallic collecting brushes are pressed by means of springs and screws. These brushes are composed of hard copper wire, electro-plated with silver, and their object, by forming a rubbing contact between the internal and the external circuits of the ring, is to offer a passage for the currents induced in the coils of the armature, so as to enable them to be used for doing external work.

When such a ring is mounted on a spindle passing through its axis, and rotated in its own plane in a magnetic field produced by two opposite magnetic poles diametrically opposed to one another, electric currents are induced in its coils which, if no external circuit were provided, would be equal in strength and opposite in direction, and would therefore neutralise one another, but by placing collecting conductors in suitable positions the two opposing branches of the current unite and flow out together, forming an external current of their combined strength. We have already considered the theory of the Gramme machine; it may, however, be useful to examine a little more in detail the manner in which the currents are generated. In Fig. 125 N and S are two magnets having their dissimilar poles opposed to one another, and producing a magnetic field between them, within which is rotated a ring of iron which is wound with an endless coil of insulated wire. If now it be supposed that those convolutions which are on the outside circumference of the ring, are stripped of their insulating covering, and against the denuded wires so produced, two fixed springs or brushes X and Y are caused to press at points A and B at opposite ends of a diameter which is perpendicular to the line passing through the axes of the magnets and the centre of the ring, and if wires forming part of the external circuit be attached to X and Y, then the following phenomena will take place. In the first place that portion of the iron ring to which the north pole is presented will assume, by induction, south polarity, while at the opposite side to which the south pole is presented, north polarity will be induced.   The ring may therefore be regarded as made up of two semi-circular horseshoe electro-magnets with their similar poles joined together in a line with the axis of the magnets, and as these magnets are fixed, the polarity of the ring is also fixed with regard to space, whether the machine is in motion or not.   The effect however of rotating the ring is to cause each point in succession around the ring to be brought under the inductive influence of the magnets N and S, which has precisely the same effect as if waves of magnetic force were constantly travelling round the ring in the reverse direction to that in which the ring is turning, a corresponding series of waves of

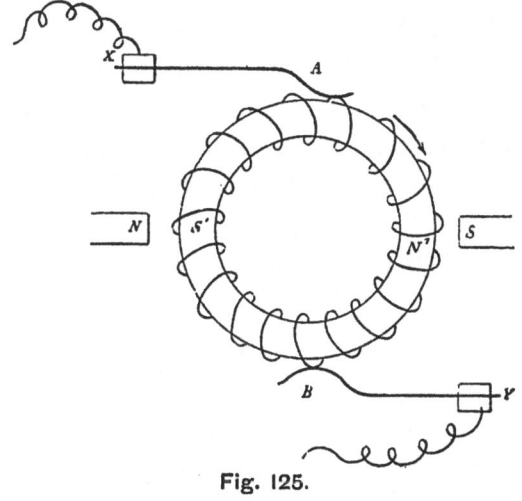

Fig. 125.

electricity would be induced in the wire surrounding it, and there would be no change of effect if the iron portion of the ring were fixed and the coils of surrounding wire were caused to run round it.   But while the iron ring plays this important part in the phenomena, it is but one of the causes at work, for a still more important action is taking place between the lines of force within the magnetic field and the convolutions of the coils of the induction ring.   As the lines of force constituting the magnetic field of the two magnets form a series of elliptic arcs joining their two poles (as we have already seen) it follows that not only does a large proportion of those lines of force with the iron ring, thereby causing it to become strongly magnetic, but the convolutions of the induction coils, attached as they are radially to the ring, are moved during its rotation within the magnetic field in directions perpendicular to a large per-centage of those lines of force, and therefore in a way especially advantageous to the induction of magneto-electric currents; this is illustrated by the diagram Fig. 126. Now when a closed solenoid or coil of wire is made to approach the north pole of a magnet, a current of electricity is induced in it, which is in the reverse direction to that induced in a similar coil approaching a south pole. In the rotation of the Gramme ring between the magnets in the direction shown by the arrow (Fig. 125) the coils surrounding one half the ring have induced in them a positive current, while those coiled on the other half of the ring are traversed by a negative current. When the external circuit between the rubbing springs X and Y is open, the two currents in the coils being equal and opposite, neutralise one another, but when external connection is made between X and Y the currents unite and flow out into the external circuit. In this respect the action of the two halves of the ring bears a close analogy to that of two equal voltaic batteries joined together by their similar poles. Fig. 127 represents two voltaic batteries $c \ z$ and $c^1 \ z^1$

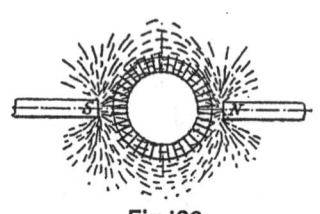

Fig 126.

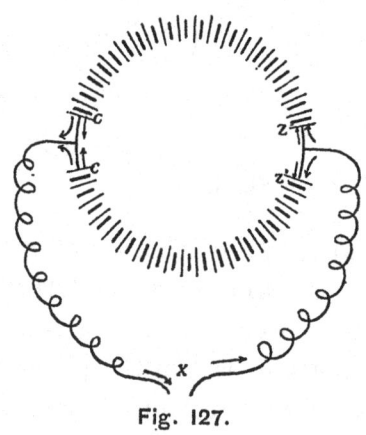

Fig. 127.

of equal power connected together in this manner, and, for the sake of making the analogy to the ring more close, they are shown as if the cells of each battery were arranged in a semi-circle. As long as the external circuit remains open as at X, the current produced by one battery will neutralise that of the other, and no work will be going on, but the moment that the circuit is closed at X, then the currents from the two batteries unite so as to produce an external current equal to the sum of the two.

Fig. 128.

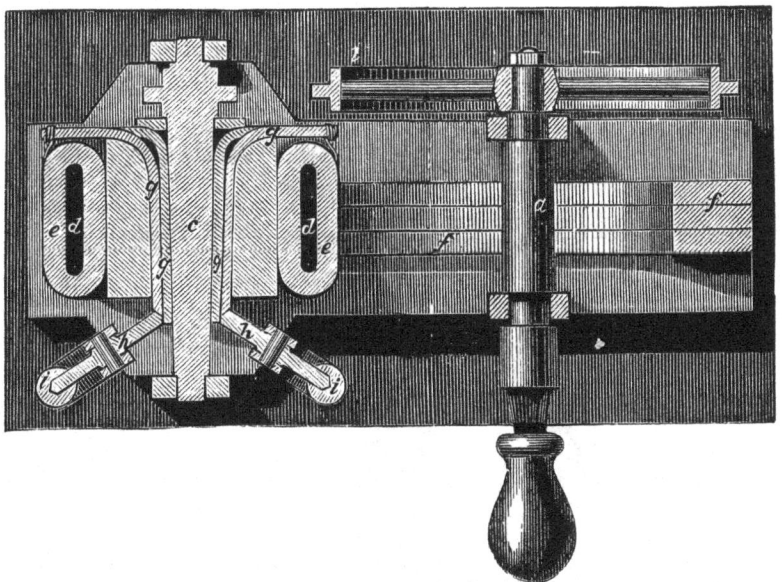

Fig. 129.

In the earliest rudimentary machine (Figs. 128 and 129) the field was made by three permanent magnets f; e is the ring armature rotating between two circular pole-pieces; h h, metallic discs which press against

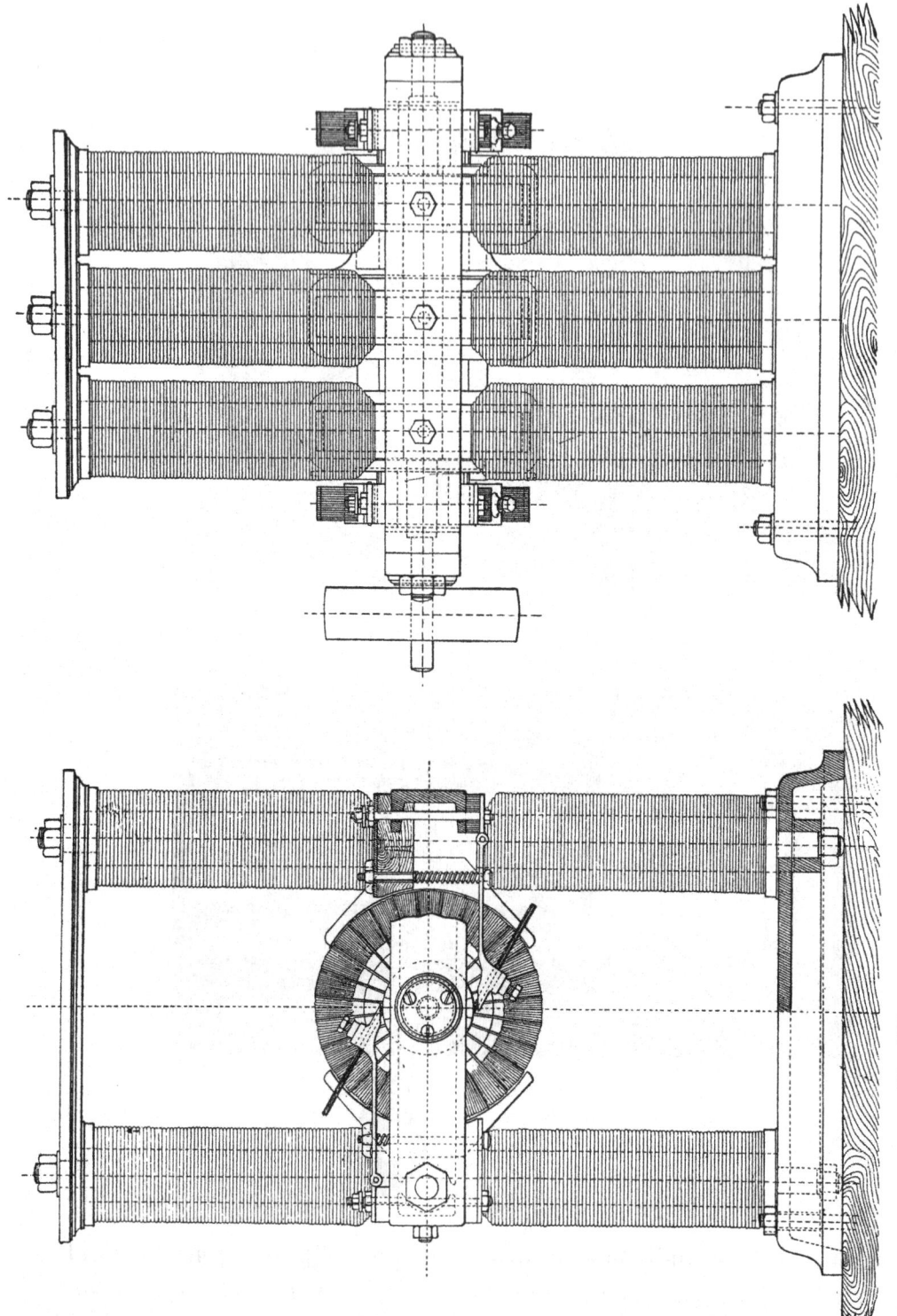

Fig. 130.

Fig. 131.

Gramme Generator used in 1873 for the signal light of the Westminster Clock Tower.

the surface of the spindle and collect the current from the axial conductors, and *i i* are the terminals of the machine.

In the larger machines, made somewhat later (1872), the permanent were superseded by electro-magnets. Figs. 130, 131, and 132 represent one of these with six vertically-placed cylinders of soft iron having similar poles opposite each other but not in contact. This arrangement was adopted to facilitate the insertion of three armatures (Fig. 132), one of which was used to excite the electro-magnets, whilst the other two supplied the current for external use.

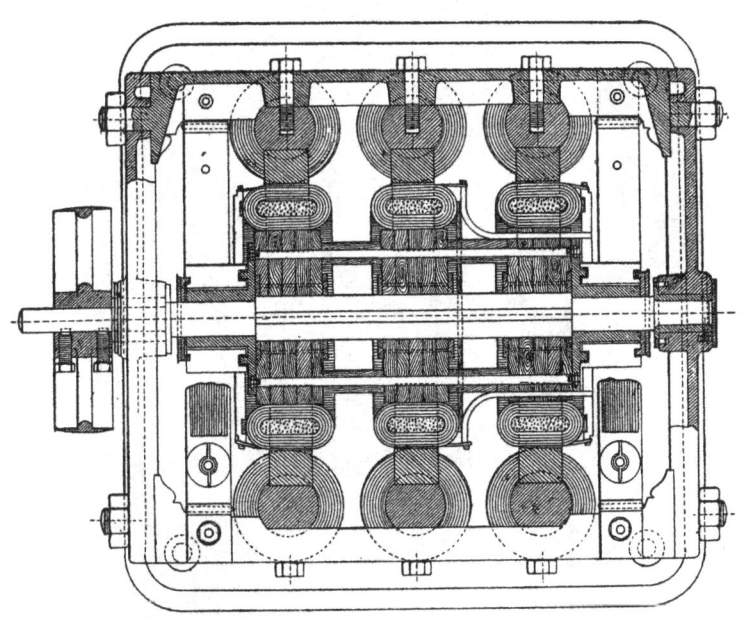

Fig. 132.

This machine was brought to London in 1873 and used in the production of the electric light which, during several months' competitive trial with Mr. Wigham's gaslight, shone out from the lantern of the Clock Tower of the Houses of Parliament. Some interesting experiments illustrative of the vast energy of the new machine were made under the direction of Mr. Conrad W. Cooke in presence of many leading scientific men. A platinum wire 18 B. W. G. and 15 feet long was raised to vivid incandescency, whilst a length of 32 feet of highly conducting copper wire was rendered red hot. A round file, half an inch in diameter and four inches long, was fused in five minutes, and a small diamond instantly volatilised. When the armature was driven at a speed of 300 revolutions per minute, the intensity of the light was estimated as equal to that of 900 Carcel burners.

Another of the early forms (1873) of the machine, designed especially for electro-metallurgical purposes, is shown in Fig. 133; B B

Fig. 133.

are the electro-magnets, four in number, and they had the wire so coiled that poles of the same name are opposite each other. H is one of the armatures fixed on the axle of the driving-wheel E; D D a pair of silver brushes to collect the current. A similar arrangement is placed near E to collect and transmit the current from the second armature round the soft iron cylinders, thus converting them into a powerful magnetic magazine. The wire wound round the electro-magnets weighed 270 lbs.; that round the two armatures 80 lbs.; the whole machine 920 lbs. The normal speed was 300 revolutions per minute and the current developed was equal to that furnished by 32 Bunsen cells.

These forms of the Gramme machine were characterised by a plurality of armatures; but in the newer and far more powerful machines this arrangement is superseded, there being but one induction ring, and the coils of the electro-magnets are included in the circuit, the whole of the current traversing them. Fig. 134 represents a more recent form of

Fig. 134.

a modern Gramme generator, such as is used for lighting purposes. The ring is mounted on a horizontal shaft, which is revolved at a speed of about 900 revolutions per minute within the magnetic field of the four horizontal electro-magnets shown in the drawing. The two upper magnets are united to a common pole-piece, which embraces about three-eighths of the circumference of the ring, and the two lower magnets are connected to a similar pole-piece opposite in polarity to that attached to the upper magnets.

Smaller Gramme machines have been made by M. Breguet for use in physical laboratories. They are all of the magneto-electric type, the field being made either by the ordinary horseshoe, Fig. 135 and 136, or by Jamin's laminated magnet, Fig. 137. They are constructed to be driven either by hand or by steam power.

By coupling together two of these machines in such a way as to send the current developed by one, through the armature of the second, the latter rotates. It thus appears that the Gramme machine may be

Fig. 135.

used either as a generator or as a motor. By spending energy in driving the armature we obtain an equivalent of the work done in the form of an electrical current; by sending a current from an independent source, a voltaic battery or another machine, we convert the electricity into energy of motion. The Gramme is, therefore, a reversible machine. This property of reversibility is among the most valuable discoveries of contemporary science, enabling us to complete the theory of electric-motors, and opening up new avenues for the application of electricity. Figs. 138 and 139 show two standard types of generators as made in 1874.

Fig. 136.

Fig. 137.

In the former of these the vertical electro-magnets were arranged in two triangles, instead of being parallel. In this form the two armatures were retained, and the current was used for one or two lights simultaneously;

or the current from one armature was used to excite the electro-magnets. This machine weighed 1,540 lbs.; the wire on the electro-magnet weighed 396 lbs., and that of the armature, 88 lbs. It could feed one light of 500 Carcels, or two of 150 Carcels each.   Fig. 139 is one of the earliest forms

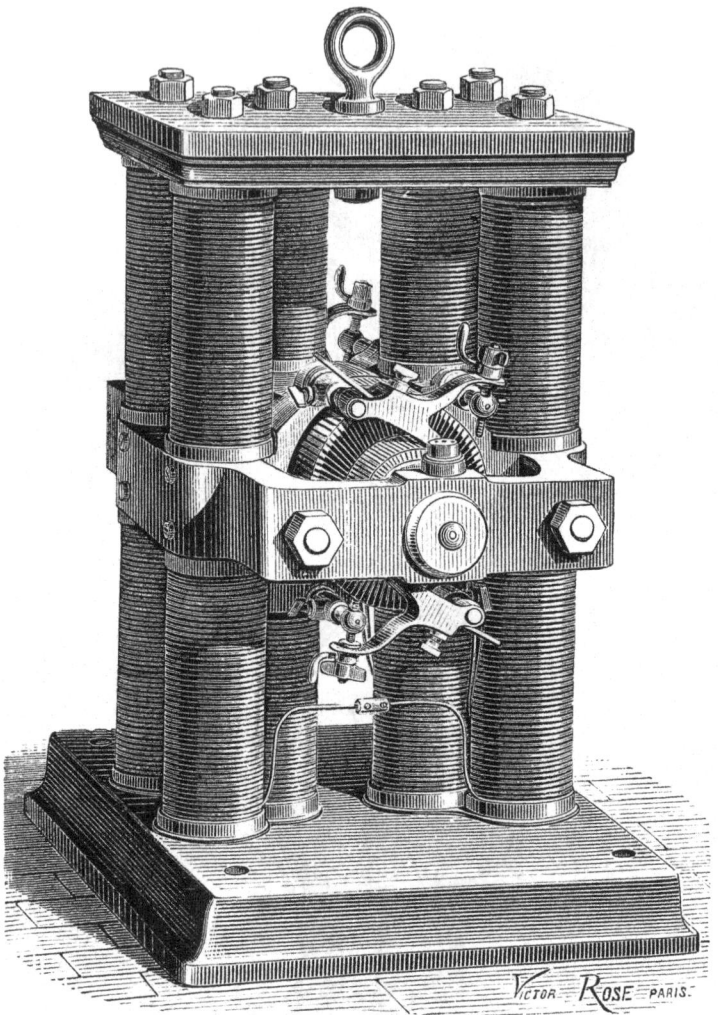

Fig. 138.

with horizontal electro-magnets; the arrangement will be clearly under-stood from the drawing, when it will be noticed that there are two sets of collectors on each side of the poles, in order to obtain two distinct currents, which, by means of the longitudinal cylinder at the base of the machine, could be coupled for quantity or tension.

Generators similar to that illustrated in Fig. 138 were used on board the "Suffren" and the "Richelieu" of the French Navy; of the "Livadia" (not the more modern Imperial yacht of that name) and the "Peter the

Fig. 139.

Great" of the Russian Navy. The results obtained were good, but the cost of the machines was high, and the luminous power given was not very intense. Out of Figs. 138 and 139 was developed the type (shown in Figs. 140 and 141) which gave much better results. In this machine the two vertical cast-iron frames were connected by four horizontal rods of steel forming the coils of the electro-magnets.   The ring, instead of being wound with one wire connected by equal fractions to a common conductor, was formed of two wires of the same length rolled parallel, and connected to two collectors for taking off the current.   The pole-pieces of the field magnet embraced

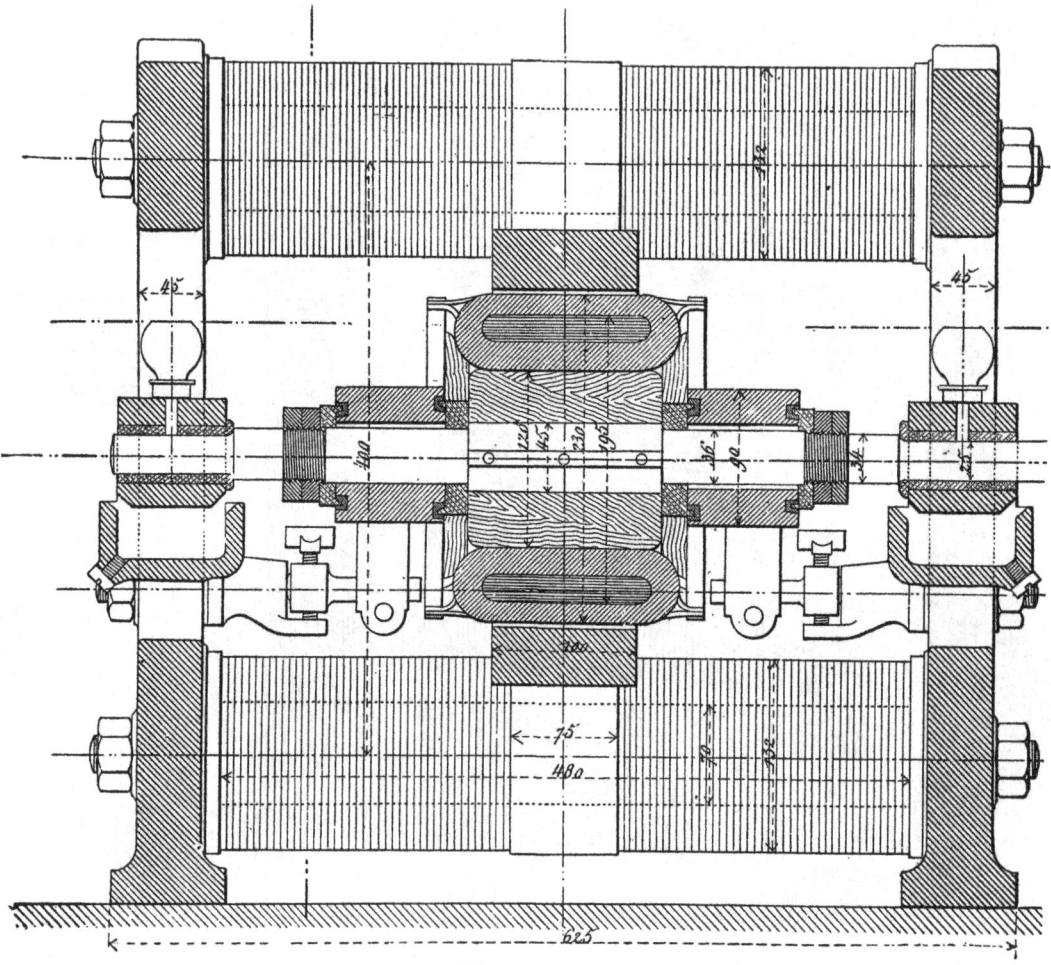

Fig. 140.

the greater part of the ring; the electro-magnets were placed in the circuit.   The double coil on the ring was connected to 120 collecting bars, 60 on each side.   The following are some of the leading particulars of the generator, which was adopted by the War Department in France as well as in other countries :—

Electro-Magnets :

Diameter of the core of electro-magnet	2·75	in.
Length	15·90	,,
Diameter of coil	5·19	,,
Diameter of wire	·0125	,,
Weight of copper rolled around each electro-magnet	52·8	lbs.

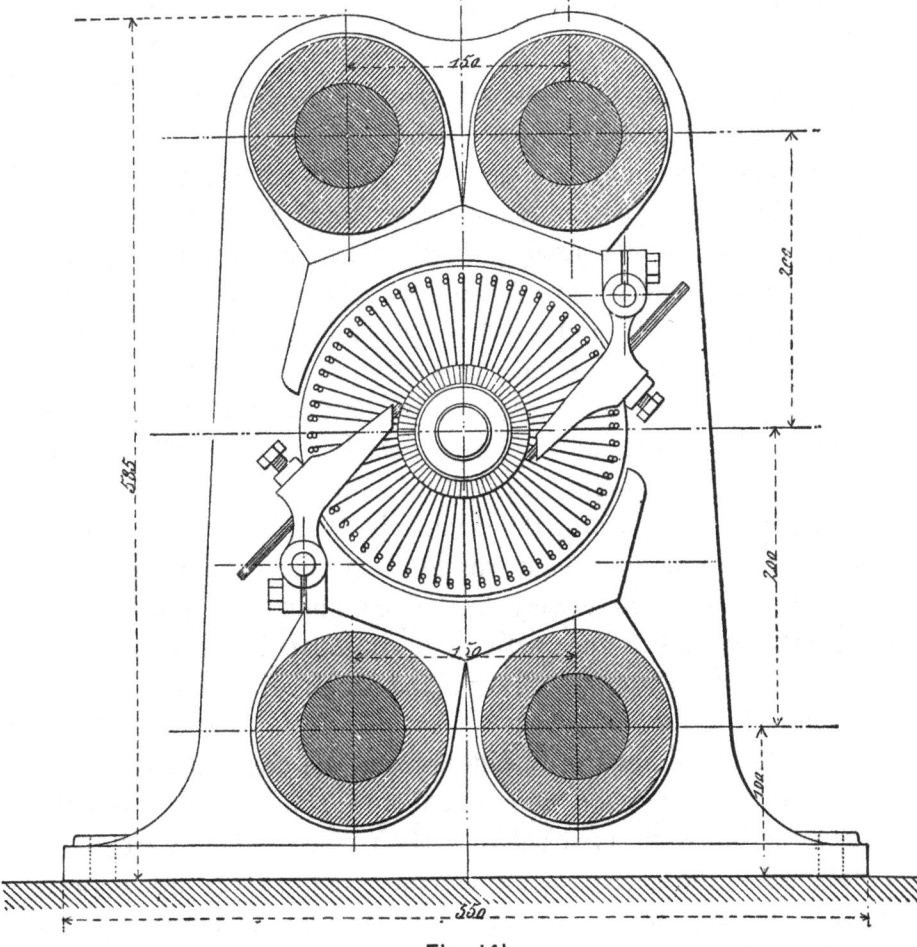

Fig. 141.

Armature :

Outside diameter of soft iron ring	7·67	in.
Inside	6·18	,,
Width of soft iron ring	4·68	,,
Outer diameter of armature	9·05	,,

M

Inner	....	....	....	....	....	....	4·72	in.
Diameter of wire		....	....	....	....	·01	,,
Total weight of wire	....		....	...	....	308	lbs.
Diameter of conducting cylinders			....	....	3·54	in.
,, ,, lamp wire....		....	....	....	03	,,

Machine :

Total length, including pulley		....	....	31·5	in.
,, height	....	....	....	....	....	20·03	,,
,, width	....	....	....	....	....	21·65	,,

Figs 142 to 146 illustrate the type of machine employed in 1878 for working the Jablochkoff light in the Avenue de l'Opera, and of which a large number have been made. It consists of a ring of soft iron similar to that used in the earlier continuous-current machines of the same inventor. This ring is, like its predecessors, wound with coils of insulated copper wire, but unlike those of the earlier machines the direction of winding is alternately right and left handed, the wire being wound in one direction, so as to cover one-eighth part of the circumference of the ring, then changing its direction it is wound in the contrary way over the next eighth part, and so on round the ring. Each of the eight sections is wound in the reverse direction to that in which its two contiguous sections are wound. Thus, while the ordinary Gramme ring might be described as an electro-magnet bent round in a circle, and joined to itself, the ring of this machine may be looked upon as eight curved electro-magnets placed end to end, with their similar poles in contact so as to form a circle. This ring is rigidly fixed in a vertical position to the solid framing of the apparatus, the inductive electro-magnets revolving within it. Here, again, it differs from the continuous-current machine in which the magnets were fixed, and the ring was rapidly rotated between the magnets fixed between their poles. There are eight electro-magnets fixed radially to a central boss revolving upon a horizontal steel shaft running in suitable bearings attached to the framing; an external pulley enables the machine to be driven by a band from a steam engine or other motor. The radial electro-magnets are alternately right and left handed in the direction in which their coils are wound, so that if they be numbered respectively 1, 2, 3, 4, &c., up to 8, those represented by even numbers will have one polarity when a current is sent through them all together, while those whose numbers are uneven will have an opposite polarity. The poles furthest from the central boss, in all the magnets, are spread out so as to increase the area of the magnetic field by which electric currents are induced in the coils of the ring.

In the early machines of Gramme, which we have already described, the electro-magnets were excited by the current from a special Gramme ring mounted on the same shaft as the two rings which produced the current for the external circuit, and the current

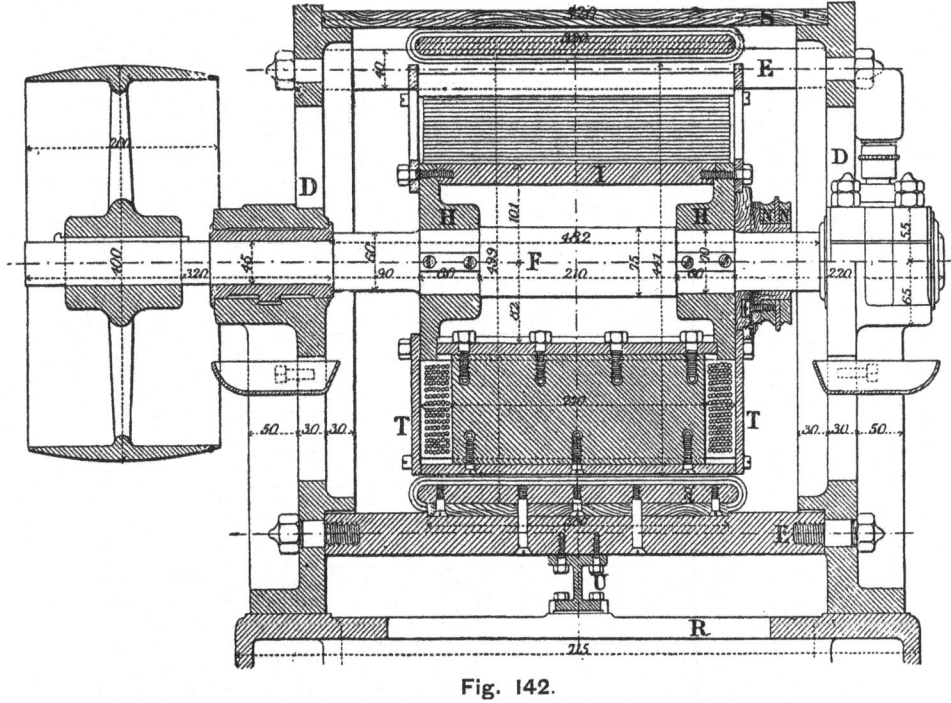

Fig. 142.

from the first ring was exclusively devoted to exciting the electro-magnets. In further improvements upon his continuous-current machine, we have seen how M. Gramme dispensed with an extra and special ring for charging the magnets, and devoted a portion of the current from a single ring to that purpose, two coils being wound on the same ring. In the machine now being considered, however, there is no self-contained apparatus for producing the current by which the electro-magnets are charged, although there is nothing to prevent an ordinary Gramme machine being attached to the same shaft. It is found more convenient, however, to use this machine simply as an apparatus for dividing the current and for giving to it an alternating character, and therefore an altogether separate exciter is employed. The old device of an ordinary voltaic battery of sufficient power would have answered the purpose, but in practice a small and separate Gramme machine of the continuous-current type was employed, driven by a separate strap, and the current from the machine being caused to traverse the coils of the rotating radial electro-magnets, they were magnetised to saturation with but a small expenditure of motive power.

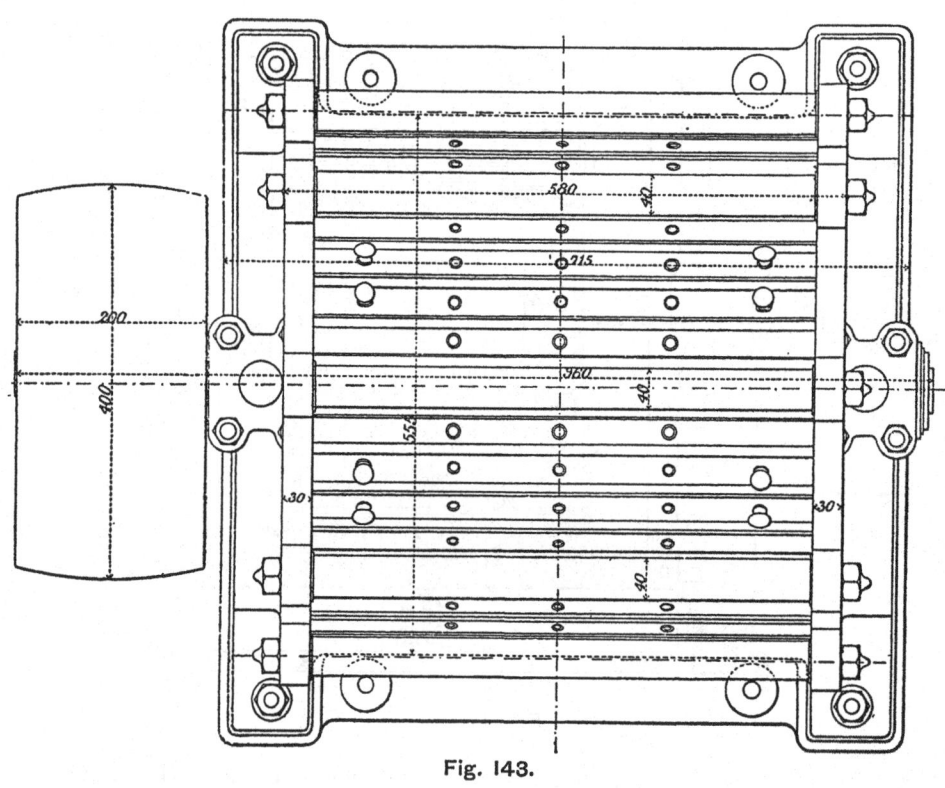

Fig. 143.

Each of the eight sections of the induction rings, is made up of four sub-sections, *a, b, c, d, a, b, c, d,* &c. (see Fig. 144), all of which in any one section are wound in the same direction. By coupling the coils of these sub-sections in various ways, a division of the current may be made into 32, 16, 8, or only 4 circuits, and this is one of the most valuable qualities of the system. From a consideration of the above description it will be seen that all the sub-sections or coils marked *a* are influenced by the rotating magnets in precisely the same manner, for the influence of a north pole upon

a coil wound in a positive direction, is precisely the same as that of
a south pole upon a coil wound in a negative direction, and similarly

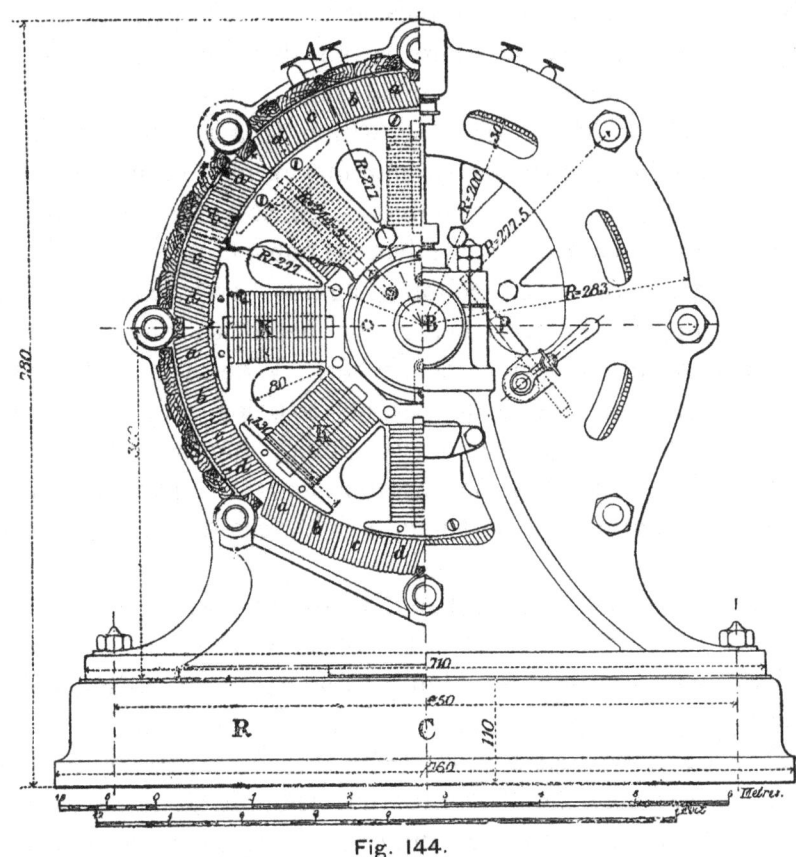

Fig. 144.

the currents induced in all the coils marked *b* are similar in direction
to one another, no matter what may be the position of the rotating

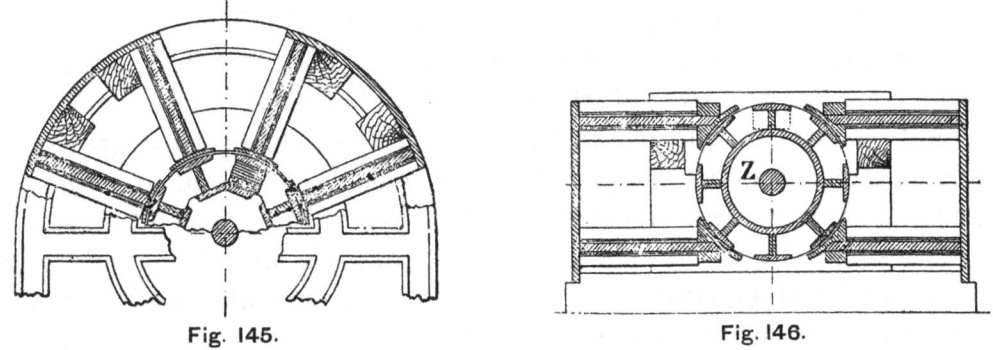

Fig. 145.   Fig. 146.

magnets.   And in order to obtain four currents from the machine
by which four lamps may be illuminated, all that is necessary is to

connect together in series all the coils marked *a* for one circuit, all the coils marked *b* for a second circuit, all the coils marked *c* for a third circuit, and all the coils marked *d* for a fourth circuit. As the induction rings are fixed, there is no difficulty in tapping the currents from the machine; the apparatus is fitted with terminal screws on the exterior of the rings by which this can be effected, and by which variations in the method of coupling-up may be made.

The current from the small machine is led to the rotating magnets through flat brushes composed of silvered copper wire fixed to the framework of the machine and rubbing against two insulated drums or cylinders of copper, the first being connected to one end of the magnet circuit, and the second to the other. When the shaft carrying the magnets is rotated, the wide magnetic poles pass close to the coils of the induction ring, and in so doing, induce within them currents of electricity, the strength of which depends upon the intensity of the magnetic field brought over them, and increases with the speed of rotation.

The generator represented in the figures is capable of supplying the current to 16 Jablochkoff candles when worked at a speed of 600 revolutions per minute, and, doing this duty, it absorbs about 16 horse-power.

Figs. 147 to 152 show a later form (1879) introduced by M. Gramme. This machine differs considerably from the Gramme generator brought out in 1878, both in the greater amount of light it is capable of producing per horse-power, and in its lower first cost.

By reference to the engravings it will be seen that the arrangement of the machine, which is self exciting, is as follows: On a cast-iron foundation are fixed two plates of the same metal almost circular, connected together by six square bolts and provided with bearings for the main shaft. One of these plates is furnished on the inner side with a circular rib on which are mounted the electro-magnets of the exciting portion. As in the earlier models, the coils for the alternating currents rest on the square bolts connecting the end plates with packing pieces of hard wood. The six movable electro-magnets are fixed radially around a hexagonal sleeve which is bolted to the shaft. The armatures are held by screws as shown in the longitudinal section. The shaft carries on one side the small exciting coil, and on the other the induction coil. The bearings are wide, and in the larger machines a system of automatic lubrication is employed.

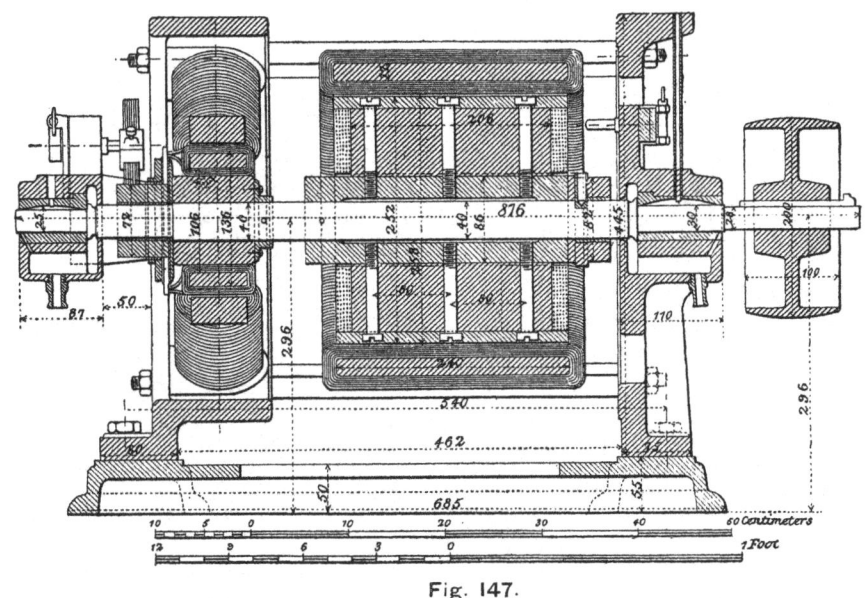

Fig. 147.

An arm carrying a wire brush, shown in the section, Fig. 147, serves to place in communication the coils of the moving electro-magnets with

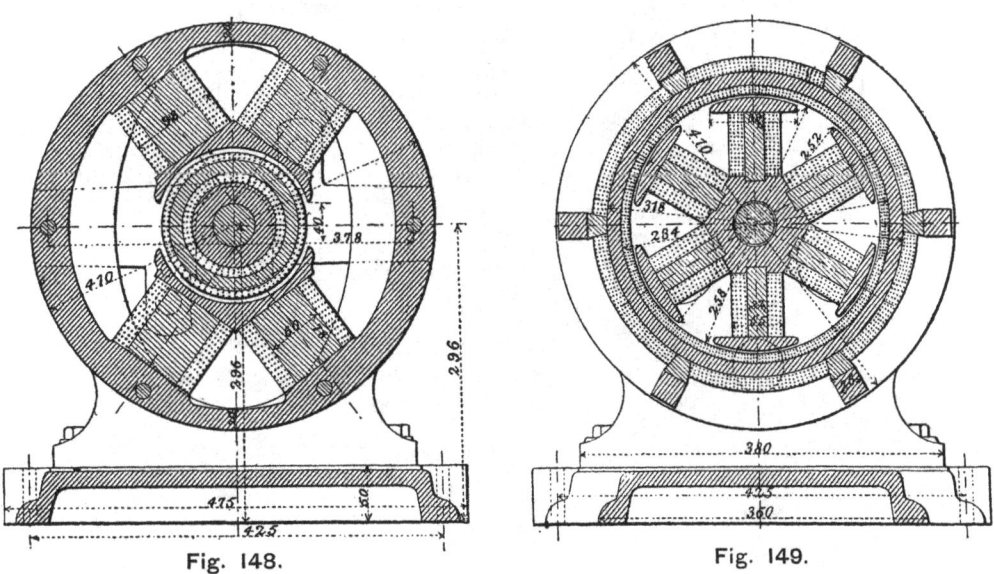

Fig. 148. Fig. 149.

the exciting circuit. The current is collected and transmitted by small brushes of silvered copper wire. The brushes are moved by means of a small endless screw. For regulating the power of the machine, a copper wire, the length of which can be varied at will, is introduced between

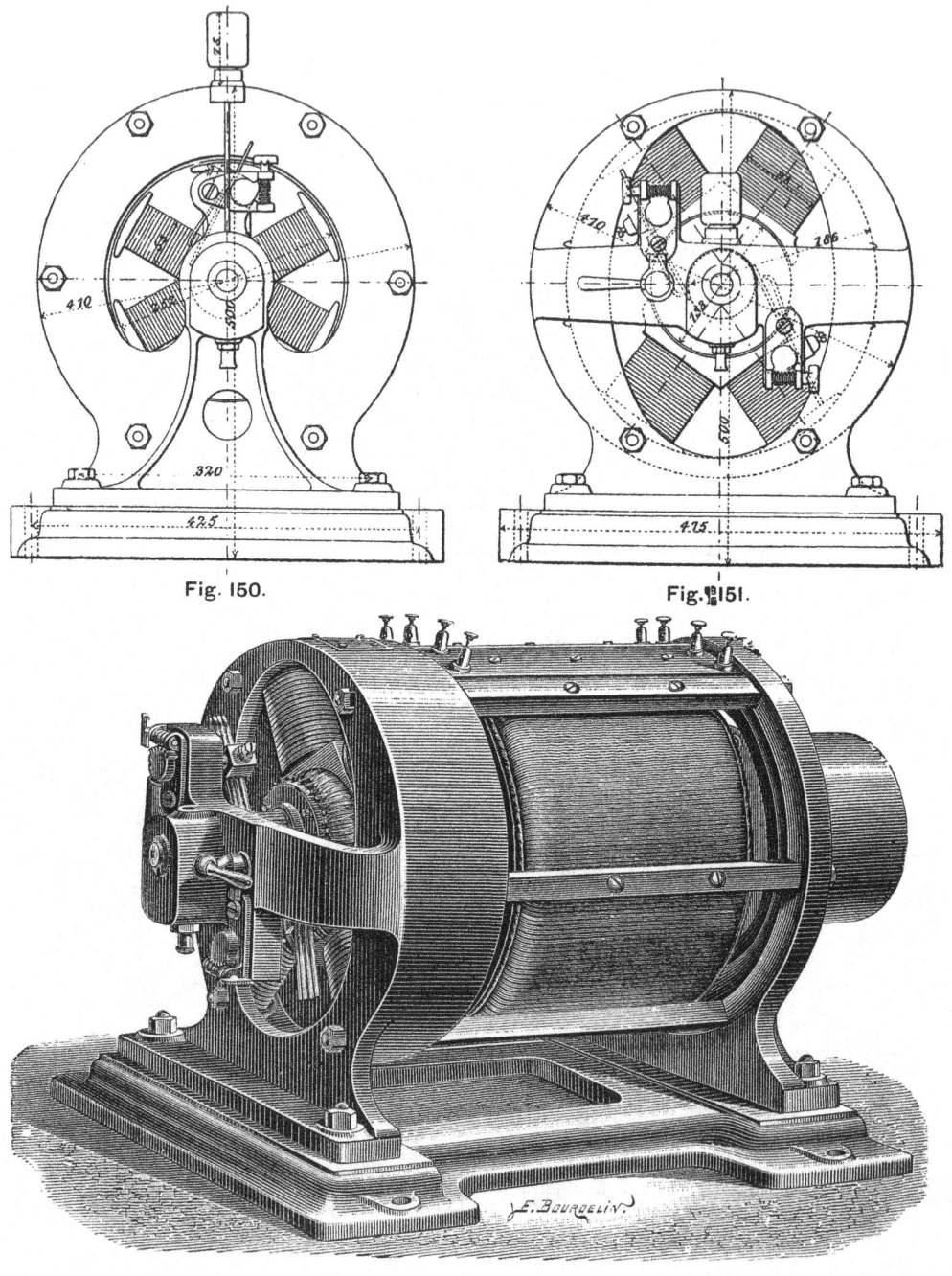

Fig. 150.

Fig. 151.

Fig. 152.

the excitor and the electro-magnets. The method of coiling the wire is the same as that adopted in some of the other machines; instead of one wire two are coiled, as in the generator shown in Figs. 140, 141, in order to obtain by this mode of coupling, tension currents for small lights, or quantity currents for large ones. Two types of this machine were manufactured. The smaller weighed 616 lbs. and supplied 12 Jablochkoff candles of from 20 to 30 Carcel burners, or 8 candles of from 40 to 50 burners. The larger machine weighed 990 lbs., and furnished power of 24 candles of 20 to 30 burners, or 16 of 40 to 50.

The following Table contains the results of some experiments made to ascertain the performance of this machine :—

Number of Revolutions per Minute.	Horse-power Expended.	Number of Lamps.	Power of each Light in Carcel Burners.
1,400	5	12	28·5
1,425	6	8	43·0
1,200	4	6	48·5
1,000	13	16	48·0
1,020	13	16	51·5
1,200	14	24	31·0

With a machine of this class, especially arranged for small lights, there were obtained with a speed of 1,250 revolutions, 14 lights of 20 Carcel burners each, with an expenditure of power of 4·66 horse-power. The candles employed had carbons 3 mm. (·12 in.) in diameter. In all the experiments made a much steadier light was obtained than that given by the machines employing an independent exciter.

Figs. 153 to 157 illustrate the last new type of the Gramme dynamo-electric generator, adapted for feeding five arc lights. Several examples of this machine were shown at the late Paris Electrical Exhibition, for two, five, ten, and twenty lights, and which were employed for lighting in the nave and three saloons on the first floor. These generators were exhibited by the Gramme Company, by MM. Sautter, Lemonnier and Co., and by the Spanish Electrical Company. The machine consists of two cast-iron standards, in the centre of each of which a bearing is formed for the driving shaft on which the ring is mounted; near the top and bottom of each frame is a face with a transverse groove made in it (Fig. 153) for the reception of the cores of the electro-magnets. The shape of the cores is shown in

Figs. 153 and 155; they are flat rectangular bars 11·93 in. long by 1·26 in. thick. These cores, besides resting on the grooves formed in the

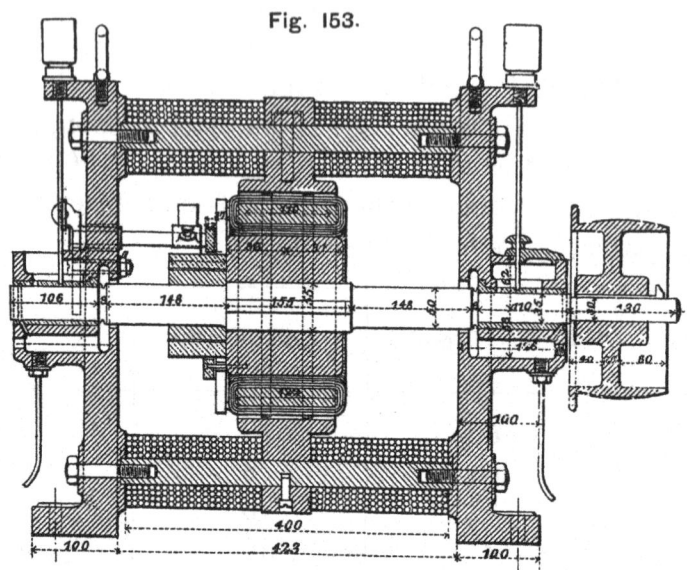

Fig. 153.

frame, are secured by bolts passing through, three at each end of the core (Fig. 154). The wire coiled around these cores brings them to the

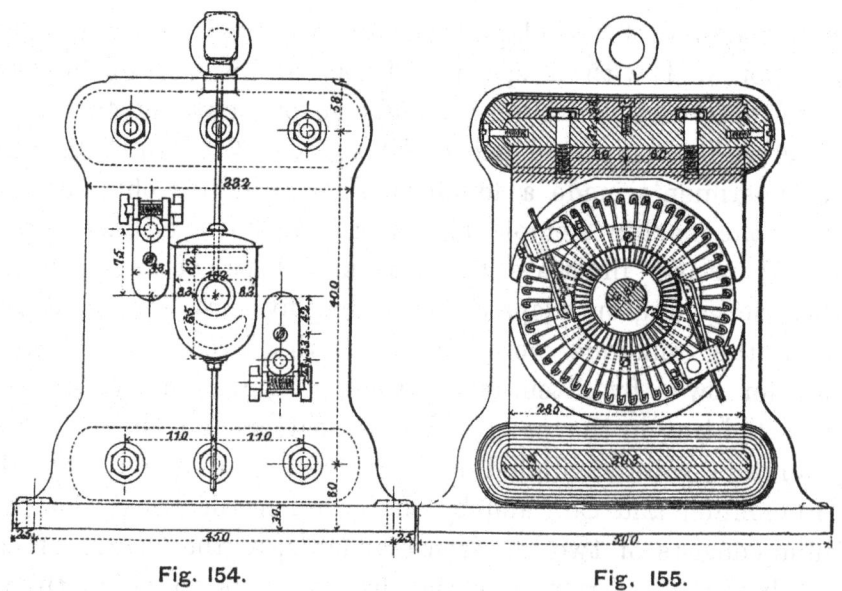

Fig. 154.   Fig. 155.

form and dimensions indicated on the drawings. In the middle of each magnet is a pole-piece, 4·72 in. thick and held by screws to the core (Fig. 155); the pole-pieces, rather wider than the ring (11·22 in.) and the inner faces are semi-circular, so as almost to embrace its complete circumference.   The central shaft carrying the revolving armature is extended at one end beyond the frame to take the driving pulley, 7·87 in. in

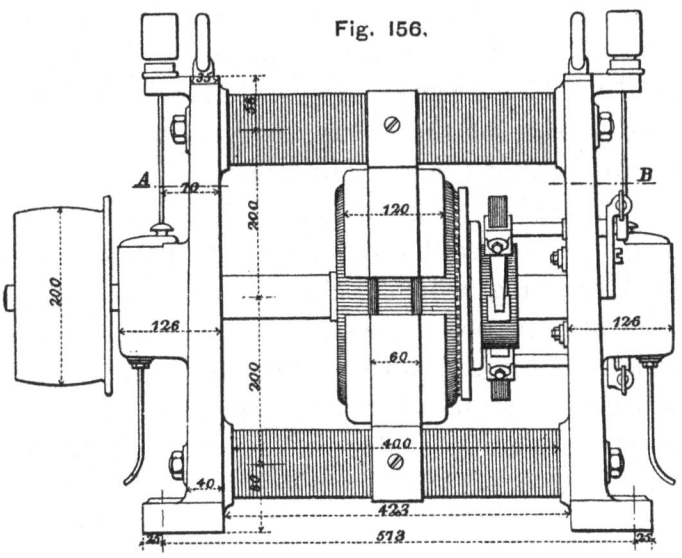

Fig. 156.

diameter; the size of the shaft varies from 2·17 in. in the centre to 1·13 in. at the end.   The ring is carried on a wooden core, 5·94 in. wide and of such a diameter as to fit tightly within the ring, which is wound, as

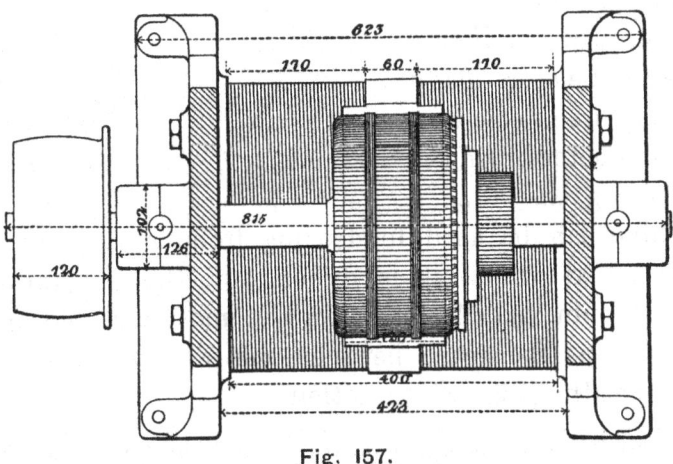

Fig. 157.

shown, upon a core 4·76 in. wide.   The outside diameter of the finished ring is 10·26 in.; Figs. 153, 155 and 156 show the relative positions of this ring and the pole-pieces, and it will also be seen that the ring is strengthened by the addition of two bands or hoops.   The collector is shown clearly in the same figures; it is mounted on the wooden core by means of a copper ring, through which screws are passed at intervals into the core. Between the radial plates of the collector, on which the brushes bear, and the central shaft, a metal sleeve is interposed, as shown in the section. The detailed arrangement of the collector is indicated in Fig. 156. The brushes are controlled by an endless screw operated from the outside of the frame, so that their position may be regulated to secure the best effect. The great simplicity and excellent design of this machine give it quite a special character, and amongst the novel details, may be indicated that of forming the bearings in the frames and the mode of lubricating, which is clearly shown in Fig. 153. It will be noticed that the bearings are of unusual length—4·32 in.—to enable the machine to run at a high speed with perfect steadiness. The following are some of the leading features of this generator :—

Total length including pulley	32·09 in.
Width	19·69 ,,
Height	21·19 ,,
,, of axis or shaft above base	11·02 ,,
Diameter of pulley	7·87 ,.
Width of belt	3·94 ,,
Total weight	792 lbs.
Resistance of ring	1·5 ohms.
Weight of wire on ring	24·2 lbs.
Resistance of electro-magnet	3 ohms.
Weight of wire of electro-magnet	140 lbs.
Outside diameter of ring	10·26 in.
Width of ring	5·90 ,,

This machine can be used for feeding from one to five lamps by varying the speed at which it is driven and the resistance of the conductors. The annexed Table shows the most suitable speeds and resistances, as well as the space between the carbons during work and the length of arc at the point of extinction :—

Number of Lamps.	Number of Revolutions per Minute.	Resistance in Ohms of Conductor.	Normal Distance apart of Carbons.	Length of Arc at point of Extinction.
			mm.	mm.
1	500	1·00	·25	·60
2	700	2·00	·25	·57
3	975	3·00	·25	·55
4	1,125	4·10	·25	·55
5	1,300	5·50	·25	·55

LONTIN.

The system of electric lighting devised by M. Dieudonné François Lontin, of Paris, received some years since, considerable notice in France, where for a considerable time it found a large and varied application for lighting open spaces, and is indeed considerably used at the present day. In 1875, M. Lontin patented in England an arrangement by which the whole of the electricity produced in the revolving armature of a machine is transmitted into the exciting coils of the field magnets, instead of only a portion, as had hitherto been done. This of course renders the magnetic field very powerful in a short time, and the mechanical resistance to the rotation of the bobbin increases in a few moments to such an extent that it is almost impossible to overcome it. The circuit is then broken by an automatic commutator, and the current passes into the external circuit, in which useful work is to be done, such as the production of gas, the deposition of metals, &c. The action of the machine then becomes normal and continues so, the superabundant energy of the current being utilised in chemical decomposition. Lontin reserved to himself all the applications which might result from this new mode of employing the current from any sort of dynamo-electric machine, whether in the production of chemical, magnetic, heating, or lighting effects. He also proposed to utilise this class of generator, in which all the current is returned into the inducing electro-magnets, or, in other words, which have a single circuit, for the production of brake power for locomotive or other purposes. When the circuit is open the armature revolves without any difficulty, but as soon as the circuit is closed, it encounters considerable resistance to rotation; a machine weighing 100 kilogrammes, for example, requires only a few pounds to work it when the circuit is open, but six to eight horse-power will be insufficient to overcome the resistance to turning, when the circuit is closed.

In 1876 (Patents Nos. 386 and 3,264), Lontin patented in England further improvements of magneto-electric generators. One defect in machines of this nature is the great heat developed in the rotating armatures, which is due to over frequency of induction. To overcome this objection Lontin multiplies the number of coils on the revolving armature. He constructs the armature in the form of a wheel provided with a central boss and spokes of soft iron, and mounted on a shaft to which rotary motion can be imparted.   This wheel is shown in side and front elevation and plan by Figs. 158, 159, and 160.   Each soft

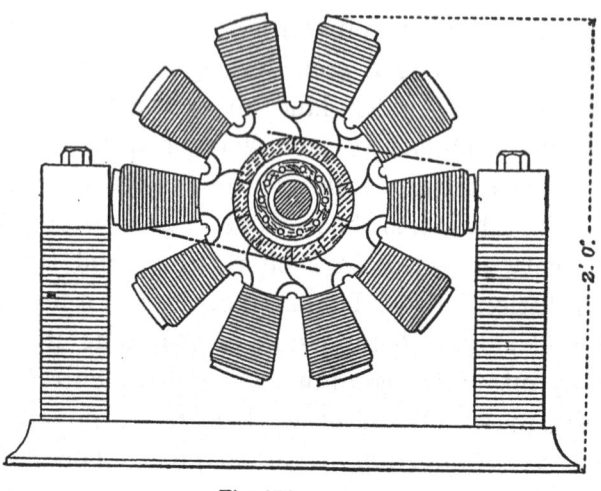

Fig. 158.

iron spoke of the wheel has a coil of insulated wire around it, and forms an electro-magnet, which becomes a source of induced electricity when the wheel is revolved between the poles of the fixed vertical electro-magnets

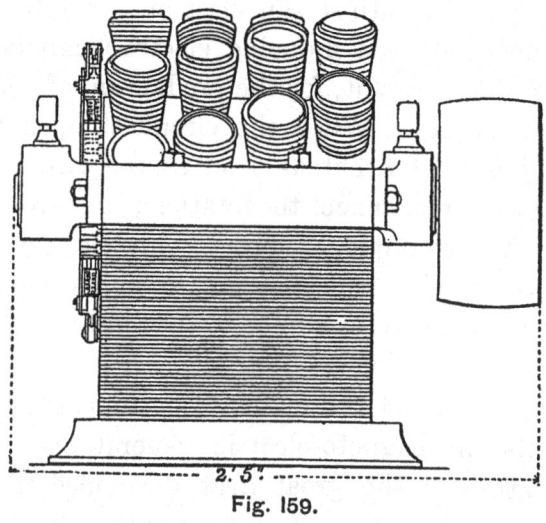

Fig. 159.

between which the armature is mounted on a horizontal shaft running in suitable bearings and having a driving pulley at one end.   The residual magnetism of the cores of these field magnets is sufficient to generate at first a feeble current in the coils when the armature is revolved, and a portion of this current, kept in one direction by a commutator, is diverted in the usual manner into the electro-magnets, in order to intensify them so that they may in turn induce yet more powerful currents in the bobbins. Lontin reserved the right of applying one or several of these induction wheels on the same shaft, and placing them opposite one or more series of electro or permanent magnets. In the course of his specification he states that if two wheels are fixed to the same shaft, one of them can supply currents exclusively

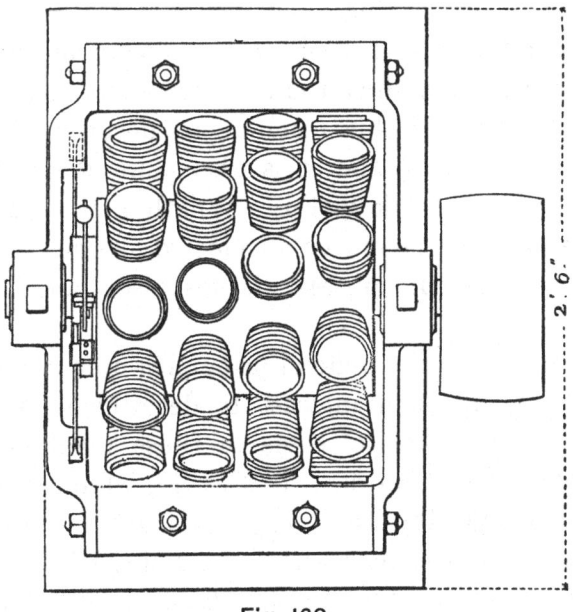

Fig. 160.

for feeding the electro-magnets, and the currents from the other can be used for external work. When there is only one wheel on the shaft, as a portion only of the current generated is employed to feed the magnets, the remainder can be turned to other purposes. If these currents are required to be invariable in direction, a commutator or a collector is used, otherwise the electricity may be collected by friction or contact rings. When commutators are used, one for each coil, or at most each pair of coils, is placed on the shaft, and to each is attached the two extremities of the wire of the corresponding coil or pair of coils. When collectors are used all the coils on the wheel are connected up in series so as to form a completely closed circuit. The result is that all the bobbins which are approaching a pole of the electro-magnet are inversely electrified to those receding from the same pole. A metal strip or rubber is placed opposite this pole of the electro-magnet to collect by contact the electricity generated in the bobbin at the instant that its polarity becomes reversed, a similar rubber being applied at the other pole of the electro-magnet to form the second pole or electrode of the machine by which the induced current is led away. It will be seen that if the arms of the wheel are sufficiently numerous the induced current will be continuous and even equable. Commutators are liable in powerful machines to rapid destruction by the oxidising and volatilising heat of the sparks emitted. M. Lontin avoids this oxidising effect in presence of air by enclosing the commutators in a bath of non-drying oil, an arrangement brought forward as a novelty since the date of his patent.

Various modifications of this machine are described in the patent to which we have been referring, and it is also pointed out that it can be efficiently utilised as an electric motor. Whether the wheel be placed between the poles of a permanent or an electro-magnet, if actuated by a battery or a dynamo-electric current, it possesses considerable motive power.

Having thus indicated the general principles upon which the Lontin generator depends, it will be of interest to describe in somewhat more detail the actual construction of some of these machines which have done, and are doing, useful work. In Fig. 161, which is an elaboration of the diagram Fig. 160, the revolving armature is in the form, as already described, of a star-shaped wheel consisting of a central boss P, into which are fixed ten or more radial bars of soft iron, circular in section but slightly conical in form, marked in the drawing D D D. Each of these radial spokes is wound with a coil of insulated copper wire, the ends of which are connected together in series, and to a cylindrical commutator. The armature is fixed on a shaft and revolves in the magnetic field of two powerful electro-magnets A A, fixed vertically into a base-plate of iron by which they become the two limbs of a horse-shoe electro-magnet. The *C* ores of the radial magnetic inductors as they revolve, approach very close to, and recede from, the poles of the inducing magnets A A, being at their point of closest proximity when their axes pass the line X X, Fig. 161. As has been already explained, a coil approaching a north pole of a magnet has a current of electricity induced in it in one direction, and when receding from the same pole,

the induced current is in the opposite direction; a coil receding from
a north pole has a current induced in it in the same direction as a
similar coil approaching a south pole, and *vice versa*. Bearing these
facts in mind, and referring to Fig. 161, it will be seen that when the
machine is revolving in the direction of the arrows, all the radial bobbins
above the horizontal line X X are receding from the left-hand or south
pole of magnet A A, and approaching the right-hand or north pole, while
those below the line X X are doing just the very reverse, that is to say,
they are receding from the north pole and approaching the south pole;
the currents induced, therefore, in all the bobbins above the line X X are

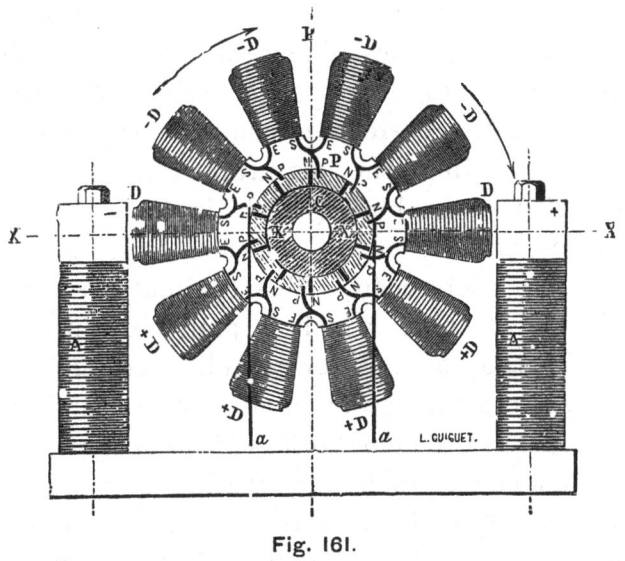

Fig. 161.

similar in direction to one another, but inverse to those below the
line X X. The coils in the arrangement shown in the figure are coupled
together in series as above stated, and each pair of contiguous coils is
connected to a sector P of the cylindrical commutator, there being as
many sectors as there are coils, insulated from one another by thin
strips of ebonite N N N. Against the surface of this commutator,
collectors *a a* are pressed by means of springs, the one taking off the
positive current and the other the negative, and the currents so
produced are transmitted through the coils of the large vertical electro-
magnets. An ordinary type of the Lontin generator is provided with
four armatures fixed on the same shaft, each carrying 10 coils. In
such a machine the ends of the iron cores of the 40 radial coils

revolve close to the horizontal pole-pieces of the two powerful electro-magnets, which are flat in section, and fixed in a vertical position as already described. In such a machine the bobbins are attached to the central boss in such a way that those of one wheel pass the poles of the magnets a little in advance of those of the next wheel; in other words they are mounted helically on the shaft. The object of this arrangement is to obtain greater uniformity in the distribution of the mechanical resistance to the rotation of the machine—no two coils being at their maximum points of resistance at the same time—and to insure greater regularity in the strength of the current produced by the machine. In some of the more recent machines on this principle, M. Lontin has introduced several improvements; one of these is to construct the pole-pieces of the inducing magnets in such a manner that their distance from the revolving armatures is capable of being adjusted at will, so as to increase or diminish within certain limits, the intensity of the current produced. He also carries the conducting wires connecting the armatures with the commutator through the axis of the machine (which is for that purpose made hollow for a certain portion of its length); this is a revival of an old device already referred to. The commutator may be placed altogether outside the machine, which is preferably inclosed in a pro-tecting casing, while the commutator is quite accessible. M. Lontin makes his collecting contact-pieces of an alloy of lead and tin, and they are maintained against the commutator with a constant and adjustable pressure by means of a counter-balanced lever.

The machine described above is the continuous or single-current generator, and it may be used with a single lamp in the place of any other single-current machine, such as the Siemens or Gramme, both of which have a higher efficiency The special feature, however, of M. Lontin's system of electric illumination consists in his method of dividing the current and distributing it over a series of different lamps, producing at the same time an alternating current in each circuit, by which the carbon pencils in the electric regulators are consumed at a more uniform rate, and their extremities remain pointed, two very decided advantages in an electric lamp. In the Lontin, as in several other distributing systems, two distinct machines are employed, one being devoted exclusively to producing the current by which the electro-magnets of the second machine are excited, and the other for inducing from the electro-magnets so excited, a series of currents of electricity, and distributing them into a number of illuminating circuits. The feeding or exciting machine used under the Lontin system is that just described, and the currents of electricity which are generated by it are led by conducting wires to the second or distributing machine which is illustrated in Figs. 162, 163 and 164. This machine consists of a revolving cylinder of brass *a*, Fig 162, around the circumference of which are mounted radially a number of flat electro-magnets A A A. The coils of these magnets (shown by Fig. 162, and

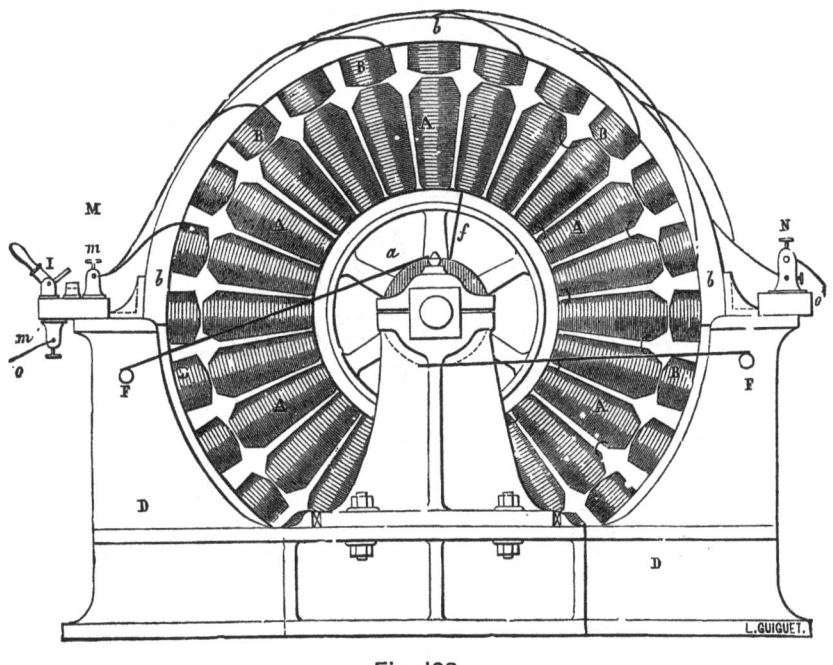

Fig. 162.

which represents the elevation of a machine having 24 inductors, and side-wise in Fig. 163) are connected in series, but in such a manner that when the current from the exciting machine is transmitted through them they become a series of powerful magnets, of which every alternate magnet has one polarity, the polarity of the intermediate magnets being in a reverse direction. That is to say, if the revolving magnets were numbered consecutively around the circumference from 1 to 24, all the even numbers would have a north pole at their outer extremity, and all the odd numbers would have a south polarity. Enveloping this revolving series of electro-magnets and fixed to the annular frames *b b b* is a crown of flat electro-magnetic coils B B B, having their ends presented towards the axis of the machine, and therefore towards the radial electro-magnets attached to the revolving wheel. There are as many of these induction coils as there are radial magnets, and it will be seen from the construction of the machine that when the apparatus is set in motion a series of alternating currents of electricity will be induced in each of the fixed coils, as the revolving magnets, alternating in polarity, approach to and recede from it. The ends of the fixed coils are connected by suitable conductors to the terminal screws *m* of the manipulator

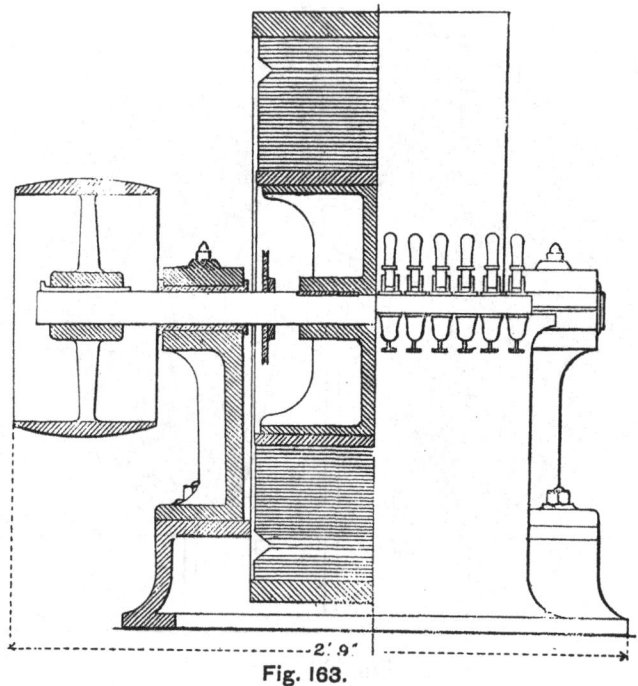

Fig. 163.

or commutating switching apparatus M, which is best shown in the perspective view of the machine, Fig. 164. By this portion of the apparatus the current from all or any lesser number of the outside induction coils can be diverted into a number of different circuits; the number of such external circuits depending upon the size and construction of the machine, and upon the number of the induction bobbins that are capable of producing the currents required. In the machine, as shown in Fig. 162, the coils are coupled up in pairs, so that 12 circuits may be fed by the 24 bobbins. Attached to the bracket, on which the commutator switches are fixed, are 24 terminal screws, 12 above it marked *m*, and 12 below it marked m^1, and by manipulating the 12 bell-crank switches I, any number of the upper terminals may be connected to any number of the lower. Thus, as any of the bobbins may be connected by wires to any of the upper terminals, and the lower terminals may be connected to any of the outside circuits, the arrangements of the connections of the bobbins, and the coupling up of the circuits, is under perfect control, and any circuit may be completed or thrown out of the machine instantaneously, by a movement of the switches. The wires *o* (Fig. 164), attached to the terminals m^1, lead

Fig. 164.

to the different lamp circuits, and the wires o^1, attached to the terminals N, are the return wires from the same circuits. To the terminals F F are attached the wires leading from the exciting machine, the current of which is transmitted to the coils of the rotating electro-magnets A A by the two pieces *a a*, which are maintained in close contact against the contact cylinders, one of which is shown in the figure, and to which the coils of the electro-magnets are connected.

The Lontin system was employed at the St. Lazare terminus, in Paris, of the Western Railway of France, for illuminating the goods' platform, waiting hall, and front of the station by the electric light in 1875, and it supplied the light to every alternate light round the balcony of the Paris Hippodrome, when it was tried, in 1878, against the Jablochkoff system, by which the intermediate lamps were illuminated. It was also the Lontin machine that was employed for the electric lighting of the exterior of the Gaiety Theatre in London, and is interesting for that reason as having been the first practical application of electricity to street lighting that was ever made in this country. Further reference to this system will be made under the head of "Installations;" meanwhile it may be mentioned that the Lontin system is being worked in this country by the Electric Light and Power Generator Company.

WALLACE-FARMER.

Amongst the many systems of lighting by electricity which within the last few years have claimed public attention, the Wallace-Farmer, although it appears already to be forgotten, in this country at all events, requires a somewhat detailed notice. The Wallace-Farmer system was introduced into England from the United States, by the enterprise of Mr. W. Ladd, towards the end of 1878; it presents some salient points of interest both in the method of generating and in the manner of distributing and applying the electric currents. The Wallace-Farmer machine was, as its double name implies, the product of more than one brain, and its inventors have the reputation of being amongst the most able electricians on the other side of the Atlantic. Mr. Moses G. Farmer is professional electrician to the United States Government Torpedo Department at Rhode Island. It was he who first demonstrated in the United States, that it is possible to put into one circuit the armature of the dynamo-electric machine, its field magnets, and the "work," without introducing separate branches or duplicate coils for the exciting of magnetism in the stationary field magnets. It need hardly be added that this is a fundamental principle in the machine, which was brought to its developed form some few years after the demonstration above referred to. The Wallace Brothers, of Ansonia, Connecticut, are also well known as pioneers in electrical science, and leaders in the manufacture of dynamo-electric machines for industrial purposes, such as the deposition of metals, and the production of the electric light. The lamp usually employed with this machine, and of which a detailed description is given in another portion of this volume, is the invention of Messrs. Wallace Brothers, though some improvements were introduced by Mr. Ladd, after its importation into England.

The Wallace-Farmer generator, of which a general view is shown in Fig. 165, consists of four powerful fixed electro-magnets, between the

Fig 165.

poles of which is rotated upon a horizontal axis a large double armature comprising a number of separate coils. The four field magnets are placed horizontally, two above and two below the rotating axis, and are bolted to the frame. Their shape is peculiar, being flat on the sides turned towards the axis, as is shown (partly in dotted outline) in the vertical section given in Fig. 166. The object of this form is to obtain as powerful a field as possible with as little unnecessary material as is compatible with this condition; hence the iron cores, which are also of the same flattened shape, are made broad in order to give a wide extent of field, but extend neither above nor below the corresponding iron cores of the bobbins carrying the separate coils of the armature. The wire of the field magnets is stout and well insulated, and in each of the four there are about 450 or 500 turns. The length of these field magnets is 9 in., and their breadth about as much; cheeks of light brasswork serve to support the coils at the ends. The double rotating armature, which is an essential feature of the machine, differs from the armature of almost every other electric generator, having elements in common with several of them, but not precisely resembling any. Perhaps the nearest approach to it in Europe was the armature devised by M. A. Breguet. In many

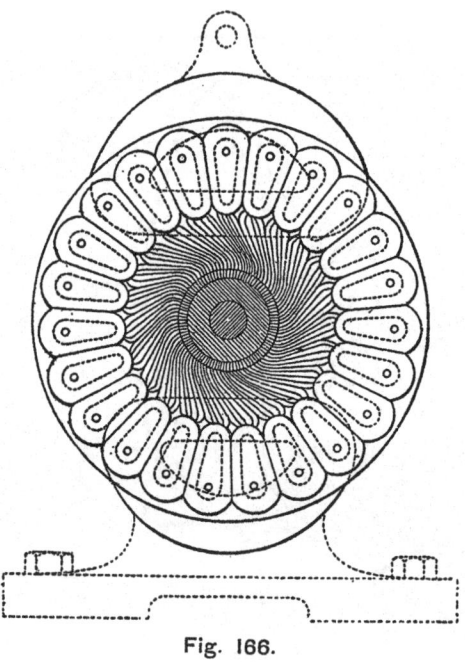

Fig. 166.

of the earliest magneto-electric machines, the armatures consisted of bobbins, usually two in number, carrying coils of wire and fixed upon solid iron cores. The Siemens dynamo machine employs, as is well known, a long narrow armature wound with coils from end to end longitudinally; whilst the armature of the Gramme machine is a rotating ring of iron, overspun with coils of wire arranged in separate sections. In the Wallace-Farmer armature, the coils are wound upon 50 bobbins of peculiar shape (see Fig. 166), arranged in two sets of 25 each, around the circumference of a circle, being attached to two plates of soft iron each nearly an inch thick, and about 16 in. in diameter, and having a clear space of about an inch between the two plates. The cores of these separate bobbins are also of peculiar form, not circular, but of a rounded wedge-shape, so as to allow of their being packed closely together around the circle. They are all pierced with a small hole, to allow of circulation of air, to reduce any possible heating effect in the armature. The wire of which the separate coils of the armature are constructed, is somewhat thinner than that of the field magnets, and it is carefully insulated. From every one of these separate bobbins of the armature, four strands of wire run to the commutator fixed on the axis between the upper and lower field magnets. There are indeed two commutators, one on each side of the double armature, and connected with their respective set of bobbins. In point of fact the machine is duplex, the two halves being set symmetrically about the vertical middle of the machine. In practice, however, it is found advisable to work the machine as one, for which purpose the two halves, which can upon occasion be employed on separate circuits, are united into one continuous circuit by the stout conductor of copper which is seen joining the two middle pair of the four terminals shown on the base of the machine in Fig 165. The commutators consist, as in the machines of Gramme and Siemens, of strips of copper arranged on the circumference of a cylinder of non-conducting material, and are each pressed, above and below, by a pair of metallic brushes to lead away the currents that have been generated in the armature. The cores of the separate bobbins of the armature as they rise through one half of their journey round the axis, are passing from one polarity to the reverse, and as they descend through the remaining half of a rotation, are having their magnetism again changed back. Hence by the laws of electro-magnetic induction during half the rotation, while the coil is ascending it is supplying, say, a positive current to the upper brush, and while descending it supplies a contrary current to the lower brush, of the commutator; and as the separate coils are numerous, and a similar action is continually going on in all varieties of phase in the separate coils, the currents produced are, as in the Gramme machine, practically continuous. The advantages sought to be obtained in this form of armature over the ring armature of the Gramme machine, were many. The trouble of winding the coils in the latter generator is very great, as the wire must of necessity be threaded through the ring separately for every turn of the coil. Here, as there are separate bobbins, it is a simple matter to wind on the wire. The iron of the ring armature of the Gramme machine is moreover always separated from the poles of the field magnets by at least the thickness of the coils. Here, as the iron cores of the separate coils are placed at right angles to the plane of the ring, or parallel to its axis, they may be made to approach as near as possible to the field magnets, thereby increasing the magnetic induction. Lastly, these iron cores, being short, there is not the same tendency to form back currents. In all those dynamo machines in which an iron portion must have its magnetic polarity reversed rapidly, whether that part be a moving one as in the annular core of Gramme, or fixed, as are the magnets of Lontin's distributor, there is a practical limit to the speed at which the machine can be driven (and therefore to its efficiency), owing to the residual magnetism of the iron cores, which takes a definite time to vanish. Now the time taken by this residual magnetism to vanish appears to be, though the function is an obscure one, dependent *inter alia* upon the length of the magnet. That is to say, a long magnet cannot have its magnetism effectively reversed as rapidly as a short magnet. Makers of induction coils know this, and they have attempted to hasten the vanishing of the residual magnetism, as well as to avoid secondary currents, by employing cores consisting of bundles of iron wires. In the Wallace-Farmer machine the cores of the separate bobbins are relatively short, and ventilation is secured by the central hollow of the armature. With a speed of 600 revolutions per minute, which for most purposes is sufficient, there is not much risk of heating, but if a higher speed is employed, the importance of this arrangement is found, for the air which issues above is perceptibly warm, and in rising draws up cold air, thereby keeping the moving parts cool. The driving power required, varies with the work which the machine is doing. Two 8-in. pulleys, one at each end of the axle, are served with bands from a motor, though there is no necessity to drive at both ends. The experience of practical use showed the amount of power required per light to be not very dissimilar from that required for driving the Gramme or the Siemens machine, namely, about one horse power per light of 800 candles' power. We have described this machine in the present tense, as if it were still being constructed. We do not know whether this is so in the United States or not. In this country, certainly it must be regarded as historical, for although the results obtained with it were tolerably satisfactory, it was soon found that the Brush machine presented many great advantages, and, as both systems were controlled in this country by the same interests, the more valuable system of course replaced the other.

DE MERITENS.

One of the most interesting and remarkable machines for the induction of electric currents is the magneto-electric generator of M. De Meritens, of which several illustrations are given, Fig. 167 representing one of the earlier type. This apparatus is interesting because it embodies and combines in its construction the principle of action of the early magneto-electric machines of Clarke and

Fig. 167.

Holmes with that of the modern dynamo-electric machines of Gramme and Brush; and it is a remarkable machine on account of its very high efficiency as an apparatus for the conversion of motive power into electricity, by induction from a magnetic field produced by permanent magnets. On this account it stands alone among all magneto-electric generators as having been able, not only to hold its own for producing the electric light, in a field which is otherwise monopolised by dynamo-electric generators, but to displace them in some of the most important installations.

The distinctive feature of the De Meritens machine lies in the construction of its rotating armature which, like the Gramme armature, is in the form of a ring, and wound in a similar way, but it differs from it in being made up of a number of segments, each constituting a separate arc-shaped electro-magnet with expanded poles, which are joined end to end with a thickness of insulating material between them, so as to form, mechanically speaking, a continuous ring. It may, for the sake of illustration, be looked upon as a Gramme armature cut up into a number of short lengths and built up again into a ring, each length being magnetically insulated from the other. This ring forms the periphery of a wheel, which is mounted on a horizontal spindle, and is rotated on it in a vertical plane, within a crown of compound horse-shoe permanent magnets, and just underneath their polar ends, the distances of which from one another are equal to the length of the armature segments, which is the same as the pitch of their poles around the ring; and the armature coils, although all wound in the same direction, are so connected up that while the current sent into the external circuit of the machine is alternating in direction, the currents generated in the several parts of the armature are at any one moment all of the same sign, and therefore in no way tend to neutralise one another.

In the machines as originally constructed, the magnetic crown or battery consisted of eight horseshoe compound permanent magnets, arranged as shown in Fig. 167, and of the construction figured in detail at E, Fig. 168. These magnets are rigidly attached to two circular frames of brass firmly bolted to the base-plate, and within the crown so formed, the armature or induction wheel is caused to revolve on the horizontal spindle, its electro-magnetic inductors being so arranged that their poles in revolving, pass in succession as close as possible to the magnet poles without actually touching them. The cores of the induction helices (see Fig. 168) do not consist of a solid mass of iron, but are composed of a large number of thin laminæ punched out of sheet iron one millimetre in thickness, and laid flat one upon the other, until a sufficient thickness is obtained. In the machine shown at the Exhibition which was held in the Albert Hall, in the year 1879, there were as many as 50 such thin plates composing each segment of the armature ring. The advantages of this form of construction are twofold; first, the laminated character of the armature cores, assist their rapid magnetisation and demagnetisation, breaking up and otherwise preventing cross currents by which their magnetic capacity is reduced; and second, the building up of the cores by means of superposed plates of the shape of its longitudinal section offers great facilities for construction. While each segment of the ring is thus built up, the ring itself is, as we have before pointed out, also compound, as will be seen by reference to Fig. 168. The segments after having been wound with insulated wire are arranged end to end around a brass wheel, with strips of copper between them, so as to break up the magnetic continuity of the ring, making each segment

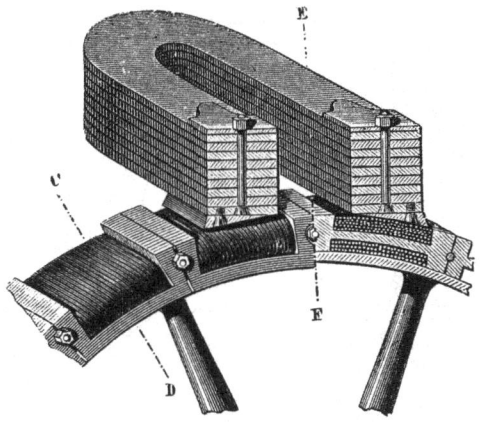

Fig. 168.

a complete and separate magnetic inductor, and they are firmly held in their places and together, by a bolt passing through each joint, which is tightened up by a nut, as is shown on the diagram. The advantages of this method of construction are obvious to any one who can appreciate the slow and inconvenient process of winding a Gramme ring, which being a continuous circle can only be covered with wire by hand, the spool containing the wire to be wound, having to be passed over and back through the ring for each convolution of the wire. By making the ring in segments, however, no such tediousness is experienced, for each piece can be put in a lathe and wound as an ordinary coil, and when so wound all can be joined up together with insulating pieces between them—a method of construction possessing especial advantages both electrically and constructively.

The action of the apparatus can best be understood by imagining an extremely simple case of a machine (illustrated in the diagram, Fig. 169), having an armature ring built up of only four segments, one of which is shown in section marked A B, and revolving within the magnetic field of two horseshoe magnets, N S, N S, placed with their poles at equal distances around the ring, and so disposed that any point in the ring will during its revolution pass them in succession, assuming thereby alternately north and south polarity.

The expansion of the poles A and B, so as to become flush with the exterior surface of the fully wound coil, besides facilitating the operation

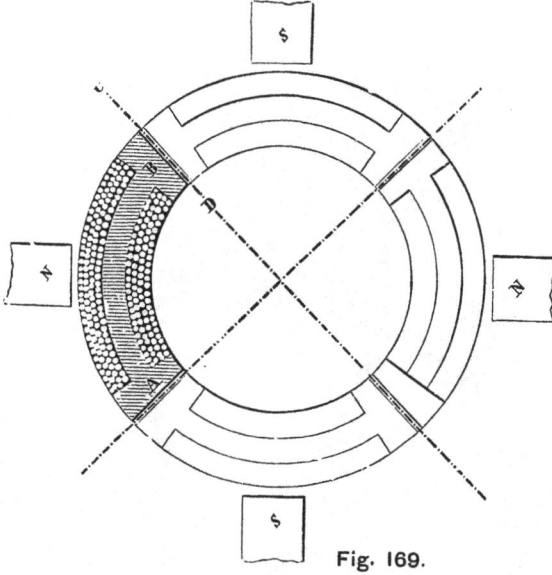

Fig. 169.

of winding, plays a still more important part, and it is to this that a great proportion of the high efficiency of the machine must be attributed. By this construction, which, as we shall see hereafter, is also taken advantage of in the very beautiful dynamo-electric machine of Mr. Brush, the iron of the cores of the several segments of the ring, is brought at its polar extremities into very close approximation to the inducing magnets N S in its passage beneath them, and it is for that reason in the best possible condition for being influenced in passing through their magnetic field.

The action of the machine illustrates, perhaps more clearly than any other apparatus, the principle of that group of phenomena which is known as Lenz's law. Thus, supposing the armature to be rotating in

the direction of the hands of a watch, the pole-piece A approaching the
north pole N of the permanent magnet, will cause an electric current to
be induced in the coil A B in a certain direction, while the opposite pole-
piece B will, in receding from the same magnet pole N, cause the
induction of a current in the coil A B, in the same direction
as is produced by the approach of the pole-piece A; the two effects
are, therefore, superposed or rather added together, and at the same
time the current is still further increased by the convolutions of the
coil of the same core passing from left to right through the magnetic
field of the magnet N. As the coils of all the segments are wound in the
same direction, it follows that the next segment which passes below
a south pole at the same moment as the segment A B is passing below
a north pole, will have induced in-its helix a current of electricity in the
reverse direction to that developed in A B, which would be opposed to
that current, and would, therefore, neutralise it, were it not for the fact

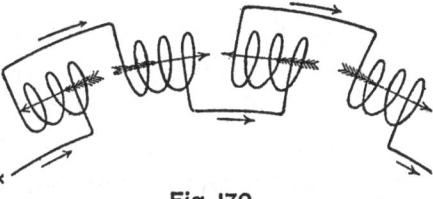

Fig. 170.

that the successive coils are connected together in the manner shown in
the diagram, Fig. 170. It will be seen that the outside end of one helix is
connected to the outside end of the coil which is in front of it, while its
inside end is connected to the inside end of the coil which is behind it,
from which construction it follows, as will be seen in the diagram, that
although the currents induced in any two contiguous helices are opposite
to one another in direction, the several helices are so connected together
that the alternating currents transmitted into the external circuit by
each of the coils are all at any one moment in the same direction, in
other words, they are all, so to speak, either pulling or pushing together.
The two ends of the entire circuit of all the electro-magnetic helices
(which is not, as in the armatures of Gramme, Siemens, and some
others), closed within the machine, are connected respectively to two
disc wheels of copper attached to the axis of the machine and revolving
with it, but insulated from one another, and upon these wheels stout
copper collectors or rubbers are maintained with a certain frictional

pressure, and are connected to the other two terminal screws of the machine. By this arrangement all complicated commutators are dispensed with, and the construction of the apparatus is thereby much simplified. By variations in the method of coupling up the helices of the armature they may be connected in series, or more or less in parallel circuit, or may be grouped in combinations of the two, and by this means the character of the current generated by the machine may be determined for the work to which it is intended to be applied.

In 1879, the Comte Du Moncel placed on record the statement, that by this machine as many as three Jablochkoff candles could be maintained burning steadily with an expenditure of motive power of not more than one horse power, and without appreciably heating its armature or its coils when running at a speed of 700 revolutions per minute, but there does not appear to have been any photometric measurement taken by which the luminous intensity of these candles so maintained could be compared with the light emitted by the same candles illuminated with the Gramme generator. M. De Meritens, it is true, invented a candle of his own, which differed from the ordinary Jablochkoff bougie only in having a thinner carbon pencil placed between the other two, from which it was insulated in much the same way as in M. Jablochkoff's arrangement. The two outer carbons were connected to the machine, and the arc passed between them, the intermediate carbon acting merely as a bridge or stepping-stone to enable the current to pass across under variations of strength. M. De Meritens claimed for this form of construction the merit of greater steadiness of the light. It is, however, only fair to state that candles of this identical form had been constructed and worked by M. Nysten several months before, and it should be added that the device appears to have fallen short of success, and is never heard of at the present time.

The more recent development of the De Meritens magneto machines formed one of the most important and interesting of the exhibits in the Paris Exhibition of Electricity of 1881, where they were shown in the form in which they are now to a great extent replacing dynamo-electric generators in many of the lighthouses of this country as well as in those of the French administration.

The first thing that strikes any one who examines the machines of M. De Meritens is the thoroughly sound mechanical design of the apparatus looked at merely in the light of a machine, apart from its electrical properties, and the next most striking feature is the high-class workmanship and finish of every part, characteristics which must contribute very largely to the excellent results obtained by M. De Meritens' system.

The machines now made by M. De Meritens are of three distinct types: (1) the generator adapted to the illumination of lighthouses; (2) the form for use in workshops and small installations of electric light; and (3) the continuous current machine. The first two types are illustrated by Figs. 171 to 186.

Referring to the illustrations, Fig. 171 is a transverse section of the machine, and Fig. 172 a longitudinal section taken through the axis so as to show, in both views, the armature ring and the position of the field magnets with respect to it, while in Figs. 175 to 177 are shown the details of the armature bobbins marked H, the iron core-pieces $h \, h$, and the projecting pole-pieces, which form enlarged ends to the latter, and are marked g. In Fig 175, which represents a section through half of the ring, the method of attachment and of coupling up is very clearly shown. On reference to Fig. 171, it will be seen that each armature ring G is built up of 16 flattened oval bobbins H, separated from one another by the projecting pole-pieces g, and around each ring are fixed radially to the framing of the machine, eight very powerful compound permanent magnets, each composed of eight laminæ of steel. The distance apart of the two limbs of each magnet, as well as the distance between the north pole of one magnet and the south pole of the next, is precisely equal to the distance apart or pitch around the armature of the pole-pieces and the coils. In referring to Fig. 171, the machine, for the sake of clearness, has been treated as consisting of one armature ring and one crown of magnets; but, as will be seen in Fig. 172, there are in the lighthouse machine five of such single machines all working together on the same shaft; there are, therefore, 40 horseshoe permanent magnets composed of 320 laminæ of steel and producing magnetic fields through which the 80 induction bobbins are being driven at a high velocity.

The details of the magnets and their method of adjustment and attachment to the framing of the machine are clearly shown in Figs. 173 and 174. Each magnet is built up of eight laminæ of steel, 10 millimetres in thickness, and are held together tightly by the bolts and nuts $c \, d$, the whole being attached to the brass frames F, which are fixed to the framing of the machine in radial slides by which the distance from

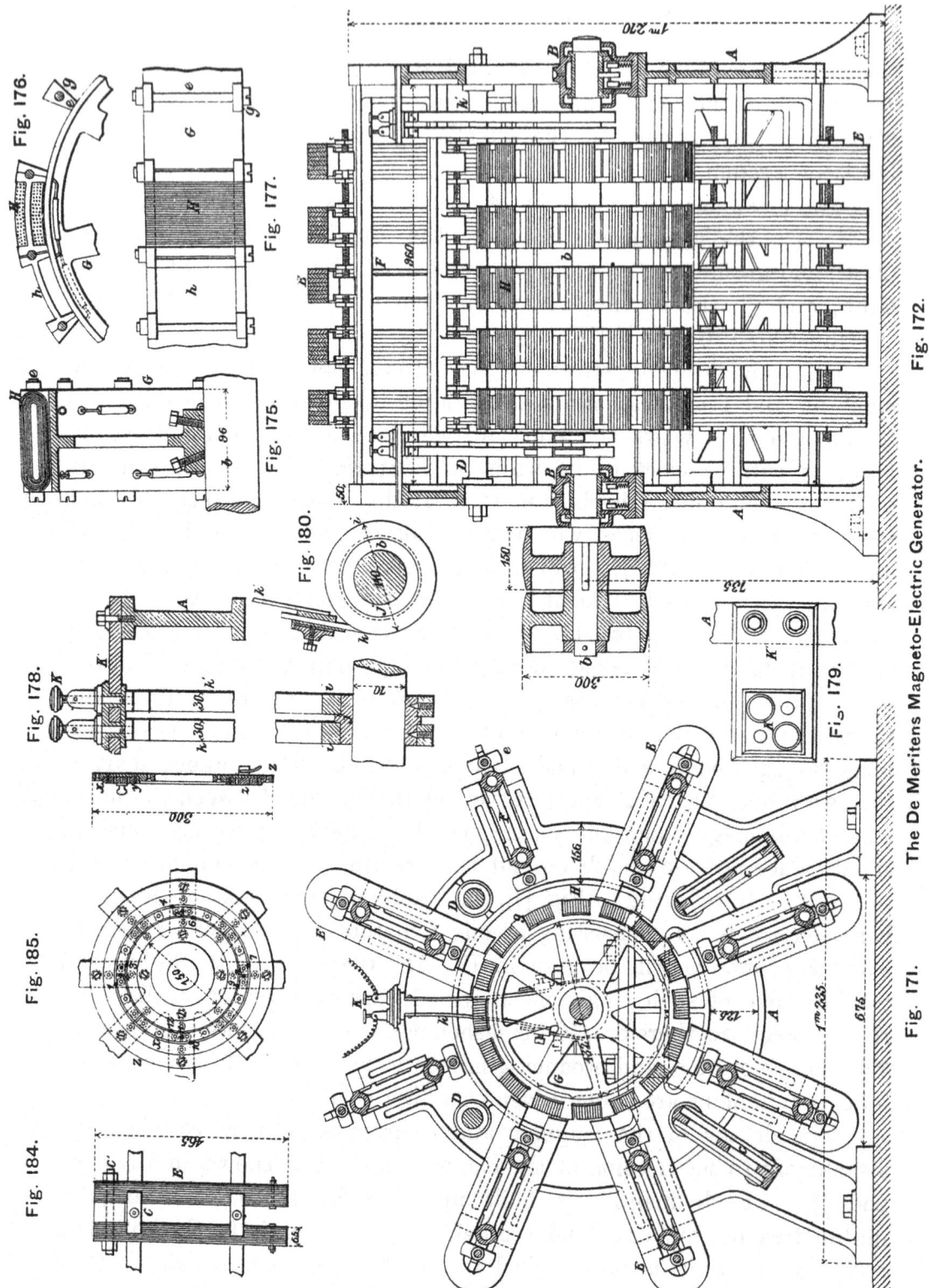

Fig. 176.

Fig. 177.

Fig. 175.

Fig. 172.

Fig. 180.

Fig. 179.

Fig. 178.

Fig. 185.

Fig. 184.

Fig. 171.

The De Meritens Magneto-Electric Generator.

the armature ring can be adjusted with great accuracy. The total weight of the 40 magnets is about one ton.

The currents from the five armatures are brought together in two groups to the four brass collecting discs *i*, which are mounted in pairs on an insulated bush J fixed to the principal shaft of the machine; the details of the collecting apparatus are shown to a larger scale in Figs. 178, 179, and 180. Against the discs *i* are pressed by means of springs the four collecting plates or brushes $k^1 k^1$, which are in metallic connection with the attachment screws *k k*, of which there are two pairs, one at each end of the machine, as shown in Fig. 172.

The construction of the armature of this machine is almost identical with the armature of the earlier machine described on a previous page, the induction coils shown at H, Figs. 175 to 177 being composed first

Fig. 173.

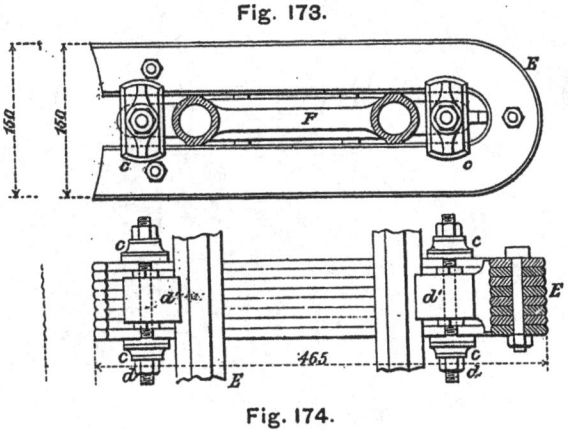

Fig. 174.

of a flat spool or bobbin of iron of the form marked *h* in Figs. 176 and 177, and this is wound in a lathe with insulated copper wire 1·9 mm. in diameter, and of which the total weight in the whole machine is from 120 lbs. to 130 lbs. The iron cores of these coils are built up of 80 thicknesses of soft sheet iron 1 mm. in thickness, and stamped out by a machine. The coils are wound and attached to the armature wheel by a set of bolts marked *e* in the figures, which pass through the projecting lugs *g* of the wheel, and through the cylindrical hole formed by the semi-cylindrical grooves in the ends of the iron core-pieces when abutting the one against the other.

The method of coupling the armature coils is the same as that adopted in the earlier machine, and which has already been illustrated in the diagram, Fig. 170, and described in connection therewith; thus the combined current is alternating in direction, because all the reversals take place at the same moment ; but, at any one instant all the bobbins are feeding into the external circuit, currents in the same direction. The general appearance of this machine is shown in Fig. 181.

Of the De Meritens lighthouse machine there were no less than six specimens in the Paris Exhibition, all of which were at work. One of

Fig. 181.

these supplied the currents to illuminate the great lighthouse in the middle of the nave, and two others illuminated the two smaller light-house apparatus in the west gallery of the building, and which formed part of the exhibit of Messrs. Sautter, Lemonnier and Co. One illuminated two large arc lamps in the south gallery, and at the same time, by another circuit, a large arc lamp in the Pavilion of the Ministère des Postes et des Telegraphes. Another machine of the same size supplied the light to the Berjot lamps in the same pavilion, or it could be switched on to a series of Jablochkoff candles; it being capable of illuminating 30 Jablochkoff candles or 40 Berjot arc lamps. The sixth machine was used sometimes in connection with Jablochkoff candles and sometimes with arc lamps. Each and all of these machines displayed much beauty of design, workmanship and finish, and worked with the greatest regularity and steadiness.

Another very interesting form of machine designed by M. De Meritens is a smaller type of apparatus for the illumination of factories and such-like installations. In this machine, illustrated in Figs. 182 and 183, the permanent magnets are disposed horizontally as in the earliest

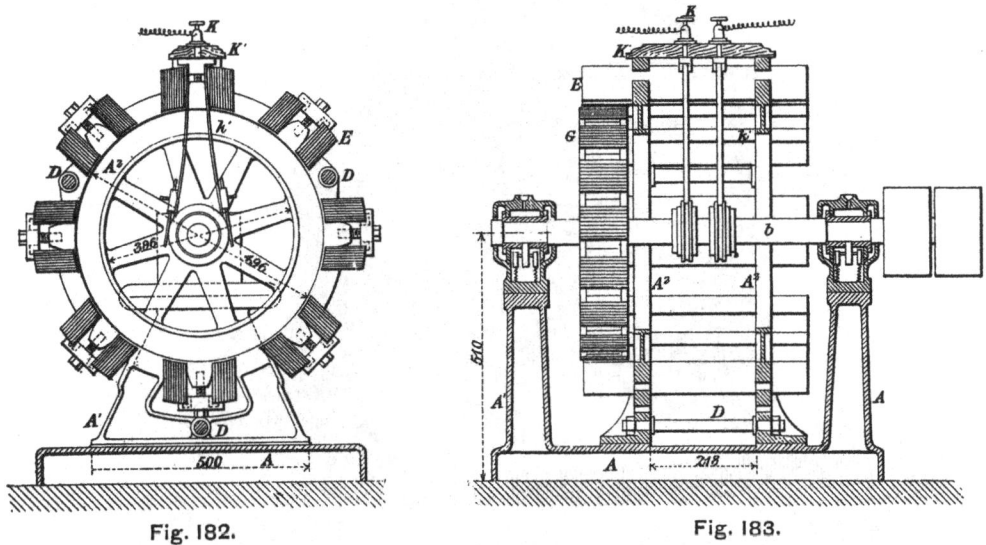

Fig. 182.   Fig. 183.

form of the apparatus, but the eight compound horseshoe magnets of the earlier form are, in the new machine, replaced by 16 compound straight magnets E E arranged in pairs, as the inventor has found that with this arrangement he obtains a set of more powerful magnetic fields. Each compound magnet is built up of 12 layers of steel about 18 in. long, 2½ in. wide, and 4½ mm. in thickness; the total weight of the magnets on the machine is about 3 cwt. Fig. 184 (page 194) shows a pair of these compound magnets which are united at their upper end by an iron block between them screwed up by the bolt and nut c', so as to convert the whole into a sort of horseshoe magnet.

The armature, of which in this machine there is only one, is of precisely similar construction to those of the lighthouse machine, and is built up of 16 induction bobbins, which are wound with insulated copper wire of a diameter of 1·9 mm., giving a total weight of wire on the armature of from 24 lbs. to 26 lbs. The bobbins are connected up in the same way as are those of the larger machine, and the collecting apparatus is very similar. With a mean speed of 1,000 revolutions per minute the machine absorbs about three horse-power while supplying the currents to four Jablochkoff candles.

One of the most interesting features of the machine is the commutating arrangement attached to the face of the armature disc, and

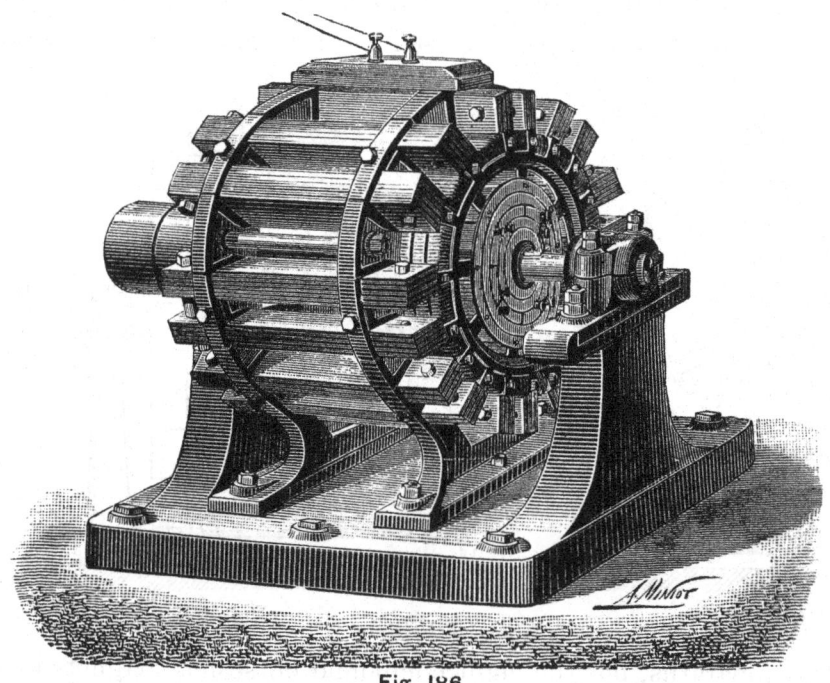

Fig. 186.

illustrated in Fig. 185. By the very simple commutating plate the 16 bobbins of the armature can be coupled up either in series or in parallel circuit, and with various modifications of both. This apparatus consists of a flat disc of wood or other insulating material, on the face of which are attached by means of screws, two circular bands of brass x and y, connected respectively with the two terminals of the machine, and between them are fixed eight segments of the same metal z, insulated from them and from one another, but capable of being connected with each other, or with the circular bands, by the insertion of small screw plugs in screw holes provided for them, and which are numbered in the figure from 1 to 12. In order to couple up all the bobbins of the armature in series, all that is necessary is to insert the plugs in the holes marked 1, 3, 5, 8, and 11, and with this arrangement the currents transmitted into the external circuit are capable of illuminating four Jablochkoff candles or six Berjot lamps of about 180 candle power each. If, however, it is required to couple up half the number of the bobbins in series, the plugs are inserted in the holes 1, 3, 5, 7, 9, and 11, under which arrangement the machine can feed two Berjot lamps of from 380 to 480 candles power each. And, again, should it be desired to couple up one-fourth of the bobbins in series, by placing the plugs in the spaces 1, 3, 4, 6, 7, 9, 10, and 12, the apparatus will supply the current capable of maintaining a single arc light of 960 candles. The general appearance of this machine is shown in Fig. 186.

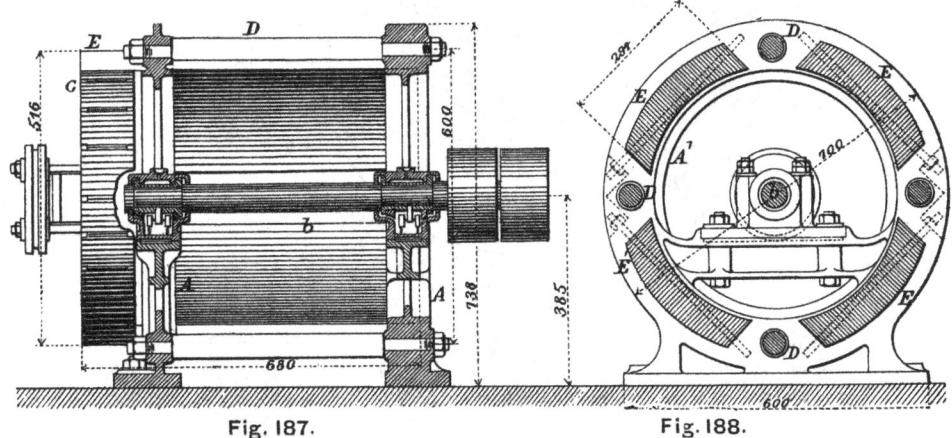

Fig. 187.   Fig. 188.

M. de Meritens also exhibited at Paris a machine of the old type, which we have already described and illustrated, but fitted with the commutating arrangement just referred to, and also two continuous-current magneto-electric machines, which calls for detailed notice.

In the De Meritens machine for producing direct and continuous currents—a perspective view of which is given in the engraving in Fig. 192, and which is illustrated in detail in Figs. 187 to 191—the magnetic fields are produced by four compound permanent magnets E E¹ (best seen in Fig. 188) arranged with alternating poles around the two circular frames of the machine, so as to lie parallel with its axis, and to form parts of a cylindrical surface within which the armature G rotates. Each of these compound magnets is built up of 64 laminæ of fine steel

1 millimetre in thickness and about 27 in. long, which are held together
and to the brass framing of the machine by three bolts, the position of
which is indicated by the dotted lines in Fig. 188. The whole of the
magnetic portion of the apparatus weighs about 7¾ cwt.

Fig. 189 is a diagrammatic view of the armature ring which in
general construction and form differs but little from the armature of the

Fig. 191.

Fig. 189.

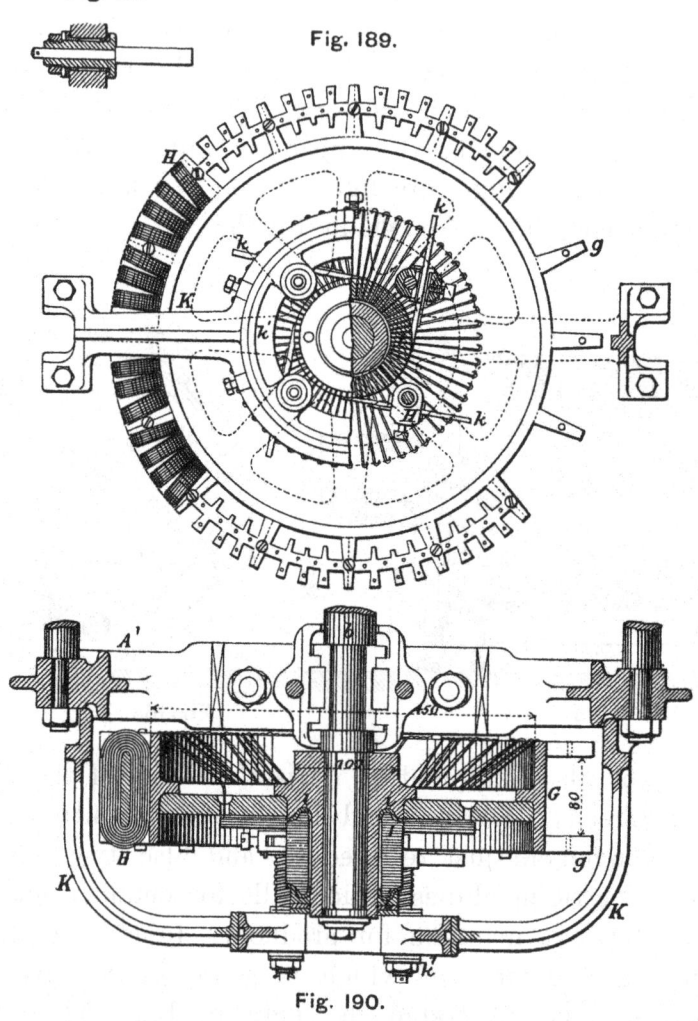

Fig. 190.

alternating-current machines described already; in the same figure are
also shown the commutator and collecting brushes. In Fig. 190 is
shown a section of the armature from which its relative position to the
framework of the machine and the commutator will be understood.
The armature in the continuous-current machine consists of a ring like

that of the alternating-current machine, but built up of 64 bobbins H H, each of which is wound upon a compound iron core of the form shown at *h*, each core being made up of 80 laminæ of soft sheet iron 1 millimetre in thickness. Each of these cores is divided into four parts by the projecting pieces shown in the figure, and in the recesses so formed are wound coils of insulated copper wire, the outer surfaces of the coils, when finished, being flush with the outer surface of the teeth which separate them; this is shown at H (Fig. 189). Each compound bobbin is attached to the brass wheel exactly in the same way as in the armature of the alternating machines, being fitted

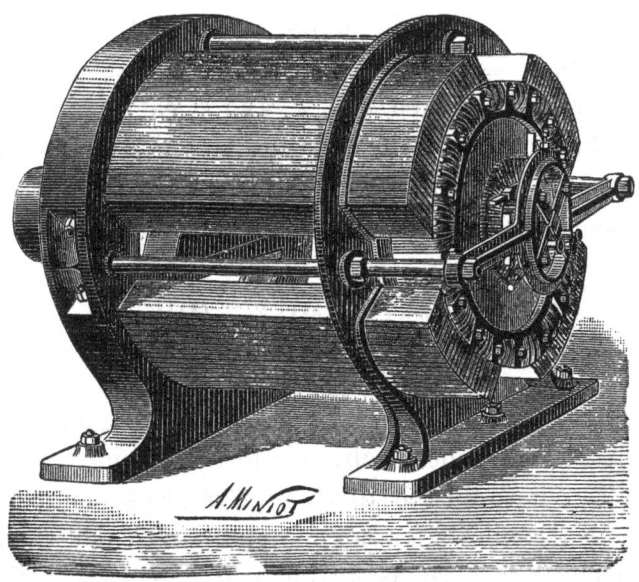

Fig. 192.

between the projecting pieces *g* of the brass wheel and held on in the same way by screwed bolts passing through the projecting pieces *g* and through grooves at the ends of the bobbin cores. The principal difference between this armature and those of the alternating machines lies in the method by which the coils are connected up. It will be remembered that the coils in the armature of the alternating machine are connected up in such a way that although the currents in the bobbin are in opposite directions, their combined current in the external circuit is in one direction (see Fig. 171). In the machine now described all the bobbins on the ring are connected in series as on those in the Gramme armature, and in Fig. 193 is shown the method of connection.

Each of the 64 bobbins is wound with copper wire 1·9 millimetres in diameter, and the total weight of wire on the armature ring is about 44 lbs. The junction between each contiguous pair of bobbins is connected to its corresponding section of the commutating cylinder, which in principle of construction and form is identical with that of the Gramme and Siemens direct-current machines. The commutator, which is shown at I (Fig. 189), is built up of 64 strips of copper insulated from one another by layers of silk, and forming a hollow cylinder which is attached to the armature shaft of the machine. The method of attachment, as well as the disposition of the commutator with repect to the armature, is shown in the sectional plan Fig. 190.

As there are two pairs of inducing magnets there are two pairs of magnetic fields, and, therefore, two pairs of collecting "brushes;" these are shown in Fig. 189 marked k, k, k, and k, each being mounted on an insulated pin (shown in detail in Fig. 191), and maintained in contact with the commutator by spiral springs, shown in Fig. 190. The four pins upon which the brushes or collectors are pivoted are attached to a

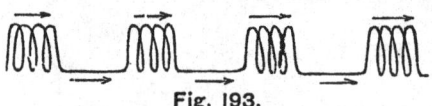

Fig. 193.

ring of brass, which can be rotated through a certain angular distance around the principal axis of the machine, and by this means the position of the collectors can be adjusted to that of greatest efficiency. This method of mounting also allows for the complete reversal of the collectors, and when the machine is used as an electric motor, the direction of rotation—as in the electric hoist of Dr. Hopkinson—can be determined by altering the position of the brushes with respect to the magnetic field.

The collectors are not strictly brushes, but consist of a bundle of strips of hard elastic brass, alternate collectors being connected together so as to form one pole, the two intermediate collectors being also united so as to constitute the opposite pole of the machine. Thus, if the collectors were numbered around the commutator 1, 2, 3, and 4, 1 and 3 would be united, so as to form the positive pole, and 2 and 4 would be connected together and would constitute the negative pole. The two pairs can, however, if required, be used separately, so as to transmit the currents into two distinct circuits.

M. De Meritens claims special advantages for this description of machine over those in which the dynamo-electric principle is involved, for electro-plating purposes or for charging secondary batteries, on account of the impossibility of reversing the polarity of the magnetic fields by the back currents generated in galvano-plastic troughs, and in such batteries as the Planté, Faure, De Meritens and Sellon accumulators; for, as the magnetisation as well as the polarity of the field magnets are entirely independent of the direction of the current in the coils and in the external circuit, the polarity of the magnetic fields is perfectly permanent and independent of external influences.

In the Paris Exhibition there were two of M. De Meritens' direct-current machines, one forming part of his fine collection of apparatus not shown at work, and the other driven by a small vertical steam engine of very peculiar form, and supplying the currents for illuminating electric lamps fixed in another part of the building.

The thought suggests itself that the special form of field magnets adopted in these machines would be especially suitable to the introduction of a second armature ring at the opposite end of the shaft, for in the machines as at present constructed there is a magnetic field in every respect equally powerful to that in which the armature rotates, and yet this latent store of energy is absolutely useless. The addition of a second armature ring would obviously be more economical, both in construction and working, than the making and driving of two separate machines, the symmetry of the machine would be much improved, and the strains and working would be more evenly distributed.

The construction and finish of the machines of M. De. Meritens are as perfect as they can be, and must contribute largely to the great success of his system. The alternating-current generators seem to be especially successful in lighthouse installation, as is proved by the fact that both the Adminstration des Phares, in France, and the Corporation of the Trinity House, in England, have adopted the De Meritens machine for the electric illumination of lighthouses; thus the celebrated first-order lighthouse at Planier (Bouches du Rhône) is now lighted by two De Meritens generators. And two of the large lighthouse machines which were at work at Paris have been erected in the lighthouse at Dunkerque, while two other similar machines also in the Exhibition are installed in the lighthouse at Baleines (Seine Inférieure). It is the intention of the Administration des Phares to establish De Meritens machines in no less than 42 lighthouses around the coast of France, and to replace the Alliance m acines at Cape Grisnez and La Hêve by the same magneto-electric system. In this country the two most important double lighthouses on the south coast, that is to say, the Lizard and the South Foreland, have recently been fitted with De Meritens generators, and the Corporation of the Trinity House have ordered similar apparatus for the Macquarie lighthouse in New South Wales.

BRUSH.

The Brush system of electric lighting which during the last four years has, in the United States of America, been developed into a very important commercial industry, is one of the most interesting systems of illuminating by electricity that has yet been before the public ; it has also become widely appreciated in this country, and has indeed taken a leading position among the lighting systems of Europe. It has for some years been employed for illuminating the terminus of the Great Eastern Railway at Liverpool-street, and it has been adopted by the Great Western Railway Company for their Paddington Station, at the Charing Cross terminus of the South Eastern Company, the Waterloo terminus of the South Western Railway, the South Kensington Museum, and too large a number of public buildings to be mentioned here. It has also been applied to light a considerable area of the City of London, in competition with the Siemens and Weston systems, each of which has had a district of similar area of the City to illuminate, so that the public have been furnished with an excellent opportunity of comparison.*

The Brush, like the Siemens system, includes both a special generator and lamp, and may therefore be looked upon as complete in itself, differing in that respect from all those so-called systems which are represented by either a particular form of machine or a special

---

* The contract for the Blackfriars district, which was entrusted to the proprietors of the Brush system, has at the date of writing been extended for another twelve months.

arrangement of lamp, and like the proprietors of the Siemens system, the Brush Company manufacture their own carbons, making the whole installation still more peculiar to themselves.

The Brush machine, a general view of which is given in Fig. 194, possesses some points of resemblance to the dynamo-electric generators of Gramme and of Wallace-Farmer, as well as to the magneto-electric machines of M. De Meritens, while at the same time its principle of construction differs from that of all others, and confers upon it at once its special excellence and its originality.

Its point of resemblance to M. Gramme's generator consists in the fact that its armature is of annular shape, a form of construction first adapted to the purpose by M. Pacinotti, in 1860. The Brush armature differs, however, from the Gramme ring in the arrangement and disposition of the helices of wire with which it is wound, as well as in the way in which the several coils are connected with one another. In the Gramme armature the coils are wound contiguously to one another, so as entirely to envelop the iron core, hiding it completely from view, and they are permanently connected to one another in consecutive series. In the Brush armature the diametrically opposite bobbins are alone permanently connected together, and a current generated in one pair of bobbins does not necessarily traverse all the others, and in fact never does, as will be pointed out further on. Again, the individual coils are separated from one another by a considerable sector of the iron ring, which is of larger sectional area between the coils (see Fig. 195), so that in the revolution of the armature within the magnetic field, the coils alternate with masses of iron which from their enlarged section are brought into as close proximity to the poles of the field magnets as are the outsides of the coils themselves. It is this difference from the Gramme ring which constitutes the similarity between the Brush armature and that of M. De Meritens, but the Brush generator differs in all other essential respects from the latter machine, in the disposition of its coils, in its mode of connection, in the method and arrangement of its magnetic field, and in the continuous nature of its current.

The Brush machine has nothing whatever in common with the Gramme generator in the disposition of its field magnets and the relative positions of the revolving helices and the magnetic field. In the disposition of its magnets the Brush more nearly resembles the Wallace-

Fig. 194. The Brush Dynamo-Electric Generator.

Farmer generator, but their form is entirely different, and the armature of the Wallace-Farmer bears no sort of resemblance to the Brush ring, its coils being more allied to those of M. Lontin's machine than to any other of the characteristic generators. Having thus pointed out the position occupied by the Brush machine with respect to other typical generators, showing wherein it approaches and wherein it departs from the machines of other systems, its construction may be described.

The most characteristic feature of the Brush machine lies in the form and construction of its armature, which consists of a cast-iron ring, the cross section of which is generally rectangular, but in the direction

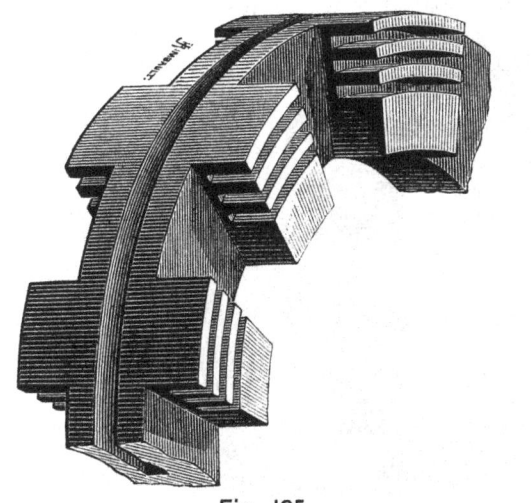

Fig. 195.

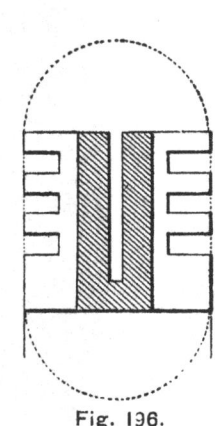

Fig. 196.

of its circumference it is alternately wide and narrow, as shown in the sketch, Fig. 195, which represents a portion of the iron armature ring and explains its construction better than can be given by description. On reference to this figure, it will be seen that the ring is divided up into as many sectors as there are bobbins to be wound, by a number of rectangular depressions or grooves; in these the coils of insulated copper wire are wound until the groove is filled up and the flat converging recesses become flush with the face of the intermediate thicker portions or pole-pieces by which they are separated from one another. Fig. 196 is a cross section of one side of the ring taken through one of these portions, and it will be observed on reference to both figures that the intermediate thicker portions of the ring are grooved out by a series of deep concentric grooves, the object of which

is partly to reduce the mass and lessen the weight of the revolving
armature, partly for the purpose of ventilating the ring and thus carrying
away a portion of the heat generated by the working of the machine,
but chiefly for the localisation and isolation of currents generated by
induction in the iron, and which would tend not only to reduce the
efficiency of the machine by diminishing the magnetic capacity of the

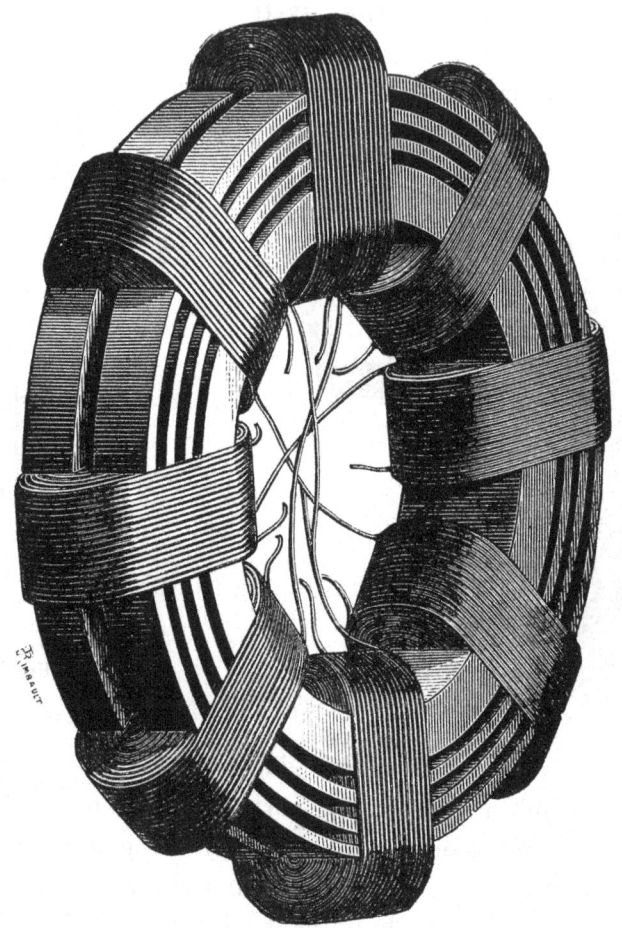

Fig. 197.

armature, but also to produce a heating of the ring, and therefore of the
coils, whereby a portion of the current would be lost through their
resistance being increased.   For a similar reason the periphery is grooved
out deeply so as almost to sever the ring; by this means all cross
currents are effectually cut off and induction currents are compelled
to flow in directions which are not detrimental to the efficiency of
the machine.   This again increases the area of radiating or cooling

surface, and consequently helps to prevent the armature becoming overheated.

In what is called the 16-light machine, which may be regarded as the normal size of the Brush generator, the armature ring is 20 in: in diameter, and it is wound with eight radial coils of cotton-covered copper wire of No. 14 B. W. G., whose middle planes radiate from the axis of rotation, being distributed round the ring at equal angular distances apart of 45 deg. Each coil contains about 900 ft. of wire, weighing about 20 pounds, and is wound in the rectangular grooves or spaces shown in Fig. 195, filling up the space completely so that the vertical faces of all the coils are flush with the vertical faces of the thicker portions of the iron ring. Fig. 197 is a drawing of the armature ring with all its coils wound, but it must be looked upon rather as an explanatory diagram than as an illustration of the proportions of this part of the apparatus. The two sides of each groove, and therefore of each coil of wire, are parallel to the centre line or radial plane of the coil, and by the adoption of that form of bobbin one of the practical difficulties in the winding of annular armatures of the ordinary form is avoided. All the coils are, like those in the Gramme generator, wound in the same direction.

Fig. 198 is a diagram illustrative, not only of the distribution of the coils around the ring, but of the method by which the connections are made; the inner end of each of the coils is connected by a wire to the inner end of the corresponding coil, at the opposite end of the same diameter of the ring, and the outer ends of all the coils are brought through the shaft of the machine, and are connected to corresponding portions of the commutator, where the currents are collected by suitably placed copper plates or brushes. Referring to the diagram, it will be seen that the inner end A^1 of the coil 1 is connected to A^5, which is the inner end of the coil 5; A^2 is connected to A^6, A^3 to A^7, and so on round the ring, and the outer ends B^1, B^2, B^3, &c., are all connected to the commutator by conducting wires insulated from one another. The two free ends of each pair of diametrically opposed coils are, after passing through the shaft of the machine, attached respectively to two diametrically

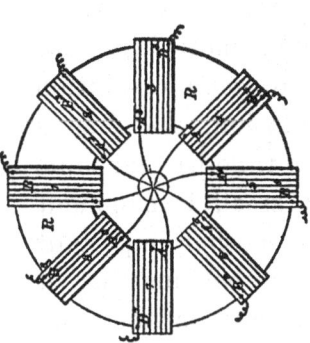

Fig. 198.

P

opposite segments of the same commutator, which segments are insulated from one another and from any other pairs of coils. The commutator, which is attached to and rotates with the driving shaft of the machine, consists of a set of separate copper rings or flat cylinders, of which there are as many on the shaft as there are pairs of coils on the armature. Each of these cylinders consists of two segments insulated from one another on each side of the shaft by two small air spaces about one-eighth of an inch wide, and by a piece of copper separated from the segments. The arrangement is clearly shown in Fig. 199, in which A and B are the two segments connected

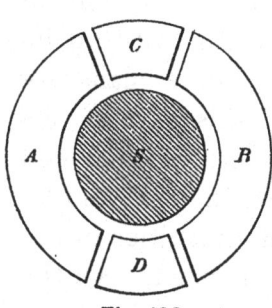

Fig. 199.

respectively to corresponding coils on opposite sides of the armature, and separated by an insulating material from the shaft S; C, D are the copper insulating pieces, the object of which is to separate either of the flat copper brushes or collectors, which press upon the periphery of the commutator, from either of the segments during the interval occupied by one pair of coils passing the vertical, in other words, through the neutral portion of the magnetic field; this occurs twice in each revolution of the armature, and therefore of the commutator. At the time when any pair of bobbins is in this way cut out of the general circuit, their own circuit is open, so that no current can circulate or be induced in them. By this most ingenious arrangement, each pair of coils has in succession in each revolution a period of rest equal to one-quarter of a revolution, and has a current passing through it for only 75 per cent. of the time the machine is running; to it is, in a great measure, due the very small development of heat in the Brush machine, and it presents also another important element of efficiency namely, that each pair of bobbins as it passes the neutral portion of the magnetic field, and is therefore incapable of doing work and contributing electromotive force to the general current, is itself cut out of the circuit, and thus two causes operating against the efficiency of the machine are eliminated. The first cause is common to most armatures which have, like that in the Gramme machine, a permanently closed circuit, namely, that the currents generated in the bobbins have two routes open to them, the one through the conductors and commutators to the brushes, and the other through the idle bobbins, and thus by a species of short-circuiting they rob the external circuit of some of its current. The other cause of inefficiency which is avoided in the Brush machine is the reduction of its internal resistance by an amount equal to the resistance of two of the bobbins, for just as a certain amount of resistance is an item of efficiency when belonging to coils which are doing work and contributing electromotive force to the general current, so does it become an element of inefficiency when belonging to cores which are idle, for in that case it diminishes the current supplied by the active coils while at the same time contributing no current of its own in compensation. By the arrangement of commutator described above, Mr. Brush has, therefore, got rid of two considerable drawbacks to the efficiency of the generator.

What is, however, one of the important features of the Brush machine is the arrangement of the magnets by which the magnetic field is produced, and by which the armature coils are, during their revolution almost continually passing through a very intense magnetic field. Upon reference to Fig. 194 it will be seen that the armature ring is closely embraced by the large horizontal electro-magnets the poles of which are expanded so as to be presented to three of the armature coils on each side, leaving one pair of coils free from their direct influence, and this is the pair which is passing through the neutral region of the magnetic field. For the sake of illustration, the disposition of the magnets towards the armature and towards one another, may be described as two horseshoe electro-magnets placed opposite one another in a horizontal position, their similar poles being presented towards one another, and having a small space between them in which rotates the armature ring. This space in the machines we are describing is however so nearly equal to the thickness of the armature that there is hardly any clearance between them, the high-class workmanship of the apparatus allowing such close working with perfect safety.

The methods of exciting the field magnets in dynamo-electric machines are very different under different systems. In the early Gramme machines a separate armature was set apart for producing the current which by traversing the magnet coils produced the magnetic field between their poles, and the current from this armature had no other work to do, the useful or external current being taken direct from one or more other armature rings attached to the same shaft. The exciting of the magnets in one machine by a current supplied from another, or from

P 2

a special armature of the same machine, was long since suggested and employed by Mr. Wilde, of Manchester, whose name with that of Mr. Holmes must ever be associated with the early history of the production of electricity by mechanical means. In the Brush generator, as in the later form of the Gramme, as well as in the ordinary Siemens direct-current machine, the whole of the current from the armature is transmitted through the magnet coils, they forming with respect to the armature, a portion of the external circuit. In several other systems, such as the Siemens alternating-current machine, &c., the field magnets of one or more generators are excited by the current from a special machine which has no other work to do; and in what is known as the shunt machine (a special form of the Siemens generator), the magnet coils form a shunt circuit to the armature, by which arrangement the machine becomes to a great extent its own regulator.

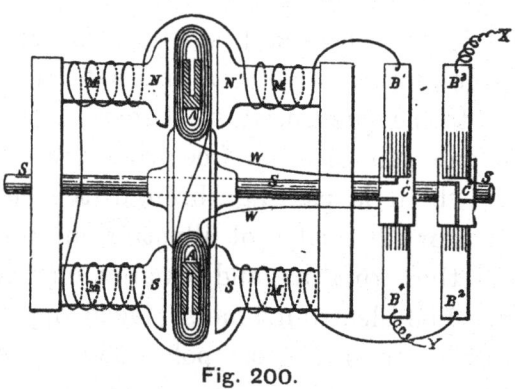

Fig. 200.

Fig. 200 is a diagram illustrating the connection between the armature bobbins and the magnet coils in one position of the armature revolution. Referring to this diagram, M M and M M¹ are the two magnets having their similar poles presented towards one another on opposite sides of the armature coils A A¹. Thus the coil A is under the influence of a magnetic field produced by the two north poles N N¹, while at the same time its corresponding bobbin A¹ is under the influence of the two south poles S S¹; a current is therefore induced in the pair of bobbins A A¹, which is transmitted by wires W W passing through the shaft S to the commutator C¹, whence it is collected by the brushes B¹ and B⁴, at the same time the other portions of the commutators (which are in connection with the other working armature bobbins) are in contact with the brushes B² and B³. At X and Y are the terminals of the external circuit, and to the brushes B¹ and B² are connected the two ends of the field wire. The field magnets are connected together as shown in the figure.

The commutators form a most beautiful piece of apparatus, and are, as we have said, peculiar to the Brush system; the arrangement by which the position of the collecting brushes can be adjusted so as to place them in contact with the commutators in any desired angular position with respect to the possession in the magnetic field occupied by the several bobbins to which the commutators are connected, is exceedingly simple and ingenious.

Attention may be called here to the fact that the mechancial, no less than the electrical details of the machine are thought out and executed with much care and excellence; the armature rings are accurately turned and centred, and are balanced with great accuracy, so that the weights are symmetrically distributed around the axis of rotation, which is absolutely coincident with the axis of gravity. The bearings in which the shaft runs are exceptionally long, and ample provision is made for preventing any end motion of the shaft. All such mechanical refinements increase the efficiency as well as the permanency of the machines, enabling the poles of the fixed magnets to be brought almost to touch the rotating armature, and therefore putting them into the best possible position to produce the maximum inductive effect upon the bobbins.

From the foregoing description the operation of the commutating apparatus will be clearly understood, and it will be seen that by the simplest of mechanical means it collects and distributes the currents from the active armature coils, sending them into the magnet helices and into the "line," and cutting out of circuit the armature coils one by one as they pass through the neutral region between the poles. The whole commutating apparatus is fairly represented on the extreme right of the general view of the machine (see page 206, Fig. 194), and consists of two pairs of rings of the form shown in Fig. 199, attached to and revolving with the main shaft, so that their position is fixed with respect to the revolving armature of the machine. Rubbing tangentially upon the circumferences of these rings are two pairs of copper collecting brushes, one pair pressing above and the other pair pressing below; a line forming the points of contact being a diameter of the ring. The copper " brushes," as they are called, are flat strips of elastic copper about 2 in. wide cut at the ends which press against the rings into eight tongues, so as somewhat to resemble a grainer's comb, and each comb or brush is wide enough to cover or be in contact with two armature rings; in this way, although two of the coils are insulated twice in each revolution, the main circuit is never interrupted. The disposition of the brushes with respect to the commutators will clearly be understood by comparing Fig. 194 with Fig. 200.

For the sake of adjusting the brushes so as to make contact with the commutators at the most effective angular position with respect to the magnetic field, they are mounted to the opposite ends of two rocking levers, which are capable of oscillation concentric with the driving shaft and can be fixed in any desired position by means of a set-screw, which clamps a stout wire rising from the base of the machine. The currents are conveyed from the brushes by wide strips of thin sheet copper, shown in the general view, and in order to allow for the variable distance of the free ends of the brushes from the base of the machine they are made undulating or wavy, doubling up as the distance is shortened and stretching out when it is increased.

The average total resistance of the sixteen-light machines as now constructed is about 11 Ohms, to which the four coils of the field contribute about 6 Ohms, that is 1·50 Ohms each, and the magnet coils about 6 Ohms or 1·5 Ohms for each helix—the resistance of the connections, conductors, contacts, &c., within the machine being inconsiderable.

A very interesting series of experiments was made with the Brush machine at the works of the Telegraph Supply Company, at Cleveland, Ohio. The size of generator under investigation was that known as No. 7, *i.e.*, the ordinary 16-light Brush generator, and with it 16 Brush lamps of the usual construction were employed through a circuit of 200 ft. of No. 10 copper wire. The machine was driven at an average speed of 770 revolutions per minute, and careful comparative measurements were made for determining the difference of potential at the two terminals of the lamp by the following simple and ingenious method devised by Mr. Brush, who was assisted in his experiments by Mr. G. H. Wadworth, the chief electrician to the Western Unions Telegraph Company at Cleveland. The 16 lamps were connected together in series and with the machine, and when the latter was running at its proper speed they were adjusted so as to produce as nearly as possible arcs of equal length; the two terminals of each lamp in succession were then placed in circuit with a galvanometer, and with a Daniell battery, fitted with a simple commutator by which any number of cells up to 48 could be instantly put into circuit with the galvanometer and lamp under examination, the connections being so made that the voltaic current and the dynamo-electric current through the coils of the galvanometer were in opposite directions.

Fig. 201 will explain the general arrangement, in which M represents the machine, B a Daniell battery of 48 cells, G a galvanometer, and L^1 L^2 L^3 L^4 and L^5 represent 5 of the 16 lamps under observation. From the above arrangement it is clear that if the difference of potential between the terminals of the lamp be greater than the difference of potential between the two ends of the battery, a current will be forced through the battery in opposition to its own current, deflecting

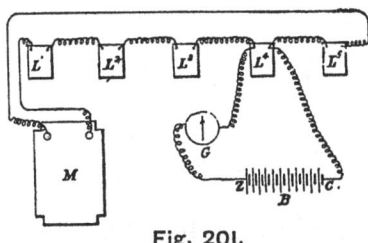

Fig. 201.

the galvanometer in one direction; but if the potential difference between the battery poles is greater than that between the lamp terminals, the current through the galvanometer, and therefore its deflection, will be in the opposite direction, and it can only be when the difference of potential between the battery poles and between the lamp terminals are equal that there will be no deflection of the galvanometer. Mr. Brush, therefore, adjusted the commutator of the battery so as to include in the circuit of the galvanometer and lamp just as many cells of the battery as would balance the current from the machine, bringing the galvanometer needle to zero, and when that balance was obtained, the appearance of the arc as well as the speed of the machine were carefully noted. A similar series of observations were made with each of a number of lamps all at work together; and, as the results obtained for the measurements of the various lamps agreed with remarkable closeness, a very correct average was obtained, and the difference of potential between the terminals of the average lamp was found by this means to be equal to that of 42·46 Daniell cells when the machine was running at 770 revolutions per minute.

By a very similar method of working, Mr. Brush determined the resistance of the average lamp. To do this, one of the lamps was removed, and its place was occupied by a resistance of nearly 700 lbs. of thick wire, its length being so adjusted that the current of 42 cells of the battery exactly balanced the difference of potential between the two ends of the resistance wire, while the machine was running at the same speed as in the former experiment. The arrangement of connections is shown in Fig. 202, in which M represents the machine, L^1, L^2, L^3, and L^5 the lamps, and R the resistance wire substituted in the circuit for the lamp L^4. G and B are the galvanometer and battery respectively, arranged as in the first experiment. The current from the machine

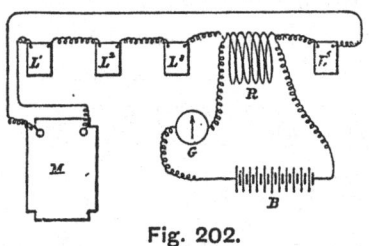

Fig. 202.

increased the temperature of the coil slightly, and before any appreciable loss of temperature took place, its resistance was measured by the ordinary method and found to be $4\frac{1}{2}$ Ohms, and from those figures the resistance of the average lamp, whose difference of terminal potential we have seen was balanced by rather more than 42 cells (*i.e.*, 42·46), was easily calculated, being somewhat more than $4\frac{1}{2}$ Ohms, viz., 4·56 Ohms, and this multiplied by 16 gives a total resistance for all the lamps of 72·96 Ohms.

After a series of careful measurements the resistance of the generator between brush and brush, at the average temperature which it assumes after settling down to its work, was found to be 10·55 Ohms; and this added to the resistance of the lamps (that of the conductors being inconsiderable) gives a total internal and external resistance for the whole circuit of 83·51 Ohms, of which 87·36 per cent. is in the lamps alone, so that it would appear from these data that no less than 87·36 per cent. of the current developed in the Brush machine is utilised in the external circuit, showing a remarkably high degree of efficiency.

In order to determine how much energy is absorbed in the 16 arcs by conversion into heat and light, Mr. Brush placed the upper and lower carbons in each lamp in firm contact with each other, and by the method shown in Fig. 202, he measured the total resistance of the 16 lamps, including all connections, magnet-coils, carbons, &c., and found it to be 2·1 Ohms, which, deducted from 72·96 Ohms (the total resistance of lamps when burning), leaves 70·86 Ohms as the resistance of the 16 arcs, from which it was estimated by Mr. Brush (allowing for the 1 per cent. of current shunted out of the current of each pair of carbons by its fine wire regulating helix) that the production of the arcs involves an expenditure of energy equal to 84 per cent. of the total electrical energy developed by the machine.

While the above electrical measurements were being made, a series of experiments were conducted to determine, by means of indicator diagrams taken on the engine, the driving power absorbed by the machine when working at its normal speed, and the results of these experiments—which are given in the Table annexed—taken in conjunction with the electrical measurements to which we have already referred, show that in the Brush system 81·89 per cent. of the motive power absorbed, is converted into electrical current; taking into consideration the fact that 84 per cent. of the total electrical energy is utilised in the arcs, it follows that 68·79 per cent. of the entire power absorbed in the production of the current is represented by the heat and light produced in the 16 arcs. In the following Table will be found a summary of the results of the various measurements to which we have referred :—

Resistance :

Resistance of the machine between terminals	10·55 Ohms.
,, external circuit	72·96 ,,
,, whole ,, ...	83·51 ,,
,, the 16 arcs	70·86 ,,

Current :

Electromotive force of current	839·02 Volts.
Volume of current	10·04 Ampères.

Power :

Total driving power required	15·48 H.P.
Driving power absorbed in production of current	13·78 ,,
Energy of current expressed in horse power	11·285 ,,

Per-centages :

Per-centage of current available for external work	87·36
,,	current appearing as heat and light in the 16 arcs	84·00
,,	gross power converted into current	72·90
,,	absorbed power converted into current	81·89
,,	gross power appearing in arcs	61·24
,,	absorbed power appearing in arcs	68·79

The figures given in the above Table are highly interesting, and indicate in almost every point a very high degree of efficiency both in the generator and in the whole system. Since those measurements were made, however, several improvements have been effected in the machine, and it would be a matter of the greatest interest if a new series of careful scientific measurements were taken in this country and with the improved apparatus, for there can be no doubt that it is by careful dynamometrical and electrical measurements carried out under scientific management, that the real practical efficiency of any system of electric lighting can alone be determined. Photometrical measurements, except as supplementary, are in the highest degree unsatisfactory, and the possibility of reading a newspaper so many feet from a lamp, or the lighting of a given floor-space to the intensity of lunar illumination in autumn, involves too much of the element of moonshine in it to be of much value as a standard of photometrical measurement, although for certain special purposes it may probably be approximate enough. Thus, at the very interesting installation at the harbour of Montreal, there are 17 lights worked by a single Brush machine through a circuit of nearly three miles, the light of each lamp enabling a newspaper to be read with ease at a distance from it of over 160 ft., and in this case the current is conveyed from the machine station through a copper wire No. 9 B. W. G.

BÜRGIN.

The Bürgin generator is the invention of M. Emile Bürgin, of Bâle, in Switzerland, and it was at first designed as a single-light machine.

Mr. Crompton, of Messrs. Crompton & Co., Arc Works, Chelmsford, however, saw that it was capable of being modified so as to yield an electromotive force sufficient to supply several lights, while at the same time its small size and simplicity of construction would lend themselves to portability and cheapness. The generator is illustrated by Figs. 203 to 208, where Fig. 203 is a general view of the side on which the commutator is placed, Fig. 204 is a similar view of the other side showing the driving pulley, Fig. 205 is a view of the armature, Fig. 206 is a part longitudinal section and part elevation, and Fig. 207 a corresponding transverse section.

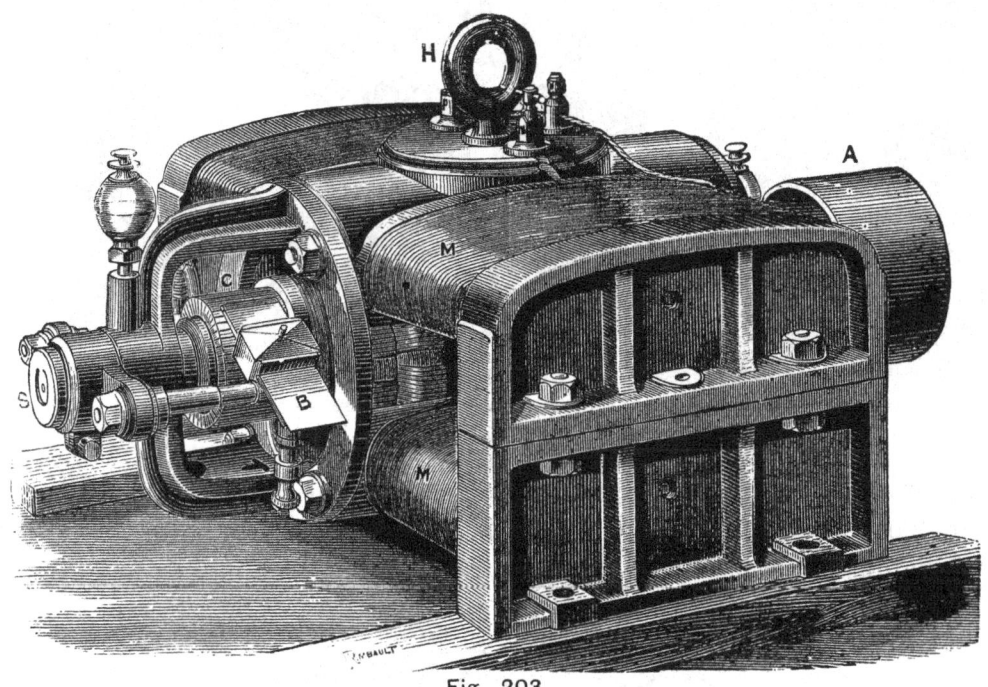

Fig. 203.

To get a high electromotive force, a high surface velocity of the coils of the armature is necessary, and the ordinary Gramme ring is not perfectly adapted for this, because the wire of the coils is closely confined and wound so thickly that it is apt to heat with the current. The coils of the Bürgin ring, on the other hand, are wound as shown in Fig. 205, which represents the armature of a three-light machine. Instead of a single ring as in the Gramme machine, it is composed, as will be seen, of several individual rings R, mounted on the same spindle S, so as to form a kind of barrel. The coils of these rings are, however, all connected in series to the sections of the commutator C, so that the current is collected in the ordinary manner by a pair of copper "brushes." Instead of being a circular hoop as in the Gramme,

Fig. 204

each core is a hexagonal frame of soft charcoal iron wire built up into a square bar. Each of the six sides of the core, after being covered with varnished tape, is wound with a coil of wire in such a manner as to leave

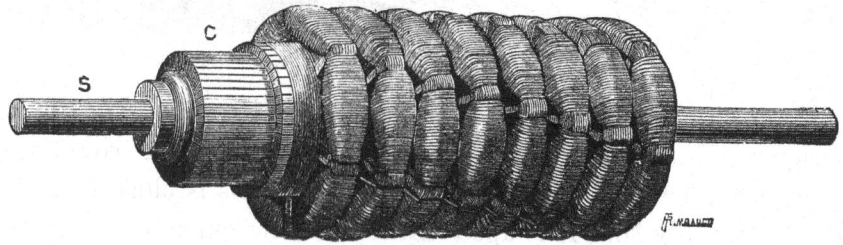

Fig. 205.

the iron corners bare and bring the general outline of the coil circular. This arrangement will be readily understood by reference to Fig. 206, where R is the ring, I the bare corners of the core, and W the coil of wire. It offers several practical advantages ; for example, the bare angles of the core passing close to the poles of the inducing magnets give rise to an intense magnetisation of the ring. Moreover, the

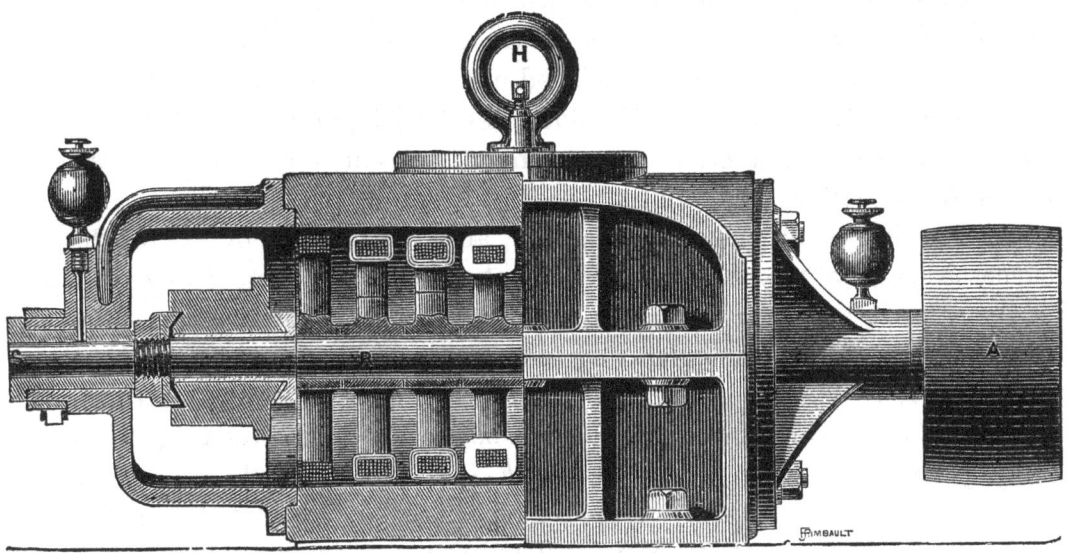

Fig. 206.

simplicity of the winding renders the ring easy to make, and the small number of layers in each coil, together with the free access of air

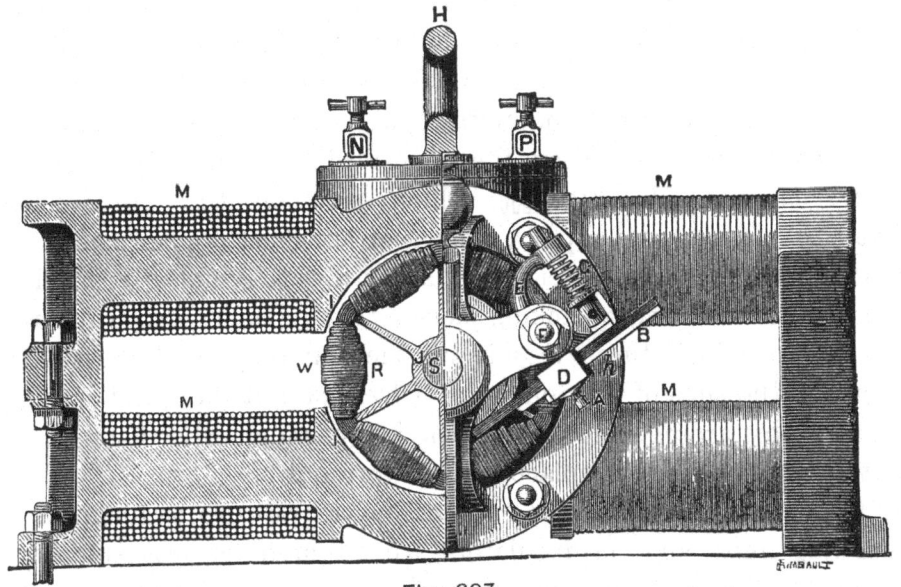

Fig. 207

to the exposed surfaces of the wire prevent the coils from being unduly heated by the current. In the Gramme machine there are usually 14 layers of wire in a coil, whereas in the Bürgin there are only five, the result being that a high tension current has no ill effect on the armature.

As the six coils on one ring are not sufficient to provide the current required for three lights, eight rings are arranged side by side on the same spindle, and each ring is displaced one-twenty-fourth of a revolution in advance of the last. Fig. 205 illustrates this ingenious construction, which favours the production of a continuous current, for the coils of the barrel which follow one another at this fraction of a revolution are connected in series one after another and their ends joined to the successive slips of the commutator. That is to say, the successive coils of the same ring are not connected in series as in the Gramme armature, but the coil of one particular ring is connected in series to any coil on the next ring which follows it at a distance of one-twenty-fourth of a revolution. This arrangement will be better understood by consulting Fig. 208, where a, b, c, d, e, f, g, h, i are the coils which properly

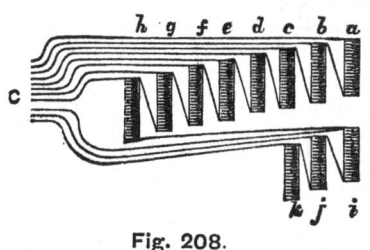

Fig. 208.

succeed each other in the length of the barrel. They are connected in series, and wires proceed from each to the slips of the commutator c. The last coil on the left, h, is properly followed by the first coil i of the next set on the right, and i is connected to j, k, &c., as before. This method of joining the coils obtains throughout the whole ring.

The contact slips of the commutator are made of phosphor-bronze, a metal which wears better than the copper which is usually employed. The two "brushes" B, Fig. 206, which draw off the positive and negative currents from the slips C, consist of six sheet copper combs laid one above another. Each is carried by a box d, into which it is fixed by an adjusting screw a, and a packing piece h; the whole being supported by a bracket e which turns loosely on the spindle f, and presses the brush against the slips of the commutator by means of the spiral spring g. One merit of this arrangement is that the brushes and other parts of the commutator can be readily replaced when worn out.

In the Gramme machine the circular ring requires to be mounted on an insulating hub; and, as insulators are all more or less subject to atmospheric influences, the efficiency of the machine is not entirely independent of the climate or country in which it may be placed. But the hexagonal rings of the Bürgin armature enable it to be mounted on the spindle S by a hub and spokes of metal as seen at J, in Figs. 206 and 207. The whole armature is light and accurately balanced, and is capable of a high velocity. It comprises 48 coils, each containing 48 ft. of cotton-covered copper wire 0·065 in. in diameter, and the total weight of this wire is only 25 lbs. The electric resistance of the whole armature between the two brushes is 1·6 Ohms, and the electromotive force or difference of potentials between them for a speed of 1,500 revolutions per minute is 195 Volts. For 1,600 revolutions per minute, the electromotive force is about 206 Volts. The speed of the mean diameter of the coils through the magnetic field corresponding to a velocity of 1,500 revolutions per minute is 2,550 ft. per minute. The driving pulley is shown at A in Figs. 203 and 204.

In the Siemens generator the cores of the electro-magnets are built up of separate bars of wrought iron, but in the Bürgin they form one solid iron casting, which is bored to admit the armature and is fitted with a cylinder cover to carry the brushes. This mode of construction, besides being inexpensive, gives considerable steadiness of working. Moreover, the mass of iron seems to intensify the magnetic power, and the passage for the armature being truly bored, the bobbins run concentric with the magnetic poles. The cores of the magnets are of oval section, and the heavy pole-pieces are 2 in. thick and crescent-shaped. They are wound, as shown in Figs. 203 and 204, at M M, with 190 ft. of cotton-covered copper wire 0·141 in. in diameter, and the total weight of wire on the four coils is 140 lbs. They are, as will be seen from the figure, coupled in opposite pairs, so as to form two poles of opposite name, between which the bobbins of the armature revolve with great velocity. The electric resistance of these magnet coils is only 1·2 Ohms.

The electro-magnets of the Bürgin machine can either be separately excited by the current from an auxiliary continuous-current machine, or they can be self-excited by the current from the armature. In this case the magnet coils are connected in circuit with the armature coils by suitable terminals; and the total internal resistance of the machine is then 2·8 Ohms, or 1·2 Ohms for the magnets added to 1·6 Ohms for the armature. By separately exciting the machine the internal resistance of the machine is simply the 1·6 Ohms due to the armature, and hence a higher electromotive force can be obtained.

MAXIM.

A most interesting system of lighting by electricity, the invention of Mr. Hiram S. Maxim, was introduced into this country, in the year 1881, by Mr. N. J. De Kabath, the financial representative of the United States Electric Lighting Company of New York.

This system, the various apparatus of which is illustrated by Figs. 209 to 213, is, like the Siemens, the Brush and the Lontin, a complete system, having its own special generators, as well as lamps peculiar to itself, and it would be difficult to say whether the Maxim machine or the Maxim lamp is the more characteristic and original feature. Of the two, the generating apparatus appears to bear the credit of the greater novelty, on account of the arrangement by which the strength of the current is regulated to the work it has to do, this regulating apparatus being designed upon an ingenious automatic principle.

The generating apparatus of the Maxim system, like the Siemens and Gramme alternate current machines, as well as the dynamo-electric generator of M. Lontin, consists of two distinct machines : (1) an electric generator whose magnetic field is produced by powerful electro-magnets, and whose current is utilised in the lamp circuit; and (2) a smaller machine on the dynamo-electric principle, the current from which is utilised exclusively for exciting the field magnets of the larger machine. The Maxim system has moreover, in addition to these, a third form of apparatus, viz., a current regulator by which the strength of the current in the lamp circuit is automatically controlled and adjusted to the number of lamps in action.

The Maxim machine shown in Fig. 209, bears (on account of the disposition of its field magnets, and the arrangement of its magnetic field) a remarkable resemblance in general appearance to the small Siemens exciting machine used for the alternate current generators, or to one of the more recent upright Siemens machines, in which the magnet helices are fixed in a vertical position. The machine, however, differs entirely from the Siemens machine in the construction of the rotating armature, which, instead of being in the form of a solid cylinder having its various wire sections passing over its ends from one side to the other, is in the form of a cylindrical ring or tube, its coils being wound longitudinally inside and outside the ring as in the Gramme armature, of which it is a modification; but it differs from the Gramme armature in the fact that its length is considerably greater than its diameter, in other words, that it is rather in the form of a tube than of a ring. It also differs from the modern Gramme armature in having two commutating cylinders, one at each end of its axis, and the coils of the armature are

Fig. 209.

connected alternately to these two commutators; that is to say, if the wire sections were numbered consecutively 1, 2, 3, 4, &c., all the even numbers are connected with one commutator and the odd numbers with the other. The Maxim machine is in physical principle a combination of the Gramme and Siemens generators, and may, in that sense, be described generally as a Gramme armature working in a Siemens magnetic field, but it has, in constructive detail so many points of

Q

originality as to give to it a separate and distinct position among electric generators.

Referring to the engraving, Fig. 209—which represents the principal machine connected up as a dynamo-electrical generator, exciting its own magnets, and requiring, therefore, no separate exciting machine—it will be seen that the field magnets consist of two sets of iron bars of rectangular cross section, surrounded at the upper and lower ends by rectangular coils of insulated wire. These are so connected together that the current traverses them in series, and in such directions as to produce a line of north polarity along the middle of one set of curved iron bars, and a similar line of south polarity along the corresponding medial line of the opposite set of bars, thus producing an intense magnetic field between them, within which the armature cylinder is rotated at a speed of about 900 revolutions per minute. This armature consists of a quantity of soft iron wire coiled into the form of a cylindrical tube about 15 in. long and about 12 in. in diameter, and wound after the fashion of a Gramme ring with a set of 16 sections or helices of insulated copper wire, each helix consisting of four wires, the ends of which are connected to the 64 sections of the commutators. The system of winding, as well as the disposition of the electro-magnets, is best shown in the illustration of the smaller machine, Figs. 211 and 212, the ring and magnetic field of which are very similar in form to those of the large machine. The armature, however, of the latter differs from that of the exciter in the fact that its contiguous coils are connected to different commutators, eight coils being attached to the commutator at one end of the shaft while the eight alternate and remaining coils are coupled to the other. The ring may thus be regarded as two armatures in one, each producing a distinct set of currents, which by a very ingenious commutating switch shown at the top of Fig. 209 and in plan in Fig. 210, can be united either in parallel circuit or in series. Referring to the latter figure, this switching apparatus consists of four pieces of brass A, B, C, and D, separated by channels shown in the figure, while at M, N, and P are peg-holes into which conducting pegs can be inserted when required. On each of the brass pieces A, B, C, and D is a binding screw, which can be connected by wires respectively to the top or bottom brushes of either commutator or to

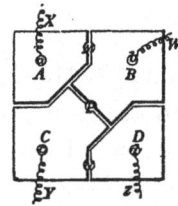

Fig. 210.

the lamp circuit. By inserting conducting pegs in M and N, A is connected to B, and C to D. In inserting a peg in P, B is connected to C, but both are insulated from A and D, and by this means the machine

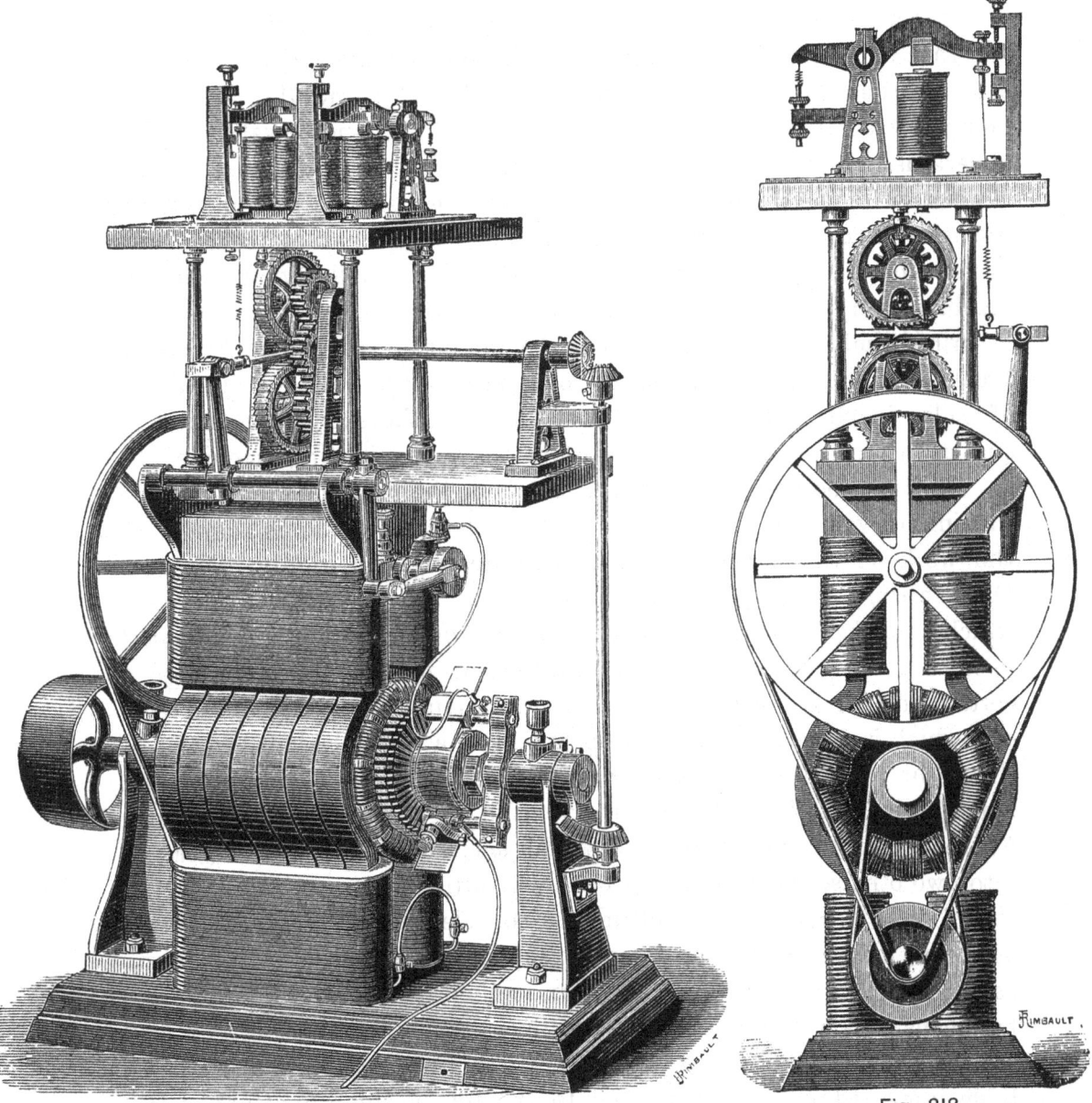

Fig. 211.   Fig. 212.

can be arranged to supply currents of high electromotive force to circuits of high resistance, or by combining the two currents of the machine in parallel circuit, it can be arranged to produce a

Q 2

great quantity of current to a single arc.   Fig. 209 represents the
machine arranged as a generator for a single-arc light, the whole
of the current being caused to traverse the coils of the field magnets;
but, for incandescence lamps, or for circuits in which the work is
at all varied, the current from the machine flows direct into the
external circuit, and the magnets are excited by the current from the
second machine and regulator, shown in Figs. 211 and 212, the former
being a general perspective view and the latter an end elevation.   This
apparatus, which, with its regulator, more than anything else, is the
characteristic feature of the Maxim system of generating currents, as his
incandescence electric lamp is the distinguishing characteristic of his
system of utilising the current, consists of a small dynamo-electric
machine, very similar in construction to the larger machine just
described, but fitted with but one commutator and a single pair of
brushes, and having its armature cylinders wound with a single set
of coils; in all other respects, except in size, the exciting machine is
similar in principle and general construction to the lighting machine.

The method by which the current from the machine, and therefore
the intensity of the magnetic field of the large machine, is adapted to
the work on the lighting circuit, is interesting and highly ingenious.
On referring to Fig. 211 it will be seen that the collecting "brushes,"
which consist of a bundle of flat copper springs or plates, are mounted
upon a rocking frame, which is capable of being rotated through a
certain angular distance upon the armature spindle by means of a bevel
arc, which is geared into by a bevel pinion on a vertical shaft shown on
the right of the figure, and this shaft is actuated from above by a
horizontal shaft through the intervention of a pair of bevel pinions.
At the opposite end of the horizontal shaft is a spur-wheel gearing
into two others, which are respectively pivotted to two ratchet wheels,
shown in Fig. 211, but more clearly in the elevation, Fig. 212, a portion
of the framework being cut away to show the position of an oscillating
pawl between them, to which a reciprocating motion is communicated
by means of a rocking lever worked by a small connecting rod and
crank, clearly shown in Fig. 212.   Above this ratchet gearing is a small
table supported on four pillars, and to the top of this are attached two
electro-magnets; the coils of one are of considerably higher resistance
than those of the other, but it is with that of the higher resistance that
we have to do at present.   Above the upper pole of this magnet is a soft

iron armature supported on a lever, which, by means of an adjustable tension spring, shown on the left of Fig. 212, can be balanced, not only against the weight of the pawl, to which its opposite end is attached by an electric connecting rod, but also against any required magnetic attraction of the electro-magnet.

The mechanical action of the apparatus is as follows: during the working of the machine a reciprocating motion, by means of a small crank and rocking bar, is communicated to the pawl, and in the normal position to which the upper level is adjusted simply moves backwards and forwards between the ratchet wheels without touching either. If, however, the magnetic intensity of the electro-magnet diminishes, its attractive force is overpowered by the tension of the adjustment spring, the lever rises, and the pawl is lifted so as to gear into the upper ratchet wheel, and the whole train of counter-shafts will be set into motion in one direction, the collecting brushes moving around the commutator towards the point of minimum efficiency; the current sent into the magnet coils of the large machine will, therefore, be weaker, thus dimishing the intensity of the magnetic field, and therefore of the current induced in the armature rotating within it. If, on the contrary, the magnetic intensity of the electro-magnet of the regulator increases, the reverse action will take place, the pawl will engage in the teeth of the lower ratchet wheel, and the brushes will be moved towards the point of maximum efficiency, and the magnetic field of the larger machine will, therefore, be intensified. The mechanical details are so proportioned and arranged that the brushes can be moved through a range which is limited on one side by the zero of efficiency, that is to say, the neutral plane of the magnetic field, and on the other by the point at which the maximum current is collected by the brushes, and between these two extremes the efficiency of the machine can be varied by stages corresponding to the motion of the brushes caused by the rotation of ratchet wheels through an angular distance equal to the pitch of their teeth.

In order to make the work on the lighting circuit control the smaller machine, and therefore the strength of its own current, the regulating electro-magnet is placed in a shunt circuit derived from the lighting circuit of the large machine. This arrangement will be readily understood by reference to the diagram, Fig. 213, in which M represents the large machine, and R the regulating portion of the smaller machine; P and N are the wires leading from and to the lamp circuit, and x and y are fine wires forming, with the electro-magnet to which

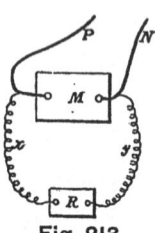

Fig. 213.

they lead, a shunt circuit of comparatively high resistance to the main circuit of the large machine. As the current from one terminal of the large machine to the other has two paths open to it, one through the lamps and the other through the regulating magnet of the smaller machine, it follows that the strength of the current flowing through the latter, depends upon the relative resistances of the two circuits. And, as the resistance of the magnet circuit is fixed, while that of the lighting circuit is variable according to the number of lights in action, it further follows that the addition or subtraction of one or more lights from the circuit, immediately affects the strength of the derived or magnet circuit, and so adjusts the brushes to collect just sufficient current to give to the magnetic field of the large machine an intensity which is sufficient and no more than sufficient for its work.

From the arrangement of circuits just described, it is obvious that should a large number of lamps be suddenly cut out, through the accidental rupture of the wire or from any other cause, there would be great danger of the lamps still in circuit being destroyed by the whole of the current being divided over a smaller number, and moreover, there would be a great risk in such a case of the machine itself being injured. To protect the system against one or other of such results is the object of the second magnet, which is shown at the top of Fig. 211, and which, while being of lower resistance than the regulating magnet which actuates the pawl, is included in the same circuit. This second or guard magnet is provided with an armature similarly mounted to that of the regulator, but its lever, instead of actuating a pawl, is in fact a magnetic relay or switch, forming, when the current is at its normal strength, a part of the circuit of the helices of the field magnets of the small machine. The moment that the strength of the magnet circuit increases beyond that to which the switch armature is adjusted, the armature of the second magnet is attracted, and the field magnets of the small machine are thereby immediately cut out of the circuit, the exciting current ceases, the intensity of the magnetic field of the large machine is thereby reduced to zero, and the main current is for a moment suspended, causing a momentary extinction of the lights but saving them from destruction, and the machines from injury.

WESTON.

The name of Mr. Weston has long been before the public as the inventor of a very successful dynamo-electric machine for electro-plating, and we believe, at the present time, the greater bulk of the largely increasing industry of nickel-plating is performed by Mr. Weston's machines, which in general principle are very similar to the dynamo-electric machine of M. Lontin. The Weston lighting

Fig. 214.

generator has, however, nothing in common with the plating generator of the same inventor, for while it has points of resemblance in certain of its details with the machines of Siemens, Gramme, De Meritens, Brush, and Maxim, it is distinctly original and cannot be confounded with any one of them.

The general external appearance of the machine, which is shown in

Fig. 214, on account of the distribution of its field magnets, bears a slight resemblance to one form of the Gramme machine. These magnets consist of 12 horizontal cylindrical electro-magnets, six above the axis of rotation of the armature, and uniting in a single upper pole-piece, and six below the armature uniting in the lower pole-piece, the two polar extensions having opposite polarity. The coils of these magnets are connected together in series, and the whole current generated by the machine is transmitted through them, thus producing an intense magnetic field, within which the armature is rotated at a speed of about 900 revolutions per minute. Those parts of the pole-pieces which embrace the armature, instead of being solid and continuous as in the Gramme machine, are cut into a series of tongues. These are shown marked T T T, T^1 T^1 T^1, &c., in Fig. 215,

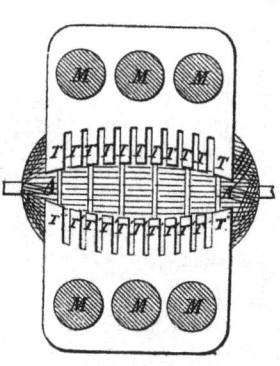

Fig. 215.

which is a diagram of the disposition of the pole pieces, as would be seen from one end of the machine if the magnets M M M, and back plate were removed. It will be noticed also that the space between the two pole-pieces is wider in the middle than at the two ends of the armature A A, the length of the tongues measured from the magnets decreasing from the centre towards the ends by a sort of elliptic curve. Mr. Weston states that by this arrangement greater regularity of current is insured, the inductive effect of the magnets upon the coils of the armature during its rotation taking place from the centre towards the ends and *vice versâ*, instead of simultaneously throughout the whole of any one section of the armature wire. The object of the slits in the pole pieces, whereby the latter are divided into a number of tongues, is twofold. In the first place it effectually prevents cross currents being induced in the pole-pieces by induction from the armature; and in the second place, the arrangement materially assists the ventilation of the armature and magnetic field, and thus prevents the machine from becoming overheated.

The armature of the Weston machine is of very peculiar construction and although it bears somewhat of a resemblance, when wound and in its place, to the Siemens armature, the construction of its iron core and method of winding are very different. The core of the armature, like that of the Maxim generator, is compound, being built up of a number of iron discs of a form peculiar to itself. One of these discs is shown

Fig. 216.

in Fig. 216; it is shaped something like a spur pinion, having around its circumference 16 projecting teeth. The core of the armature is constructed by threading 36 of these discs upon the driving shaft of the machine with small separating washers between, so that there is an air space between each washer and its neighbour, equal in thickness to the thickness of the washer. When the core is completed by the addition of solid end-pieces of a diameter equal to that of the discs, measured across the bottoms of the notches, the whole presents generally, the form of a cylinder with hemispherical ends and having a series of 16 longitudinal grooves cut in its circumference parallel with its axis, and a number of annular transverse grooves between the discs, and perpendicular to the longitudinal furrows. Into the 16 longitudinal grooves are wound as many coils of insulated copper wire, the method of winding being similar to that in the Siemens armature, that is to say, it is laid along one groove, across one end, and back by the corresponding and diametrically opposite groove on the other side. By this peculiar method of construction, it is obvious that the ventilation of the armature is exceptionally well insured, for not only is it hollow, but it is provided with 16 openings from the inside to the circumference of every disc of which it is composed, or 576 openings in the whole armature; and, as the ends are also provided with openings, a continuous stream of air is, during the working of the machine, constantly being driven through the orifices of the armature and between the spaces of the poles into the outer air.

The commutator which is shown in Fig. 217 is similar in principle to

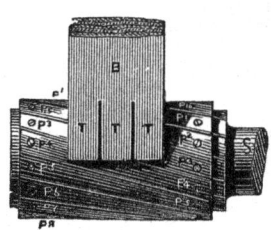

Fig. 217.

the commutators used in the Siemens and Gramme machines, but its construction differs from them in the fact that the various segments of which it is composed are arranged helically on the shaft, and are separated by air spaces instead of by solid insulating material. The object of this helical disposition of the copper strips of which it is composed is, that the rubbing edges of the collectors or " brushes " B are, in all positions of rotation of the commutator, in contact with two of its segments, thus insuring greater uniformity to the strength of the current.

The collectors or "brushes," of which there are two (diametrically opposed to one another), are made up of ten or a dozen thin curved plates of elastic copper cut by slits into three tongues, and so held against the commutator that the edges of the whole bundle press against the copper segments rotating beneath them, and the brush holders are like those of the Maxim generator, mounted upon a swing frame by which the strength of the current and of the magnetic field can be adjusted by setting the collectors by hand to position of varying efficiency. This arrangement is very clearly shown in the general view of the machine, Fig. 214. The Weston system is a complete one in itself, that is to say, it comprises a special generator and lamp, the description of which latter will be found in another part of this volume.

HEINRICHS.

The Heinrichs generator, a general view of which is given in Fig. 218, and of which a cross section taken through its axis of rotation

Fig. 218.

is shown in Fig. 219, possesses certain points of resemblance to the dynamo-electric machines of both Siemens and Gramme, while at the same time in some of its essential features it differs from either and from every other electric generator. Its point of resemblance to the Siemens machine consists (as will be seen on reference to Fig. 218) in the method by which the magnetic field is excited, and in the general disposition of the field magnets; while its point of resemblance to the Gramme machine lies in the fact that its armature is of annular form, and is wound, Gramme fashion, with insulated wire forming a closed

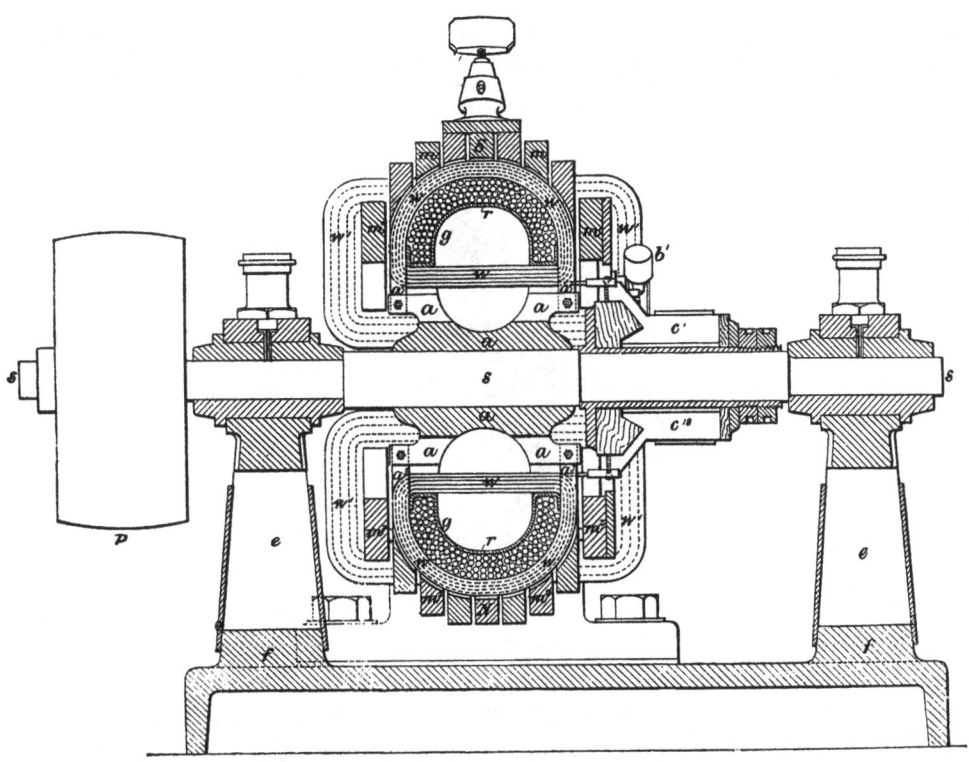

Fig. 219.

circuit. The armature differs, however, from the Gramme ring in its totally different cross section, which, instead of being rectangular, is U-shaped, as seen in Figs. 219 and 220, which latter represents a portion of the ring partly wound and having the U-shaped core-ring R and its coils of insulated copper wire C C. The construction of the armature ring will be best understood by referring to Fig. 220, while the general arrangement of the machine is very clearly shown in Figs. 218 and 219. Referring to the latter figure it will be seen that s is a shaft driven on a horizontal axis by the belt pulley p. To this shaft is keyed the boss *a a a* carrying the armature ring *r r* of the section indicated, and having wound upon it ⊔-shaped helices, which during the working of the machine are revolved within an intense magnetic field produced by the electro-magnet extension pieces m^1 m^1 m^1, &c., m^2 m^2 m^2, &c. These are excited by the powerful rectangular magnetising helices *w w* that form part of the main circuit of the machine. The various armature coils communicate, as in most machines of the Gramme type, with a set of copper segments *c c c*, &c., forming a cylindrical commutator, containing as many segments as there are wire helices on the ring; on to these commutators are pressed a pair of collecting brushes by which the currents generated in the armature are transmitted into the magnet coils and external circuit.

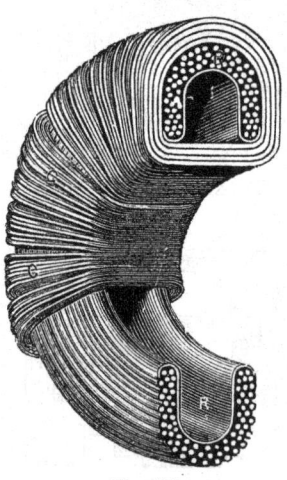

Fig. 220.

It will be observed that a special feature of this machine is the lateral disposition of the polar extensions of the field magnets, which are so placed with respect to the sides of the armature as to embrace and be very close to every portion of its external convex surface, and it is this feature which, while considerably increasing the efficiency of the machine, constitutes the essential difference between the field magnets of Mr. Henrichs' machine and that of Messrs. Siemens Brothers and Company.

The construction of the armature ring, more than anything else, is a characteristic feature. The iron core is, as remarked above, and as is clearly shown in Figs. 219 and 220, of ⊔-shaped cross section. It is constructed by coiling upon a thin cast-iron mould or shell *r r*, a quantity of soft charcoal-iron wire, which in the three-light machine, illustrated in Fig. 218, is of No. 10 B. W. G. When completed, the core consists of a thick iron ring, convex on its outer circumference, but having a deep grooved channel around and within its inner surface.

On this ring-core is wound, after the manner of a Gramme armature, a number of D-shaped helices of insulated copper wire which in the three-light machine are of No. 16 B. W. G. (*i.e.*, about ·065 in. in diameter) wound in 36 sections, their ends being connected to as many copper segments forming the commutator, and constituting a closed circuit, and the ring is held on to the boss by six brass loops, of which two are shown in Fig. 219, marked $a^1 a^1$, $a^3 a^3$. These loops pass over the iron wire core, one at every sixth section, and their ends are attached by screws to the boss $a\,a$. Besides holding the ring-core in its place, these brass strips form so many openings from the inside of the ring into the channel, and thus, by ventilating the armature, help to keep it cool.

The high efficiency of Mr. Heinrichs' form of armature as an apparatus for inducing currents of electricity from a magnetic field, will be understood from the following considerations, and by a comparison of the typical forms of armatures illustrated in the diagrams Figs. 221 to 224.

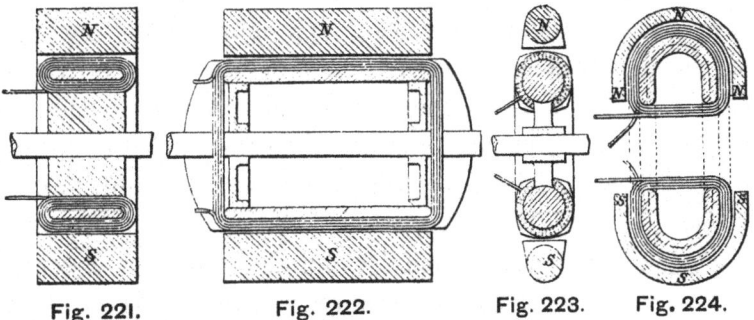

Fig. 221.   Fig. 222.   Fig. 223.   Fig. 224.

The first is a diagram of a Gramme ring within its magnetic field, while Figs. 222, 223, and 224 represent respectively the armatures and magnetic fields of the Siemens, Pacinotti and Heinrichs machines. In all generators of this general type there must necessarily be a certain proportion of the wire upon the armature which is idle, that is to say, which does not contribute, under the influence of the magnetic field in which it moves, as much current to the main circuit as it takes from it by its own resistance, and it is clear that the greater the proportion of active wire on the armature, to dead or idle wire, the greater, other things being equal, will be the efficiency of the machine. As we have already seen, M. Antoine Breguet showed that one action of the iron ring in a Gramme machine is to screen off, almost entirely, lines of magnetic force from crossing the interior of the ring, and from this it follows that all the convolutions of the armature coils which pass across the ring over its inside surface, contribute no current, for they cut no lines of magnetic force, while at the same time they add to the total resistance. Referring to Fig. 221 it will be seen that the quantity of wire traversing the ring on its inside is equal to that on the outside. Thus, if the ring be 6 in. wide, in every convolution of the wire there will be 6 in. of wire outside the ring under the influence of the magnetic field, and 6 in. inside the ring protected from the lines of magnetic force, and therefore having no useful effect. In a Siemens armature (see Fig. 222) 10 in. long and 6 in. in diameter, there are about 10 in. of useful line to about 7 in. of wire, which, by passing across the ends of the armature, does not in its revolution cut through any lines of magnetic force, and therefore does not contribute to the current generated by the machine. In the Pacinotti machine (treating it as a generator) the amount of wire under the immediate influence of the magnets N and S, Fig. 223, is very small compared to the rest of the current of the armature. In the U-shaped section of the armature core (see Fig. 224), surrounded by D-shaped helices, and these embraced all over their exterior surfaces by U-shaped magnet poles N N N, S S S, the maximum quantity of the armature circuit is exposed to the inducing influence of the magnetic field with the minimum of idle wire, and therefore the minimum of useless resistance is included in the circuit; thus if the ring shown in Fig. 224 be 4 in. wide there will be but 4 in. of idle wire in each convolution, to 10 in. of conductor fully exposed to the inductive action of the magnets which so nearly surround it. Figs. 225 and 226 are two other forms, suggested by Mr. Heinrichs, of his channeled ring, in which the screening action of the iron core is made use of to eliminate back-currents induced in the idle wire by the action of the magnetic field upon those portions of the armature which are more remote from the inducing magnets.

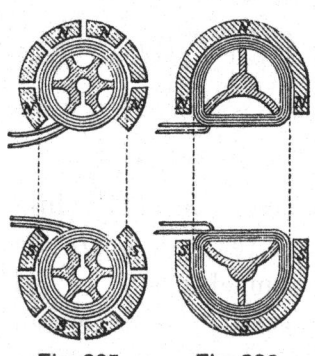

Fig. 225.   Fig. 226.

It is a curious fact that Mr. Heinrichs was not led to the adoption of the ∪-shaped ring for the reasons to which we have just referred, valuable as they undoubtedly are, but from theoretical considerations which are highly ingenious and decidedly interesting, and Mr. Heinrichs attributes the efficiency of his machine more to these theoretical considerations than to the causes already referred to. It is, of course, well known that if the north pole of a magnet N, see Fig. 227, be approached to the end of a soft iron bar *s n*, magnetic induction will take place, and the end of the bar which is nearest to the magnet will assume south polarity, while the opposite end will become a north pole, there being a neutral plane *l l* about half-way between them ; and if an insulated wire be wound round the iron bar and its ends joined so as to form a closed

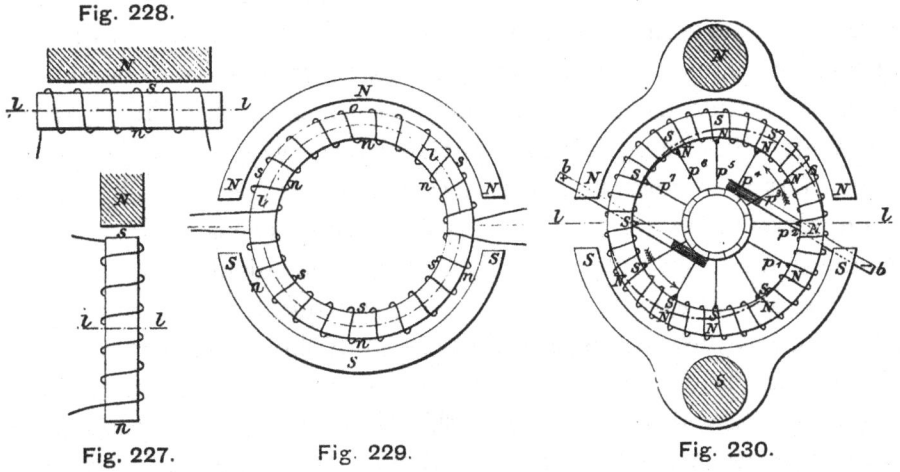

Fig. 228.

Fig. 227.　　　Fig. 229.　　　Fig. 230.

circuit, a current of electricity will be generated in the wire every time the magnetic intensity of the bar changes through a movement of the magnet N, or a change in the magnetic intensity of its field of force. Arguing from this illustration, Mr. Heinrichs goes a step further, and assumes the case of a broad poled magnet N approached to the same iron bar and coil, but this time not against its extremity but towards its side, (see Fig. 228). The neutral plane *l l* of no polarity will now lie through its axis, and the half cylinder above that neutral plane will assume south polarity, while the lower half will become north. The effect of vertically withdrawing the magnet from, or approaching it to, such a system, would have but insignificant inductive effect upon the helix of wire, for the reason that the lines of magnetic force lie so nearly parallel to the planes of the convolutions of the wire that they are not cut by them in their movement, and therefore the only currents generated in the wire, are produced by the slight aberration of parallelism which may exist between the coils and the lines of magnetic force when either the magnet or the coiled bar is moved in a vertical direction.

In Fig. 229 we have a condition of things which may be looked upon as a special case of Fig. 228. In this case the bar is bent round into a continuous ring, and the magnet poles N N N and S S S are curved concentrically with respect to the ring. Here again, for the reasons given in reference to Fig. 228, but small inductive effect will be produced upon the wire helices by the vertical movement of the two pole-pieces N N and S S, although rather more than in the case illustrated in Fig. 228, for the reason that as in the position shown in Fig. 229 the lines of magnetic force are, like the convolutions of the helix, more or less radial to the ring, a vertical motion of the pole-pieces will cause certain lines of magnetic force near their outer extremities to be cut by a few of the convolutions near them, and electric currents will, therefore, be generated. But such a movement is in no way analogous to the action of a Gramme machine or apparatus of that type, which depends upon the rotation on its principal axis of the armature ring within the magnetic field, whereby a lateral or angular displacement of the convolutions of the armature takes place, cutting through the nearly radial lines of magnetic force in every portion of their revolution.

Mr. Heinrichs' theory of the action of the Gramme machine is illustrated by the diagram, Fig. 230, which represents a Gramme armature with its commutator and brushes revolving in a direction indicated by the arrows within a magnetic field produced by the magnetic pole-pieces N N and S S. Assuming that in a solid iron ring of the Gramme form, the influence of each of the magnets N and S is to induce polarity opposite to itself on those parts of the ring nearest to it, and polarity of its own sign on the parts more remote, it would follow that if the ring were not in motion, its polarity would be distributed as is shown in Fig. 229; that is to say, the outer circumference of the upper half of the ring and the inner circumference of its lower half would have south polarity, while the inner circumference of the upper half and the outer circumference of the lower half, would assume north polarity, and the neutral region of the ring would be a cylindrical surface passing through the circular axis of the iron ring. The moment, however, that the ring is rotated upon its principal axis (according to Mr. Heinrichs' theory) a different kind of polarisation is set up. For the purpose of explanation, this may be called the magnetic influence of N upon *s n*, in Fig. 227, *longitudinal* polarisation, and that in Fig. 228, *transverse* polarisation, the motion of the magnet in both cases being in a vertical direction. If, however, the magnet N in Fig. 228 be laterally displaced, that is to say, if it be moved in a direction parallel to *l l*, then, from the fact that all iron takes a certain period to become polarised—the parts nearer the inducing magnet being influenced an appreciable time before those parts which are more remote—the influence of the magnet upon the bar will be something between longitudinal and transverse polarisation, and may be termed for the purpose of this argument, *oblique* polisaration, the polarity of the more remote parts dragging, so to speak, behind that of the others and causing the neutral plane to become more or less diagonal. Now, just as Fig. 229 represented a special instance of what was described generally under Fig. 228, on the assumption of a *vertical* displacement of either the magnet or bar, so Fig. 230 may be regarded as a special instance of Fig. 228, on the assumption of a *lateral* displacement of either of the principal parts. In this case the neutral surface will become a spiral (dividing the ring into two parts), whose obliquity with respect to the circumference of the ring will be dependent upon several causes, the principal of which are the speed of rotation, the strength of the magnetic field and the magnetic capacity, or rather the rapidity of magnetic induction of the iron ring.

According to this theory that portion of the ring opposite p^1, which has been for a considerable portion of its revolution under the influence of the south pole S S, will be almost entirely of north polarity; but immediately after coming under the influence of the north pole-piece N N, opposite P^3, its polarity will somewhat slowly be changed from its outer surface inwards, and its magnetism will not become wholly reversed until it has been some time under the influence of N N, as it will have been when it reaches the point opposite p^7, after which the process will be repeated again under the influence of the south pole. On the same theory Mr. Heinrichs accounts for the fact that the most effective angular position of the collecting brushes in machines of this type depends upon, and varies with, the speed of the machine as well as with the intensity of the magnetic field.

From the considerations involved in the above hypothesis—by which

R

it may be shown that the proximity of the iron core to the convolutions of the armature helices traversing the inner surface of the armature, are positively detrimental to the efficiency of the machine—Mr. Heinrichs was led to the channelled form of ring, in which the iron core is well removed from the inner convolutions of the armature coils, which, moreover, by forming so many bridges across the channel with nothing but moving air around them, are not only kept cool themselves, but by conduction considerably assist in carrying off heat developed in the working portions of the coils.

Whether or not Mr. Heinrichs' theory be a correct explanation of the phenomena taking place in the working of generators of the Pacinotti or Gramme type, there can be no doubt that he has produced a machine of much efficiency and simplicity. The large surface of armature wire exposed to the influence of the magnets, the small proportion of idle wire, the embracing of the armature ring all round and on all sides by the magnetic pole-pieces, and the small weight and thorough ventilation which the channelled form of armature core insures, are sufficient, independent of more speculative reasons, to account for the very excellent results which Mr. Heinrichs has obtained with his machine.

The mean diameter of the armature ring of Mr. Heinrichs' three-light machine is 8 in., which at a speed of 800 revolutions per minute gives a circumferential velocity of about 1,600 feet per minute, or half that of the 16-light Brush machine. The four magnet coils contain about 80 lbs. of insulated copper wire 14 in. in diameter (about No. $9\frac{1}{2}$ B. W. G.), and as there are $16\frac{1}{2}$ lbs. of No. 16 wire on the armature of the three-light machine, an expenditure of wire of $5\frac{1}{2}$ lbs. per light is involved as against 10 lbs. per light in the Brush machine.

The following figures give the results of a series of measurements of the Heinrichs machine taken by Mr. W. H. Preece, F.R.S., and Mr. Augustus Stroh :—

Mean number of revolutions of armature per minute....	850
Number of lamps in circuit	3
Resistance of machine between terminals and including magnet coils	1·83 Ohm.
Resistance of armature between brushes	·85 ,,
Resistance of each arc	1·24 ,,
Electromotive force	130 to 150 Volts.
Current	33 to 38 Ampères.

Annexed are the results of a series of measurements made upon the same machine by Mr. H. R. Kempe :—

Number of Lights in Circuit.	Number of Revolutions per Minute.	Current.	Diameter of Carbons.
		Ampères.	Millimetres.
1	700	40	13
2	800	37	13
3	1,000	30·8·	13

Carbon resistances were then substituted for the lamps, and the following measurements were taken :—

Carbon Resistances.	Number of Revolutions per Minute.	Internal Resistance of Machine.	Current.	Electromotive Force.	Calculated Horse-Power.
			Ampères.	Volts.	
2·1	700	1·83	36·4	143·3	7·1
2·6	800	,,	33·7	149·3	6·7
4·3	900	,,	26·3	160·0	5·6
7·3	900	,,	15·7	143·3	3·0
7·3	1,000	,,	17·7	161·6	4·0

The working of Mr. Heinrichs' system is exceedingly satisfactory, the apparatus is simple and easily managed, and although the machine, during the trials above referred to, was driven by a very inefficient engine, it maintained three perfectly steady arc lights, showed no tendency after long working to becoming heated, and was evidently absorbing but little power. It is reasonably to be expected that this very promising generator, which is now being further developed in Paris, will before long play an important part in electric lighting.

JABLOCHKOFF.

Most of the leading electric light companies own one or more distinct types of dynamo or magneto machines, as well as distinctive forms of lamps. When the Société Générale d'Electricité was established it possessed only the patents relating to the Jablochkoff candle and its accessories. Not having any alternating dynamo-electric generator of their own, they were obliged to use at first the Alliance, and afterwards almost exclusively the Gramme machines. As a consequence of this they could not compete favourably with other companies more fortunately situated. It was for this reason that the alternating-current dynamo-machine of M. Jablochkoff was designed. The excellent results obtained with Gramme machines under such different circumstances speak more strongly than anything else to the practical value of these generators. The experience of the Compagnie Générale appears to amount to this, that while the Gramme machine in competent hands leaves little to be desired, when ignorantly or carelessly treated, accidents and breakdowns are not unfrequent. It was to the avoidance of these accidents that M. Jablochkoff's attention was especially devoted, and he has aimed at producing a generator which shall work successfully and without breakdown under the varied and trying conditions that necessarily attend the popular employment of the electric light. It is unnecessary to recall here the construction of the Gramme machine, but we may remind our readers that the main feature of the alternate-current generator is a cylinder open at both ends, and around the inner and outer faces of which, the wire is coiled. This cylinder is in close contact with six electro-magnets. Fig. 231 explains the arrangement, where A A is the induction coil, and B B the electro-magnets. This arrangement is subject to the inconvenience common to all dynamo-electric machines, that of heating when at work, which as one result reduces the magnetisation, and consequently the inductive action. Another tendency is to expand the wire forming the coil, and to swell or soften the insulating material, giving the wires a tendency to lift from the cylinder along which they are wound and to sag in the middle. This, although of comparatively small importance on the outside of the cylinder, is very serious on the inside, on account of the slight clearance between the coil and the faces of the electro-magnets, the two surfaces sometimes coming in contact. This danger may also arise from another cause; the clearance being so small a very slight wear of the bearings causes the shaft to sink, which brings the magnets in contact with the coil either along its whole length or on one side. Careful watching prevents such an accident, but without care it may easily happen that the wear of the bearings goes on unnoticed, till contact takes place, and the coil is torn. These causes for binding are always present, and however slight either of them may be, they can easily result in accident on account of the minute distance between the coil and the magnet. Now the tearing of a coil in conjunction with the high speed of the machine often produces very serious

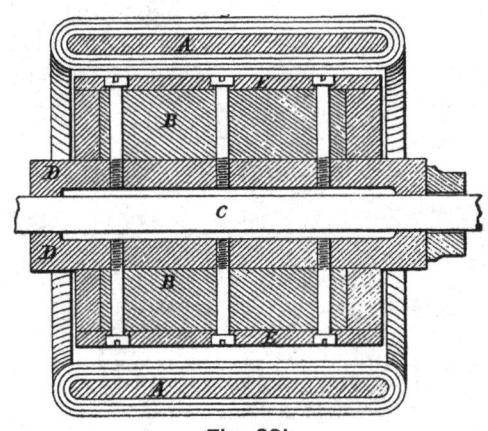

Fig. 231.

accidents, the minimum damage being the destruction of the coil; and it is only a reasonable precaution for the employer to keep by him a spare coil in case of accident, which of course involves an extra first cost.

Accidents of this kind it must be remembered are rare, owing to the admirable workmanship of the Gramme machines, which not unfrequently run in constant use during three or four years without requiring any repairs, but it is none the less a fact that they possess certain inherent elements of danger, and it is with the view of avoiding these, and obtaining a machine capable of resisting rough usage, that M. Jablochkoff has designed a special arrangement of dynamo-electric generator.

The first patent for this machine was taken out in France on the 25th September, 1878, by M. Jablochkoff, for "Improvements in the construction of magneto and dynamo-electric machines," and the specification begins thus: "The modifications we have introduced in the construction of magneto and dynamo-electric machines have for their object, the improvement in the working of such machines, the increase of their duty, or the reduction in the motive power they absorb, and to facilitate their maintenance and repairs." The generator to which this patent refers, after having been studied

Fig. 232.

theoretically by M. Jablochkoff, was brought to a practical form only a long time after, when the inventor had returned to Paris from a protracted stay in Russia. It is shown in Figs. 232 to 236, which represent the general arrangement of the first type constructed for feeding 16 lamps. It will be seen from these engravings that it comprises eight electro-magnets of a helicoidal form. To avoid the serious difficulties that would be met in the construction of these spiral electros, if they had been formed in a single piece, they are made in a series of stars of plate iron, from three to four millimetres thick, with eight branches, and pierced with a hole at the centre; these are mounted on an axis, one against the other, and to each a slight angular

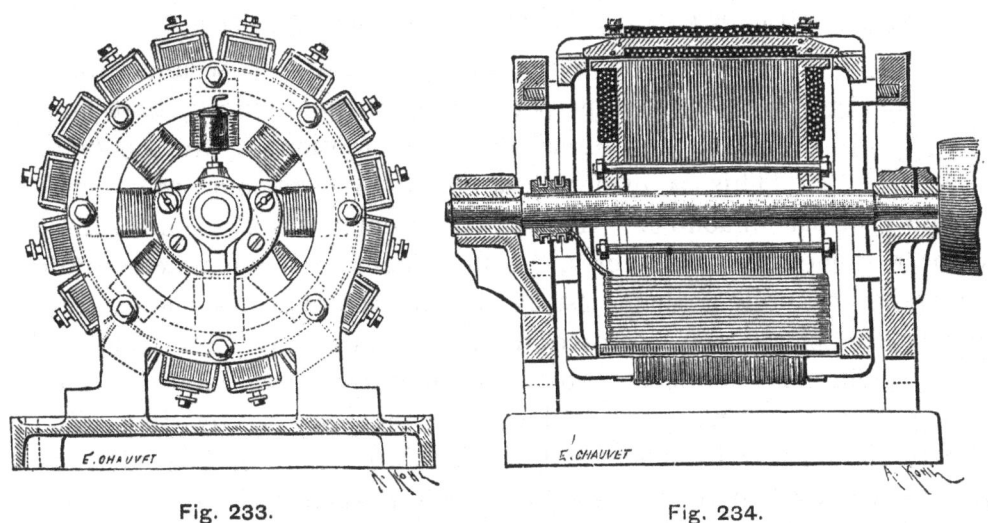

Fig. 233.   Fig. 234.

advance over the preceding one is given; by this means the desired spiral surface is obtained. The eight branches are connected by locking bolts passing through the stars, and the steps formed by the

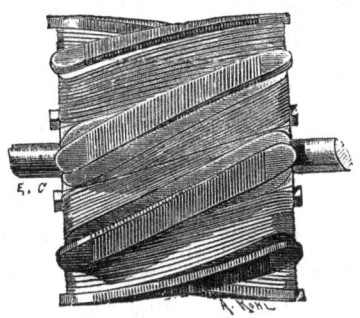

Fig. 235.

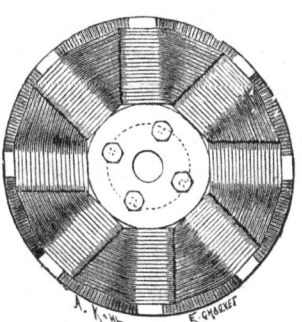

Fig. 236.

angular advance, are reduced to a continuous surface by filing, or other mechanical means; the core formed in this manner is externally solid and compact, and not likely to be injuriously affected by the action of centrifugal force. We shall moreover see, further on, that the speed at which the Jablochkoff machine is driven is sensibly less than that of many other generators. On the inclined surfaces of the core, coils are wound in which the current of the exciting machine circulates. The fixed coils receiving the induced current, are formed in a somewhat similar manner to the inductors, being made of a series of iron plates one millimetre thick, placed together so as to form a flat core. The annexed sketch, Fig. 237, shows the form of these plates, which are held together by four clips. The wire is wound around these cores perpendicular to their length, that is to say in the direction of the motion of the magnets.

Fig. 237.

This is evidently a good arrangement as regards the safety of the machine. If any wire from the fixed coils should become detached from the cores, the movement of the magnets would tend rather to protect than destroy it In the core of the electro-magnets, the plates are separated from each other by thin pieces of paper, which prevent the creation of currents injurious to the proper working of the machine, and reduce the tendency to heating. The magnets, the form and arrangement of which are shown in Figs. 233 and 234, are fixed by two copper armatures held at their ends by two rings attached to the frame. Their dimensions, and that of the wire covering them, are so calculated that each of them corresponds to a normal Jablochkoff candle of four millimetres diameter, so that a generator with 16 magnets is also a 16-light machine. If in the course of working, one of these magnets is disabled, it is very easy to remove it and replace it in a few minutes by a spare one, an advantage peculiar to this machine. On account of the special characteristic of having each magnet to correspond to one lamp, it is possible to make the exchange when the generator is at work, by simply cutting out the lamp corresponding to the magnet being shifted, and without extinguishing the other lights.

The following are some general particulars of the 16-light machine of the first type that was experimented with at the works of the Jablochkoff Company, 61, Avenue de Villiers, Paris. The generator, arranged as described, was furnished with a current from a Gramme exciting machine revolving at 750 revolutions per minute, and furnishing a current of about 48 Ampères. The 16 magnets were connected in four series of four magnets each. Each circuit, therefore, supplied four Jablochkoff candles of an average intensity of 45 Carcels; the resistance of each magnet was ·3 Ohm, the total resistance from this source being thus 1·2 Ohm in each of the four circuits. The most favourable speed for driving the Jablochkoff machine is 720 revolutions; at this velocity the light produced is very steady. The fixed magnets are grouped as indicated in Fig. 238, which represents the development on the same plane of the upper series of eight. The helicoidal branches of the revolving magnets are shown in dotted lines; the successive electros having different polarities, the relative magnetisation in the magnets will be as indicated. The four magnets, 1, 3, 5, 7, are coupled in tension to form a circuit; magnets 2, 4, 6, 8 are similarly coupled to form the second circuit. About 14 horse-power are required to drive the generator when furnishing 16 lights.

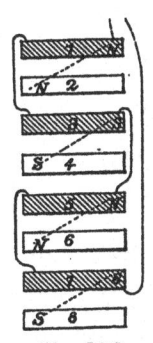

Fig. 238.

At the Paris Exhibition of Electricity two types of the Jablochkoff generator were shown, one for 16 lights constructed by M. Breguet, and differing only in a few minor details from the arrangement just described, and the other a 32-light machine made by MM. Sautter and Lemonnier, of Paris. The former was placed on a cast-iron frame 24 in. long and 19·7 in. wide; with its pulley, which was 8·6 in. wide over all, it occupied a space 32·3 in. × 19·7 in., its height being 25·6 in. The magnets had a useful length of 8·46 in. and a width of 2·76 in. The wire in the coils was ·08 in. in diameter, and the length rolled around each was about 144 ft. The clear space between the magnets is ·008 in. and the diameter of the cylinder to which they are secured, is 16·5 in. The eight electro-magnets are covered with wire ·146 in. in diameter. This machine gives practically the same duty as the one described above. It will be seen it is small and compact; the weight is 776 lbs.

In general arrangement the 32-light machine does not differ from the 16-light generator. Like it, it is formed of a series of electros of helicoidal form, the polar surfaces of which, revolve before the magnets. There are 16 of these branches to the armature, and each one is covered with 7 layers of wire ·15 in. in diameter. The magnets are fixed somewhat differently to those in the 16-light machine, and they are covered with 3 layers of wire ·087 in. in diameter. Their exterior surface is protected by an envelope of copper pierced with holes to provide a passage for the air, which is driven out rapidly by the movement of the helicoidal armature. We may note in passing that this constant draught produced by the machine is extremely useful in keeping all the parts cool. The leading dimensions of the generator are as follows :—

Length of generator	48 in.
,, shaft with pulley	66 ,,
Height	61 ,,
Diameter of circle described by the magnets	33·08 ,,
Clear distance between magnet and armature....	·04 ,,

Such are the two first industrial types of the alternating-current generator of M. Jablochkoff. They will doubtless be made largely, and with improvements, by the able manufactures who produced the models we have described, and which give extremely satisfactory results. Indeed, the machine seems likely to take an important part in electric lighting. It is solid, simple, efficient, and easily maintained and repaired; and by having a supply of spare magnets it may be used without risk in districts far removed from the manufacturer. The cost price appears low, so that it can be sold at a very moderate sum. All these points would seem to show that the Jablochkoff generator possesses a strong chance of success in electric lighting industry.

LACHAUSSÉE.

The very interesting electric generator of M. Lachaussée, of Liège, constituted, at the Paris Electrical Exhibition, a striking feature of the highly important exhibit of the Compagnie Générale Belge de Lumière Electrique of Brussels. In general appearance, and to some extent in principle and construction, the Lachaussee machine bears some resemblance to the Siemens alternating-current machine as well as to the Wilde generator. The distinguishing characteristic, however, of the Lachaussée machine lies in the fact that the induction portion of the apparatus or armature is fixed, while the field magnets are rotated, a form of construction possessing several features of convenience, and which it shares with the Jablochkoff, alternating Gramme, and Gerard machines, the latter of which it more nearly resembles than any other.

The rotating magnetic field is produced by two sets of electro-magnets, each set consisting of a crown of 12 electro-magnets, and attached to the shaft in such a position that the free ends of the one set are directed towards the free ends of the other set, but separated by a space within which is fixed the armature. The coils of these field magnets are connected together in series, in such a way that the direct current from a small Gramme machine in traversing the series, gives to

Fig. 239.

the magnets, polarities alternating in direction around each crown, and at the same time produces opposite polarities in the corresponding magnets in the two crowns which are presented towards one another. These two sets of magnets are mounted on two discs of iron, which are keyed on to the driving shaft of the machine, and are rotated with it in the usual way by a belt driven from the engine. The current from the exciting machine is conducted to the rotating magnets (one set of which is shown in the illustration, Fig. 239), through a pair of metallic brushes pressing against cylindrical contact pieces forming part of the driving shaft, but at the same time being insulated from it. The position of these brushes is shown to the left of the figure.

The induction portion of the machine, or armature, consists of a central disc fixed in the open space between the two opposed crowns of electro-magnets, and to which are attached 12 elliptical helices of insulated copper wire of the form shown in Fig. 240, arranged around the disc at a distance from the centre equal to that of the field magnets, with which they correspond. Each helix consists of an elliptical coil of insulated copper wire F wound round a hollow core L composed of thin sheet iron rolled into a short oval tube ; each of these helices is attached to a small block of wood P, which is inserted into a rectangular hole in the cylindrical wooden casing of the machine, being fixed therein by suitable screw attachments. The free ends A A[1] of each coil pass through its respective block and terminate in small attachment blocks of brass

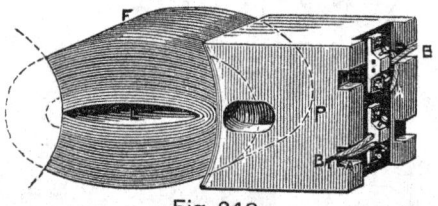

Fig. 240.

shown on the right of Fig. 240, and to which the conductors B B[1] are connected. Each coil is therefore with its block and attachment screws complete in itself, and can be removed or exchanged at pleasure without interfering with the working of the machine. The conducting wires B B[1] are led from their respective bobbins around the circumference of the drum to which they are attached, and below the casing, to a table at the top of the machine (shown in Fig. 239), to which are attached two sets of 12 binding screws, each coil in the armature having for its terminals one pair of the binding screws on the upper table. By this arrangement it is obvious that the 12 bobbins in the armature can be coupled up in several different ways, according to the nature of the work the currents generated by the machine are required to do. Thus 12 distinct circuits can be fed with alternating currents, or several may be grouped together so as to combine the currents from two or more bobbins, so as to form one large current. In this case alternate bobbins are connected together in series, and if the intermediate helices are added, their corresponding binding screws on the table are so connected to those of the first set, that all the currents transmitted into the circuit are in the same direction.

With one of these machines, six Soleil lamps in the picture gallery of the Palais de l'Industrie were worked in series, each having an arc of 20 millimetres, absorbing for that installation about 24 horse-power.

Passing over the mechanical differences which exist between the Lachaussée machine and the alternating-current machine of Messrs. Siemens Brothers and Co., the two generators differ most essentially in the fact that in the Siemens machine there are no iron cores in the induction bobbins as there are in the Lachaussée apparatus. The absence of iron, especially in an alternating-current machine, undoubtedly

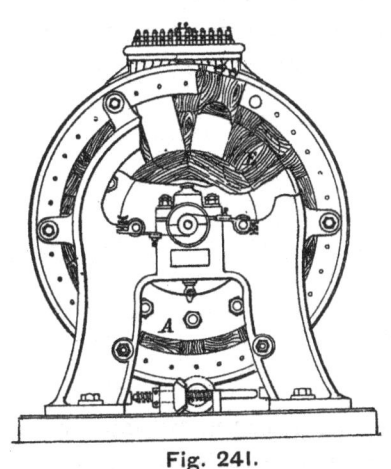

Fig. 241.

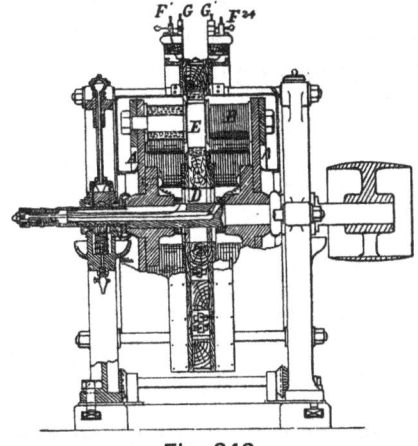

Fig. 242.

reduces the liability to the heating of the cores, and it also permits a greater speed of rotation being given to the machine ; on the other hand, the lines of force in the magnetic field are more concentrated by the presence of iron in the coils, and from some experiments recently made with the Lachaussée generator, it appears to have been proved that a higher efficiency is obtained when iron cores are employed. The making of the iron cores of small mass, and with a hollow space in the centre, greatly reduces the tendency to heat under the influence of rapidly reversing currents, and at the same time permits of very complete ventilation of the machine. It will be remembered that the arrangement was recognised as advantageous among the early magneto-electric generators.

Figs. 241 and 242 show a similar form of generator to that just described with a few later modifications.   The characteristic feature, of course, is that all the coils are independent of each other, and are so arranged that each of them may be put into and taken out of the machine without interfering with the others.   The rotating plates A A carry the field magnet coils B B, the two sets of which are separated by the fixed drum D, in the thickness of which are placed any suitable number of the metallic spirals E of small thickness, but of a diameter sensibly equal to that of the electro-magnetic bobbins.   Fig. 242 shows the construction of the drum D, which is made of wood; one bobbin E is in its recess, while the other recesses are shown empty.   The ends of all the coils are connected to small terminals F^1 to F^{24}, placed parallel to the large terminals G and G^1 that serve to connect these to one or more general or return wires, and allow of the grouping of the currents in tension or in quantity, even when the machine is working, as already explained.

EDISON

Mr. Edison is one of the more recent workers in devising mechanical means for the production of electric currents, one of his first patents for a dynamo-electric machine having been obtained in 1879. Since that time, however, he has laboured with his characteristic energy and enthusiasm, and of the two classes of generators which to-day stand associated with his name, one type is remarkable for its enormous dimensions, and both are, more or less, original in their conception, and admirable for their excellence of detail.

The first generator proposed by Mr. Edison in 1879, involved a utilisation of the often-patented idea of vibrating magnets. It was the exact converse of Helmholtz's electro-magnetic tuning fork, which it somewhat resembled, and bore something the same relation to that instrument, which the Gramme machine bore to Pacinotti's electro-magnetic engine. It consisted of a tuning fork about $6\frac{1}{2}$ feet long kept in vibration by a small steam, water or gas engine, or other source of mechanical power attached to the free end of each prong; the driving power thus resolved itself into a pair of single-cylinder engines working isochronously, their speed being regulated and controlled by a large tuning fork, which took the place of a fly wheel. To the outer surface of each prong were attached two or more electro-magnets, which were, when the fork was in vibration, carried backwards and forwards in front of two other corresponding electro-magnets, fixed with their poles opposed to those of the magnets attached to the fork; the ends of the wires forming the helices of these electro-magnets were connected to an exceedingly simple commutating apparatus by which the direction of the currents was reversed at every semi-vibration of the fork. If we imagine that the magnets on the forks possessed so much residual magnetism as to produce opposite their poles a magnetic field no matter how feeble, an electric current would be induced in the coils of the fixed magnets every time the vibrating magnets approached to or receded from them, the current being in one direction in the one case, and in the opposite direction in the other; and as the coils of one set of magnets were in circuit with the other, a current of electricity, first in one direction and then in the opposite, would be transmitted through all the coils of the apparatus as well as into the external circuit of the machine. When it was required to produce a current in which the direction was constant, the commutator above referred to was employed, and as this mechanically reversed the direction of the external circuit at the moment that the internal circuit was reversed by the action of the machine, it followed that a constant direction was given to the external circuit by the process of the double reversals. Mr. Edison made little more practical use of this idea than other inventors had done before him, and quickly abandoned the vibrating magnets for more practical forms. Fig. 243 shows the type which he devised in 1879 and 1880, in which the two vertical electro-magnets *a a* are wound with coarse wire, and are provided with pole-pieces N S that are recessed as shown to admit the horizontal armature. The current was collected by brushes and led to the external circuit by the wires W W.

Early in 1880 we find Mr. Edison patenting certain devices intended to obviate the induced current which circulates in the body of rotating armatures made of a solid mass of metal, or of large rings of metal. To effect this object he made the armature of very thin discs or rings, one-thirty-second to one-sixty-fourth of an inch thick, secured together and slightly insulated from each other by sheets of tissue paper, and to avoid sparkling at the commutator, the commutator brushes were arranged to bear obliquely upon the face of the commutator, preferably at an angle of about 30 deg. with the axis.   In his specification, Mr. Edison refers to a means for securing even wear over the commutator, by the aid of a disc forming the armature of an electro-magnet, attached to the frame of the machine and fixed to the end of the shaft.   When the magnet is charged,

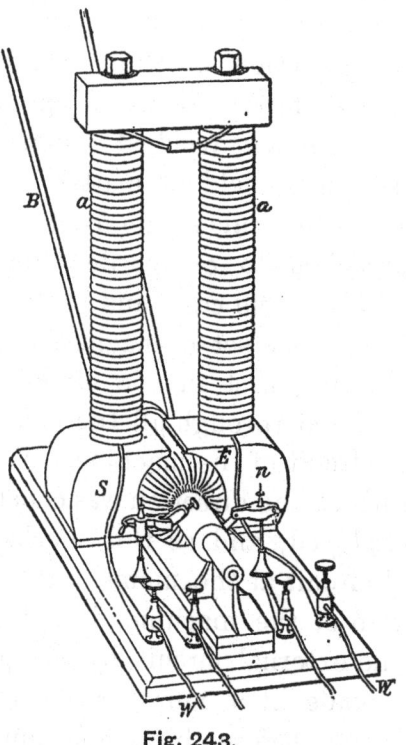

Fig. 243.

its armature, together with the shaft is attracted, a spring forcing them back when the magnet is inactive, or an iron pulley fixed to the shaft in the vicinity of the field magnets taking the place of the spring, the attraction between it and the magnets serving for the recoil of the shaft.   The magnet is alternately charged and discharged by means of a contact breaker consisting of a spring bearing upon the periphery of a rotating disc, one part of the periphery being of insulating and the other part of conducting material.   Fluctuations in speed consequent upon variations of load or work, are corrected by a centrifugal governor driven by the machine shaft, the governor acting upon and controlling an adjustable circuit lever which acts by either making or breaking the circuit to the motor. To prevent too sudden fluctuations, the main shaft is provided with a heavy fly wheel, and in order to get rid of the large sparks occurring when the main circuit is broken in one place only, the circuit is broken at several places simultaneously by an arrangement of contact breaking levers, operated by the governor either directly, or by means of an auxiliary electro-magnet and armature which the governor connects with a local batttery. The disposition of the various parts is clearly shown in Figs. 244 and 245.

Fig. 244.　　　　Fig. 245.

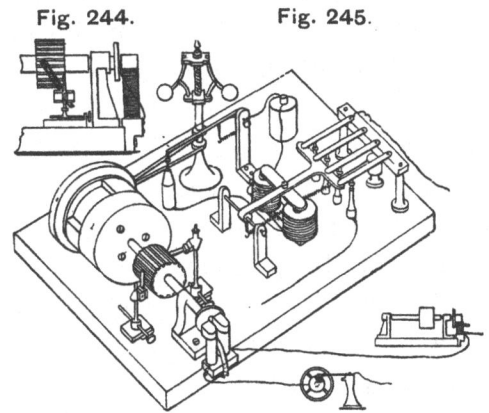

Before passing on to describe the great generator exhibited by Edison at the Paris Electrical Exhibition of 1881, we may examine a patent taken out by him in this country on July 6th in that year. The object of the first part of this invention is the production of an electric generator in which the iron core of the armature will not be necessary, and the loss of power caused by the heating of the same will be avoided, only the inductive portion of the armature being passed between the polar extensions of the field magnets. Another object is to construct a machine of this character so that it will generate a continuous current of high electromotive force in the same direction without the use of " pole changers," all the inductive portions of the armature being constantly in circuit, and the internal resistance of the machine being very small. This object is accomplished by constructing the armature in the shape of a disc divided into radial sections 1 to 16 (Fig. 246). These sections, which form the inductive portion of the armature, are naked copper bars joined edgewise by non-conducting material, shown by solid

s

black radial lines in the drawing. The radial bars are turned outwardly at their inner ends (Fig. 247), and are each separately

Fig. 246.

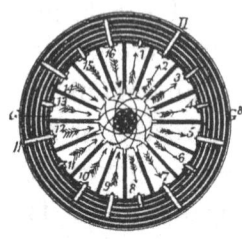

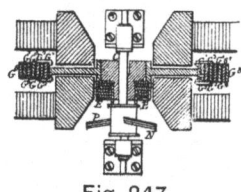
Fig. 247.

connected to one of the insulated circular plates E surrounding the hub. Each plate has a tongue, which is turned outwardly and secured in a groove of the commutator cylinder. "The bars nearest to diametrically opposite positions are in this way connected together in pairs, and with a commutator between them," as is diagrammatically shown in Fig. 246. The bars revolve between magnetic poles (Figs. 247 and 248), and outside such poles are surrounded by copper concentric rings G^1 to G^8 insulated from each other. Each ring is connected to two of the metal bars as shown diagrammatically in Fig. 246. The number of exterior rings and of plates on the boss is severally half that of the radial bars. Each ring connects two radial bars, the terminal bar of one opposite pair being thereby connected to the initial bar of another pair, so as to make a continually closed circuit through all the bars. Calling the inner rings No. 1

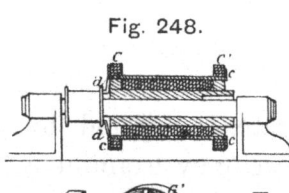
Fig. 248.

Fig. 249.

and the outer rings No. 8; No. 1 will connect the bars 1 and 11; No. 2, 8 and 15; No. 3, 6 and 13; No. 4, 4 and 10; No. 5, 3 and 9; No. 6, 7 and 16; No. 7, 5 and 14; and No. 8, 2 and 12. The bars are connected together at their inner ends, and to commutators in pairs as follows: 1 and 8; 2 and 11; 3 and 10; 4 and 13; 5 and 12; 6 and 15; 7 and 14; 9 and 16. The concentric rings are outside the polar field, and excite no electromotive force. The neutral line extends vertically through the centre of the armature, while the commutator brushes P N make contact at the end of a horizontal diameter, the bars next the neutral line being connected with the central side commutator bars as shown. On the left-hand side of Fig. 246, all the currents run inwardly, while on the right side they all run outwardly. In the position shown, one portion of the current will pass from the negative to the positive commutator brush, *viâ* 1 G^1 11; 2 G^2 12; 5 G^7 14; 7 G^6 and 16, and the other portion, *viâ* 8 G^2 15; 6 G^3 13; 4 G^4 10; 3 G^5 9; the proper plates E and straps H being included in the circuits. There are two ways of giving extra strength to the disc, neither of which is shown in the drawings: (1) A supplementary strengthening disc can be added, formed by winding a thin strip of iron around the hub with a strip of paper interposed in spiral convolutions. Around this disc there is shrunk an iron ring, and radial bolts are passed through holes in the disc and screwed into the hub. This strengthening disc from its spiral shape does not cut the lines of force, but becomes a detached portion of the magnet which revolves with the armature. (2) Another way is to make the copper bars double with a spiral strengthening disc between them. Fig. 248 illustrates the remaining portion of the invention, and relates to the application of concentric rings for making the multiple arc connections of the copper bars in an electric generator or motor, having an armature of cylindrical form. This appears to apply more nearly to the Edison's generator shown in the Paris Exhibition, wherein, as we shall presently see, the armature consisted of a cage of copper bars united at their ends in pairs by discs. At each end of the cylindrical armature is arranged a series of insulated copper rings C C¹, placed outside the polar extensions of the exciting magnet. Each ring is provided with two projecting fingers c which are turned inwardly and connected with the ends of the proper longitudinal bars. Midway between such fingers at the commutator end of the machine each ring is connected with a bar of the commutator cylinder by means of an angular piece d.

The bar armature is more clearly explained in a subsequent specification of Mr. Edison, illustrated by Figs. 250 and 251, and of which the

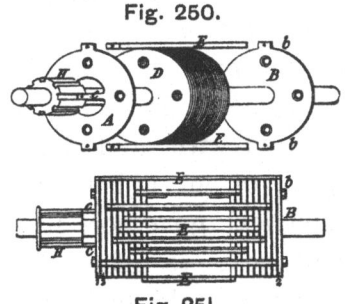

Fig. 250.

Fig. 251.

summary is as follows:—The object is to furnish means, for easily removing and replacing one or more coils of the armatures of dynamo or magneto-electric machines. For the generative portions of the armature, coils of wire or naked bars are employed, electrical connection from each bar to the corresponding bar on the opposite side of the armature being made through discs. If each longitudinal set of wires, or each bar, be considered as one coil, a series of discs equal in number to half the coils is used at each end of the armature. Referring to the illustrations, which are diagrammatic in

s 2

their character, A (Fig. 250) represents a disc at the commutator end or the armature, and B a disc at the other end. Each disc A has a hole punched out of the centre, and a finger *c* bent out at right angles to the commutator block H. In Fig. 251, two only of such fingers are shown for sake of clearness, connected to discs Nos. 1 and 3. D is a core, not shown in Fig. 251, of insulated discs of soft iron. Supposing the upper bar E to be moving through the polar space of the magnetic field, a current would traverse it, say, from left to right, enter the disc 2 by the lug *b*, cross the disc, and return by a bar not shown, but immediately behind the lowest bar E, to the disc 3, and then two paths would be open to it, one by the tongue *c* to the commutator, and the other through the second bar E attached to the disc 3. The same would occur in each pair of bars and discs, the course of the current being similar to that in a Siemens low tension dynamo-machine for plating.

This leads us naturally to a description of the gigantic dynamo-electrical machines with which Mr. Edison proposes to illuminate a certain district of the city of New York, and for which installation, underground conductors have been laid for conveying the currents from the central driving station to the various houses and warehouses where the light may be eventually adopted.

One form of the complete apparatus, illustrated by Fig. 252, which was exhibited at the Exhibition of Electricity at Paris, in 1881, comprises a horizontal steam engine of 125 horse-power, and a dynamo-electric generator of enormous size, the whole being attached to a single bed-plate so as to constitute one machine. Some idea of its size may be gathered from the fact that the whole apparatus weighs no less than 17 tons, of which 10 tons belong to the field magnets alone, and the rotating armature weighs upwards of $2\frac{1}{2}$ tons. The shaft of the engine is co-axial with that of the machine, the two being coupled together by a pair of pin discs and a drag link between them. By this arrangement, as in that adopted by Brotherhood, all driving belts are dispensed with, and the sudden cessation of the light through the slipping off of a belt is avoided. The engine is fitted with a very symmetrical governor, in external form similar to a fly-wheel, on the main shaft, so as to offer but little beating resistance to the air, and acting immediately on the eccentric, increasing or diminishing its throw or eccentricity, and thus controlling the engine by varying the length of traverse of the slide valve.

Fig. 252.   Edison's Large Dynamo-Electric Generator, 1881.

This dynamo-electric machine was one of the first of the colossal type made by Mr. Edison, and although the principle has been preserved in later machines, we shall see further on that several alterations have been made in what may for the time be called Edison's standard type. The magnetic field is produced by 8 very long cylindrical electro-magnets (about 8 feet long) fixed in a horizontal position, the coils of which form a shunt or derived circuit to the principle circuit of the machine. Of these field magnets three are attached to the lower pole-piece, while five are fixed above them to the upper pole-piece. From this construction it is obvious that this special machine must be regarded, as an experimental apparatus, for the two extra upper magnets were possibly added after the machine was constructed, on account of the six magnets not producing so strong a magnetic field as they were expected to do. The unequal distribution of the field magnets with respect to the neutral plane in the magnetic field, could not but tend to produce an unsymmetrical disposition of the lines of force within the field; this arrangement has been modified in subsequent generators.

Another point that must strike electricians as extraordinary, is the great length of the field magnets, which are covered with insulated wire from end to end. It is claimed for these long electro-magnets that they confer on the machine greater economy in working, and that by their use copper is saved in the construction of the machine at the expense of iron, but it would appear probable that but very little magnetism is added to the magnetic field by the magnetisation of that half length of the magnets, which is more remote from the pole-pieces, and certainly not sufficient in inductive effect on the armature to compensate for the greatly increased quantity of wire necessary for their construction.

The arrangement of the armature of Mr. Edison's machine has been already referred to. In principle it belongs to the Siemens direct-current type, but the connecting up is very different and more nearly resembles that of the Gramme machine. The armature is of cylindrical form and revolves at a speed of 350 revolutions per minute in a hollow cylinder formed by the interior surfaces of the two massive pole-pieces to which the magnets are attached. There is no wire used in its construction, the inductive portion of the armature being built up of a number of straight copper bars or prisms, trapezoidal in cross section, fixed longitudinally around the circumference of the laminated iron cylindrical core and insulated from one another by prepared brown paper. At each end of the armature, and strung over the central shaft are a number of flat copper discs or washers, there being as many of these discs as there are bars on the armature, and they likewise are carefully insulated from one another. Each copper disc is connected at its edge to one end of one of the induction bars, and the corresponding and diametrically opposite bar on the armature is connected to the same disc diametrically opposite to the junction of the first bar. This latter is connected in a similar manner to the edge of one of the discs at the opposite end of the armature, and this disc is connected to the end of the bar which is next to the bar which we originally started with; the opposite end of the third bar is attached to a fresh disc at the end of the armature, and so on until all the bars have been included in one continuous electrical circuit. It will thus be seen that, electrically speaking, the iron core is coiled with a metallic conductor, a current transmitted to it passing up one bar across a disc to the opposite side, along the opposite bar across a disc at the other end, and through the contiguous bar to that in which it commenced. By this simple arrangement, the resistance of the armature, and especially of the inactive portion of it at the ends, is reduced to the lowest possible amount, and that awkward complication of wires involved in the wire armatures on the Siemens principle is avoided. In consequence of the fact that each succeeding bar is connected to the disc next in order to that to which the previous bar was attached, it follows that the line of junctions between the bars and the discs has at both ends a helical path. The bars are all, however, of precisely the same length, those that project most at one end doing so less at the other. In the machine we are describing there are 138 of these bars, and as many conducting discs, the bars being about 3 ft. 6 in. long, of which length about 3 ft. 3 in. lies within the magnetic field.

On the end of the shaft furthest from the engine is a cylindrical commutator of the Gramme and Siemens type, made up of 138 insulated copper sectors, respectively connected to the copper discs, and against which the collecting brushes are pressed by means of their elasticity, the pressure and position being regulated by adjusting spindles worked with worn and worm-wheels. The commutator cylinder is about 9 in. long, and there are two broad collecting brushes, each made up of eight separate bundles.

It is stated that the electromotive force of this machine at a speed of

350 revolutions per minute, is 103 Volts, and at the Paris Exhibition the current was supplied to 2,000 "half lights," or 1,000 "whole lights," which were distributed as follows : Over the grand staircase at the north central entrance to the Palais de l'Industrie were suspended 10 horizontal frames of wood ; across each of these was stretched a trellis of wires, to which were hung the lamps. There were 25 lamps on each frame, or 250 in all, and in addition to these the staircase was illuminated by 148 lights attached to the central chandelier, making a total of 398 whole lamps.

There was a further number of 200 lamps working in the room devoted to the Edison exhibits, and a number of other lights were illuminated with the same current in various parts of the building.

The conductors were solid rods of copper segmental in cross section, and embedded in a resinous material enclosed in iron gas tubing, which was laid across the Exhibition below the soil ; these conductors were of the form and size employed in the large main conductors which will form part of the Edison installation of New York.

The resistance of the armature of this machine is almost inconsiderable, being only 008 of an Ohm., while that of the magnet coils which form a derived circuit to that of the armature, is 30 Ohms, and it is calculated that the resistance of the external circuit when all the lamps are at work is about ·32 of an Ohm, or 40 times that of the armature.

We may notice here an arrangement of commutator patented by Mr. Edison in July, 1881, and illustrated by Fig. 253. The object of the

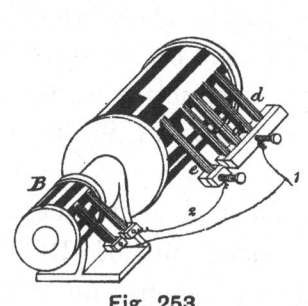

Fig. 253.

invention is to minimise the spark produced by the brushes and the commutator bars in their revolution, by dividing it up, since "the spark at each part is reduced about as the square of the number of points at which the circuit is broken." In carrying out the invention, the insulation (shown in black) is widened and the conducting bars narrowed at one end of the commutator, and upon this portion a single brush *e* is arranged to bear, noticeably behind the ends of the main brushes *d d*. This brush, called the "isolated brush," is not connected with the main brushes directly, but is connected with a series of breaking points on a "breaking cylinder" B, which has conducting bars and insulating spaces corresponding with those on which the "isolated brush" bears. This cylinder may be end to end of the commutator cylinder, or be fixed to the end of the spindle. In working, the local circuit between two bars, and a portion of the main circuit, are continued through each "isolated brush" after the main brushes have left each commutator bar, so that no spark is produced at the points of the main brushes. When each "isolated brush" leaves a commutator bar, the current passing through it is also broken at a number of points on the breaking cylinder simultaneously with the breaking of the current on the commutator cylinder by the "isolated brush." The connections are clearly shown on the drawings, 1 being the main lead. There is a corresponding set of brushes at the opposite side, but they are not visible.

Since the Paris Exhibition of Electricity, Mr. Edison has introduced several modifications into his large type of generator, although probably they still remain in a transition state. Two such generators were recently brought over to London, and are erected at the Offices of the Edison Electric Light Company on the Holborn Viaduct, for working a central station there. In this installation no less than a 1,000 full size, or 16-candle incandescence electric lamps can be maintained constantly in operation. All these lamps derive their current from one of the two machines which are fixed in the basement of the Company's Offices, the other machine, which is very similar, being kept in ressrve in the event of an accident occurring to the working generator, as well as to take its place during periods of repair.

Fig. 254 is a general view of one of these interesting machines, which consists of an enormous dynamo-electro generator, similar in general design to the great machine at the Paris Exhibition, but differing in several important details of construction, placed side by side upon the same bed-plate and foundation with a horizontal high-pressure Porter engine of 130 horse-power, the main shaft of which is a prolongation of the axis of the armature of the generator.

Upon referring to the illustration it will be seen that this generator is of the Edison horizontal type, the magnetic field being produced by a battery of 12 horizontal electro-magnets disposed in three rows of four magnets each, two rows being attached by their ends to the upper pole-piece, and one row to the lower pole-piece, the further ends of all being united by a massive heel plate, seen at the back of the figure. The armature is driven at a speed of 350 revolutions per minute, within a cylindrical space bored out of the very massive pole-pieces shown in the front of the illustration, and which are built up of 12 heavy blocks of cast

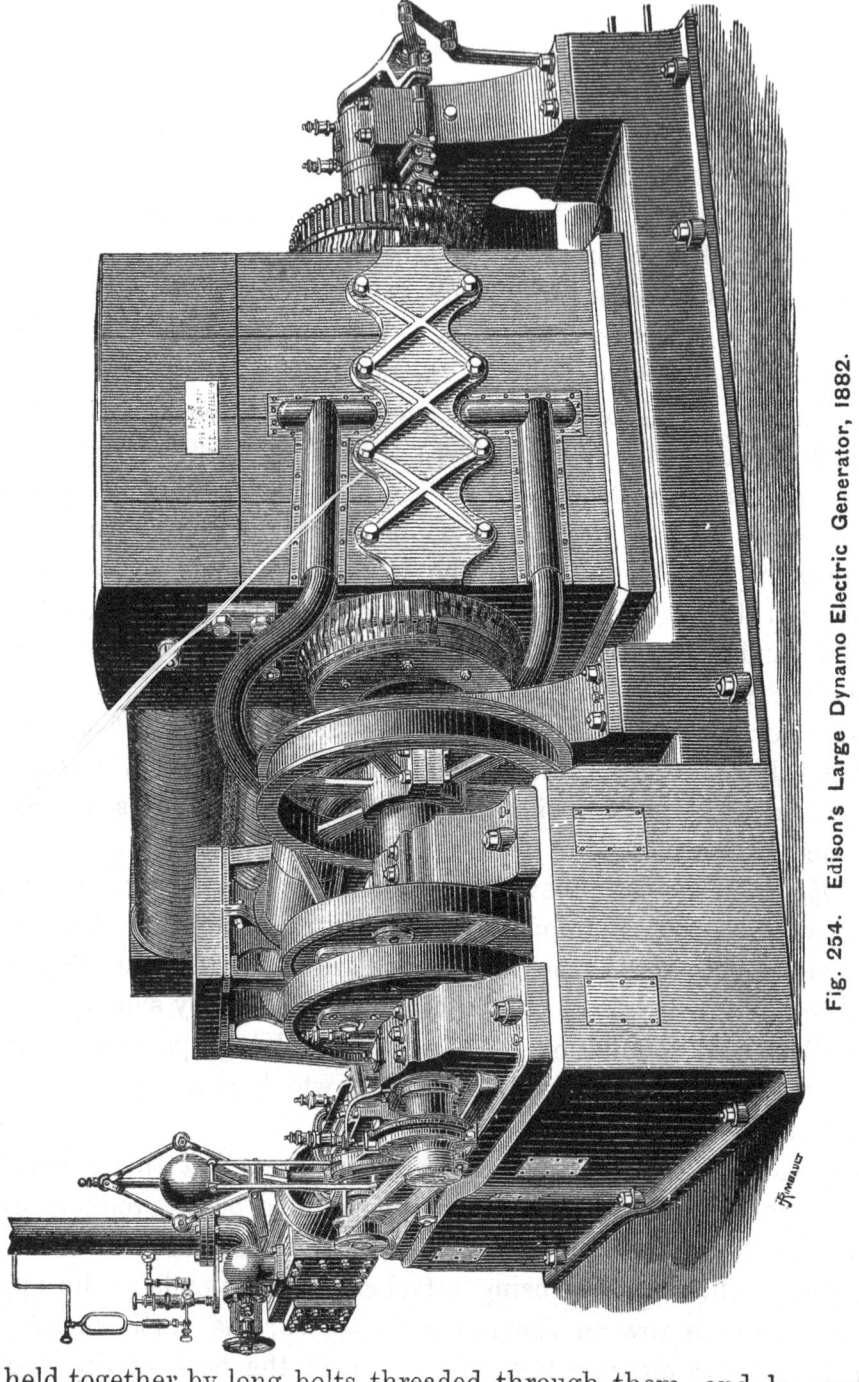

Fig. 254. Edison's Large Dynamo Electric Generator, 1882.

iron held together by long bolts threaded through them, and by surface
coupling plates screwed on to three of their faces. The armature, like

that of the Paris machine, is of cylindrical form, and is composed, first of a core built up of no less than 2,200 discs of very thin sheet iron, alternating with discs of tissue paper, while at every foot, measured in the direction of its length, is a disc of thick iron to give stability and rigidity to the armature ; the whole is bolted together by eight longitudinal bolts passing through all the discs, and through a pair of end plates by which pressure is applied, and the core is bushed with a tube of wood, so as to insulate it from the rest of the machine. The induction portion of the armature is composed of 108 straight longitudinal thick bars of nearly pure copper of trapezoidal cross section, arranged at equal distances around the core and insulated from it. These 108 copper bars are united at alternate ends to as many copper discs, half of which number are strung on to the shaft at one end of the armature and the other half at the other ; all the discs are insulated from the shaft and from one another and the bars are so connected with them that the bars and discs together form a continuous and closed circuit, wound as it were longitudinally around the core. This coupling up is effected as has been already described, and it will be seen that for all electrical purposes the bars and discs together may be looked upon as making up a thick copper coil of extremely low resistance, wound longitudinally over an iron cylinder in a single layer, the various convolutions of which are at equal angular distances apart, around the circumference of the cylinder.

In the disposition of the inductive portion of this armature it is identical with the armature of the ordinary Siemens direct-current machine, but the coupling up is simpler, and the course taken by the currents produced is in consequence somewhat different.

The diameter of the armature when complete is 28 in., its length is 5 ft., and its weight is over four tons. When it is remembered that this mass is revolved within the hollow cylindrical space between the pole-pieces at a speed of 350 revolutions per minute, it will readily be anticipated that if the heavy bars of copper were attached to the armature only at their ends, which are between 4 ft. and 5 ft. apart. and revolving as they do at a circumferential speed of 43 ft. per second, the armature would be speedily destroyed by the bars flying out under the influence of centrifugal force, and coming in contact with the iron pole-pieces which embrace them. To prevent this, the bars are held together at short distances along the length of the armature by coils of steel pianoforte wire bound tightly round the bars over bands of mica, by which they are insulated from them, and some idea may be formed of the high class workmanship and fitting together of this finely constructed machine, when we state that although the diameter of the revolving armature is 28 in., that of the cylindrical space within which it revolves is only 28¼ in., thus allowing but one-eighth of an inch clearance between the induction bars and the pole-pieces. It is needless, of course, to point out that this accuracy of construction adds very considerably to the efficiency of the machine, by enabling the armature to revolve in a more intense magnetic field than if, through less accurate workmanship, the magnetic poles, to insure the safety of the machine, had to be farther off.

The electro-magnets by which the magnetic field is produced consist of 12 horizontal cylindrical bars of iron about 8 ft. long, and coiled throughout their whole length with thick insulated copper wire. The coils of these magnets are connected together in two parallel circuits of six coils each, and the resistance of the circuit so arranged, and which form a derived or shunt circuit to that of the machine, is 21 Ohms. The resistance of the armature, as will be apparent from a consideration of its construction is practically inconsiderable, measuring only about $\frac{5}{10,000}$ of an Ohm (·00049 Ohm).

The commutator is a cylinder built up of a number of insulated copper sections as in that of the Siemens and Gramme machines, there being as many copper segments as there are induction bars on the armature, and are connected to them by as many radial copper rods attached in such a manner that owing to a slight elasticity at the junction, there is no tendency for them to be sheared off in the starting and stopping of the machine, a defect which showed itself in the machines of this type which were first constructed. Upon the cylindrical surface of this commutator are pressed two sets of metallic brushes or collectors mounted in spring fittings attached to a rocking arm—shown to the extreme right of the illustration—by which the angular position of the points at which the brushes make contact with the commutator, can be adjusted with respect to the neutral plane of the magnetic field so as to obtain the maximum efficiency of the machine.

The motive power is a horizontal engine of the Porter type, of 130 horse-power nominal, fitted with a Porter governor and expansion gear, and working with a steam pressure of 120 lbs.; it drives the armature which is mounted on the crank shaft at a velocity of 350 revolutions per minute, the steam being supplied by one of Messrs. Babcock & Wilcox's compound tubular boilers.   The weight of the machine with its engine and bed-plate, which is common to both, is over 20 tons.

In order to keep the armature cool, thereby protecting its insulation

Fig. 255.

and keeping down its resistance, there is a small blowing fan driven by the engine, and from this blower three pipes are led which communicate with three air channels cut through the pole-pieces, so as to maintain three jets of air constantly impinging on the middle of the rotating armature, which, escaping right and left are, during the working of the machine, continually bathing the induction bars with air. The efficiency of this arrangement is proved by the fact that streams of perceptibly warm air are, when the apparatus is in action, continually issuing from the two ends of the hollow cylindrical space within which the armature is rotating.

The smaller type of generator employed by Mr. Edison for smaller installation is somewhat similar to that illustrated by Fig. 243, but has been modified since that date. The present class of generator is indicated by Fig. 255. It consists of a longitudinally wound armature of low resistance revolving in a magnetic field produced by two long vertical cylindrical electro-magnets, their upper ends being united by a rectangular block of iron, while their lower ends terminate in massive polar extension pieces, whose opposed faces are bored out so as to form a hollow cylinder, within which the armature is rotated at a speed of about 1,200 revolutions per minute, at which speed its electromotive force is 110 Volts. Fig. 256 is a diagram of one of these machines, showing

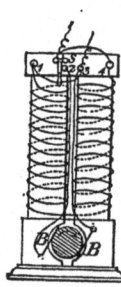

Fig. 256.

the way its connections are made with respect to the four attachment screws numbered 1, 2, 3, and 4, which are fixed to the upper plate of the machine. It will be seen from this figure that the screws 1 and 4 are the terminals of the magnet coils, while the screws 2 and 3 are connected to the collecting brushes, and may therefore be regarded as the terminals of the armature helix. In order to explain clearly the mode of producing currents with this generator, and the method of employing them, we will devote some little space to describing this installation of machines at the Crystal Palace, at Sydenham.

The electric currents for working all the Edison lamps in the building were supplied by 12 of these machines, erected in two parallel rows on the basement floor, close to the principal garden entrance of the Palace. From the extensive nature of this machine installation, forming as it does two long rows of generators, worked by three engines through the intervention of six contershafts, as well as from the beautifully simple and well-arranged system for making the connections and regulating the current to the work it has to do, it must be commended as a thoroughly well-arranged mechanical installation. The system is clearly indicated by the diagram, Fig. 257 and 258.

This figure shows three 25 horse-power semi-portable engines by Messrs. Robey and Co., of Lincoln, marked on the diagram E^1, E^{11}, and

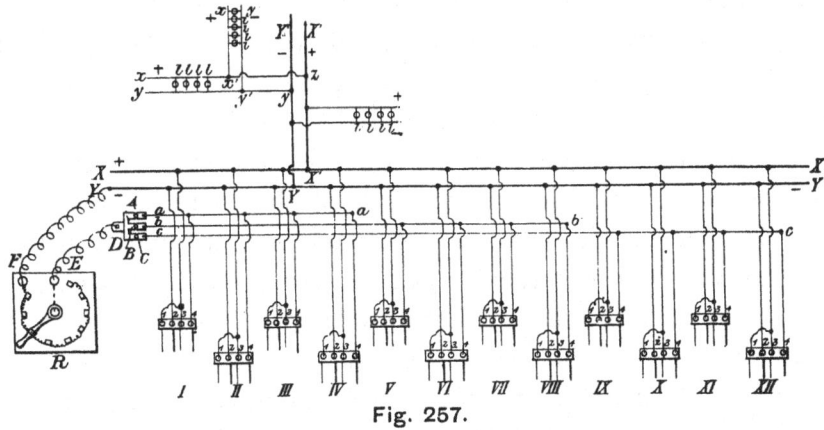

Fig. 257.

E^{111}, and on the crank shaft of each are fixed two driving wheels P^1 and P^2, the one on one side of the engine and the other on the opposite side.

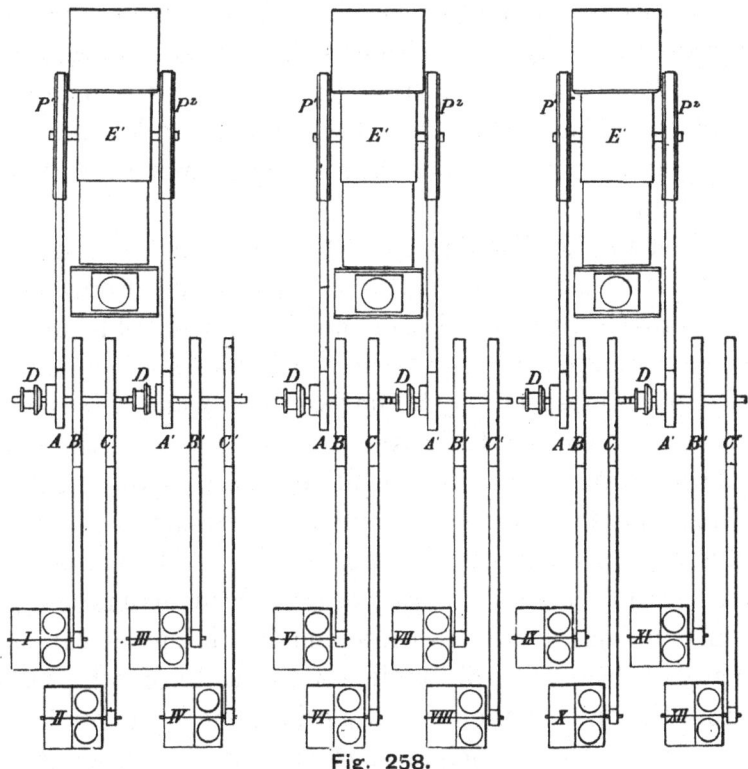

Fig. 258.

Each of these pulleys drives, by means of a belt, a separate countershaft, which can in a moment be thrown into or out of action by a conical clutch or coupling D, actuated by a hand-wheel. To each of these countershafts are keyed three pulleys, one smaller than the other two, which is driven by the engine belt, and two larger pulleys B C, B^1 C^1, &c., but equal to one another, each of which drives the belt by which it is connected to the pulley of one of the generators, which are shown in the diagram marked by Roman numerals I. to XII.

The electrical arrangements, as well as the very simple and interesting method by which the strength of the currents is regulated, and by which the whole set of 12 machines is under complete control, are as thoroughly carried out as any other part of the system, and form a particularly interesting feature of the installation.

The arrangement of the conductors and regulators will be understood by reference to Fig. 257, in which the 12 generators are represented at the bottom of the diagram, marked I., II., III., IV., &c., up to XII., the terminal screws of each machine (see Fig. 256) being numbered 1, 2, 3, and 4. Stretched along a board fixed in a vertical plane above and behind the machines are five copper wires, two of which for collecting the combined currents of all the 12 machines shown in the diagram by thick lines at X X and Y Y, and transmitting them to the various circuits in distant parts of the Palace, while the three others marked *a a*, *b b*, and *c c*, are employed in connection with the regulating apparatus to be presently described.

Referring first to the distribution of the currents generated by the machines and their transmission to the various lamp circuits, it will be seen that the arrangement adopted is of the simplest possible description. The two principle terminal screws of each machine (marked 2 and 3 in Fig. 256), are connected respectively to the two wires Y and X in the manner shown in the diagram, so that the positive terminals of all the machines are connected with X X while all the negative terminals are in connection with Y Y. From these two wires any number of branch mains X^1 Y^1 can be connected in a precisely similar manner to that in which the machines were connected with them, and from these branch mains branches x and y can be led off similarly, and from these, smaller branches can again be led. Across any of these pairs of conductors, so as to offer a passage for the current from the one conductor to the other, one or more lamps can be placed, and it is on this principle that all the Edison lights at the Crystal Palace are connected up, and it is the system adopted throughout all the installations of the Edison Company. In the diagram the position of lamps is indicated by the letters *l l l*, &c.

The method of regulation by which all the machines and lights are maintained under complete control is equally simple, although at first sight it may, by the repetition of parts necessary for the working of a dozen machines, appear a little complicated. The arrangement is, in principle, the same for all the machines, so that a description of one will suffice for all.

The mode by which the regulation is effected is identical with that by which the Swan lights at the Savoy Theatre are controlled, that is to say; the main current is regulated by increasing or diminishing the intensity of the magnetic field of the machines, by varying the strength of the current transmitted through the coils of the field magnets, and this is effected by throwing more or less resistance into the exciting circuit of the machines. In all Edison's generators the exciting helices of the electro magnets form a shunt or derived circuit to the lamp circuit of the machine, and this arrangement in itself tends to regulate the current. The strength of this derived current is, however, increased or diminished by means of the variable resistance box or regulator shown at R in the diagram. This regulator consists of a table, below which are a number of resistance coils, consisting of iron wires stretched on skeleton bobbins, so as to have a free circulation of air around and between their convolutions. The quantity of this iron wire thrown into the circuit of the magnet coils is regulated by a pivotted lever attached to the centre of the top of the table, and making contact in its rotation with a series of contact pieces corresponding to various lengths of the iron wire. The terminals of the regulator are connected respectively with the lower thick collecting wire Y, and with the branch piece D, which can be placed in metallic connection with any or all of the three thin wires *a*, *b*, *c*, by the insertion at A, B and C of contact pegs such as are used for resistance boxes.

Upon reference to the diagram, it will be seen that the terminals of the magnet coils marked 1 and 4 on each machine are coupled in the following way: No. 1 is connected with the wire leading from No. 3 to the thick wire X, while No. 4 is connected to one of the three thin regulating wires. The object of employing three wires instead of one is to divide the 12 machines into three separate groups of four machines each, any of which groups can be thrown in or out of action by the insertion or removal of one or more of the contact pegs at A, B and C. Thus each machine of the group marked I. to IV. is connected to

T

the upper wire *a a*, and is controlled by the peg A; the next group from V. to VIII., being connected to *b b*, is controlled by B; and the last group, IX. to XII., is connected to the lower wire *c c*, and is thrown into or out of action by the contact pin C. By this very beautiful arrangement it is possible to throw out of action any group of machines without perceptibly affecting the strength of the main current or the illuminating power of the lamps, and at the same time the expenditure of motive power, and therefore, the consumption of fuel, is proportioned to the work actually being done. In the same way the throwing of a large number of lamps out of circuit can instantly be compensated, and no variation in the intensity of those left in action can be detected.

In permanent installations an electromotive force indicator is provided at the central station, and an assistant keeping his hand on the lever of the regulator, and his eye on the index of the indicator, can maintain the strength of the current within very narrow limits, notwithstanding what may happen to the lamps throughout the different circuits and their branches.

The Edison machines, of the form illustrated in Fig. 255, are made in two sizes, one known as the "A" machine, capable of maintaining 75 full lamps or about 120 half lamps, and a second, of similar bulk, called the "B" machine, which, however, is only applicable to the smaller or half lamps, of which it is capable of maintaining also 120 in circuit. The two machines differ, only in the internal resistance of the armature, and in the coupling up of the coils of the field magnets. Thus in the "A" machine the resistance of the armature is ·14 of an Ohm, while that of the coils of the field magnets, which are connected in series, is 60 Ohms, each of the magnet coils having a resistance of 30 Ohms; and this machine, when running at a speed of 1,200 revolutions per minute, has an electromotive force of 110 Volts. The "B" machine has an armature resistance of ·035 of an Ohm, while that of the magnet coils (which are exactly like those of the "A" machine) is only 15 Ohms, each coil having a resistance of 30 Ohms, but as they are connected in parallel circuit the resultant resistance is half instead of being double that of each coil, as in the "A" machine. The electromotive force of the "B" machine when driven at 1,200 revolutions per minute is 55 Volts. In both generators, however, the resistance of the magnetising coils is rather over 400 times that of the armature, a proportion which might probably be increased with advantage.

SIEMENS.

Next to the Gramme machine, and in fact almost contemporary with it, is the dynamo-electric generator of Messrs. Siemens Brothers and Company, which from the moment that it was introduced until the present time has occupied a leading position. This position it has maintained in the face of a large number of more recent competitors, and is still the most powerful generator of electricity in existence, whether estimated by a comparison of its light-producing power with its bulk and weight of material, or with its absorption of motive power or cost of working.

The most remarkable characteristic of the Siemens machine is its extraordinarily small size and weight in proportion to its power, and nothing shows more strikingly the great advances that have been made in the construction of machinery for the induction of electric currents during the last 20 years, than a comparison between Holmes's machine, which was first used at Dungeness lighthouse, and the modern Siemens dynamo-electric apparatus, which was installed at the twin lighthouses on the Lizard promontory; thus, while the Dungeness machine occupied no less than 484 cubic feet, the Siemens machine takes up but little over 4¼ cubic feet, and the corresponding weights of the two machines are 5¼ tons for the Holmes, to 3¾ hundredweight for the Siemens; again, while the Holmes machine produced a light equal to 670 candles, the light from the Siemens apparatus is equal to 3,620 candles; and with regard to the question of economy, while the apparatus at Dungeness had an illuminating power of 244 candles per horse-power absorbed, that of the Lizard machine was 1,034, and the cost of working per unit of light is 1,294 at Dungeness to ·0,147 at the Lizard. We have thus the following proportions between the apparatus at the two stations of Dungeness and the Lizard respectively. Bulk, 114 to 1; weight, 28 to 1; light produced, 1 to 5·4; light produced per horse-power, 1 to 4·24; and cost per unit of light produced, 1 to 0·113. The figures are interesting, comparing as they do the apparatus which was employed in the first application of Faraday's transcendent discovery in the illumination of the coast, with one of the latest and most perfect apparatus which modern science has developed from that discovery.

As Faraday has demonstrated, when part of a closed electrical circuit is caused to move within a magnetic field so as to cut through the lines of magnetic force in a path more or less perpendicular to their direction, a current of electricity is induced in the circuit, the strength of which depends upon the combination of a variety of causes, of which the principal are, the intensity of the magnetic field and the favourable presentation to its inductive influence, of those parts of the circuit which are moving within it. All dynamo or magneto-electric generators, whether their currents be induced from permanent or from electro-magnets, or from a magnetic field excited by their own action and by the circulation of their own currents, depend, as has been already pointed out, for their action upon the great principle first discovered by Faraday and investigated so fruitfully by Ampère, Arago, Lenz, and others, and the efficiency of all such machines is determined by the perfection with which those laws and the forces which depend upon them are economised. A perfect machine would be one which, producing a magnetic field of the greatest possible intensity, is so constructed as to cause convolutions of insulated wire forming a circuit, to move rapidly within that field in such a manner as to be always cutting through the lines of magnetic force to the best possible advantage for magneto-electric induction, the length of the circuit being so proportioned to its conductivity as to present the largest number of convolutions to the inductive influence of the magnetic field without diminishing the strength of the current by unnecessary resistance, and it is the degree to which these theoretical requirements are fulfilled, that constitutes the efficiency of a machine. How closely the Siemens dynamo-electric generator approaches to the ideal standard may be judged from the experiments of Dr. Hopkinson, F.R.S., which showed it to be capable of converting mechanical power into electricity with a loss of energy of about 10 per cent., that is to say, 90 per cent. of the power exerted on the driving shaft of the apparatus is converted by it into energy of electricity, a transformation of one form of energy into another with an exceptionally high efficiency.

The Siemens machine, which is illustrated by the figures accompanying this description was devised in March, 1872, by Mr. Von Hefner Alteneck, chief constructor to the firm of Messrs. Siemens and Halske of Berlin, and was first publicly exhibited in this country at the loan collection of scientific apparatus, which was held at South Kensington in the year 1876.

In a more elementary form, however, it had been shown at the Vienna International Exhibition of 1873, where it arrived nearly two months after the exhibition had opened; in this arrangement the armature revolved around a fixed core, as will be presently described in more detail. It may be mentioned in passing that Messrs. Siemens and Halske showed at the same Exhibition another electric generator, illustrated by Fig. 259, which will indicate the advance they had made up to that date since 1867, when as we have already seen they had discovered the principle of dynamic action. In this machine, a Siemens armature revolved horizontally between the pole-pieces of the electro-magnets, which were excited by a smaller and similar generator. As will be seen,

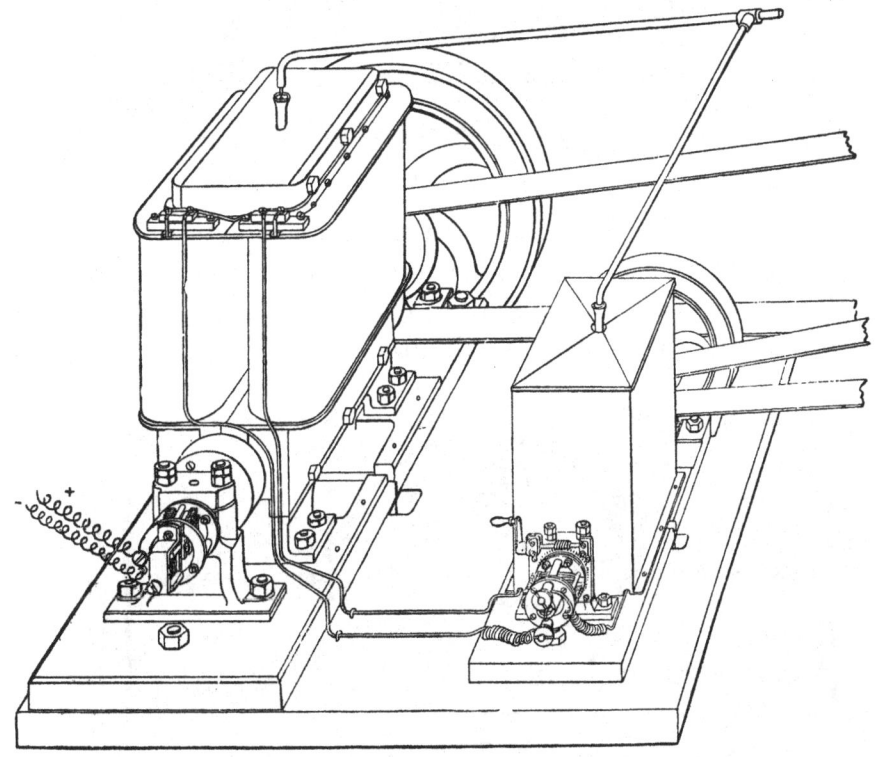

Fig. 259.

both machines were enclosed in casings through which water circulated to keep the moving parts cool. The small armature was driven at a speed of 1,600 revolutions, and the larger one at 800 revolutions per minute, and with 2 horse-power a 2,500 candle light was produced, which was employed during the time the Exhibition was open, to light up at night the Imperial Crown that decorated the culminating point of the Rotunda.

Figs. 260 and 261 illustrate an early form of the Hefner Alteneck machine, which possesses considerable historic interest. On the fixed shaft C and within the armature is fixed the iron core $s\ s_1,\ n\ n_1$. The armature drum $a\ b\ c\ d$ is made of a very light cylindrical shell of German silver, closed at the end with coned castings terminating as shown in cylindrical sleeves that take their bearings in the standards F^1 and F^2. Eight independent insulated coils are wound

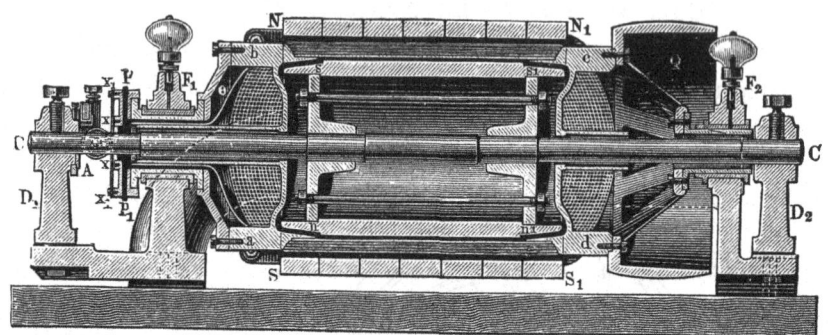

Fig. 260.

upon the armature body; the cores of the field magnets $N\ N^1$, $S\ S^1$ are curved above and below the armature in the usual way before extending horizontally to receive the coils $E\ E^1$. The shaft C and the core $s\ n$ are stationary; the latter consists of a cylindrical casting, and two end discs

Fig. 261.

with long bosses by which they are keyed to the shaft; the long through bolts shown convert the whole into a rigid cylinder. It will be noticed that on the left-hand side of the longitudinal section (Fig. 260) an annular space is left between the sleeve on the coned end of the drum armature and the fixed shaft C, and that this communicates with the disc $P\ P^1$, through which suitable passages are formed; this disc is the collector and revolves with the armature. On the right-hand end of this latter is bolted the pulley Q, by which motion is given to the machine. This modified form of armature was designed by Mr. Von Hefner Alteneck, with the object of avoiding several difficulties experienced in their ordinary form of armature for this class of machines. By using a light German silver body, there is no mass of iron necessary, no induced currents are set up in the armature, and no trouble from heating is experienced. The commutator attached to this generator calls for some detailed description. As stated above, the drum armature is wound with eight independent sections of coils, and the circular commutator has a corresponding number of radial divisions separated from each other with narrow spaces, the whole forming part of the disc P that rotates with the armature. The eight metallic sections are faced with insulating material, on which the covering disc is placed so that each section is insulated from the others. The wire terminals from each coil pass through the annular passages between the sleeve on the drum and the shaft C, and the terminals 1 1^1 or 2 2^1 of each of the coils, leaving the drum at the opposite sides of a diameter, are connected to two corresponding radial parts of the disc P, so that while one wire is in the strongest positive field, the other is in the strongest negative field.

The generator illustrated was the largest size of this type made by Messrs. Siemens and Halske, and when driven at a speed of 450 revolutions per minute, and absorbing 6 horse-power, it supplied a light of 14,000 candle-power, or maintained at red heat a copper bar 4 in. in diameter and 39 ft. 5 in. long. The machine is of small dimensions, being 43½ in. long, 18½ in. wide, and 12½ in. high.

At the loan collection at South Kensington the machine attracted very considerable attention on account of its extremely small size, standing not more than 9 in. from the ground, and producing a light which far exceeded in power that developed by its much larger competitors. Fig. 262 is a perspective view of this type, Fig. 263 is an end elevation with the front framing removed, so as to show its principal working parts, and Fig. 264 is a half sectional elevation, the section being taken in a vertical plane close to its axis of rotation.

The principal distinctive feature of the Siemens machine lies in the peculiar method adopted to form the generating coils of the revolving armature. The moving conductors, in which the electric currents are induced by the action of the machine, consist of lengths of insulated copper wire wound in several sections longitudinally over a cylindrical core of soft iron.

Fig. 262.

The longitudinal coils of wire completely envelop the iron cylinder, all sections being wound at equal angular distances from one another

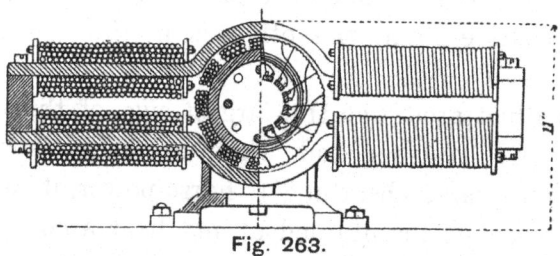

Fig. 263.

around the surface of the cylinder, and parallel to planes passing longitudinally through its axis, there being as many planes as there are sections to be wound.

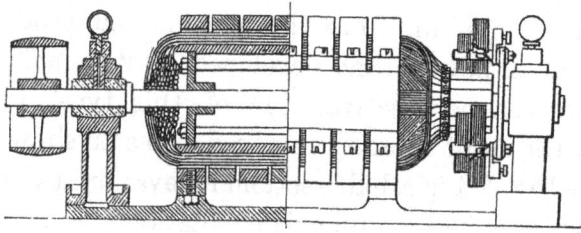

Fig. 264.

Each of the sections of wire coiled upon the cylinder consists of two separate coils, each having two ends, that is to say, there are double as many ends as coils. These ends are brought two by two to a number of copper sectors insulated from one another, which, together build up a cylindrical commutator (Fig. 264) much resembling that of the Gramme machine, and rigidly attached to the armature spindle; and the currents are collected in a similar manner by conducting brushes and pressing against the commutator as it revolves.

Simple as is the winding of the armature cylinder, the connecting up of the ends of its coils is not so easy to explain even with the aid of diagrams. The attachment of the ends of the coils to the sectors of the commutator is partly shown in Fig. 263, and more clearly in Fig. 265, in which, for the sake of simplicity, an armature of only eight coils is shown, there being the same number of commutator plates. In the diagram the two ends of the same coil are numbered 1 and 1^1, 2 and 2^1,

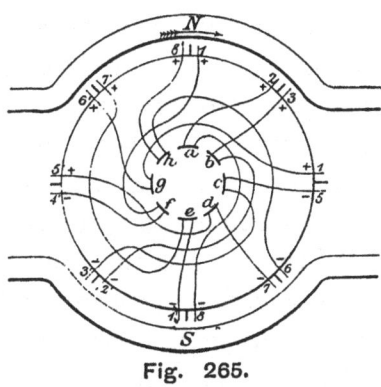

Fig. 265.

&c., and they are connected to the sectors of the central commutator in the order shown. If these connections be traced, it will be seen that all the coils are united into a continuous circuit, the commutator sectors being traversed in succession; the signs plus and minus indicate the direction of the current induced at any particular spot in the position of rotation shown in the diagram.

The advantage claimed for this form of winding the inducing coils, is that all the wire is brought into the magnetic field, with the exception only of those portions crossing the ends of the cylinder, and in order to reduce this idle resistance to a minimum, the armature is made long as compared with its diameter.

In the earliest machines, as we have seen, the core of the armature consisted of a stationary iron cylinder, around which the various coils of copper wire were wound upon a separate shell, which was constructed to revolve in bearings surrounding the fixed spindle carrying the core. This form of construction obviously involved structural complication, and in the later machines the stationary cylinder has been replaced by a revolving one, sometimes built up of coils of iron wire wound at right angles to the copper conductors. It is understood that this modification, while obviating much awkwardness of construction and cheapening the production of the machines, does not detract from their efficiency.

This system, consisting of the iron cylinder with its enveloping coils, is surrounded by a powerful magnetic field produced by a series of electromagnets, the coils of which are included in the circuit of the rotating armature through the brushes. The curved bars shown at N. S., Figs. 263 and 264, are of soft iron of rectangular section, and are the prolongations of the cores of the powerful electro-magnets, the coils of which are long flat bobbins wound with insulated wire, which are best shown in the general view, Fig. 262. Upon reference to Figs. 262 and 264, it will be seen that the magnet cores, instead of being flat continuous plates are divided longitudinally into several bars having air spaces between them, the object of which is, firstly, to prevent cross currents being induced in the magnets; secondly, to maintain the lines of magnetic force parallel to the length of the bar; thirdly, to permit of a circulation and escape of air between the revolving armature and the magnets, so as to reduce the accumulative heating of the machine; and lastly, for convenience of manufacture. Of the curved portions of the magnetic cores N. S., Fig. 263, each surrounds two-sixths of the entire circumference of the induction cylinder, so that two-thirds of it are embraced by the magnets, and the coils of each set of magnets are so wound as to produce a consequent point or pole in the centre of their length, a north pole being produced in the upper set of magnets, and a south pole being produced in the lower set of magnets. Thus a very intense magnetic field is formed within the cylindrical space included between the upper and lower sets of magnet bars, and within this space is revolved at a high velocity the induction cylinder or armature which has been described.

The action of the apparatus is as follows:—When the machine is set in motion by a steam engine or other motor, the permanent or residual magnetism within the magnet cores induces in the wire on the revolving armature a current of electricity initially weak. This being led through the commutator brushes to the magnet coils increases their magnetic intensity, and consequently that of the magnetic field in which the armature revolves. The effect of this action is to induce in the revolving conductor a stronger current, which, being again transmitted through the coils of the magnets further augments their power, and these by their reaction induce a still stronger current in the conductors. This mutual action continues until the magnetic limit of the soft iron of the electro-magnets is attained.

The maximum inductive effect of the magnets upon any one convolution of the rotating coils takes place when it passes through the middle of both magnetic fields, *i.e.*, approximately when it reaches a position perpendicular to the magnet bars, and the least effect is experienced when the convolution is at right angles to that position, or when it lies parallel to the direction of the magnet bars.

By Lenz's law a coil starting from the position of least effect (the neutral point) on one side of the axis, towards the north pole of the electro-magnet, would be subject to a direct induced current, and that portion of the coil on the opposite side of the axis would be traversed by a current of opposite direction. By following the winding of the coils, and in order to do this more readily the reader is referred to Fig. 265, it will be found that the currents in the two portions of the coil, although of relatively opposite directions, make one stream round the coil.

The ends of the armature coils are so connected to the sectors of the commutator, and the collecting brushes are so placed with regard to it, that they take off the currents generated by the machine, at the position giving the least spark at the commutator, the absence of "sparking" being an indication that the machine is working at its greatest efficiency. There is to every machine of this construction one position for the brushes to bear upon the commutator, in which the current is strongest, and where sparking is reduced to a minimum, and, in order to facilitate the adjustment of the brushes to this position, they are sometimes mounted upon a frame, shown edgeways in Fig. 260, which is capable of being moved through a certain angular distance around the principal axis of the machine. By this means the position of the brushes at which sparks disappear at the commutator, or are reduced to a minimum, is very readily determined, and the frame being fixed in that position by set screws, the adjustment of the apparatus is complete. Except, however, for experimental purposes the brush holders are made fast to the frame in the most advantageous position, as otherwise inexperienced attendants shift them to the prejudice of the machine.

The brushes are sometimes arranged in two pairs, so fixed that one brush of each pair is slightly in advance of its fellow, the object of which is to render the current more uniform by preventing the occurrence of a break in the circuit, and causing one brush to compensate for any imperfect contact of the other. By a comparatively slight alteration of the brush holders and of the wire connections, the armature can be made to revolve in the opposite direction to that for which it has previously been set, and the machine driven to the right or to the left, as found convenient.

The only wearing parts of the Siemens machines are the brushes and the commutator, which are not only always rubbing together, but are subject to be more or less burnt under the influence of the current when sparking is taking place. The renewal of the brushes is a matter of only a few minutes, involving a very trifling cost, but the renewal of the commutator was, until the later improvements made by Messrs. Siemens, a much more serious business, and could only be done by the maker, and at considerable cost of time and money. In some forms of the Siemens machines, each sector of the commutator cylinder is separate and renewable, being fixed in its place by a single screw; by this means the removal and renewal of any one plate, or even, if necessary, of the whole commutator, is a very simple and insignificant matter, occupying only a few minutes instead of several days.

A modification of the ordinary Siemens dynamo machine has been designed by Dr. C. William Siemens, and was introduced by him to the notice of the Royal Society, in a paper read on March 4th, 1880.

The ordinary dynamo generator labours under the disadvantage that the intensity of the magnetic field, in which the armature revolves, varies very considerably if the electrical resistance of the external circuit be disturbed. For any given speed the current furnished by the revolving wires rises rapidly as the external resistance is diminished, and on the other hand if this resistance be increased, the intensity of the magnetic field diminishes, causing a corresponding diminution of the induced current. It is thus evident that when a disturbing influence is at work in the outer circuit, whether occasioned by an ill-regulated lamp or by a short circuit, the power of the machine increases when it is least wanted, and decreases when the demand upon it to overcome increased resistance is greatest, resulting in an aggravation of the original disturbance. This instability of the dynamo machine proper is liable to effect in practice an injury to the insulation of the inducing wires of the armature by excessive currents, in all cases where the resistance of the external circuit is exposed to comparatively large fluctuations.

Following in the steps of Professor Wheatstone, who, in his communication to the Royal Society, in February, 1867, suggested that only a portion of the entire current should be allowed to pass through the electro-magnet coils and produce the magnetic field of a dynamo machine, Dr. Siemens constructed a generator in which the wire of the coils surrounding the field magnets constitutes a shunt to the main circuit, *i.e.*, the electric current flowing from the revolving armature divides, a small portion traversing the electro-magnet coils, and returning direct to the armature, while the other and larger portion passes to

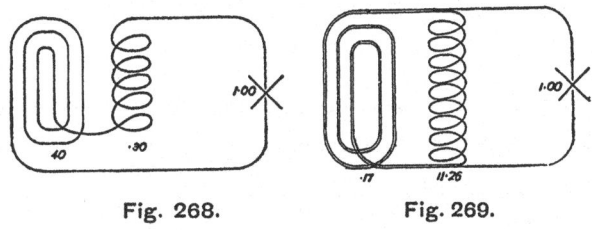

Fig. 268.   Fig. 269.

utilise its energy in the external circuit. Figs. 268 and 269 illustrate the difference existing between the two modes of winding.

When the machine is in action and a disturbance, such as has been described, occurs in the outer resistance, it will be seen that the machine adjusts itself to the new condition; for supposing the disturbance takes the form of an increase of resistance, the resistance of the electro-magnet coils virtually diminishes in relation to the total resistance in circuit and more current is forced through them. The magnetic field is, therefore, intensified and the power of the current is strengthened to overcome the additional hindrance thrown in its way. Similarly a diminution of the external resistance causes more current to flow through the external circuit and less through the magnet coils, the effect of which is to weaken the electromotive force of the machine correspondingly to the diminution of impediment it is required to resist.

The essential features manifested in the working of each type of machine, the " direct " and " shunt " wound, are clearly depicted in the accompanying diagrams, Figs. 270 and 271. The abscissæ represent the resistance in Ohms introduced into the circuit external to the machine, while the ordinates give the electromotive force in Volts, the current

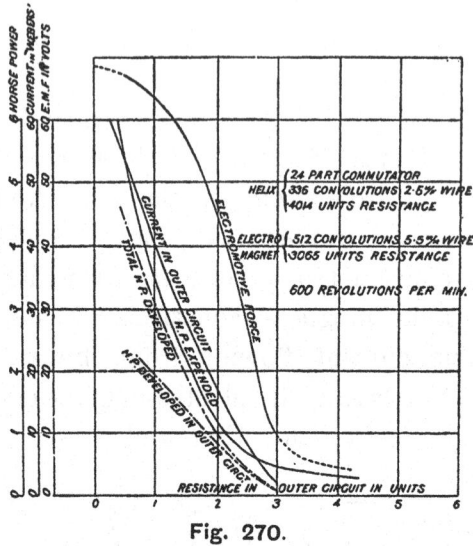

Fig. 270.

in Ampères, and the horse-power, corresponding to the various plotted curves, which latter explain themselves.

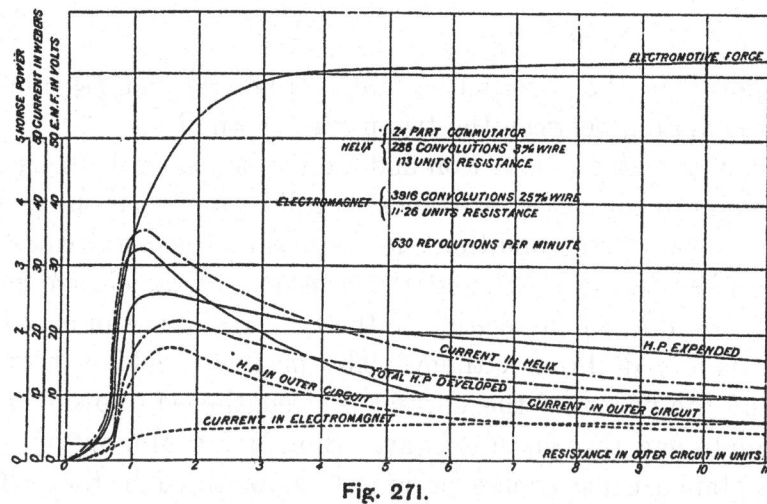

Fig. 271.

In experimenting with the direct form of winding, the results of which are shown in diagram Fig. 270, a machine was made use of having a commutator with 24 sections, the armature being wound with 336 convolutions of copper wire of 2½ millimetres diameter, and having a resistance of approximately 0·4 Ohm.   The electro-magnet coils, which were in direct circuit with the inducing coils, consisted of 512 convolutions of 5½ millimetres wire with a resistance slightly greater than 0·3 Ohm.   The most noticeable feature is the very rapid falling away of the electromotive force and of the induced current of the machine as resistance is added, until, when the latter amounts to 3 Ohms, the electromotive force sinks as low as 10 Volts.

The second diagram Fig. 271, illustrates the effect of the "shunt" winding upon a machine of the same size as that just described, the details and mode of winding *only* being different.   Upon the revolving cylinder were coiled 288 convolutions of copper wire 3 millimetres in diameter, the resistance of the complete helix being 0·173 Ohm.   The electro-magnets comprised 3,916 convolutions of 2½ millimetres copper wire, having a resistance of 11·26 Ohms.   It is interesting to note that in this diagram the electromotive force of the machine, rising quickly at first, never ceases to rise to meet the increase of resistance opposed to it. The flow of the current, on the other hand, also increasing rapidly, and attains a maximum when the resistance reaches 1¼ Ohms.   It then commences to fall, but more gently, and never disappearing, is maintained at a practicable and useful value through resistances which the previously described machine could not face.

Naturally these illustrations are merely intended to explain the respective advantages which may be secured by the two different modes of winding, and they do not pretend to define the limits of the capabilities of either machine, as it is evident that by varying the dimensions of the framework, the size and length of the copper wires constituting the winding, and the velocity of rotation, almost any desired effect can be produced with the Siemens-Alteneck machine, whether "direct" or "shunt" wound.

The shunt wound generator lend themselves very readily to winding with different sizes of wire, and they possess therefore the quality, shared by the others in a less degree, of being able to generate currents widely differing in intensity and force.

For many purposes, such as burning incandescence lamps, all of which are not invariably alight together, this type of machine is endowed, we are led to believe, with peculiar qualifications, while for laboratory use it doubtless possesses advantages over its older brother on account of its more greatly varied capacity.

We now pass on to consider the second type of machines for the production of electric currents manufactured by Messrs. Siemens; this is the alternate-current generator, of which a very large number are now in use. To the great majority, even of the scientifically educated public, electric lighting was nothing more than an abstract idea until they visited the Paris Exhibition of 1878. The sight of the Avenue de l'Opera brilliantly illuminated throughout its whole extent, came upon them like a revelation, and on their return home the Jablochkoff candle furnished them with a never-ending subject of conversation. There were at that time but two electric generators in this country that were well known, the Siemens and the Gramme dynamo machines, and neither was adapted for producing more than one light. The general feeling was that the electric light needed to be divided, if it were to be applied with useful effect in every-day life, and consequently inventors and manufacturers began at that time to turn their attention to apparatus for burning lights in series. Those who have had practical experience of the difficulties of getting one regulator lamp, as they were then constructed, to burn satisfactorily on a circuit, will not wonder that the early attempts were directed to the production of a generator suitable for use with candles of the Jablochkoff type, which of course require alternating currents to ensure the two pencils burning at the same rate. In 1878, Gramme had solved this problem with great success, as we have already seen, and at the Exhibition of Electric Light Apparatus, which took place at the Albert Hall in 1879, Messrs. Siemens Brothers showed a generator adapted for the same purpose, the latest development of which is illustrated in Fig. 272. Upon the base plate are bolted two circular frames united by stays, and upon these frames there is fixed a number of electro-magnets, each consisting of a stud encircled by a coil wound on a bobbin and turned down at one end to fit a hole in the frame, in which it is held by a nut. There is an equal number of these magnets, which constitute the field magnets, on each frame, and they are so wound that the free end of each magnet has a pole of opposite name at either side and in face of it. To render the magnetic field between each pair of opposite magnets more intense they are provided with polar extensions arranged radially.

The coils or armatures in which the currents are generated are arranged round the circumference of a wheel mounted upon a central rotary shaft. These coils are wound upon wooden cores of slightly elongated form, which allows of a greater length of wire than if they were circular, and they are equal in number to the magnets on each frame. The ends of the coils are held by discs of German silver or brass, pierced with holes to prevent the circulation of currents in them. Even with this provision it is necessary that the plates should be kept as thin as possible and made from metal of comparatively low conductivity, as from their immediate proximity to the field magnets and their large sectional area, as compared with the wire of the coils, they would otherwise be capable of absorbing an enormous amount of power and applying it to heating the stream of air thrown off by the machine

Fig. 272.

when in action. In the early generators of this type the currents from a part of the coils or armature, were led to a commutator on the shaft, and after being turned so as to flow in the same direction, were employed to excite the coils of the field magnets. This plan, however, was soon discarded, and a separate dynamo machine (Fig. 273) of small size employed for the purpose.

As all the armature coils are in the same relative positions with regard to the magnetic fields that they are entering or leaving, it is evident that they may be connected together either in series, or in multiple arc, according as currents of high or low potential are required.

U

The currents are led directly to the external circuit without the intervention of a commutator, all that is needed for their conveyance from the rotating spindle to the stationary brushes which form the terminals of the leading wires, being simple cylindrical contact surfaces. The number of contact rings required depends upon the number of circuits that the generator is intended to feed; generally speaking, there are two rings for each circuit. Thus, if 16 armature coils were arranged to supply four circuits, they would be divided into four groups of four each, and one end from the first and last bobbins of each group would be connected each to a separate ring, while the adjoining ends of the four bobbins would be connected together. In a machine with eight

Fig. 273.

armature coils it is usual to connect the first and last bobbins to independent rings, and to bring a conductor from the wire connecting the fourth and fifth bobbins to an intermediate ring. This arrangement permits of two dispositions of external circuit. In the first, the leading wire commences at the first ring and terminates at the third, and the current is due to the electromotive force of all eight bobbins. In the second there are two external circuits, one beginning at the first ring and ending at the intermediate one, and a second beginning at the intermediate ring and ending at the third. The first arrangement may be illustrated by the letter D where the straight part represents the internal and the curved part the external circuit; while the second arrangement is like the letter B, and has two external circuits.

The armature coils may also be connected in multiple arc when currents of low potential are desired, as in the case of incandescence lighting, or part of them may be connected in series to supply a circuit of arc lamps, and part parallel to feed a number of Swan lamps.

The subjoined table gives the principal data required to calculate the power of these generators, and by its aid the best method of connecting the coils in any particular installation can be determined. It must be remembered, however, that two alternate-current generators cannot be coupled together unless they are driven positively, and that even the use of a common return wire will produce unsteady working, except in the case where it bears no appreciable proportion to the total resistance in circuit :—

SIEMENS' ALTERNATE-CURRENT GENERATOR.

Distinguishing Mark or Size.	W3	W6	W2	W1
Number of arc lamps, of 400 candle-power each, that the machine will feed	8	12	18	28
Current circulating through the lamps, in Ampères	12	12	12	12
Current required to excite field magnets, in Ampères	17	19	21	24
Number of coils in armature	8	10	12	16
Resistance of each coil, in Ohms	0·27	0·27	0·29	0·32
Electromotive force of each coil, in Volts	40	47	57	67
Revolutions of armature per minute	800	700	650	630
Horse-power required	4½ to 6	9	11	20

Figs. 274 and 275 show a recent form of alternating-current generator made by Messrs. Siemens and Halske, of Berlin. As will be seen the machine consists of a cast-iron bed-plate A, and two circular vertical side frames B B, around which are arranged the circles of electro-magnets C C. There are eight magnets on each side, secured to the frames by bolts, their inner and opposing ends carrying soft iron pole-pieces of a wedge shape as indicated in Fig. 274. These magnets are arranged to present alternating polarities, and their pole-pieces are placed as close together as is practicable, to obtain a magnetic field of high intensity. The narrow parallel space thus left, is occupied by

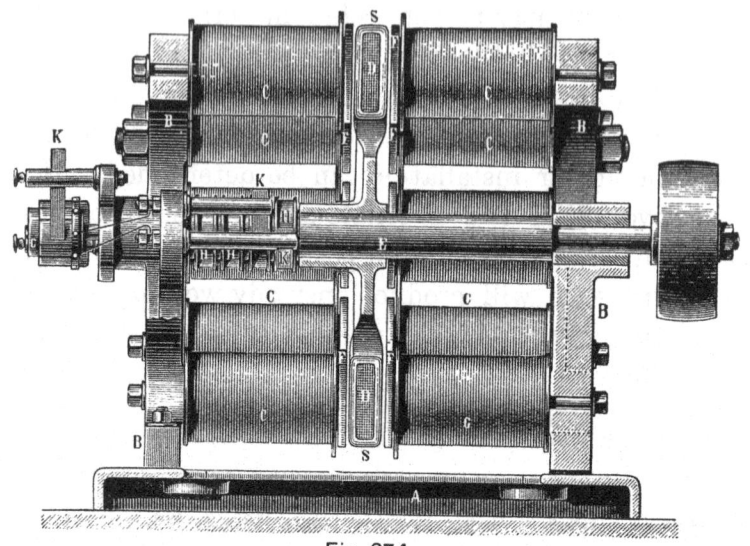

Fig. 274.

an armature formed as shown in the drawings. It consists of a ring, either made solid, or built up of thin plates or wires, and carrying eight

Fig. 275.

rectangular shaped coils, the insulated wires on which are wound radially as indicated in both figures. This ring is connected to a disc having eight radial forked arms, which hold the ring so that the arms lie between the coils. Each of the field magnets has a solid iron core, over which a sleeve is placed insulated from the core, and upon this sleeve the wire is coiled, the inner end of the wire being electrically connected to the sleeve, for convenience of coupling up. All the magnets are wound either in one circuit or in any desired combination, and the ends can be led to the collector G, which is mounted, outside the frame, on the shaft E. In the generator illustrated, two of the armature coils D are employed to excite the electro-magnets. To distribute the current produced by the revolution of the armature through the external circuit, a number of contact rings H H are placed on the shaft E, insulated from it, and from each other; the coils are connected with these rings, the current being taken off by the brushes K K. A large number of these machines have been made by Messrs. Siemens and Halske.

WILDE.

We have seen how important a part Mr. Wilde has played in the development of machines for the production of electric currents. Fig. 276 illustrates the latest form of his dynamo-electric generator, a type employed by the Compagnie Parisienne d'Eclairage par l'Electricité, in France, and which has found considerable employment in the British Navy. As will be seen, this generator has a striking similarity to several others now in use; its prototype will be found illustrated by Figs. 117 to 119, on page 143, illustrating Mr. Wilde's patent of 1868. It consists, as will be seen, of an armature carrying a series of coils, revolving between the free ends of a number of electro-magnets, arranged in a circle on each side, and secured at one end to the frame of the machine. The armature coils are provided with iron cores, differing in this respect from the Siemens alternating generator. One of the coils is set apart for exciting the electro-magnets. There are two commutators in this machine, one for transferring the work-performing current into the external circuit, the other for diverting the current from the exciting coil to the field magnets. If a number of these machines be worked together, it is preferable to set apart one of them to excite the others, so that the whole of the current from the other generators may flow into the external circuit. The commutators of this machine are so arranged that the current flowing into the external circuit may be either continuous

Fig. 276.

or alternating. With a velocity of 1,250 revolutions per minute, this machine supplies about eight Wilde candles connected in series.

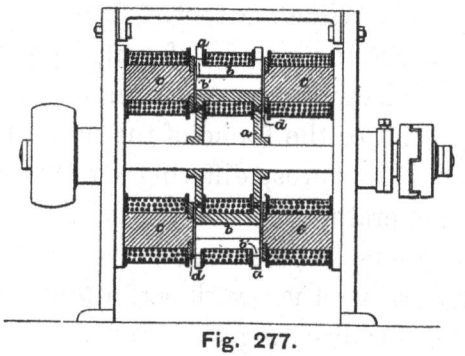

Fig. 277.

Fig. 277 shows a somewhat modified form of the Wilde dynamo-machine. In this arrangement *a a* are wheels or discs made of brass or other non-magnetic metal, and slit in a radial direction; *b b* are the armatures armed with pole-pieces $b^1 b^1$, the slits in which coincide with the slits in the wheels. The pole-pieces are fitted in recesses in the brass wheels *a; c c* are the electro-magnets armed with pole-pieces *d* slit radially and about one-half larger in diameter than the cores, while the pole-pieces $b^1 b^1$ are about one-sixth larger in diameter than the armature cores. These slits are intended to prevent cross currents, and enable the electro-magnets to be more readily exicted by their own residual magnetism. In order to prevent destructive sparking on the commutator when the direct current is used for external work, and the circuit from any cause is interrupted, an electro-magnet excited by the main current is employed to actuate a switch which shunts the current through the machine till the external circuit is re-established.

ANDREWS.

Figs. 278 to 283 illustrate Andrews' dynamo-electric generator. It has two massive field magnets, the cores of which are of cast iron, very heavy as shown; they are hollowed out at the back, and curved in front to admit the revolving armature. Wire is coiled round each of the field magnets, or rather wire is coiled on to a thin shell the form and shape of the magnet, this shell being afterwards placed upon the magnet. The complete armature is star-shaped in section, and is built up of a number of thin plates similar to the Jablochkoff and several other generators; these plates are mounted on a shaft, and pieces of paper are placed between each to insulate them, and prevent the setting up of cross currents. Insulated wire is wound longitudinally round each radial arm thus formed, as shown in Fig. 280. This armature is mounted on suitable bearings to revolve between the pole-pieces of the field magnets. There are 12 arms to the armature, and the opposite coils are joined up to so as to form six pairs, each pair being connected to the double

commutator shown in Figs. 281 and 282.   Each section of the commutator consists of two discs divided circumferentially and transversely,

Fig. 278.

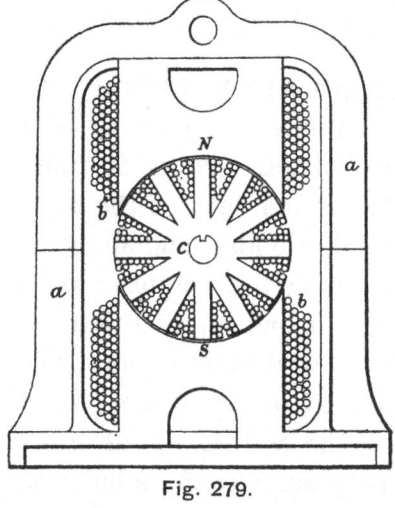

Fig. 279.

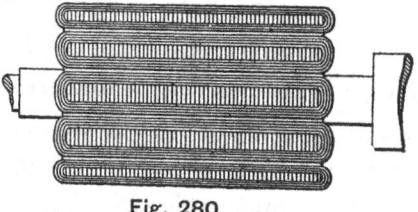

Fig. 280.

this part being insulated as shown, and the two placed together so as to
form a complete ring. These rings are mounted on a wooden drum, through
which 12 holes are formed at equal distances apart, from end to end, and

parallel to the axis of the shaft. The 12 wires from the six pairs of coils are passed through these holes, and each pair is joined up to its special

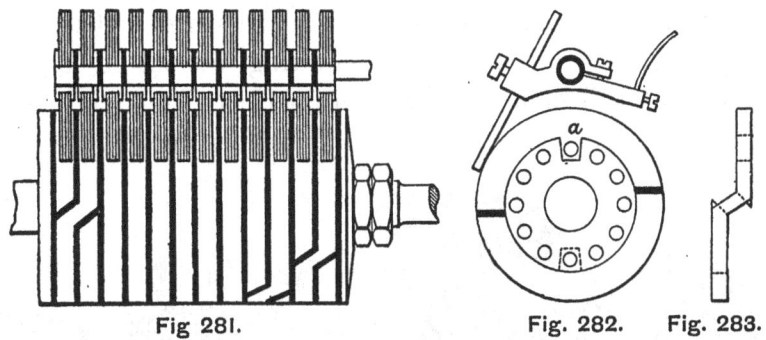

Fig 281.   Fig. 282.   Fig. 283.

commutator rings by the projection in the latter indicated at *a* in Fig. 282. It is intended with a generator of the kind illustrated to work six separate circuits, of which one is used to excite the field magnets.

GÜLCHER.

Fig. 284 shows an interesting dynamo-electric machine devised by Mr. R. J. Gülcher, of Biala (Gallicia). In this machine the magnetic field is produced by eight electro-magnets of oval section arranged in two sets of four, with their free ends directed towards one another and at their opposite ends, attached to two vertical circular frames carrying at their centres the bearings in which the main shaft of the machine runs; the opposed ends of each pair of magnets are connected together by a U-shaped pole-piece, the open jaw of which is directed towards the centre of the armature ring. Between the two sets of magnets and within the pole-pieces, is rotated a flat ring armature, the coils of which are wound Gramme fashion, but with spaces between, through which the iron wire core is exposed. Mr. Gülcher's machine occupies in principle an intermediate position between the Gramme and the Schuckert machines, as the pole-pieces not only embrace the lateral, but also the circumferential convolutions of the armature helices. The commutator is of the Gramme type, being cylindrical, and divided into as many sections as there are coils on the armature. It is, however, of extraordinary length, consequently requiring collecting brushes of great width, and by this construction more perfect electrical contact is insured, and the chances of "sparking" are reduced. A curious

Fig 284.

peculiarity of this machine consists in the fact that the magnet coils are wound with a double strand of copper wire, consisting of two thick wires twisted together; this arrangement is adopted probably for convenience of winding, but there must be a considerable loss of efficiency on account of there being no part throughout the whole of the exciting circuit in which the electric current is transmitted in a direction perpendicular to the axes of the magnets or of the fibre of the iron of which they are composed. On another page will be found a description of Mr. Gülcher's lamp, and of his system of lighting.

SCHUCKERT.

Figs. 285 to 288 are illustrations of a dynamo-electric generator used considerably on the Continent and in this country, amongst others by the Brush Electric Light Company. This machine, constructed by Mr. S. Schuckert, of Nürnberg, consists of two vertical frames A, between which are fixed the cores of the electro-magnets M M. These cores are cylindrical except at the middle of their length, where they are widened as shown, and formed with a central groove to receive the

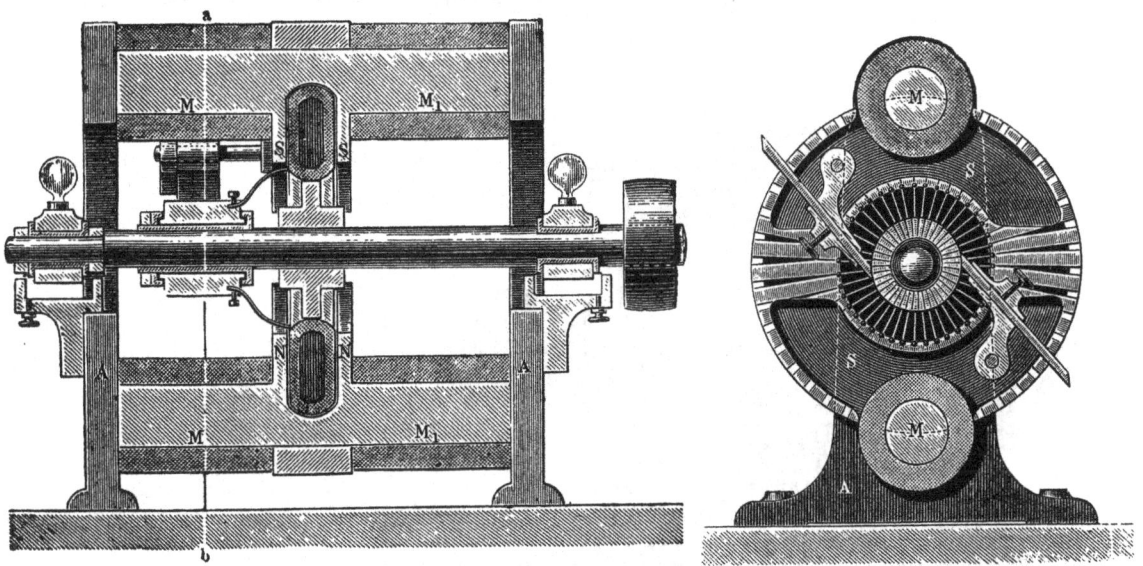

Fig. 285.   Fig. 286.

revolving armature. The latter consists of a ring built up of a number of thin discs, insulated from each other, and covered with coils as shown in the section. This ring is nearly enclosed by the semi-circular pole-pieces before mentioned as being formed in the middle of the field magnets, and which approach each other within a short distance, near the horizontal axis of the ring. The object of this special arrangement is, of course, to bring as large a number of the convolutions of the armature coils as possible within the influence of the magnetic field. The commutator of this machine consists of a number of metal segments insulated from each other, these segments being connected respectively to the sections of the armature coil. The number of these sections varies from 10 to 100, according to whether the machine is intended for low or high tension, and the end of each section is connected to its corresponding segment by a screw, so as to be easily removed for repair or renewal. In the generator illustrated, the commutator is shown within the frame, but in the latest pattern it is placed outside the frame, so as to be more readily accessible. The arrangement of coils on the armature, and their connections with the commutator, are shown in Figs. 286 and 287, which also illustrate the way in which the brushes are attached to the pole-pieces. It will be seen that they are each pivotted

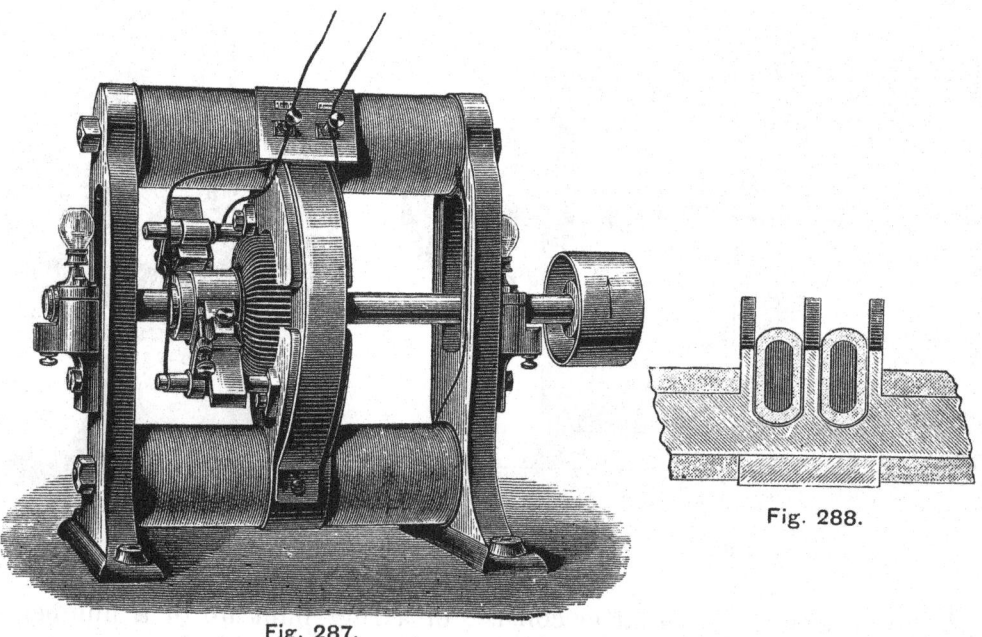

Fig. 287.

Fig. 288.

on a pin, and are free to move, the brush passing through a recess and being fixed in position by set screws so as to be capable of exact adjustment.

As already stated, this generator is in considerable use, and appears to give very good results. It is constructed after several different types, varying with the size, and the purpose for which they are specially intended, such as for the deposition of metals or the production of the electric light. In one modification these machines are made self-exciting, with commutators on both sides of the armature, about one-third of the current being taken off to excite the field magnets. In another form two rings are placed side by side on the shaft, the flat form adopted, lending itself to this arrangement, which, of course, involves the use of two commutators. Although the machine is driven at a high speed in ordinary working there is little or no trouble from heating.

This generator has probably the highest electromotive force of any in practical use, and for this reason it shares with the Brush machine the advantage of being especially applicable to installations in which the driving station is at a considerable distance from the area to be illuminated, but it of course requires proportionately increased care in handling, and in the manipulation and fixing of the conductors.

JÜRGENSEN.

Figs. 289 to 291 show an interesting form of dynamo-electric machine, the joint design of Professor C. P. Jürgensen and Dr. P. L. V.

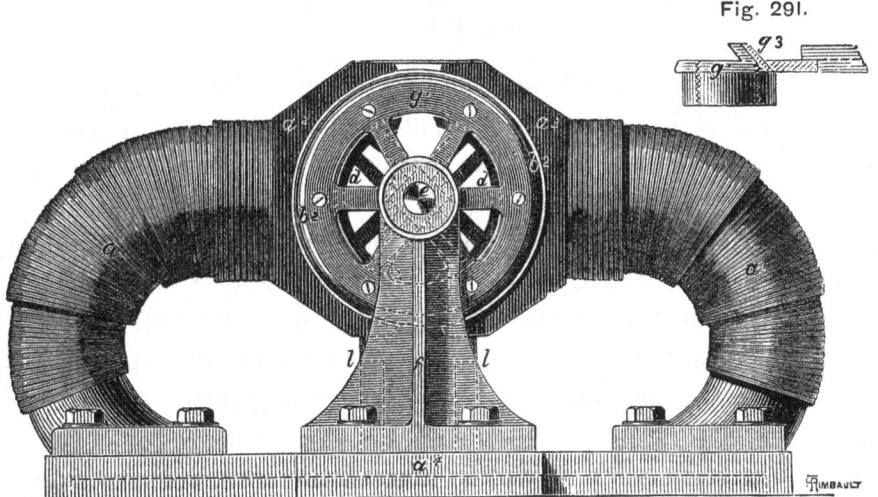

Fig. 291.

Fig. 289.

Lorenz, both of Copenhagen. This generator embodies several special features in its construction, the principal one being the introduction within the armature of field magnets intended to utilise the inner

portion of the coils on the ring.   The ring consists of a number of thin
sections b^1 placed side by side, and insulated from each other to prevent
the creation of induction currents; by this arrangement very wide rings
can be constructed.   The sections are held together by bolts passing
through them at intervals as shown.   Around the ring thus formed the
wire is coiled, as indicated in Fig. 290, to complete the armature, which
is mounted as follows : To the outer sections of the ring studs b^3, they are
secured by means of insulated rings $g^1 g^2$ provided with spokes, but each ring
is of a somewhat different form.   Thus g^1 has a small boss, to which is

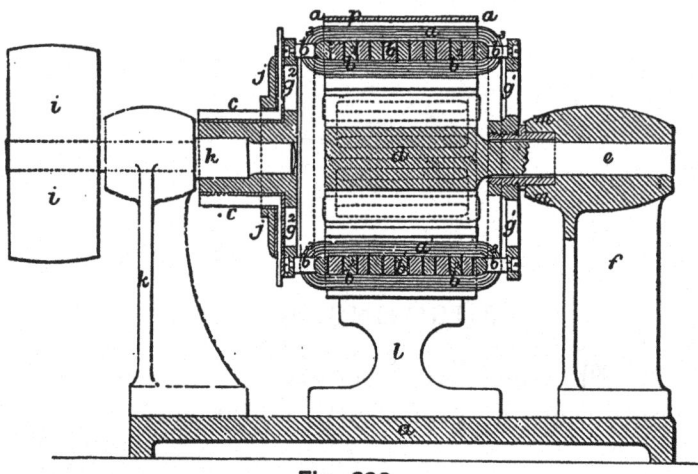

Fig. 290.

screwed the sleeve n, which enters the bearing m of the main standard f.
The other boss g^2 is formed with a long sleeve passing over and secured to
the driving shaft h.   The spokes connecting the ring g^1 with the boss are
provided with wings g^3, shown in Fig. 291, for the purpose of creating an
active current of air, and keeping the various parts of the machine cool.
Motion is given to the armature by the pulley i mounted on the shaft h,
which as will be seen terminates flush with the inner face of the ring g^2.
The outer electro-magnets $a\ a^1$ are of the curved form shown in Fig. 289,
and are bolted to the base plate a^4.   Each magnet terminates in a wide
pole-piece $a^2\ a^3$, the faces of which embrace the armature for nearly the
whole of its circumference.   It will be noticed the coils in these magnets
increase in volume up to the pole-piece, the object of this being to obtain
a more powerful magnetisation with less resistance.   In order to check
vibration, and increase the steadiness of the machine, the pole-pieces
are held together at the top by an insulated plate p, and at the bottom

by two insulated standards *l l*. So far then the machine consists of a cylindrical armature, carried at the end of a short shaft, and situated between the widespread pole-pieces of the two curved electro-magnets. To complete the machine the space within the cylindrical armature is filled with a second electro-magnet, as shown in Fig. 290. This magnet is fixed at the end of the shaft E, which passes through the standard F and the sleeve N which revolves round it. It is formed of two bar magnets at right angles to each other, as shown in Fig. 289, the method of coiling being indicated in the section Fig. 290. At the ends of these magnets are the pole-pieces *d* curved so as to present their surfaces to nearly the whole inner side of the ring. The north and south poles of this magnet are opposite the similar poles of the outer magnets.

The commutator of the machine is shown at *c*, and consists of a series of copper L-shaped bars, one face of which rests on the sleeve to which the driving shaft is secured, the other face butting against the ring g^2; the section illustrates two of these commutator bars. They are, as shown, insulated from the sleeve and ring, and also from the large washer *j*, by which they are fastened to the ring g^2. The collecting brushes are not shown, but they are of the usual form, and are placed on the bracket *k*; they are capable of adjustment. The current to the inner and outer electro-magnets may be either taken from the main current of the machine, or from a part of the armature reserved for that purpose, or a separate exciting machine may be employed.

The special features which are claimed by Messrs. Jürgensen and Lorenz, are: the application in dynamo-electric machines of an electro-magnet within the revolving armature, with or without the use of external electro-magnets; the use of annular armatures made up of a series of sections or narrow rings, insulated from each other; the special mode of ventilating and cooling the machine by means of the fan or wings attached to the ring at the end of the armature; and the special form of outer electro-magnet. It is to be regretted that no figures are available as to the duty of this machine. Messrs. Jürgensen and Lorenz claim that for its size it is the most powerful dynamo-electric machine yet made, but there appear to be no data on the subject.

ARAGO.

Among the numerous dynamo-electric generators shown at the Paris Electrical Exhibition were two machines of American origin, which, under the name of the "Arago Disc Dynamo," were exhibited by the Whitehouse Mills Company, of Hoosac, New York. Figs. 292 and 293 show both these machines, the former being the simpler of the two. It comprises, as will be seen, two pairs of electro-magnets E E, placed horizontally, and attached to the frame of the machine so that their polar extremities are opposite each other. These magnets are so wound

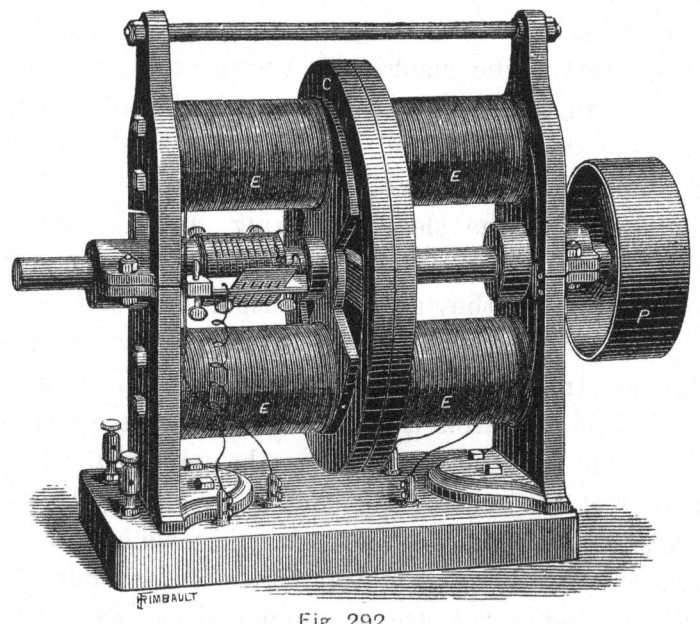

Fig 292.

that they present alternate polarities, and the polarity of each of the pair of magnets facing one another is of an opposite nature. The magnets are provided, as shown, with sector-shaped pole-pieces, which collectively form the greater part of a disc ; these are so placed as to leave a narrow

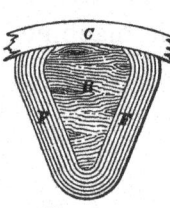

Fig. 294.

parallel space between them. Within the magnetic field thus formed, revolves a wheel or disc, in which is mounted a series of six sector-shaped bobbins. These are clearly shown in Fig. 294, where B is a wooden core, F the coil, and C the rim of the disc. All six of these coils are wound in one direction, but are coupled up in such a way that a current passing through them in series would circulate through them in opposite directions. The six coils are held in position by two plates on the driving shaft, and by the double copper ring C, that forms the periphery of the wheel.

The second and larger machine is shown in Fig. 293. It is constructed on the same principle, but involves more complication. In

Fig. 293.

this generator there are six pairs of electro-magnets, arranged as in the smaller machine, and furnishing six alternating magnetic fields. The revolving disc is similar to that already described, only it has eight coils with wooden cores. The commutator consists of 24 plates. The means employed for rendering the current continuous is the same in principle as that used in the Hefner Alteneck machine. The wire connecting each adjacent pair of coils of the disc is connected to three plates of the commutator, which is formed of 24 plates arranged in three groups of eight plates, each plate having a plate from each of the other groups on either side of it.   The two brushes are always in contact with two plates, corresponding to the line that divides the induction coils in two equal parts, traversed by opposite currents.   The armature of this machine has a special arrangement which does not obtain in the smaller one; each of the coils is in two parts; the induced current set up in the larger of these parts, is thrown into the external circuit, while the induction produced in the other portion creates a current used for exciting the electro-magnets.   On account of this disposition the collector is divided into two distinct parts, each corresponding to one of the circuits just mentioned.   The brushes L L are mounted in as shown; their supports are fixed on a ring A beyond the frame of the machine, and the brushes themselves are carried by springs fixed to the ring.   From the foregoing description it will be seen that the Arago disc dynamo is a self-exciting generator, and that when constituted according to the latter modification it involves the use of two sets of brushes and two commutators, one connecting the magnets with the exciting coils, and the other connecting the induction coils and the outer circuit.   The frame on which the brushes are mounted can be turned through a considerable angle, so as to adjust the brushes to the position of highest efficiency.   It is stated that the larger of these machines, driven at a speed of 1,000 revolutions, can feed eight arc lights of average intensity, consuming about 6 horse-power, but accurate data as to its capacity are wanting.

FEIN.

M. Fein, of Berlin, patented in May, 1880, a modification of the Gramme machine illustrated by Figs. 295 and 296, and the details of which are described by the inventor substantially as follows: In dynamo-electric generators constructed on the Pacinotti-Gramme principle, the poles of the electro-magnets exert their inductive action only on the exterior part of the ring, whilst the other parts of the wire on the armature are scarcely affected and serve only to increase the resistance of the circuit, which gives rise moreover to a very disadvantageous heating. In order to avoid these inconveniences some generators have been constructed with a flattened ring instead of the more common cylindrical form, in such a way that the coils can be subjected on both sides to the action of the electro-magnets. By this arrangement the wires, instead of being parallel to one another, are wound radially; from this it follows that in order to have on a flat ring, groups of coils of the same total length, and the same number of turns as with a cylindrical ring, it is necessary to give a much larger diameter to the ring. But this increase in diameter produces an increase of resistance, and in order to drive the flat ring more power is required than for the circular armature.

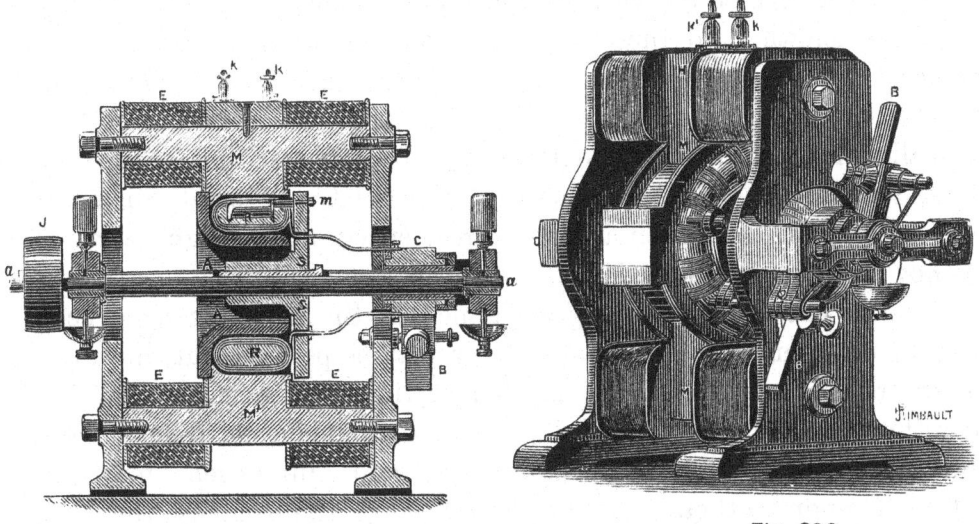

Fig. 295.   Fig. 296.

The way by which M. Fein proposed to obviate this inconvenience, and at the same time to secure the most efficient action of the electro magnets, is shown by the form of generator in the illustrations, of which one is a perspective view, and the other a longitudinal section. In the latter the ring is shown at R; it is fixed by means of screws m to the body S, which is keyed fast on the shaft A, on which is the driving pulley J. The ends of the wires from each group of coils on the ring, pass through insulated holes in the body S, and are led to the commutator C which is placed outside the frame, and revolves in contact with the two brushes B B[1]. To avoid cross currents in the ring, and consequent heating, the core is formed of a number of thin plates insulated one from the other. The cores of the electro-magnets E E are connected to the pole-pieces M M, which surround the outside of the ring, and have, bolted to them, auxiliary pole-pieces A A, which envelop the inner side of the ring, as shown in the longitudinal section. In this way almost the whole surface of the ring is exposed to the action of the electro-magnets, there being only the small part adjacent to the body S which does not assist in the production of electro-motive force. It is stated that a very simple experiment illustrates the efficiency of this machine; after measuring the intensity of the current produced in its normal state, then, if the inner pole-pieces A A be removed, it will be found that the intensity is hardly one-half that which it was before. The advantages claimed by the inventor for this arrangement are: (1.) The coils on the ring are subject to inductive action for almost their entire length, and therefore the resistance of those parts which produce no electromotive force is very slight. (2.) The mode of constructing the ring is such that the formation of cross currents is practically prevented; consequently the loss of power and production of heat are small. (3.) On account of the relatively small diameter of the ring it can be driven at a very great speed, and a high degree of efficiency attained, in proportion to the power expended.

Whether all these advantages claimed by the inventor have been tested by experience does not appear; it may be pointed out, however, that several other makers have arrived at the same result of utilising nearly the whole surface of the ring, in what would seem to be a more practical manner—Schuckert, Heinrichs, and Jürgensen may be mentioned among others. As for building up the core of the ring of a number of thin plates insulated from each other, that has been done also by many inventors.

———

In concluding this review of what has been done during nearly half a century, to produce mechanically, electric currents, we may mention briefly a certain number of devices that have been recently proposed, but which have not been, and possibly may never be, in their published forms, subjected to actual experience. The record of electric-current generators would not however be complete without them, and we have therefore, in grouping them together, confined ourselves only to such brief descriptions as shall render the proposed devices comparatively clear.

Fitzgerald.—This arrangement is of the Gramme class of generators, and aims at the substitution of comparatively small electro or permanent field magnets, in place of the large ones generally employed; these magnets are disposed in the plane of the ring and nearly surrounding it and its coils. Fig. 297 is a sectional elevation, in which A is the ring, carried by a disc on the shaft F. The coils of wire B have their extremities brought out and connected to the commutator; I, I¹, I² are the field magnets, and their essential characteristic is that they are always curved in the direction transverse to the axis of the magnet, whilst they are also, sometimes, curved in the direction of this axis. When electro-magnets are used, they are surrounded by coils of wire,

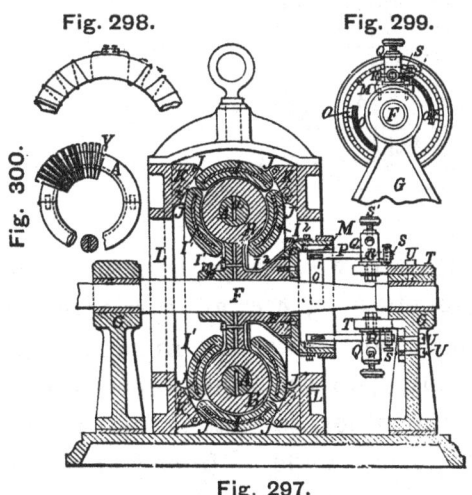

Fig. 297.

every turn of which is partly convex and partly concave. The winding of the wire is reversed at two points diametrically opposite (see Fig. 298), to produce opposing poles on each side of the machine. Owing to the form of the ring A, interstices are often left between the different coils at their outside edge, and these are filled up by soft iron wedges or blocks V (Fig. 300) cast, or slipped, on to the ring. The ends of the coils are attached to plates M (Figs. 297 and 299), and the currents are collected by springs o o^1, fixed to the bar P carried in the terminal Q.

Perry.—This proposed dynamo-electric generator is shown by Figs. 301 and 302. It is based on the use of angularly disposed coils, on a movable armature or ring, so arranged that the plane of each winding makes an angle between 90 deg. and zero with the direction of motion.

Magnetic poles of opposite name are placed so as to break joint on the opposite sides of such a coiled armature. In the illustration, the ring is shown rotating between the field magnets, and built upon a core, which may be of steel or phosphor-bronze, or, in some cases, may be dispensed with entirely. The coils are first wound on a peculiarly-shaped bobbin, then soaked with insulating material, and when dry are taken from the bobbin and slipped on to the core at an angle to it, as shown in Fig. 302. The bobbins and core are carried on a wood centre fixed on the shaft *a*,

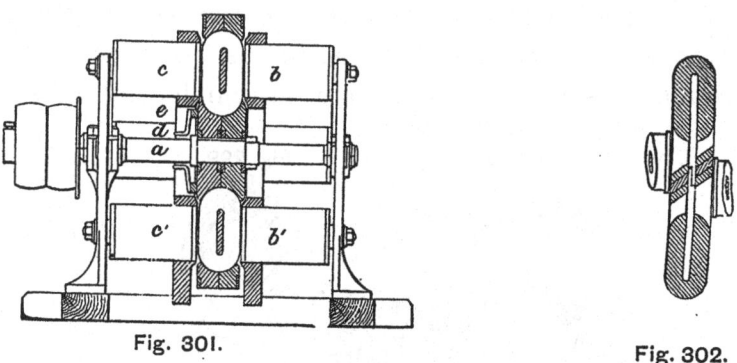

Fig. 301.   Fig. 302.

and are further held by two wood clip rings screwed together on the outside. The field magnets *b b¹*, *c c¹* are carried at their bases in cast-iron frames, and at their ends in wooden rings, shown in section. The opposite magnets are not upon the same centre line, one system being turned in advance of the other, so that a line joining the centres of corresponding poles makes an angle of from 20 deg. to 70 deg. with the direction of motion of the armature. The commutator is formed of cranked or curved rods *d* attached to an insulating ring *e*, and separated by air spaces.

Zipernowsky.—Mr. K. Zipernowsky, of Buda Pest, has patented a generator designed for producing currents of high potential. For this purpose a cylindrical armature, wound lengthwise with coils of insulated wire is caused to revolve within an electro-magnet, consisting of an iron ring also wound over sections of its circumference with coils of insulated wire, so as to produce a number of successive polar fields of alternating polarity. The spaces on the ring between the coils are formed, as shown in Fig. 303, with inwardly extended pole-pieces which embrace the greater part of the circumference of the armature. The coils of the armature are wound in the form of rectangles with their long sides parallel to the axis, the width of each coil being such that its two sides are at the same time approaching and leaving two of the polar fields. The core of the armature, for very strong currents, is made of wrought-iron plates whose continuity is broken; and for producing continuous uniform currents, it is formed of spiral-shaped rings of insulated iron wire wound round a cylinder of non-magnetic material. The currents are collected by a cylindrical commutator, Fig. 305, which, in order to diminish wear

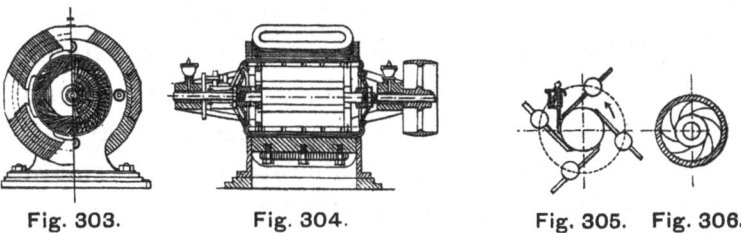

Fig. 303.   Fig. 304.   Fig. 305. Fig. 306.

and tear of the plates, is lubricated by pure petroleum, drawn from a tank placed above it, through a porous wick, which extends from the tank in such a manner as to bear tangentially against the commutator. The armature is made hollow and has openings in its ends and periphery, so that the air, which is put in motion by internal oblique fans Fig. 306, may keep the machine cool. The drawings without further reference show the disposition of the various parts.

Hussey and Dodd.—The object of this arrangement is to produce a compact, and powerful generator. In the Figs. 307 and 308, A is

Fig. 307.

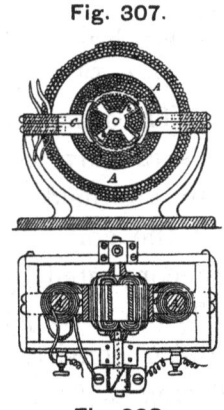

Fig. 308.

the field magnet, which may be either a ring or a hollow cylinder, and is furnished with polar extensions C C. The rotating armature is of the section shown, and is provided with air passages *b b b*; it may be wound either by the Siemens or the Gramme method, or the coils may surround each of the ribs *b b*. The commutator presents no feature of novelty.

Müller and Levett.—The object aimed at in the construction of the generator, illustrated by Figs. 309, 310 and 311, is the production of separate currents, one of which may be used to excite the field magnets, and at the same time perform work in the external circuits, whereas the other currents perform work in the external circuits only, and these different currents from the same machine may be of various degrees of intensity. Referring to the illustrations, the armature rotates between two sets of field magnets, of which there are six in each set. The magnet cores are connected by brass plates O^1 to O^6, and are coupled together in one circuit by means of a wire L, which proceeds

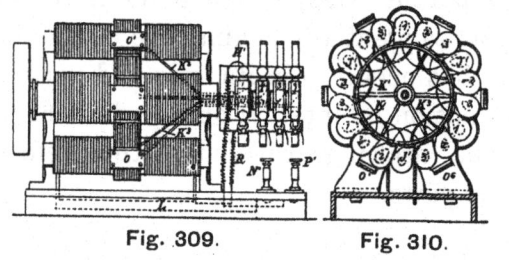

Fig. 309.   Fig. 310.   Fig. 311.

from the binding post N^1, traverses the whole of the magnets in succession, and goes in the form of a coil R to the lower brush of the first commutator I^1. From the other brush H^1 of the same commutator a wire extends to the binding post P^1. These two posts form the terminals of the external circuit. The armature wheel has eighteen coils, designated J^1 to J^3. These coils are divided into three groups of six each, that is, as many as there are field magnets, in the following manner: A wire K^1 connects all the coils $J^1 J^1 J^1$ for each group, while the coils J^2 for each group are connected by a wire K^2, and the coils J^3 are connected by a wire K^3. These wires pass to the respective commutators $I^1 I^2 I^3$. Each commutator consists of six segments, arranged concentrically within a flanged ring *a* (Fig. 311); the segments 1, 3, and 5, are connected with each other, and the segments 2, 4, and 6 are also connected with each other. The wire K^1 connects the segments 1, 3, and 5 of the commutator I^1, and the coils $J^1 J^1 J^1$ of an armature coil group, with each other, and passes back to the commutators and connects the segments 2, 4, and 6 of this commutator I¹. In like manner the wire K² connects the segments 1, 3, and 5 of the commutator I², the coils J² J² J², and the segments 2, 4, and 6 of the commutator I², and so on. The sides of the armature coils are so arranged that in the moment that they leave one field magnet one-eighth of their cross section covers the next adjoining magnet, as shown in Fig. 310.

Higgs.—The generator proposed by Mr. P. Higgs, of New York, is of the Gramme type. Fig. 312 is a perspective view, and Fig. 313 is a detailed view, explanatory of the construction of the coil. The ring is made up in short segments, practically identical with the method of M. De Meritens.

Fig. 312.

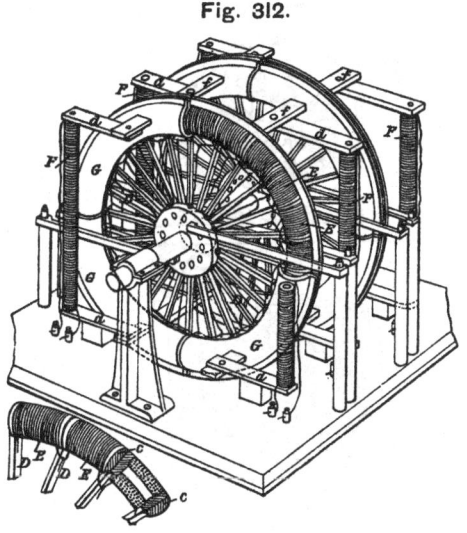

Fig. 313.

Then the whole of these segments are built up into a continuous ring by connecting their end flanges to the arms or spokes projecting from the boss. Plates of mica *c*, Fig. 313, are interposed between the bobbins to prevent the circulation of induction currents. The field magnets are formed of soft iron bars F F, around which the exciting coils are wound. These coils may be extended around the parallel pieces *d*, connecting the cores with the common pole-pieces *f*, which join the magnets F at their corresponding poles, so that the whole constitutes one electro-magnet consisting of two cores and helices F, four connecting pieces *d*, and two pole-pieces *f*. At each of the ends of the pole-pieces is a slotted circular tube marked G, and when four of these are placed together they form a hollow ring in which the armatures revolve, the slot being left for the passage of the spokes D. The armature coils may be all of one quality of wire, and be connected to a single commutator, or they may be alternately of two or more qualities and be connected to two or more commutators, and supply currents circulating in independent fields. The commutator consists of a cylinder of insulating material pierced with holes for rods of copper, which, as they wear under the destructive action of the spark, can be set up afresh to the face of the cylinder.

Hussey and Dodd.—Figs. 314 and 315 show a second arrangement of generator devised by Messrs. Hussey and Dodd, of New York, in

Fig. 314.

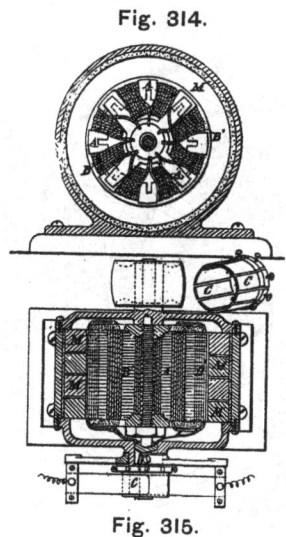

Fig. 315.

which M is a permanent ring magnet enclosed in a case in which it is embedded in plaster-of-paris. The armature A is composed of a number of plates arranged side by side, "and having alternate inward and outward projections." Between the outward projections are spaces for coils of wire B^1, which are connected together, and to the commutator plates C (Fig. 315); the latter pass through a collar of insulating material around the journal. The magnet is provided with poles or consequent points, and according to the inventors "the armature travels not only before the poles, but also before the whole length of the magnet, and through the entire field of force; hence electric currents are generated throughout its entire revolution." The magnet sections are magnetised by being placed in contact with the poles of a powerful stationary electro-magnet. The parts of the ring " where north and south poles, or consequent points, are desired to be formed" are placed opposite the south and north poles of the said electro-magnet. A second " electro-magnet, having its poles furnished with inward extensions, is arranged so that its poles will correspond with those of the first electro-magnet, and it is then repeatedly moved over the side of magnet section from one portion of the periphery towards the stationary electro-magnet, and from a diametrically opposite portion of the periphery towards the first electro-magnet." The section is then reversed, and the opposite side treated in the same way. As shown in the inventors' specification drawings, the electro-magnets are each wound, so that both their poles are of the same name.

Lane-Fox.—Besides having worked with marked success in the manufacture of incandescence lamps, Mr. St. G. Lane-Fox is the patentee

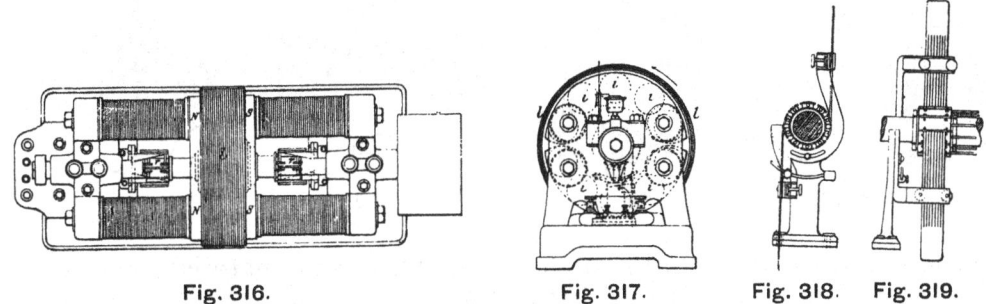

Fig. 316.   Fig. 317.   Fig. 318.   Fig. 319.

of a dynamo-electric generator possessing several points of interest, and which is illustrated by Figs. 316 to 319. Fig. 316 is a plan and Fig. 317 an end elevation of this generator. There are four field magnet coils on each side, or eight in all. Each pair with poles of like name is fitted with a common pole-piece N N or S S, and consequently there are two pole-pieces of opposite signs (as shown in dotted lines in Fig. 317) at each side of the central disc. The armature is made of a disc of hard wood bolted to a metal nave, and provided with bobbins $i\,i$ seated in recesses in its periphery. They are held in place by wedges, and the whole served with cord l. The two ends of each coil are carried to separate commutators $r\,r$. Currents are generated in the bobbins as they pass from the upper to the lower pole-piece, and *vice versâ*, and these currents are in opposite directions at the two sides of the machine. The commutators may be coupled up in parallel circuit or in series, as desired. Figs. 318 and 319 show the construction of the commutators.   Upon the shaft is a cylinder of baked wood, and upon this, segmental brass strips. The working faces of the strips are covered with light removable copper plates to take the wear.   The brushes can be set at an angle to the horizontal plane.   There is no connection between adjacent coils, each commutator strip being insulated from everything except one end of its own coil, and the brush when it happens to pass under it.

Moffat and Chichester.—In this generator, the poles of the field magnets   are   arranged   in such a manner that they will envelop a large portion of both the inner and outer surfaces of the armature which is a rotating cylinder, having the greatest length of cross section parallel with its axis.   Referring to Figs. 320 to 323, the armature core is formed of pieces B grouped together to constitute a ring   (Fig. 323).

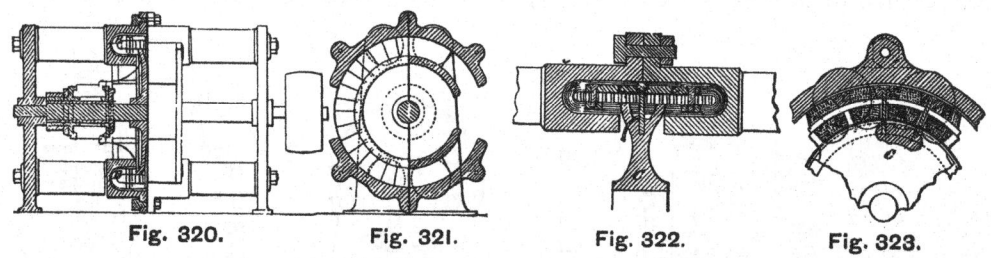

Fig. 320.   Fig. 321.   Fig. 322.   Fig. 323.

C is the hub provided with arms of non-magnetic material; $i\ i$ are small lugs, ribs, or fillets of iron which are secured between the bobbins.   They have metallic contact with the core pieces B, and serve to convey the magnetic force to them from the field magnets. The wires are wound round the core pieces B as shown, the ends being brought out to the commutator.

Harling and Hartmann.—In this generator the ring armature is brought into close proximity with the field magnets by means of projecting pieces constructed in the following manner :   The projecting parts have formed on them flat soft iron plates, attached to the sides, or to the peripheries, as the case may be, of a number of soft iron parallel rings.   A space is left between the end of each plate and the outer side of each projection, but the plates may also be made to overlap.   Between the projections $p\ p$ the coils e are wound.   The magnetic field is so arranged that the point of greatest magnetic intensity is not situated in the middle of the field, but near the end where the coils leave, so that when one of the above-mentioned projections and its generating coil, pass from one neutral point to another it is withdrawn suddenly from the magnetic influence. N S represent the magnets, and P^1 to P^4 the pole-pieces. The strength of the magnetic field is varied by a regulator or keeper. Two soft poles of opposite polarity are arranged to attract a bar of soft iron moving in close proximity with the faces of the field magnets, but never coming in actual contact therewith. The more the regulator is attracted the more it will cover the faces of the magnets, and so cloak or disguise their effect on the rotating coils; d (Fig. 324) is one form of the regulator. Any further attraction of d in the direction of the arrow will lessen the space between d and P, the attraction being exercised in opposition to the weight s suspended from the eccentric f. The circuits are arranged

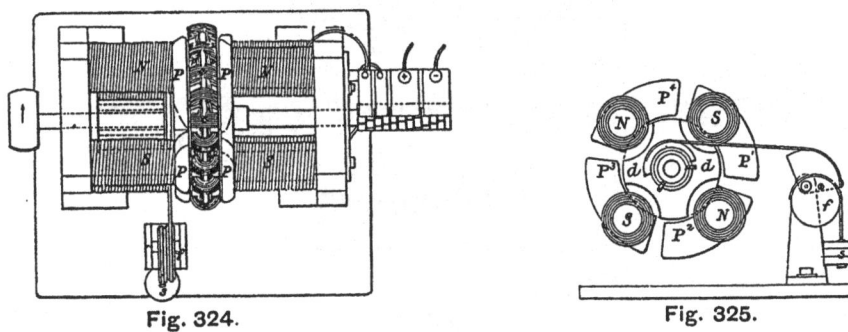

Fig. 324.   Fig. 325.

as follows : Considering four diametrically opposite coils, the first end of coil 1 is connected to the first end of coil 2, the last end of the latter to the last end of coil 3, and the first end of coil 3 to the first end of coil 4; the last ends of coils 1 and 4 are connected to their respective segments of the commutator in such a manner that each end is in contact with four diametrically opposite segments, which do not, how- ever, form part of the same cylinder. The segments 1, 3, 5, and 7 of the commutator are all connected with coil 1, while the segments between the same are connected to coil 4. When the coil is under the influence of the beginning of the pole-pieces P, its current is conducted through the field magnet coils; but as soon as it approaches the cores of the magnet, the current is sent through the working circuit.

Möhring and Bauer.—This dynamo-electric generator, the joint design of Mr. H. G. Möhring, of Frankfort, and M. G. Bauer, of

Stuttgart, is illustrated by Fig. 326. As will be seen it possesses
very little claim to novelty, but is of neat and compact design, being
contained within a cylindrical casing, to the side of which the
various parts are attached. The six electro-magnets H H are fastened
to one end of the cylindrical casing H by the bolt K. These
electros are of an ordinary form and present alternate polarities.
A shaft D, revolving in a bearing in the end plate K, and
carrying the driving pulley E, has fixed to it a disc, to which are
secured the six coils G which compose the revolving armature. A
brass sleeve L is keyed on one end of the shaft, and this sleeve
carries the commutator, outside the casing; this commutator is formed

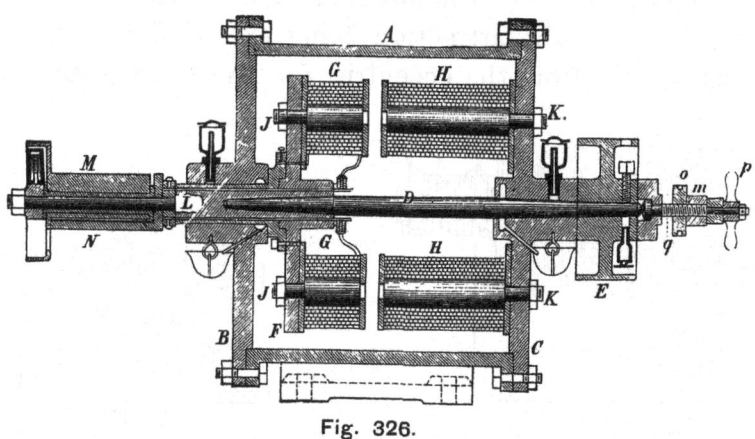

Fig. 326.

of two parts M N insulated from each other. While in the Weston
generator the fixed electro-magnets form one circuit, in the Möhring-
Bauer arrangement they are placed in parallel circuit with the coils G.
The current flows from one of the commutator brushes to an insulated
stud, from which six wires are taken, one to each of the electro-magnets
H, which, as stated above, are coiled so as to present alternate polarities.
The drawing illustrates a very neat arrangement for adjusting the
distance between the armature and the electro-magnets, so as to alter
the width of the magnetic field. By means of the nut o and the wheel
p, the bolt q and the shaft D can be shifted endways, which brings
the magnets and the armature closer together, or sets them further
apart; currents of varying intensity may thus be generated.

SECTION III.

—·◆·—

VII.

CONDUCTORS.

THERE is a general tendency in the applications of modern industry to domestic use, to proceed in crowded centres, by the creation of large works situated in suitable localities, and connected to the houses of consumers by a system of canalisation. This system is thoroughly carried out for the supply of gas and water in the majority of large towns. In Paris too, a very curious example of the principle is found in the distribution of compressed air, employed in a large number of houses, to drive several thousand pneumatic clocks. In a similar manner the general transmission of power and the production of electric light appears to be only a question of time, and of quite a short time.

In all these different methods of the distribution of various physical agents, a study of the canalisation, that is to say, of the *réseau* by which the production of the central works is delivered at the houses of the consumers—gas, water, time, or electricity—is, it will be readily understood, a study of the greatest importance, because it must combine the primary conditions of economy, simplicity in establishment, completeness and durability.

During the early days of electric lighting, canalisation had but very little importance, on account of the proximity of the electric generator to the single lamp that it supplied. It is scarcely eight years, since a separation of 150 or 200 yards was a maximum, and under these conditions the conductors employed were always satisfactory both as regards conductivity and insulation. But gradually the divisibility of the electric light, in separating the lamps fed from the same generator, resulted in the necessary consequence of extending the radius of operations of one installation, and consequently the *réseau* of conductors became more and more extended. Although, up to the present time, the principle of general electric lighting distributed from great central factories, has not been practically applied, a large number of private and public installations may be cited which present on a small scale the same general condition, and which involve the installation of thousands of yards of conductors. Thus, for example, we have in London the lighting of the City and of the Thames Embankment, and Edison's personal venture on the Holborn Viaduct; in Paris, the installation of the Avenue de l'Opera, the Louvre, the Hippodrome, the Place du Carrousel, the Grand Hotel; at Havre, that of the Docks; at Antwerp, that of the railway stations, and so forth; installations which have certainly involved the laying of some hundreds of thousands of yards of cable. These examples, not to quote others, show the ever-increasing importance of the rôle played in electric lighting installations by the canalisation, and fully justify the detail with which we propose to enter upon the subject.

As is universally known, electric currents flow with varying facility through different bodies, which, on account of the property for allowing the passage of the current, are called conductors. Metals especially are the bodies possessing this characteristic in the highest degree, although in very variable proportions. This will be seen from the following list, in which are ranged in the order of their decreasing conductibility the various metals, silver, the best conductor, being taken as unity :—

Silver	....	....	100	Palladium	...	...	12·5
Copper	....	....	80	Platinum	....	....	10·5
Gold	....	....	55	Lead	....	....	7·8
Zinc	....	....	27	Antimony	....	....	4·3
Tin	....	....	17	Mercury	....	....	1·6
Iron	....	....	14	Bismuth	....	....	1·2

The foregoing list shows at a glance that the high price of some of the metals named, and the want of ductility and strength of others, restrict within very narrow limits the choice of metal to be employed as electric conductors. The first on the list—silver—is too costly for any but exceptional use. It is employed, however, in the construction of instruments of precision, in which a very high and unvarying degree of conductibility is necessary. It possesses, moreover, the special property that its oxyde is a conductor, while that of copper is an insulator. Mercury is sometimes employed in spite of its relatively feeble conductibility, on account of its fluidity. But except these two metals, the utilisation of which is exceptional, conductors are almost wholly made of red copper, of iron, and of certain alloys, such as phosphor-bronze, of late largely used for telephone lines.

The employment of iron and of steel is almost wholly reserved for telegraph and telephone conductors, though, as we shall see further on, it has even found a certain application for electric lighting, where, especially in overhead installations, considerable tensile strength is required in the wires, and a lower degree of conductibility is sufficient. But for this purpose red copper has hitherto been almost solely used for conductors, either as single wires, or in cables formed of several wires twisted together, and either bare, or protected by insulating envelopes of various materials. For transmission to a distance which requires conductors of a relatively large diameter, and especially when this is insulated or placed in not easily accessible positions, the use of single wires has been almost entirely abandoned, for the following reasons :—

(1.) The conductors necessary for the passage of the currents for the Jablochkoff, Siemens, or Brush lamps, in a word for lamps producing the voltaic arc, are generally wires of 3 or 4 millimetres diameter, for moderately short distances. For incandescence arc lamps, such as those of Werdermann or Reynier, the diameter necessary is much more considerable. The dimensions requisite would render single wires very difficult to handle when it was desired to twist or bend them sharply, while wires that have been already used, cannot be easily straightened out and employed satisfactorily a second time. Ropes, on the other hand, formed of wire of ·7 mill. to 1·15 mill. in diameter twisted together to form a flexible conductor lend themselves to all the requirements of installation.

(2.) Another advantage is the facility with which such cables can be connected together at junctions. If it be desired to join together the two extremities of a thick wire of 3 or 4 mill. diameter,

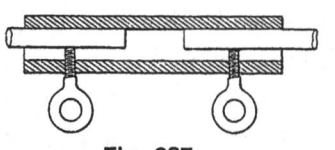
Fig. 327.

Fig. 328.

the junction is generally made by means of small brass sleeves, within which the ends of the wire are placed and fixed by means of two screws, after the surfaces of the wire have been cleaned with

Y 2

a file to remove the oxydation. This mode of fastening, which is illustrated by the sketch, Figs. 327 and 328, is also applicable to cables, but junctions in the latter are preferably made by means of splicing, which is effected by untwisting the strands of each end and weaving them together again in the ordinary manner. This has the advantage with insulated wires, that the splice can be protected with an insulated covering, which is very difficult to do efficiently with the sleeve and locking screw connection. Sometimes the splice is rendered further secure by immersing it in a bath of solder. But as a matter of fact, the passage of the current destroys the contact of the solder, and consequently makes it useless.

Such are the principal considerations that have led to the abandonment of single conductors in most lighting installations, their application being chiefly reserved for overhead and air insulated connections, and of course for small service leads.

Electrical conductors used in underground canalisation, or wherever they are placed against a wall or within reach, are always carefully insulated. There are two reasons for this. The first is that electricity, especially in currents of high tension, has the tendency to flow out at any points that give it an easy passage. The dampness prevalent on walls offers a great facility for leakage in conductors placed against them, the smallest contact with any moist object allowing it to escape; the hygrometric condition of the atmosphere that bathes non-insulated wires, also promotes the discharge of electricity into the air. These losses manifest themselves in oscillations of the light, in its more or less important diminution, and sometimes in the disappearance of the illuminating power. The second cause refers to the danger that exists in leaving conductors within the reach of persons ignorant of the elementary characteristics of electricity. Evidently this is a danger that may be easily exaggerated; every industry in operation presents danger, and, as Sir W. Thomson has said, there is more danger in a circular saw running at full speed in a workshop, than in a dynamo-electric generator of high electromotive force. But it is none the less true that prudence dictates the necessity of precaution, and insulation of conductors is of the first necessity.

The larger proportion of electric conductors are therefore protected by an insulating covering. The finer wires are enveloped in silk, the larger are either surrounded with gutta-percha, caoutchouc, or other compounds. There is also the very useful system of enclosing conductors in lead, as patented by Wheatstone as long ago as 1845, and revived lately by MM. Berthoud and Borel, whose process of manufacture will be referred to further on.

Gutta-percha is somewhat less costly than other insulating materials, but this advantage is counterbalanced by several serious defects. Gutta-percha becomes softened under exposure to very moderate heat, which tends to throw the wire out of the centre of the envelope, and reduce the insulation. Exposed to cold, on the contrary it hardens, then cracks, and sometimes flakes off, allowing damp to penetrate, and thus opens a passage to the current. Conductors protected with caoutchouc are much more efficient. They are usually made with an envelope of this material over the wire or cables, and are then wrapped with two coverings of rubber cloth, wound in opposite directions, in order that all the joints may be thoroughly covered. It is an ordinary practice to tin the wires before insulating them, in order to prevent any sulphurisation of the copper by the action of the vulcanised rubber. The only objection to this mode of protecting conductors is its price, which has greatly increased of late years, owing to the rise in the cost of caoutchouc.

Before we pass on to the question of cable installations we may briefly examine the physical conditions which control their usefulness, commencing with a glance at the laws that determine the flow and distribution of currents, in doing which, we must pass rapidly over some of the same ground we covered at the commencement of this volume.

The first and fundamental law of electric currents is that discovered by Dr. G. S. Ohm, and known as Ohm's law. It sums up the two causes which affect the strength of an electric current in the following statement: *The strength of a current is directly proportional to the electromotive force that tends to drive the current through the wires of the circuit, and it is inversely proportional to the resistance which the whole circuit offers to the passage of the current.* The law may be illustrated in the following fashion: Suppose either a dynamo-electric generator, or a battery of voltaic cells, be employed to send a current through a long line of wire and a series of lamps which offer a certain considerable resistance to the flow; to keep this flow going in spite of the resistance, requires a continual steady pressure, as it were, behind. The name of "electromotive force" is given to this particular power of the generator or battery, by virtue of which it tends to urge electricity through the circuit.   If through the same circuit it is desired to send a current of double strength, then twice as great an electromotive force must be applied, and we must double the existing electromotive force either by driving our dynamo-electric generator at a double speed, or by using a larger and more powerful generator, or in the case of a voltaic battery, by employing twice as many cells as before.   The standard by which electromotive force is measured is called "one Volt;" it corresponds in practice very nearly to the electromotive force of a single Daniell's cell.   The dynamo-electric generators in use for electric lights urge the currents forward with electromotive forces that vary from 50 Volts to 2,000 Volts, according to their construction, &c.   Within certain limits the electromotive force for a machine of given construction, is proportional to the speed at which it is driven.

The standard by which electric resistance is measured is, as we have seen, denominated "one Ohm," which is approximately as great a resistance as that offered by about 100 yards of ordinary telegraph wire. The arc or flame in an electric lamp may offer, according to circumstances, a resistance of from one to ten times as great.   The resistance of incandescence lamps, such as those of Edison and Swan, ranges from 130 Ohms in the former to 35 Ohms in the latter.   The unit or standard in which currents of different strengths are expressed is the "Ampère" (formerly called the "Weber").

An electromotive force of 1 Volt, when applied to drive electricity through a circuit whose resistance is 1 Ohm, produces therein a current whose strength is one Ampére.   The currents used in telegraphy may be as small as the hundreth part of an Ampère.   For producing an arc light the current must not be less than about 10 Ampères, and may be as great as 50 or more Ampères.   The current in incandescence lamps is usually a little more than 1 Ampère in strength.   A Swan lamp gives the best result with a current of 1·35 Ampères.

Ohm's law may be expressed exactly as follows : *The strength of the current in Ampères is found by dividing the number of Volts of electromotive force by the number of Ohms of resistance in the circuit.*

It follows from Ohm's law, that if an electric arc light begins to exert a higher resistance by its carbons burning away and leaving a gap between them, increased resistance must necessarily produce a decrease in the number of Ampères of current that flow through the circuit to the lamp, which must therefore shine less brightly unless some means are adopted either to lessen the resistance or to increase the electromotive force.

Another consequence of Ohm's law is that if a circuit branches into two routes through either or both of which the current can flow, the greater part of the current will pass through the route of lesser resistance. In fact, the strengths of the currents flowing in the two branches will be inversely proportional to the relative resistances of the two branches.

Ohm's law is the key to the distribution of electric currents, and is therefore of extreme importance to a clear understanding of the management of electric lighting.

Wherever an electric current meets with resistance to its flow, a part of the energy of the current is used up in overcoming the resistance, and this energy makes itself evident by the heating which always takes place at the point where the resistance is met with. A thin wire offers more resistance than a thick one, there being, as it were, less room for the currents to flow in. A short length of thin wire interposed in a circuit rapidly becomes hot, and may even be melted by the current. Now all wires, however thick, and of however good a conducting material, offer *some* resistance to the current, and therefore are to some extent heated. But with stout copper wires this is usually imperceptibly slight. The arc or flame between the two carbon pencils of an electric lamp offers a great and concentrated resistance and therefore is the seat of an intense development of heat. The thin filaments of carbon used in incandescence lamps grow hot because of the great resistance they offer to the current. Now as we do not want to have heat developed anywhere except in the right place, it is of great importance to make all other parts of the circuit—the coils of wire in the generator and its magnets, the conducting wires, and the coils of the magnetic apparatus that regulate the feed of carbon in the lamps—of stout copper wire, so as to offer as little resistance as possible, and to waste in useless heating as little as possible of the energy of the current. Edison has utilised the heating effect of resistance, by introducing into his system of lighting "safety catches" to prevent risk of fire by overheating. These safety catches are merely pieces of lead wire of a thickness suitable for the probable strength of current that must be carried, interposed in the circuit. If the current through them becomes by any accident too strong, the safety catch fuses and so breaks the circuit and stops the current. The amount of heat developed in any given resistance depends both on the resistance and on the strength of the current. The greater the resistance, the greater is the quantity of heat developed by a current of equal strength. Stronger currents develop more heat, the effect being proportional to the strength of the current multiplied by itself. A current of 1 Ampère running through a resistance of 1 Ohm, is found to develop therein about 15 calories or Gramme-units of heat every minute. A current of 2 Ampères, running through the same resistance, will produce four times as much heat; while a ten-fold current will yield a hundred times as much heat.

This important relation between the resistance, the strength of the current, and the heat produced is known as Joule's law. It may be stated thus: *The amount of heat developed by a current in passing through a resistance is proportional to the resistance, to the square of the strength of the current, and to the time that the current lasts.*

Recently some very careful investigations have been carried out by a physicist whose name is well known in France—Professor Joubert. The investigations of this gentleman to which we allude are connected with the conductibility, insulation and capacity of cable conductors. The specimens with which he experimented were of various types of manufacture. Two of them (A and B) were insulated with caoutchouc; a third (C) was covered with tarred thread; and a fourth type D was encased in lead, and made by MM. Berthoud and Borel.

The conductibility of a cable is the most important characteristic. It is measured, as we have seen, by the inverse of the resistance, which increases, as is known, with the length of the conductor, and varies in an inverse ratio to its cross-section. Resistance varies not only with the nature of the metal, but in conductors of the same metal it may alter considerably according to its purity; it also varies with the temperature, rising with the increase of the latter. The resistance of a wire of pure copper, 1 mill. in diameter, and 1 kilometre in length, is 20·57 Ohms at 0 deg. Centigrade, and that of a wire n mill. in diameter, at the temperature t is

$$\frac{20 \cdot 57}{n^2} (1 + 0 \cdot 0038 t) \text{ Ohms.}$$

Thus at 20 deg. Centigrade, a wire 3 millimetres in diameter and 1 kilometre long, of pure copper, has a resistance of 2·46 Ohms. It was at this temperature that the experiments of M. Joubert were made, with the following types of conductor :—

A ; Cables formed of 5 strands of 1·14 mill. diameter insulated with caoutchouc. Three samples of this were tried. B ; a similar type to the foregoing ; two samples tried. C ; cables of 7 strands, 1·06 millimetres diameter, insulated with tarred thread. D ; this was a conductor enclosed in lead by the Berthoud-Borel process. It was a single wire 3 mill. diameter, and insulated with a mixture of resin and paraffin.

Designation.	Resistance per kilometre in Ohms.	Difference from Pure Copper of 3 millimetres diameter.	Specific Conductibility : that of Pure Copper being 100.
A[1]	2·63	0·17	94
A[11]	2·54	0·08	97
A[111] (not insulated)	2·50	0·08	98
B[1]	2·54	0·08	97
B[11]	2·50	0·06	98
C	2·84	0·38	87
D	2·48	0·02	99

From the foregoing experiments it may be deduced, that a length of 400 metres of such conductors, which is the average length of a lead in a lighting installation on the Jablochkoff system, has a resistance equal to 1 Ohm.

To measure the resistance of the insulating material, the experiment consists in charging the cable at one of its extremities, the other being insulated, and the envelope remaining in contact with earth; the time required for the cable to lose one-half its charge is then recorded. This value being determined, the resistance R of the insulating material will be obtained by the formula $R = \dfrac{t}{\log. 2}$. The insulation of the cable D given above is very perfect; during an hour the conductor lost only $\frac{2}{100}$ of its charge; the resistance was in fact so high that it could not be expressed, for the length of 16 metres, with which the experiment was made. This class of conductor would therefore be perfect, if the great advantage of its excellent insulation was not counterbalanced by several inconveniences. Amongst others, a deficiency in elasticity, and the ready deterioration of the leaden envelope, would naturally bring about faults in the insulation.

The insulation value of the A and B cable was almost identical· They lost one-half of their charge in about 12 minutes. Their capacity being ·15 microfarad per kilometre, the resistance of the insulation for this same length was 6,000,000,000 Ohm's, or about 6,000 megohms. The insulation of the cable C was very bad, the charge having been lost in an inappreciable time, and the resistance of the insulant was measured by another process. It was found to be 125,000 Ohms per kilometre, or 40,000 times smaller than those of cables A and B. This insulation is, however, sufficient. Indeed, the resistance of the metal core being 2·5 Ohms per kilometre, that is to say, 50,000 times less than the envelope, the current would be divided between the conductor and the insulant, in the proportion of $\frac{49,999}{50,000}$ passing through the former, and $\frac{1}{50,000}$ lost through the latter. The capacity of a conductor is the quantity of electricity it takes when it is in communication with one of the poles of a Daniell's cell, the other being in connection with earth. This capacity may become very high, as in the case of condensers, when the conductor is separated by a very thin insulating piece, from another conductor connected with the ground.

With continuous currents the influence of capacity is insignificant; when once the cable is charged it becomes nothing. But this is not the case with alternating currents; at every reversal, a certain quantity is necessary to give to the conductor the new charge in the opposite direction; that is to say, that at every reversal an absolute loss takes place of a quantity of electricity equal to double that which corresponds to the capacity. If we assume 500 changes in direction per second, the loss is one thousand times the charge corresponding to the capacity, or that corresponding to a capacity one thousand times greater.

From this point of view, the Berthoud-Borel conductor, with its lead sheathing, ought to possess marked disadvantages. Its capacity has been fixed at ·273 microfarad per kilometre, which is largely above that of a kilometre of transatlantic cable (·18 microfarad). Conductors insulated with caoutchouc have a capacity of ·15. This result appears surprising, because an insulated cable without a conductive envelope has usually an insignificant capacity, as in the case of an ordinary caoutchouc covering. But the conductors in question are, as we have seen, covered with a double sewing of rubber cloth, which to a certain extent may act as a conductor. The capacity is less than in the Berthoud-Borel conductor, because the insulation is thicker. As to the practical importance of the foregoing results, In taking ·20 microfarad as a mean capacity, and 500 as the number of reversals per second of the alternating current, the quantity of electricity lost each second, and for each kilometre of cable, will be that corresponding to a capacity of 200 microfarads ; this is one-third the capacity of the French Trans-atlantic cable ; one-third also of the capacity of a sphere as large as the earth. But, as compared with the quantity of electricity that traverses the cable every second to produce a useful effect, this amount, which appears enormous, is absolutely insignificant, being only one two-thousandth part.

The influence of capacity is felt also in another manner. Let us assume that we pass through a cable, a current of short duration. This current starting from zero, arrives at a maximum, and returns to zero. The curve of intensities is more pointed as the duration of the current is shorter, and its maximum intensity greater. This electric wave is propogated with extreme rapidity, but constantly altering in form, which is always represented by a tooth-shaped curve, the base of which widens, and the maximum ordinate shortens. On arriving, the current, instead of being instantaneous as at the point of departure, has a duration which varies, all other things being equal, as the square of the length of the cable traversed. If a wave in the opposite direction follows the first too closely, they partially neutralise one another on arriving, and give an effect equal only to their difference. The question to be solved is, therefore, as follows : With a given conductor, at what intervals should the pulsations of an opposite nature succeed each other, in order that they should not counteract one another, and thus partially reduce the useful effect ? or reversely, with alternating currents succeeding each other at given intervals, what is the maximum length that can be given to the conductor, without the natural interference of the pulsations ? Admitting, as we have assumed already, and which is practically the case with the dynamo machines employed, 500 reversals of the current per second, a capacity of ·20 microfarads, and a resistance of 2·5 Ohms per kilometre of conductor, it will be found by calculation too long to reproduce here, that the effect in question only commences to show itself in a circuit of 58 kilometres.

We cannot within the limits of a general review like the present enter into the details of the manufacture of electric conductors, but must confine ourselves, in order to show the importance of this new industry, to giving a short notice to some of the industrial establishments, whose production now reaches an enormous amount.

Among the French manufacturers of conductors, we should first mention MM. Rattier & Cie., whose business has recently been acquired by the Société Générale des Telephones. This establishment dates from 1828, and during the first 26 years of its existence was occupied solely in the manufacture of caoutchouc. In 1854, that is to say almost as soon as the valuable qualities of gutta-percha were realised, the Rattier establishment added to its caoutchouc the preparation of the newer material, and was the first in France to manufacture telegraph wires with gutta-percha. In 1858 it made the first underground telegraph cable laid in Paris, and in 1861 constructed the first submarine cables that were employed to connect France and the various outlying islands. Later, when the development of science brought about the employment of electricity for lighting purposes, MM. Rattier & Cie. were among the first to manufacture as a matter of every-day trade, cables specially for that purpose, insulated either with gutta-percha or caoutchouc, and since that date have continued to supply these conductors on a large scale.

Besides the Rattier establishment, may be mentioned that of M. Menier, which is of great importance, and the works of MM. Berthoud, Borel and Co., who make a special kind of conductor, patented in 1845 by Sir Charles Wheatstone, re-invented and patented some years later, and very recently brought forward again with much success.

The early experiments of M. Borel, who revived this system of conductors, were devoted to the production of a cable with a conducting wire of lead or tin, made as shown in the annexed sketch,

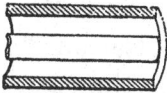

Fig. 329.

Fig. 329, where a lead or tin rod was placed within a leaden tube; the annular space between the two was filled with an insulating material such as sulphur or resin. The structure thus formed could be drawn out to any desired degree of fineness, and would produce a conductor in which the initial proportions would be preserved, while the insulation was sufficiently good, the protective material not being destroyed, even in bends of the conductors. With colophane as an insulator it was found that a conductor on this system lost only one three-hundreth of its electric charge in four hours, and that, charged with static electricity, it retained sufficient after 10 days, to affect the gold leaf of an electroscope. However, it was soon realised that this arrangement was impracticable, partly on account of the low conducting power of lead, and partly on account of the insulating material being reduced to dust, and thus brought into an unstable condition. Consequently M. Borel, while preserving the same principle, arrived gradually at a type of cable widely different.

In the form now manufactured by MM. Berthoud and Borel, at their works at Grenelle, Paris, and Cortaillod (Canton de Neuchâtel, Switzerland), the conductor is formed of one or several copper wires, covered with several thicknesses of cotton wrapped in opposite directions; that is to say, if the first covering is wound from left to right, the second will be from right to left, and so on; in this manner the spaces between the wires are well closed. This first operation is effected by means of a special machine. The covered wire thus obtained, is rolled on a reel and plunged into a bath containing a melted insulating material, which is generally a mixture of paraffin and colophane, kept at a temperature of 392 deg. Fahr., in order to remove the moisture from the cotton, and to make the insulant penetrate the structure of the latter to improve the insulation. The wire or wires thus protected are then encased in lead. Figs. 330, 331, and 332 show the general arrangement and details of the machine employed at Cortaillod. A lead ingot is compressed by a hydraulic cylinder, formed into a tube and drawn over the wire coated with the cotton and paraffin as already described. The lead ingot is placed in the chamber P, which it fills completely; it has an axial hole equal in diameter to the exterior diameter of the tube G. The piston F is forced upwards by the hydraulic press placed below the apparatus; its diameter is equal to that of the ingot. The tube G is connected firmly to the piston F by cross-pieces. At its lower end is a conical piece (Fig. 331) which determines the interior diameter of the envelope. The tube G, which receives the conductor, is surmounted by the recipient A, in which is a bath of melted insulating material, which is allowed to enter freely the interior of the tube by means of the opening *h*. The stopper *i* serves solely for emptying the recipient when the operation is stopped. The vessel B is filled with hot oil, which maintains the temperature of the insulating material and preserves it in a liquid condition. From the recipient B, it descends into the envelope E by a tube, and thus heats the lead chamber, flowing off by an overflow opening. The chambers C and D are filled with coke dust, or with some other bad conductor of heat. The process of manufacture will be easily understood.

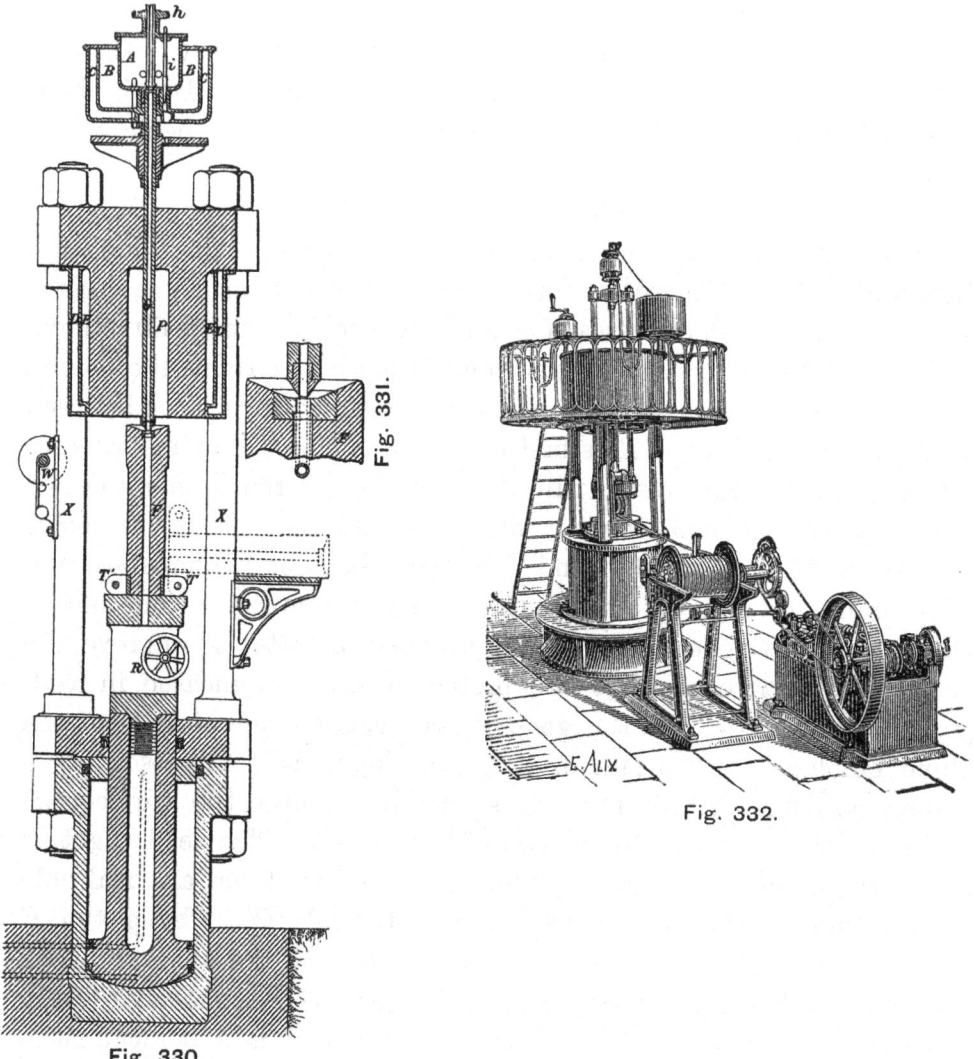

Fig. 331.

Fig. 332.

Fig. 330.

The wire covered with cotton and rolled on its reel is plunged into the bath of paraffin and colophane; it is then slowly unrolled, and passes down the tube G, where it receives a second charge of melted insulating compound. The piston F forced upwards, compresses the lead cylinder, which can escape only through the narrow annular space between the exterior part of the tube G and the diameter of the hole which traverses the piston. It is thus drawn into a tube which exactly covers the cable. The apparatus is so arranged that the speed with which the cable is unrolled is equal to the speed with which the drawn lead flows from the compression chamber. The finished product passes over the small pulley R, and is then rolled upon drums.

The lead is heated, as described, to facilitate drawing. Experiment showed that when cold this operation required a pressure of nearly 57,000 lbs. per square inch, but that when raised to a temperature of 250 deg. Fahr. it required only from 20,000 lbs. to 40,000 lbs., according to the thickness of tube produced. To facilitate placing the lead cylinder in the compression chamber, and changing the matrices required for different kinds of cable, the piston F can be turned on the bearings T so as to be placed horizontally on a bracket, as indicated in the section by dotted lines. To replace the piston, a small winch W is used, fixed to one of the columns of the press.

The conductors thus completed can only be used in the air. If it is intended to place them in the ground, precautions have to be taken to protect the lead from deterioration and also to secure it from the attacks of certain boring insects. For underground work the cable, made as above described, is enclosed into a second casing of lead, and tar is forced in the space between the two.

The speed with which this class of conductor can be made varies with the type; for example, a cable with three wires ·5 mm. diameter within a lead envelope 4 mm. (·16 in.) exterior diameter, is completed at a speed of 50 ft. a minute in regular working, though a maximum of nearly 150 ft. can be made. At the Cortaillod works one machine is in operation, and two are being constructed; at Paris there are two machines in operation. The perspective view Fig. 332, shows the general arrangement of the machine with its platform, from which the whole series of operations can be watched, with the reel on which the cotton-covered cable is rolled, and the three-cylinder pumps which work the compressor. These pumps are driven by a small 4 horse-power engine, 2½ horse-power being sufficient for the operation of drawing the lead tube.

MM. Berthoud, Borel and Co., make 17 different types of cable for telegraphic, telephonic, bell, and lighting purposes; differing in the nature and arrangement of the conductors and the thickness of insulating material. They are all protected in the same manner by an outer tube of lead separated from the inner by a thickness of tar. The

cables for bellwork, however, are not enclosed in the outer casing. In
the first series the inner lead tube has a standard thickness of ·0295 in.,
the outer tube is ·04 in., but these thicknesses can be varied at will. The
length of cable produced in one piece is limited only by the size of the
compression chamber of the machine. Where joints are required they
can be very easily made; the ends of the wires are bared, then joined
and covered with cotton, which is saturated with insulating material. A
thin piece of lead is then wrapped around and soldered along its edge
and at the ends to the enclosing tubes.

The following is the classification of sections made by this process,
and the annexed diagrams, Fig. 333, show the disposition of the conducting wires :—

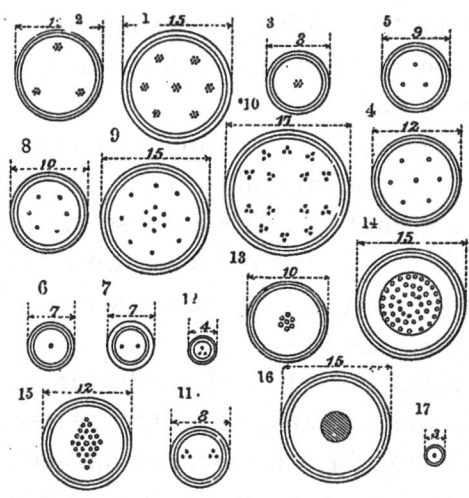

Fig. 333.

No.	1. Telegraph conductor,		7 cables of 7 wires each of		·5 mm.
No.	2.	,,	,,	3	,,	7	,,	·5	,,
No.	3.	,,	,,	1	,,	7	,,	·5	,,
No.	4.	,,	,,	7 single	,,	·9	,,
No.	5.	,,	,,	3 ,,	,,	·9	,,
No.	6.	,,	,,	1 ,,	,,	·9	,,
No.	7. Telephonic	,,	2 ,,	,,	·7	,,
No.	8.	,,	,,	6 ,,	,,	·7	,,
No.	9.	,,	,,	14 ,,	,,	·7	,,
No.	10.	,,	,,	14 lines of 3	,,	·5	,,
No.	11.	,,	,,	2 ,, 3	,,	·5	,,
No.	12.	,,	,,	1 ,, 3	,,	·5	,,

No. 13. Lighting cable   7 strands   each of   1·14 mm.
No. 14.   ,,   ,,   37   ,,   ,,   1·14   ,,
No. 15.   ,,   ,,   12   ,,   ,,   1·14   ,,
No. 16.   ,,   ,,   1   ,,   ,,   5·12   ,,
No. 17. Bellwork line   1 conductor

The prices of these conductors vary from about £10 a mile for the last-named section, to 17 times that sum for a cable like No. 14.

The properties of this class of conductors call for a few remarks. The insulating material employed is more perfect than either rubber or gutta-percha, an insulation of over 30,000 megohms per kilometre being easily obtained at an ordinary temperature. The process of manufacture proves that it can sustain a high temperature without damage, so that it can be laid with safety under tropical suns. All these facts, combined with the comparative cheapness of the conductor, should assure for MM. Berthoud, Borel and Co. a large demand for their speciality. At the recent Exhibition of Electricity, held at the Palais de l'Industrie, about 17,000 yards of their cables were utilised for the transmission of currents for the production of force and of light. A cable with a number of conductors, connects, through the sewers, the Ministère des Postes et Telegraphes with the Exhibition. The General Telephone Company has also some miles of cable with 14 conductors laid in the Paris sewers.

Reference should here be made to some of the English systems of conductors, and of their manufacturers. Copper, as we have already seen, is practically the only material employed for electric light conductors; but even now Mr. Lane-Fox is employing round iron rods, an inch or more in diameter, insulated in longitudinal wooden troughs filled with pitch; and Mr. Edison, we understand, proposes, scarcely with seriousness we should imagine, to utilise old iron rails for the purpose should the half-round copper bars which he now employs, prove too expensive.

The ozokerited core manufactured by Mr. W. T. Henley's Telegraph Works Company, at North Woolwich, is a departure from the time-honoured gutta-percha core adapted not only for telegraph and telephone cables, but for electric light leads. In practice, ozokerited core is made by curing the ordinary india-rubber core in melted ozokerit or paraffin under pressure, or by super-heated steam, and then placing it in a heated chamber, still under pressure, until all moisture has evaporated. Next, it is heated in hot ozokerit which is forced through the various coats of rubber on the core, stopping the pores and rendering it impervious to moisture. The insulation resistance is very high and runs to 10,000 or 12,000 megohms per nautical mile at 75 deg. Fahr.; gutta-percha being 800 or 1,000 ordinarily. The following tests taken for Mr. W. H. Preece, electrician to the Post Office, will serve as a comparison. For ten miles of ozokerited core supplied to the department by Mr. Henley, the results were :—

Conductor resistance	23·723 Ohms.
Capacity	·281 microfarads.
Insulation resistance	8,196 megohms or millions of Ohms.
Temperature	75 deg. Fahrenheit.

For ten miles of similar core insulated with gutta-percha the results were :—

Conductor resistance	22·36 Ohms.
Capacity	·273 microfarads.
Insulation resistance	860 megohms.
Temperature	75 deg. Fahrenheit.

Its imperviousness to moisture fits it for underground positions, and it is less liable to become soft by heating than gutta-percha. The core has given satisfaction as a submarine cable, and there appears no reason why it should not prove equally successful as an electric light conductor, especially as it is cheaper than gutta-percha. This cable has been used by Mr. R. E. Crompton in fitting up the s.s. " Cotopaxi " with incandescence lamps, and appears to give no trouble to the electricians employing it.

The " Nigrite " core invented by Mr. Price, and now manufactured by Messrs. Latimer Clark, Muirhead and Co., of Regency-street, Westminster, is another combination of ozokerit and india-rubber (or gutta-percha); but in this case the union is more intimate. The ozokerit or black earth wax is, in fact, melted and kneaded with the rubber into one mass, which can be coated on the wire like gutta-percha by means of a die or draw hole. Nigrite core has a high insulation resistance and is cheaper than gutta-percha.

A third application of ozokerit exists in the new electric light leads manufactured by the United Asbestos Company, of 161, Queen Victoria Street. These consist of copper wires and strands covered with a mixture of asbestos and ozokerit. The asbestos is not only an excellent insulator but a non-conductor of heat. Wires entirely insulated with asbestos in the form of a braid are also manufactured by them; and other kinds have a skin of india-rubber or tape over the asbestos braid. A fourth variety consists of asbestos braided wire wound with india-rubber, and again covered with a braid of asbestos. The lining of asbestos tends to keep the undue heating of the conducting wire from spoiling the insulation. The United Asbestos Company also make a very useful insulating tube of asbestos and ozokerit for enclosing electric light cables. Altogether there appears to be a promising field for this new insulator.

Messrs. Walter T. Glover and Co., Booth Street, Manchester, the well-known makers of insulated wire for the Post-Office and the telephone and railway companies, supply very excellent electric lighting conductors as well as indoor leads of every kind. A novel and highly serviceable feature in their cores is a covering or skin of hemp and preservative pitch over the insulator. In addition to this there is also a longitudinal warp added. This covering gives a lengthened life to the core and guards it from injury when coming against the corners or sharp edges of pipes in drawing it through. Moreover, even if it should be cut it does not ravel off like the tapes usually employed to cover cores. The sizes most in demand by the principal electric lighting companies are the following :—

Number.	Size. B. W. G.	Resistance per Statute Mile of 1,760 Yards.
		Ohms.
1	7 strands No. 18	3·62
2	7 ,, ,, 16	2·22
3	19 ,, ,, 22	3·15
4	19 ,, ,, 18	1·20

The first two types are chiefly used by the Anglo-American Brush Electric Light Corporation employing the Brush and Lane-Fox lights.

The third type is mostly used by the British Electric Light Company (Brockie system), Messrs. R. E. Crompton (the Crompton system), and for some Swan installations; while the fourth type ($\frac{19}{18}$) is also used for the Brush lamps, the Swan lamps, and by the Electric Power and Generator Company in connection with their Weston and Maxim lamps. This type was employed in lighting up the Pleasley and Risca collieries.

Other makers may also be mentioned: the Gutta-percha Company, of Wharf Road, City Road; Mr. Armand Levy, of Goswell Road; the Electric Lighting Supply Company, of Queen Anne's Gate; Phillips Brothers, of Macintosh Lane, Homerton; and Messrs. Siemens Brothers, of Charlton. The conductor made by Messrs. Phillips is of vulcanised rubber, the finer service leads being covered with paraffined cotton. Messrs. Siemens Brothers manufacture a great variety of lighting conductors insulated with gutta-percha and india-rubber. Strands of 7, 14, and 19 wires are insulated with tape and india-rubber, or with two layers of gutta-percha sheathed either in oakum and tape, or iron wires and compound, according to the requirements of the user. The iron-sheathed conductors are of course better calculated to withstand violence than those protected simply by tape or yarn.

Stoneware spigot and faucit pipes for containing electric light conductors when laid underground, are. made by Messrs. Doulton and Co., of Lambeth, and manholes are provided for getting at the wires for repairing purposes. Light, strong and rigid pipes of steel are also manufactured for the same purpose by Messrs. Exton, Berridge and Partners, of Page Street, London. These pipes have a joint which any labourer can quickly detach in order to get at the wires to effect repairs; while they are so tight-fitting that they can be used for conveying gas, water, or steam. The box street curb of Mr. W. Reddal, of South Street, Finsbury, is also worthy of remark. It is designed to utilise the space ordinarily taken up by a solid curbstone for holding electric light leads. The curb box is of iron, and the lid can be readily taken off so as to obtain access to the wires threaded through the boxes placed end to end.

We may now pass on to examine the practical conditions under which electric conductors are ordinarily laid, with a few preliminary words on the different classes of circuits.

When more than one electric lamp is to be worked by the current some mode of grouping the lamps and of connecting them with conducting wires must be adopted. The two principal methods of grouping in common use are respectively denominated " simple circuit " and " compound circuit."

The lamps are said to be arranged " in simple circuit " or " in series " when the current is led (see Fig. 334) to one lamp, then through another, then through a third, and finally return to the generator.

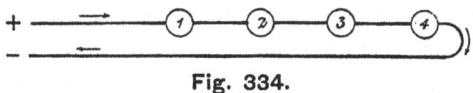

Fig. 334.

The lamps are said to be arranged " in compound circuit " " in parallel arc,"* or in multiple arc when they are placed as bridges across the circuit from a main leading wire to a main return wire (see Fig. 335). This system, which may be said to resemble the circulation of blood in the human body having a main artery and a main vein with many fine

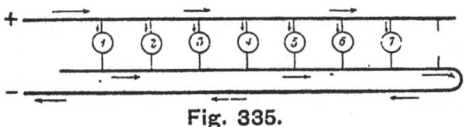

Fig. 335.

capillaries, is adopted for incandescence lamps such as Edison's, each lamp being placed across the circuit. To work this arrangement satisfactorily there should be an equal amount of current flowing through each of the parallel arcs or lamps. This can only be attained by making the resistances of all the lamps equal, a result easily practicable with incandescence lamps, but more difficult with arc lamps where the resistance in the arc increases as the carbons burn away.

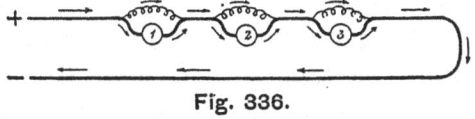

Fig. 336.

A third method of grouping is frequently adopted, in which the lamps are connected " in derived circuit," or " in derivation." English electricians sometimes denominate this grouping as an arrangement " in shunt circuits." Fig. 336 shows the general mode of carrying out the

---

* Or " in quantity;" a foolish expression and unmeaning, that has survived from the days when electricians did not understand Ohm's law.

idea. The current passes, as in simple circuit, first to one lamp and then to another; but instead of the whole current flowing through the lamp, a portion only is sent through it, another portion being led past it by a thin wire forming another route of higher resistance. The lamps may be regarded as being each set in loop lines *derived* or branched from the main leading line. A loop-line or branch of this kind is in telegraphy called a "shunt." It would be equally correct to regard the lamp as being placed in a shunt from the main line, or to regard the lamp as being placed in the main line, and the thin wire as a shunt of high resistance. When a current has thus two routes before it, it divides, the amount of current going by each route being, as stated above, inversely proportional to the resistance of that route. If the resistance of the thin wire shunt be equal to that of the lamp, half the current will flow by each route. If the resistance of the thin wire shunt is ten times as great as that of the lamp, ten-elevenths of the whole current will flow through the lamp, and one-eleventh through the shunt. This is more nearly the kind of proportion observed in practice. The advantage of this mode of grouping is that the failure of one lamp does not involve the extinction of the rest, as there is still a route open for the current. Another advantage is that the regulating apparatus of the lamp—the electro-magnet or solenoid which controls the mechanism by which the carbons are fed into the arc—may be placed in the shunt part of the circuit, the result of this arrangement being as follows: If the resistance in the lamp becomes too great owing to the carbons burning away, and leaving too long a gap between them, a larger proportion than before of the whole current will take the shunt route, and will at once bring the regulating mechanism into play and re-adjust the lamp. In the Siemens, Brush, Crompton, Weston, and many other systems of electric lighting this method of grouping is adopted.

Where more than one generator is used to supply a circuit, the generators must be connected together in such a way as to cause them all to drive their currents into the leading wire. The question as to how they should be most advantageously grouped must be determined by their construction, and by the circumstances of the case. It is possible to group generators either in simple circuit or in parallel arc,* the former

---

* That is either for "intensity" or for "quantity," as the older electricians used to say before they understood Ohm's law. The former case gives currents of great electromotive force; in the latter the internal resistance is small. Since an "intense" current and a "strong" current mean the same thing, the old terms should be allowed to die—the sooner the better.

method being the right one where a great electromotive force is required to overcome a great resistance (as of a lot of lamps in series), the latter being right where no great electromotive force is required, but where the resistance of the external circuit is already small. In general it is safe to say that the generators should be grouped in the same manner as the lamps in the circuit. If the lamps are "in series," either "simple" or by "derivation," then group the generators "in series." If the lamps are "in parallel arc," bridging across from the leading main wire to the return main wire, then set the generators also "in parallel arc" across the circuit.

We have already said that under certain circumstances uninsulated wires may be employed. This has no other advantage than that of economy, and the wire may be laid in the same manner as ordinary telegraph wires, with similar posts and porcelain insulators. When it is possible, higher posts should however be employed, in order to protect the wires from all imprudent or malicious handling. At the points where the wires are near the generators or the lamps, and in fact, wherever they are easily accessible, it is necessary to replace them with insulated conductors. Naked wires should never be used in damp climates, for the moist air surrounding them causes a continual loss of electricity along the whole length of the line. From the foregoing it will be seen that the applications of uncovered wire for electrical lighting are very restricted, and that almost universally, conductors clothed in some kind of insulating material should be used. A cable commonly used for regulators and electric candles, is composed of 7 strands of copper, 1·14 mill. in diameter; for incandescence lamps burning in air 34 or 35 strands of similar wire are necessary. For the small incandescence lamps, such as those of Maxim, Swan, Edison and others, single wires of from 1 to 2 millimetres diameter are used. These sizes are approximate for average distances ; it is evident that they cannot be absolute, and that they would have to be increased or diminished according as the length of the circuit is great or small, and according as the lamps are arranged in parallel arc or in series. As a rule the resistance of the leads should not exceed ten per cent. of that of the lamps. Ten Swan lamps in series have, when in action, a resistance of about 350 Ohms, consequently the two leads may be about 35 Ohms with a very wasteful expenditure of power. Ten similar lamps in parallel arc would have a resistance of 3·5 Ohms, and their leads a resistance of ·35 Ohms.

From the conditions that the resistance of a Swan lamp is 35 Ohms, and that the current required to raise it to the proper temperature is 1·35 Ampères, it follows from Ohm's law, $1 = \dfrac{E}{R}$, that the differences of potential at the two ends of the filament must be 46·25 Volts. These data being determined it is a simple matter to calculate the particulars of the current required for each special installation.

Insulated cables may be carried, like naked wire, as we have said, on posts and insulators, or they may be fixed to any convenient support, the greater portion of the cable hanging free. This is the ordinary workshop system: in internal installations, the conductors are laid side by side along the walls, secured by staples at convenient intervals, exactly in the same manner as ordinary gas pipes. Care has to be taken not to drive in the staples with too great force, lest the curved portion should cut through the insulation, and establish a contact between the conductor, the hook and the wall, and so cause an important waste of current. Staples covered with an insulating film of enamel are manufactured; they are, however, more costly, and have not hitherto been largely used.

It is generally found preferable to arrange the wires parallel to each other, and not to employ cables containing two or a larger number of separate conductors; the single wires are more expensive, but the use of compound cables is accompanied with the risk of induction between the wires. When several conductors, traversed by alternating currents from independent generators, are employed in one building they must be kept at considerable distances apart, or the main lead and return lead from each machine must be twisted together. The reason is easy to see: at one moment the currents from two generators will be pulsating in unison; at the next moment, in consequence of the slight variation in speed that always exists between machines driven by belts. the alternations will be in opposition, and the character of the inductive effects entirely changed. Such variations are distinctly visible on incandescence lamps. When it is necessary to lay conductors in the ground, several precautions have to be taken. If, as in Paris, it is possible, under municipal authorisation, to make use of the system of sewers for placing the conductors, the problem is relatively a simple one, since the wires can be secured against the sides or top of the conduits, in the same manner as in buildings. But these facilities are exceptional, and most frequently it is necessary to lay the conductor in trenches made in the ground. Wires insulated only with caoutchouc, last but a very short time under these conditions. Even by surrounding them with a protecting envelope of lead, only a comparatively short existence can be relied upon, as the lead frequently corrodes rapidly, and exposes the insulation. This has happened in several instances, as at the Docks Station, in Antwerp, during the early stages of the installation with ordinary conductors covered with lead; and at Havre, with the Berthoud-Borel cables. In some cases a sheathing of iron wire somewhat similar to that used for submarine cables, is employed. The current is transmitted at the outer port in Havre by such a conductor, and has hitherto answered very well.

Again the wires may be protected within earthenware pipes, as already mentioned, securely jointed together; insulating discs are placed within these pipes at intervals of about 18 in., and the wires are passed through them, as shown in the sketch (Fig. 337). The necessary

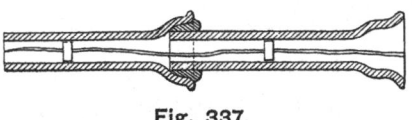

Fig. 337.

connections in the Avenue de l'Opera, in Paris, are made in this manner, and have now been down more than four years. Joining up the two ends of a cable may be done either by means of a coupling, or by splicing, as we have already indicated. In both cases it is necessary to cover the point of junction with a serving as strong as the insulating covering. If the joint is made with a sleeve coupling, the screws are turned as tightly as possible with a pair of pliers, and then the heads are filed down, so that there may be no projection on the sleeve to interfere with laying on the serving properly. Fig. 338 shows a T-shaped sleeve coupling of this nature. A very ingenious coupling used by the Brush Electric Light Company, is illustrated in Figs. 339 to 342, and is especially useful for joining up short lengths of conductor. The positive coupling has a small pair of trunnions formed on the outside, and carries a plunger that is pressed forward by a spiral spring; the negative coupling is provided with a pair of hooks and a recess in the body to receive the end of the positive plunger. In joining up two lengths, the positive end

Fig. 338.

is held slightly inclined, and the plunger is pressed against the negative end until the trunnions clear the hooks; when released the spiral spring forces the plunger forward and presses the trunnions against the hooks.

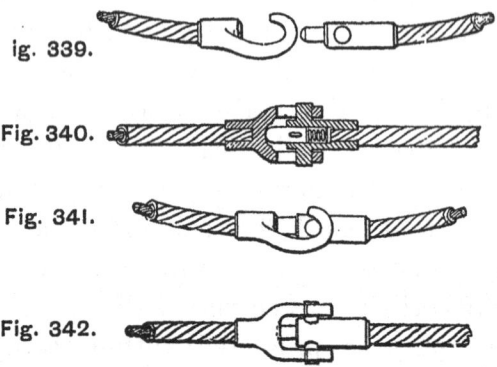

ig. 339.

Fig. 340.

Fig. 341.

Fig. 342.

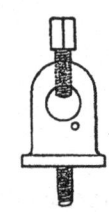

Fig. 343.

Connection of the wires to lamps, generators, &c., is effected by means of binding screws or terminals, of which a general type is shown in Fig. 343. They are secured to the piece to be connected electrically by the lower screw; the end of the conductor, previously stripped of its insulating envelope, and cleaned with emery paper, is passed through the opening o, to which it is held fast by the screw v, which has either a milled head for turning by hand, or a square head, as shown, to be turned by means of a key.

Generally the lamps are placed in closed and distinct circuits, commencing at one terminal of the generator and ending at the other. Sometimes with a view to economy, one return wire is employed for several circuits. Successful experiments have also been made of completing the circuit by leading the cable to earth. To do this it is necessary that the extremity of the circuit, and the end of the cable attached to the negative terminal of the generator, are in thorough contact with metallic surfaces buried in the ground or placed in a pit.

A very favourable example of this method of conduction is to be seen at Holyhead. The new harbour, the property of the London and North Western Railway Co., is lighted by five large lamps fed from five dynamo generators situated in the engine-house in which the pumps for actuating the whole of the hydraulic machinery of the cranes, turn-tables, &c., are fixed. From this building, high-pressure mains branch in several directions, and three which lie along the quay sides are utilised as the return leads from the lamps. The distance from the generators to the farthest light is more than half-a-mile. A similar plan is often adopted in steamships lighted by incandescence lamps. One terminal of the generator is connected to the nearest angle iron on the vessels skin, and from each lamp a return wire is led to a conductor which is put in electrical contact with the metal work of the ship at frequent intervals. This arrangement is followed in the "City of Rome," the "Alaska" and the "Orient," in all of which alternating currents are employed.

The Lyons Station in Paris has been lighted for a long while by the Lontin lamp in this manner, the extremities of the wires being soldered to the iron columns carrying the roof. In the Jablochkoff installation on the Thames Embankment, experiments were carried on during a considerable time, to complete the circuit by means of the river. In certain trials made by the Jablochkoff Company in Paris, the return was effected by large cast-iron pipes, sunk about 45 ft. into the ground, or some 5 ft. below the top of the water bearing level. So far, however, this method has met with no general application, but it is of sufficient importance to merit further and careful investigation, because a successful solution would permit of a large economy in conductors, which will form so large an item in first cost, when central stations for distribution of current are established.

We may notice here a form of American conductor recently devised, which will doubtless come into extensive use. It is patented by Mr. Edison, and is illustrated by Fig. 344, which shows the conductors

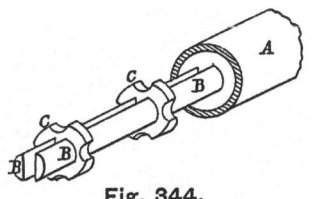

Fig. 344.

employed and the method of laying them ; B B are two D-shaped rods of copper held apart by washers C C of Manilla pasteboard and enclosed within the iron tube A. The vacant space is run up with asphaltum.

To complete this very general study of conductors for electric lighting, some necessary apparatus connected with the installation of lamps may be referred to. In the first place, there is the commutator, distinct from the device of the same name which forms so important a feature in the electric current generator. The commutator for conductors plays the same part in the distribution of currents, as do the taps of water and gas mains, with this distinction, that while the electric current flows when the commutator is closed, the water or gas passes when the tap is open. Assume a circuit A B, C D consisting of

A_____B C_____D

two wires, one being a prolongation of the other, and that it is desired to establish or interrupt at will the continuity of the circuit. This can be done by introducing between B and C a commutator arranged, for example, like that of which a drawing is given in Fig. 345. It consists of

Fig. 345.

a circular block of wood carrying a brass key which is in connection with the conductor bringing the current. This key, extended through the centre of the disc, is provided with a pivot that can be turned, and takes in its movement a tongue-piece that can be brought at will into connection with the various keys communicating with the wires, through either of which the current can flow, when the connection is thus completed. Effectual contact is secured by a steel spring between the movable tongue and the keys. Generally, the commutator carries an intermediate key, on which the tongue-piece rests when the current is interrupted. The figure explains all these details quite clearly. It will be seen

Fig. 346.

that the number of keys, that is to say, the number of directions in which the current may be sent, is only limited by the size of the commutator. Fig. 346 is a two way commutator, in which the current arrives by the contact A; B, B[1] are the two direction contacts, and C is an insulated stop. Some other forms of commutators will be found illustrated and described in another part of this volume.

Lastly, we may refer to the resistances which it is sometimes necessary to introduce into electric circuits, either for modifying at will the total resistance, or to establish at certain times resistances equal to those of lamps that may be cut out of the circuit. Thus, when alternating currents are used, it is often useful to vary the resistance of the circuit between the exciting machine and the distributing generator; such a variation is produced by means of a device such as is shown in Fig. 347. With this it is possible to modify,

Fig. 347.

while the machine is at work, the length of the intermediate circuit, and so to vary the extent of magnetisation of the field magnets. The commutator may be introduced at any point of the circuit, by fixing the ends of the wire in the terminals A and A. The terminal B is connected to the terminal 0 by a very short wire of low resistance; then between the contact 0 and the contact 1, a resistance R is introduced, usually formed of 2 metres of insulated copper wire, of 2·2 millimetres diameter, for the auto-exciting Gramme generator No. 1. (This is the type for feeding 4, 6, or 8 Jablochkoff candles). For a No. 2 machine feeding from 16 to 20 candles, the length is 3 metres. As will be seen from the figure a similar arrangement is introduced between each of the contacts 1, 2, 3, 4, and 5. The working of the apparatus will be easily understood. If the movable contact A^2 be placed on the contact *o* of the commutator, the minimum of resistance, and the maximum intensity are obtained. If the movable contact be placed on 1 the resistance is a little greater, and the current somewhat less intense, and so on up to contact 5, at which the resistance is a maximum and the intensity a minimum.

It will be seen that this little apparatus is susceptible of many modifications; the copper coils can be replaced, for example, with wire of other metals, and a large range of resistance obtained. By this arrangement a lamp in circuit, that has been extinguished can be at once replaced, either by hand or automatically, by an equivalent resistance, so that the electrical balance is not affected in the generator, the conductor, or the remaining lamps. Resistances are also employed, formed of carbon pencils fixed in metallic terminals and placed in a frame.

VIII.

CARBONS.

THE manufacture of carbons for the purpose of electric lighting has already grown into an industry of great dimensions, and is one that appears to be steadily on the increase. Several of the principal electric light companies manufacture their own carbons, such as the Brush, Messrs. Siemens & Co., La Compagnie Générale d'Electricité, &c. Then there are independent manufacturers of carbons on a large scale, chief of whom are Carré and MM. Sautter, Lemonnier & Co., of Paris. The process of production has received almost as great a development as has the generation of currents, since 1810, when Davy showed his historical experiment with the rapidly wasting charcoal pencils. Long, however, before the construction of any practical machine for producing currents, the pioneers of electric lighting had set to work to elaborate the manufacture of suitable carbons and this, like all else they did, was little short of marvellous in its anticipation of the practice of to-day. To give precise details of the mode of producing carbons, followed now by some of the principal makers, is out of the question, as the secret—if there be one—is jealously kept. We will, however, trace very rapidly what there is of historical interest in this matter, and give such data about current practice as are available; in one of the important instances these are fortunately very complete.

W. Greener and W. Edwards Staite patented, in February, 1846, a mode of manufacturing electric carbons, as follows:—They took lamp-black or powdered purified coke, which they digested in a weak solution of nitro-muriatic acid and afterwards washed thoroughly in distilled water. The powder was dried and moulded into sticks, prisms or cylinders, solid or perforated as might be desired; they were then subjected to intense heat during 24 hours.

W. E. Staite, in July, 1847, patented the application of melted sugar to a mixture of coal and patent coke, secured by Church in 1845. This coke was purified from sulphur by passing a powerful electric current through it, while it was in process of manufacture in the oven. The coal and coke when pulverised were intimately mixed, and were subjected to great pressure in wrought-iron moulds, which were plunged into melted sugar while hot, this operation of heating and immersing being continued alternately as long as was desired. Mr. Staite in a patent taken out the following year, says "In some cases I prepare electrodes for use by wrapping them tightly in several thicknesses of metal foil, such as tin, for the purpose of giving them greater strength and making them better conductors.

M. Le Molt, patented in 1849, a mixture of two parts of retort carbon, two parts of wood charcoal or coke, and one part of liquid tar. This compound, after having been carefully cleansed, was thoroughly blended, and by means of a mixing mill was reduced to a homogeneous paste which was moulded into form under pressure. The pieces thus moulded were soaked in a saturated solution of sugar, and then heated in a retort for about 24 hours.

C. T. Pearce.—In the same year, C. T. Pearce obtained letters patent for a method of manufacturing carbon rods by joining short lengths together with carbon dowels let into holes bored in their ends.

T. Slater and J. J. W. Watson, in October 1852, patented a mixture of pulverised coke thoroughly purified, beech charcoal, powdered coal and gas tar. This mixture having been combined, reduced to a paste and pressed into form, was soaked in caustic lime, brought to a white heat and placed in a solution of alum. It was then again brought to a white heat, immersed in molasses and again heated. In some cases for powerful lights, soluble glass was used instead of the alum solution.

W. Binks, in January, 1853, secured, by provisional protection only, the idea of covering metal bars with a coating of carbon for the purpose of electrodes.

W. E. Staite, in March, 1853, described in the course of a specification a process of preparing carbons by boiling them in oil or other fatty matters, and then baking them.

C. W. Harrison patented, in February, 1857, the process of placing rods of metal in gas retorts, and so covering them with a deposit of carbon; and also a mixture of spongy metal and plumbago, for use as an electrode.

MM. Lacassagne and Thiers, in 1857, introduced some modifications in the process of manufacturing carbons. These were cut into sticks from the carbon adhering to the sides of gas retorts, and placed in a bath of melted potash or caustic soda. The carbons were kept in this bath at a red heat for a considerable time, and were afterwards washed in boiling water. They were then enclosed in small porcelain retorts, kept at a red heat for several hours, a current of chlorine being passed through the retort. The object of the preliminary bath was to convert into soluble soda, the silica contained in the carbon; and that of the chlorine, to drive out the other earthy matters which the potash had not attacked. This process does not appear to have had much practical value.

B. Burleigh and F. L. Danchell, in December, 1857, filed a specification for making carbon suitable for electrodes by moulding it, preferably under percussive action, either unmixed or combined with resinous, oleaginous, saccharine or other suitable cementing matter. The electrodes thus formed were baked in a close retort.

E. Hunt, February, 1858, describes a mode of utilising the residuum of tar or pitch. Ground to an impalpable powder, and mixed with tar, it was then moulded into the desired form, subjected to a red heat, and then to a series of heatings and immersions in tar till the desired density was obtained.

M. Curmer,* introduced a process, in which lamp-black, benzine and essence of turpentine were amalgamated and moulded into form. The pencils thus produced were subjected to great heat, which drove out all the volatile matter, leaving the stick porous; the pores were filled up by repeated immersions in saccharine solutions or in resinous substances, followed by repeated heating. In this way a high degree of homogeneity was obtained.

*This and the five following processes are described in detail in *L'Eclairage Electrique* of M. Fontaine, from which book the information is obtained.

M. Jacquelain, chemist at the Ecole Centrale, conducted a series of experiments having for their object the imitation of the natural process in the formation of carbon in the gas retort. The necessary conditions involved the contact with the incandescent sides of the retort, of very dense hydro-carbons, of which a portion is volatilised and the remainder is decomposed, depositing on the sides of the retort a film of carbon. In the gas retort, these hydro-carbons carry with them all the impurities contained in the coal, and M. Jacquelain therefore made use of distilled tar, from which he obtained hard deposits of carbon that could be cut into shape by a saw, and which when used gave a very uniform and steady light, whiter and more intense than that obtained with ordinary carbons. This mode of production, very satisfactory in the laboratory, did not answer in practice. The process was a costly one, and the material was so hard that it could be sawn in sticks only with difficulty and with considerable waste.

M. Peyret, a doctor at Sourdes, made some carbons from elder pith saturated in a sugar solution, and subjected to heat. This process he repeated as often as necessary to produce density. No practical results attended this mode.

M. Archereau, about 1876, made electric carbons by the agglomeration of powdered carbon and magnesia, moulded into form and baked. The alleged advantage of the magnesia was to increase the steadiness and intensity of the light. M. Fontaine found some of these superior, and others inferior to ordinary retort carbons.

M. Carré, who is among the largest manufacturers of electric carbons, has conducted a great number of experiments in impregnating retort carbon with various materials. The general feature of the process consisted in impregnating the porous carbon, and then subjecting it to prolonged boiling in concentrated solutions. From these experiments M. Carré deduced the following :—

(1.) That a solution of potash and soda, doubles the practicable length of the voltaic arc, renders it silent, combines with the silica, and drives it out when burning in limpid liquid globules, which accumulate near the incandescent points, and increases the intensity of the light in the proportion of 1·25 to 1.

(2.) That lime, magnesia and strontian increase the light as 1·40 to 1, and impart different colours to it.

(3.) That boric acid renders the carbons more durable by covering them with a vitreous envelope, without, however, adding to the power of the light.

(4.) Lastly, that the impregnation of pure and uniformly porous carbons with various solutions is a cheap and simple way of producing their spectra, but that it is better to mix the compounds at once with powdered carbon.

The composition, according to M. Fontaine, that M. Carré employs for his carbons is powdered coke, calcined lamp black, and a syrup formed of 30 parts of cane sugar and 12 parts of gum. In January, 1876, M. Carré patented the following proportions :—

Very pure coke, reduced to almost impalpable powder .... 15 parts.
Calcined lamp-black .... 5 ,,
Special syrup .... 7 to 8 ,,

This mixture is intimately combined and from 1 to 3 parts of water are added, according to the desired consistency of the mixture, and to compensate for the losses by evaporation. The coke is made from selected coal, previously ground and carefully washed. When the paste has been properly mixed, it is put into a press and forced through a die of the form to be given to the pencils.

The carbons thus made are placed in a crucible and subjected to heat for a considerable time, when they are ready for the various subsequent processes. In the first stage the carbons are placed horizontally in a cast-iron crucible, on a bed of powdered coke, a sheet of paper is laid on them, then a second layer of carbons, and so on. When the crucible is nearly full, a layer of powdered coke and then a layer of sand is added, and the cover is fastened down. After four or five hours baking at a cherry heat, the carbons are removed, and kept during two or three hours in a boiling solution of sugar, with two or three intervals of cooling to facilitate the penetration of the syrup into the pores. The carbons are finally washed in boiling water to clean their surfaces. After having been dried the carbons are subjected to a second, third, or more processes similar to that just described, until the required density is obtained. The final drying is effected very slowly in stoves, at a temperature of about 170 deg. and lasts about 15 hours, the carbons being placed in V-shaped troughs to maintain their shapes. The Carré carbons enjoy deservedly a very high reputation, and the defects observable during the early stages of their manufacture have been practically eliminated. It will be observed that the process of making them is based on the earlier methods mentioned above.

M. Gaudoin, of Paris, has also made numerous experiments with carbons containing other materials, such as bone phosphates, chloride of calcium, borate of lime, silicate of lime, pure precipitated silica, borate of magnesia, phosphate of magnesia, alumina and silicate of alumina. The quantity of foreign matter thus introduced was about 5 per cent. of the finished carbons, and with a direct current from a Gramme generator a length of arc varying from 10 to 15 millimetres could be maintained. When the negative carbon was placed beneath, M. Gaudoin observed the following results :—

First.—A complete decomposition of the phosphate of lime, under the combined effect of electrolytic action, heat and the reduction of the carbon. The calcium in the negative carbon coming in contact with the air, burnt with a reddish flame. The lime and phosphoric acid passed into the air, producing considerable smoke, and the intensity of the light was about twofold that resulting from ordinary retort carbons.

Second.—The chloride of calcium and the borate and silicate of lime were also decomposed, but the boric acid and silica appeared to be volatilised. These bodies yielded less light than the phosphate of lime.

Third.—Silica diminished the light, melted and volatilised without being decomposed.

Fourth.—Magnesia, the borate and phosphate of magnesia were decomposed, vapour of magnesium emanated from the negative carbon and burnt with a white flame in contact with the air ; magnesia, and boric and phosphoric acids were developed in the form of smoke. The increase in the light was less than with the salts of lime.

Fifth.—The alumina and silicate of alumina were decomposed with a powerful current, the aluminium passing from the negative carbon like a jet of gas, and burning with a bluish flame.

The smoke which was a constant attendant upon these mixtures, proved an obstacle to their utilisation, and M. Gaudoin abandoned further experiments in this direction, turning his attention to making agglomerated carbons, a process which he patented in July, 1876. Like several previous inventors, M. Gaudoin employed instead of coke, which is always more or less impure, or lamp-black which is costly, the residue from tar, bitumen, and several other substances, which he distilled in close retorts of plumbago, provided with outlet tubes to carry off the lighter products, some of which were burnt below the retorts. The mass of coke left in the retorts after a sufficiently long exposure to heat, contains always a certain amount of lamp-black. This carbon having been reduced to an impalpable powder and mixed with solutions to convert it into a paste, is placed in a stee cylinder, and under a high pressure is forced out of openings in the cylinder, not at the bottom, but at the sides, these openings being inclined so that the issuing carbon takes an angle; as it comes out it is supported in tubes to prevent its breaking.

M. Fontaine in his book gives the results of experiments he conducted with various carbons, as follows:—

Class of Carbon.	Dimensions.	Speed of Generator, rev. per min.	Wear of Negative Carbon per hour.	Wear of Positive Carbon per hour.	Mean Wear.	Remarks.
			Mill.	Mill.	Mill.	
Retort Carbon	9 mill. sq.	800	19	36	63	Light irregular, sparkling intermittant, and some disintegration.
,, ,,	9 ,, ,,	920	23	48		
Archereau	10 ,, diam.	800	20	60	85	Tolerably regular. Considerable cinder, containing oxyde of iron. White, well formed light.
,, ...	10 ,, ,,	920	30	60		
Carré	10·4 ,, ,,	800	18	60	92	Slight disintegration and sparks; more cinder than the previous one, red for a greater length.
,, ...	10·4 ,, ,,	920	26	80		
Gaudoin	11·3 ,, ,,	800	20	38	73	No disintegration, nor sparks; less cinder than Carré or Archereau. Light very regular.
,, ...	11·3 ,, ,,	920	38	50		

The light produced by the retort carbons was equal to 103 Carcels; that from the Carré and Archereau carbons varied from 120 to 180

Carcels, and from the Gaudoin from 200 to 210. The wear of the latter carbons took an intermediate place between Archereau and Carré, for an equal section, but was much lower than any for the same amount of light, the wear being—

For the Gaudoin carbons 35 millimetres per 100 Carcels.

  ,,   Archereau ,,   44   ,,   ,,   ,,

  ,,   Carré   ,,   51   ,.   ,,   ,,

  ,,   Retort   ,,   49   ,,   ,,   ,,

In April, 1877, M. Gaudoin introduced some further modifications in the manufacture of his carbons. Instead of carbonising wood, reducing it to powder and then agglomerating it, he formed the pencils in hard dry wood, and approximately of their definite shape; these he afterwards thoroughly carbonised, and saturated them in the manner already described. Before being heated, however, the wood was soaked in acids and alkalies to remove impurities.

The Jablochkoff Candle.—We have said that although there is a great degree of secrecy observed by manufacturers of electric carbons, as to the details of the processes they follow, we are enabled to describe with some minuteness, the manufacture through all its stages, in the production of the carbons required for one of the important systems of electric lighting to which we shall again refer at considerable length. At the same time it will be convenient to follow throughout the manufacture of the Jablochkoff candle, in which the preparation of the carbons is only one part.

The Société Générale d'Electricité, of Paris, some years since established works able to produce from 3,000 to 4,000 carbons per day, and the Compagnie Générale, which absorbed the other association, continued the same manufacture. It would be very difficult to describe with absolute precision the various operations involved, because the whole merit of their success lies in the special skill employed, and in certain details, the secrets of which are guarded. The material used for the fabrication of the pencils is retort carbon as free as possible from silica. The coke resulting from the distillation of petroleum is purer than that of coal, but it is considerably dearer, and a mixture of the two products is usually employed. The coke, coarsely broken, is subjected to the action of stamps, which reduce it to an almost impalpable powder; it is then combined with a certain proportion of gum and water and the mixture then undergoes a prolonged trituration in a granite mill. On leaving the mill, the mixture is cohesive and elastic; it is worked up in small lumps and then passed on to the drawing machines. These are of the various sections it is desired to impart to the sticks—cylindrical, elliptical, or otherwise. Sometimes, the carbon has to be made annular; in this case, the opening in the forming plate is provided with a small frame with three or four arms, in the midst of which is fixed a needle to form the central hole as shown in Fig. 348. The formers for

Fig. 348.

the various sections are placed at the bottom of vertical cylinders into which the mixture is put; a piston is then forced down, pressing the carbon through the openings provided for it. The pencils fall vertically and are removed as soon as the desired length has been obtained; the material even in this early stage possesses a remarkable amount of consistency. When detached from the cylinder, they are placed one against the other on plaster surfaces, and as soon as a certain number are collected, a steel rod is placed on each side of the row; flat iron strips are also laid across the carbons to prevent them from bending during the drying process. The sticks thus held in all directions cannot get out of shape; the plates carrying them are then placed in a drying stove, kept at, a low temperature but with a rapid circulation of air. Here they are left for a considerable time, until, in fact, they part with a great deal of their moisture, a certain amount of which is absorbed by the plaster. Before being removed from the stove, the carbons have acquired a sufficient degree of hardness to enable them to be handled with ease; they are then subjected to a series of bakings and saturations in a saccharine syrup; afterwards they are gathered up in bundles, bound round with iron wire, and placed in cylindrical retorts, either of iron or refractory clay, the spaces left between the bundles being filled with coke dust. A number of such retorts, when they have been charged, are placed on the hearth of a reverberatory furnace, and subjected to a sufficiently high temperature, and for such a time until the whole of the water is evaporated and the gum contained is decomposed; but now the carbon sticks are sensibly porous, and this porosity has to be removed by saturation under pressure with concentrated syrups; the sugar penetrates into the pores and in a subsequent baking process becomes decomposed, leaving within the pores a deposit of carbon. A series of three or four of these double operations of saturation and baking, imparts to the sticks a remarkable compactness and strength; the baking process may be performed very economically in a Hoffman kiln. The Compagnie Générale d'Electricité recently constructed a furnace on this system, at their works in the Avenue de Villiers. Carbons properly made ought to have a high degree of resistance, a smooth surface, be almost as sonorous as steel, and absolutely straight. Before the sticks, manufactured as has been described, are passed forward for a subsequent process, they are subjected to the following tests: They are first examined as to their exterior qualities, their solidity, straightness, and to a less degree for the smoothness of their surface. Then their lengths are measured, each stick having to be long enough to make the two carbons of one candle, the effective length of which is about 9 in.; the sticks must therefore be about 19½ in. long, including those portions which are placed in the sockets. The cylindrical carbons are then sorted out according to their diameters. This measurement is effected by means of an iron plate, in which square recesses are cut, the successive widths of which are 4·1, 4·2, 4·3 millimetres, &c. These plates are fixed on a small table at which the workman is employed in calibring. He tries one stick after another in the different recesses, as shown in Fig. 349, and is thus able to separate the whole into distinct

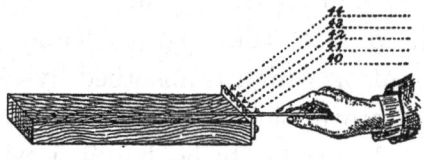

Fig. 349.

groups. The sticks from 4 to 4·3 millimetres are used in the manufacture of the 4 millimetre (·16 in.) candles, it not being found advisable to go above or below these limits. With a smaller diameter the carbon would have a higher resistance and would consequently burn with a longer cone; a larger diameter would give a greater length to the voltaic arc and would render it less steady. The electric resistance is then measured, a certain per-centage being selected for this purpose at random; with a Carré carbon, the specific resistance for 1 metre of length, and 1 millimetre diameter, varies between 40 and 50 Ohms. A workman can gauge about 2,000 metres of carbons per day, or sufficient to make 4,000 candles.

The operation of impasting is now abandoned; it consisted in introducing the ends of a carbon, into two cylindrical copper sockets; it was then cut in the centre and the two portions were placed in a little device which kept them at the desired distance apart. Then by means of a wooden palette knife, a slip of glass or asbestos cardboard, covered with a mixture of sand, kaolin, or silicate of potash, was placed between the sockets. More recently the glass or asbestos was dispensed with, and the sockets were connected with a mixture of 400 parts of kaolin, 100 of sand, 200 of powdered asbestos, and some silicate of potash; this mixture did not splinter off like the preceding one, but it shrunk in drying, reducing the distance between the carbons, and making from 25 to 30 per cent. of the candles defective. After November, 1878, the following mixture was substituted : refractory clay 550 parts, kaolin 450, silicate of potash 720. This mixture was more homogeneous and easy to work, and the proportion of bad candles was reduced to 12 per cent. ; it was, however, defective on account of its hygrometric properties, which caused a passage of the current from one socket to the other, sometimes lighting the candle at the wrong end. It was then that the paste was replaced by distance strips of wood, hard rubber or horn. All these strips had some defect; those of wood were found to be irregular, those of rubber produced a disagreeable smell in heating, and those of horn were too costly; however, the system of fastening was a good one, and it only remained to find a suitable material. M. Minet, engineer of the candle factory, succeeded with a mixture of kaolin, six parts; magnesia, two; sulphate of lime, two; water containing 5 per cent. of gum, four parts. This mixture is easily amalgamated, hardens rapidly in the air without contraction, and is sufficiently strong to resist the pressure of the sockets at the moment of lighting. In making these distance-pieces, the materials are combined in a mixer (see Fig. 350) for 10 or 15 minutes; the paste obtained is placed in the cylinder of a hydraulic press (Fig. 351), in the base of which is an opening (see Fig. 352) of the same form as the distance-piece, and a length is driven out through the press, is received in guides and laid to dry upon a marble table. Before the paste has attained a high degree of consistency, that is to say, about an hour after it has been forced from the cylinder, it is cut up into lengths of about 4 centimetres each (see Fig. 353); 1,400 grammes of the mixture are worked up at each operation; this quantity is sufficient to make 500 distance-pieces, and two men can turn out about 5,000 a day. The copper sockets are made of small rectangular

leaves of metal rolled up in such a manner as to convert them into
cylinders open on one side (Fig. 354); the width of this opening is 2·2
millimetres, and it serves to give passage to the piece which connects the
two ends of the distance blocks; these sockets are 4·5 centimetres high.

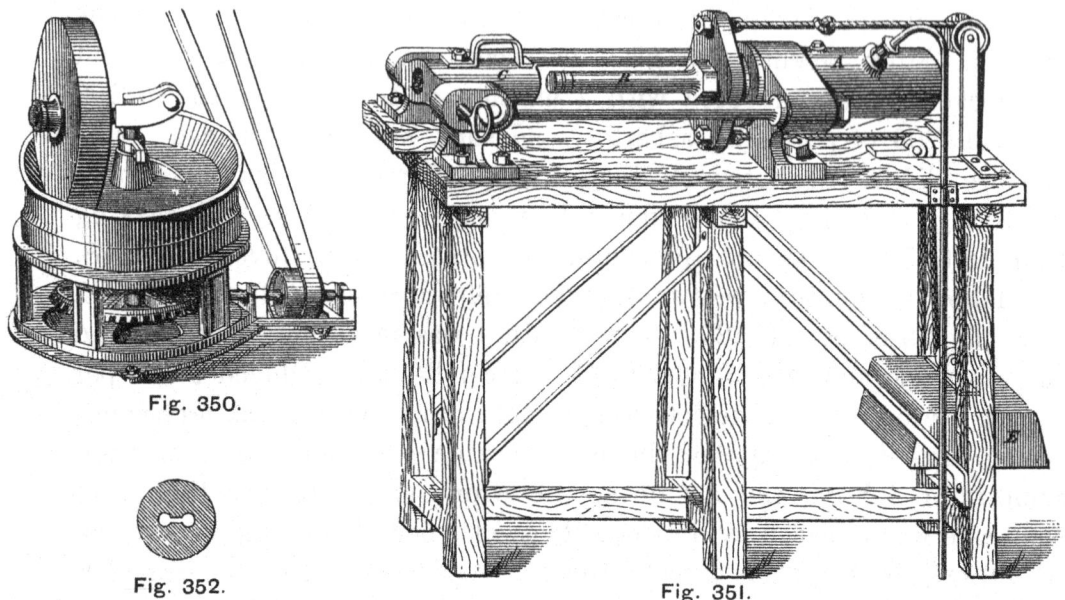

Fig. 350.

Fig. 352.

Fig. 351.

The workman takes two sockets and introduces them into the two
recesses A and B of a guide, Fig. 355, in such a manner that their
openings are opposite to each other; with his left hand he places a
distance-piece on the two sockets, and with the right hand he brings

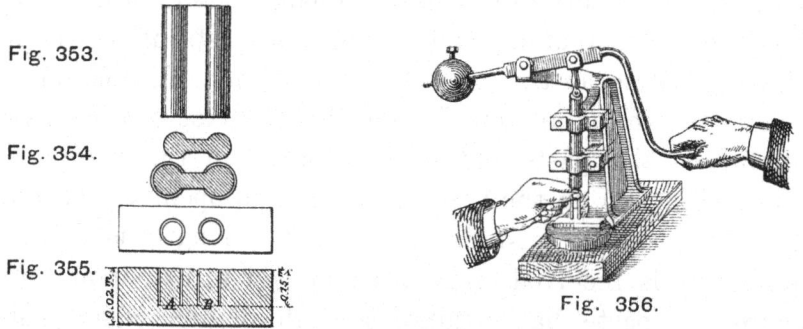

Fig. 353.

Fig. 354.

Fig. 355.

Fig. 356.

down a press, which forces it into the sockets. From the dimensions
given, it will be seen that the distance-piece is below the level of the
sockets leaving sufficient space for the carbons to enter to a distance of
1·5 centimetres. Fig. 356 shows the whole of the apparatus and the

manner in which the operation is performed; a workman can place 2,500 distance-pieces into their sockets daily. The two sockets thus united are parallel to each other, but the ends are placed in a conical guide to destroy their parallelism and give a convergence to the two carbons in order to lock them against the "colombin" and prevent it from falling out. The stick of carbon is then taken and cut in the middle; the freshly fractured ends are used to form the top of the candle, the original extremities being introduced into the sockets. It is an invariable practice to make each candle from the halves of the same carbon rod, in order to secure as far as possible complete homogeneity, and equal conductivity. It is then necessary to ascertain if the distance between the two carbons is correct; this detail fixes itself, provided that the carbons are well made, the gauging correct, and the size corresponds exactly with that of the sockets. If it be found that either of the carbons be slightly curved, it will be necessary to place it as shown in Fig. 357, in order to make it bind tightly against the "colombin;" if placed as shown in Fig. 358, the result would be entirely unsatisfactory; with such a bend as this it would be necessary to turn the carbon upside

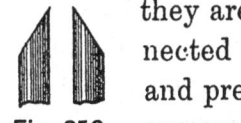

down, so as to make it conform somewhat to the carbon shown in Fig. 357. This process is tested with a standard "colombin." One workman can place 900 candles per day into their sockets, a second workman can test 5,000 a day.

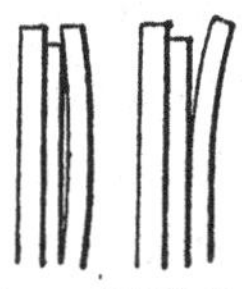

Figs. 357 & 358 The upper ends of the carbons are tapered (Fig. 359) in order to facilitate lighting; this cutting is performed by emery wheels 15 in. in diameter, and having a speed of 1,800 revolutions per minute;

they are covered with a plate-iron hood in which a tube connected with a ventilator draws off the particles of carbon and prevents them from entering the lungs of the workmen;

Fig. 359. one man can cut about 2,000 candles a day.

The manufacture of the insulating strip or "colombin" is one of the most important operations in the manufacture of the Jablochkoff candles. As already stated, the value of the "colombin" was warmly disputed by electricians who investigated the subject; some maintained that it was useless because certain lamps, such as that of Wilde, are composed of parallel carbons; others maintained that it was disadvantageous because it absorbed a certain portion of the light. Both assertions are equally erroneous, and it is indisputable that the "colombin" does assist the steadiness of the voltaic arc. In candles without this insulator, the arc is frequently shifting, and it sometimes disappears either from a current of air or from the irregularity in the working of the dynamo machine; in fact, air is only a conductor when it is brought to a high temperature, and a given temperature corresponds to a tension necessary for a current supplying the lamp; from this it results that if a current of air cools the arc, the tension is no longer sufficient, and the light goes out; it may also be insufficient when the air preserves its temperature, and consequently its conductibility, if the speed of the dynamo machine is reduced. These inconveniences are much less pronounced in candles provided with an insulating strip, indeed, which volatilises in the arc giving up particles brought to a high temperature and so increasing the conductibility of the arc; there results also from its use a diminution in the necessary tension, and consequently a somewhat less power is required than without the "colombin." As the arc remains in a more conductive medium, the air has less action upon it, and if the tension falls on account of a reduction in the speed of the machine, the arc leaves the points at the end of the carbon, and plays on the surface of the "colombin" without disappearing. As to the light produced, it is the same with or without the distance-piece, a fact which is very easily demonstrated. To sum up, the "colombin" insures a steadier light, and diminishes the amount of motive power required. The only inconvenience it presents is as follows :—In electric lamps with opposing carbons the points may be brought together as closely as may be desired, and consequently, with regulators adapted to the division of the current, a considerable number of lamps may be placed in the same circuit. In lamps with parallel carbons without "colombins," the minimum distance apart of the points is equal to the diameter of the carbon; with the Jablochkoff candle the thickness of the "colombin" increases this minimum distance. Now the thickness of the "colombin" cannot be reduced beyond certain limits without rendering it very fragile; the divisibility of the light, although favoured by the action of the "colombin" in the flame, is reduced on account of its thickness; in a word there are two properties to be considered. If, as may be hoped, manufacturers will arrive at casting the carbons with an insulating material between them, this latter may be reduced to a very slight thickness, so as not to hinder the divisibility of the current; and improvement in this direction may be looked for in the manufacture of the

candle N P. A large number of materials have been experimented with in
the production of the " colombin ; " as has been already said, kaolin was
at first employed, but this produced a sensible diminution in the light ;
different refractory earths also presented serious drawbacks ; they
volatilised easily when formed of fusible silicates, or they did not disappear
at all and curled down the side of the candle. The mixture definitely
decided upon is sulphate of barytes, one part, and sulphate of lime, two
parts. This mixture does not reduce the intensity of the light,
disappears as the carbons are consumed, and very seldom curls down the
side of the candle. The materials are very carefully screened in a fine
sieve, and are then moistened with water so as to form a semi-fluid
paste. This paste is placed at one end of a long marble table previously
coated with oil, and the workman is provided with a kind of comb, the
hollows of which are formed to the contours of the " colombin." By

Fig. 361. Fig. 360.

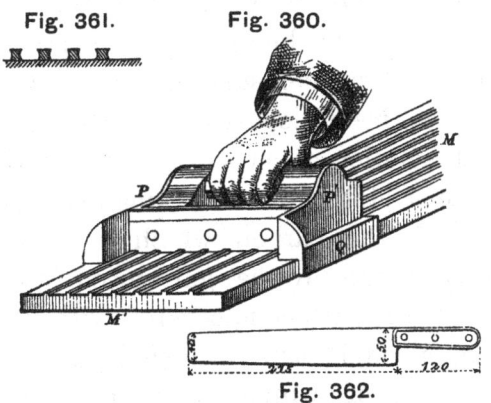

Fig. 362.

means of this comb he drives before him the paste, which leaves upon
the marble a number of strips corresponding to the form of the comb ;
this operation is repeated perhaps a dozen times ; after each pass, the
comb is carefully cleaned, dried, and covered with a little oil. By the
time the " colombins have taken their ultimate form they are nearly dry,
on account of the rapid setting of the plaster ; they are then detached
from the marble by means of an extremely thin knife-blade, are cut into
the proper lengths and placed upon small plaster tiles (see Figs.
360 to 362), which are kept in a stove for three or four hours at
a temperature of 212 deg. This heat drives off the water, and prevents
the " colombin " from losing its shape ; 4,500 grammes of the mixture
are sufficient to make 1,050 " colombins," and two men can turn out
8,000 of them a day. The " colombin " is placed between the two carbons

the sides of which it embraces, the slight convergence of the carbons helping to keep them in place. The top face of the " colombin " is then cut until it is about a quarter of an inch below the points of the carbons, which are dipped into the fuse material, composed in equal parts of gum-water (three parts of water and two of gum), and a mixture containing three parts of coke and two parts of plumbago ; one workman can fix 3,000 " colombins " per day, and another can fuse 3,000 candles.

As we shall see hereafter, a large number of the Jablochkoff candles are electro-plated, which makes them more durable, but at a certain expense of steadiness and uniformity of colour. In many cases, and especially where industrial lighting is concerned, economy is the first point to be considered ; since the month of August, 1879, when the process of galvanising the carbons was practically introduced, four-fifths of the total production have been thus coated. This operation is performed before fixing the " colombin ; " it is preceded by a careful washing, which removes from the surface of the carbons any greasy matter which may have been deposited on them. The carbons are placed in a 10 per cent. solution of caustic potash, and they are afterwards carefully wiped with clean cloths ; one workman can wash 3,000 candles per day. That part of the carbon which is in contact with the " colombin " does not require to be galvanised, and to prevent the adherence of the copper, the surface is covered with varnish prepared by dissolving gum lac in alcohol ; this varnish is laid on by means of a wooden spatula, the end of which is covered with velvet. It is sufficient to pass this spatula between the two carbons in order to make the required band. One workman can perform this operation on 2,500 candles a day. Galvanisation is effected by electrolytic decomposition in a bath, by means of a small direct-current machine, similar to the Gramme exciter for four Jablochkoff candles. This machine is driven at 1,200 revolutions a minute, and produces a current of 40 Ampères when coupled to the bath ; three baths are employed of the following dimensions : length, 24 in. ; width, $10\frac{1}{2}$ in. ; height, 12 in. The candles are fixed 12 at a time, points downwards, to frames provided with copper pincers (see Fig. 363) ; one bath contains seven such frames for 84 candles, and eight soluble anodes of red copper. These anodes and the rows of candles are arranged alternately ; the current reaches the bath at one end, and passing all the soluble anodes simul-

taneously, leaves the bath by means of the seven frames; it then goes
on to the second, and afterwards to the third bath. By this arrangement
in the same receiver, the anodes are coupled in quantity and the baths
in tension; the copper plates are 11 in. long and 9·8 in. wide. By means
of these three baths arranged as explained, about 380 candles can be
galvanised per hour; this result conforms to theory. Approximately a
96-Ampère current deposits ·032 grammes of copper per second, or 115
grammes per hour, therefore the current of 40 Ampéres will give a deposit
of 48 grammes per hour; each candle requires ·35 grammes, which
corresponds to 137 candles plated per hour and per bath, or 401 for
the three baths, approximating very closely to the 380 actually
treated; three men can galvanise 4,000 candles per day. The electrolytic liquid employed is a solution of sulphate of copper containing
15 per cent. of salt; this solution is generally acidulated by means
of dilute sulphuric acid, the quantity of which varies with the size of

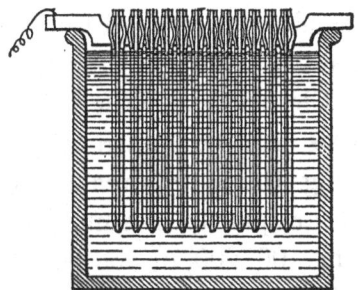

Fig. 363.

the anodes and their distances apart. Occasionally from 5 to 6 per cent.
of soda is added. Practice and careful examination alone can enable the
workmen to decide whether the process is going on well or badly; if the
bath appears too weak, a small portion of it is drawn off and replaced by
a more concentrated solution of sulphate of copper; if the reverse, a few
drops of sulphuric acid are added. The galvanised candles on coming
from the bath are dried in sawdust, and then varnished by immersion in
a metallic varnish, which preserves the copper from oxidation; one man
can treat 6,000 candles per day. After this has been done, the
"colombin" is introduced and the fuse added. Summarising the various
operations which we have described, the work of each man per day, after
the actual manufacture of the carbons has been completed, will be seen
at a glance:—

					Candles per Day.
Gauging....	....	....	....	....	4,000
Distance-pieces	....	....	....	....	2,500
Putting sockets on distance-pieces	....		....	....	2,500
Fixing carbons	....	....	....	....	900
Testing ,,	....	....	.	....	....	5,000
Cutting ,, for fuses	....		....	....	2,000
Manufacture of " colombin "			....	....	4,000
Fixing " colombin "	....		....	....	3,000
,, the fuse	....	....	....	....	3,000
Washing	....	....	....	....	3,000
Galvanising	....	....	....	....	1,500
Varnishing	....	....	....	....	6,000

From these figures, assuming an output of 5,000 candles per day, we shall find the following staff to be necessary :—a foreman, a storekeeper, a packer, a fitter, and 27 workmen, of whom some are employed in several operations; 5,000 candles correspond to an output of 150,000 per month, or 1,800,000 per annum, which is very nearly the production of the Compagnie Générale d'Electricité. When finished, the Jablochkoff candles are not extremely fragile objects, the packing them requires some special precautions. They are put up in small wooden boxes, each containing 100, 200, or 500, and the candles are all laid in little cardboard frames placed one on top of the other. These frames are shown in the annexed

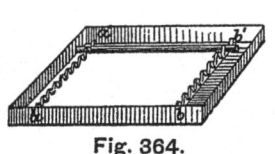
Fig. 364.

sketch, Fig. 371, where *a b* are two bands of felt, folded on themselves, and between which the candles are held, so that they cannot shift; each frame contains 10 candles.

It is quite essential in the manufacture of Jablochkoff candles that the two carbons of the same bougie should have exactly the same electrical resistance, in order that they may burn equally. It is also necessary that the various candles should have a similar resistance so that their maximum duration can be guaranteed, and that on any given circuit the different candles may give an equal light. The first requirement is fulfilled by making each candle from the same carbon rods which, as has been already explained, are made long enough for that purpose, and experience has shown that each stick of carbon is sufficiently homogeneous to have practically the same resistance throughout its length. As to the second point, it is necessary to have recourse to a system of selection, by which the rods can be classified in groups of similar resistance, and those carbons which are unsuitable may be rejected. The operation is based on a principle analogous to that of the Wheatstone bridge, and a few words on the theory of the subject will be of interest. The Wheatstone bridge may be represented by the figure A B C D (Fig. 365); in the diagram P is a

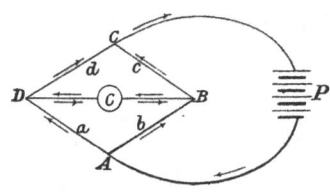

Fig. 365.

battery the current from which flows to A, divides between the routes A B C and A D C, and returns to the battery from C. Between the points D and B is introduced a connection which includes on its line the galvanometer G. In order that the needle of this galvanometer should remain stationary, it is necessary that no current should flow from D to B, which condition will exist as long as the potentials D and B are equal. We may examine the conditions necessary for this to exist.

If P be the potential at A, P^1 the potential at C, X that at D, and Y that at B, it is necessary that X and Y should be equal. Now in the branch A D C, the intensity of the current being equal, we shall have $\dfrac{P-X}{a}=\dfrac{X-P^1}{d}$. In the branch A B C we shall have for the same reason $\dfrac{P-Y}{b}=\dfrac{Y-P^1}{c}$. If X=Y it is necessary that $\dfrac{a}{d}=\dfrac{b}{c}$. From this simple formula is deduced a method of measuring resistances which in the particular case under consideration, can be applied very con-

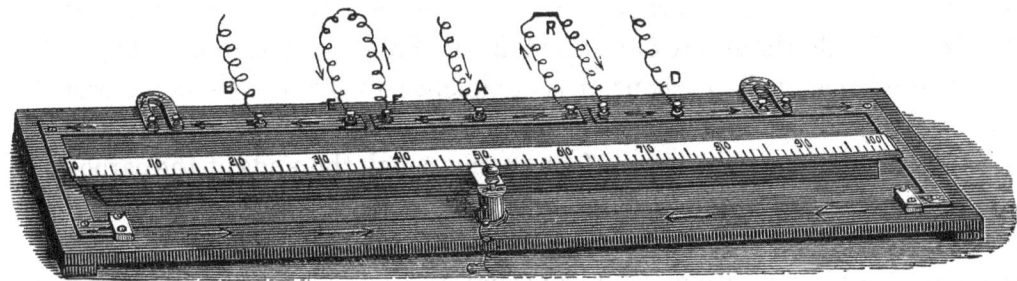

Fig. 366.

veniently and correctly within one-hundredth of an Ohm. The apparatus, the principle of which is shown in Fig. 366, is composed of a rectangle, three sides of which are made of a metallic strip, the resistance of which may be neglected; the fourth side is made of German-silver

wire. This rectangle is carried on a wooden table, on the longer axis of
which is a graduated scale. The current from the battery flows to the
rectangle at A in the middle of the metal strip; it returns from C by a
shifting terminal on the German-silver wire, the length of which can be
divided at will. A pointer carried by the terminal lies on the division of
the scale, and indicates the proportion in which the German-silver wire
is divided. Two terminals D and B receive the wire from the
galvanometer. The copper band is interrupted in two places; in one is
interpolated a small coil with a resistance of 3 Ohms; in the other is the
resistance that has to be measured.

It is necessary to move the index C until the galvanometer needle
remains at zero, and the measure of the desired resistance will be found
from a reading and a proportional deduction. Thus, if x is the resistance
to be measured, if we assume that the index was stopped opposite the
number 35 of the graduated scale, this latter being 1 metre long, we
shall have $c = 35$ and $d = 65$, and the formula will be $\dfrac{3 \text{ Ohms}}{x} = \dfrac{65}{35}$
whence $x = 1\cdot63$ Ohms. For the special purpose for which this apparatus
is intended, is is necessary that it should be used with great rapidity.
it is essential, that the operator should replace the carbons one after
the other very quickly, and that the galvanometer should come
to rest at zero with but few oscillations. The first requirement is
fulfilled by means of a small device with two metal supports. The
operator places the rod of carbon on these two supports, then by
means of a pedal he lowers two metal clips, which are in contact with
the wires E and F The carbons, for a constant length, are thus brought
into circuit, while the index once fixed at the point corresponding to the
normal resistance of the carbons, it is sufficient to see if the galvanometer
remains always at zero, with each carbon placed in the apparatus. The
operator remains seated with one foot on the pedal, the ends of the carbon
rods in his hands; he has before him the clips in which he places
successively the pieces to be measured. In this way he is able to work
with considerable rapidity as well as accuracy.

The galvanometer employed is a Marcel Deprez mirror galvanometer, formed with a small movable frame, that oscillates between the
branches of a vertical magnet, combined with a hollow cylinder of
soft iron. A small mirror, 1 centimetre in diameter, is fixed on the
wire carrying the frame, throwing on a graduated scale, a luminous line,

the reflection being obtained from a small lamp. There is scarcely any appreciable oscillation with this galvanometer, so that the reflection becomes stationary almost at once, enabling the operator to complete his test in the time required to fix a carbon in place.

In this way the selection of carbons of a given normal resistance is rapidly carried on. Two persons are easily able, in one day, to test a sufficient number of carbon rods half-a-metre long, for the manufacture of 10,000 carbons.

M. Napoli, of Paris, who has done much to develop the capabilities of the Werdermann lamp, has been also successful in producing electric carbons of high quality, and on a large scale. The materials he employs are gas tar, and the residue of coal slowly distilled at a dull red heat. The coke is first reduced to an impalpable powder in a Carr's disintegrator, and is afterwards carefully screened in the apparatus shown in Fig. 367. This consists of two rapidly vibrating sieves placed in a wooden

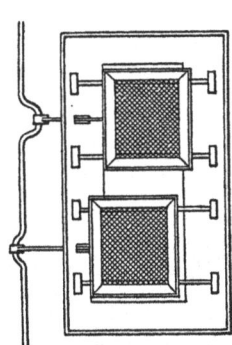

Fig. 367.

chamber which is kept covered. The screened material falls through an opening in the bottom of the chamber, and is transferred to a mixing mill with edge runners, where it is combined with the tar, the proportions being 75 per cent. of coke and 25 per cent. of tar. The usual charge for the mill is about 65 lbs., and the time required for grinding is four hours. When the mixing is completed, the compound is very compact and solid, requiring considerable pressure to force it into its ultimate form. The arrangement employed for this purpose is shown in Figs. 368 and 369; it consists of a hydraulic press, the ram *a* carrying a table on which is the cylinder *b* containing the mixture of carbon and tar. Fig. 369 is a section of this cylinder. Within it is the ram *c* bolted to a heavy timber attached to the head of the press. It will be seen that the cylinder *b* terminates in a base *b*¹, which contains a continuation of the cylinder but curved round at right angles, and reduced gradually in diameter. The end of this extension is closed by a die plate *e*¹ containing the die *e*, through which the mixture of coke and tar is forced. The cylinder *b* is jacketted, and steam is led to it by the pipe *d*. This was found to be an absolute necessity, as the carbon would not flow through the die with satisfactory results when the mixture was cold. An inclined table *f* is

attached to the head of the ram for receiving the stream of carbon as it
flows from the die. The action of the apparatus will be at once
understood. As the ram is raised, carrying with it the cylinder containing the carbon compound, the latter is forced out in a continuous stream
through the die *e*, and is received upon the inclined table *f*, being divided
into any desired lengths as it comes out. During the earlier stages of this
manufacture, considerable difficulty was experienced in lowering the
hydraulic ram, on account of the film of carbon that was squeezed
between the piston and the wall of the compressing cylinder, which
caused the whole of the apparatus to adhere firmly together. After

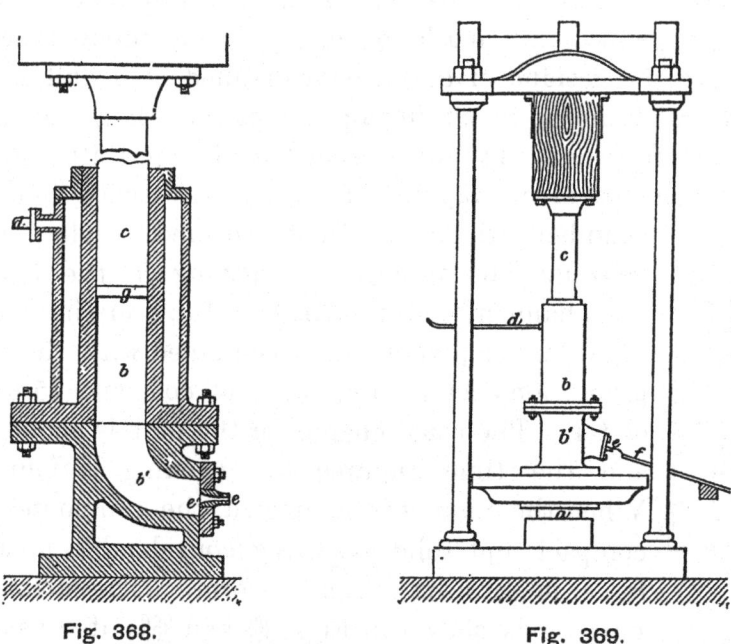

Fig. 368.   Fig. 369.

several trials this trouble was obviated by using a piston of smaller
diameter than the cylinder, and making the joint with a bronze disc *g*
about 2 centimetres thick. With this arrangement the ram, table and
compressing cylinder fell easily, and the disc was removed by means of a
key inserted in a hole formed in its centre. The proper form to be
given to the die was also a matter of considerable experiment; at first it
was impossible to produce straight and uniform carbons, but ultimately,
when dies shaped as in Fig. 368 were employed, this difficulty was
effectually surmounted.

In baking the carbons, M. Napoli is careful that in the first stage they should not be exposed to a higher temperature than that by which the component parts have been obtained; that is to say, that they should not be brought beyond the dull red heat at which the tar has been distilled from the coal, and the coke produced. The carbons placed in a retort, are gradually brought to this temperature, and after some time the heat is slowly increased, so as to facilitate the escape of gases without damaging the structure of the carbons. When the latter are raised to a cherry red, they are maintained at that degree during several hours, and then slowly cooled. The slight proportion of tar used by M. Napoli as a combining material plays an important part in this stage of the process, which is always attended with shrinkage of the carbons to a greater or less extent according to the amount of tar or other binding substance, employed. The shrinking is often accompanied by a rupture of the carbons, and by their disintegration, but in the compound used by M. Napoli, the shrinkage is so slight as not to cause inconvenience. After the first baking and cooling, the carbons are subjected to a second and similar operation, only this time they are raised to a white heat, which drives out the final traces of gas, and leaves the carbons of great hardness, and with a fine fracture like that of crucible steel.

In this condition, and without any further treatment, the pencils can be used for lighting purposes; they burn with perfect steadiness, and give a pure white light. It is claimed moreover, that the relative consumption as compared with the Carré carbons of similar size, and in the same time, is as 1 to 3·3.

Before the carbons are baked for the first time, and as they are taken off the delivery board in issuing from the die, they are laid in cast-iron plates, fluted with semi-circular grooves large enough to hold the sticks. The plates are grooved on both sides, and laid on top of one another, so that the soft carbons are held securely in all directions, and cannot become deformed in the process of baking. When a sufficient number of charged plates are piled together, they are placed in a muffle-furnace, and heated as described above. For the second baking, the carbons having become dense and hard, no such precautions are required: they are simply fastened in bundles, and placed in refractory clay crucibles packed with powdered coke. The covers of the crucibles are carefully luted down, a number of them are placed in a second muffle-furnace, and they are heated as already described. The use of the powdered coke, which is also employed during the first baking, is to absorb the gases thrown out by the carbons when heated.

As stated above, the carbons after the second baking are in a fit condition for use; but for several classes of lamps, especially for incandescence-arc lights, M. Napoli finds it advantageous to add the supplementary processes common in the production of electric carbons, that is to say, he increases their density by saturation in saccharine syrup, and subsequent baking. A special apparatus, illustrated in Fig. 370, is used for the saturation. It consists of a cylinder *a* surrounded by a steam jacket, supplied by the pipe *b*, and having a blow-off cock at the bottom; this cylinder is mounted on trunnions *c*, and the cover can be easily removed and replaced. At the top of the cover is a pipe connection *d*, which can be connected either to an ejector for making a vacuum, or to a steam boiler for producing pressure within the cylinder; at the bottom is a blow-off cock *e*. This apparatus is used as follows :—The pipe connections are first unfastened, the cylinder is turned horizontally on its trunnions, and the cover removed. A wire basket charged with carbons from the second baking, is placed in the cylinder, the cover of which is replaced, and the pipe connections made good. Steam is turned into the jacket surrounding the cylinder, until it and its contents are thoroughly heated,

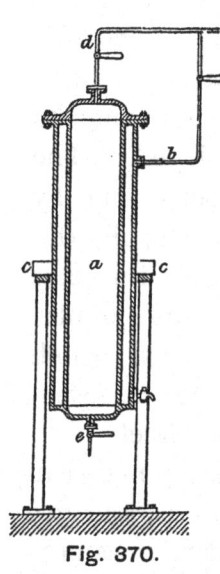

Fig. 370.

then communication with the steam ejector is opened, and a partial vacuum produced in the cylinder, a large proportion of the minute particles of air imprisoned in the carbons being thus set free. The tap *e* at the bottom of the cylinder is then placed in connection with a vessel containing syrup, the cock is opened, and the syrup rises forciby in the cylinder, penetrating the pores of the carbons. After this operation is completed, the syrup is allowed to run out of the cylinder, leaving the carbons saturated, and high-pressure steam is then turned on, with the result of completely washing the carbons, and removing the superfluous syrup from their sides, which, had it remained, would have impaired the surface of the finished pencils. The cylinder is then opened, the carbons removed, and placed in the furnace. A second similar operation follows, when the carbons become too dense to absorb more syrup, and all as perfect as the process of fabrication admits.

Messrs. Mignon and Rouart, of Paris, have lately designed a special apparatus, which is indicated in Figs. 371 and 372, for the manufacture of carbons covered with composition. In the former figure the carbon is placed in the central tube and the covering in the annular outer chamber. The two are forced out together into the nozzle T, the central part by

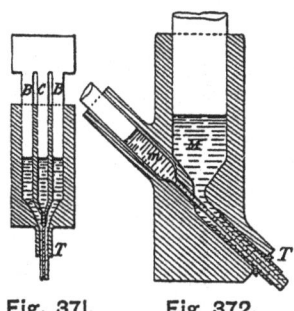

Fig. 371.   Fig. 372.

the ram C, and the external part by the annular or tube-like ram B. Fig. 372 illustrates another apparatus for the same purpose. M is the outer composition and *m* the inner. The two rams can be geared together with change wheels to produce cores of different diameters. By a modification, hollow pencils can be produced, or a hollow pencil and a core can be made separately and threaded together. As a protective, the rods are covered with enamel or glass.

SECTION IV.

—•♣•—

X.

ARC LAMPS.

LTHOUGH the voltaic arc was produced by Davy in 1810, and the experiment was repeated in numerous laboratories in this and other countries, there was, if we may take the silence of the Patent Office Records as evidence, no hasty recognition of the discovery as a practical means of illumination. The invention of the Grove battery in 1836 and the Bunsen battery in 1842, provided, however, generators of electricity that would produce a current sufficient to maintain a light for some considerable time, and almost immediately the subject of electrical illumination seems to have emerged from a state of scientific to one of practical experiment, and to have engaged the attention of one worker after another with regular succession until the year 1859. From 1840 to 1859, many patents for electric lamps, chiefly controlled by clockwork mechanism, were taken out, but about the latter date inventors seemed to have come to the conclusion that the subject was not one that had an extended field of operation, and with a few exceptions they abandoned it. Yet the prospects then had never been brighter. It had been practically proved that ample currents could be economically generated by magnetic induction, and one of Professor Holmes' machines was soon to be set to work at Suta Point, where it has done good service ever since. In spite of all this, however, for 12 years no improvements on existing lamps were patented, although in the meantime great steps had been taken. Wilde's generators had been followed by the dynamo machine, and there was fierce contention among the re-discoverers of the principle as to whom belonged the honour of the discovery, yet every one seemed satisfied with the Serrin and Duboscq lamps, which it must be admitted gave good results and still retain their place.

It is our purpose in the next few pages to take a rapid survey of the early arc lamps, and to give our readers a brief account of each that appears to be worthy of notice. Our investigations have been largely directed, in this part of our subject, to those cases in which experimentalists have taken out patents for their inventions, for after the long interval that has elapsed it is often difficult to obtain trustworthy and detailed descriptions from other sources. In the present part of this volume, however, we have not by any means exhausted the subject, and the reader is referred to the abstracts of patent specifications in the latter part of the book.

Wright.—The first arc lamp that we find is one patented by Thomas Wright, of Thames Ditton, in 1845.* The electrodes were made in the form of discs rotated by clockwork with a slow continuous motion, a method of feeding which, though not successful in practice, has often commended itself to the experimentalist when contending with the

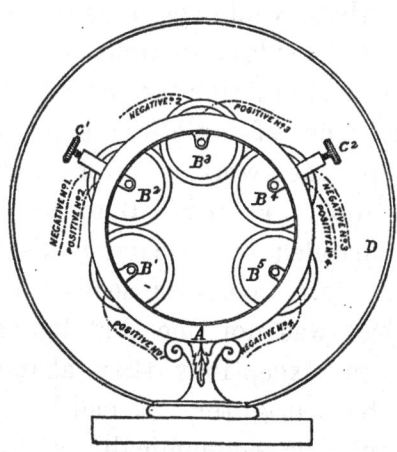

Fig. 373. Wright.

difficulties of an intermittent action. Fig. 373 illustrates the arrangement very clearly. Within the glass globe D is a frame carrying five discs of carbon, or plumbago, held between metal plates. To begin with, all the discs are in contact with one another, but when the current is established B^2 and B^4 are drawn outwards by the screws C^1 and C^2, and an arc is established at each side of them.

Staite.—During the next few years the subject seems to have fallen into the hands of W. E. Staite, who, in 1847, after some attempts in which the light was produced by the incandescence of carbon in air,

* Leon Foucault, of Paris, had the year before, produced a hand regulated lamp, in which the carbons were formed from retort coke sawn into sticks.

aided in certain cases by minute arcs, brought out a lamp in which the length of the arcs was regulated by mechanical devices*. He employed carbon rods as his electrodes, and arranged them vertically one over the other, making the feeding of them dependent on the current

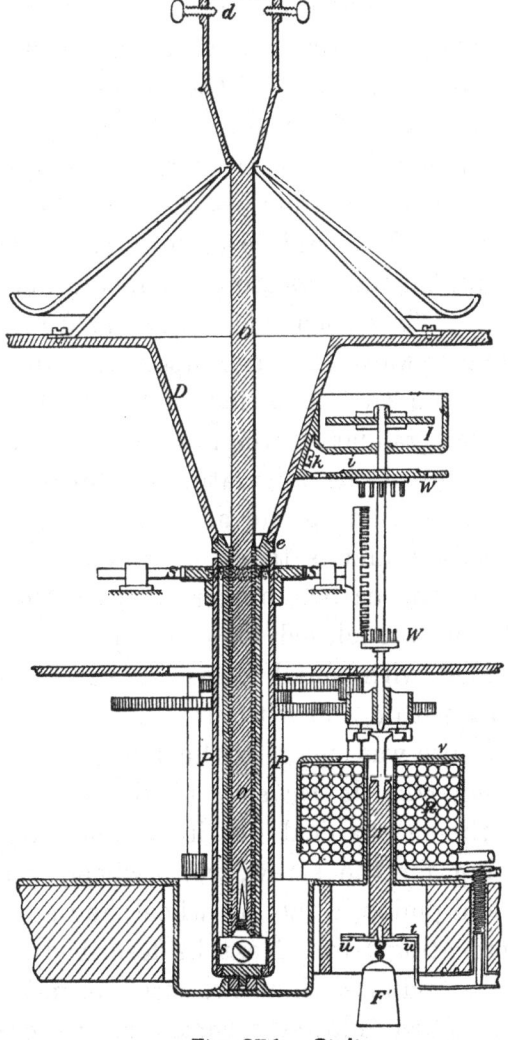

Fig. 374. Staite.

traversing the circuit. Fig. 374 illustrates the feeding and regulating mechanism that was applied to the lower carbon; the upper one was

* The French word "*regulateur*," applied to this class of lamp, has unfortunately been adopted for a similar definition in this country. The English word "regulator" is not only ill chosen for the purpose, but becomes confusing, as the same term is necessarily and properly applied to devices for regulating the flow of currents.

stationary. The holder of the lower electrode d is mounted upon a screw O, which works through a fixed nut e. The lower end of the screw O bears a cross-piece s, which serves to connect it to the slotted rotatory tube P, after the manner of a feather key. The tube P takes a bearing round the nut e at the top, and rotates on a silver-plated footstep at the bottom. It is driven in either direction according to the strength of the current by a worm and worm-wheel s, actuated, according to the direction of motion required, by a crown wheel and one of the pinions W W. The vertical pinion shaft is driven constantly in one direction by clockwork, and is supported by the iron movable core r of the solenoid R. In its mid-position the shaft is prevented from rotating by the arm i coming in contact with the stop k. When raised or lowered, by the action of the solenoid upon the core, beyond this stop, its revolution is controlled by the centrifugal brake I. The solenoid R is included in the lamp circuit, and exercises a varying upward attraction upon its core dependent on the resistance of the arc and the condition of the battery. Its power is regulated by means of graduated weights, such as F^1, and needless oscillations and reversals of the apparatus are prevented by the loose weight t, which is dropped on to the adjustable shelf u when the core falls below its mid-position, and is taken up when it rises above it. A saucer is placed round the carbon holder to catch the sparks, and the upper part of the lamp is enclosed in a transparent globe (not shown in the figure) provided with small holes or valves to maintain the equilibrium of pressure within and without.

Within a year, Staite was again at the Patent Office, and this time he offered quite an *embarras de richesses* to the public. First, he had an improvement in the mechanical details of the lamp we have just illustrated, substituting for the screw, worm wheel, worm crown wheel and pinions, a rack and pinion, a ratchet wheel, and a double ended pawl on a constantly reciprocating arm. Next, he described a lamp with a disc for the upper electrode, and after that, two forms of incandescence lamps with iridium bridges. Then followed several arrangements of mechanical lamps of which we illustrate the best (Fig. 375). The carbon holder B^{11} is constantly impelled upwards by a cord and weight not shown in the drawings, and not very greatly heavier than itself. This difference is maintained constant, notwithstanding the consumption of the electrode by its picking up a chain as it rises, equal in weight, inch for inch, to the carbon pencil. The action of the solenoid A is to draw down its core, and by means of the spring pawl *g* attached to the core, the carbon and its holder, until the current is so far diminished by the increased resistance of the arc that an equilibrium between the electric force and force of gravity is obtained. As the carbons consume, the core is gradually drawn up again, and the spring pawl *g* is forced by the incline H out of the ratchet teeth, when the rod B¹¹ is free to rise until again arrested by the action of the increased current. No one, we think, can fail to see in this invention much that has been produced as new within the last two or three years. Among the most successful lamps to-day are those in which the regulation is effected by a solenoid rather than a

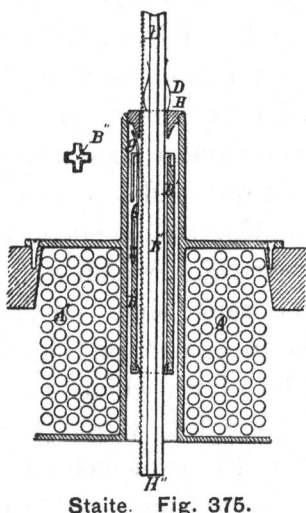

Staite.   Fig. 375.

magnet. In them, as in the lamp before us, the actual work of liberating the catch or the gripper, and feeding the carbon, is effected by gravity, and the influence of the coil is only employed to establish or lengthen the arc and regulate the motion. It is probable that this lamp did not work satisfactorily in practice, but it is nevertheless evident that it only required good mechanical knowledge, and not inventive ability to make a success of it.

Foucault.—We have seen that Foucault, so early as 1844, devised a crude means of maintaing the electric arc by hand regulation, employing however gas retort coke for his carbons. In 1849 he schemed an ingenious but cumbrous piece of mechanism for the same purpose. The following is his own description of the apparatus: "The two carbon holders are pressed towards each other by springs; but they can only be brought into contact by the action of a train of clockwork, the last wheel of which is controlled by a detent. It is at this point that the action of the electro-magnet is introduced. The current producing the light traverses the coils of an electro-magnet, the energy of which varies with the intensity of the current. This electro-magnet acts on a piece of soft iron, normally drawn in an opposite direction by an opposing spring. On this movable piece of soft iron is mounted the detent which checks the clock train, or allows it to move, and the direction of the movement of the detent is such, that it presses on the wheel train when the current is strong, and releases it when it becomes weak. Now, as the current grows stronger or weaker precisely as the space diminishes or increases, it will be understood that the carbons become free to approach each other at the moment when their distance begins to increase, and that this approach can never end in contact, because the increasing magnetisation which results, soon opposes an insurmountable obstacle which is however again removed as soon as the interpolar distance is once more increased." In this arrangement the carbon holders were mounted on small travelling carriages, which were pushed towards each other by springs that at the same time served as conductors. The reduced speed of the carriage to compensate for the lesser wear of the negative carbon was effected by a lever which acted direct on the carriage of the positive carbon, and through a system of cord and pullies on the carriage of the negative carbon, the position of this latter was regulated by a small hand winch. The electro-magnets through which the current passed was provided with an armature, the action of which could be regulated by a spring lever. A connecting rod passed from the armature to the clock train, described above. A current regulator was also employed; it was formed of two immersed plates of platinum placed parallel, and 1 millimetre apart, the two plates being insulated. The current passed from one plate to the other traversing the liquid—a solution of sulphate of soda—and by varying the depth of immersion, the intensity of the current was increased or reduced.

Archereau.—In this lamp, devised by M. Archereau, in Paris, as early as 1848, a current traversing a hollow electro-magnet was employed to regulate the length of the arc. The upper carbon was held by a bracket carried on a vertical post; the lower carbon was placed in a holder formed of a rod of soft iron, which descended into the hollow magnet. The holder and the carbon were carefully balanced by a counterweight passing over a pulley on the post, carrying the upper carbon holder, this weight always tending slightly to raise the lower carbon. When the current passed, the magnetisation of the solenoid drew down the iron bar within it, separating the carbons until the strength of the current was sufficiently reduced to allow the counterweight to pull up the lower carbon again into contact. This arrangement was as simple as it was ingenious, but apart from other practical drawbacks, the focus of the light was always shifting with the consumption of the carbon.

Pearce.—Passing the next two lamps, which, although interesting as evidences of the activity with which the subject was pursued, have no other value, we come, in the next year (1849), to one constructed by C. T. Pearce, in which there is a novel expedient for restoring the arc when it has been blown out, or otherwise accidentally destroyed. At one end of a lever is a short metal tube, in which is fitted a piece of carbon shaped like a wedge. To the other end of the lever is connected a soft iron armature to which is attached a spiral spring. The lever is kept in position away from the electrodes by an electro-magnet, so long as the current is passing. The instant the light is blown out, and the circuit is broken, the magnet loses its power, and the spiral spring moves the lever so as to bring the carbon wedge between the electrodes. The circuit is by this completed, the magnet regains its power, and, moving the wedge back again, the light is restored.

The same year *Le Molt*, who was working hard in this field, in France, fell back on the general idea of the Wright lamp, and designed one in which two discs of carbon were kept revolving in contact, by a suitable train of clockwork. His patent is curious because he claims the use of all kinds of carbonized material for electrodes, especially gas retort coke, and also a means of rotating two discs, and of varying the distance between their peripheries.

Roberts.—In 1852 an attempt was made by M. J. Roberts to dispense with the feeding mechanism of arc lamps. In his designs, one electrode was fixed and the other was maintained in one position until the arc failed, when the two carbons came together and were then separated by the length of the arc, and held until the arc again was broken. In one

c c

form of lamp, the upper carbon was pushed by a weight through a tube with a platinum-lined spring mouth-piece. Two clips or tongs passed through slots in the tube and gripped the carbon. When the current failed, the pencil slid through the tube and the clips until it rested on the lower pencil and completed the circuit. Immediately an electro-magnet, included in the lamp circuit, attracted its armature and by a system of rods and levers tightened the clips on the pencil, and at the same time raised both the tube and the clips the length of the arc.

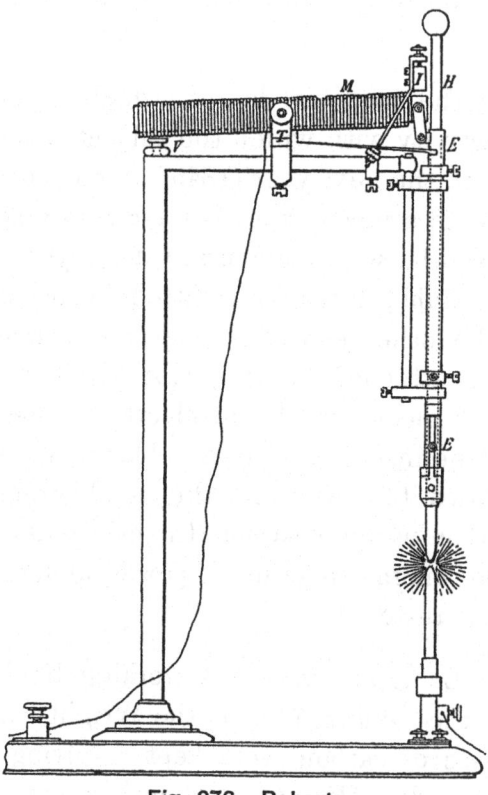

Fig. 376. Roberts.

A second lamp by the same inventor is shown in Fig. 376. The tube E acts as the carbon holder, the pencil being normally retained by a slight frictional contact, so long as the weight of the brass-faced rod H is not allowed to rest upon it. This weight, when free, is sufficient to force the pencil through its holder. The tube E can slide up and down in guides, and is connected by a link to the electro-magnet M. When the current begins to circulate, the magnet attracts the weight H, and, rotating on the studs or trunnions T, rises towards the fixed armature I until stopped by the set screw V. When the current fails, the magnet rotates in the opposite direction, the weight H is released, and the carbons come together, whereupon the magnet is again excited and the arc is formed.

Slater and Watson.—The next lamp that calls for notice is remarkable for the fact that the carbon holder passed through an eye in a lever which acted as a clip or gripper. When the lever was horizontal the holder was free to slide through it, but as soon as its outer end was raised

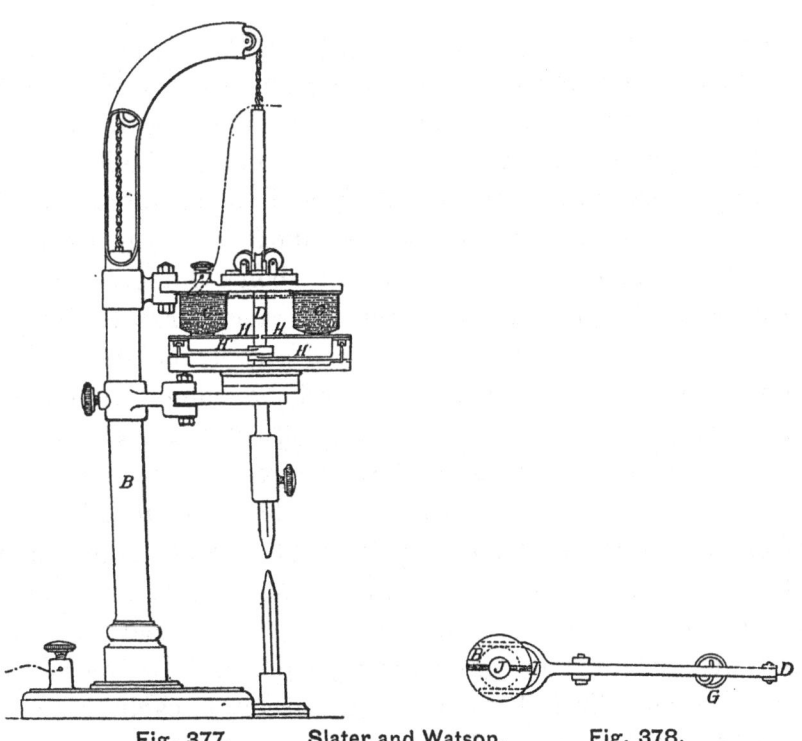

Fig. 377.   Slater and Watson.   Fig. 378.

it bound or locked the holder, and stopped its motion. Our readers will, we think, recognise the lamp as an early type of some that are now working very successfully. Referring to the illustration, C C is an electromagnet carried by a bracket on the column B. The armatures H H are attached to brass levers H¹ H¹, fitted with eyes, through which the carbon holder D passes freely when the magnetic induction ceases. When the current traverses the magnet coils the outer extremities of the levers are raised, and their eyes simultaneously grasp the rod, and retain it until from the increased resistance of the arc the current diminishes or ceases.

In a modified form of this lamp the electro-magnet was placed in the base, and its armature connected to a rod passing up the hollow pillar to a gripping lever shown in plan in Fig. 378. The forked end of this lever embraces the collar B, which consists of two semi-conical pieces hinged together at I, and nips it firmly to the rod, so long as the armature is attracted to the magnet. When the current fails the spring G raises the end D¹ of the lever, and releases the conical clip from the grip of the forked end, whereupon the upper carbon slides down to the lower one, and then the arc is again formed.

Banks.—In January of the following year (1853), Christopher Banks filed a provisional specification in which were several ideas that have since been often revived by inventors, but seldom with any practical result. He proposed, in the first instance, to carry the lower electrode on a floating stem, immersed in liquid, and so balanced that as the carbon consumed, the stem would rise. In another arrangement a carbon pencil was to be placed within a hollow carbon cylinder, and the arc to be formed across the intervening annular space ; or a carbon rod was placed beside a second, and gradually moved past it, the arc being formed between the point of the stationary one and the side of the moving one. In another instance a stream of mercury constituted one electrode. The principal feature of the invention, however, consisted in an arc lamp wherein the usual constant or intermittent feed was abandoned, and in place of it the electrodes were caused to alternately approach to and recede from each other (short of actual contact) by a rapid movement imparted to one or both of them by any mechanical agency. The movement was obtained in various ways, such as "by the vibration of wires," or of a metallic fork carrying one carbon on each branch ; by an eccentric wheel, or by the revolution of two discs on a common axis, but revolving in contrary directions and carrying carbon arms or radii, acting as the electrodes and crossing each other scissorwise; by the rapid rotation of two carbon wheels with serrated edges, placed edge to edge, or by rings placed concentrically "with the light-emanating positions serrated or made angular." An apparently uniform emission of light was also to be obtained by bringing the electrodes into actual contact and separating them with great rapidity. A rotatory movement could be combined with the reciprocatory one to the end that the several independent flashes might be blended together into an apparently uniform ring of light. This specification is not illustrated, and so far as we know the invention was not put into operation.

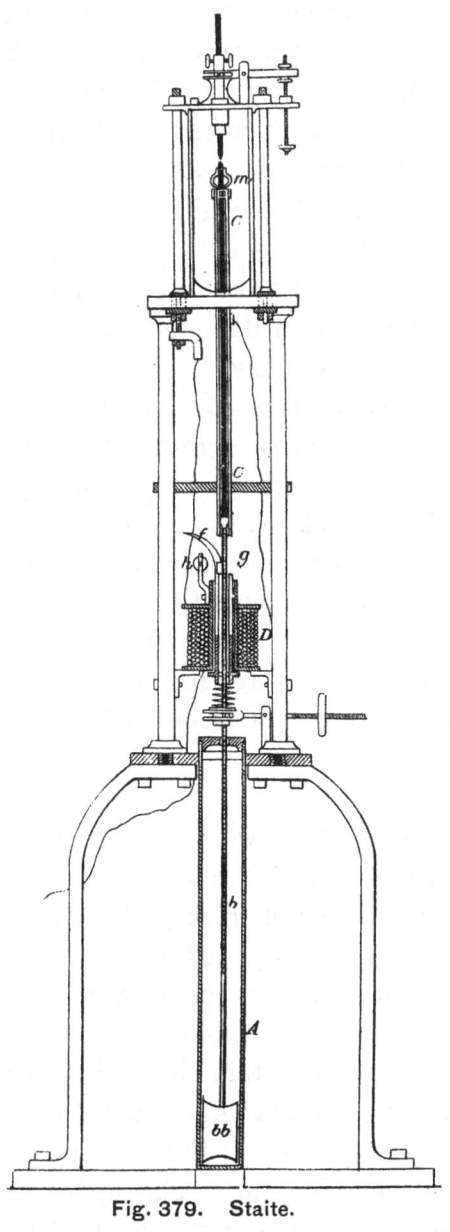

Fig. 379. Staite.

Staite.—Within a few months of the last named patent, W. E. Staite, patented a lamp in which the idea of raising the lower electrode by a float was reproduced. This lamp is illustrated in Fig. 379, where *b b* is a float attached to the stem *b*, and moving within the vessel A, which is filled with liquid. The weighting of the float is so regulated that its tendency to rise causes the lower electrode to be very slowly pushed up, and only as fast as it is consumed. D is a helix wound on a metal reel, and provided with a hollow soft iron core, capable of free motion up and down within it The weight of the core is borne by the spiral spring and lever below it. When the normal current is flowing around the helix, the core is drawn downwards until the spring click *f* is pressed against the roller behind it, and so forced to engage with the teeth of the rack on the stem. As the arc lengthens the core rises until the click comes out of gear with the rack and so releases the stem, when the buoyancy of the float carries the carbon upwards until the length of the arc is re-adjusted, whereupon the core is drawn down and the feed arrested. The specification goes on to say, that in place of the ratchet and click it is preferred to attach " to the electro-magnet, at the top, a friction clip, which, as the electro-magnet " (core?) " is pulled down by its attraction towards the centre of the coil, clasps the rod *b* and holds it firmly to the magnet, so that it cannot ascend until such diminution in the power of the magnet be produced as shall allow it to rise to a position at which the clip is detached from the ratchet." Unfortunately this arrangement is not illustrated; it would be very interesting to know exactly what was in the inventor's mind, and how nearly it resembled the frictional lamps that are now in vogue.

Chapman.—During the next two years, no improvements of any importance were patented in electric lamps, but early in 1855 we find a specification bearing the name of Henry Chapman, in which five regulators are illustrated. The lamps, which are all modifications of the same design, do not present any features of great novelty or originality, but they have the merit of being thoroughly practical. The inventor appears to have endeavoured to attain his end rather by the exercise of care and good mechanical knowledge than by the aid of what is popularly called " inventive genius." Fig. 380 illustrates the simplest and most typical form of the apparatus. The upper electrode G is fitted into a socket H, and slides through a split tube D, divided longitudinally and held together by springs. Attached to the socket is a weight I suspended from a chain, which passes over pulleys, and is attached at its other end to the barrel K, the motion of which is controlled by a brake

wheel L on the same axis.   This wheel is embraced by a shoe on a lever
which carries an armature situated over the electro-magnet O.   The
attraction of the magnet is opposed by an adjustable spring *d*, which is
overpowered when the arc is of the proper length.   As the carbons
waste, however, the spring raises the lever and allows the brake wheel

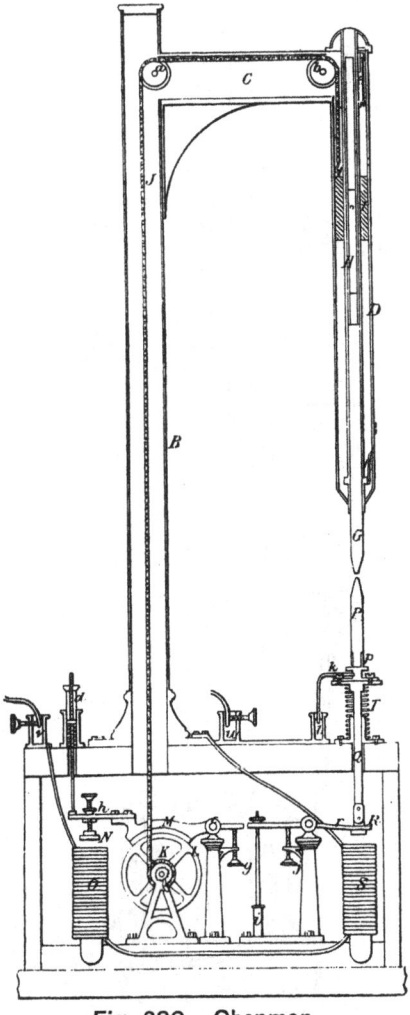

Fig. 380.   Chapman.

and barrel to rotate, and the upper electrode to approach the lower one
to adjust the distance between them.   The lower carbon P is fixed in a
socket on the rod Q, which is connected to the armature of the electro-
magnet S, and supported in opposition to the pull of the magnet by the
springs T and *i*.   As soon as the circuit is established this armature is
attracted and the arc formed.   The specification illustrates a focussing

lamp, a horizontal lamp, and a lamp in which both the feeding and the establishing of the arc are effected by the same electro-magnet.

Lacassagne.—The next lamp we meet is remarkable as embodying in the clearest manner the differential principle now almost universally

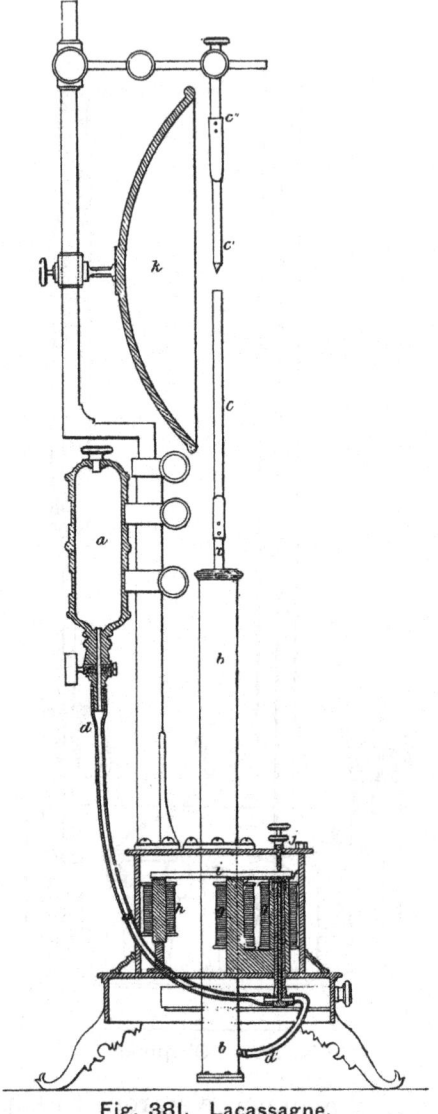

Fig. 381. Lacassagne.

adopted where several lights are burnt in series. It was patented by Joseph Lacassagne, Chemist, and Rodolphe Thiers, Manufacturer, both of Lyons, on the 20th of October, 1856. The general arrangement of the lamp is clearly shown in the annexed illustration, Fig. 381. The lower

electrode is mounted on a piston or float in the cylinder *b b*, and is fed
upwards by the intermittent introduction of mercury from the reservoir
a into this cylinder. The liquid passes by the flexible tube *d* to the
cylinder through the regulating passage *e*, in which it rises up one of the
vertical ducts, flows under the india-rubber valve *f*, and down the second
duct. The valve is controlled by the oscillating armature *i*, one end of
which is over the electro-magnet *g g*, which is in the lamp circuit, and the
other end over the magnet *h*, in a shunt or derived circuit. The
differential action is described in the specification in these words :—" The
armature, which shuts the valve by the action of the current producing
the light, is attracted in an opposite direction by a spring, or by a second
electro-magnet receiving magnetic action from the derived current, the
intensity of which is determined by the resistance offered to it by a coil
interposed for that purpose, consequently the passage of the mercury from
the reservoir to the cylinder tends to establish itself under the action of
the derived current, which opens the valve, and to be interrupted under the
action of the principal current, which closes it ; the resistance offered to
the derived current is invariable. The resistance offered to the principal
current increases by the widening of the distance, or space, between the
two electrodes, and diminishes by their coming nearer to each other. If
the distance between them increases, the principal current, meeting with
a greater resistance, diminishes in its intensity, which also diminishes
the magnetic power of the electro-magnet due from the principal current
which closed the valve ; the intensity of the derived current (the resistance of which is invariable) increase sin the same ratio, and tends also by
its action on the second electro-magnet to open the valve." * * * *
" It is then evident, that after having determined the resistance of the
derived current by the length of the wire of the resisting bobbin or coil,
there will be established between the two currents a kind of equilibrium,
which will maintain a passage for the exact quantity of mercury that
should necessarily flow from the reservoir to the electrode cylinder to lift
and regulate the space between the electrodes." It would be very
interesting to know how many persons have imagined themselves to be
" the first and true inventors " of the differential method of regulation
during the last four years.

Way.—At this time inventors appear to have conceived the idea of
avoiding the difficulties of an intermittent feed, by the use of flowing

electrodes of melted metal, or mercury, and during the years 1856 to 1859
there was a constant succession of patents for apparatus based on this
principle. For the sake of its historical interest, as illustrating one of
the steps in the progress of electric illumination, rather than for any
practical value, we give a drawing of one of these lamps invented by
J. T. Way, in 1857. In Fig. 382, *a* is an iron tube communicating at
its upper end with a reservoir of mercury, and carrying at its lower end a

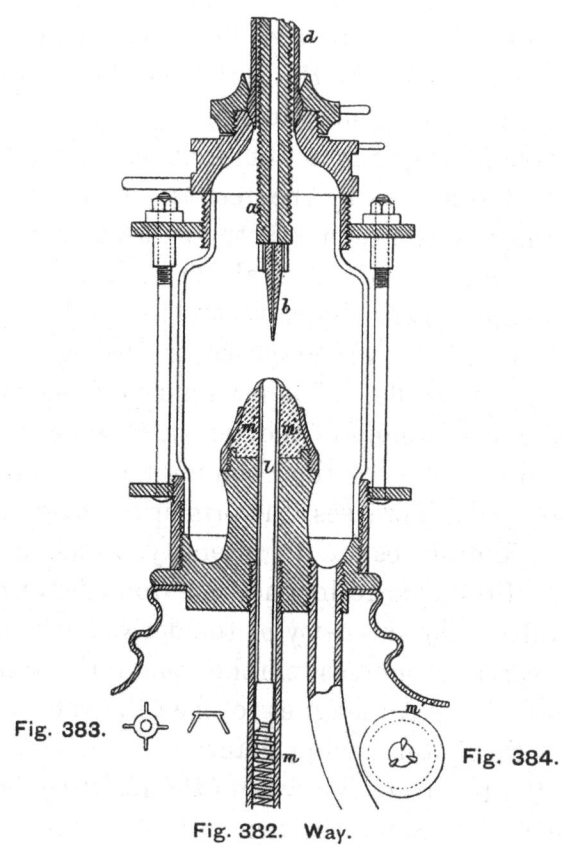

Fig. 383.

Fig. 384.

Fig. 382. Way.

jet of fire-clay cemented into it. It is so held that its position can be
adjusted in every direction, *l* is the lower electrode which is pressed
constantly upwards by the spiral spring *m*. Its higher end is prevented
from rising above its proper position by a cap of fire-clay, m^1, carrying a
ring of iridium of smaller internal diameter than the carbon rod. The
ring is shown separately in Fig. 383, and it will be seen that it is furnished
with four projections, which are embedded in the fire-clay cap before it is
burnt. An alternate method of manufacture is to imbed a series of

separate points in the clay projecting inwards, as shown in Fig. 384. The mercury may be heated before using, and the breaking point of the jet be regulated by altering the lead of fluid, which when once set may be made constant by an arrangement acting on the principle of the bird fountain.

Serrin.—In 1857, Serrin took out his first patent. His invention marks the close of the early era of electric lighting, and is the only one

Fig. 385.

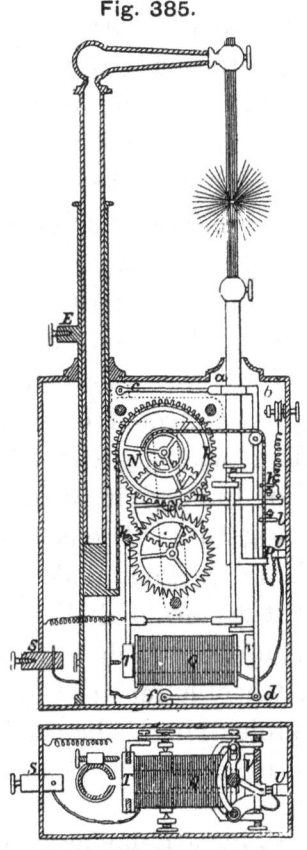

Fig. 386. Serrin.

that maintained an existence across the interval of 17 years that elapsed before the production of the Jablochkoff candle, again called the attention of the public to the powerful means of illumination that had so long lain unused and unappreciated. From the illustrations, Figs. 385 and 386, it will be seen that this lamp has been considerably improved since it first appeared, although the main features still remain. The carbon holders are sustained by two cords, each of which passes round a drum fixed to the wheel k. One of these drums is double the diameter of the other, to compensate for the difference of consumption of the two electrodes. The wheel k forms part of a train of which the last member is provided with ratchet teeth, and revolves or is stationery accordingly as the pawl m is in or out of gear with it. The arc is struck, and the feed regulated by an oscillating system which forms the particular feature of the invention, but which is not very clearly shown in the drawings. It is composed of two horizontal bars, $a\,e,\,f\,d$, and one vertical bar, $b\,d$, pivotted together and to the frame. The vertical bar carries an armature V, which is attracted downwards by the electro-magnet Q when the current is flowing. The cord which supports the lower carbon holder passes round a pulley N on the vertical bar, and when the latter descends, the cord, pulley and carbons all fall with it, thus forming the arc. The pawl m is carried on the same system, and can oscillate between the fixed stops h and l. As soon as it touches the upper one it is raised out of gear with the ratchet wheel, and allows it and the train of wheels to revolve, paying out the two cords and allowing the electrodes to approach each other. T is a second armature opposite the other end of the magnet. It is suspended from an arm, and is connected by a rod to an arrangement of mechanism on the lower holder, whereby when the armature is attracted the electrode is slightly rotated to break it off from the upper one in case the two have become welded together. The course of the current is from the binding screw S, round the magnet coil, through the chains U P, and the two electrodes to the terminal E. The specification also includes a simpler form of lamp, in which there is no torsional arrangement, and where the ratchet wheel is replaced by a fly that is alternately engaged and released by a projection on the oscillating system. A rack and pinion is also substituted for one of the drums and cords.

Marçais and Duboscq.*—This arrangement, which was first schemed in 1858, and some years afterwards brought before the public, merits description as an ingenious arrangement rather than a practical device. The lamp has a cylindrical body A (Fig. 387), filled with oil up to a certain height, above which is a piston D. This piston is carried by a rack, around which is a strong coiled spring R; on the top of the rack is a standard, carrying the upper carbon holder H. On the vertical arm E

---

* See " Les Applications de l'Electricité," Vol. V. By the Count du Moncel.

G of this standard, which is square, moves a slide E I carrying the lower carbon holder J. To this slide is connected a lazy tongs device, which moves about the axis K, and the motion of which is controlled by a stud

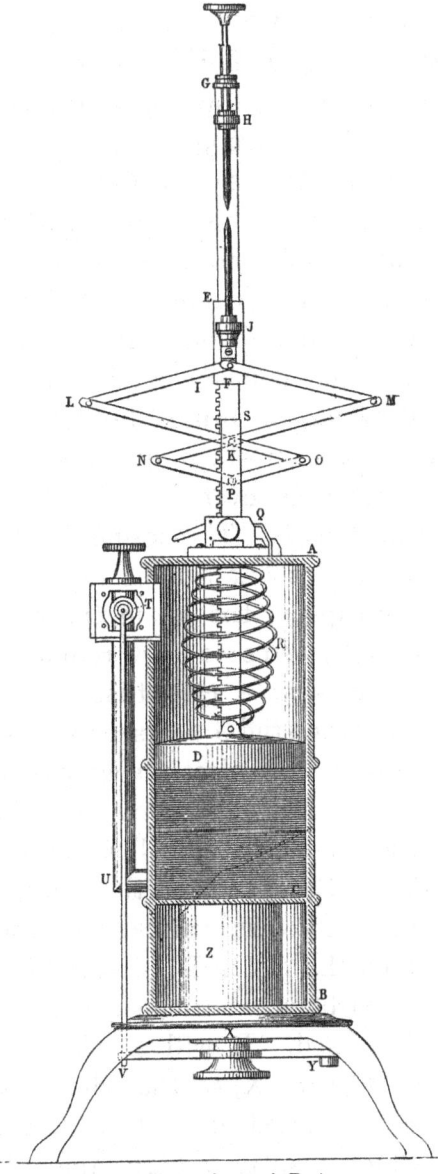

Fig. 387. Marçais and Duboscq.

fixed on the rack, and acting on the joint P. The pin K is not fixed on the rack, but on a rod fastened to the body of the lamp. From this arrangement it will be seen that as the rack falls it re-acts at the same time on both the carbon holders, lowering the upper one H which is in one with it, and raising the other J, by the opening out of the lazy tongs; but as the lozenges of this latter are of different sizes and so calculated that their sides are in the ratio of the wear of the carbons, it follows that when the lozenge K N O P is opened a certain amount a, the lozenge K L F M is opened a greater amount b, which bears the same ratio to a as the wear of the positive carbon does to the negative carbon. By this arrangement the position of the light remains unchanged, and the necessary action is obtained by means of the travel of the piston D. The second part of the arrangement refers to the method by which the movement of the parts is made dependent on the intensity of the current, which furnishes the voltaic arc. To the outside of the cylindrical base is attached a bent tube U, which is open to the top and bottom of the cylinder; at the top of this tube is a box T having an orifice furnished with a rubber diaphragm. On this diaphragm rests a stud connected to a long lever N, articulated to the horizontal lever V Y, which carries the armature X of a tubular electro-magnet Z. The spring R always tends to drive down the piston D, but its descent can be regulated according to the provision made for the escape of the oil in the cylinder. As long as the pipe U is open, the oil is free to pass round to the upper side of the piston, which then falls at a rate that varies with the size of of the opening; but when the stud J bears upon the rubber diaphragm, which it does by the action of the electro-magnet Z, the passage for the flow of the oil will be more or less obstructed, according to the energy of the magnet, and the motion of the piston will be correspondingly modified. As the action of the electro Z depends on the energy of the current producing the light, which in its turn varies with the distance between the carbons, the motion of the piston D is placed in direct relation with the consumption of the carbons, and consequently they are brought together as they are consumed. It will be observed that the electro-magnet Z has a screw armature X, so that the pressure to be brought upon the stud T can be regulated accurately. This lamp belongs to that numerous class which display a high amount of ingenuity, but have failed to establish a position in practice; it must not be confounded with another form of Duboscq lamp, in which the movement of the carbon was controlled by clockwork, and which for years was very successfully employed.

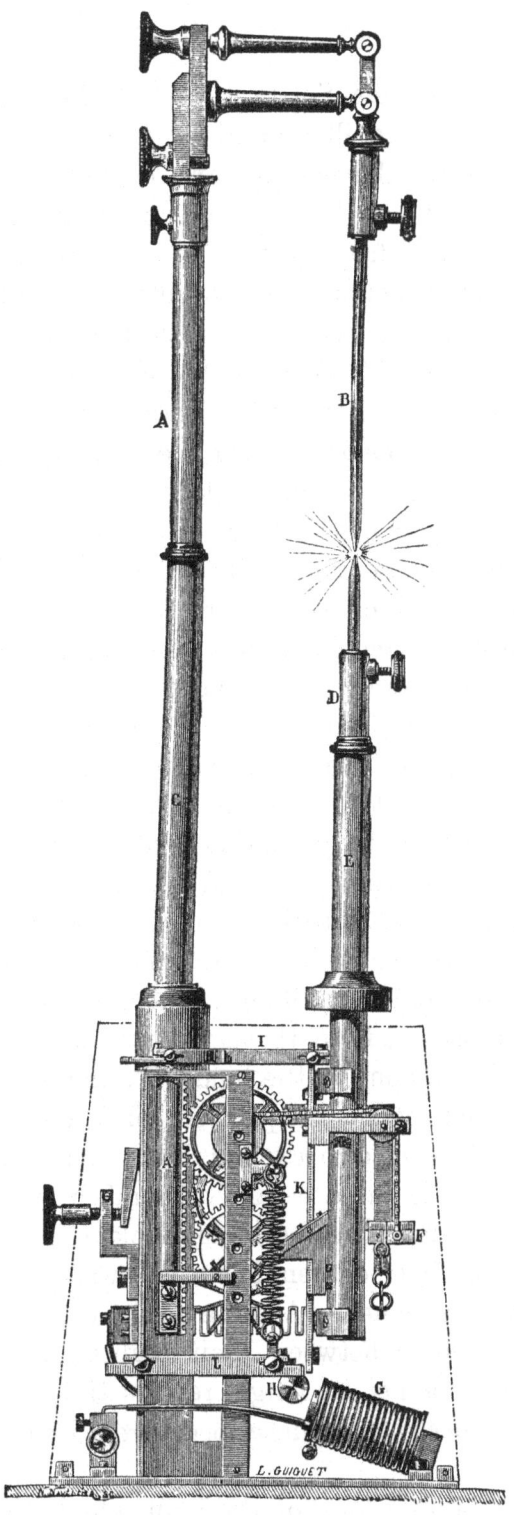

Fig. 388.   Serrin (Suisse Model.)

Serrin (Suisse Model).—The Serrin lamp, of the type known as Suisse, and devised in 1859, is illustrated by the annexed engraving, Fig. 388, where A and D are the positive and negative carbon holders respectively. At the top of the lamp is a horizontal connecting link which may be traversed to and fro by means of an adjusting screw. The stud immediately below is fast on a small eccentric, and by turning the two the upper carbon holder can be moved in both directions, so as to adjust its point exactly opposite that of the lower carbon. The vertical rod A holding the upper carbon slides in the vertical standard C; as will be seen, this rod is extended below C into the body of the lamp, the lower part being provided with a rack; it is this rod which, by its gravity, forms the motive power of the lamp. Gearing into the rack is the first wheel of a clock train and on its spindle is a drum having a pitched chain, the end of which is fixed to the bracket F and lower carbon holder. The last of the train is a star, held by a finger on the rectangular articulated frame J, which is free to oscillate, and which serves to regulate the feed of the carbons. The spring K balances the weight of the movable pieces connected to the oscillating parallelogram, and pulls against the action of the armature H of the electro-magnet G. The tension of this spring can be regulated by a lever from the outside of the lamp casing and stud. When the current does not pass, the electro-magnet C is inactive, and the oscillating rectangle is held up by the springs; the star wheel is liberated, and the carbons by the action of the clockwork and the falling rod A, approach each other. As soon as the current passes, it flows through the upper carbon holders and descends through the carbons to the holder E, and thence through the zig-zag plate at the back of the lamp to the electro-magnet G, and back to the generator. The electro excited by the current attracts its armature, and with it the articulated parallelogram J, drawing down the lower carbon, while at the same time the detent, locks the clock train and prevents the upper carbon from descending; the two carbons are thus separated and the arc is established. As soon as the consumption of the carbons increases the distance between their points, and the current grows weaker, the electro-magnet no longer retains the armature in its lowest position, and the process above described is repeated.

Gaiffe.—The Gaiffe lamp is one of those in which the action of a coiled spring supplies the motive power instead of the falling weight of the upper carbon holder. It is illustrated by Fig. 389, where H and H¹ are respectively the positive and negative carbon holders. The former is

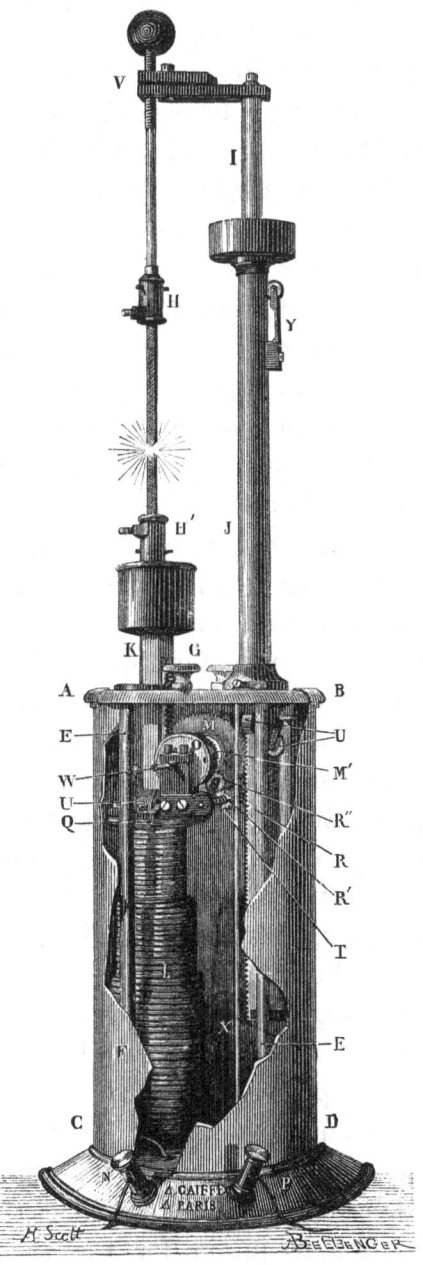

Fig. 389. Gaiffe.

suspended from a horizontal bar V at the top of the vertical rod I, which descends into the casing of the lamp, and is furnished at its lower end with a rack. Above the casing it slides in the hollow standard shown The bar K supporting the negative carbon holder is rectangular in section; it also is furnished with a rack, and a tail below enters the hollow core of the electro-magnet L, the coils of which increase in thickness from the top to the bottom. The feed of the two carbons in the proper ratio of 1 to 2, is secured by a coiled spring in a box O, on the spindle of which are the two toothed wheels M M^1, the latter being half the diameter of the former; these wheels are insulated from each other, and gear in the racks of the positive and negative carbons respectively. The force of the spring in the box O can be regulated by a key, the square end of the spindle W projecting beyond the case of the lamp. The carbon holders can, however, be raised or lowered independently of the mechanism just described, by means of the pinions R, R^1, R^{11}, which are mounted on axes that can be moved horizontally so as to gear in to the racks of I and K. By means of a key turning the axes of these pinions, the holder can be raised or lowered without affecting the rest of the mechanism; a coiled spring withdraws these pinions to their normal position out of gear as soon as the key is removed. The rods K and I can also be locked by turning the lever G on top of the case, to engage in the rack. The current is supplied from the terminal P (the negative terminal is shown at N), and proceeds by the vertical rod X to the standard of the positive holder J, a perfect contact with the rod I being secured by means of a roller mounted on the spring Y; the roller passes through a slit in the column J, and presses against the bar I. Guiding pulleys U U in the casing of the lamp, press against the rods I and K and prevent their turning.

The action of this apparatus is as follows :—So long as no current passes, the tension of the spring O holds the carbons in contact, but when the circuit is established, the electro-magnet L draws down the bar K against the action of the spring, and of course in so doing, turns the axis W and the toothed wheels M^1 and M; through the latter the rod I is drawn up, the carbons are separated, and the arc established. As soon, however, as the wear of the carbons weakens the attraction of the electro-magnet L, its effect is overcome by that of the coiled spring O, which draws the carbons together and again re-establishes the arc at its normal length. The exact tension required for the spring can be easily ascertained by trial.

Tchikoleff.—In order to overcome certain faults in the lamp of

MM. Foucault and Serrin, M. Tchikoleff, Chief of the Electric Lighting Service of the Russian Artillery, devised, in 1869, a use of the "differential action of derived currents" for that purpose, and in 1871 he communicated the fact in a paper read before the Society of Naturalists at Moscow. In 1874 he designed the lamp shown in the diagram, Fig. 390, and this also was exhibited to the same Society. In his arrangement,

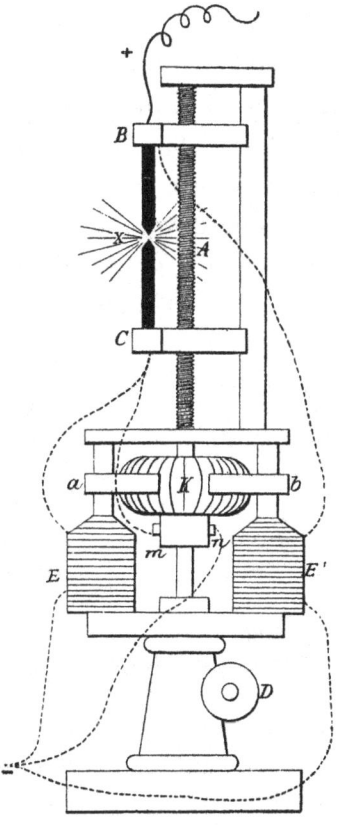

Fig. 390. Tchikoleff.

E E[1] are electro-magnets disposed like those of some other systems, and having on the poles *a b* circular plates like the pole-pieces in the Gramme machine. K is a Gramme or Siemens ring, the rotation of which, will cause the carbons to move by the action of a right and left screw, and the two nuts B C, which serve as carbon holders. The screw D is used to raise or lower the luminous point.

The current passes from the positive pole of the generator to the negative by three derivations, one of which, after having traversed the arc, reaches the ring by the brushes *m n*; a second, which after having also passed the arc, excites the electro-magnet E (or the two electro-magnets at once, but in a given direction); and finally a third which, without passing across the arc, excites the electro-magnet of fine wire E^1 (or the two at once in a direction contrary to the last), but in such a manner that its action on the ring is contrary in direction to that of E.

As a result of this disposition, with a normal resistance of the arc x, the influence of the electro-magnets E E^1 on the ring K will be about nil; but if the resistance of the arc augments, the action of the electro-magnet E becoming more feeble will be overpowered by that of the electro-magnet E^1, and the ring K will move in such a manner as to bring the carbons nearer together. The inverse effect will naturally take place if the resistance of the arc diminish.

Figs. 391 to 393 show a recent development of this system, patented by MM. Tchikoleff and Kleiber, both of St. Petersburg. In this device the controlling of the arc and the continuous adjustment of the carbons

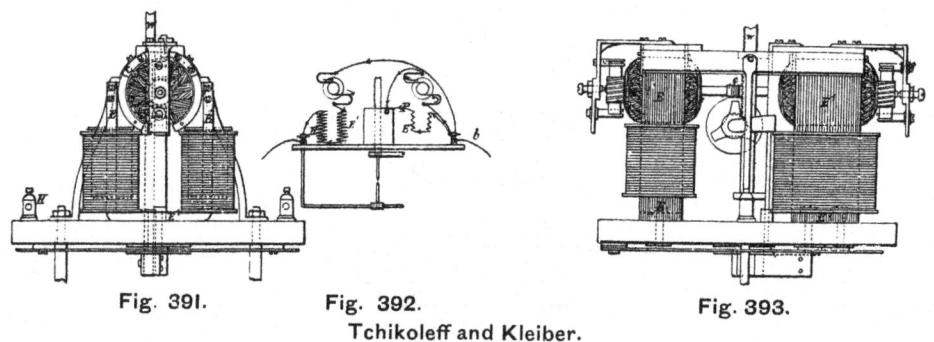

Fig. 391.   Fig. 392.   Fig. 393.
Tchikoleff and Kleiber.

is effected by two small electric motors combined with the usual mechanism of a lamp. These motors oppose each other, one tending to increase, and the other to decrease the arc, the former being placed in the main circuit, and the latter in a shunt circuit between the lamp terminals, following the differential principle. The motor which transmits the main current is wound with coarse wire, while that on the shunt circuit is wrapped with fine wire to reduce the amount of current that leaks past it, and also to enable sufficient convolutions of wire to be used in a reasonable space to give it power to operate the lamp mechanism. Figs. 391 and 393 are two elevations of the mechanism. E and E^1 are two small Siemens electro-motors, each having two field magnets with extended pole-pieces surrounding its armature. Both armatures are fixed on the same spindle on either side of a worm *e*, which, through intermediate wheels, raises and lowers the rack W attached to the carbon holder. The leading wire *b*, as shown in Fig. 392, is divided into three branches, the first of which traverses the field magnet coils, and the second the armature coils of E, both meeting at P and going through the carbons and across the arc to the other lamp terminal H. The third branch leads in succession through the armature and magnet coils of E^1 to the same terminal H, the connections being so made that the two armatures tend to rotate in opposite directions. When direct currents are employed, one motor need only be used with the high and low resistance coils wound side by side on its field bobbins.

Von Hefner Alteneck.—This lamp is regulated in its action by an electro-magnet, operating a ratchet wheel and gear. Fig. 394 illustrates the arrangement ; the positive current enters at K, and flows to the upper carbon holder by two routes, one of which includes the coils of the electro-magnet E, while the other is direct. After crossing the arc, the current returns to the generator at Z. The two carbon holders are geared to two wheels on the same axle, one of which is half the diameter of the other, to compensate for the unequal consumption of carbons. The action of the lamp is as follows :—If no current passes, the two carbon points will run together by gravity, the speed being somewhat retarded by a fan geared to the wheel train. As soon as the current circulates through the lamp, the armature A is attracted to the electro-magnet and the lever T, pivoted at *k*, partly rotates the last member of the wheel train by means of the ratchet M, and slightly separates the carbon points. By its oscillation the lever T comes in contact with the clamp screw *c*, and the current now passes direct through the lever T instead of through the electro-magnet ; the armature A is released and the lever T is pressed back by the spring *f* into its normal position against the adjustable stop *d*. This process is repeated and a regular oscillation set up in the lever T by the make-and-break contact A, until the two carbon points are so far apart that the arc is established. When the current passing round the electro-magnet becomes too feeble to overcome the action of the spring *f*, which is in electrical connection with the upper carbon holder, the lever T remains for a while at rest, and, as the carbons consume, it gradually falls back against the set screw *d*, until the pawl *m* is lifted out of gear by the stop *n*, allowing gravity again to act on the wheel train and bring the carbons closer together.

Von Hefner Alteneck Differential Lamp.—Fig. 395 illustrates the Von Hefner Alteneck differential lamp, the principle which controls it being indicated in Fig. 396, where R represents a coil covered with thick

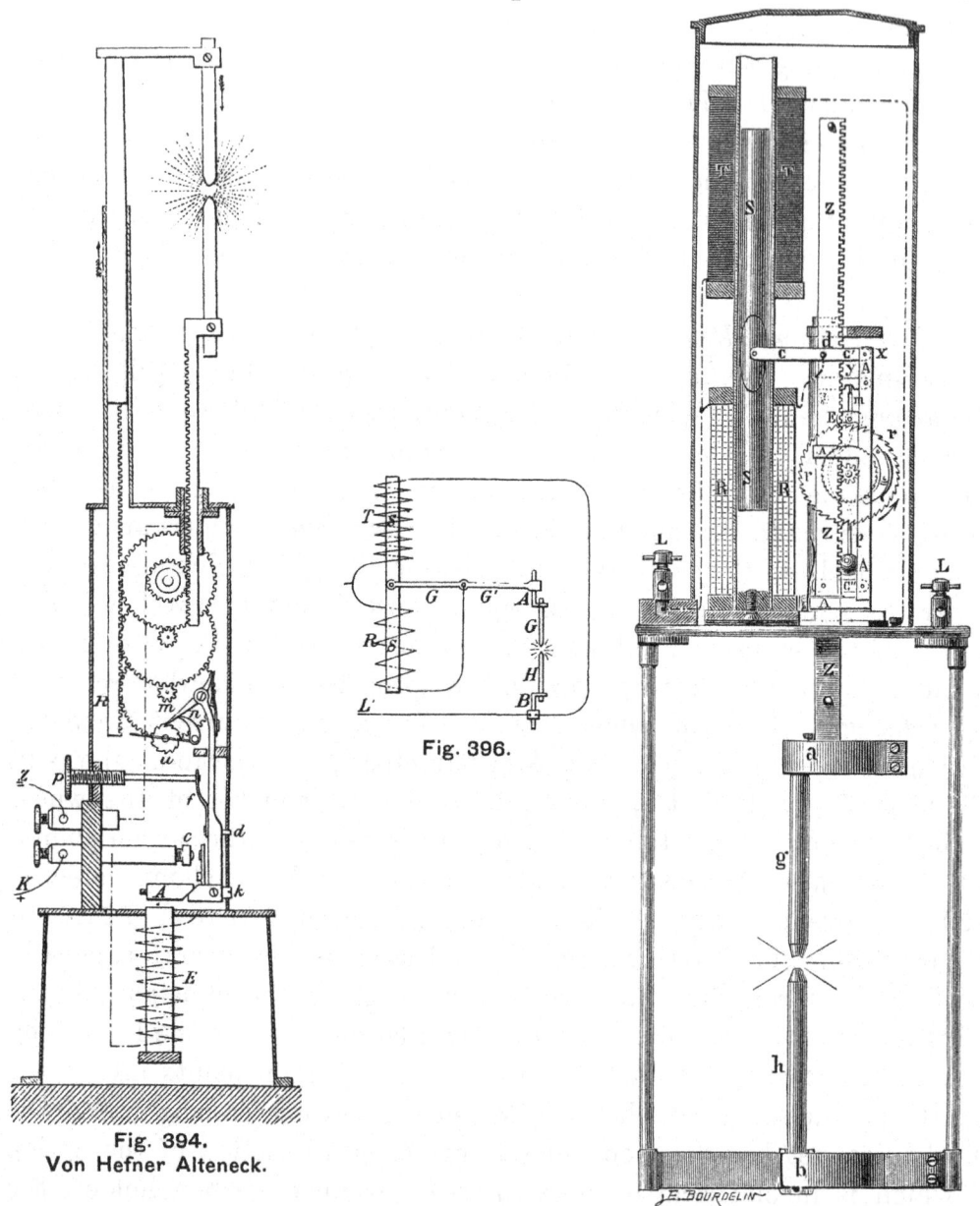

Fig. 396.

Fig. 394.
Von Hefner Alteneck.

J.E. BOURDELIN

Fig. 395. Von Hefner Alteneck Differential Lamp.

wire, and T a high resistance coil wound with fine wire. Both solenoids are fitted with a core S S, connected by the lever G G¹ with the upper carbon holder A, while the lower carbon holder B is stationary.

Following the course of the current which enters at L, it will be noticed that it is divided, part flowing through the coil T, while another part, passing through the arc, goes to the coil R and the return wire. The rise and fall of the core S S¹ depends therefore entirely on the difference of magnetic attraction in the two coils.

Assuming the current to enter at L while the carbon points are far apart from each other, then all the current passes through the upper coil T, the other line being broken between the two carbons. The solenoid core then rises, bringing the carbons closer together, and the carbon holder G¹ is depressed. In this position the carbon holder A is automatically detached from the bar G¹; the upper carbon drops upon the lower, and the current now passes through the carbons to the lower coil R, thus raising the lever, when the carbon holder A is again attached to the lever G and the arc established. A steady arc is thus maintained until a certain consumption of carbon has taken place, when the same series of operations is repeated, and an automatic feed is thus produced. A change in the external intensity of the current does not affect the action of the lamp, since the current is proportionately affected in both coils. From the foregoing, which describes the principles controlling differential lamps generally, the construction and action of the Von Hefner Alteneck lamp, which is made by Messrs. Siemens and Halske, of Berlin, and illustrated by Fig. 395, will be easily understood, the arrangement of the mechanical details only requiring a few words of description, which refer chiefly to the detaching gear for the upper carbon, consisting of a rack and pinion, with an escapement wheel controlled by a pendulum p, the latter being arrested in its oscillation by a small bar y fixed to sliding piece A, that is coupled to the solenoid. It will be seen on reference to the drawing, that the rack Z carrying the upper carbon is not directly connected to the solenoid bar G G¹, but the sliding bar A is pivotted to the bar G G¹ at d, and, by a short parallel motion bar C¹¹ at the bottom, is free to move vertically only. The rack Z acts on the pinion and ratchet wheel r, the rotary motion of which is governed by the escapement bar E, fixed to the pendulum p. The pendulum bar is extended upwards beyond its pinion, and this extension m arrests the pendulum by the notch in the cross-bar y when the sliding bar A is in the position shown; when this bar descends, the cross-bar y is lifted up by a fixed pin shown in the engraving; the escapement is then free to act and the carbon holder descends, until the difference in magnetic force, due to the distribution of current, again lifts the bar A. It will be readily understood that a number of these lamps may be disposed in one circuit. By a simple arrangement this lamp is also made self-focussing ; the lower carbon is contained in a metallic case and is forced upwards against a copper ring of a diameter slightly less than the carbon itself, by a spiral spring, only that part of the carbon which is already partly consumed passing through the copper ring. These rings do not waste as rapidly as might be expected, and they can be easily replaced. The generator used for these lamps is the Siemens alternate-current generator already illustrated and described.

Girouard.—This lamp, which was designed in 1875, was brought under the notice of the French Academie des Sciences in January, 1876. The lamp consists of two distinct apparatus connected by two circuits,

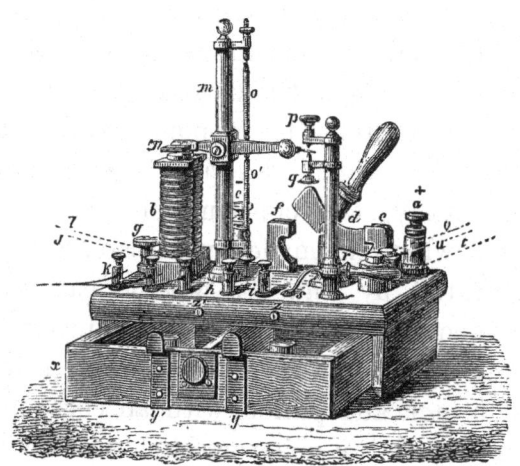

Fig. 397. The Girouard Relay.

one of high intensity which produces the arc, after it has passed through the relay constituting one apparatus, the other, a feeble current used only to operate the clockwork of the lamp. Fig. 397 shows the regulating relay, which consists of a coarse wire electro-magnet *b*, the armature *n* of which is placed on the end of a lever, and held by the two springs *o o¹*. The other end of this lever has a small contact plate situated between the contact screws *p* and *q*, which are coupled to the electro-magnetic systems that control the lamp carbons. Under normal conditions this plate remains midway between the contact screws, but if the current becomes too powerful or too weak, contact is made with one or other of the screws, and the clockwork is operated to make the carbons approach or recede from each other. When the carbons are in contact, and the full strength of the current passes through, then the attraction of the armature n causes the contact to be made with the screw p, and actuates the gearing in the lamp to separate the carbons : if on the other

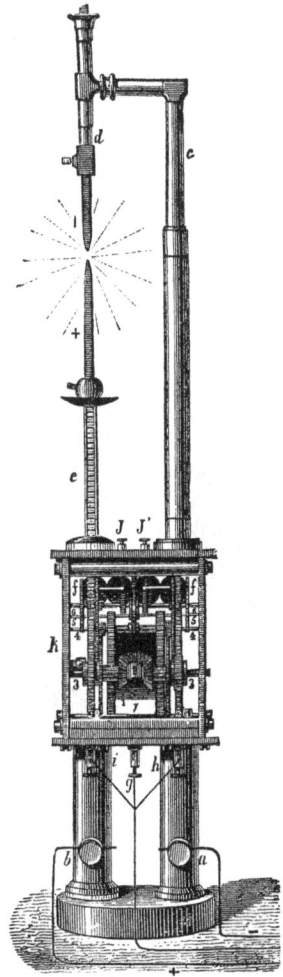

Fig. 398. Girouard.

hand, the distance between them becomes excessive, the spring o^1 draws down the lever until the plate at its end is in contact with the screw q, and the reverse action takes place, drawing the carbons together. The lamp, which is illustrated by Fig. 398, has the peculiarity of two distinct trains of clockwork, one for each carbon holder, each being connected to an electro-magnet corresponding to the screws p and *q*, Fig. 397 of the relay by means of a detent working into the last wheel of the train. Both systems are actuated by the same spring barrel; the batteries working the electro-magnets of the lamp are placed in the base of the relay.

Wallace.—The Wallace lamp, or rather its modification as it was used in England, is shown in Figs. 399 and 400; and, as will be seen at

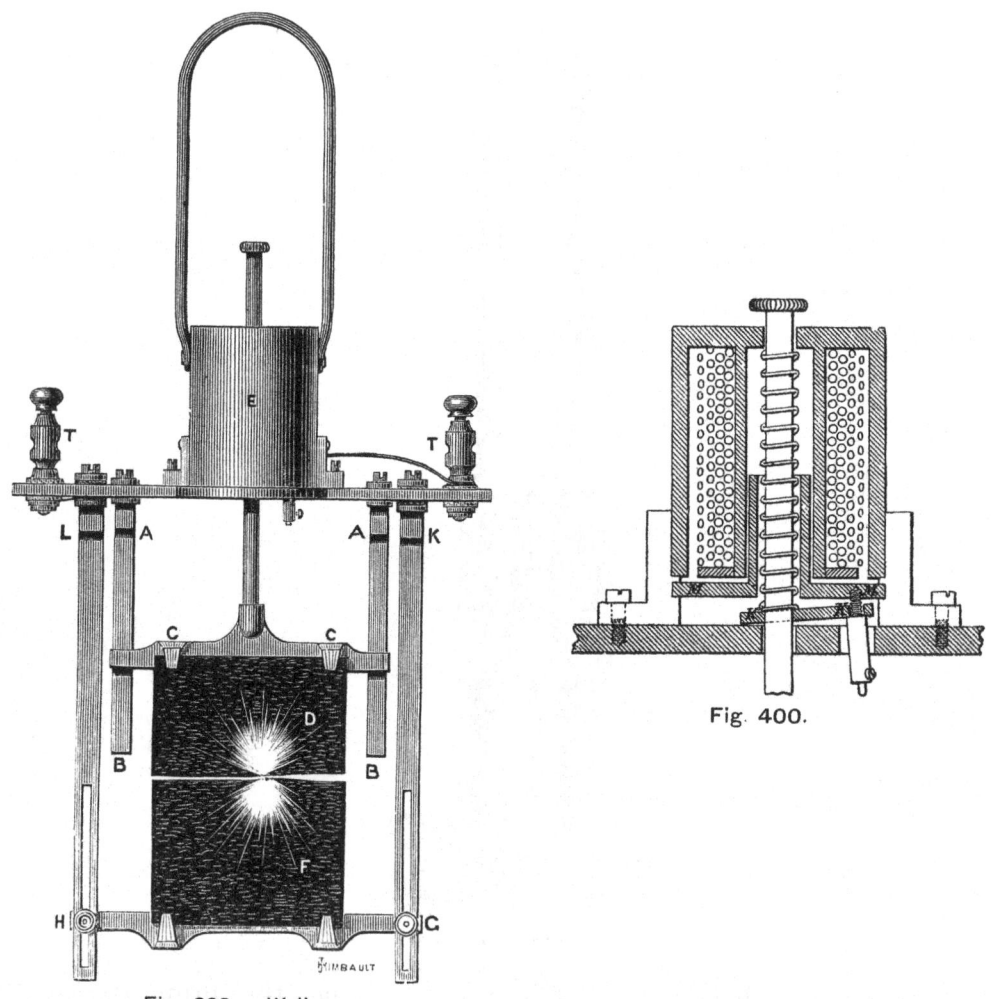

Fig. 399. Wallace.

Fig. 400.

a glance, its characteristic feature lies in the fact that instead of having rods or pencils of carbon, the arc is produced between flat plates of carbon set in an appropriate frame. The object here is to attain that durability of the light which constitutes so great a difficulty in the application of the electric light to many purposes. The carbons of the plate lamp properly adjusted, lasted for 20 hours or more, and in the older form of lamp used in America, which was broader and more clumsy than the pattern illustrated, carbons have been employed which would last without attention or renewal for a hundred hours of work. Another interesting feature of the lamp is the simplicity and ingenuity with which the automatic adjustment of the arc is both attained and maintained, without the employment of any of the delicate or complicated clockwork contrivances of many arc lamps. This automatic adjustment will be understood by reference to the drawings. Fig. 399 gives a general view of the lamp, and shows the carbons D and F clamped to a light framework of brass, by which also the lamp is suspended at any desired point. The lower carbon is fixed, but the upper carbon is capable of motion vertically, its holder C C being attached to a smooth rod of brass, which slides centrally through a brass box containing the regulating electro-magnet; its motion up and down is confined by guides A B, which also convey the current to it. The electro-magnet above is of tubular form, having the coils wound round in the space between an inner and an outer cylinder. The mechanical device next to be described provides for the automatic maintenance of the arc. The double difficulty to be overcome is first to move forward the carbon as fast as it burns away, and, secondly, when so moved up, to draw apart the carbons to the requisite distance. This double requirement is met in a very ingenious manner by the contrivance at the base of the electro-magnet shown in Fig. 400. The brass rod which holds the upper carbon slides loosely through a hole in a small steel bar K K. This bar is itself loosely traversed by a pin screwed into a flat annular disc of iron M M, which is nothing else than the movable armature of the electro-magnet, and which, though heavy enough to fall by its own weight, can readily be lifted by the electro-magnet when the current is once established. When no current is passing through the lamp, this armature M M and the clamping piece K K alike fall loosely down, and the brass rod is free to descend, thus allowing the upper carbon to rest against the lower. So soon as a current passes, the armature is attracted upwards, and in rising, lifts the clamping piece K K obliquely, thus jamming it on to the brass rod; the armature goes on rising about the eighth of an inch further, lifting clamping piece, brass rod, and carbon plate with it, and establishing the arc. Should the arc by any means break, the armature is released, the upper carbon falls and re-establishes contact, once more to be clamped and lifted to the right height. The advantage claimed for the plate form of carbon was its greater durability, since the arc has to eat away a considerable width of the carbon pole before any further adjustment was needful. It is for some purposes a disadvantage that the arc should thus range along the line between the edges of the plates of carbon, travelling from a point where the nick has become widened by burning, to a narrower point. But the disadvantage of itself is slight in those cases of practical application for which this lamp was designed. Enclosed in a light lantern, either of clear glass or of opal, it was especially intended to illuminate a factory, shed, or open space. Clear glass was preferred in England, where it was for some while in use at the Liverpool Street Station. For a considerable time the proprietors experienced a great difficulty in procuring carbons of suitable quality, the ordinary plates sawn from blocks of crude retort carbon being unfit for the purpose, and liable to split and blaze unpleasantly. It was stated that the cost of carbons did not exceed one penny per hour per lamp.

In the Wallace-Farmer system six or eight lamps could be connected in series, and up to the limit of six lights, apparently without any detriment to the amount of light evolved at each lamp, the only difference being the amount of power required to drive the machine. It will be remembered that the Wallace-Farmer machine is duplex, and can be employed in two circuits having five lamps in each, though it was found preferable where possible, to work the machine as one, a result which agrees with the experiments made by the authorities of the Trinity House at the North Foreland, with two Alliance machines, whose united power was more than double that of either alone. The power required to drive the machine was roughly proportional to the speed at which it must be driven, this in turn depending upon the work of the circuit; for feeding five lamps a speed of about 500 revolutions was required, and this was estimated to consume 5 horse-power. A considerable number of factories in the United States have been illuminated by this system, and in this country the well-known tobacco factory of Messrs. W. D. and H. O. Wills, of Bristol, was furnished with the Wallace-Farmer light; the system also received extended trial at the Liverpool Street Railway Station in Shoreditch, though here the experiments were delayed and spoiled for lack of a supply of good carbons. The machine was also tried at the Avonmouth Docks, where it was proposed to employ it in illuminating the dock gates and on the quays for unloading vessels. The English patents are in the hands of the Anglo-American Electric Light Company, and at one time the system appeared to have a promising future here. After, however, the Company named acquired the Brush patents, it was not considered by them expedient to push a system they considered less perfect than the Brush; this is probably the reason why the Wallace-Farmer generator and lamp have not been heard of here for a long while.

Crompton.—Mr. R. E. Crompton, Arc Works, Chelmsford, has been remarkably successful in developing a thoroughly efficient arc lamp. Taking the Serrin lamp as the type to work from, Mr. Crompton, looking at it as a mechanical engineer rather than as an instrument maker, modified one part after another, introducing simple devices in the place of those of a complex and expensive character, solving the problem as far as possible in a mechanical and workmanlike manner. Thus lathe work has been substituted, wherever practicable, for vice work; the rods, guides and protecting case are of cylindrical instead of rectangular section, and the expensive linkwork parallel motion has been replaced by the more obvious and simple device of guide rods and slides.

Referring to the figures (of which Fig. 401 is a sectional elevation, Fig. 402 a sectional plan taken through the line A B, and Fig. 403 a view in plan of the lower carbon holder), it will be seen that Mr. Crompton, taking advantage of the facility which his simplified form of construction has given him, has made a lamp capable of being used with equal facility either suspended from a ceiling or standing on a table. In the figure the apparatus is shown as a suspension lamp, the carbons being below the box containing the mechanism; but in order to convert it into a lamp of the ordinary form, all that is necessary is to take the carbon holders from the lower ends of the lamp rods, putting them on to the upper ends, and interchanging them, so that what was the lower carbon holder in the first position becomes the upper carbon holder in the second; but this does not of course involve any change in the disposition existing between the lamp rods and the upper and lower carbons respectively.

The action of this lamp does not differ in any important particular from that of the Serrin. The rod to which the upper carbon is attached

is connected by a train of wheelwork with the holder of the lower carbon.
so arranged that the former, falling under the influence of its own gravity,

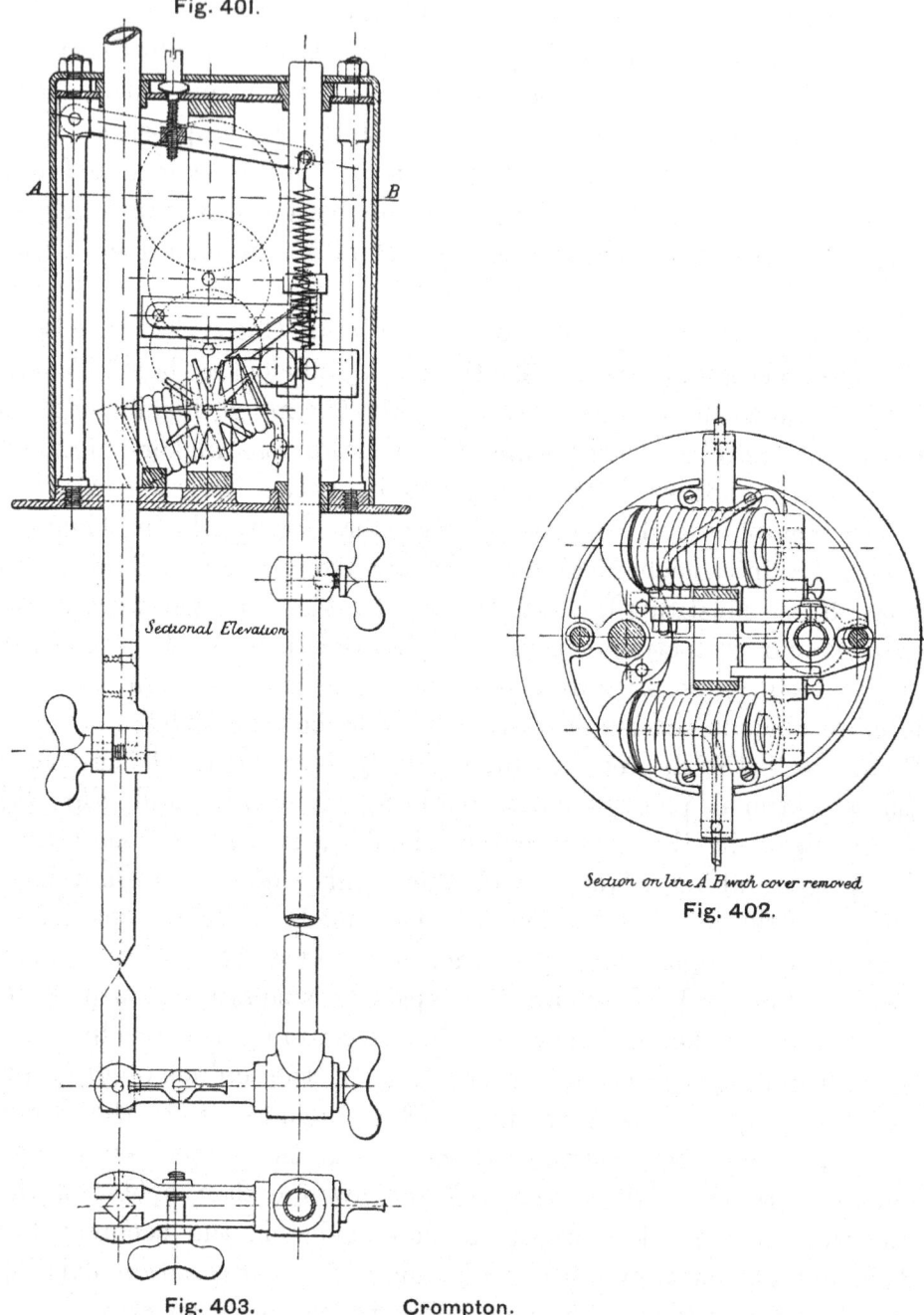

Fig. 401.

Sectional Elevation

Section on line A B with cover removed

Fig. 402.

Fig. 403.   Crompton.

raises the latter at half its own speed, that is to say, the lower carbon
rises through a distance of 1 in. for every 2 in. through which the upper

carbon falls, which is approximately the proportion of consumption of the two carbons respectively when used on a continuous current circuit. Connected by multiplying gearing to the train of wheels by which this is effected, is a small star-shaped checking wheel, which, as long as the carbons are running together, and no current is passing, is rapidly rotating; the moment, however, that the electric current is transmitted through the apparatus, an electro-magnet, included in the circuit (and shown in Figs. 401 and 402) draws down an armature attached to the lower carbon-rod, separating the carbons to a certain distance depending on the strength of the current, the downward pull of the magnet being opposed by an adjustable spiral spring tending to draw the rod in the opposite direction. The action of drawing down the rod, causes a detent attached to it to engage in the radial arms of the star-shaped wheel, thereby stopping its rotation, and with it that of the train of wheels; and consequently the running together of the carbon pencils; and as the position at any moment, of the lower rod, is that of equilibrium between the upward pull of the adjustable spiral spring which is constant, and the downward drag of the magnet acting on the armature, which drag is variable with the strength of the current, it follows that the length of the arc is continually being automatically adjusted to the strength of the current, and the light is maintained uniform.

Mr. Crompton has in some of his lamps modified the method by which the carbon pencils are held in the holders, making use of the conical clip arrangement adopted by M. Faber in his pencil-cases.

Rapieff.—In the Rapieff lamp the single vertical carbons, common to most arc lights, are replaced by two, inclined to one another at an angle of about 20 deg, and meeting so as to form the letter V; between the point of intersection of the upper pair of carbons and that of a similarly arranged pair below, the electric arc is produced. Each carbon pencil is free to move between guide rollers or within tubes in the direction of its length, and is drawn with a uniform pressure, by means of a cord and weight, towards the apex of the V, the feed of each being stopped only by the two butting against each other. A Rapieff lamp of this construction is shown in Fig. 404, in which the four carbons are shown in the position they would occupy when no current is flowing, that is to say, the lower pair is held in contact with the upper pair, by a light spiral spring in the base acting through a vertical rod which passes freely up through

the right-hand vertical pillar shown in the illustration.   To the free end
of each of the carbon pencils is attached, by means of a screw clip, a silk
thread, which, passing over suitable pulleys, enables a rectangular

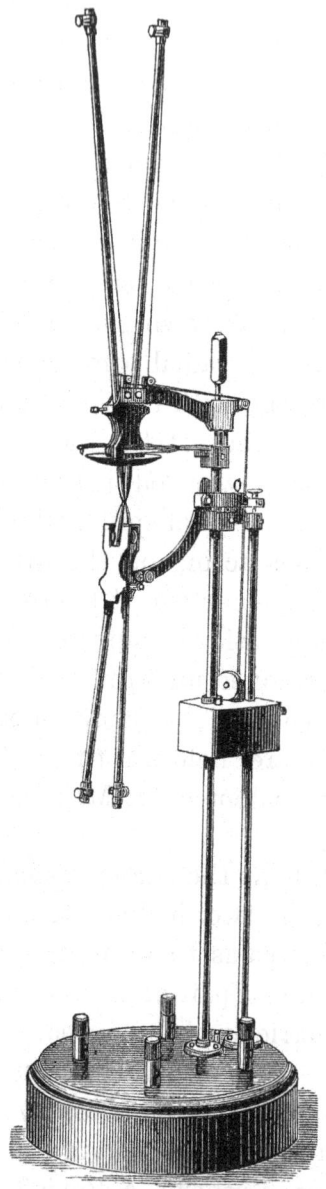

Fig. 404. Rapieff.

weight, which slides vertically up and down the two pillars, to draw the
carbons towards their common point of intersection.   When the lamp is
placed ·in circuit and an electric current sent through it, the lower pair

of carbons drops away from the upper pair to the distance appropriate for the formation of the arc, and this distance remains constant as long as a current is flowing, but if there occur any interruption—due to the stoppage of the machine or to any accident—the two pairs of carbons are instantly brought into contact again, and the light is here re-established, or the lamp is in a condition to start again the moment that the current is renewed.

The automatic arrangement by which the carbons are separated is concealed below the circular mahogany base of the lamp, and is shown in Fig. 405. It consists of a pair of electro-magnets (shown about the middle of the circular cavity) one of which is fixed while the other is hinged or pivotted at that end which is to the left of the figure, in such a

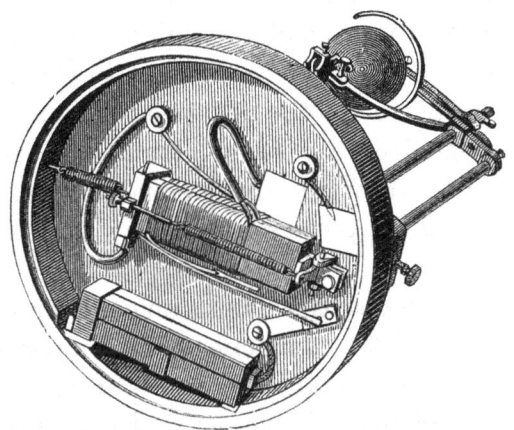

Fig. 405.

way that the magnetisation of the magnets causes the pivotted magnet to approach the fixed one, and in so doing to lift the sliding rod which passes up the vertical pillar of the lamp, and thus to separate the carbons. A spiral brass spring, shown in the figure, re-establishes the connection when no current is passing to excite the magnets.

Below the stand, and also shown in the drawing, is fixed an apparatus, which is an essential feature of the Rapieff system for sub-dividing the electric light, and by which the extinction of one lamp in a series is prevented from rendering the others brighter, or otherwise affecting them. This consists of an automatic shunt which throws into the circuit a resistance about equal to that of the burning lamp; the moment that, through any cause, the arc becomes extinguished, the current ceases to pass through the coils of the electro-magnet, and the

E E

armature thus released is withdrawn by a short brass spiral · spring fastened to the rim of the stand; to this armature is attached a long brass arm carrying a contact piece, which, when the armature is released, presses against a second contact piece, connected to one end of a carbon resistance (shown at the lower part of the figure). The further end of this resistance box is connected to the terminal of the lamp to which the negative pole of the machine is attached. It will thus be seen that there are two courses open to a current entering the apparatus, according to whether the circuit through the carbon pencils is broken or complete. In the latter case its course is as follows :—From the positive terminal through the magnet coils to the upper pair of carbons, thence through the arc to the lower pair which are connected to the negative terminal. When, however, the circuit through the carbons is broken its course, is as follows :—From the positive terminal, through the contact piece attached to the armature, through the artificial resistance to the negative terminal, which may either be connected to the negative terminal of the generator, or to the positive terminal of the second lamp of a series, and as the two courses through one lamp are about equal in resistance, it follows that the extinction of that lamp cannot have any appreciable effect upon others in the same series.

This system was adopted at the *Times* office, as long since as 1878, and is still in use there. At the commencement, six lamps were placed in one circuit, but with more powerful currents that number could be doubled, and the extinction of one, two, four, or even five lamps does not increase the illuminating power of those that are left.

Fig. 406 represents an extremely convenient form of M. Rapieff's lamp, in which the two pairs of carbons, instead of being disposed the one vertically over the other, are placed side by side, making angles with one another, so as to form the four corners of an acute-angled pyramid, at the top of which the arc is formed. The carbon pencils are in this lamp drawn upwards, in a similar manner to that described in reference to the first lamp, by means of cords passing over pulleys and attached to the cruciform weight shown in the figure.

The Rapieff system possesses two special features, to which attention should be called : the first is, that as the position of the arc is determined by the geometrical intersection of two fixed lines, and not by the rate of consumption of the carbon pencils, or by any regulating clockwork, it follows that the same lamp is equally well adapted for continuous and for alternating currents, for although in the one case the one pair of carbons would consume at twice the speed at which the other pair would be burnt, while in the other, the two pairs would disappear at equal rates, still the position of their points of intersection would remain constant, and all that is necessary to do in the case of a continuous current is to have

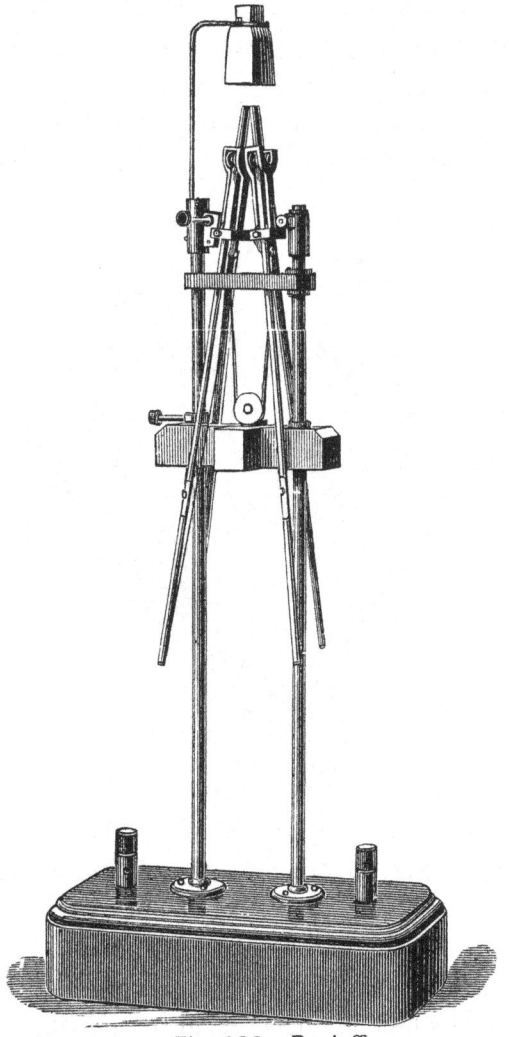

Fig. 406. Rapieff.

longer pencils in the one pair than in the other. The other advantage is also a very material one. It is well known that one principal cause of fluctuation in the ordinary electric light is due to imperfections in the carbon employed, small fragments of silica, as well as cracks and fissures, causing the pencils to split or to break away, producing a flicker or variation in the light. By the employment of two or more carbons instead of one, the effect of such imperfections is reduced to a minimum, for the continuity of the one makes up for the defects of the other, and a steady performance is insured.

Hedges.—Mr. Killingworth Hedges' lamp belongs to the Rapieff type; it is illustrated by Figs. 407 and 408.

Referring to Fig. 407, A and B are two troughs, rectangular in cross section and inclined to one another so as to form the letter V. Within these troughs slide two carbon pencils of circular

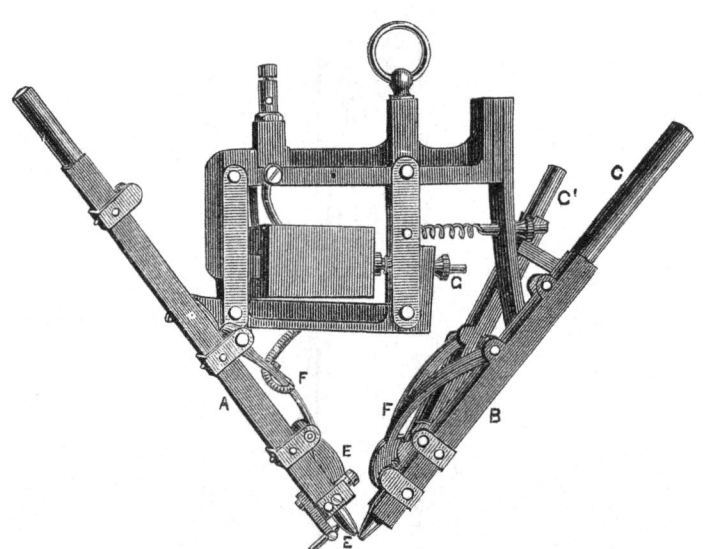

Fig. 407. Hedges.

cross section, meeting, when no current is passing, at the lower point E. The carbon holder B, to the right of the illustration, is rigidly attached to the framing of the lamp, but the trough A, which carries the negative carbon, is attached to the framing by a pivot shown in the figure, and on this pivot the carbon holder can rock, its motion being controlled by the position of the armature of an electro-magnet above it, the coils of which are included in the circuit of the apparatus. By this means, the moment the current is established through the lamp, the armature is attracted, and the points of the two carbons are separated, thus forming the arc. The positive carbon C is held from sliding by the gentle pressure against it of the smaller carbon rod C¹, which also lies in a trough or tube fixed in such a position that the point of contact between the two rods is sufficiently near the arc for the smaller rod to be slowly consumed as the other is burnt away; the latter in this way is permitted to slide gradually down the trough as long as the lamp is in action. The negative carbon holder A is provided with a little adjustable platinum stop E, which by pressing against the side of the conical end of the negative carbon holds the latter in its place, and prevents it sliding down the trough except under the influence of the slow

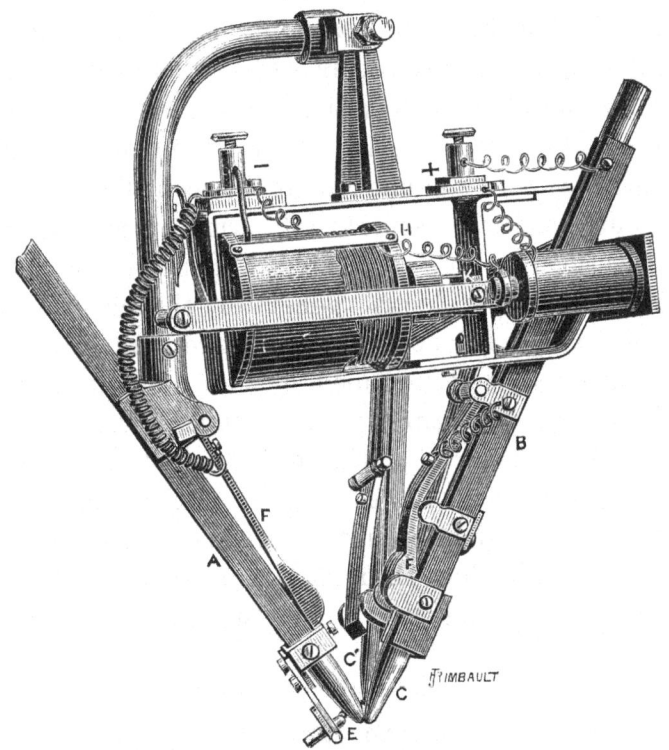

Fig. 408. Hedges.

combustion of the cone resulting from the wasting action of the arc. The position of the stop with respect to the conical end is determined by a small adjusting screw.

In order to maintain good electrical contact between the fixed conducting portions of the lamp and the sliding carbons, Mr. Hedges fits to each carbon holder a little contact piece F F hinged to its respective trough at its upper end, and carrying at its lower or free end a relatively heavy little block of brass grooved out to fit the cylindrical side of the carbon, against which it bears with an even pressure. By this arrangement the length of that portion of the carbon rods which is traversed by the current is always the same, notwithstanding the shortening of their total length by combustion; the resistance of the carbon electrodes is therefore maintained constant, and, for the reason that the contact pieces press against the rods very near their lower ends, that resistance is reduced to a minimum. The length of the arc can be determined by the adjustment of the screw G, by which the amount of movement of the armature is limited.

Fig. 408 represents a modified form of Mr. Hedges' lamp, designed for installations when it is desirable to burn a number of lamps in series. In this arrangement the carbons are separated by the attractive influence of a solenoid upon an iron plunger, to which is attached (by a non-magnetic connection) the armature of an electro-magnet. The coils of the magnet, which are of fine wire, form a shunt circuit between the two terminals of the lamp, and so disposed with respect to the armature as to influence it in an opposite direction to the solenoid. When the circuit of the lamp is completed with the electric generator, the carbons are drawn apart by the action of the solenoid on the plunger, and the distance to which they are separated is determined by the difference of attractive forces exercised upon the armature by the solenoid and the magnet; but as the latter forms a short circuit to that of the arc, it follows that should the resistance of the arc-circuit increase, either through the arc becoming too long, or through imperfection in the carbons or contacts, a greater percentage of current will flow through the magnet coils, and the arc will be shortened, thereby reducing its resistance and regulating it to the strength of the current.

Brush.—The regulating electric lamps designed by Mr. Brush are quite as much an element of the remarkable success of his system of electric lighting as are the generators themselves. They can be constructed to burn continuously for 8, 16, 24, or, indeed, for any desired number of hours, and the mere hanging of a lamp in its place puts it thereby into the circuit of the other lamps of the series, and with the generator; while, on the other hand, the extinction of one or more lamps or, indeed, their removal altogether, does not affect the others except (if the machine be driven at the same speed) to make them brighter, but by driving the engine slower, and therefore using less steam, the same initial degree of illumination can be maintained in each of, say, 14 lamps as if all the series are in operation. In the Brush lamp the feed is actuated by gravity alone, while it is controlled solely by the influence upon a bar of iron, of a magnetic field, the intensity of which varies with the strength of the electric current passing through the lamp circuit. The general external appearance of three forms of Brush lamps is shown in Figs. **409** to **411**, the first representing the single carbon or

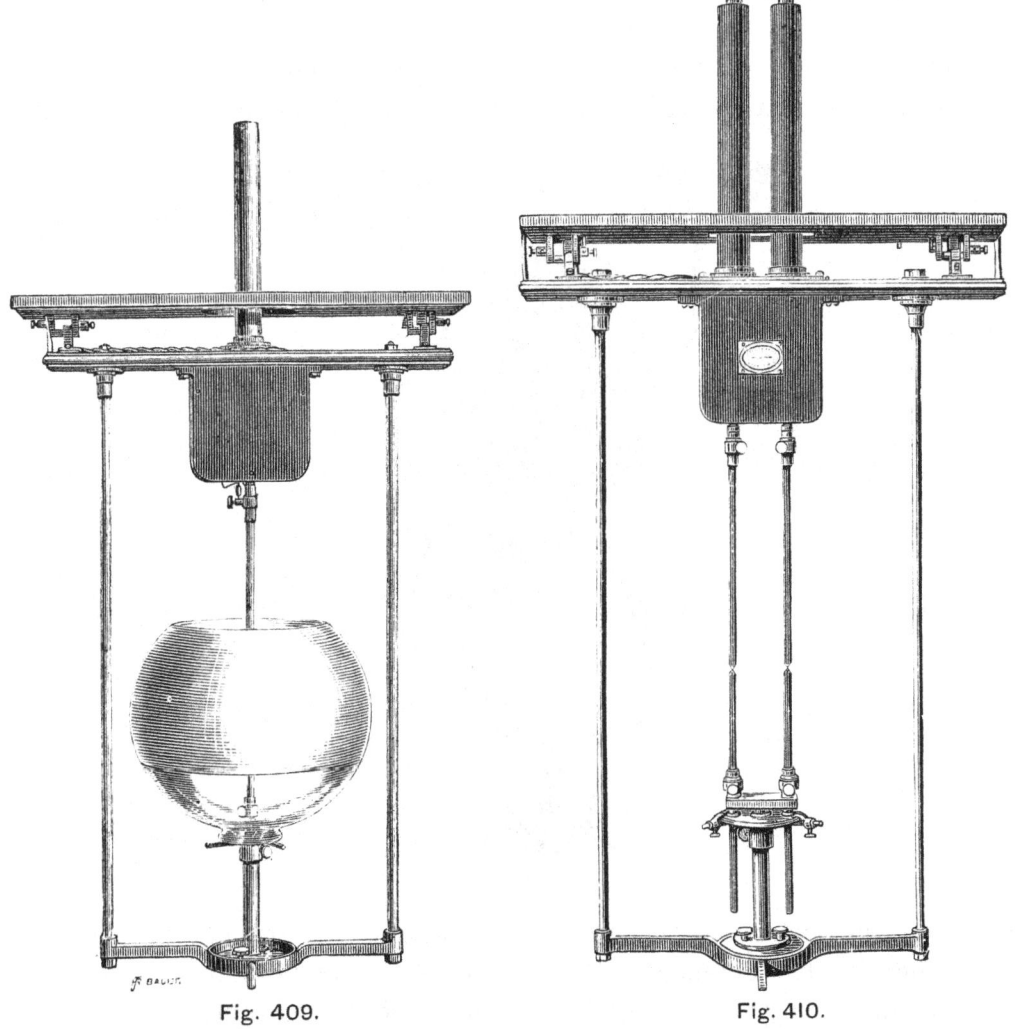

Fig. 409.   Fig. 410.

8-hour lamp; the second, the double-carbon, or 16-hour lamp, with its glass globe removed; and the last being a sketch of a more ornate form of the double-carbon lamp, the controlling apparatus being enclosed in a nickel-plated case. In all these lamps, which are designed for general illumination, the lower carbons are fixed, and as they burn away the upper carbons follow them, so the arc gradually descends, but for certain special installations, such as lighthouses or for purposes of projection, a focus-keeping arrangement has to be applied.

Like most of the modern electric lamps, the upper carbon descends by its own weight until it touches the lower carbon, and the circuit is thereby completed ; the effect of this in the lamps we are describing is to cause a soft iron plunger to be drawn to a greater or less extent within a hollow coil, or sucking magnet as it is sometimes expressively called, and through the intervention of a lever and an ingenious annular clutch surrounding the rod of the upper carbon like a washer, the upper carbon is lifted

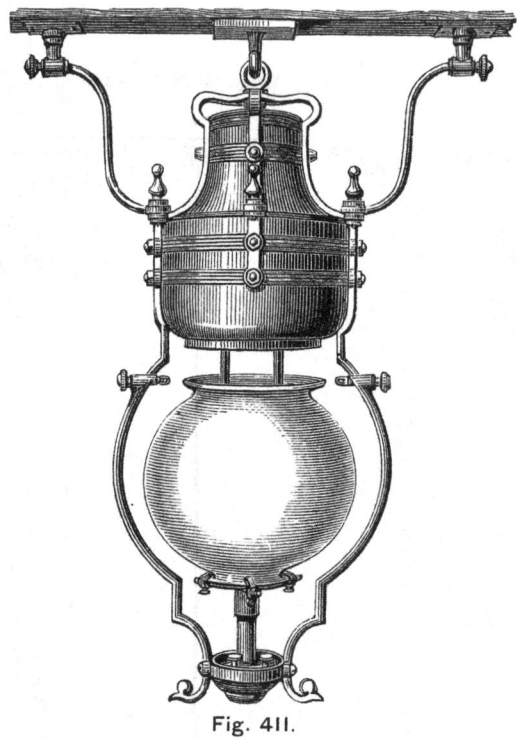

Fig. 411.

away from the lower and the arc is established. As the carbons burn away, the arc has a tendency to become longer and its resistance to increase, and this, by reducing the strength of the current, diminishes the supporting power of the magnetic helix, allowing its plunger to descend, and in so doing to lower the carbon and shorten the arc until the proper strength of the current is restored, when the rising of the plunger once more holds the carbon in position. In the above very general description the influences have been spoken of as *tending* to produce the effects, for as a matter of fact so sensitive is the controlling apparatus to the smallest variation of current, that there is practically no reciprocating action such as the above description might suggest, but the normal condition of the upper rod is to be slowly sliding through the clamp as the carbons become shorter, but if by any chance the rod slides a little too far, the clutch immediately raises it again and the carbons are adjusted to their proper distance.

The manner in which each lamp in a circuit is enabled to control itself independently of the action of all the others in the circuit, and yet dependently on the strength of the current in that circuit is as follows :— Each of the bobbins of the controlling magnet is wound with two distinct coils of wire, the first consisting of a number of turns of thick wire, through which the current is transmitted to the arc, and the second of a much larger number of convolutions of fine wire, which forms a secondary or shunt circuit of high resistance between the terminals of the lamp, and which being always closed is independent of the arc. The connections are made in such a way that the electric current shall pass through the shunt circuit in a direction opposite to that in the primary or thick wire circuit. It will readily be understood that, as the current flows in opposite directions around the magnet, the influence of the fine wire circuit will be to neutralise or weaken the attractive influence of the thicker helix, but the number of the convolutions of the two coils, as well as their respective resistances, are so proportioned to one another that the attractive influence of the primary helix (when the arc is of its normal length) shall overcome the influence of the secondary circuit. Owing to the greater resistance of the latter, not more than 1 per cent. of the main current is transmitted by the fine wire helix, but its magnetic influence is rendered considerable by its greater number of convolutions.

From the above description it is clear that since the electric current has two routes from one terminal of the lamp to the other, the one through the arc and the other independent of it, it follows that should the arc become too long, its resistance will increase and a larger proportion of the current will be shunted through the fine wire helix, while the strength of that in the thicker helix will diminish, the resultant magnetic influence on the plunger will be reduced, and the upper carbon will be brought closer to the lower; on the other hand, if the arc becomes shorter than its normal-length, its resistance is reduced, more current flows through the primary helix and less through the secondary, and the carbons are drawn further apart.

Fig. 412 is a diagrammatic sketch showing the course of the primary circuit, and illustrating the general principle by which the arc is controlled; it also shows the short-circuiting contrivance by which any accident to one lamp, or irregularity of working, cuts it out of the general circuit, and does so without exercising any influence on the other lamps in the series. In this diagram X and Y represent the two terminals of the lamp, in most cases consisting of hooks, which by being dropped over pins attached to the ceiling—and which are in connection with the line circuit—place the lamp in circuit with the machine. The current entering at X is transmitted through the two hollow bobbins H and H[1] in parallel circuit, the outcoming ends being joined together and connected to the upper carbon holder N; if the

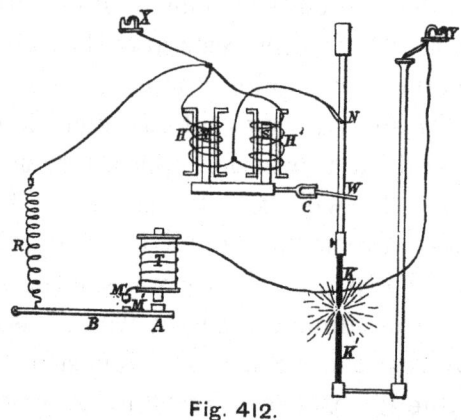

Fig. 412.

carbons are in contact the current flows through them, and by the vertical rods of the lamp to the terminal hook Y; the effect of this is to convert the solenoids H and H[1] into magnets, which by drawing into themselves the two iron plungers N and S, lift one edge of the washer clutch W; this by its oblique action seizes the carbon rod (much in the same way as a tent-rope tightener grasps its cord) and lifts the upper carbon until its influence is balanced by that of the fine wire helix, which, it must be remembered, surrounds the thicker coil.

The short-circuiting apparatus shown to the left of the carbons (see Fig. 414), consists of an electro-magnet T wound with a thick and a fine wire coil similar to those of the regulating solenoids, but both wound in the same direction. When the thick wire circuit of this magnet is complete, it forms a shunt of low resistance between the terminals, and therefore short-circuits the lamp, transmitting the current to the main circuit. Now, the fine wire wound upon this magnet is in circuit with the fine wire of the regulating solenoids; it follows therefore that if through a failure of the arc, or through its becoming abnormally long, or through any considerable increase in its resistance, a larger proportion of current be diverted through the fine wire circuit, the attractive force of the electro-magnet T would be increased, and its armature A, which is attached to the pivotted lever B, would be attracted, the contact pieces M and M¹ would thereby be brought together, and the terminals would be short-circuited through the thick coil M, and the resistance spring R. By thus short-circuiting the terminals through a route altogether independent of either the fine or the thick wire solenoids, their magnetic action ceases, and either the upper carbon is dropped, or it is burnt out; if from any other cause the arc is not established, the contact pieces M and M¹ are held firmly together by the attraction of the magnet T, and the current flows past the faulty lamp to the others in the series, its extinction calling attention to it, while the other lamps become proportionately brighter.

In installations in which it is desired frequently to vary the lights burning on one circuit, and it is inconvenient to alter the speed of the machine, the automatic regulator shown in Fig. 413 is employed. By this apparatus the amount of current generated by the machine is automatically controlled to suit the number of lamps in action at any one time. This regulating apparatus is of a very simple and effective character, and depends for its principle on the fact that the electrical resistance of a series of carbon plates or laminæ varies with the degree of pressure to which they are subjected. In the arrangement shown in the figure, four sets of carbon plates are connected with two terminal screws on the generator, so connected up that the regulating apparatus forms a shunt circuit across the electro-magnets. From this it is evident that if a high resistance be placed in the shunt, the greater portion of the current generated in the machine will be transmitted through the coils of the electro-magnets, and the strength of the magnetic field will be proportionately increased; while, on the other hand, if the shunt be of low resistance, a greater portion of the current will be diverted by it from the magnet circuit, and the strength of the magnetic field will be reduced. In the apparatus to which we refer, the columns of carbon plates constitute the variable resistance, the pressure upon them being controlled by a lever actuated by a pair of solenoids included in the main circuit of the machine. Thus, if any of the lamps in that circuit be extinguished, a larger amount of current passing through the solenoids causes the two iron plungers to be drawn into them, the lever is thereby lifted, and the carbon plates are pressed closer together, thus reducing the resistance of the shunt and causing a greater amount of current to

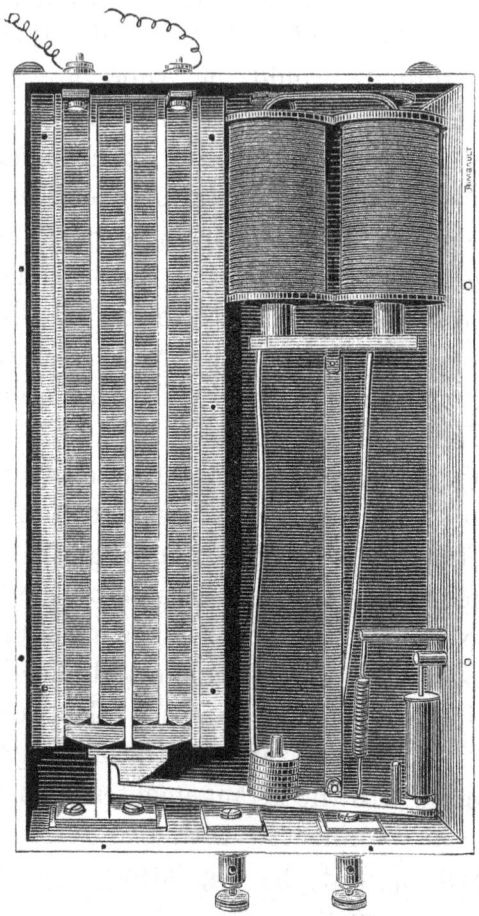

Fig. 413. The Brush Regulator.

be diverted from the magnet coils. The magnetic field being thereby reduced, the machine absorbs less motive power, and when the lamps are turned on again the operation is reversed, and thus the power required to drive the generator is proportioned to the work it is required to do. This apparatus will automatically control a No. 8, or 40-light, Brush machine, so that the number of lamps in one circuit may be varied from 40 down to 2 without altering the speed of the generator.

The carbon pencils consumed under the Brush system are, before being used, a foot long, and are electro-plated with a thin covering of copper; they last for about eight hours, during which time about 9½ in. of the positive and about 4 in. of the negative carbon are consumed. When, however, the lamps are required to burn for a longer period double carbon lamps are employed, fitted with two pairs of carbon holders, as shown in Fig. 412, and each is controlled in the same way as in the single carbon lamp, but the controlling apparatus is not duplicated. When the lamp is put into action the arc is formed between one of the pairs of carbons, and when they have burnt out the other pair automatically starts

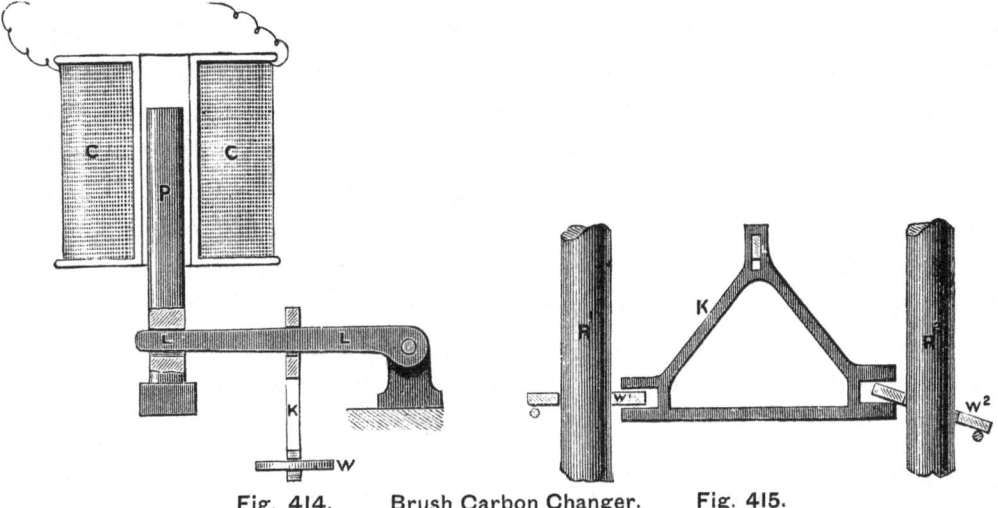

Fig. 414.   Brush Carbon Changer.   Fig. 415.

into action and continues to burn until it is in turn consumed; in this way the time of burning is doubled, and in order to still further extend the period of burning it is only necessary to employ a lamp with a third pair of carbons, when it will be capable of illuminating for 24 hours, each additional pair of carbons extending the period of burning for another eight hours.

The change from the one pair of carbons to the new pair is effected by purely mechanical means, and by a contrivance the ingenuity of which is only measurable by its extreme simplicity. This little apparatus will be understood by reference to the explanatory diagram, Figs. 414 and 415, which also more clearly illustrate the action of the washer-clutch than the preceding figure. C C is the hollow solenoid or "sucking magnet," which exercises a magnetic influence upon the soft iron plunger P. The rising or falling of this plunger raises or lowers the little frame K (shown in both figures) by the intervention of the lever L. R¹ and R² are the carbon holders which pass vertically through the casing of the lamp, and W¹ and W² are the clutch-washers which surround them. One side of each of these washers passes between a pair of jaws forming part of the little frame K, and when the latter is drawn upwards, its action is first to tilt the washers and then to cause them to clutch the rod on diagonally opposite corners. By the very simple device of making one pair of jaws a little higher than the other it takes its grip before the other begins to act, and therefore lifts its corresponding carbon higher than its neighbour; the consequence is that only one arc is established, viz., across the lesser distance, and, in all the subsequent feeding and controlling, the pair of carbons first started are alone affected, because although both carbons are raised and lowered together, the lower end of the reserve carbon is always higher than the other by the difference in the height of the two pair of jaws on the frame K. A time arrives, however, when, owing to the shortening of the consuming carbons, they can no longer meet when the frame is dropped, and the current by which they are again separated can only be transmitted by the reserve carbons coming into contact; the circuit is thereby completed, the new carbons are separated, and the arc between them continues to be controlled by the magnet and clutch in the same way as the first arc was.

Heinrichs.—Figs. 416 to 419 illustrate an automatic lamp devised by Mr. Heinrichs, whose dynamo-electric generator has already been described in these pages. The lamp belongs to that class in which the position of the arc is determined by the coincidence of the axes of the carbon pencils, of which the earliest type, that of Rapieff, will be found in another part of this chapter. In the arrangement now being considered, the arc is formed between the point of contact of one pair of circularly curved carbons and the corresponding point of another similar pair below it, the planes of the two circles being perpendicular to one another, as shown in Fig. 416; the upper pair are both positive, being connected to the positive pole of the generator, and the lower pair are negative, being in electrical connection with the other terminal of the machine. The feed of the carbons requires no mechanical apparatus such as clockwork or gearing, but is determined solely by the mechanical wasting away through the consumption of the carbon rods at their point of contact, the rods falling together by their own weight, as their extremities are dissipated under the action of the current.

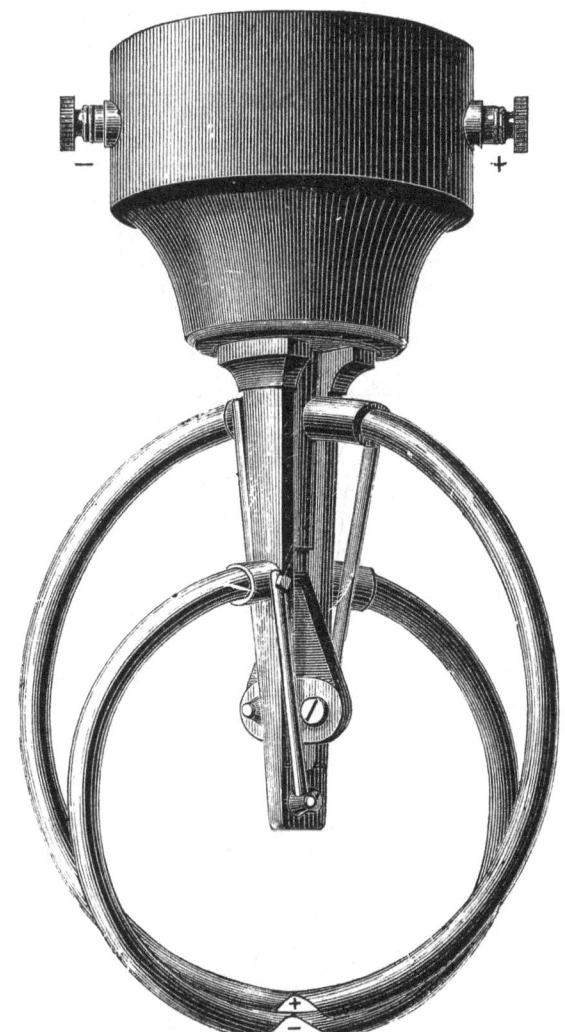

Fig. 416. Heinrichs.

While the feed of the carbons is determined in this way, the regulation of the length of the arc is effected by electrical means, the strength of current determing the degree of approach or of separation of the carbon rings. On referring to Figs. 417 and 418 it will be seen that the upper or positive pair of carbon holders a^1 and a^2 are pivotted at the points s^1 and s^2 to two small equal-sized pinions p^2 and p^1, shown in the detailed plan Fig. 418, which by gearing into one another, render the swing of the arms a^1 and a^2 equal and symmetrical on each side of the vertical axis of the lamp. A similar method of pivotting is adapted to the arms a^3 and a^4, which are the holders for the pair of lower or negative carbons. The geared pivots of the upper pair are attached to the rod f,

Fig. 417.

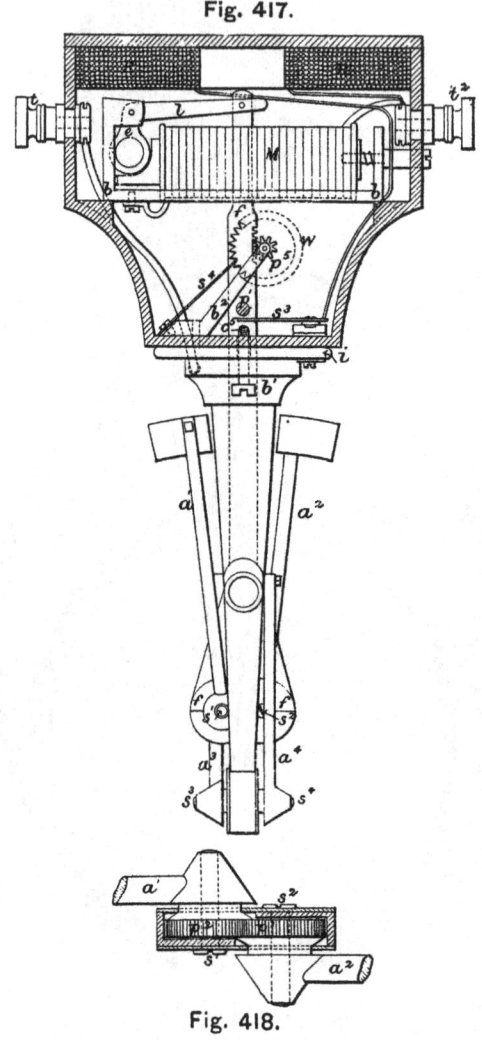

Fig. 418.

which is in its turn suspended from the long arm of the bell-crank lever l, of which the short arm is the armature e of an electro-magnet M. When the current passes through the lamp this arm by attracting the armature e lifts the rod f, and with it the upper pair of carbons separating them from the lower pair, and thus establishing the arc. To prevent any sudden lifting, Mr. Heinrichs has adopted the checking apparatus shown at p^5.

This consists of a small rack forming part of the rod f, and gearing into the small pinion p^5, on which is the large notched tooth-wheel w, against the teeth of which the light click spring s^4 presses, preventing any sudden vertical movement of the rod f, to which the upper carbons are attached. The lower or negative pair of carbon rods are attached to and in electrical connection with the framing b^1, which, while insulated from the case of the lamp and positive terminal t^2 by the ebonite washer i, is connected by a wire to the negative attachment screw of the lamp t^1 seen to the left of the figure.

At R R is a resistance coil of insulated German-silver wire through which the current is shunted, if from any cause it ceases to pass from one pair of carbons to the other, and, as the resistance offered by the coil is practically equal to that of the arc which it supplants, the strength of the current throughout the rest of the lamp circuit is not affected by the interruption. It also enables one of the lamps to be extinguished without affecting the others of the same series. The arrangement by which the circuit of R R is substituted for that of the arc when the latter is interrupted is shown in Fig. 417. On the rod f is fixed a projecting pin p^1 which, when the rod drops, brings the contact spring s^3 against the contact screw c^5, which is connected to b^1 and by the wire b to the negative terminal. The same action brings the two pairs of carbons together, and in this position a current has two courses through the lamp open to it, one through the resistance R R, and the other through the carbons and through the magnet cells M; the latter portion of the circuit causes the carbon to be separated by the lifting of the rod f, the arc is thereby established, and the pin p^1, rising with the rod f, releases the spring s^3 from the screw c^5, thereby cutting the resistance R R out of the circuit and allowing the whole current to pass through the carbons and the arc.

Fig. 418 is a diagram showing the very simple and ingenious method of constructing the pair of pivotted joints of the radial carbon holders a^1 and a^2. In this figure it will be seen that the two pinions p^2 and p^1, which are geared together, are rigidly attached respectively to the radial arms a^1 and a^2, turning on the pins s^1 and s^2, which are fixed to opposite sides of the little framing. The pinions are simply pushed over the pins from opposite sides, and are so proportioned that when pushed against their respective stops they are in gear and the rods are in their proper position, the axes of their carbon holders lying in the same vertical plane.

F F

Mr. Heinrichs has constructed his carbon pencils by horizontal compression between dies, the diameter of the rings before they are cut in half varying from 10 in. to 14 in. With carbons of circular cross sections Mr. Heinrichs makes the positive carbon rod 13 millimetres diameter and the negative 11 millimetres, and the length of time which such carbons will burn varies from 20 to 30 hours, according to the diameter of the rings, or in other words to the length of curved carbon rod employed—a continuance longer by many hours than can be required in actual practice except in the most unusual applications.

The latest form of carbon rod adopted by Mr. Heinrichs (see Fig. 419) is of rectangular cross section, measuring 14 millimetres by

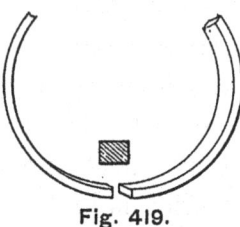

Fig. 419.

12 millimetres, the sectional area being about equal to that of the carbon formed of circular section. This figure will show better than any description the mutual positions of the longer and the shorter side of the cross section of Mr. Heinrichs' rectangular carbon; this form offers several facilities in manufacture.

In our description of the Heinrichs generator, we have referred to some tests made in which the lamps employed were of the type just described, the result obtained being, as will be remembered, highly satisfactory.

Weston.—The Weston lamp devised by the same inventor whose generator has been already described, gives very satisfactory results, regulating itself with accuracy according to the strength of the current, and producing a steady white light. Fig. 420 is a general outside view of the lamp, and Fig. 421 is a diagram illustrative of its very interesting and ingenious regulating mechanism, which is contained within the small dome-covered box shown in the upper part of Fig. 423. Referring to Fig. 421, R represents the rod carrying the upper carbon, which, when no current is passing, slides freely through the clutch lever C, and therefore allows the upper carbon to rest on the lower. M M is an electro-magnet of peculiar construction, and A·A is an armature capable of motion to and fro in a vertical direction only, being supported by the two parallel flat steel springs N and O, which are rigidly attached at their further extremities to the framing of the lamp. It will be observed that in its position of rest the upper end of the armature is just a little below the lower edge of the upper pole of the magnet, so that the action of a current through the magnet coils would be to cause the armature A to be attracted vertically thereby lifting the end of the clutch lever C, to which it is attached by a link. This lever is so constructed that when it is raised beyond a certain point it seizes and lifts the rod R, much in the same way as the lifting washer in the Brush lamp lifts the rod of the carbon holder. S is a tension spring attached, at its upper end, to the top of the armature, and hooked, at its lower end, to the arm of the bell-crank lever L fixed in such a position that the tension of the spring S can be regulated by an adjustment screw shown in the figure The

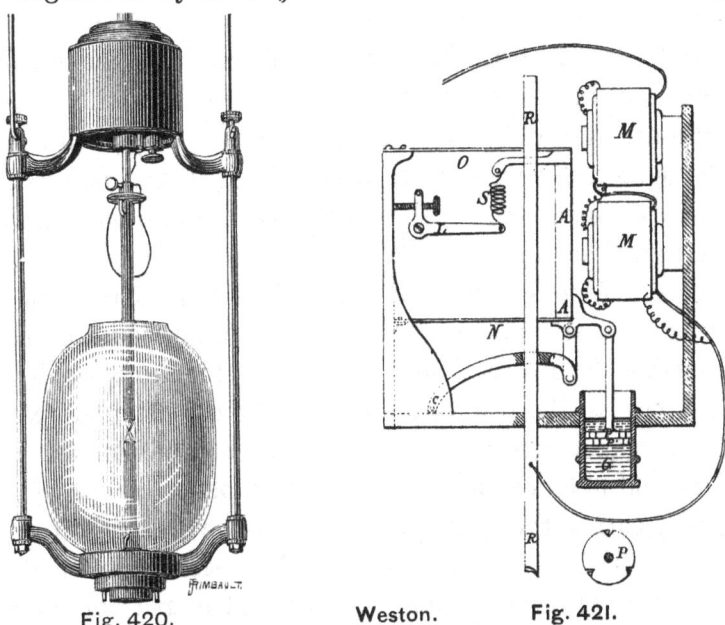

Fig. 420.   Weston.   Fig. 421.

movement of the armature therefore is controlled by the varying attractive force of the magnet M M drawing it upwards, and the adjustable tensile force of the spring S pulling it in the opposite direction; in order to prevent a sudden and violent variation in the attractive influence of the magnet from setting the lamp "jumping," the lower end of the armature is attached to a small piston rod and piston, which works in a little cylinder or "dash pot" G containing glycerine, in which the piston is immersed, and the controlling action of this little contrivance can be adjusted within certain limits by allowing more or less glycerine to pass from one side of the piston to the other.

This adjustment is effected by a very simple device shown in the figure. The piston of the glycerine cylinder is composed of two discs, the one fixed to the piston rod and the other capable of being rotated through a small angle on its centre, which is coincident with that of the other disc as well as with the axis of the piston rod. Around the edges of both discs are cut three notches at equal angular distances apart of 120 deg., and the notches can be made to coincide with or pass by one another, and by altering the angular position of the movable disc with respect to the fixed disc, the rapidity with which the glycerine can pass from one side of the piston to the other can be regulated with great nicety. The discs are shown in elevation in the diagram, Fig. 421, marked P¹ and P², as well as in the plan immediately below.

The magnet M M is of very peculiar construction, inasmuch as each limb is surrounded by no less than three coils wound over one another, although there are but two circuits which traverse the magnet in opposite directions, and it is by the difference between the magnetic inductive influence of the two circuits that the regulation is effected.

It is scarcely necessary to recall the fact that if two electric currents circulate around an electro magnet in opposite directions they will tend to neutralise the magnetic effect of each other, and that just as the inductive influence of the one or the other circuit predominates, either through the strength of its current or the magnetic efficiency of its coil with respect to the iron core, so will the direction of polarity as well as its intensity be determined. Now, the magnet in Mr. Weston's lamp is a differential magnet, one of its circuits being composed of thick wire, and forming part of the principal circuit of the machine and arc, and the other a fine wire shunt circuit wound in the opposite direction, and independent of the arc; and as the strength of current in shunt circuits, other things being equal, is determined by the ratio that exists between the resistance of shunt circuit and that of the circuit to which it is a shunt, it follows, from the construction of Mr. Weston's lamp, that the magnetic intensity of the field of the regulating magnet depends upon the resistance of the main circuit of the machine which is inclusive of the arc between the carbons. Fig. 422 is a sectional diagram of one of the limbs of the magnet, and shows an iron core surrounded, first, with a bobbin of fine wire, over which is wound a helix of thick wire and over this again is a second coil of fine wire, the two coils of fine wire being wound in the opposite direction to the thick wire, but in the same direction as one another. Both limbs of the magnet are alike, and the two helices of thick wire are joined together and in circuit with the lamp circuit, while the four fine wire bobbins are united in series, and included in the fine wire shunt circuit above alluded to.

The various connections of the lamp will be better understood by reference to Fig. 423, which represents the general distribution of the essential parts. The current entering the positive terminal at A has two channels open to it, one through the thick wire of the magnet coils to the upper carbon, down through the lower carbon, and through the left-hand suspension rod E to the negative terminal B. This causes M M to become a magnet, the armature is raised

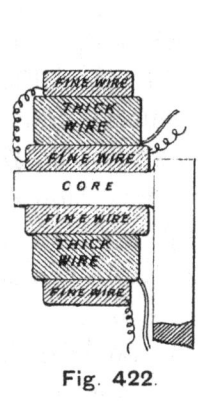

Fig. 422.

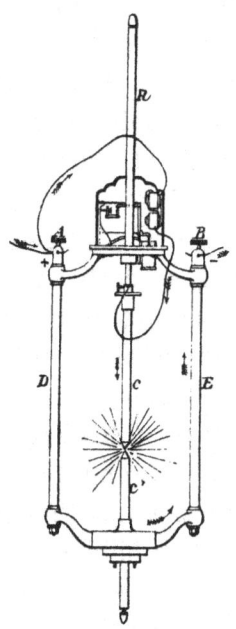

Fig. 423.

and the carbons are separated. Another course for the current is from the positive terminal through the fine wire of the magnet, whose strength it reduces, to the negative terminal of the lamp. The action of this arrangement is as follows: The moment M M becomes a magnet its armature is lifted, and the carbons being separated, the arc is established between them, its length being determined by the strength of the magnet in relation to the tensile downward pull of the spring S to which it has been adjusted. If through any cause, such as the shortening of the carbons, the arc increases in length or in resistance, more current is shunted into the reverse coils of the magnet, which is thereby reduced in magnetic intensity, the armature sinks, and the arc is diminished in length, thereby decreasing its resistance and allowing less current to pass into the shunt. If the arc circuit from any cause becomes interrupted, the armature drops altogether, bringing the carbon into actual contact and the current is re-established. All these operations, which take long to describe, are performed automatically with ease and promptness by the apparatus itself, and the result is a steady light under perfect control.

De Mersanne.—At the recent Paris Electrical Exhibition a part of the nave of the Palais de l'Industrie, and Salle 19-on the first floor, were lighted very successfully with the De Mersanne lamp, one of that numerous class in which the carbons are actuated by clockwork mechanism. Examples of two different types were exhibited, horizontal and vertical, both of which are illustrated by Figs. 424 to 428. Figs. 424 and 425 show the former arrangement, Fig. 424 being a general view of the lamp and Fig. 425 an enlarged drawing of the mechanism. Fig. 427 is a section showing an earlier modification, and Fig. 428 is a section of one of the boxes, serving at the same time as a holder and as a feed for the carbons. The motive power of the lamp is a coiled spring in the box A, which drives, through gearing, the horizontal shaft in the lower compartment of the lamp. This shaft, which is divided into two and coupled by a clutch, has at each end a bevel wheel which gears into another wheel on the vertical shaft b, Fig. 428. This spindle passes through the leg carrying the carbon holder, and has at its lower extremity a second bevel pinion gearing into a wheel on a spindle which carries a friction wheel e, driving the two small friction rollers $g\ g$. These press on the upper side of the carbon, the under side being in contact with two other rollers $h\ h$, the pressure of which against the carbon can be regulated by a spring and adjusting screw i. This mechanism is similar for both of the carbon holders. The train of clockwork is controlled by the electro-magnet B, the armature n of which is held by the spiral spring o and rests against the screw v. The armature controls a star wheel, the last of the clock train. The electro-magnet B is placed on a branch circuit of the main current, and consequently only acts when the current meets with a sufficient resistance to pass freely through the derived circuit. When this is not the case, as when the ends of the carbons are in contact or only a normal distance apart, the electro-magnet C, in the main circuit, acts on

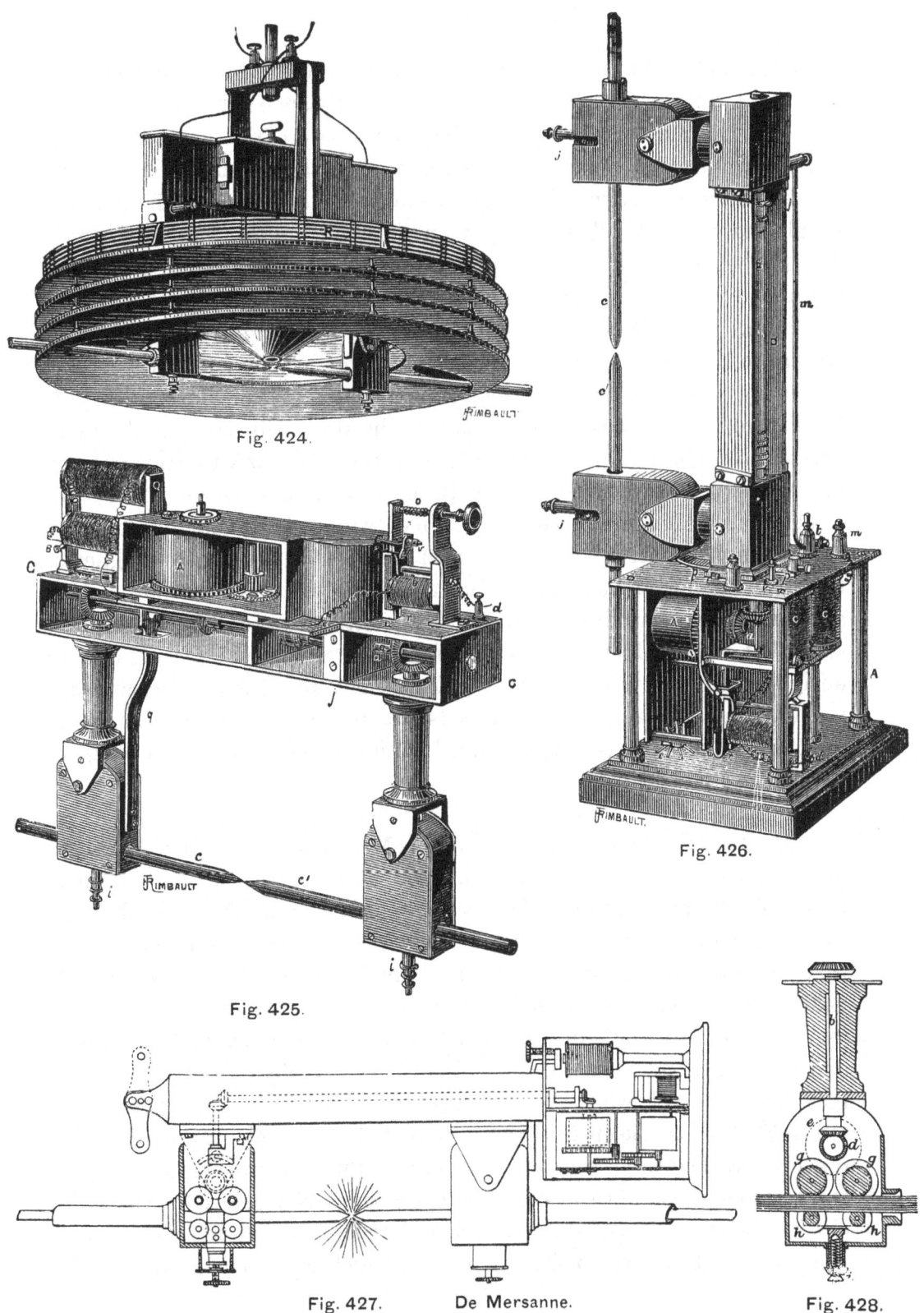

Fig. 424.

Fig. 425.

Fig. 426.

Fig. 427.   De Mersanne.

Fig. 428.

the rod *q* and pushes back one of the carbon holders which are each hung on a pin as shown. This action increases the distance between the carbons *c c¹*. At the moment of lighting, when the two carbons are in contact, the electro-magnet separates them, and when their distance apart is too great the electro-magnet B brings them together, the combined action regulating the arc. The arrangement for diffusing the light is shown in Fig. 424. It consists of a series of circular louvre-like reflectors, placed one above the other as shown, and a central reflector coned in the reverse direction. Fig. 426 shows the vertical De Mersanne lamp, which is similar to the one just described, except so far as changes are rendered necessary from the carbons being vertical instead of horizontal. In this figure, A is the base of the lamp, D the standards carrying the upper carbon holder, *a* is the shaft, passing from the clockwork to drive the feed mechanism; A is the mainspring, B the electro-magnet on the branch circuit, *r* its armature, *e* the star wheel which is liberated when the resistance increases beyond the proper limit; C is the electro-magnet on the main circuit, and *q* is the spring actuating the lower carbon holder, and tending to withdraw it when the resistance is too low. It will be seen that the De Mersanne lamp offers several original features in its design; it appears to work with regularity, and from the peculiar arrangement of the carbon holders and feeds, carbons of any convenient length may be employed, so that a prolonged illumination without renewing the carbons can be effected.

Mackenzie.—The electric lamp of Mr. Mackenzie first attracted public attention in the summer of 1880, at the Glasgow Exhibition of Gas and Electric Lighting Apparatus, where it was shown by Messrs. Strode and Co., of London, the manufacturers. This lamp belongs to that class which automatically readjust the length of the arc to a normal value. So soon as the waste of the carbons has widened the arc too much, the current fails in the regulating electro-magnet, and the keeper of the latter being released, the carbons are thereby permitted to come into direct contact, when the full strength of the current is immediately re-established, the keeper is again attracted by the electro-magnet, and the original length of the arc is restored. Such a process of stopping and starting afresh must of course cause a temporary eclipse of the light, but if it be effected almost instantaneously the break is only momentary; and not to be distinguished from the flicker ordinarily caused by inequalities in the carbons themselves. A good lamp of this kind ought therefore to act promptly, else the light which it gives will not appear steady; but if this condition be fulfilled the simplicity of the mechanism required for its regulation on this plan is greatly in its favour for practical use, especially in private works where much attention cannot be given to it.

As will be seen from the diagram, Mackenzie's lamp consists of a base P supporting an upright standard C, which carries a projecting lever

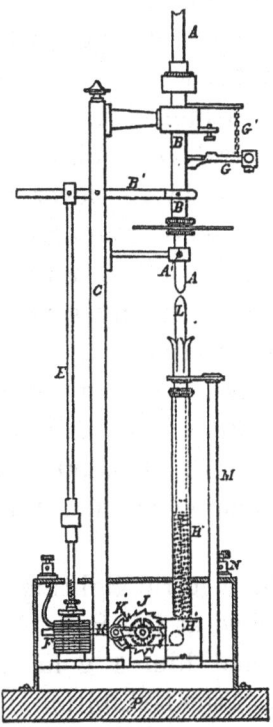

Fig. 429. Mackenzie.

arm B[1], supporting the tubular holder B of the upper or positive carbon A. The carbon itself is maintained or clamped in the holder by the pressure of the toe, or cam-like end of the weighted lever G, which is pivotted to a bracket fixed to the holder. A vertical rod E, passing downward from this lever B[1], bears at its lower end the soft iron armature or keeper F of an electro-magnet, and, when the current is passing through the arc and the electro-magnet, this keeper is held down by the poles of the latter. The depression of the keeper in this way raises the tubular holder B and maintains the arc between the upper and lower carbons.

When, owing to the consumption of the carbons, the arc is

excessively widened, and the current is proportionately weakened, the keeper is released, the lever E rises, and the tubular holder B descends. The small weighted lever and cam G being held up by the chain G¹ allows the upper carbon to drop on the lower one, thus making contact, which at once re-establishes the full current, pulls down the keeper once more, and raises the holder, so as to withdraw the carbons and restore the arc. The length of the arc is determined by the adjustment of the lever G, which at a certain point in the travel of the holder again falls and holds the carbon tight in the tube.

The lower or negative carbon is supported in an upright tube having a spiral spring H, which propels the carbon upwards. This spring is held down by a chain H¹ which is attached to it by one end. The other end of the chain is wound on a barrel to which an escape wheel J is fixed, and which is released tooth by tooth by the lever K and pallets K¹ every time the keeper of the regulating electro-magnet (which is also attached to the lever K) moves up and down. M is a rod conveying the current to the lower carbon, and N N are binding screws for connecting the lead wires from the generator to the lamp.

It will be seen that Mackenzie's lamp preserves the arc, or source of light, always at one point. Such a one as that illustrated is designed to burn five hours, and give a light of 1,600 candles with about 4½ horse-power. But it can be constructed to take in much longer carbons so as to burn for a greater time.

Jaspar.—The system of electric lighting, devised by M. Jaspar, of Liége, which was distinguished at the Paris International Exhibition of 1878 by the jury, who awarded him a gold medal, attracted also much attention among electricians at the Exhibition, in the Palais de l'Industrie, of 1881, on account of the simplicity of the lamp, its efficiency in working, the softness of the light produced, and the ingenuity of the whole system. The arrangement consists in placing the apparatus producing the light, at the bottom of a cylinder with opaque sides, which conceal it from view. The luminous rays escape from the upper and open end of the cylinder, and strike, at a distance above, a large white umbrella-like reflector, which diffuses them, and distributes a general illumination, very soft and pleasant to the eye. The Jaspar lamp belongs to the numerous category of apparatus in which the weight of the positive carbon holder acts as the motive element. The system is monophotal, that is to say, it is not adapted to the devisibility of the voltaic arc, and only one lamp can be placed in circuit. Fig. 430 is a side view of the lamp, from which a part of the enclosing box has been removed in order to show the mechanism. Figs. 431 to 434 are respectively two vertical and two horizontal sections, and from these the description will be readily understood. The frame of the lamp consists of two platforms A A¹, the first

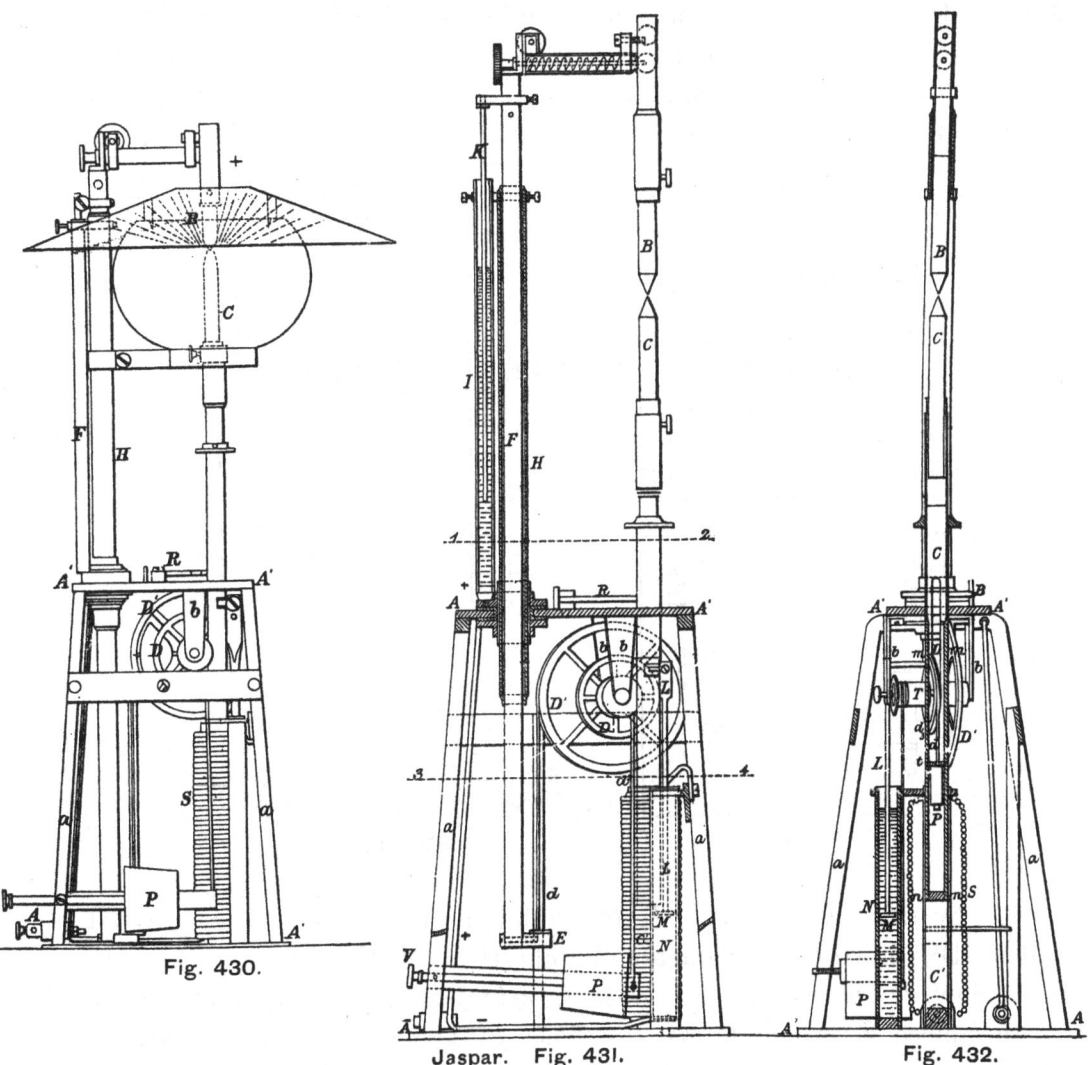

Fig. 430.

Jaspar. Fig. 431.

Fig. 432.

larger than the second, and connected by four bars a making a cage, formed like a truncated pyramid. The sides of the frame are closed by varnished iron plates, which can be removed at will, and which protect the mechanism against injury and from dust; the apparatus can be suspended from two screwed rings. The negative carbon holder C is formed of a copper tube traversing the upper platform, and sliding at its lower part in a second tube C¹ fixed on the base A. This lower portion of the negative carbon holder is formed of a hollow cylinder of soft iron. In the middle, the tube C is pierced with a longitudinal slot wide enough to allow a part of the circumference of the grooved pulley D to pass; around this pulley is wound the cord d, which descends in the axis of the tube C, and carries a weight P that keeps it stretched. This weight abuts upon a transverse piece t in such a way that when the cord is wound up, the weight in rising lifts the carbon holder. The grooved pulley is fixed on a small drum T,

Fig. 433.

Fig. 335.

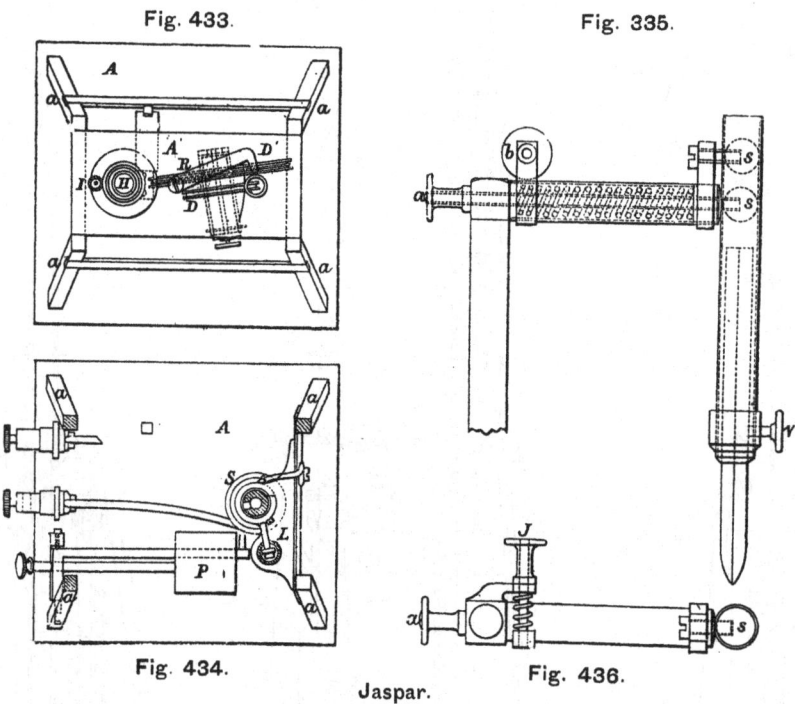

Fig. 434.

Jaspar.

Fig. 436.

carried by the two brackets b, on which is also placed the grooved pulley D¹, having a diameter double that of the pulley D. On the pulley D¹ is wound the cord d^1, attached at its lower extremity to the piece E, which forms the base of the positive carbon holder. It will be seen that by the direction in which the cords d d^1 are wound, that the weight of the positive carbon holder tends to raise the negative carbon holder, and that this will move through only half the distance of the former. This difference corresponds to the unequal wear of the carbons. On the drum itself is wound a third cord d^{111}, to which is attached a weight P¹, free to be moved longitudinally on its rod by means of the screw V placed outside the case of the lamp. This weight helps to counterbalance that of the positive carbon holder, which it always tends to raise. Its position is regulated in such a manner that its influence is greater in proportion as the current is more feeble. By this means the lamp can be used with very varying intensities of current. The positive carbon holder is formed of a copper tube F furnished with a clip to which the carbon is attached. Figs. 435 and 436 are detail views showing how, by aid of the studs *a* and *b*, the regulation of the position or direction of the upper carbon can be effected in two planes, perpendicular movements being rendered possible by the spherical articulation S. The carbon is held securely in place by the locking screw *v*. The tube F slides in a tubular guide H fixed in the upper platform of the lamp, and insulated electrically by an ebonite ring. Beside the tube H, and parallel to it, is a smaller tube I containing mercury (about 260 grammes). Into this dips the pendent rod K, which is fixed to the positive carbon holder, and transmits the current from the electric generator. The negative carbon holder also carries a pendent rod L, terminating in a small piston which can move within the tube N filled with mercury (about 350 grammes). The diameter of the piston is less than that of the tube, so that when the piston moves down by the motion of the carbon holder, the mercury passes from one side to the other of the piston by the narrow annular space. This acts as a very efficient brake, and prevents any sudden action which would show itself in irregularity of the light. The arrangement is somewhat similar to that employed in the Brush lamp, but we do not know to whom the priority of the idea is due. On the tube C^1 is wound an insulated copper wire forming a solenoid, and intended to produce the separation of the carbons and to counterbalance the effect resulting from the weight of the positive carbon holder. We may now proceed to examine the working of the mechanism we have described. The positive pole of the electric generator is in communication with the tube I, the mercury it contains, the pendent rod K, and the positive carbon holder. The negative pole of the generator is connected to the solenoid, the tube N, the pendent rod L, and the negative carbon holder; the contacts, secured by the immersion of the rods in the mercury, are excellent and always certain. The points of the carbons are in contact when the current passes; the solenoid exerts an attraction on the iron rod which forms the end of the negative carbon holder, and tends to draw it down. The weight of the positive carbon holder, regulated by means of the counter-weight P acts in an opposite direction, making equilibrium between the two opposing influences—the weight and the magnetic attraction—and the separation of the points would remain unaltered if the action of the solenoid were constant. In reality this is not the case, it varies with the position of the iron rod, and the distance between the carbons tends always to increase. To counterbalance this augmenting action of the solenoid it occurred to M. Jaspar to place between the weights of the large pulley, a counterweight acting in the opposite direction, and the moment of which increases at the same time as the effort of attraction of the solenoid; the distance between the carbons consequently remains constant all the time that the lamp is in action. The spring R is placed outside the box, the object is to lock the apparatus when it is not at work, and to facilitate replacing the carbons, and taking the lamp apart.

It will be seen that the Jaspar lamp, which is very simple, and not liable to get out of order, combines three striking characteristics: (1) The counterweight P, allowing the employment of the lamp with currents of varying intensity; (2) The counterweight Q, compensating automatically the variable action of the solenoid; (3) The pendent rods acting as brakes, and preventing any sudden variation in the movement of the carbons.

Krizik and Piette.—The electric lamp known as the " Pilsen " is the invention of two Austrian engineers, Messrs. Franz Krizik and Ludwig Piette, but takes its name from the town of Pilsen, in Bohemia, where it was devised. It is an arc lamp with two solenoids and a movable core of soft iron as the regulating part; but the special feature in it is the peculiar shape of the core and its suspension. When a cylindrical iron core of the ordinary form is placed inside a solenoid, through which a current is sent, the core takes up a balanced position in the coil, and when forcibly moved from it always tries to regain it. Thus with two solenoids M S, Fig. 437, the cylindrical core will take up a position proportional to the power of one coil over the other; and when the current alters in one or other coil the core will take up a new position, always tending however to return to the old one. By using a spindle-shaped core, however, such as is shown in Fig. 441, there is no tendency to take up a particular balanced position. It has a longer range of free movement, and can be utilised to feed the carbons of the arc directly, that is to say, without the intervention of clockwork, such as is necessary in lamps having cylindrical cores. Figs. **438** to **440** represent the core at different points of its range, and in each of these positions there is no tendency to move up or down. Figs. **442** to **445** show other forms of core which possess the same stability, and may be employed in the Pilsen lamp. In each the greatest mass of metal is in the middle, and the least at the

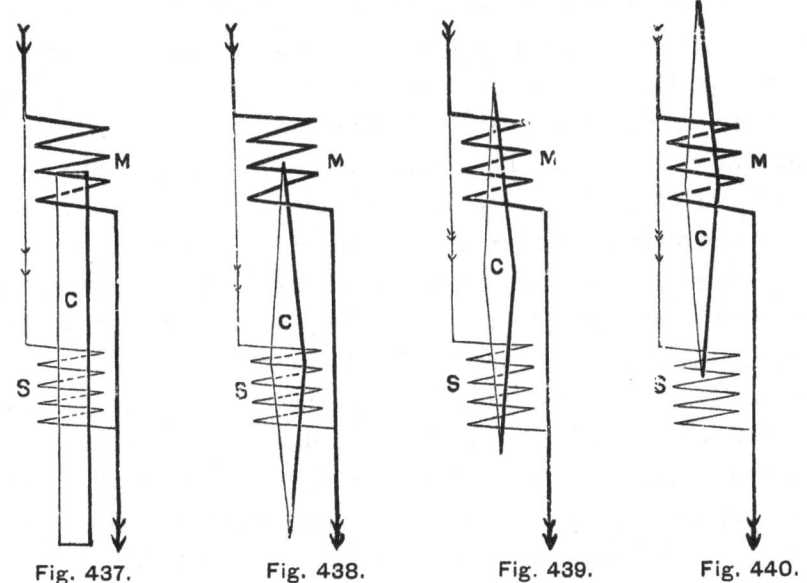

Fig. 437. Fig. 438. Fig. 439. Fig. 440.

extremities. Whether it be placed in a vertical or horizontal position this arrangement operates with the same uniformity of action.

Fig. **446** shows the arrangement of the lamp as now constructed. The wire from the negative pole of the machine is connected to the

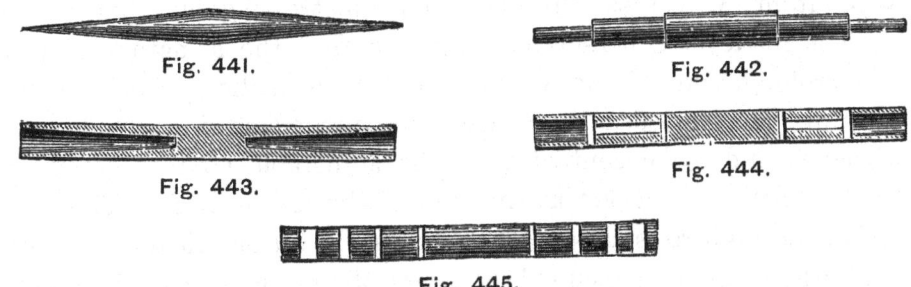

Fig. 441.

Fig. 442.

Fig. 443.

Fig. 444.

Fig. 445.

terminal on the top at the left, and goes from thence through the main coil M by means of the sliding contact spring S. After passing through M the current splits, part going through the small coil A and the rod D^1, which is insulated from the frame of the lamp, then to the contact roller *r* and lower carbon holder H. The other part goes through the resistance coil E to the other rod D, which is also insulated from the frame, and thence to the lower carbon holder by the contact roller *r*. The positive current enters at the right-hand top terminal, and goes to the frame and upper carbon. The shunt circuit G is of thick German-silver wire connected to the iron frame of the electro-magnet A. From thence a fine coil of German-silver wire of about 130 Ohms resistance is wound round the outside of the lower solenoid M^1, and the end of this coil is connected to the contact screw *n*, and to a comparatively thick copper wire of low resistance coiled inside the German-silver wire on M^1. The other end of this copper wire passes to the metal frame of the lamp, and thus completes the shunt circuit.

The object of this supplementary resistance G is to make up a total resistance, together with the copper coils inside M^1, equivalent to the lamp when burning; and to avoid heating when the main current all passes through that shunt or partial shunt circuit.

Among the numerous ingenious details in the Pilsen lamps the following device for altering the length and number of turns of the wire on the main solenoid M seems to be novel. After the wire is all coiled on M (see Figs. 446 and 447) the insulator is scraped off a strip ½ in. wide, leaving a ridge of bare copper wire; and connection with this is made through the contact spring S, which slides up and down the fixed rod, as shown. In this way the pull of the coil M can be adjusted, and the length of the arc regulated.

The use of the small electro-magnet A is to automatically short-circuit the high resistance coil M^1. As before described, a split wire from M passes through this electro-magnet. The iron frame of A is made of a bell-crank shape, so that the armature at the top is a prolongation of the bottom pole, thus utilising both poles. This armature is pivotted in the centre, and on the free side is weighted to overbalance and make contact with the adjustable contact screw *n* fixed to the insulated bracket at the side of the frame of A. Thus by means of A the high resistance wire on M^1 can be automatically cut out when the carbons are all consumed, or when there are none in the lamp, and the main current is then passed through the thick supplementary resistance wire and inside copper coil on M^1. This is effected as follows: One end of the fine wire shunt coil on the outside of M^1 is connected to the wire of G, and to the metal frame of A at bottom; the other end being connected to the copper shunt coil and insulated bracket and contact screw.   When the lamp is burning the main current passes through

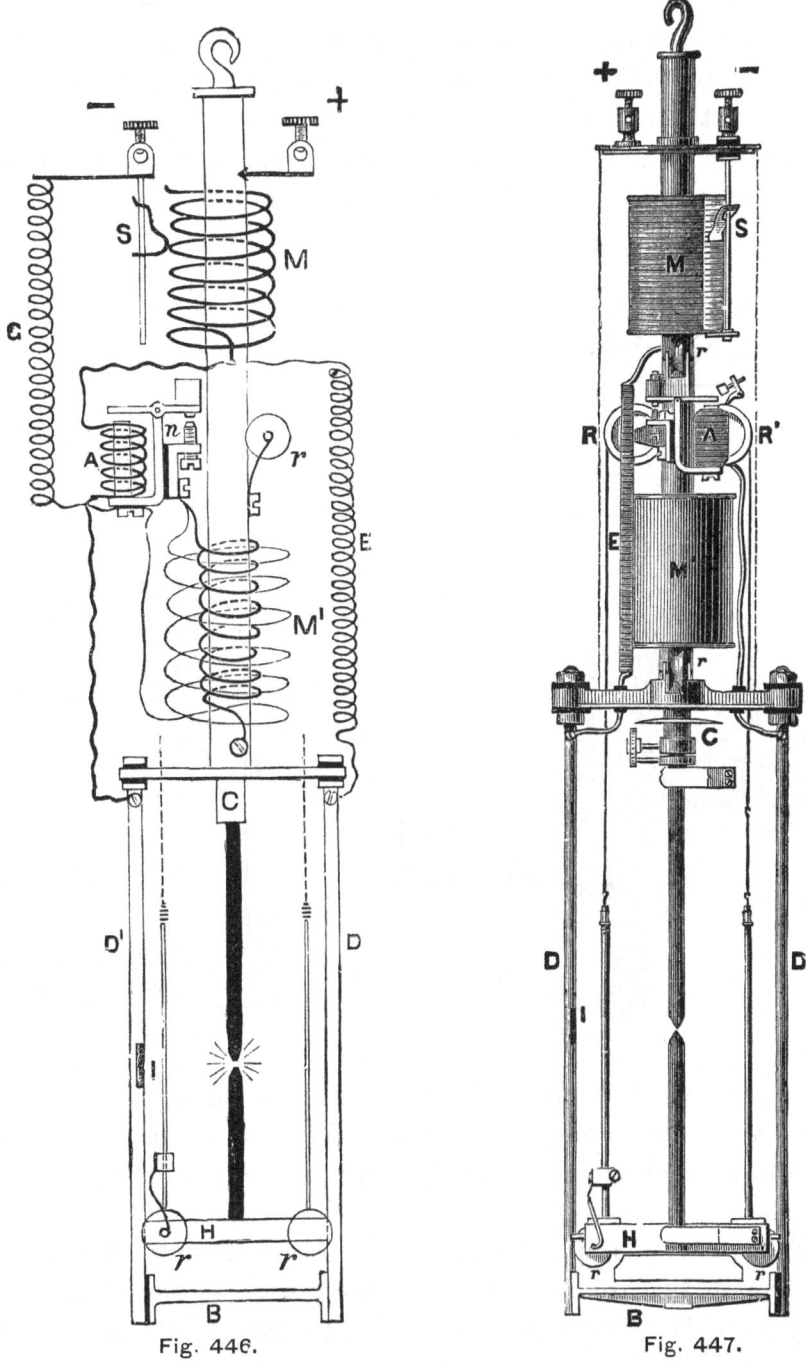

Fig. 446.

Fig. 447.

A and attracts the armature retaining it, and thus breaking the conta actt
n and adding the fine wire coil of M¹ to the shunt circuit, which then

balances the lamp. When the carbons are quite consumed this electromagnet A automatically cuts out the lamp. This is done by inserting a small insulated piece in the guide rod D^1 at I, so that when the bottom holder H rises to this insulated piece I the roller r on that side becomes insulated from the rod D^1, and the electro-magnet A is cut out of the circuit, the armature is released, and the weighted side remakes the contact and short-circuits the German-silver coil on M^1.

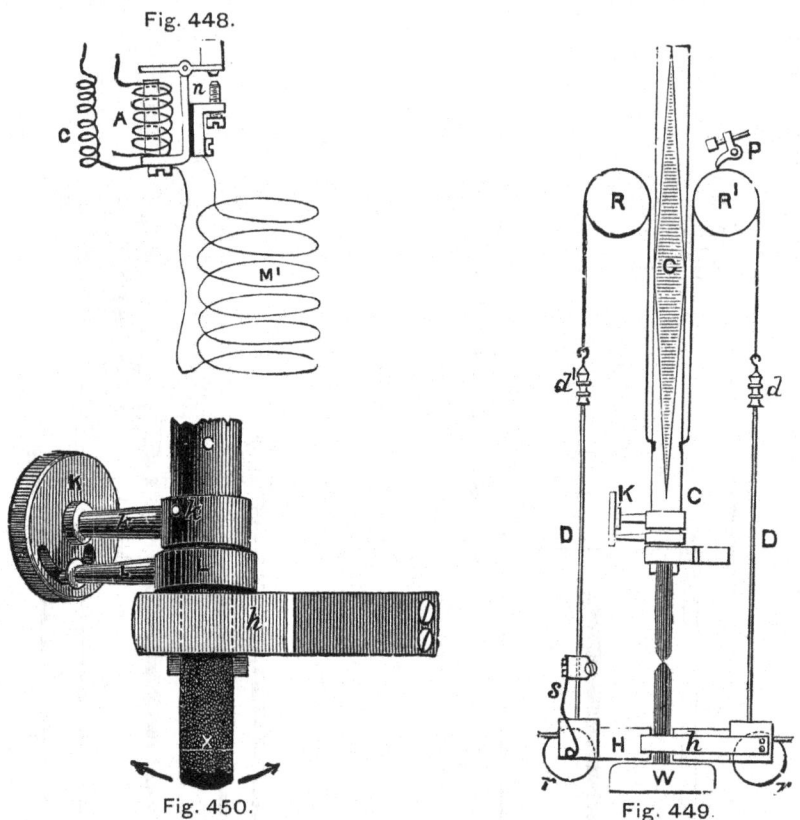

Fig. 448.

Fig. 450.

Fig. 449.

The upper coil then preponderates over the lower one, and pulls the top carbon up, thus completely separating the two carbons and breaking the arc; the current then passing through the thick shunt coil and wire A.

The frame of the lamp consists of a brass tube mounted between two brass plates with two iron rods D^1 D, joined across at the bottom by the stretching piece B. The iron core C (Fig. 449) fits loosely inside a brass tube which again fits loosely inside the tube of the frame and is guided by contact rollers B B^1, so that it moves with very slight friction. The weight of the iron core is counterbalanced by that of the bottom carbon holder by means of cords joined to the rods D^1 D at d^1 d and passing over the pulleys R^1 R. The metal rods D^1 D are used instead of cords near the arc to prevent scorching. The carbons are fixed as shown, one to the core C and the other to the lower tube holder H.

If a lamp with fixed focus be desired, the pulleys R^1 R are made double, one being twice the diameter of the other; the cord that passes over the larger pulley being connected to the top electrode whilst that passing over the smaller pulley is connected to the bottom electrode. The different sizes of the pulleys allow the carbons to feed together in proportion as they are consumed. One of the pulleys R^1 (see Fig. 449) has very fine teeth cut in its outer rim, and a weighted pawl or click P gears into the teeth. The object of this device is to restrain the cord which must rub over the pulley R^1 in one direction, and thus serve to arrest the parting of the carbons, when there are any fluctuations in the current. The carbons are fitted into the spring clips h h, and the bottom holder can be partly adjusted by altering the lengths of the two side cords at d^1 d. The top carbon can be adjusted to any position by the device shown in Fig. 450, where the spring clip h is attached to the collar and arm L, and both are pivotted at k and free to move laterally. The collar k is fixed to the core tube C. On one side of k is an arm carrying a milled head free to turn, and having an eccentric slot cut in its face with a pin projecting from L inserted in it. By turning the milled head, the arm L and the carbon clip h are moved, the latter being shifted backwards or forwards as required. The core tube C can also be moved round. The lower carbon holder H runs by means of the contact rollers r and r between the guide rods D^1 and D, the contact being maintained by the spring S. The top of the lamp with the coils is completely encased in a zinc box. The globe is put on from below in the usual hanging lamp, and is supported on the ash-pan which slides up and down the rod at the bottom. This is very convenient for putting in fresh carbons.

Fig. 451 represents a form of ornamental hanging lamp suitable for the interior of theatres; a factory lamp made to burn 100 hours and having plate carbons, as in the Wallace-Farmer lamp, is shown in Fig. 452. The electrical connections are the same in this lamp as those shown in Fig. 446, but the parts are made stronger. Figs. 453 and 454 represent the Pilsen horizontal lamp; in the former figure, C is the iron core, *r r* the contact rollers, and M M¹ the solenoids as before. One carbon is fixed to the end of the tube C and the other to the pivotted arm

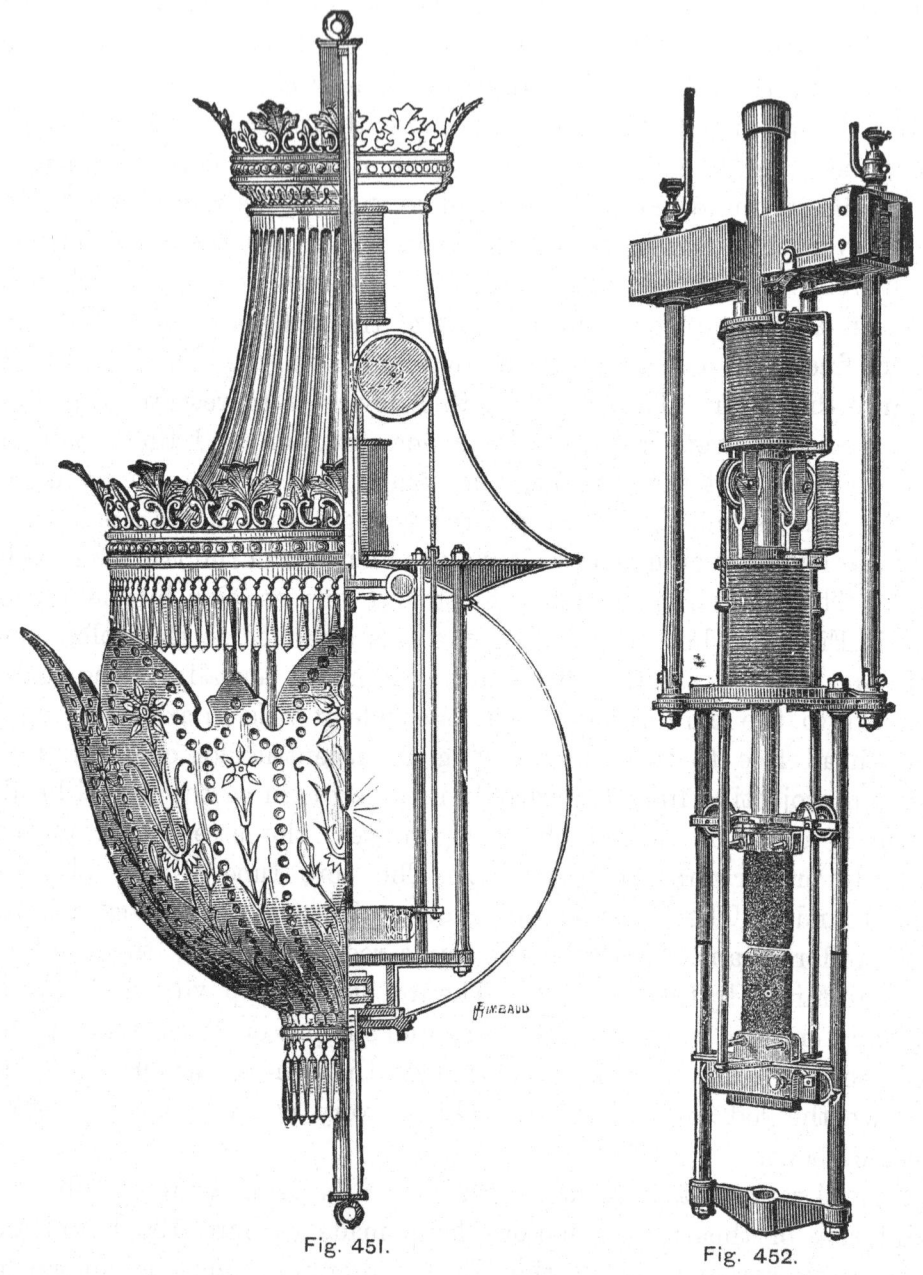

Fig. 451.

Fig. 452.

D. Such a lamp can be used with advantage in low-roofed factories, close to the ceiling. Fig. 455 shows the Pilsen lamp adapted to street lighting, in which the counterbalances of the iron core are shown reversed, and the coils are placed at the bottom beneath the lamp. The side rods D¹ D, the counterweight W, holder H, and rollers R are the same as in the hanging lamp, but have their cords reversed. This arrangement permits the regulating coils to be safely protected and avoids shadows on the globes.

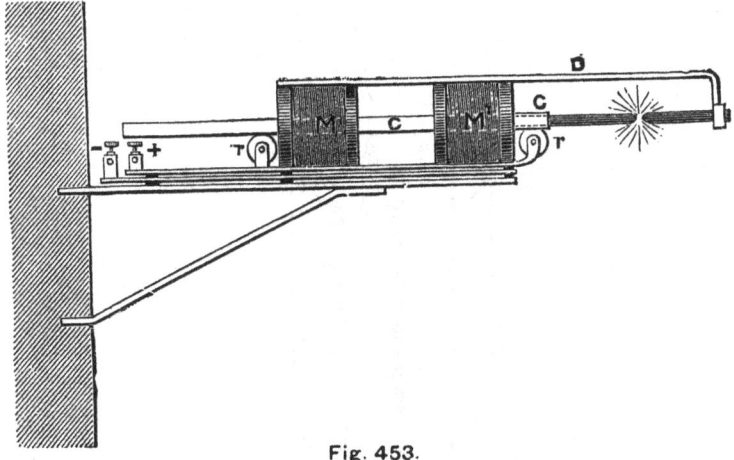

Fig. 453.

The Pilsen lamp has the great merit of simple construction and remarkably steady working. The feeding movement is continuous, the electric resistance of the arc is sensibly constant, and it can be worked

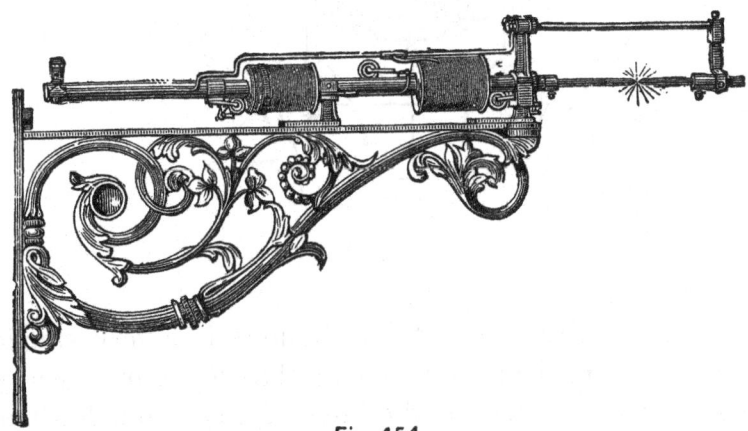

Fig. 454.

with almost any kind of engine, as the absence of either clutch or releasing gear enables it to quickly adjust itself to variations of engine speed. At the Electrical Exhibition at the Crystal Palace, the Handel Orchestra was lighted by six Pilsen lamps hanging from the roof, and all connected in one circuit. They were fed by the current from a Schuckert dynamo-electric machine, driven at 850 revolutions per minute, by a Ruston and Proctor single-cylinder portable steam engine with the ordinary governors; the power of each lamp is that known as 2,000 candles. Messrs. Rouatt and Fyfe, London, are the owners of this system in England.

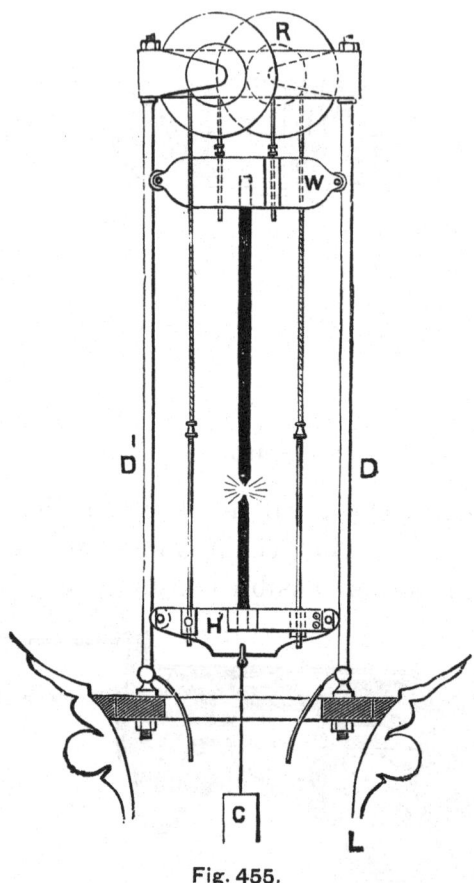

Fig. 455.

Gramme.—The electric lamp of M. Gramme is a very simple and ingenious piece of mechanism; it is one of that numerous class which depend for their action upon a derived current passing through a fine wire electro-magnet, which comes into play when the distance between the carbons increases.

Of the drawings, Figs. 456 and 457 are respectively side and front elevations of the lamp; Figs. 458 and 459 are side and front elevations to an enlarged scale of the auxiliary magnet for feeding down the upper carbon.   Figs. 460 and 461 (see page 457) are an elevation

Fig. 458.

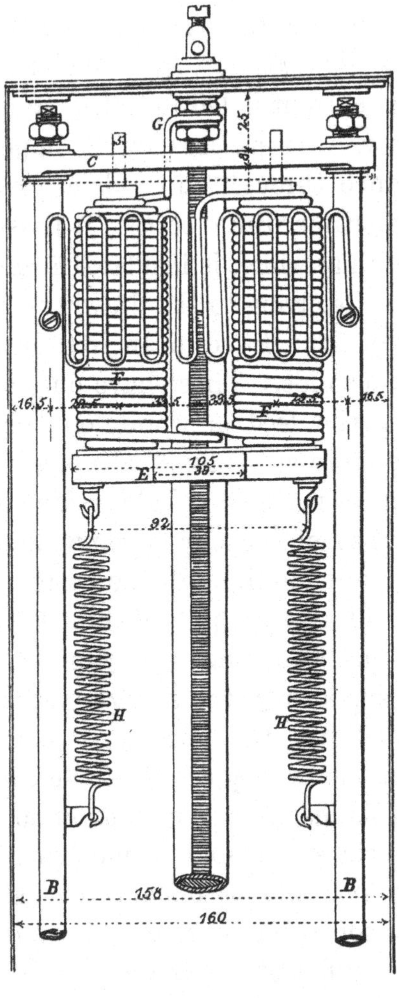

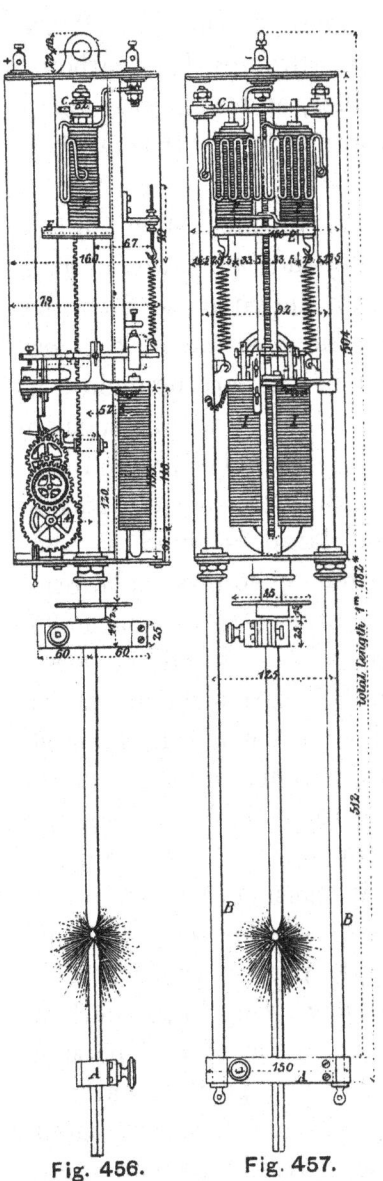

Fig. 456.　　Fig. 457.

Gramme.

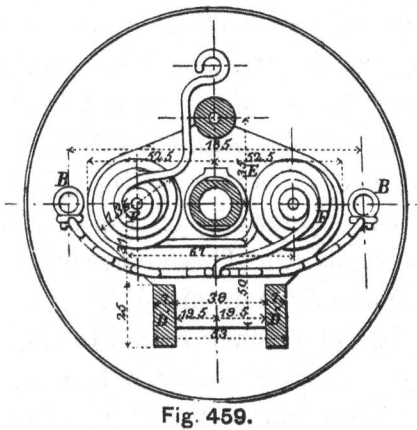

Fig. 459.

and plan of the coarse wire electro-magnets for moving the lower carbon, and Figs. 463 and 464 are plans of the upper and lower carbon holders. The current is led from the generator to the terminal on top of the lamp, whence it passes by the upper carbon holder, through the carbons, thence through the rods carrying the lower holder, and, after traversing the coarse wire electros, it passes back to the generator by the negative terminal which is insulated from the lamp.

Referring to the drawings, it will be seen that the mechanism of the lamp is enclosed in an iron case, below which depend the lower carbon holder rods. These have only a very slight motion imparted to them, so that the feed is wholly on the upper carbon; the position of the arc therefore is not fixed, although in a modification of the arrangement, it can be made so. Taking first the lower carbon holder, it will be seen to consist of the clip A carrying the carbon, the two rods B B, which are free to slide through the bottom of the case, and the cross-piece C at the top of the lamp. Within D^1 the case of the lamp is fixed a post D, and two other uprights D^{11}, to which is secured by screws a platform E, supporting the electro-magnet F. The form, arrangement and mode of coiling the magnet is shown in Fig. 458, where it will be seen that the cross-piece C of the carbon holder rods B B forms the armature of this magnet; the mode of connection with the rods, which is adopted to secure freedom of movement, is also shown. The connection with the negative terminal G is also shown. The rods B B are held up by coiled springs H H, hung to the platform E. These springs are sufficient to maintain (when no current is passing) the armature C clear of the electro-magnet F, in other words to hold up the negative carbon. But when the current passes through the electro, it attracts the armature and forces down the lower carbon through the range given to it. Passing on now to the second portion of the mechanism, that for controlling the feed of the positive carbon; the electro-magnet I (Fig. 462) coiled with fine wire is the main agent in this operation; on top of it rests the platform K, having two small standards K^1. These carry by screw trunnions the balanced lever L, which is normally maintained in a horizontal position by the coiled spring N, hung to a bracket bolted to the post D. This spring can be delicately adjusted to set the lever L in its right position. At one end of this lever is the armature M; M^1 is a check screw to prevent its rising suddenly. At the other end of the lever L is the adjustable screw P, which presses

Fig. 460.

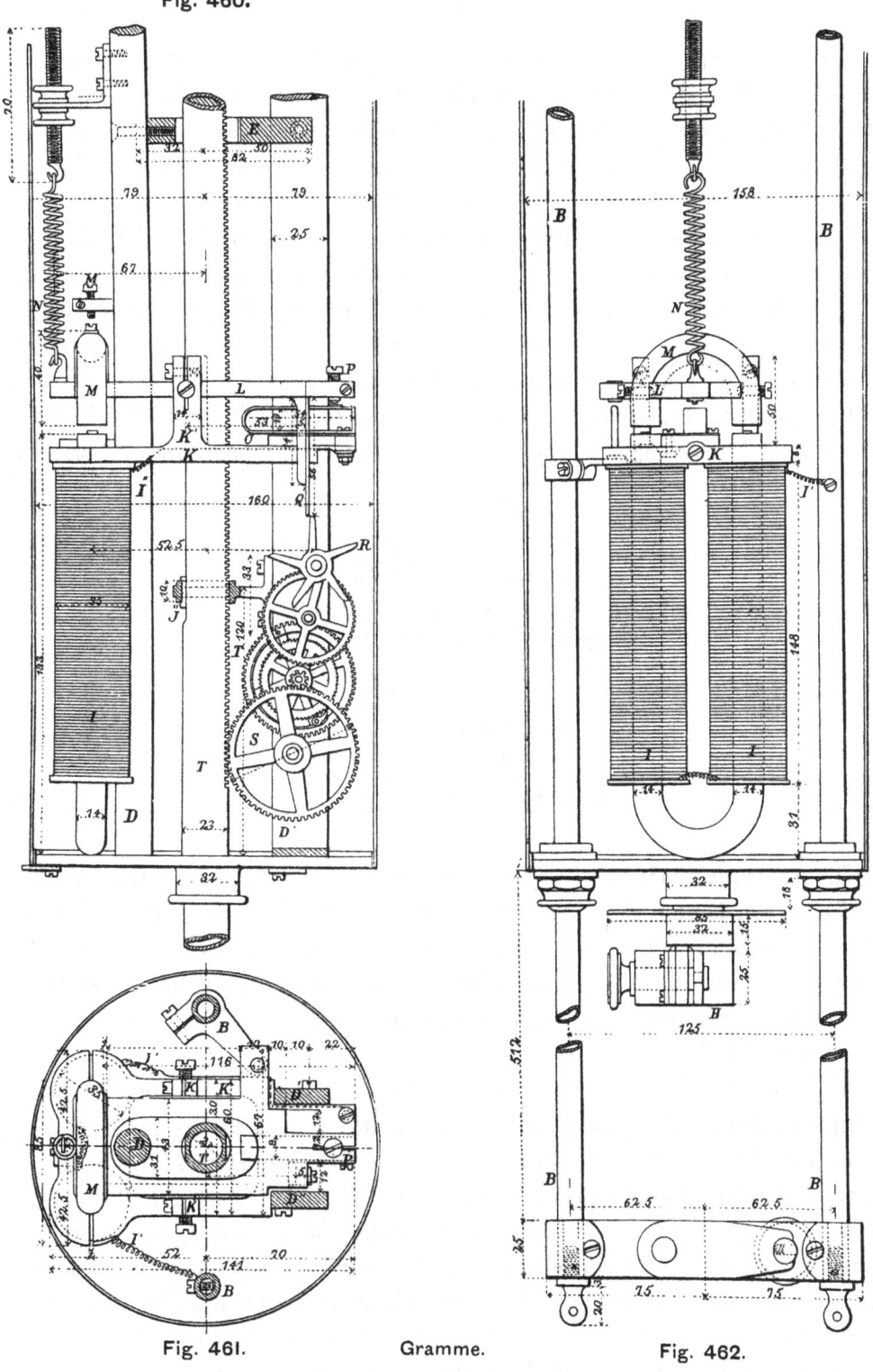

Fig. 461.   Gramme.   Fig. 462.

upon the bent spring O fastened to the frame K on top of the electro-
magnet. From the lever L depends the arm Q carrying the spring Q^1.
This spring engages in the star-wheel R, which is the last of the train of
clockwork, the first wheel S of which gears-in the rack T^1 on the side of
the positive carbon holder T. The derived current to the electro I is
taken by a connection from the positive pole to the insulated spring O,
whence it passes, when the arc is of normal length, through the lever L,
and its trunnions K^1 to the coils, which are coupled at their other
extremity to one of the negative rods at I^1. The upper carbon
holder T is tubular as shown and is guided by a socket in the
bottom plate of the case, and also by the guide T^{11} screwed to the
standard D^1. Fig. 463 is a plan of the positive carbon holder shown

Fig. 463.

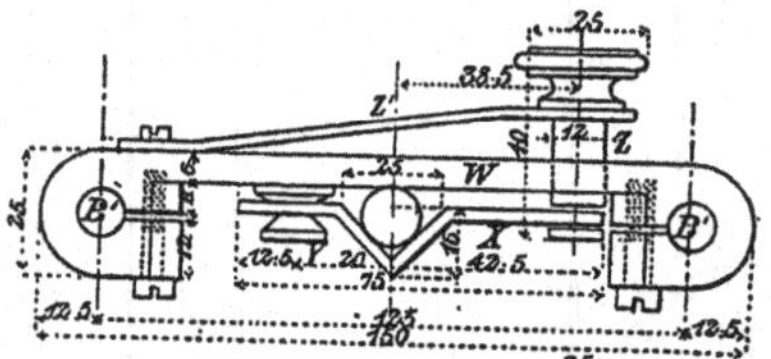

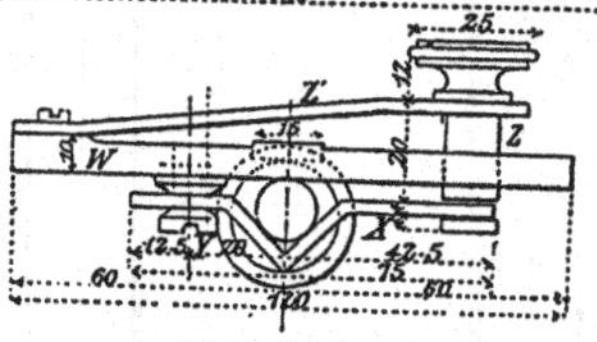

Fig. 464.

in side elevation at H, Fig. 462. The plate W is attached to the rod T,
and the carbon is clamped by a recess in the bar X which swings on the
pivot Y, and is held up tight against the carbon by the stud Z and the
spring Z^1. The lower carbon holder is very similar, except that the
plate W is larger, and is provided with spring openings B^1 to receive the
ends of the rods B.

The action of the lamp may now be described. So long as the
principal current maintains a certain intensity, that is to say, as long as
the arc is not of too great a length, the spring Q^1 holds the star-wheel R
and prevents the positive carbon from descending by its own gravity. But
as soon as the arc lengthens, and the resistance is increased, the electro-
magnet I comes into action, and attracts the armature M, the movement
of which draws down the lever L on which it is mounted, and lifts the
spring Q^1 clear of the wheel R, thus releasing the clock train, and allowing
the positive carbon holder to fall by its own weight to bring the carbons

closer together. At the same time the contact between the spring O and the stud P is broken by the tipping of the lever L, and the current ceases to pass through the magnet I. The armature M, drawn by the spring N, regains its horizontal position and the regulator resumes its action, but if the arc be not of the normal length the same movement recommences. It will be seen, at the moment when the upper carbon descends, the electro-magnet F F becomes active and attracts its armature C, which causes the lower carbon to descend also, the electro I becoming of course inactive. This stopping of the positive and lowering of the negative carbon establishes the distance necessary to produce the arc. Five of these lamps, fed with the Gramme generator for five lights, which is illustrated and described on a previous page, and each giving nearly 1,500 candles, were maintained with an expenditure of 6 horse-power.

Gülcher.—The Gülcher electric lamp, which possesses several ingenious and highly practical details, is attracting much attention at the present time by the steadiness of the light it gives, and the simplicity of its construction. Although it does not belong to that class devoid of regulating mechanism, such as the Soleil, Werdermann and others, it has no clockwork, and therefore may lay claim not only to great simplicity, but also to especial suitability for exposed situations where clockwork may become easily deranged. The regulation of the carbons is effected by a single electro-magnet, rendering the lamp very cheap to manufacture and repair, and not liable to be easily put out of order. Fig. 465 shows the construction of the regulator, in the body A of which is a chamber B containing the regulating mechanism. This consists of a pair of bearings C, in which are placed the pivots of a metal ring D. The ring carries an electro-magnet E, which is free to oscillate upon the trunnions of the ring D. At the base of the regulator is the insulated binding screw F, which is connected to the positive pole of the generator, and the current is then led by a wire to the bearings C (which are also insulated), passes to the ring, and thence by the coil to the core of the electro-magnet, whence it flows, partly through the spring G, which is in contact with the end of the magnet, and also in metallic connection with the case of the regulator, and partly to the rod H in contact with the other end of the magnet. This rod H is the holder of the upper carbon, as will be seen from the drawing. Finally the two currents unite and passing

through H, flow through the carbons, the lower carbon holder, and thence
by a wire or other conductor to the binding screw I, which is coupled up
to the negative pole of the generator.   The pole faces at the ends of the
electro-magnet E are curved, forming parts of a circle, the centre of

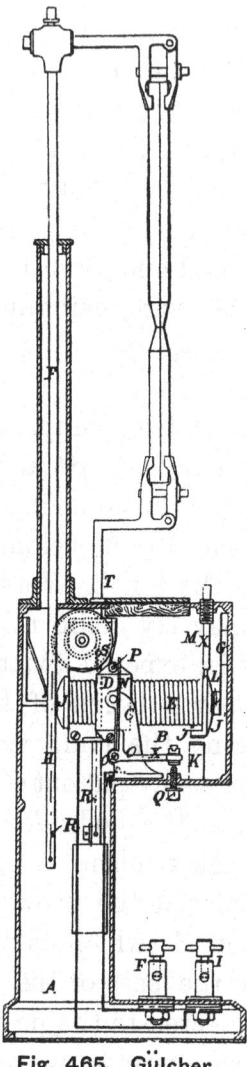

Fig. 465. Gülcher.

which is on the axis of the trunnion of the ring D.   It will be seen that
there are several parts not yet described within the box B.   One of
the poles J' is extended downwards and terminates in a projection J."
Immediately beneath this projection is a small iron block K attached to
the case of the lamp.   The lower end of the spring G is furnished with

another small iron block L, which makes contact with the pole J'. Passing through the top of the casing is the screw M, which terminates in a spindle against the lower end of which the electro-magnet rests when no current is circulating. This position of the magnet is secured by means of the spring N, which is connected to the angle lever O pivotted to the insulated bearing C at O¹. The upper end of the spring N bears against the screw P, and the position of the spring and its pressure against the trunnion ring is regulated by means of the screw Q, which is insulated from the lever O.

The operation of the regulator is as follows :—As soon as it has been connected with the generator, the electro-magnet E attracts the rod H of the upper carbon holder to the pole J, and at the same time the part J" of the pole J' is attracted to the fixed piece of iron K at the bottom of the casing, thereby causing the electro-magnet to turn on the trunnions of the ring D. By this movement the carbon holder H is necessarily lifted, because of the contact with the curved face of the armature J, the motion being far steadier and more uniform than could be imparted by a rack and pinion ; the two carbons are thus separated and the voltaic arc produced. As the carbons burn away, and the distance between them becomes greater, the resistance increases, and consequently the power of the electro-magnet E diminishes, the piece J" of the armature becomes further removed from the piece of iron K, the carbon points approach each other, and the magnet continues to turn on its trunnions until its armature J' is arrested by the stop M, the position of which has been carefully adjusted. As soon as this condition has been established the magnet E has no further movement, the carbons continue to burn until they are at that distance apart corresponding to the greatest intensity of light. As soon however as further consumption increases the distance, the attractive power of the magnet E becomes too small to enable its armature to hold the bar H carrying the upper carbon, and the bar begins to slide down, until the carbons are brought nearer together, and the power of the magnet is increased so that it again holds the bar H and prevents it slipping further. If the bar continued to fall, the power of the magnet would increase until the armature J' would be drawn down to the piece of iron M and the preliminary process above described would be repeated. In practice, however, it is found that after the lamp has been started the magnet remains stationary, the consumption of the carbons and the weight of the rod H being sufficient through the intermittent action of the magnet to secure the automatic working of the lamp. The face of the pole J and the rod H are covered with thin brass to diminish the attractive power of the magnet and to facilitate the sliding of the latter when the carbons require to be fed down; this detail makes the action of the regulator much more soft and checks irregularity in the light. The block of iron L attached to the spring G, and acting as a brake, is also covered with thin brass as well as the pole J'. This block serves as a brake and damps down the action of the electric magnet when the lamp is started and secures a soft and easy movement. As is shown in the drawing, both carbon holders are made movable, so that the position of the light may be kept unaltered. One method of securing the necessary motion is shown in the drawing, where threads R, R¹—the former to the upper, and the latter to the lower carbon-holder T—are wound around the drum S, and both actuated by the movement of the rod H.

Such is the construction of the Gülcher lamp; it now remains to be seen how the peculiarities of its arrangement render it suitable for being used in groups in the same circuit, and how one lamp acts as the regulator of the second, the first and second as the regulator of the third, and so on. This regulating action depends on the positive motion of the electro-magnet, and upon the quantity of current flowing through the circuit. To take first the simple case of two lamps put in parallel circuit with the generator: If a lamp 1 be first lighted by connecting the branch wire from the trunnion of the electro-magnet, to the main conductor from the positive pole of the generator, the light will establish itself, as has been already described, then if the connection with lamp 2 be completed the current is divided, and lamp 2 is lighted. At first the part of the current flowing through lamp 2 is stronger than that passing through lamp 1, because the carbon points of the former are in contact, and the resistance is less The result of this inequality is that the electro-magnet of lamp 2 is set in motion, the carbons are separated, and the resistance is increased; at the same time that this takes place, the current passing through lamp 1 is strengthened, and the carbons are brought somewhat closer together. From this it will be seen that if in either lamp the carbons become too far separated, the preponderance of current in the other will tend through the action of the magnet to raise its carbon holder, increase the distance between the points, and so weaken the current in the first-named lamp; then the magnet will cease to maintain the carbon holder, which will slide down till equilibrium is established, or the same process is reversed, and so on, one lamp regulating the other, and consequently itself. With three lamps in parallel circuit, as shown in Fig. 466, the same processes take place; when all three are lighted, lamp 1 regulates lamp 2 and 3; No. 2 regulates 1 and 3, and No. 3 regulates 1 and 2.

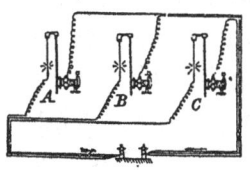

Fig. 466.

Mr. Gülcher claims in his patent specification that these. lamps, "without any addition of special mechanism, can be used for the division of the electric light by means of a simple arrangement in parallel circuits. The advantages of a good arrangement of parallel circuits, and of the application in connection with it of strong quantity currents are (in opposition to the division of the electric light hitherto attained by currents of high tension and differential lamps placed one after another) of great importance, and consist essentially in the following: That the electric light produced by this process is thoroughly white (without the violet or bluish tints in lights produced by currents of high tension); that the number of lamps which can be fed by a single source of electricity is nearly unlimited; that the power used by the dynamo-electric machine is proportional to the number of lamps which at one time are fed by it; that further, large and small lamps (of the latter a great number) can be provided for by one and the same machine; and finally the lamps can be attended to while at work without any danger, and even the contact with uncovered portions of the conducting wires by unprotected parts of the body is free from danger." It must be admitted that the experience gained at the Paris Electrical Exhibition with the Gülcher lamp went far to sustain the greater portion of these claims.

The manner in which Mr. Gülcher arranges his conductors, for working a number of lamps on the same circuit, now remains to be described. Knowing the number of lamps to be fed and the distance of the most remote lamp from the source of power, he ascertains the cross section of the main conductor required to transmit the total current. He then constructs a cable having the same number of strands as there are lamps on the circuit, and each of such a diameter that the total cross section is equal to the required area of the principal conductor. This cable is joined to the positive pole of the battery, one strand being branched off to the first lamp, another to the second, and so on for the whole number, till at the last lamp the final strand is connected. In the same manner the return conductor commences with one strand at the first lamp, and gradually increasing a strand at a time with each lamp, returns to the negative pole of the generator with the same cross section as the cable from the positive pole leaves it. Following Mr. Gülcher's explanation of the arrangement of his conductors, it will be seen that each separate circuit to each lamp is of equal length, and the cross section of each conductor corresponds exactly with the quantity of current flowing through it.

To conclude this notice we may give the summary of some trials made by Mr. Gülcher with the lamps fed by the generator bearing his name, and which we have already described. They consist of a series of measurements of the photometric values of the lamps and of the power they absorb. In these trials were also included some relating to the internal resistance and the electromotive force of the dynamo-electric generator employed.

The first table shows the lighting power obtained with a given intensity of current. The trials were made with a Gülcher generator feeding six large regulators on one circuit. The generator was driven at a speed of 640 revolutions a minute, and the intensity of the standard photometric lamp was exactly 1 Carcel. The following 10 measurements were made, the photometric values being taken on the horizontal plane of the arc :—

No.	Power of Light. Carcels.	Intensity of Current. Ampères.	No.	Power of Light. Carcels.	Intensity of Current. Ampères.
1	105·6	14·6	6	105·6	14·6
2	127·0	15·4	7	112·3	14·8
3	112·3	15·0	8	127·0	15·6
4	112·3	15·2	9	112·3	15·2
5	105·6	14·8	10	112·3	14·8

The average of the foregoing results are a photometric luminous value of 113·23 Carcels for a 15 Ampère current. Under the same conditions of trial 10 further measurements were taken, the incidence of the luminous rays being 34 deg. These measurements gave the following results :—

No.	Power of Light. Carcels.	Intensity of Current. Ampères.	No.	Power of Light. Carcels.	Intensity of Current. Ampères.
1	128·8	14·8	6	128·8	14·8
2	144·0	15·6	7	144·0	15·4
3	136·1	15·4	8	136·1	15·2
4	136·1	15·2	9	136·1	15·2
5	136·1	15·0	10	136·1	15·0

The average being 136·22 Carcels for an intensity of current of 15·16 Ampères. From the foregoing it will be seen that the intensity at 34 deg. is equal to 1·2 times that on a horizontal plane. The second part of the experiments relates to the consumption of power in proportion to the luminous values. This experiment was made by introducing into the circuit a gradually increasing number of lamps. When only one was being fed the photometric intensity was about 125 Carcels. With two lamps it was about 150 Carcels each. In increasing the number from three to five lamps, an intensity of 135 Carcels was maintained; with six arcs it fell to 120 Carcels. Mr. Gülcher, however, maintains that this last result is abnormal and was due to the slipping of a belt, that the speed of the generator had fallen from 640 to 630 revolutions, and it is probable that if the speed of 640 had been maintained an intensity of 135 Carcels would have been kept up with the six lamps. The results obtained from the trials were as follows :—

Number of Lamps.	Speed of Generator.	Horse-power Absorbed.
0	645	0·51
1	640	2·49
2	640	4·12
3	640	5·62
4	640	7·04
5	640	8·58
6	630	10·05

Corresponding to about 1·66 horse-power for each lamp of 135 Carcels.

H H

The internal resistances of the Gülcher generator appear to be very low, thus attaining one of the objects of the inventor in designing this machine. Annexed are the measurements taken (1) before the generator was doing useful work, the temperature being 16 deg. Cent.; and (2) after some hours' work, when the machine was heated to 31 deg. Under these two conditions the resistances were as follows :—

—	1.	2.
	Ohms.	Ohms.
Resistance of magnets	0·126	0·129
,, ring	0·133	0·136
Total	0·259	0·265

It will be noticed that the heating was inconsiderable, and that it hardly modifies the internal resistances.

The potential differences in Volts were as follows :—

At the terminals of the machine .... 60·00
,, ends of the field magnets .... 10·22
,, brushes .... 70·22

As stated above, the intensity of this current per lamp is about 15 Ampères; all the characteristic data of the generator are thus given. From the foregoing results it was shown that the motive power absorbed in working the six lamps would be 10·25 horse-power, distributed as follows :—

2·91 absorbed by the generator
7·34 ,, ,, exterior circuit.

To this must be added ·51 horse-power required to turn the generator, giving a total of 10·76 horse-power. This amount differs from that attained at the trial above recorded, because during the time that the intensity of the current and the differences of potential were measured, the speed was 640 revolutions, and this fell to 630 revolutions when the total work was measured. The useful effect is thus equal to $\frac{7\cdot34 \times 100}{10\cdot76} = 68\cdot2$ per cent. The intensity of each of the lamps having been 136·22 Carcels, or 817 for the whole, the relation between light obtained and power expended was $\frac{819}{10\cdot76} = 76$ Carcels per horse-power.

Berjot.—The lamp consists of a casing of the form shown in Fig. 467, enclosing the mechanism, its clockwork is held in a frame, at the top and bottom of which are the four flat springs R R¹ suitably

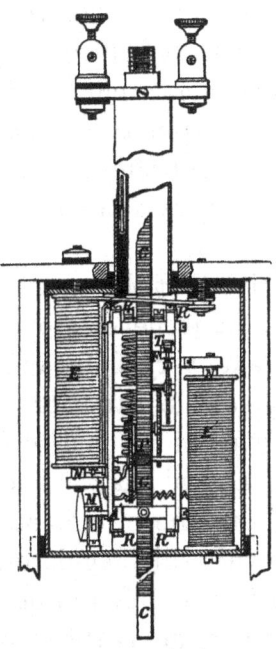

Fig. 467. Berjot.

regulated, and subject to the action, under varying conditions, of the two electro-magnets E E¹. One of these electro-magnets tends to raise the frame, the other to move the mechanism. The first one E is covered with coarse wire and is placed in the main circuit with the carbon; the other E¹ is formed with fine wire, and is placed in a derived circuit. As shown, they are straight magnets, the cores of which are cut in the middle, one-half N constituting the armature; in this manner is combined the reciprocal action of a solenoid, with the attractive action of the magnetic core. The frame containing the clockwork, carries the weight of the upper carbon holder, which is connected to a long rack C gearing with the first wheel of the clockwork, through the pinion P. A counterweight M, acting in conjunction with the upper carbon holder, tends to pull down the frame containing the clockwork; this is resisted by the electro-magnet E, which retains it in a fixed position. The carbon holder itself would always tend to fall if it were not held up by the clockwork of which the last wheel D is controlled by the brake F, fixed to the case of the lamp. As this brake may be raised or lowered by the action of the electro-magnet E¹ the rack of the upper carbon holder is free to move when the magnet E¹ is in action, the approach of the two carbons resulting from this action, which takes place when the arc is sufficiently weakened to give to the electro-magnet E¹ an excess of power over the other. The approach of the carbons is moreover effected conjointly with the locking of the case containing the clockwork, but when the carbons are so near together as to restore the strength of the current at the arc, the magnet E raises the cage A B, and the fine wire magnet E¹ releasing the brake F, prevents the clockwork from moving until the arc again becomes weakened.

The apparatus is so sensitive that the case A B is in a constant state of vibration, and as the clockwork is never stopped abruptly on account of the delicate action of the brake, very slight differences in intensity can be corrected without any appreciable variations in the steadiness of the light. When the rack has arrived at the end of its travel, that is when the upper carbon is consumed, a stud on the rack presses on a contact piece, and shunts the current into the circuit of other lamps. In order to secure a prolonged service with this regulator, without renewing the carbons, a modification is devised by M. Berjot, in which several racks are employed, one coming into action as the previous one is thrown out. As will be seen, the mechanism of the lamp is simple, its action is sensitive and certain, and its details have been very ingeniously worked out by the inventor.

Standard lamp for the French lighthouse system.—Fig. 468 shows the lamp that has recently been decided upon for adoption in the new French electric lighthouses. It is a combination of the Serrin and the Berjot lamps, both of which have been described. It comprises the two electro-magnets of the Berjot lamp, the armatures of which form an internal core, one magnet having coarse wire, and placed direct in the circuit, the other a fine wire magnet, mounted in a derived circuit. The former acts

Fig. 468. Lamp for French Lighthouses.

on the articulated frame carrying the lower carbon ; the other influences the disc brake that controls the clock train. In the drawing, S is the coarse electro-magnet acting through the lever Q, on the articulated frame or cage, the two coiled springs R and R¹ tending always to lift this cage. The tension of these springs is regulated by the bent lever L hung to the frame of the lamp, and the position of which can be altered by the screw V, that can be turned by a key passing through a hole in the casing. The fine wire coil is not shown in the drawing, but it occupies a position corresponding to that of the electro S on the other side of the clock train. The connection between the two carbon holders and the first spindle of the train, is effected by means of a steel ribbon F attached to the lower ends of the two rods *g* and *l*, the lower and upper carbon rods respectively. This ribbon is led over several pulleys and passes round the greater part of the circumference of a wheel on the first axis of the clock train, so that there should be no slip between the ribbon and the pulley. The rod *g* slides in the tube D fixed to the articulated frame, and this tube has a vertical slot formed in it to allow the ribbon to be attached. This mode of connecting the two rods supporting the carbons which replaces the chain of the Serrin regulator, allows the carbons to be set at their right height, by a slight slipping of the ribbon. Another peculiarity in the lamp is the mode of connection adopted for the internal mechanism. The current from the large cable conductor is led to the upper carbon from the rails and the non-insulated parts of the lamp. From the lower carbon it returns to the two insulated terminals H and H¹, passing partly through the articulated cap, and partly through the electro S. The connections between the pieces to which the current arrives are made with four spirals of thick copper wire, nickel-plated ; they are shown at M and N, placed above and below the coarse wire magnet S. The tube D carrying the rod *g* is not insulated from the articulated cage, but the latter is insulated from the rectangular bar by which it is carried. This arrangement is convenient in construction, because it is more easy to insulate a rectangular bar than a round piece like the tube D. An air pump T serves to deaden the motion of the cage, and prevents any violent oscillation. The ingenious details of the porous plate *v* must be noticed. When large carbons like those necessary for lighthouse work are employed, the turning piece *a* that serves to regulate the height of the arc becomes heated as well as the rods *t*¹ and *l* and the different parts exposed, often causing them to bind. To avoid these inconveniences, a porous clay shield *v* is introduced as shown. When the upper carbon holder has reached the end of its travel, it acts laterally on a small inclined rod which draws back a spring, and cuts off the connection with the fine wire electro-magnet, so that it cannot be injured by the passage of too strong a current.

Cance.—The Cance electric lamp is intended to work with a continuous current, and as the movement of the carbons to each other is regulated proportionally to their consumption, it follows that the position of the arc is always fixed. In Fig. 469, which is a longitudinal section, *a* and *b* are the two carbons fixed to the holders *c* and *d*, formed of two frames, from each of which a pair of rods rise vertically. At the top of each of the rods from the lower carbon holder are two small pulleys *e e*, around which pass the cords *f*, one extremity of each being attached to a fixed point on the top frame *g* of the regulator. The other end of each cord passes round the pulleys *h h*, and is fastened to the top frame *d* of the upper carbon holder as shown. On this frame is placed a driving weight *h* loaded with shot, and guided in its descent by the rods *i i*. The weight is formed like a nut, and fits on the central screw *j*, fixed laterally, but free to move on the pivots at the ends that turn, one in a bearing formed on the top plate *g*, and the other in a step on the lower platform *k* of the lamp. Under the effect of gravity the weight tends constantly to make the carbon descend, turning the screw, and causing, on account of the arrangement of the cords, the lower carbons to be raised ; the arrangement, therefore, tends always to bring the carbons in contact. The two platforms *g* and *k*, connected by the four guiding columns *i*, form the fixed frame of the lamp. On the lower platform of this frame are two plates *l*, carrying the coil *m*, inside which is a hollow core *n* able to move freely. The upper part of this core is extended in the same form by the core *o* made of copper, and the length of which is proportioned to the desired magnetic intensity. From the lower end of the screw *j* depends a rod *p*, passing through the axis of the core ; this rod is of copper to prevent any magnetic effect. The weight of the core of the coil can be more or less neutralised by springs. It is this coil, placed in the circuit of the main current, that constitutes the regulating organ of the lamp. As will be seen it carries above and below, cross-pieces *q r* rising and falling with it, and guided by the four rods *i i* of the frame. A regulating screw *s* in the upper piece *q*, and coming in contact

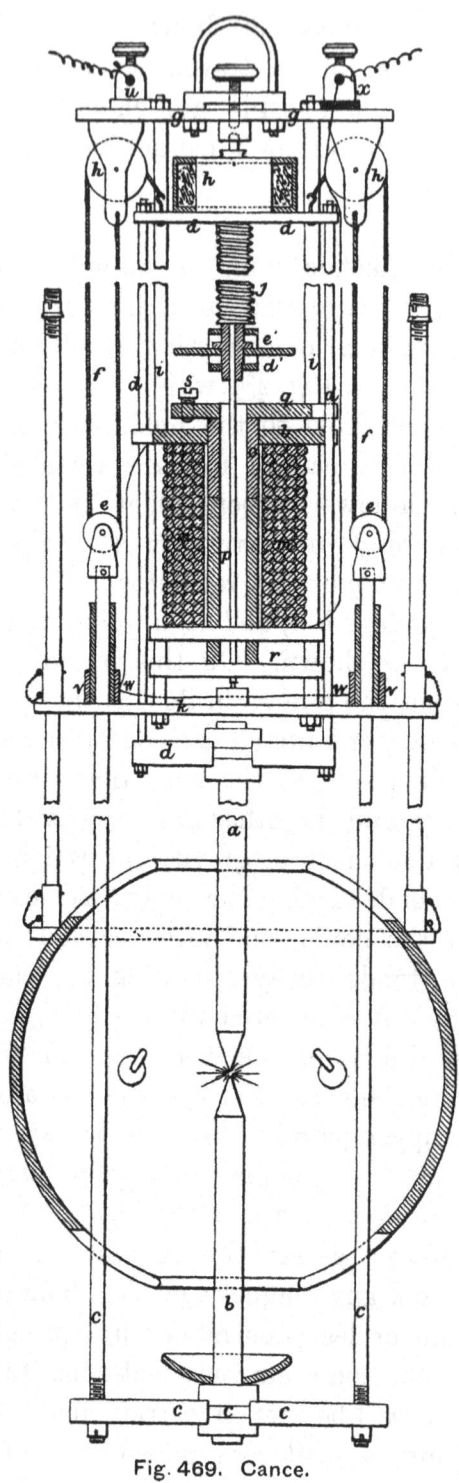

Fig. 469. Cance.

with the top of the coil, serves to regulate the position of the core within the coil, and the position of the escapement wheel, which checks the movement ; a second regulating screw at the other end of the same piece limits the separation of the carbons, and consequently the length of the arc.

The operation of the lamp is as follows :—The carbons being in contact, the current led to the lamp by the binding screw u passes through the frame to the upper carbon holder and carbon; the current then flows through the lower carbon, and the two rods of

Fig. 470.

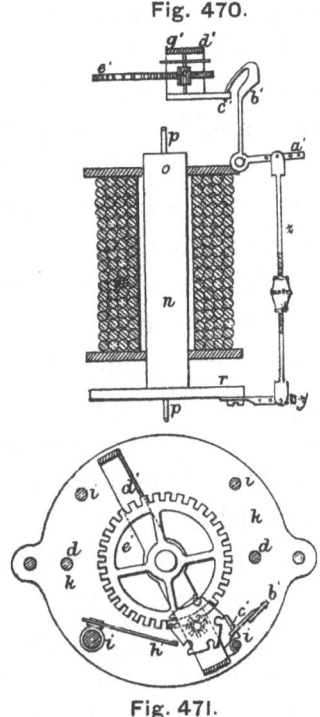

Fig. 471.

the corresponding carbon holder, which are insulated from the rest of the lamp, but are electrically connected together by the cross-pieces below. These rods slide easily in the sleeves v, electrically connected by the conductor w, and communicating with one end of the magnet coil m. The other end of this coil goes to the binding screw x on the upper platform, the return current flowing by this path after traversing the electro-magnet. In flowing through the coil of this latter, the current magnetises the core, which lifts suddenly and fixes the distance between the carbons in the following manner. The lower plate of the core carries an arm y (Fig. 470), to which is attached a connecting rod z; the length of this can be regulated by the adjusting screw as shown. The upper end of this connecting rod is linked to the arm a^1 of a cranked lever pivotted on the frame. The other arm b^1 of the lever is terminated as shown with a curved path in which engages a finger c^1, fast on the frame d^1, which is mounted loose on the extension p of the screw j. Within this frame is the toothed wheel e^1 mounted loose within the frame d^1, and fast on the rod p. This wheel, which therefore turns with the screw, gears into the pinion f^1 mounted on pivots within the frame. The spindle g^1 of the pinion carries also the escapement shown in plan, Fig. 471, which comes in contact with, or moves clear of the spring h^1 according as the carbons approach or recede from each other; the special form given to this escapement causes it to act either as a brake or a stop. As will be seen, each recess in the wheel is separated by a wide path, on which the spring h^1 bears with sufficient power to stop the movement when the magnetic variation due to the approach of the carbons has not been sufficiently great to bring the recess opposite the spring. At the moment of lighting, the core n is lifted suddenly, raising with it the connecting rod z, which tilts the cranked levers $a^1 b^1$; the curved groove at the end of b^1 pushes the frame d^1, which draws the escapement wheel against the spring b^1; the end of this spring enters one of the recesses in the escapement, and prevents this wheel and the pinion f^1, which is fast with it, from being turned by the wheel e^1. Then the pinion, always gearing in the wheel e^1, forces this latter, by its travel with the frame d^1, to move in a direction opposite to its normal motion, and as the wheel e^1 is fast on the screw j, the latter also revolves backwards, raising the weight h, and with it the upper carbon holder, while at the same time the lower carbon holder is forced to descend an equal quantity. When the arc is produced, the magnetic intensity of the coil is reduced in proportion as the carbons are consumed, the core tends to fall, and the escapement wheel retiring from the spring h^1, until it clears it, allows the carbons to advance in the direct proportion of their consumption. This operation periodically repeats itself, and the regular working of the apparatus is insured until the carbons are entirely consumed, or the lamp extinguished.

Carré.—M. Carré, of Paris, has an ingenious lamp, in the lower part of which two curved electro-magnets with hollow cores are fixed. These magnets form part of a circle, and are placed symmetrically around a common centre, at which is a pin carrying an oscillating armature, or rather two armatures, one at each end of a rod hung upon the centre pin. These armatures are curved so as to swing free to and fro in the axis of the cores of the electro-magnets. A system of articulated regulating levers and springs permits of the adjustment of an opposing force to that set up by the magnets on the armatures. Upon the bar carrying the armature there is a rod having at its upper end a pawl, engaging in a ratchet wheel, on the axis of which is gearing that drives two pinions, one belonging to the rack of the positive and the other to that of the negative carbon holders, the relative diameters being adjusted to suit the different rate of consumption. The wire is coiled upon the magnets that the two armatures are moved in the same direction by the action of the current, the varying strength of which determines the action of the pawl on the ratchet wheel, and consequently the advance of the carbons.

Sedlaczek-Wikulill.—Three very ingenious forms of arc lamps are shown in Figs. 472, 473 and 474, especially adapted for use on board ship or for locomotive head lights, the main object of the inventors being to avoid any vibration or shock to the carbons. These lamps, devised by Messrs. H. Sedlaczek and F. Wikulill, are manufactured by Mr. S. Schuckert, of Nürnberg. This system was tested at Paris during the Electric Exhibition, and more recently, further experiments have been made on the Kronprinz Rudolph Bahn, Austria, with very considerable success. In the trials on this railway the inventors first used Serrin and similar lamps. None of those lamps, however, dependent on clockwork, would answer, as a very short time was sufficient to disorganise the mechanism. Having demonstrated the unsuitability of such lamps for railway purposes, experiments were made with the form of which Fig. 472 explains the general arrangement. In two vertical cylinders connected at the bottom are two pistons a and b; a in connection with the positive carbon has only half the diameter of b, which carries the negative carbon, so that with a displacement of level in the liquid, a will move through double the distance of b. As long as the apparatus is at rest, the two cylinders are in communication through a hole at the bottom of the small cylinder, this hole being controlled by the tap e, in which is a vertical hole as shown. The piston a, though smaller, is heavier than b, and raises the piston b until both carbons are in contact. As soon as the circuit is closed, the solenoid c attracts its core and with it the small piston *d*, moving horizontally within the tap *e*, drawing it forward and cutting off the communication between the two cylinders. Thus the level of the liquid in this small cylinder is fixed, whilst the outward motion of the piston *d* increases the capacity of the larger cylinder,

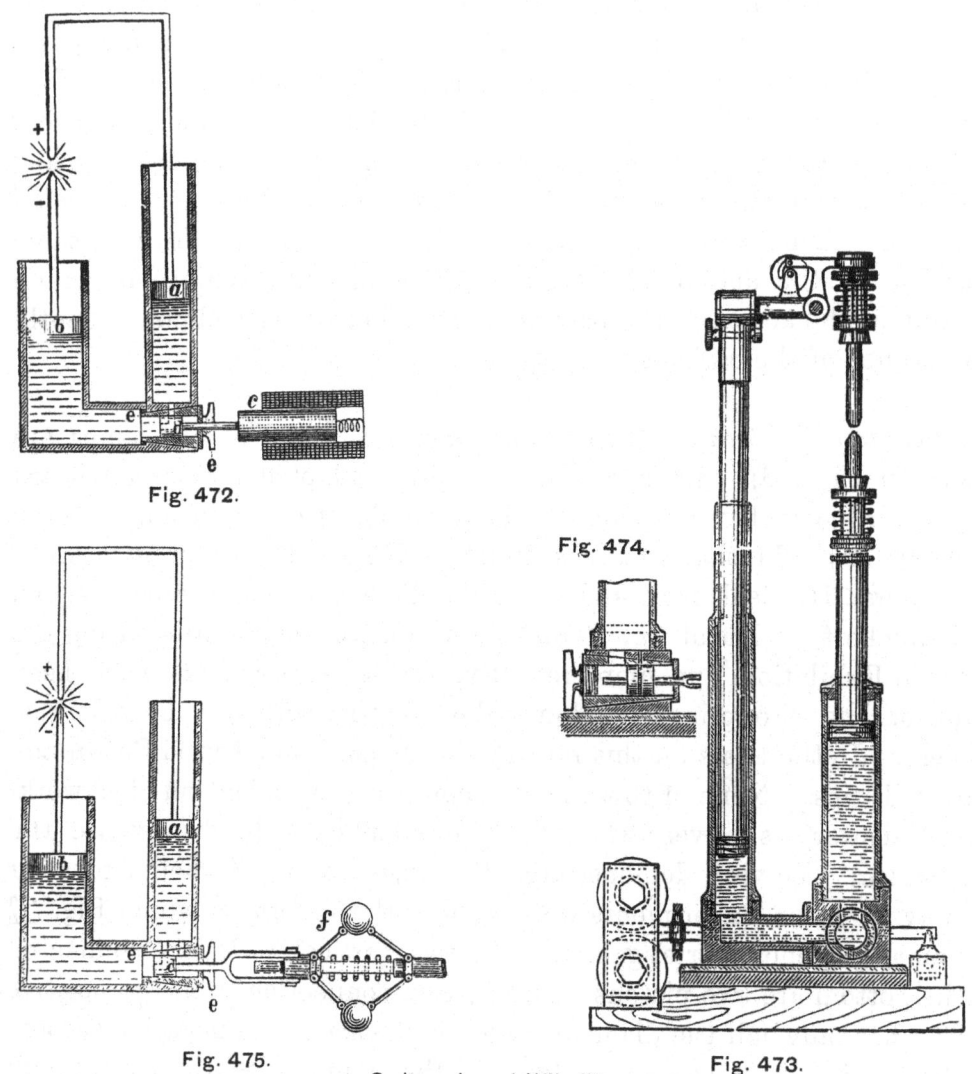

Fig. 472.

Fig. 474.

Fig. 475.

Fig. 473.

Sedlaczek and Wikulill.

and by slightly reducing the level causes the piston *b* to fall, and establish a clearance between the carbons. By these means, the proper distance for the arc between the two carbons is established. To maintain this condition a coiled spring presses against the core *c* of the solenoid. When, in consequence of the consumption of the carbons, the solenoid exerts less attraction, the spring forces back the core, and raises the piston b and the negative carbon at the same time. By turning the cock e the lamp can be cut out of circuit, for replacing the carbons, &c. Figs. 473 and 474 show one of these lamps in more detail. It weighs nearly 57 lbs. and a very strong electro-magnet or solenoid is used, so that considerable force is required to bring the two carbons nearer to one another, when the closing of the circuit has once fixed their position. During the trials above referred to, such a lamp was made to fall through several feet without the steadiness of the light being affected.

The solenoid may be replaced by an electro-magnet in which the position of the armature may be regulated by screws in such a way as closely to resemble in its working the action of a solenoid.

Fig. 475 illustrates another type, which is controlled, not by a special electric regulator, but by a governor f, actuated from the shaft of the generator. In the bottom of the small cylinder are two holes, that on the left being open to a corresponding passage in the tap d when the generator is at rest. When this is set in motion the governor f opens, withdrawing the piston d. This outward movement increasing the capacity of the cylinder b, the negative carbon falls, whilst the level in a remains unchanged as long as the correct length of the arc is maintained. The separation of the carbons by consumption causes the governor to open still further, and the piston d is withdrawn more, until the passage corresponds with the right-hand opening in the bottom plate of a, allowing part of the liquid to enter the cylinder b and raise the larger piston.

As used for a locomotive head light the lamp is mounted in a wooden case attached to the front of the smoke box, so that the bottom of the case is on a level with the highest point of the boiler. On the top of the boiler, behind the chimney, are placed a Schuckert dynamo machine, and a three-cylinder Brotherhood engine. The whole length of the apparatus is 5·feet. Such a lamp was found sufficiently powerful to illuminate a space of about one-third of a mile, and its efficiency was particularly evident in tunnels and for distinguishing signals. By a very simple arrangement the lamp can be rotated round its axis by the engine driver, so as to throw a light in various directions.

A later and somewhat modified arrangement is shown by Figs. 476 to 479. In this, as already described, the regulation may be effected by the direct action of a centrifugal governor driven by an engine, or by an electro-magnet or solenoid. In the former case the arrangement is based on the fact that as the resistance of the arc increases so does the speed of the engine, and *vice versâ*. A lamp on this principle is shown in Figs. 478 and 479, in which each of the electrodes is carried by pistons 3 and 4. In the base of the apparatus and between the two cylinders is a socket a with a brass plug b, the two being provided with ports n n^1 and c c^1. The plug is operated by a disc g and a lever connected to the sleeve of a centrifugal governor. As soon as the engine is set in motion the disc g will be caused to turn on its axis by the rod, and with it the plug will move, carrying the opening c gradually past the corresponding opening n of the

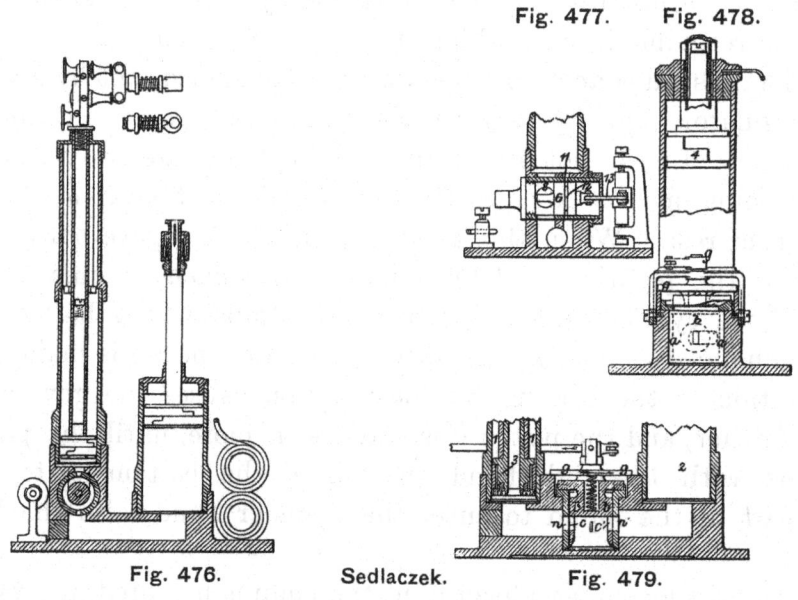

Fig. 477.   Fig. 478.

Fig. 476.   Sedlaczek.   Fig. 479.

box a until it has closed the same, and thereupon the communication between the two cylinders on the side next the positive pole is cut off, the hollow plug b being, however, during its further rotation still in communication with the negative cylinder 2 through the large opening c^1 and the passage n^1. As the rotation continues wings on the plug sliding upwards on correspondingly inclined surfaces on the case will lift the plug (Fig. 479) causing the negative electrode to fall to a corresponding extent, whereby the voltaic arc will be formed. When the velocity of the engine, in consequence of the consumption of the carbon points, reaches such a degree that the small slot of the plug between c and c^1 comes in front of the passage n, and thereby re-establishes communication between the two cylinders 1 and 2, then the carbon points will approach each other until the rotation of the engine becomes slower, and the plug is moved to break the communication. A further set of inclines is provided to come into action if the carbons break and a very rapid feed is required. Figs. 476 and 477 illustrate a lamp governed by an electro-magnet with an armature having a small range of motion balanced by a short spring acting through a long lever. As soon as a sufficiently powerful current passes through the coil of the electro-magnet and the carbons (then in contact), the piston 6, by means of the link 13, will be drawn backwards, and the holes 11 of the cock will be closed, while the space 8 at the rear of the piston will be increased, causing the negative carbon to fall and produce the arc· As the current decreases the piston returns, establishing communication between the two cylinders through the port 11, the groove 12, and the pipe below. When several lamps are used in series, the electro-magnet is replaced by differential solenoids.

Andrews.—This lamp, illustrated by Figs. 480 to 484, belongs rather to the category of electric candles than to that of regulators. The characteristic features of the Andrews lamp is the form of carbon electrodes, which, instead of being in pencils, consist of two thin plates of gas coke, fixed about $\frac{1}{8}$ in. apart, and having between them an insulated plate of plumbago $\frac{1}{32}$ in. thick, and of the same area of surface as the outer plates. The insulation is of the simplest character, consisting of nothing more than four little triangles of slate inserted between the plates, the whole being held together by the pressure of the flat spring clips *b*, by which the compound plate is held in the lamp. With plates 2 in. square, the lamp will keep in regular action for about six hours, but by increasing the size to 6 in. square and the thickness of the outer plates to $\frac{3}{16}$ in. the lamp is capable of burning for a week. The arc is formed on the extreme edge of the plates from one to the other, across the edge of the plumbago, and always maintains itself at the highest point, so that in action the arc moves slowly backwards and forwards along the upper edge, for as the carbon and plumbago are consumed the arc moves to a higher point. The use of plates instead of pencils make it resemble somewhat the Wallace lamp, but the essential difference between the Andrews plate lamp and the Wallace in the disposition of the carbons, is exactly the difference between the ordinary arc regulators and the Jablochkoff candle, or still more analogous Jamin bougie; that is to say, in the Wallace lamp, as in the ordinary arc regulators, the

carbon electrodes are placed the one above the other in or near to the
same axis, while in the Andrews lamp, as in the Jablochkoff candle, the
carbons are placed side by side and parallel to one another.   The Jamin

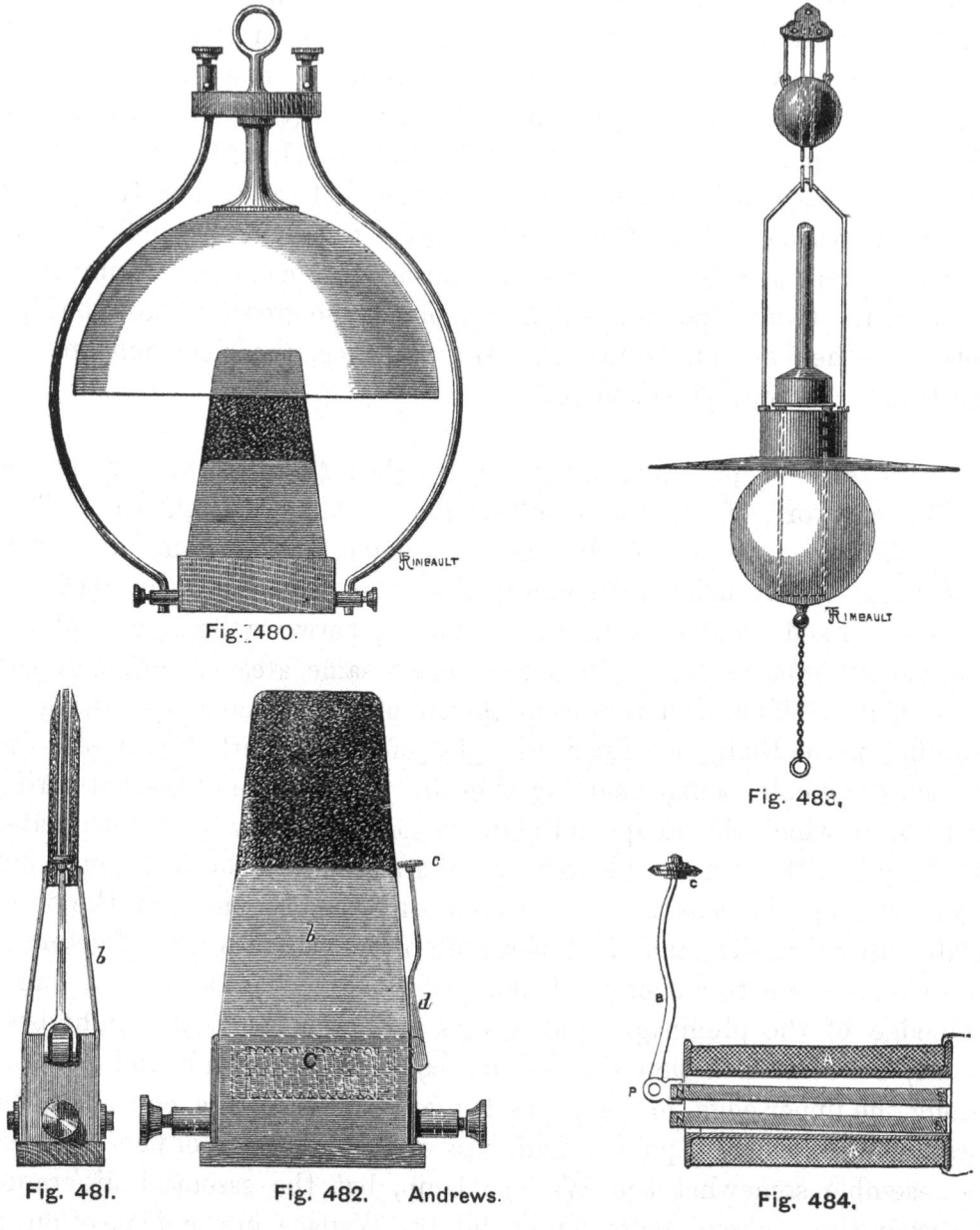

Fig. 480.

Fig. 483.

Fig. 481.   Fig. 482.   Andrews.   Fig. 484.

candle, moreover, possesses the still closer resemblance in the
disposition of its carbons to the Andrews lamp, from the fact that in
both arrangements there is an intermediary neutral carbon, the object

of which is to act in the capacity of what may be called an electrical *stepping-stone* to the arc which is formed between the two exterior electrodes. A special feature of this lamp is the method by which it is automatically relighted should it by any accident become extinguished, and by which it is lighted in the first instance. Against one edge of the carbon plates, and near to the bottom where they are held in the spring clip, rests (when no current is passing across the carbons) the edge of a little disc of gas coke C (Fig. 484), which is mounted upon a light brass arm B attached to the armature of an electro-magnet, the coils of which are included in the main circuit; this maintains a connection between the two carbons as long as no current is flowing; the moment, however, that the current is put on the lamp, the arc is established at the point where the disc touches the carbon plates, and in consequence of the current traversing the coils of the magnet its armature is moved, and the carbon disc is drawn away from the carbon plates; the arc then runs up to the top of the plates, and is, after that, maintained at their highest point.

Dornfeld.—The lamp designed by C. Dornfeld, of Essen, is in its general arrangement something similar to that of Jaspar, but differs from it considerably in detail and in the means employed for regulating the arc. In this lamp, see Figs. 485 to 487, the two carbon holders *a* and *b* are suspended by cords or light chains passing over two pulleys I and D fixed to the same shaft F. On reference to Fig. 485, it will be seen that the pulley T, over which passes the cord of the positive carbon holder *a*, is twice as large in diameter as the pulley D to which the cord of the negative carbon *b* is attached. By this arrangement the uneven wear of the two carbons is compensated, and the focus of the light is kept stationary until the carbons are consumed. To regulate the fall of the upper carbon holder *a*, which must not be made very light lest its travel should be checked by dust or oil, a fly, fixed to a shaft, carries a small pinion gearing into the toothed wheel E on the shaft F. By this arrangement the resistance of the air is employed to regulate the speed with which the rod *a* is allowed to run down. The wheel E is not keyed to the shaft F but connected to it by a ratchet clutch, so that when lifting the holder to insert a new carbon it can be raised without rotating the wheel train.

The distance between the two carbon points is regulated by a friction brake, acted upon by a solenoid, the coil of which is constantly in circuit. On the same shaft that carries the cord pulleys T and D (Fig. 486) is fixed

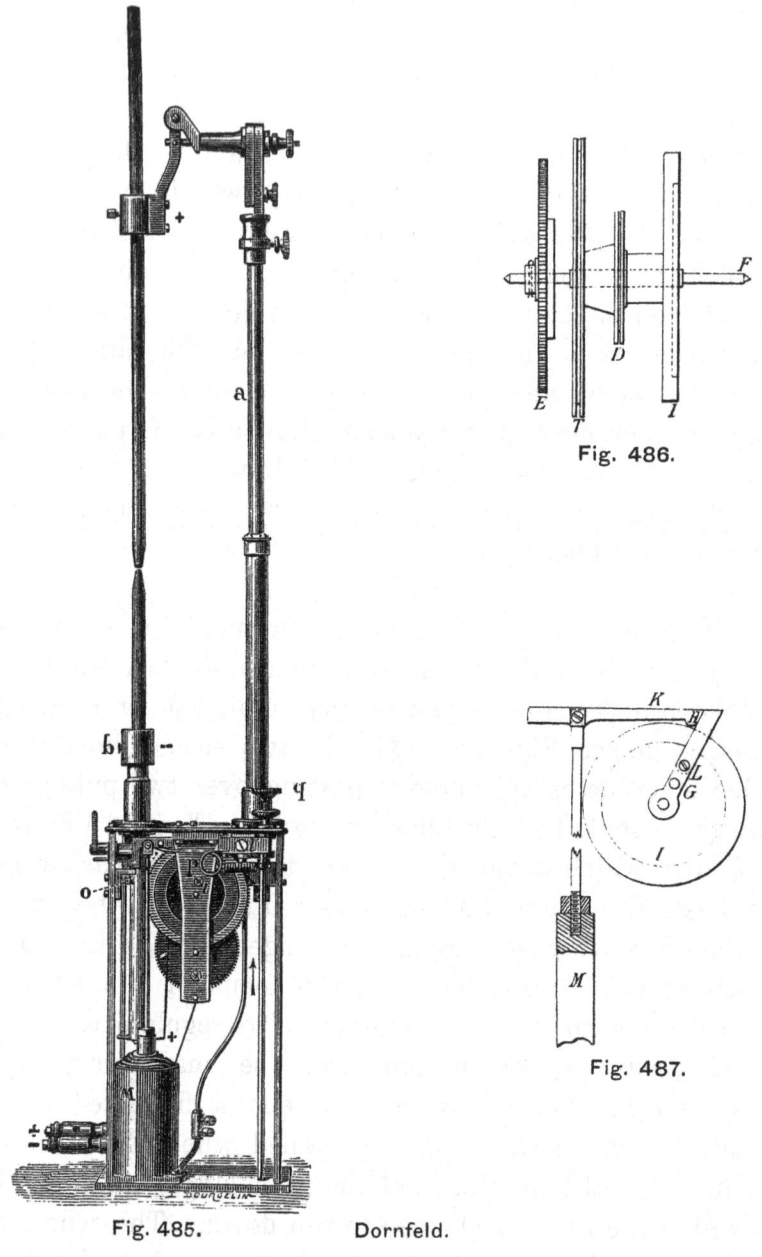

Fig. 486.

Fig. 487.

Fig. 485.   Dornfeld.

a friction disc I, acted upon by a brake (see Fig. 487) which is moved by the solenoid plunger M. When the circuit in the lamp is closed, while the two carbon points rest upon each other, the solenoid plunger M is attracted and, by means of the brake block H, actuates the friction disc I, thus separating the two points to the desired distance, which can be adjusted by a set screw O (see Fig. 485). As the carbons wear away, the current in the solenoid becomes weak, and the magnetic attraction is overcome by a spiral spring p, the strength of which can be regulated by a set screw q; the carbon points then move towards each other till the strength of the current is sufficiently increased for the solenoid to recommence its action. Several modifications of this lamp have been designed for lighting up large halls, open spaces, &c. Where a stationary focus is not necessary, the lamp is constructed with a fixed lower carbon, and pencils of such length are employed that it will burn from 10 to 12 hours without requiring any attention. In the establishment of Mr. F. Krupp, in Essen, a number of these lamps have been in operation for a considerable time, giving great satisfaction.

Fyfe.—This lamp belongs to the differential type, and has been devised by Mr. A. L. Fyfe, of London, who has introduced it with very favourable results as regards steadiness and colour of light. The construction and regulating arrangement of the lamp, which is specially designed for low tension currents, are of a very simple kind, an ordinary clock train being employed, the brake of which is placed under the influence of a shunt circuit solenoid.

The mechanism is shown clearly in Fig. 488, which illustrates a type in which the positive carbon only is movable. The upper holder is attached to a rack B geared with a pinion which forms part of the train $G^1 G^2$; on the last spindle of this train there is a fan G and a brake wheel f. In operation, the upper carbon descends by gravity until it rests upon the lower carbon; as soon, however, as the current flows through the large magnet E, the core E^1 is withdrawn into it, the bar D, pivotted at d and connected by a link to the cage carrying the whole wheel train C C^1 is deflected, the cage rises and with it the upper carbon, which is now prevented from falling by the wheel f coming in contact with the brake f^1. The arc is thus established, and the lamp burns till the distance between the carbons increases the resistance to the current to such an extent that the small shunt electro G comes into action, and by drawing up the lever g^1 and brake strap f^1, as shown to a large scale in Fig. 489, frees the brake wheel f, and allows the upper carbon to descend until the proper distance between the points is again established. The two levers D and *g* are balanced and a stop is provided at L for the latter, so that the brake block cannot act on the brake wheel when the cage carrying the wheel train is not lifted up by the large solenoid. Fig. 489, while showing clearly the action of the small solenoid, varies slightly from the arrangement in Fig. 488. the lever *g* not being attracted directly, but lifted by a hook g^4 fixed to g^1; stop *g* is provided to prevent the brake from being lifted too

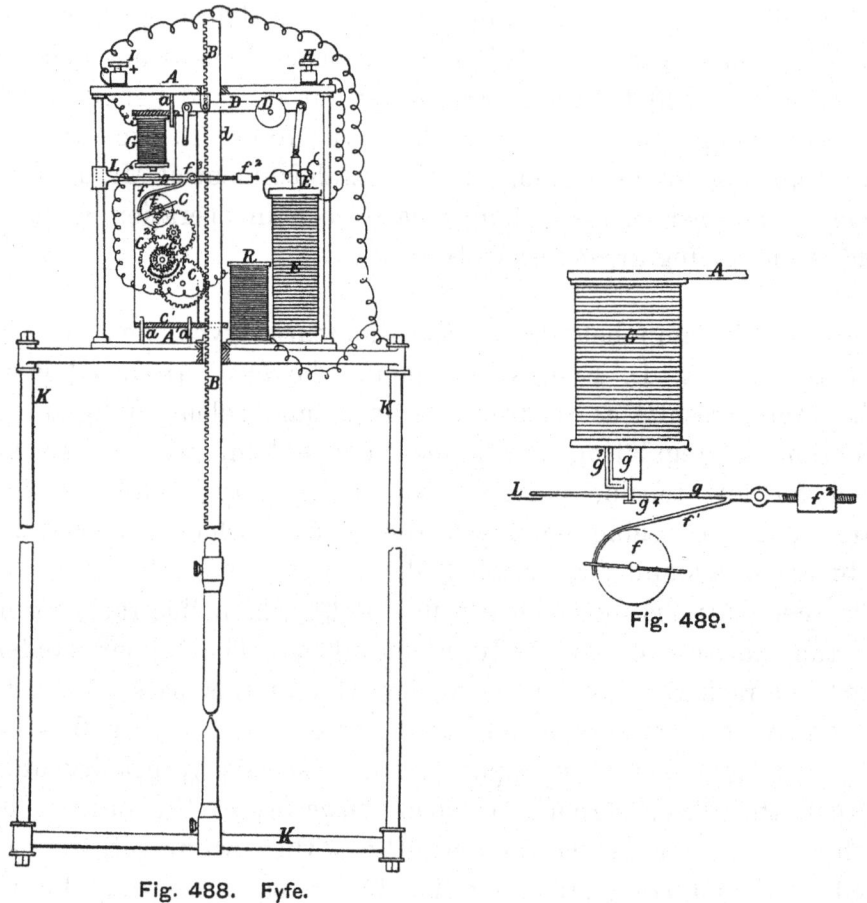

Fig. 489.

Fig. 488. Fyfe.

high. The solenoid G is fixed to the cage C, which is made to rise and fall within slight limits on the guide pins *a a*. By inserting small resistance coils R, this lamp can be readily adjusted to any current intensity, and equally good results have been obtained when the apparatus was fed by a Gramme or a Schuckert generator. In a recent installation at the Crystal Palace, six of these lamps were worked in one circuit from a Gramme generator; the power absorbed by each lamp was 1·74 horse-power.

Bouteilloux and Laing.—MM. L. J. Bouteilloux and W. Laing, Paris, have proposed two forms of lamp in one of which the arc is always maintained of the same length by allowing one of the electrodes to be pressed by its own weight, or by special mechanism, against a non-conducting or an insulated core inserted in the centre of the other, and which is gradually dissipated during the consumption of the carbons. The electrodes are held in split tubes, screwed on the outside and compressed by a nut. Fig. 490 and 491 are sections of a carbon holder, in which *g* is the carbon, *h* the split tube, and *i* the nut. Fig. 492 is an enlarged view of the carbons, in which *e* is the negative carbon made

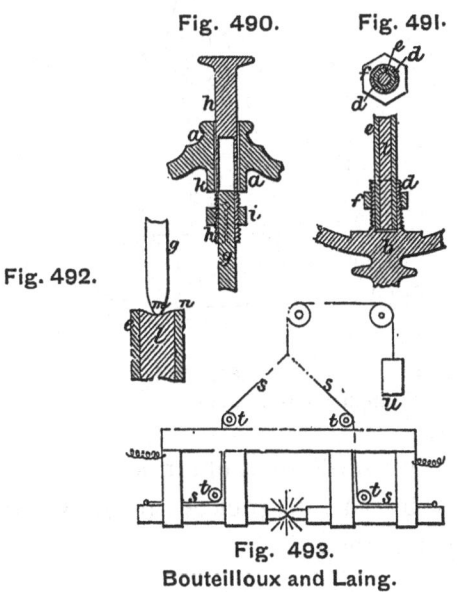

Fig. 490. Fig. 491.

Fig. 492.

Fig. 493.
Bouteilloux and Laing.

hollow to receive the non-conducting core *l*, and *g* is the positive carbon resting on it. Fig. 493 is an elevation of a lamp with horizontal carbons, arranged to maintain the luminous focus in one position. In this figure the two carbon holders are connected by a system of cords *s* passing over the pulleys *t* to the weight *u*, which maintain the carbons in contact. In a second and later arrangement of the same inventors, the idea of using a combination of carbon electrodes, of which one has a central core of insulating material for the other to bear against, is worked out in a different manner.

A third arrangement is shown in the illustrations, Figs. 494 and 495. Here two inclined pencils meet beneath the reflector S at a point below a tubular carbon with a non-conducting core, sliding within a tube *f* with a contracted orifice that prevents the cylinder from falling out. The tubular electrode is closed by a plug of carbon to facilitate lighting.

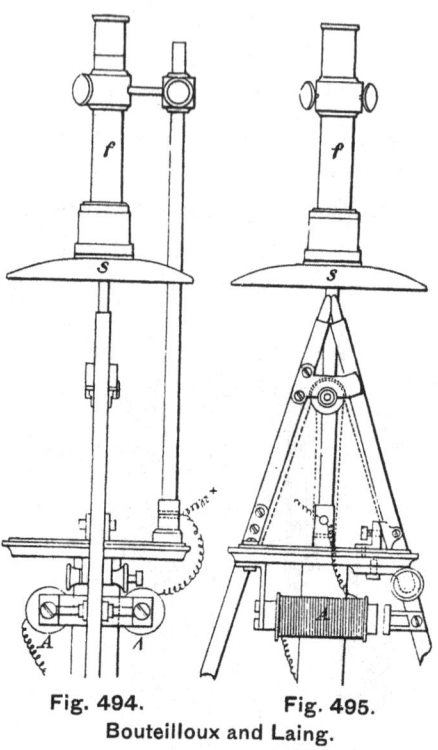

Fig. 494.   Fig. 495.
Bouteilloux and Laing.

The inclined pencils are fed upwards by a chain and a weight, and when extinguished are separated by the magnet A to allow them to rise and make contact with the upper carbon.

Gérard-Lescuyer.—One form of the Gérard-Lescuyer lamp is shown in Figs. 496 and 497. Here the electrodes are each formed of two carbons inclined towards one another in the form of a V, and bevelled at the points so as to meet. The lamp is composed of two pairs of tubular carbon holders a a^1, through which the two sets of carbons b b^1, slide. The whole of the mechanism is mounted on a plate c, to which the holders a are permanently fixed, the other holders a^1 being hinged to it at d. The electro-magnet i is placed in a shunt circuit; it serves to bring the carbons together to establish the arc, the spring g separating them to an extent determined by the adjusting screw f; m m are gun-metal rollers, which are used both to guide the carbons and to make electric contact between the holders and the carbons near their points; n n (Fig. 497) is a fan-shaped electro-magnet, through which the main current passes, curved prolongations of the cores being provided, as shown, in order to prevent the arc rising between the electrodes.

Fig. 496.

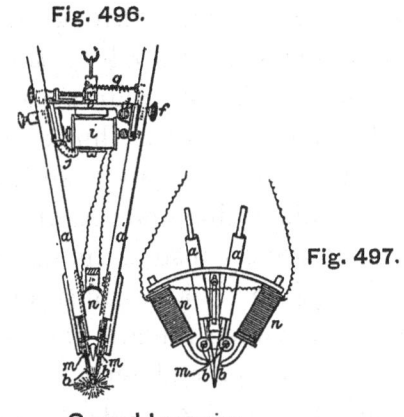

Fig. 497.

Gerard-Lescuyer.

The second form (illustrated by Figs. 498 to 500) is intended especially for the production of a small arc with a divided current. In the illustrations, A is a guide upon which slides a frame B, carrying a solenoid D of

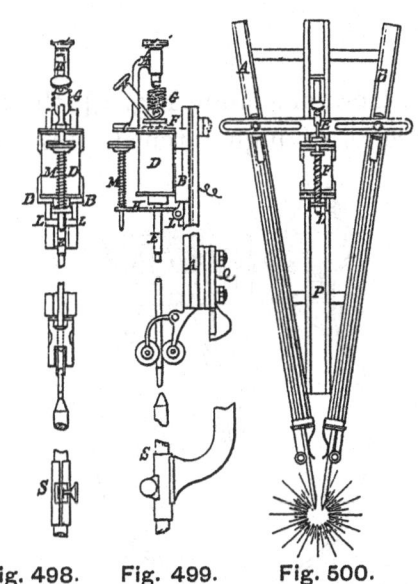

Fig. 498.   Fig. 499.   Fig. 500.
Gerard-Lescuyer.

fine wire ; through this passes freely the brass carbon holder E, the upper end being fixed to a soft iron armature F. The carbon holder E is suspended by a spring G from a small bracket H carried on the plate C, so that the armature is a few millimetres above the upper pole of the solenoid. A second armature K has a jaw L, which is caused by the spring M to bind against the guide A. The upper carbon as fixed in the holder E, is passed through guide rollers. The lower is carried by the bracket S. The current passing through the shunt circuit of the solenoid causes the armatures K F to be attracted, at the same time releasing the grip of the jaw L and allows the bracket B to descend until the carbons meet. The current is thus diverted from the solenoid, and the two armatures are withdrawn by their springs, the one causing the jaw L to again bind against the guide A, and thus arrest the descent of the bracket B, and the other causing the upper electrode to be raised to produce the arc. Fig. 500 shows a modification in which the carbons slide in two guides A B. Between them is another guide in which slides a solenoid F, having at its lower end an armature L, adapted to bind against the guide P, and at its upper end an armature for effecting the separation of the carbons. When the current traverses the shunt coil it is released from the guide and drops down the slide, carrying the carbons before it until they meet. The coil then becomes inert, the lower armature grasps the guide, while the upper one raises the bar E and separates the carbons.

Gordon.—In Mr. J. Gordon's lamp, Fig. 501, the light is produced by rapidly alternating currents of high intensity discharged between ter-

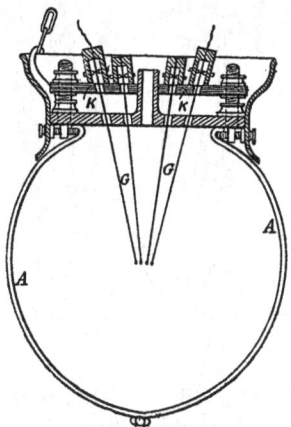

Fig. 501. Gordon.

minals of refractory metal which are thereby caused to become heated and to emit light. In the illustration, G G are wires of platinum or iridium ending in balls of the same metal, the outer ones being connected to the source of electricity, the current passing from one to another; A is a glass globe and K an insulating plate.

Siemens.—In this arrangement H, is the core of a high resistance solenoid coil E situated upon a shunt circuit. To the core are attached rods K L, supported at their lower extremities in angular beds *k l*, and armed with ratchet teeth, which ordinarily stand just clear of the ratchet wheels C D, mounted upon the axes of the carbon feed rollers *a b*. In operation, as the distance increases between the carbons A and B, the resistance to the direct circuit of the lamp is augmented, and consequently a larger portion of the current passes through the

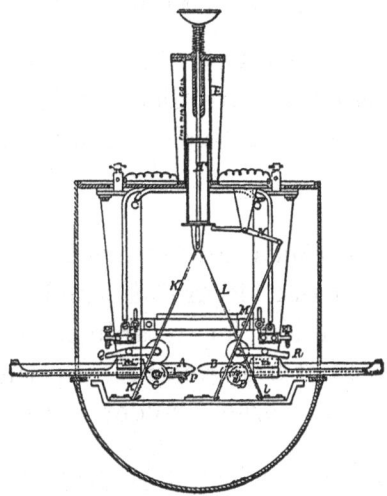

Fig. 502. Siemens.

shunt circuit of the solenoid coil. The core H is thus attracted upwards, and by its ascent draws up the two bars K and L. These, as they retire from their rests *k* and *l*, bear against the ratchet wheels C and D and partly turn them round in the direction of the arrows, thus causing the carbons to approach one another. Should the carbons be too near, the solenoid will have little power, and descending, will move the tappet lever N, raising the bar M and reversing the motion of the feed roller *b*. The lamp is focussed by the carbon A being fed rather faster than the carbon B, and being provided with an abutment stop P. When either carbon is consumed the upper feed roller drops and short-circuits the current through the contact screw R or Q.

Mandon.—M. J. A. Mandon, of Paris, proposed the arrangement indicated in Fig. 503. The lamp has two quadrant electrodes balanced by a counterweight to turn freely round a centre of rotation. The curved tube L contains mercury, and receives a float J attached to a

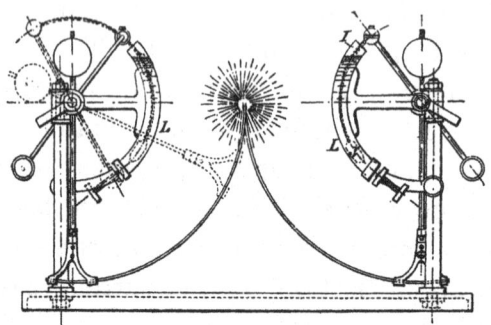

Fig. 503. Mandon.

stem I, both float and stem being so formed that as the carbons are consumed, they rise to a proportionate distance in the mercury so as to maintain a fixed length of arc. In this scheme its ingenuity is evidently the chief merit.

Hopkinson.—Dr. Hopkinson's improvements in electric lamps relate firstly to arc lights, the essential feature of which is the combination of means for utilising fluid friction for retarding the feeding of carbons without checking the motion of the armature, together with an appliance for holding and releasing the carbon holder. Fig. 504 shows the fluid

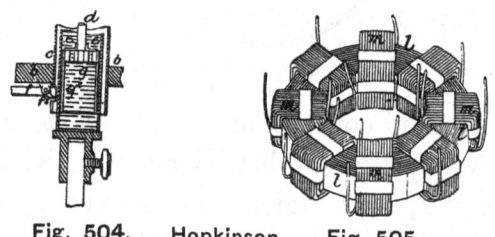

Fig. 504.　　Hopkinson.　　Fig. 505.

retarding apparatus; *b* is the armature of the controlling magnet of the lamp. In it is fixed a vertical tube *c*, rather longer than the acting length of the carbon pencil. This tube is closed at the top by a cap from which depends a rod *d*, ending in perforated piston *g*, and a short tube g^2. Between the piston and the outer tube is a cylinder *e*, filled with viscous fluid and acting, at its lower end, as the carbon holder; *f* is a bent lever provided at its outer end, which stands over, or rests upon, an adjustable stop, with a counterweight. It has a toe f^2, which bears against the cylinder *e* and jams it, when the free end is clear of the stop. When the current passes, the electro-magnet attracts the armature in opposition to a spring, lifting it and the outer and inner tubes and the piston, and thus striking the arc. As the carbons waste, the armature falls a little, and the long arm of the lever *f* coming in contact with its stop, the pressure on the toe f^2 is reduced, and the cylinder *e* descends slowly, relatively to the tube *c* and the piston *g*, and thus adjusts the arc. In a second arrangement two magnets are employed, one in the lamp circuit to form the arc and one in shunt circuit to control the feed. In this case the outer tube *c* is fixed to the framing. Fig. 505 represents a coil of wire with a considerable coefficient of self-induction, and so arranged that this coefficient may be varied. By its means an alternating current may be varied without interfering with the generator, or, if an arc lamp be used, without altering the length of the arc, and also without wasting energy by introducing a considerable additional resistance. The current may also be divided and arc lamps burnt in parallel circuit. The annulus *l* is formed of a coiled ribbon of sheet iron, round which several coils *m m* of insulated wire are wound, capable of being coupled by plugs like any ordinary resistance box. Another form of coil is made like an ordinary electro-magnet with a ∪-shaped core that can be more or less intruded into the bobbins. If either of these instruments be introduced into the circuit of an electric lamp, the current in the lamp is diminished. The specification contains two diagrams, the first showing the circular coil placed in a circuit with several incandescence lamps arranged in parallel arc, and the second illustrating four arc lamps in parallel circuit with a coil of the second form in each lamp circuit. In both cases the light may be varied by bringing more coils, or a greater portion of each coil, into operation.

Muirhead and Hopkinson.—Figs. 506 to 508 show Messrs. Muirhead and Hopkinson's lamp, the first two figures being respectively an elevation and plan. Fig. 508 shows a modified form of brake gear. The upper carbon C is carried by the holder A, upon which is cut a screw-thread B, engaging a similar thread in the boss of the wheel F. D is a key-way, which keeps the holder from revolving. In operation, as the upper carbon falls, the wheel F is caused to revolve by the screw A, and rotates the pinion G and the brake-wheel H. The brake-wheel is controlled by an electro-magnet on a shunt circuit acting against a spring, which, when the resistance of the arc increases, is overpowered and drawn away from the wheel. In Fig. 508 the toothed wheels are replaced by a brake disc provided with paddles working in a fluid.

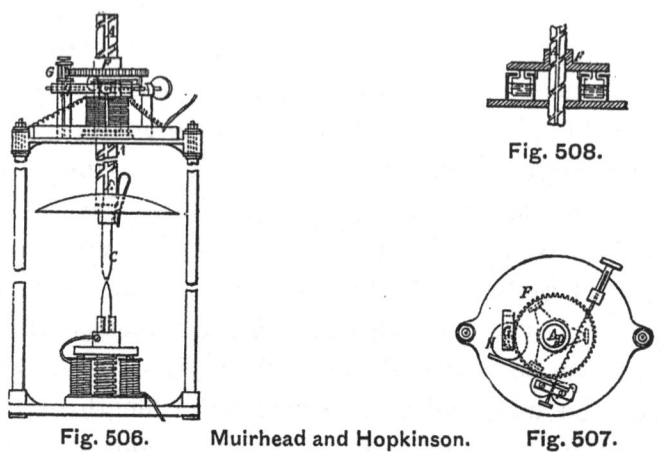

Fig. 508.

Fig. 506.   Muirhead and Hopkinson.   Fig. 507.

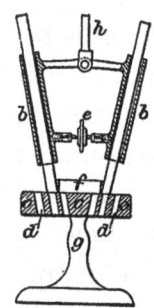

Fig. 509.
Sachs.

Sachs.—In the Sachs' lamp (Fig. 509) the two carbons are placed in tubes *b b*, and rest upon a plate *c* of asbestos, so as to be free to fall or rise as required, according to wear. The plate *c* is perforated with a series of holes *d* of smaller diameter than the carbons, which are adjusted in position over two of the holes by means of the screw *e*. A strip of graphite *f* forms continuity between the carbons; *g* is a stand for supporting the plate *c*, the tubes *b b* being suspended by the rod *h*. In another arrangement a number of carbons are arranged round a plate in a similar manner. As the carbons wear they descend by gravitation; the perforations in the plate allow the emission of the light, although not permitting the carbons to pass through.

Holcombe.—Mr. A. G. Holcombe, of Danielsonville, Connecticut, has devised a lamp shown in Figs. 510 and 511, the peculiarity of which consists in its containing no electro-magnetic apparatus. The regulation is controlled by the attraction or repulsion of electric currents traversing insulated conductors. The upper carbon *c* runs down by gravity as soon as the brake-wheel *e* is released from contact with a block carried on the lever *f*, which is actuated by the attraction of two flat coils $f^2 f^3$, one fixed and the other movable. The arc is controlled by the solenoids *m* and m^1. The carbons and solenoids form one circuit and the flat coils f^2 f^3 a second or shunt circuit acting on the well-known differential principle.

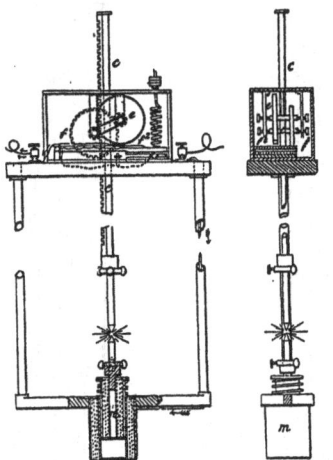

Fig. 510. Holcombe. Fig. 511.

Grimstone.—Figs. 512 and 513 are elevations, and Figs. 514 and 515 are plans on lines X X and Z Z respectively of an arc lamp devised by Mr. G. S. Grimstone, of New Cross, Kent. To the soft iron core A, working up and down in the solenoids of high and low resistance B, C, is pivotted one end of a lever D, which has its fulcrum on the fixed bracket E. To the other end of this lever is linked the suspended frame F, which has two loops F[1] F[2], through which passes freely the carbon holder G. The holder is clamped by the action of an eccentric locking piece H pivotted to the frame and weighted by an arm H[1], so as to press the carbon holder against the sides of the loops F[1] F[2]. Thus as the core A rises, owing to the increased attraction of the solenoid of high resistance C, consequent on the burning away of the carbons S, the frame F, and with it the carbon holder and upper carbon will descend until the arm H[1] of the locking piece comes in contact with the stop K, a slight further descent of the frame will cause the eccentric H to be turned by its arm H[1], so as to free the carbon holder, which will then descend by gravity through the loops F[1] F[2], its fall being regulated by the action of the loosely-fitting piston L moving the hollow carbon holder filled with glycerine. When the arc is sufficiently reduced in length the solenoid B will raise the frame and bring the arm H[1] off the stop K. N is a spring to balance the excess of weight of the solenoid. The carbon holder G has a cross-head G[1], in which are

secured two upper carbons S¹ S¹.   On first starting the lamp the current
passes through one pair of carbons, and, as that pair burn away, the downward motion of the holder causes the second pair to approach nearer than

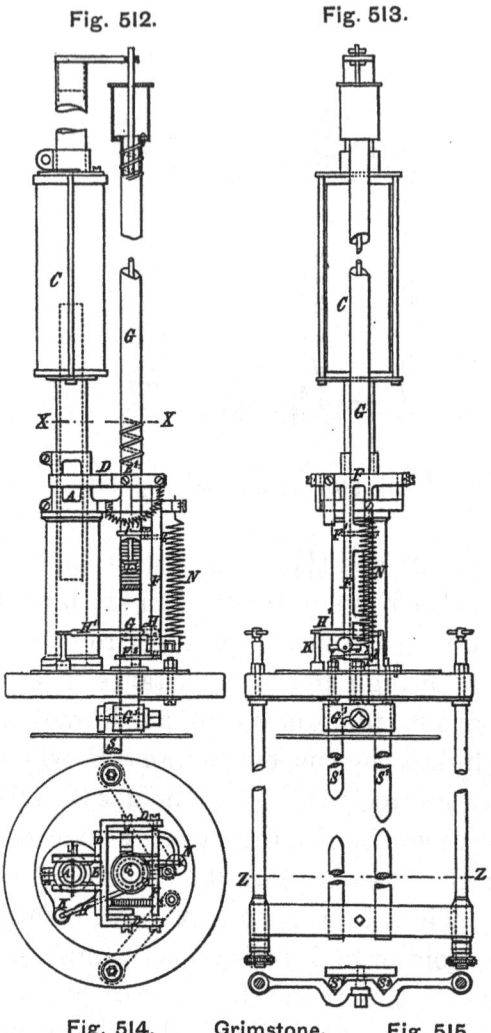

Fig. 512.   Fig. 513.

Fig. 514.   Grimstone.   Fig. 515.

the first pair, consequently the current is transferred to the latter pair.
Thus the current alternately shifts from one pair of carbons to the other
until both pairs are consumed.

Hawkes and Bowman.—Messrs. Hawkes and Bowman, of Ipswich,
recently patented an arc lamp (shown in Fig. 516 and 517), in which
the lower carbons are buoyed up by mercury, and checked in their rise by
an oscillating gripping piece operated by an electro-magnet.   Fig. 517 is a

sectional plan to an enlarged scale, of Fig. 516. B is a tube lined with glass, and containing mercury, in which the lower carbon holder is immersed, its movements being controlled or damped by a perforated leather piston. This carbon holder rises between the arms of an electro-magnet E, and also through a gripping piece F attached to the armature E^1 of the magnet. The upper carbon holder has a plate G^1, attracted by the magnet G, which is in the lamp circuit, when the circuit is closed. In order to cause the electro-magnet E to regulate the feed with great accuracy, the following contrivance is adopted: H is an electro-magnet in the lamp circuit between one pole of the generator and the lower carbon. Opposed to this magnet is a second H^1, and between the two is an oscillating

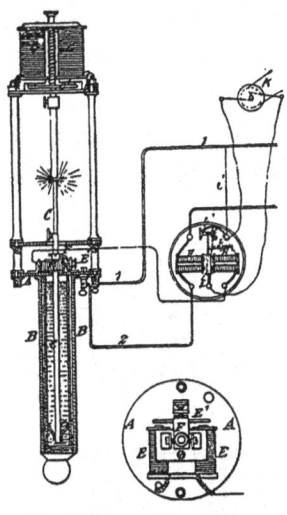

Fig. 516.   Fig. 517.
Hawkes and Bowman.

armature I. This armature connects with the electro-magnet E, and is intended to impart rapidly intermittent currents thereto for the purpose of enabling it to actuate the oscillating gripper P, and allow the carbon C to rise gradually at an almost imperceptible rate. One terminal of E is connected to the plate A, and through the wire 2 and magnet H to the generator; the other terminal of E connects with one terminal of H^1 whose other terminal is joined by the elastic coil i to the oscillating armature I. An extension of I bears against a fixed stop connected by a wire i^1 to the leading wire 1. As the arc grows longer, the current in the lamp circuit and the magnet H becomes less, and that in the shunt circuit and magnets E and H^1 greater. At a certain proportion of these

two, the armature I will be attracted by H¹, breaking the shunt circuit and
rendering inert the two magnets in it.   This will momentarily release the
carbon holder from the gripper P, and leave it free to rise until the
magnet H draws the armature I back against the contact piece again.
The armature will continue to vibrate until the arc is so reduced that the
magnet H¹ can no longer draw it away from H.   When there is only one
lamp on a circuit, the electro-magnet 4 is replaced by a spring.   If it be
desired to employ two sets of carbons in one lamp, the second set is held
out of action by a double-ended spring lever, which takes at one end into
a notch in the second carbon holder, and bears at the other end on the
plain part of the first carbon holder.   When the first pair of pencils are
nearly consumed, a notch on the first carbon holder comes opposite the
end of the aforesaid lever, which immediately enters it, and at the same
time leaves the notch in the second carbon holder, which is then free.

Wood.—The object of the device shown by Fig. 518 is to afford a
conducting path around a faulty lamp when the resistance of the lamp

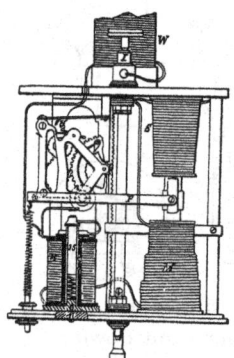

Fig. 518. Wood.

circuit rises above a regulated amount.   M and S are differential coils, S
being in the shunt circuit is as usual.   They act conjointly on a lever F,
which carries a train of wheels, of which the first pinion is in gear with
the rack on the carbon holder, and the last drives a fly or escapement,
when it is freed from a spring detent *b*, on the connecting spring H,
by the falling of the lever F.   The spring H also acts as a connecting
bar; K is a solenoid in the lamp circuit with a core 15 standing on a
spiral spring, and situated under a contact piece.   If the feeding gear
sticks, or the carbons break, the current in the solenoid K will cease and
its core will rise, making a short circuit for the current from the lamp

frame through the resistance W (equal to the arc) to the insulated terminal I. Should the carbons again meet, the current will divide between them and the resistance W, and the core 15 will be drawn from the contact piece. This arrangement was proposed by Mr. J. J. Wood, of New York.

Woolley.—Mr. L. G. Woolley, of Mendon, Michigan, proposed a lamp, shown by Figs. 519 to 521, in which the whole of the mechanism is kept inside a dash-pot, and is constantly immersed in glycerine or other lubricant. The regulating device is placed within the upper carbon holder; and the latter is made in two pieces united by easily fusible solder that gives way when the electrode gets too short. Fig. 519 shows the upper part of a lamp; Fig. 520 is an enlarged view of the dash-pot, and Fig. 521 is an elevation of the carbons and clamps. D is an axial magnet,

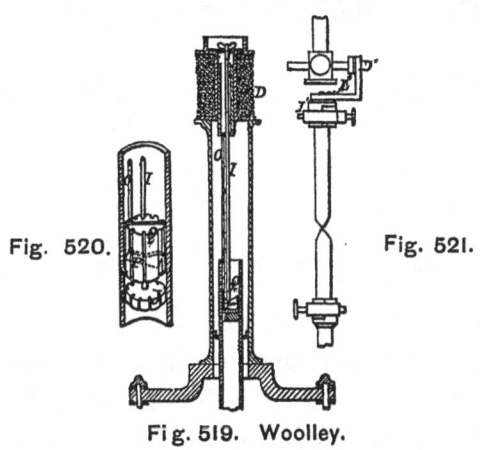

Fig. 520.   Fig. 521.

Fig. 519.   Woolley.

the core of which is connected by the rod O to the lifting lever I. This lever is pivotted in a groove in the frame Q, and when the core is raised it bites into the inner surface of the carbon holder and draws it up. When the core falls, the lever and holder drop with it and come into contact with the piston or dash J depending from the rod I, when the frictional connection is broken and the carbon holder descends by gravity until an upward motion of the core causes the lever again to bite. The clamp I^1 of the upper carbon holder (Fig. 521) is secured to the adjustable bar D^{11} by solder, which will melt when the arc approaches too closely. An insulated chain L serves to catch the clamp as it falls.

Harding.—The Harding lamp is illustrated by Figs. 522 and 523. In the former the carbon is suspended by a chain, and is regulated to descend by means of an escapement B, which is actuated by an automatic make-and-break on a shunt circuit, such escapement being arranged with a recoil action, so that the recoil is equal to any desired amount of the feed ; this serves to produce the arc by giving falls of precise measure and a separation of the carbons after contact.   D is a telescopic carbon holder, which may be replaced by a chain and guides.   The current passes through the electro and attracts the armature, which, moving the

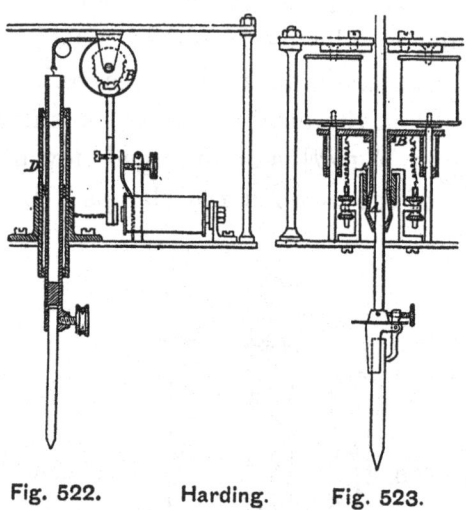

Fig. 522.   Harding.   Fig. 523.

escapement, allows one tooth to pass and the carbon to fall ; contact is then made by a platinum point on the armature feed, the escapement moves back, and the recoil is obtained.   This takes place until the carbons are in contact, and on separation by recoil the lamp is lighted.   Fig. 523 represents a clutch lamp.   The holder passes through the spring jaw or clip A, which is fastened to the armature B.   When the current fails the armature and clip descend and the holder is released, until the increased current raises the clip, and drawing the jaws within the ring E compresses them and lifts the carbon.

Chertemps.—Fig. 524 shows a form of lamp recently proposed by M. D. A. Chertemps, of Paris, in which some novelty of detail will be noticed in reference to the carbon holders.   The lower one is provided with a bracket *e*, which carries the carbon clear of the rack rod, and so permits of the use of long pencils.   The upper carbon holder is jointed in two directions, and is arranged so that by turning the screw *f*[1] the point of the electrode is moved to or from the rod k, and by turning the screw f^2 it is moved sideways. The feed mechanism is of the Serrin type. The catch g stops the wheelwork when the solenoid a is fully excited.

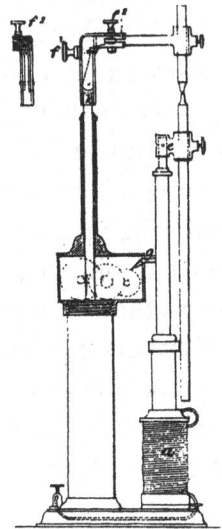

Fig. 524. Chertemps.

Harling and Hartmann.—The feeding mechanism in this lamp (Figs. 525 and 526) is allowed to revolve with greater or less velocity, or is entirely arrested by a brake, which consists of a rotating keeper revolving between the poles of two electro-magnets. One of these magnets is in the lamp

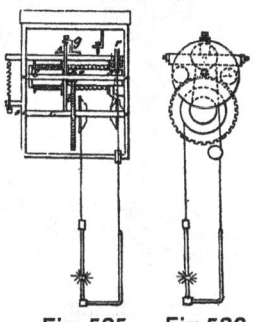

Fig. 525.   Fig 526.
Harling and Hartmann.

circuit, and when the normal current is flowing its attraction is sufficient to stop the rotation of the keeper. The other magnet is in a shunt circuit, and its power, which increases as the arc lengthens, tends to neutralise that of the first magnet and leave the keeper free to move. In

some cases a frictional magnetic brake presses against the periphery of
the keeper, its pressure varying with the force of the magnetism induced
in the said keeper.   Referring to the illustrations, it will be seen that the
two carbon holders are suspended by cords wound in opposite directions
round drums of different diameters on the same shaft.   The weight of the
upper holder having the greater leverage is able to raise that of the lower
one, and also rotate the feeding mechanism.   The cord from the lower
holder is led round a sheave *r* on the end of a lever, which is attracted
downwards when the circuit is established, thus dropping the lower holder
a sufficient distance to establish the arc.   Upon the drum shaft *c* is a
toothed wheel *t*, which gears with a pinion *p* upon a shaft which carries
the disc keeper A.   This keeper stands in front of the poles, and when
the two magnets neutralise each others' action, the keeper rotates and the
carbons are fed together; *g* is a stop that comes down upon the disc when
the current is abnormally strong.   The regulation of the lamp may be
effected by a sliding piece, that can be set to bridge the magnetic poles.

Conolly.—This lamp is intended to be worked in series, and the chief
feature it possesses is the provision for short-circuiting if the arc fails.
The current enters at the terminal *c* and divides itself, part flowing through
the magnet coils and arc, and part through the lever *e*, screw *f*, and frame A,

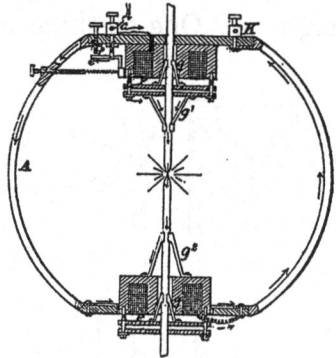

Fig. 527. Conolly.

to the terminal K.   If the carbons are in contact, or the arc is established,
the armature on the lever *e* is attracted, the connection with the screw *f*
broken, and the entire current passes through the carbons.   If, however,
from any cause the lamp does not work properly, the current goes round
it, and it is left out of circuit.   When the magnets are excited, both

the armatures F are raised, forcing the clips *g g* together and raising both carbons. In consequence of the adjustment, the upper carbon is raised more than the lower one, and the arc appears between them. There does not seem to be any provision for regulation, the feeding of the carbons occurring on the extinction of the light only ; g^1 g^2 are spring conductors.

André.—In this arrangement Fig. 528 is a section ; Fig. 529 is a sectional side elevation ; Figs. 530 and 531 are horizontal sections. The upper carbon is hung from an iron rod B, which is held up by magneto-

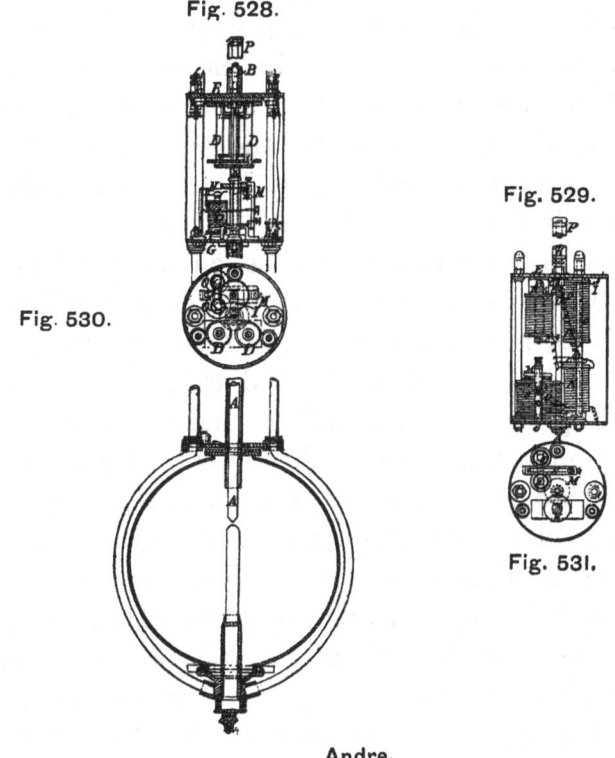

Fig. 528.

Fig. 529.

Fig. 530.

Fig. 531.

Andre.

frictional contact with pole-pieces of a magnet suspended from a pair of electro-magnets D D, so long as the arc is maintained within the determined limit of length. The greater part of the entering current is passed through the magnet D, which is attached to the plate E. A branch I^1, from the main wire I, connects to a smaller electro-magnet K, having two coils wound in opposite directions; the pole-pieces C^1 are carried out to the iron rod B, and are shaped to embrace it. " A third electro-magnet R, having similar pole-pieces C^2 C^3, is arranged on the other side of the rod B, and is wound with a coil connected to one of the coils of the electro-magnet K, and forming a second branch from the main wire I, and so that the resistance of this coil and the coil on the magnet R are together equal to that of the other coil on the electro-magnet K; the exit end of the wire is connected to a bracket M, which is insulated from the plate G and extends over an armature N, which may form electrical connection with it. The plate G carries a fourth electro-magnet O, the coils of which are wound with wire of very high resistance and form a shunt circuit past the arc. The weight of the magnet R, or rather of its core, is supported by a light spring, and its upper pole-piece is in close proximity to a peg or stop under the plate E." The action is as follows :—" When the carbons are together and the current turned on, the magnets D, K and O become excited, the main current passing through D and K; the pole-pieces C will then adhere to the rod E, and the electro-magnet D will draw up the electro-magnet K; the arc will thus be established. When the arc becomes in the least lengthened, the magnet O closes the circuit through that coil of the magnet K, which is connected to the bracket M, and through the coil of the magnet R by means of the armature N, and thus neutralises the adhesive action of the pole-pieces C, and excites the pole-pieces of the magnet R, the spring of which will be slightly compressed by the superior weight of the rod B and carbon A, and will drop on to the peg, whereby the current through the magnet O is decreased, and the circuit through one of the coils of the magnet K and through the magnet R broken. The magnet K will then resume its action, and the pole-pieces C adhere to the rod B." In a modification, the electro-magnet R is omitted, and the tube P on the top of the lamp is surrounded by a cylinder containing glycerine, and having an annular weight or easily fitting piston with two holes covered with a valve ring. The rod B is counterbalanced by chains and a weight. According to another arrangement, where a lower carbon is fed upwards, the latter is clamped to an iron cylinder fitted with a long loosely fitting float piston supported in a cylinder of mercury. In such case the magnet K has its coils wound in opposite directions, the inner being arranged to form a branch from the main conductor, and the outer coils, of high resistance, to form a shunt past the arc. Another pair of electro-magnets with coils of smaller resistance, and corresponding with the magnets D in the illustrations, are arranged underneath the first-named magnets, and have their wire coils connected, one to the positive terminal and the other in electrical connection with the carbon supporting the iron rod. These electro-magnets serve the purpose of separating the carbons when first started by drawing down the first-named, or feed magnets, as in the previous arrangements. The action is as follows :—As the arc lengthens the pole-pieces of the magnet K are demagnetised so far as to allow the rod to slide past them, impelled by the float. When the arc is adjusted the rod is again held. Instead of the branch and shunt connections, it is possible to use the Wheatstone bridge arrangements above described for the purpose of exciting the feed magnets. The whole of the framing of the lamp may consist of a tube, the upper end of the carbon only extending above it, and having a protecting casing around the magnet.

Brown.—The distinguishing peculiarity of this lamp is that the electrodes have a constant feed, less than the rate of their consumption. When the arc sensibly lengthens, this feed is temporarily increased to bring the carbons again within the proper distance, and then it falls back to its usual or normal rate until the limits of deviation, or of arc resistance, are again exceeded. Referring to the illustration Fig. 532, F is one of a

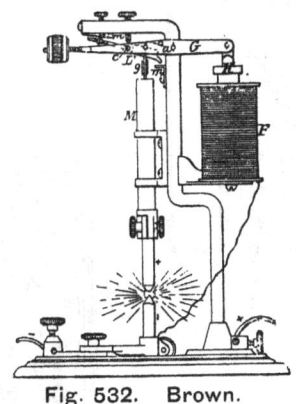

Fig. 532. Brown.

pair of coils on a shunt or derived circuit, each of which has a movable core H, attached to one and the same balance beam G, which is pivotted at *a*. From a yoke on this beam depend two rods *g* attached at their lower ends to a long piston or plunger which fits fluid tight, or nearly so, in the tubular extension of the upper carbon holder. L is a second balance pivotted at *c* to the beam G. From L depends a rod, situated between the two rods *g g*, and carrying a small plunger or cylindrical valve that fits in an axial cylindrical cavity in the piston or plunger, and controls ports or openings that form a communication between the fluid (glycerine) above and below the piston in the tubular extension M. When the lamp is burning, the upper carbon is supported by the atmospheric pressure acting on an area equal to the large piston. A slow leakage of fluid takes place through the piston, and consequently there is a steady descent of the upper electrode, which, however, is not equal to its consumption. When the arc lengthens, the increased current through the coils draws down the coil cores, lengthening the arc still further and pressing the end of the balance beam L against the stop *m.* This has the effect of tilting the beam G and raising the valve, the lower part of which is tapered, so as to allow an increased flow of fluid through the main piston, and the more rapid descent of the carbon holder. As the resistance of the arc decreases the cores rise and shut off or diminish the direct current of fluid.

Sheridan.—This is an American lamp, the patentee being Mr. H. B. Sheridan, of Cleveland, Ohio. Fig. 533 is a sectional elevation of the

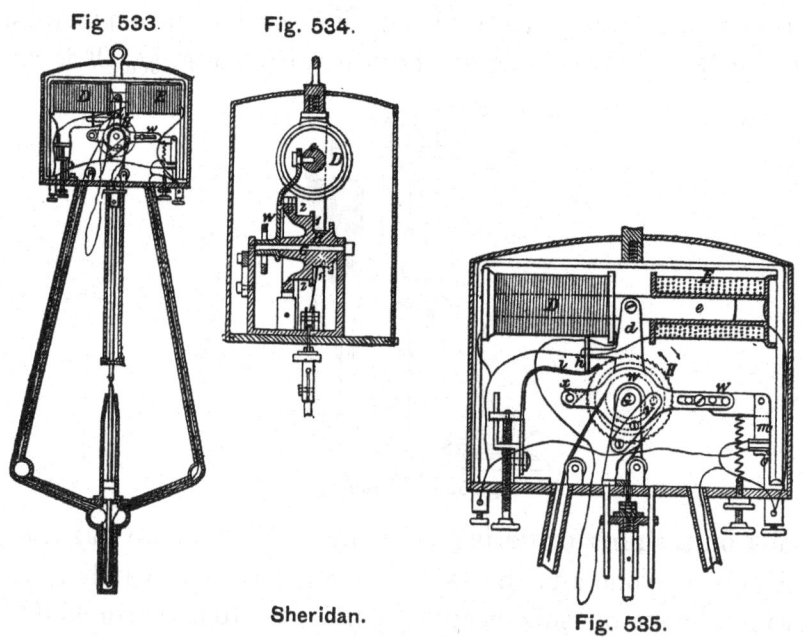

Fig 533.   Fig. 534.

Sheridan.   Fig. 535.

lamp, Fig. 535 is similar view of the regulating mechanism to a larger scale, and Fig. 534 is a section at right angles to Fig. 535. The lower carbon is raised by a cord passing over guide pulleys in the framing and wound on the groove 1 of the pulley H. The upper carbon holder depends from a cord wound in the opposite direction on the groove 2. The pulley H rides loose on a pin, and is turned in one direction by the end of the spring *i* acting as a pawl, and taking into teeth on the flange, and in the other direction by the pawl *x* on the lever *w*. The adjustable spring *i* is depressed by an arm *h* on the lever *d*, operated by the differential coils D and E, of which D is in the line circuit, and E in a shunt circuit. The lever *w* is centred on the arm *v*, and slotted out to pass the pin G; it is moved by a driving pin on an arm projecting horizontally from the lever *d*. At its outer end it carries an insulated plate *m*, electrically connected to one terminal, which is put in contact with a similar plate *o* connected to the other lamp terminal, when the carbon holders have reached the limit of their travel and the arc fails. Fig. 533 shows the position of the parts when the lamp is ready for lighting. Upon the passage of the current, the core *e* moves into the coil D, and the lever *d* presses the spring *i* on to the toothed wheel, turning the pulley in the opposite direction to the hands of a watch and separating the carbons. As the arc lengthens the core *e* moves back into the coil E, and the levers *d* allows the spring to rise, at the same time by means of the lever *w* and pawl *x* it turns the toothed wheel in the opposite direction, and causes the carbons to approach. As the levers *d*, *w*, have different fulcra, but are connected together, the pawl *x* will travel faster and further than the holding plate *h*, and the pulley will make its movement before being fully released from the spring *i*. The outer end of each coil is closed by a vibrating plate like a reed, to act as a dash-pot.

Edison.—Fig. 536 shows an arc lamp devised by Mr. T. A. Edison. To secure a steady light, either, or both, of the carbon pencils are rotated around their longitudinal axes at a speed, by preference, of two to three thousand revolutions a minute. A motor of the Pacinotti or other type in the lamp circuit, or in a shunt circuit, or clockwork, may be used to effect the rotation. Any form of feeding or regulating mechanism may be used, but the one shown in the illustration is well adapted to be com_ bined with rotating carbons. The upper carbon *c* is connected with the rod by the ball-and-socket joint *a*, and is guided near its point by the guide *b*. E is the electro-magnet of the motor, the coils of which are in the shunt circuit 3—4. The revolving armature is supported by a sleeve on the top of the frame, through which slides the rod D, the two being connected by a feather; *d e* are commutator springs touching a revolving circuit breaker. An adjustable resistance R is placed in the shunt circuit for the purpose of regulating the speed of the motor. G is a horizontal armature pivotted on the frame of the motor and playing in the fields of the two electro-magnets, of which the upper is in the lamp circuit and the lower

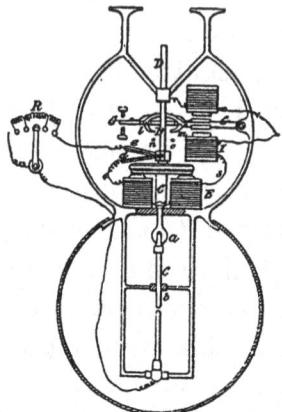

Fig. 536. Edison.

in the shunt circuit. The lever G has two spring pawls that clamp downward on the rod D. The pawls have arms *l m*, and stops *n o* are located at such points on the frame that the pawls are thrown upward away from the rod D, when the armature lever reaches a certain point in its downward movement.

Brockie.—The special feature of this lamp is that the regulation of the arc is effected not continuously, but intermittently, so that there occur periodical lapses in the continuity of the light, but of very short duration; Fig. 537 shows the construction. It will be seen that the bottom of the positive carbon holder T, which is provided with a cross-bar H, slides on the two guide rods G G; the negative carbon is fixed, so that the lamp belongs to that class having a shifting focus. The direction of the current is shown in the figure, and a derived current traverses the circuit W leading to an automatic commutator, and also the electro-magnet M, which intermittently attracts the armature K, the movement by means of a clutch, allowing the carbon to descend at such times as the armature is released by the electro. The upper carbon then falls upon the lower one, but only instantaneously, since the action of the commutator restores the action of the electro-magnet, the armature is again raised, and with it the clutch lifting the carbon, re-establishing the normal arc. The commutator upon which the successful action of the lamp depends is shown at the bottom of the figure. It is connected with the branch circuit by the terminals F F¹. The former is in electrical communication with the toothed wheel S, driven by a worm on the shaft carrying the pulley V. The other terminal is in connection with the pawl D. On the toothed wheel S is a small central disc, and a projecting pin P. The worm wheel is caused to rotate by a cord passing over the pulley V, and led to the generator shaft, and the speed can be adjusted as found most convenient. Except at intervals, the pawl D lies on the face of the central disc, and the circuit is closed. But at stated periods, say of one minute, the pin P comes into contact with the pawl and raises

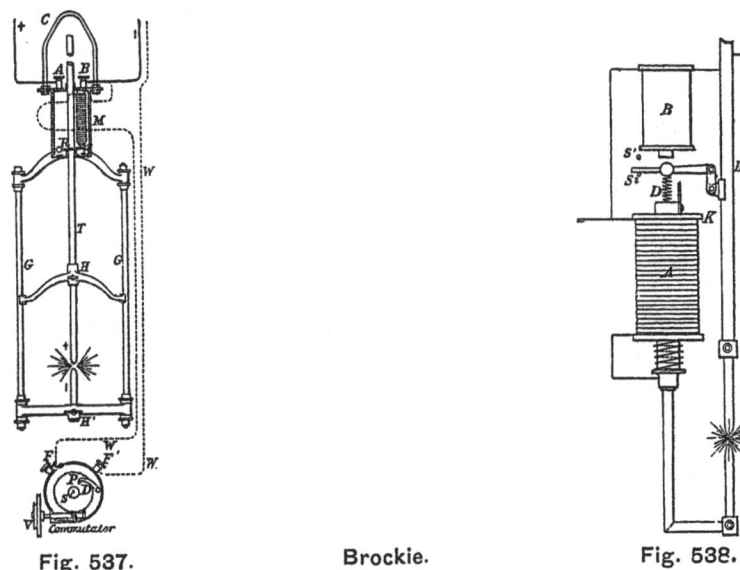

Fig. 537.   Brockie.   Fig. 538.

it off the disc; the circuit is, however, still maintained, until in its revolution the pin passes clear of the pawl, and the latter falls on to the face of the disc, the period of interruption of the circuit being that occupied by the pawl in falling on to the disc, when it is immediately restored. This very minute period suffices to allow the armature, clutch and carbon to fall as above described.

Another form of the Brockie lamp is shown in Fig. 538, where two magnets or solenoids are placed in each lamp, one in the main and one in a shunt circuit. The second magnet releases, retards or stops the feeding train, while the first magnet by its motion, tightens or slackens the spring against which the second solenoid acts, so that when a strong current exists in the main circuit, the first magnet, in addition to directly lengthening the arc (as in the Serrin lamp), will tighten the said spring. By this arrangement the lamp becomes self-regulating, and may be worked with currents of various powers without adjustment or regulation. When the solenoids are arranged as shown, and the main current from any cause becomes diminished, the solenoid A will reduce the tension of the feeding spring D, and thus compensate for the simultaneous weakening of the influence of the shunt magnet B, while at the same time it brings the lower carbon a little nearer the upper one, reducing the length of the arc and adjusting the feeding gear to the decreased strength of current. The core of the solenoid A is of iron in the upper half only, and is pulled downwards against a spring not shown. The shunt solenoid B actuates a brake K upon the upper holder H. The lever of this brake plays between two stops $S^1 S^2$, the lower of which forms an abutment which allows the lower solenoid core as it moves up and down, to diminish or increase the tension of the feeding spring D. The descent of the carbon is retarded by a dash-pot or a train of wheels.

Solignac.—A very simple and ingenious arrangement of arc lamp has been recently devised by M. Solignac, of Paris, which was recently exhibited at a *séance* of the Paris Société de Physique, when a number of them fed by a De Meritens magneto-electric generator were used with great success to light the large hall of the Society. Figs. 539 and 540

Fig. 539.

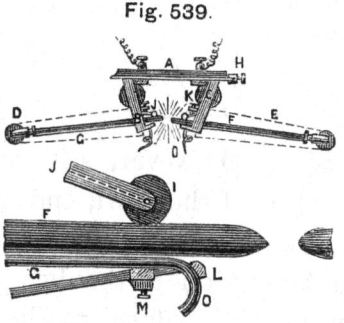

Fig. 540. Solignac.

show the arrangement of this lamp. The carbon rods F, about 19 inches long, are placed nearly horizontal and with their ends in contact; the inner ends are held in suitable guides, attached to a light frame; the outer ends are carried in sockets D with pulleys, round which a system of cords E pass over other pulleys C in the frame, and thence to a weight always tending to keep the ends of the carbons in contact. Beneath each carbon is a glass rod G similar to a laboratory stirring rod, the ends being curved and butting against a nickel stop L, the position of which can be adjusted by a screw. The current is brought to the carbon by the roller I carried at the end of the bar J attached to the frame. These details are shown clearly in the enlarged view Fig. 540. The action of this lamp is as follows : The arc being established, the incandescent cone gradually approaches the ends of the glass rods and softens them in such a way that they yield under the pressure of the weight attached to the system of ends E. Curling under the nickel stops as shown, they allow the carbons to approach and so the arc is constantly maintained. That this is the simplest system of arc lamp that has yet been schemed is beyond doubt ; whether it will stand the test of practice remains to be seen; the trials so far have been very satisfactory, but the arrangement is so new that it has not yet emerged from the experimental stage.

Schuckert.—In this lamp the arrangement is somewhat similar in principle to that of Tchickoleff. A metallic ring is mounted on an axis between the poles of four electro-magnets, placed in pairs on opposite sides of the disc (see Fig. 541); a partial rotation of this disc takes place

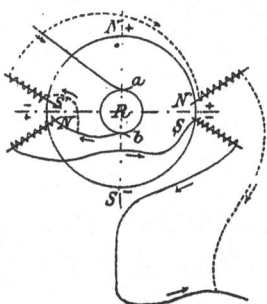

Fig. 541. Schuckert.

according to the distribution of the current, and the motion thus produced is transmitted, by means of a rack and pinion, to the carbons, which are then regulated. The distribution of the current is shown in the diagram : it flows in by a brush *a* to the ring R, leaves at *b*, and is thus sub-divided, the main current passing through the electro-magnets N S, the carbons and back to the generator, while the branch current passes through the electro-magnets S¹ N¹, and then joins the return cable. Two magnetic poles N and S are set up in the ring by this arrangement, and according to the greater intensity of one or the other, the discs will rotate in the corresponding direction.

Auxiliary Lamps.—In terminating these notices of arc lamps we
may refer to a few devices that have been introduced into practice,
having for their object the temporary continuance of the arc, if from any
cause the principal lamp has become extinguished. In some installations, such as lighthouses, practical continuity is a matter of vital, and in
all cases it is of great, importance. The principle of these lamps is that
they should be automatically lighted at a moment of emergency, and be
extinguished as soon as the necessity has passed. Fig. 542 represents an
auxiliary lamp, designed by Messrs. Siemens and Halske, of Berlin, in
circuit with the main lamp. An electro-magnet E E[1] is provided with an
armature S held in such a manner that while it rests against the lower
pole E[1] of the electro-magnet, it is held at a slight distance from the
upper pole E, by a spring *f*. To this armature is fixed the positive carbon

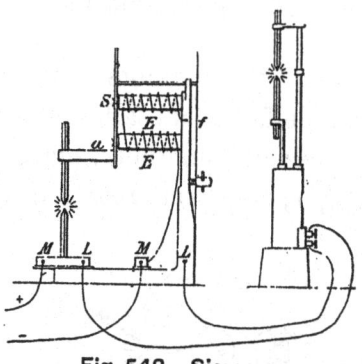

Fig. 542. Siemens.

holder *a*, and while under ordinary conditions the upper carbon rests on
the lower, a slight distance is produced between the two, as soon as the
upper end of the armature is attracted to the pole E.   As will be seen
from the engraving, the current coming from the machine is divided into
two branches, after traversing the electro-magnet of the auxiliary lamp,
one branch leading direct to the main lamp and back to the lower carbon
holder of the auxiliary lamp, while the other traverses the carbon points
of the auxiliary lamp and joins the first branch in the lower carbon
holder, and thence to the generator. In reality, however, this arrangement does not split up the current, except at the moment of starting.
As soon as the electro-magnets E E[1] are rendered active, the two carbons
are separated so far from each other, that no arc is established between
them ; the principal lamp is however lighted. But if by accident the
latter is extinguished, the carbon of the auxiliary lamp drops and an arc

is established until the carbons of the main lamp are again brought into contact, when the branch current is at once interrupted.

An auxiliary has been constructed for the Dornfeld lamp, already described, by Mr. Schuckert. The action of this also depends on the derived current, but the upper carbon *c* (Fig. 543) is raised by a split clutch *b b* forming the core of a solenoid *a*. This core is suspended to a lever *d*, adjustable by a spring *h*, and as soon as the

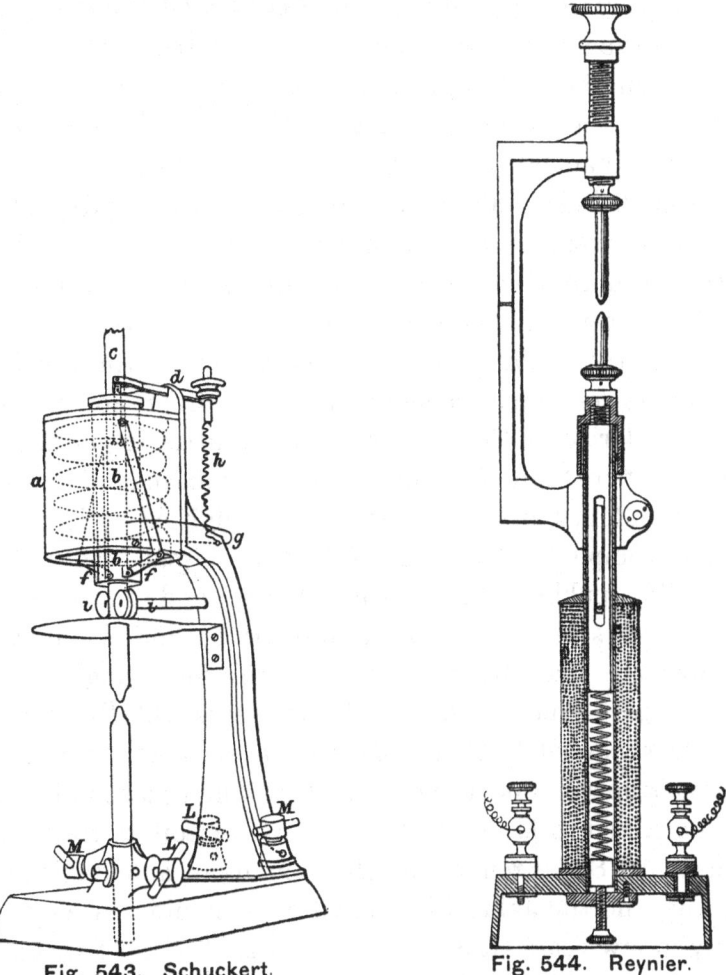

Fig. 543. Schuckert.   Fig. 544. Reynier.

current traverses the solenoid *a*, the plunger is drawn upwards, closing round the carbon, and raising it. The current is led to the carbon by the copper contact rollers *i i*.

M. Reynier has also devised a small auxiliary lamp, shown in Fig. 544. It consists of two carbon holders, the upper one of which is fixed, and the second is lower of a soft iron tube placed within a solenoid, the action of the latter being balanced with a coiled spring.

XI.

THE JABLOCHKOFF CANDLE.

THE memorable experiment which contained the germ from which has sprung the whole industry of electric lighting, dates from the beginning of the present century, but it commenced to bear useful fruit only a few years since, although, as we have seen, there was a period, more than 30 years ago, of extraordinary activity amongst inventors, who failed to arrive at any great results only for want of a practical and convenient means of generating the necessary currents. Until 1870, indeed, the electric light was familiar only to a few specialists, and existed as a scientific wonder, scarcely leaving the laboratory except for exhibition purposes, or in some rare cases where the rapid execution of public works rendered economy a very minor consideration, and where the regularity of the light was but of little importance. To-day the electric lights of many systems burns in thousands of lamps. Whole districts of London, Paris and New York are lighted by it, and hundreds of workshops employ it, to the exclusion of gas. In many railway stations, cafés, public offices and stores, and in some theatres, it supplies the sole means of illumination, and it may be confidently expected that before long it will be brought widely into private dwellings. And to whom in future years will the honour of this revolution be accorded? To no one person certainly, but amongst the crowd of scientific workers, it is evident that a first place must be accorded to M. Gramme, and to M. Jablochkoff. All visitors to the International Exhibition of 1878 will remember that Gramme and Jablochkoff made Paris brilliant, and the Jablochkoff pavilion, insignificant as it would now appear, only so few years later, contained by far the most striking exhibits in this department of industrial science.

M. Paul Jablochkoff was an engineer officer in the Russian army, and was entrusted in 1869 with some galvanic investigations at the Ecole Galvanotechnique of St. Petersburg, and afterwards he had under his charge the telegraph lines from Moscow to Koursk for the Midi Railway Company. Towards the close of 1875 he started from Russia to visit the

Philadelphia Exhibition, but went no further than Paris, where he entered the establishment of M. Breguet, and on March 23, 1876, he secured his first patent for the electric candle. A short time afterwards a Russian gentleman, M. Wyrouboff, director of the *Révue Positiviste*, introduced him to M. S. Denayrouze, and at the commencement of 1877 a group of capitalists combined in the form of a syndicate, for the investigation of electric lighting, with a capital of half a million francs. Afterwards this syndicate was transformed into the Société Générale d'Electricité (procédés Jablochkoff), and more recently into the Compagnie Générale d'Electricité, with a capital of 20,000,000 francs.*

The electric candle was presented to the Academy of Sciences by the President, M. J. B. Dumas, in the name of M. Denayrouze. On that occasion M. Dumas spoke as follows :—" I have the honour to bring before the notice of the Academy, the results of investigations by M. P. Jablochkoff on an invention which has made a great step in the problem of electric lighting. This discovery involves first the suppression of all the mechanism usually employed in ordinary electric lamps. The new luminous source is composed of two carbons fixed parallel to each other, a slight distance apart, and separated by an insulating material which is consumed at the same rate as the carbons themselves. As soon as the current commences to pass, the voltaic arc plays at the free ends of the two carbons. The adjacent insulating material becomes consumed and slowly uncovers the pair of carbons just as the wax of a candle gradually uncovers its wick. The invention in question appears to me at first sight as a vast simplification in the known processes for the production of the electric light, and in suppressing regulating apparatus. I think the advantages may be summed up as follows :—The heat from the combustion of the carbons lost in the air with regulating lamps, is utilised in the Jablochkoff candle for the combustion of the insulating material. The composition of this latter can be varied indefinitely, since a vast number of earthy materials may be used. The most refractory substances volatilise when placed in the voltaic arc, as they are placed by the arrangement of M. Jablochkoff. We have employed as insulating materials, sand, glass, lime, ground brick, &c. But the most simple, as well as the least costly, is a mixture of sand and glass." Such was the first official description of the Jablochkoff candle. M. Dumas' communication also referred to the

---

* We believe that this Company is now undergoing a fourth transformation.

ease with which the system lent itself to the divisibility of the electric
light. This was the first announcement of divisibility having been
successfully achieved; since then it has been effected in a variety of
ways.

The form and modes of construction of the Jablochkoff candle
have been changed greatly since 1876. At the commencement, M.
Jablochkoff employed direct-current generators, and, as in this case the
positive carbon burns twice as rapidly as the negative, he was compelled
to have the former twice the section of the latter, as will be seen in the
annexed sketches. Fig. 545 shows the two pencils separated with an
insulating distance-piece. Fig. 546 shows them surrounded with an
asbestos paper envelope, filled with powdered refractory material. In

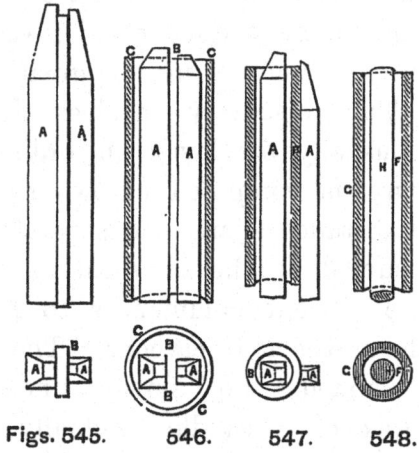

Figs. 545.   546.   547.   548.

Fig. 547, B is a tube of porcelain containing the larger carbon, and against
which the smaller one is placed. Fig. 548 shows two cylindrical carbons,
of which the larger, G, is hollow, and encloses the second, the two being
separated by a packing of refractory material. These various types have
now only an historical interest, and they have long been replaced by
forms with equal carbons, consumed under the action of alternating
electric currents.

As above stated, M. Jablochkoff arranged his earliest forms of candles
with a view to employ direct currents, which necessarily involved a consumption of the positive carbon about twofold that of the negative carbon.
This ratio is not, however, mathematically exact, and varies with the
qualities of the carbons employed. In practice it was found that these
were not consumed in equal quantities during equal periods. On the

other hand, the carbons were not symmetrical, which introduced a further cause of error; and finally, in proportion to the difference in sections, the carbons heated unequally, which added a fresh element of irregularity in combustion. By the use of alternating currents, making successively each point of the carbons a positive and a negative pole, the carbons were consumed in equal quantities in the same time. At the time M. Jablochkoff abandoned the use of unequal carbons, he also abandoned prismatic retort carbons for those of a cylindrical form made of agglomerated carbon by M. Carré. These carbons have of course been manufactured in very large quantities, and their price has been reduced in five years to one-tenth of their original cost. The Société Générale d'Electricité alone has consumed over 3,000,000 metres, or nearly 1,900 miles of these carbons. The type adopted as a standard is ·16 in. in diameter.

The early examples of the present form of Jablochkoff candle (see Fig. 549) consisted of two cylindrical carbons ·16 in. in diameter, from 4 in. to 5 in. long, and having the lower ends encased in copper sheaths *a* to secure a good contact in the holder. These two sockets were con-

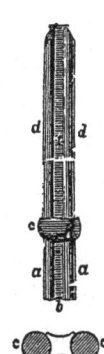

Fig. 549.

nected with a composition *b* formed of silica and silicate of potash, and were surrounded at their upper side by a fillet *c* formed of the same material. The carbons *d* were pointed at their upper ends, and between the two was worked in with a spatula, the insulating material *i*. A small piece of plumbago, tied in place by a ligature of asbestos paper, connected the tops of the carbons, and secured instantaneous lighting as soon as the current passed. The candle thus made lasted about three-quarters of an hour, but the insulating material was deficient in homogeneity, which resulted in sputtering and scattering around of particles, entirely incompatible with successful lighting. The mixture of silica and silicate of potash was then replaced by a thin hollow distance slip of porcelain from Sèvres, and then by pieces of baked kaolin, formed in steel moulds, and of the form shown in the plan Fig. 549, so that they partially surrounded the carbons. In imitation of the name of similar pieces used in faience factories, this insulating strip was called " colombin," the name it is still known by. At the same time the effective length of the carbons was increased to 9 in., which is the normal length at the present time. Later on, as we have already seen in describing the manufacture of the Jablochkoff candle, instead of the unpractical and costly mode of making the " colombin," a very simple and expeditious process was substituted.   Candles of this pattern were manufactured for several months, but it was soon found that the kaolin, melting in the voltaic arc, absorbed considerable heat, and produced a marked diminution in the light.   Experiments were then undertaken with other substances, and after many careful trials M. Jablochkoff stopped at a mixture of sulphate of lime and sulphate of barytes as pure as possible.

The question was discussed for a long time whether the " colombin " formed of this mixture increased the light, or whether the employment of carbons insulated only by air, as in the Jamin and Wilde lamps, did not give better results.   Trials made during the International Exhibition of 1878 showed that the Jablochkoff candles, with or without the insulating strips, give (steadiness apart) the same quantity of light.   But since that time it has been proved that the utility of the insulator lies in a sensible reduction in the amount of power absorbed.   It may thus be stated confidently that the " colombin " gives a definite increase of light for a given power.   The method of joining the copper sockets has also been modified in a very ingenious manner, so as to avoid completely the loss in electricity which always took place across the very impure insulating mixture connecting these two parts of the candle.   The copper bases now employed are split, and receive an insulating block which is not so deep as the socket.   The ends of the carbons are placed within them.

Fig. 550 shows the arrangement, where *a a* is one of the copper split tubes, *b* the insulator, *c d* the three parts united. Finally the lighting device, first formed of a small graphite pencil, is replaced by a sort of fuse produced by the end of the candle being dipped in a mixture of gum and plumbago. Such is the Jablochkoff candle actually employed for lighting purposes.

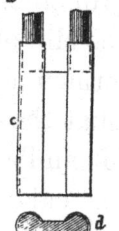

The rate of combustion in electric carbons plays a most important part in the commercial aspect of the question, and any means whereby this can be reduced must be considered (other things being equal) as an important improvement. And besides the economy resulting from slower combustion, there is another advantage—the simplication of the apparatus.

Fig. 550.   Thus, supposing it is desired to obtain a continuous light during ten hours with candles lasting two hours each, it will be necessary to have lamps with five candles, commutators in five directions, and a certain length of conductors. But if a candle lasting ten hours can be used, the necessary apparatus would become very simple, and the length of conductors largely reduced. The economy thus effected would be one of first establishment, as well as of hourly expenditure.

The first means proposed to increase the duration of the candles was to cover the carbons with a thin film of metal. The priority of this invention has been warmly contested; in France it was for a long time attributed to M. Reynier, who secured it by a French patent dated 11th October, 1875. Fortified by this patent, which he considered original, M. Reynier commenced an action against the Jablochkoff Company, who, relying on certain prior publications, electro-plated the carbons they employed without paying him any royalty. The case went to trial and was decided, on the 21st July 1881, in favour of the Compagnie Générale d'Electricité. In the course of the action certain facts, before almost unknown, were brought forward and threw a new light on the history of this invention. There were found in several English patents certain intimations, though not well defined, of the association of metals, especially of copper, with electric carbon. But there appears no question as to the priority of electro-plating carbons to be used for the electric light, since in 1868, in a supplement taken out by M. Ed. Carré for the French patent of 23rd August, 1867, under the title of "Improvements in the Means of Producing and Employing Electricity," M. Carré claims the employment of carbons coated with metals such as zinc and tin. About the same time M. Joseph Van Malderen, who contributed so largely to the development of the Alliance generator, experimented with coating carbons with copper; his trials having been made in 1867 and 1868. But these trials were not successful, and he abandoned them. Several years later more serious experiments were conducted in Russia by M. Bouliguine, lieutenant in the Imperial Russian Navy. This was in 1873. Shortly after, another Russian, M. Tchikoleff, inspector-general of lighting fortresses, &c., followed up these experiments of coating carbons, and without knowing of Bouliguine's attemps, believed himself to be the first inventor of the improvement, which he described at a *séance* of the Imperial Society of the Friends of Science at Moscow, when he soon ascertained that he had been anticipated. M. Reynier only came into the field in 1875, and in good faith, ignorant of the investigations made before his own, he believed himself to be the inventor of electro-plated carbons. The Tribunal Correctionnel of the Seine decided that he was in error, and that the process of metal coating carbons for the electric light belonged to the world. If the proprietorship of this invention has been so keenly disputed, it is because it is of real value. Coating carbons with copper or nickel, the only metals employed for the purpose, has the effect of diminishing the resistance of the carbons and preserving them from contact with the air. There results from this a considerable diminution of the incandescent cone of the carbons, and consequently a slower combustion. The following figures refer to candles of various types, coated and non-coated, burning under similar conditions :—

Type of Candle.	Length.	Time of Burning.	Difference.
	in.	h. m.	m.
Carbons, ·12 in. in diameter, non-coated	9·25	1 20	10
Carbons, ·12 in. in diameter, coppered	9·25	1 30	
Elliptical carbons, ·12 in. by ·177 in., non-coated	9·25	1 45	15
Elliptical carbons, ·12 in. by ·177 in., coppered	9·25	2 00	
Carbons, ·16 in. in diameter, non-coated	9·25	1 30	20
Carbons, ·16 in. in diameter, coppered	9·25	1 50	
Elliptical carbons, ·16 in. by ·24 in., non-coated	9·25	2 10	20
Elliptical carbons, ·16 in. by ·24 in. coppered	9·25	2 30	
Carbons, ·24 in. in diameter, non-coated	9·25	2 20	15
Carbons, ·24 in. in diameter, coppered	9·25	2 35	
Elliptical carbons, ·24 in. by ·35 in. non-coated	9·25	3 00	20
Elliptical carbons, ·24 in. by ·35 in. coppered	9·25	3 20	

Larger differences were obtained with carbons coated with nickel, but the saving is overbalanced by the increased cost of the process. Fig. 551 shows the effect of electro-plating on carbons burnt with a direct current; A indicates the result with bare carbons; B shows those coated with copper, and C those covered with nickel. The use of this coating has the inconvenience of rendering the arc somewhat less steady, and for special purposes, as for cafés, theatres, stations, &c., the carbons not coated are preferred for this reason.

Another very simple means of increasing the duration of the carbons is to enlarge their diameters. There results from this, it is true, a diminution of the light produced, since the incandescence is spread over a larger volume. But this inconvenience of itself is of no great importance, because even if there were a loss of several Carcels in the lighting power, it would be scarcely appreciable to the eye, and the resulting economy would more than compensate for it. But the objection to using large carbons is that the length of the arc, and

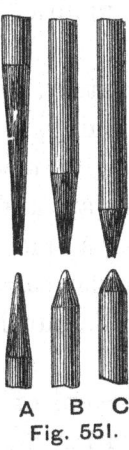

A   B   C
Fig. 551.

consequently the resistance, increases. When the device was found to be a failure, attention was directed to the section of the carbon, and after many repeated experiments it was found that the circular form could be advantageously replaced by others. If the way in which the carbons are consumed is examined, it will be seen that the cone formed by the combustion at the ends between which the arc plays, takes the form A indicated in Fig. 552 annexed. It results from this that the mean

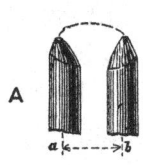

A

B
Fig. 552.

length of the arc is sensibly greater than the distance from axis to axis *a b* of the two cylindrical carbons. This difference of length of arc increases with the diameter of the carbons, and a stronger current becomes necessary to overcome the resistance; this gives rise to many difficulties with the generators, the conductors, and the candles also. There may be an advantage in giving to the carbons such a form

that their mass may be directed towards the interior of the candle, thus
for example as in B, Fig. 552, with a triangular section with the two bases
opposite each other.   Several special forms have been experimented on;
the most practical, both as regards favourable combustion and manufacture of carbons, appears to be the elliptical form, which permits at
once of an increase in section, and a closer approach of the axes.   It
will be seen indeed from A, Fig. 553, that if the radius of the carbons A B is

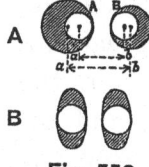

Fig. 553.

increased, the distance *a b* would become greater, whilst the
same result is arrived at by transferring the mass represented by the tinted portion, in the manner indicated in
B, the distance between the axes not being changed.
This modification of form easily adapts itself to the electro-
plating of the carbons, and the arc is somewhat steadier
with the elliptical than with the cylindrical carbons; a further advantage
is thus gained by the alteration.

In a supplement to his first patent, M. Jablochkoff claimed the use
of pencils pierced with an axial opening, in which a mixture of refractory
earth and plumbago was introduced, or else metallic salts.   This filling
gives a kind of *point d'appui* to the arc and renders it more stable.   The
combination of these three elements : electro-plating, the elliptical form
of the carbons, and the use of annular carbons filled as above mentioned,
will probably lead to a type of candle burning slowly and giving a steady
light.

The description of the employment of the Jablochkoff candle may be
usefully preceded by an examination of the special conditions that
control the electric carbons during consumption.

Such an investigation has been made with great care by M.
Joubert, general secretary to the French Physical Society, and by
M. Minet, who has especially interested himself in the manufacture
of the candles.   We will now give the results of some of the most recent
researches.

As a general rule, in burning Jablochkoff candles with carbons of
4 millimetres diameter, separated by a distance-piece or " colombin "
·28 millimetre wide, currents are employed of sufficient tension to burn
four such candles placed in the same circuit.   The work to be done
corresponds to a current of from 8 to 9 Ampères.   The fall of potential
between the two parts of the bracket carrying the candle should be from
42 to 43 Volts.

The resistance due to the Jablochkoff candle resolves itself into two parts :—

1.—The resistance of the carbons.
2.—The resistance of the other elements of the candle.

According to M. Joubert, the specific resistance of a carbon, that is to say, the resistance which is offered to the current by a cube of 1 centimetre, is 3,927 microhms (the microhm is a millionth part of an Ohm) or 2,430 times the resistance of pure copper, which is 1.616 microhm. To take another form, a cylinder 1 metre in length, and 1 millimetre in diameter, has a resistance of 50 Ohms. The following table gives the resistances for carbon cylinders 1 metre long and of varying diameter :—

Diameter of Carbon.		Resistance in Ohms.	The Equivalent Length in Copper Wire ·16 in. = 4 mm. in Diameter.		
mm.	in.		m.		yds.
1 =	·04	50·000	20,000	=	21,870
2 =	·08	12·500	5,000	=	5,466
3 =	·12	5·550	2,222	=	2,430
4 =	·16	3·125	1,250	=	1,366
5 =	·20	2·000	800	=	875
6 =	·24	1·390	555	=	606

These results were obtained at a temperature of 68 deg. Fahr., or far below that to which the carbons are raised on the production of the arc, now it is known that the resistance of metallic conductors increases with the temperature, while the reverse effect takes place in bodies of high electrical resistance.

Carbon is a substance whose resistance diminishes as the temperature increases. Between 0 deg. and 100 deg. Cent. the coefficient of reduction is $\frac{1}{8442}$ for each degree. Carbons made of gas coke have a much higher resistance, about 17 times more than the Carré carbon.

From the foregoing it will be seen that the resistance of the carbons at the temperature due to incandescence is relatively very low, and has practically no influence on the intensity of the light, which depends rather on the composition of the carbon. A metallic coating also diminishes the resistance very sensibly; under normal conditions it reduces it to about one-third of the original value.

Besides the resistance due to the carbons is that of the contact, more or less perfect, of the carbons with their sockets, and before lighting, the resistance of the lighting fuse must be added to the total. M. Joubert has carefully investigated this question, but he has not, we believe, measured the resistance due to the second cause just named, the imperfect contact of the socket and the carbons.

To measure the resistance of lighting he proceeded by differences, taking at first the resistance of the candle as it is made, and afterwards that of a candle in which he substituted for the lighting fuse a copper wire of inappreciable resistance.

The following are the results obtained with two series of candles :—

A—*Candles 9 in. long, Carbons ·156 in. in diameter.*

No.	Total Resistance.	Resistance of the Carbons.	Resistance from Lighting.
1	7·70	1·73	5·97
2	8·27	1·52	6·75
3	6·97	1·73	5·24
4	6·42	1·60	4·82
5	6·30	1·79	4·91
6	9·20	1·73	7·47
7	6·12	1·44	4·68
8	7·15	1·81	5·34
9	5·94	1·65	3·29
Average	7·12	1·66	5·34

B—*Candles 9 in. in length with Carbons ·16 in. in diameter.*

No.	Total Resistance.	Resistance of the Carbons.	Resistance from Lighting.
1	8·10	1·54	6·56
2	5·62	1·35	4·27
3	18·80	1·42	16·38
4	5·65	1·42	4·23
5	5·94	1·64	4·30
6	6·95	1·37	5·58
7	6·66	1·48	5·18
8	8·20	1·43	6·77
9	6·92	1·53	5·39
10	6·60	1·43	5·77
Average	6·74	1·46	5·27

In terms of equal lengths of conductor, the candles of the first series give the following equivalents :—

	Yards of Conductor.
Total candle with lighting	3,104
Lighting only	2,335
Candles without lighting....	726

Second Series.

Candle with lighting	2,981
Lighting only	2,305
Candle without lighting	638

Thus, the resistance due to lighting is equivalent to about 2,000 metres of conducting wire, and the resistance of the candle is only 600 metres. In a circuit of four candles, the resistance is therefore equivalent to 2,400 metres at the commencement, and metal coating reduces the amount to about 800 metres. In the course of burning, the carbons become heated, and the resistance diminishes, first on account of the rise of temperature, and also because the length of the candles becomes gradually reduced. But a new resistance also has to be considered—that due to the voltaic arc; exact experiments have not been made on this subject up to the present time. There is, lastly, another element, the resistance of which is more or less important, and which plays a constant part in the combustion of the carbons; this is the resistance of the " colombin " and of the distance-piece connecting the sockets carrying the carbons. As regards the former, no careful experiments appear to have been made, but with respect to the latter the investigations of M. Joubert furnish interesting data.

The insulating materials first used gave a resistance of 25,000 Ohms; horn insulators showed a resistance of 94,000 Ohms; and ebonite is yet a more perfect insulator. The material now used gives a lower figure than horn, but the economy in manufacture is considerable, and the insulation is sufficient for all practical purposes.

The ease with which the Jablochkoff candle can be adapted to brackets or holders for a variety of conditions, constitutes one of the striking advantages of the system. The most important parts of the bracket are the copper clips, furnished with springs, which hold the candles fast and insure close contact with the copper sockets at the base of the candles. Fig. 554 shows the first system of clips employed, and Fig. 555 indicates how it has been modified. The clip is fixed to a circular base of wood, slate, marble, or onyx; two connections serve for coupling up the wires of the circuit. This type of bracket is adapted for periods of lighting not exceeding an hour and a half, and the circuit is arranged as shown in Fig. 556, where it will be seen that the current generated by the dynamo machine M feeds the four candles A, B, C, D. As a rule, however, the period of lighting exceeds $1\frac{1}{2}$ hours, and it is necessary generally to arrange beforehand a number of candles disposed in such a manner that in burning one after the other, they last collectively for the desired period. To this end combinations of candle holders are arranged for periods of lighting varying from $1\frac{1}{2}$ to 16 hours. The special type for four candles is the one found most generally useful. Whatever may be the number of candles to be lighted one after another to afford a continuous light for a given time, it is necessary to employ a device by

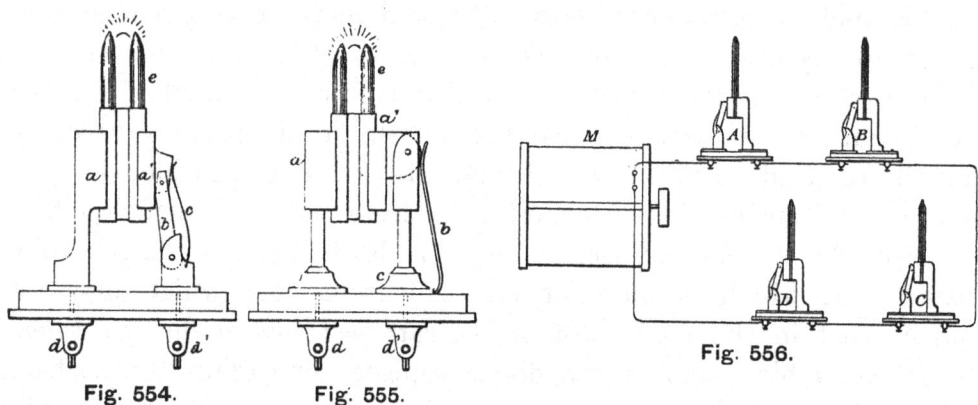

Fig. 554.   Fig. 555.

Fig. 556.

which, as soon as one candle has burnt out, the current feeding it shall be switched off to the one adjacent. This is effected either by hand, or by the use of an automatic commutator.

Two modes of arranging the circuit are used, according to whether it is desired to charge all the candles upon that circuit simultaneously, or in succession. Each of these methods requires a special type of candle bracket; for the former the bracket is circular, and for the latter it is in the form of a cross.

To explain the former type we will take as an example a bracket to carry four candles, and which is illustrated by Figs. 557, 558. It is provided with four double clips similar to those represented in Figs. 554 or 555. These clips are equally spaced round the circular base, and below the bracket is a series of eight connections corresponding with the series of clips. In the cruciform brackets (Figs. 559, 560), the double clips are placed at right angles, the fixed portion being on the inner side. All four are connected by a piece of brass and are in communication with a single binding screw. The jointed portion of the clips are on the outer side and are each furnished with a single binding screw. These two classes of brackets require a four-way commutator, Figs. 561 and 562; this apparatus

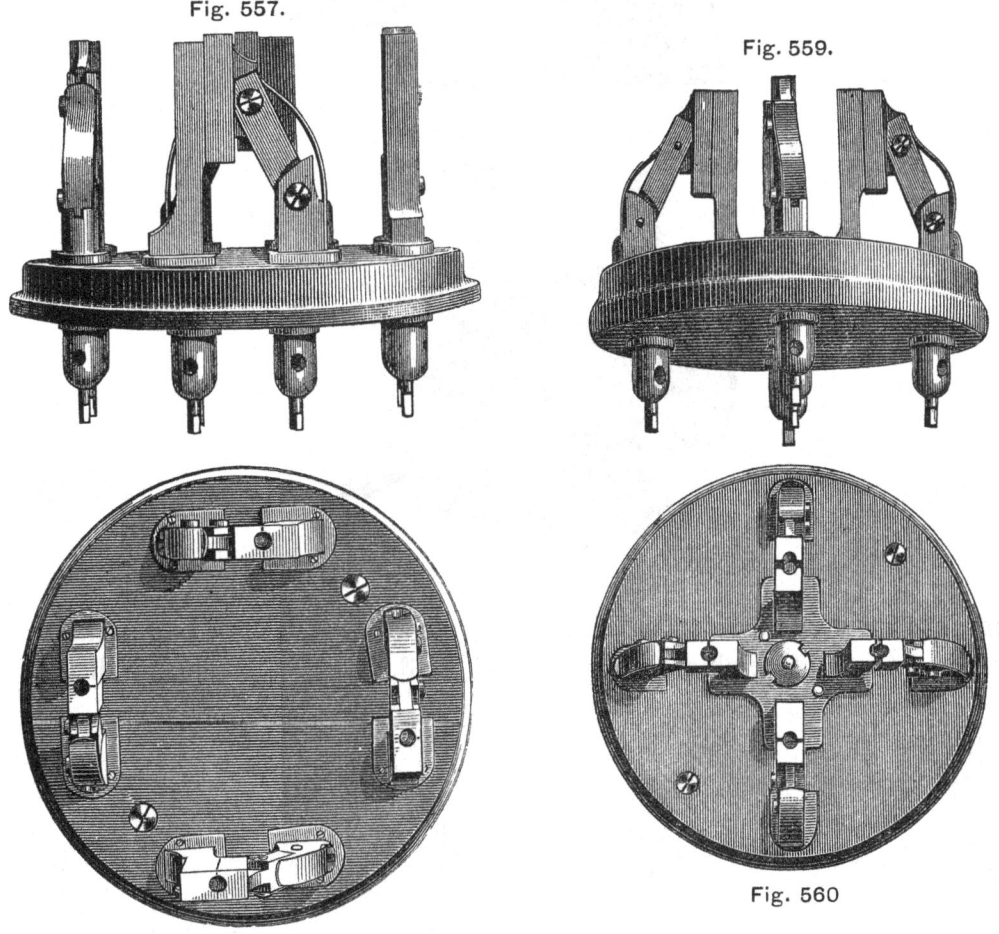

Fig. 557.

Fig. 559.

Fig. 558.

Fig. 560

consists of a wooden disc around which are arranged four copper contacts, A, B, C, D, each of which has its binding screw. The central metallic stud E is in communication on the one side with the binding screw F of the entering current, and on the other side with a movable piece G fitted with a steel spring; by means of a key, which can be placed in the square cavity of the central stud, the movable plate can be turned successively to rest on the contacts A, B, C, D. The contact H is insulated; the piece K, which carries two binding screws, can be connected electrically with the binding screw F by a metallic plug, which

Fig. 561.

Fig. 562.

fits into the space I between the metallic pieces L M. The commutator key and the contact plug are shown in Fig. 563. Figs. 564 and 565 indicate

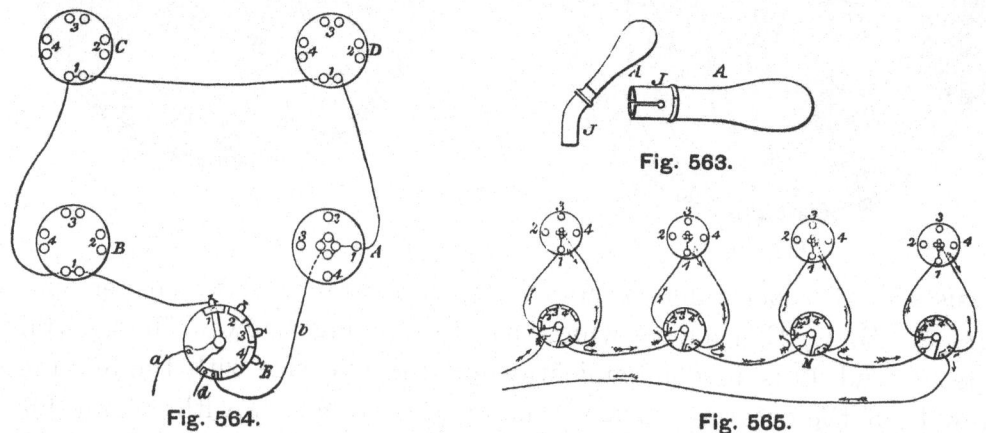

Fig. 563.

Fig. 564.

Fig. 565.

the arrangements for a circular and a cruciform circuit respectively. The installation of the circular circuit Fig. 564, requires one cruciform bracket

A, three circular brackets B, C, D, and one four-way commutator E. The first conducting wire is shown by the line connecting the clips 1, 1, 1; the current coming from the machine by *a* passes by the movable rod to the contact No. 1 of the commutator, traverses the candles 1, ignites them, and returns to the machine by the wire *b d* attached to the contact with double-binding screws. In the same manner three other conducting wires are required to join up the brackets bearing the corresponding numbers on the diagram. It will be seen that if, by means of the key, the commutator rod is brought to bear successively upon the contacts 2, 3, 4, the candles 2, 3, 4 will be ignited. The metallic contacts of the commutator are placed so close together that the movable plate does not clear any one of them before it has come into contact with the one adjacent. By this arrangement there is never any absolute interruption to the passage of the current. With such an installation a continuous light during a period of eight hours can be obtained with galvanised candles, if at intervals of two hours, the attendant operates the four-way commutator which is placed near the generator. The inconvenience of the arrangement lies in the fact that it is necessary to have four circuits 1, 2, 3, 4 between the brackets, which involve the use of a great length of conducting wire. To avoid complication the second, third, and fourth circuits are not indicated on the diagram. Fig. 565 indicates the arrangement of a circuit in which cruciform brackets are employed, and which is composed of four such brackets and of four four-way commutators placed near them. The four contacts of each commutator are connected by wires to the four binding screws of the corresponding bracket; the direction of the current is shown on the diagram by arrows, but only one circuit has been given in order to avoid complication. It will be seen that this system economises a considerable length of wire, but on the other hand, the attendant charged with the management of the light is obliged to go from one bracket to the other, which involves a loss of time and causes a waste of carbons. With this method it is possible to extinguish one of the lights without affecting the other; for example, if it be desired to suppress light No. 3, all that is necessary is to place the commutator plug in M, and the bracket will be cut out of circuit.

In some arrangements the commutator is fixed beneath the onyx or marble base of the bracket. This apparatus is technically known as a "commutator bracket," the working of which will be easily understood from the previous explanation. M. Gadot, one of the engineers of the Compagnie Générale d'Electricité, has devised an extremely simple bracket, dependent for its operation on the differences in the resistance of the lighting attachment at the end of each candle. The annexed diagram (Fig. 566) explains the arrangement. It represents four candles, 1, 2, 3, 4, placed in circuit; A is a two-way commutator by which the current can be established or interrupted; the current in passing will select that one of the four candles the lighting fuse of which offers the least resistance. As soon as this candle is almost burnt out, the circuit

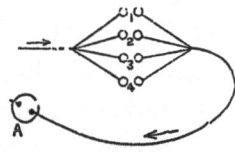

Fig. 566.

is interrupted by means of the commutator, and immediately restored. The interruption is sufficient to extinguish the candle, which now offers a higher resistance than that of the fuses on the other candles. The current will again select the easiest path, and so on until all the candles in the bracket are consumed. Ingenious and simple as this arrangement is, it offers serious difficulties in practice, which have prevented its adoption. Amongst others may be mentioned that of the current attempting to pass the four candles at the same time, with the result of destroying them all without obtaining any light. It might be possible previous to using them, to classify the candles according to the respective resistances of their fuses; but other difficulties creep in, and a bad contact in any of the various parts of the system is sufficient to completely upset the working of the apparatus. For all that, the idea has considerable merit, and may some day possess a practical value.

These various methods of producing a continuous light, require the care of an attendant to operate the commutators at comparatively short intervals, and they will be greatly simplified when candles lasting three or four hours take the place of those used at present, and which are consumed in half the time; but such method will always possess the inherent weakness of requiring the assistance of an attendant, who may forget his duty or find himself suddenly unable to accomplish it at the critical moment. It was natural, therefore, that at an early stage of working the Jablochkoff system, the engineers interested in its development devoted much attention to devising apparatus for shifting the candles automatically, and thereby dispensing with the services of an attendant. The most primitive forms of automatic brackets consist of a system of contacts, held back by threads placed near the bottoms of the candles, when, by reason of combustion, the carbons are sufficiently reduced in length, the thread is burnt, and allows the contact to fall forward and pass the

Fig. 567.

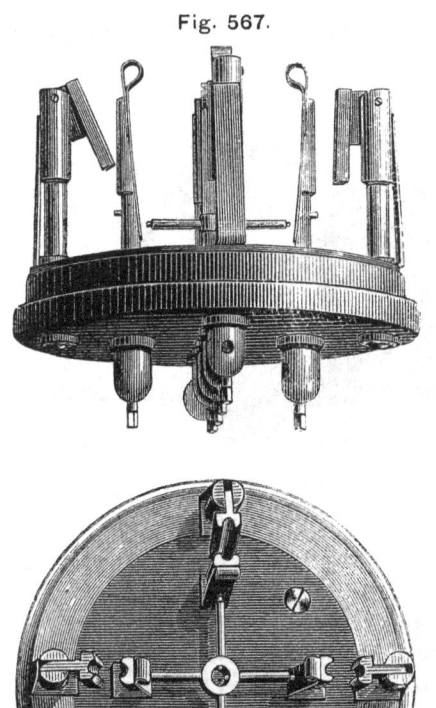

Fig. 568.

current to the adjacent carbons. This system was soon abandoned on account of the complication it involved with three or four candle brackets, and which, moreover, multiplied the number of contacts, and with it the chances of imperfect working. More recently, another and very ingenious arrangement has been devised, consisting of an expansion bracket, of an automatic mercury commutator, and of an annunciator for showing the extinction of each candle.

M M

The expansion bracket is shown in plan and elevation by Figs. 567 and 568 ; it is constructed on the same principle as the ordinary bracket, but in addition, each clip is provided with a bent compound metal strip, formed of steel and copper soldered together. When a candle is almost entirely burnt out, the voltaic arc and the incandescent portion of the carbons are brought into very close proximity to the strip, and raise its temperature. When this happens, the strip, on account of the difference in the coefficient of expansion of the two metals of which it is formed, is

Fig. 569.

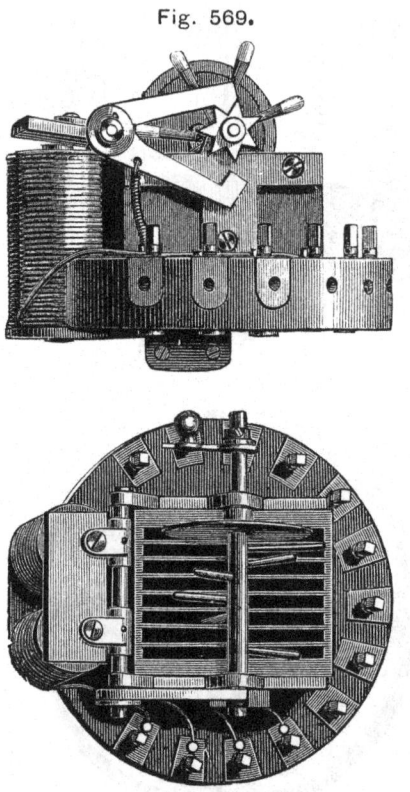

Fig. 570.

expanded differentially, and the free end curves away from the fixed clip until it touches the contact placed in the centre of the bracket. The automatic commutator consists of a hard rubber receiver divided into eight compartments containing mercury, each compartment being completely insulated from the others. The current passes to one of the outer compartments of the reservoir, and thence it is transmitted to a disc plunged in the mercury, to the spindle on which the disc is placed, and successively to seven arms which are fixed radially on the spindle. Each of these seven arms corresponds to one of the seven compartments of the mercury reservoir. Rotation is given to the spindle in order that each arm may at the proper time transmit the current through its own compartment to the corresponding candle. This movement is obtained by means of a toothed wheel and escapement, actuated by an electro-magnet. This mercury commutator rests on a wooden base, and is enclosed in a casing, on one face of which is drawn a graduated circle; an index fixed on the shaft carrying the radial arms, indicates upon the circle which candle is burning. Figs. 569 and 570 show an elevation and plan of this apparatus. The annunciator indicating the extinction of the carbons, comprises as many movable flaps, and consequently as many indicating apparatus as there are circuits in the installation. Each apparatus consists of a double electro-magnet with large wires, through which the current passes during the combustion of the candles. This electro-magnet is furnished with a movable armature, on which is fixed a rod carrying a plate divided into two parts, the one entirely blank, the other bearing the number of its corresponding circuit. A counterweight tends always to separate the armature from the electro-magnet, and at the same time to establish a contact with an insulated standard, connected with a battery and bell, in such a manner that when one of the electro-magnets is not working, the current from the battery is put into circuit with the bell.

The installation, regulation and working of these different apparatus may now be described. Near the dynamo-electric machine, and on the circuit, is placed a two-way commutator, connected to the contact disc of the automatic commutator, and consequently to the shaft and the seven arms mounted upon it (see Figs. 571 to 575). At the moment of lighting, the indicating needle 1 ought to be in a vertical position; the point 1 which is plunged into the receiver to which it belongs is itself connected to the contact 1; each of the contacts 1, 2, 3, 4, 5, 6 is connected to the fixed portion of the clips on the bracket corresponding to the 1st, 2nd, 3rd, 4th, 5th, and 6th candles; the return wire from the bracket is connected to the stud C of the commutator, and the expansion contact is connected to stud D. This understood, we may observe what passes when one of the candles is nearly burnt out. The expansion strip on becoming heated, bends, and the outer end approaches the corresponding branch of the contact in the centre of the bracket, until it touches it. The current arriving by the fixed part of the clip thus finds two passages opened for it ; one by way of the candle, where it encounters a high resistance, and the other by the expanding strip and the contact where the resistance is much less. The greater part of the current is shunted through this latter path, passing the stud D and the electro-magnet N, and thence along the ordinary return wire of the circuit.

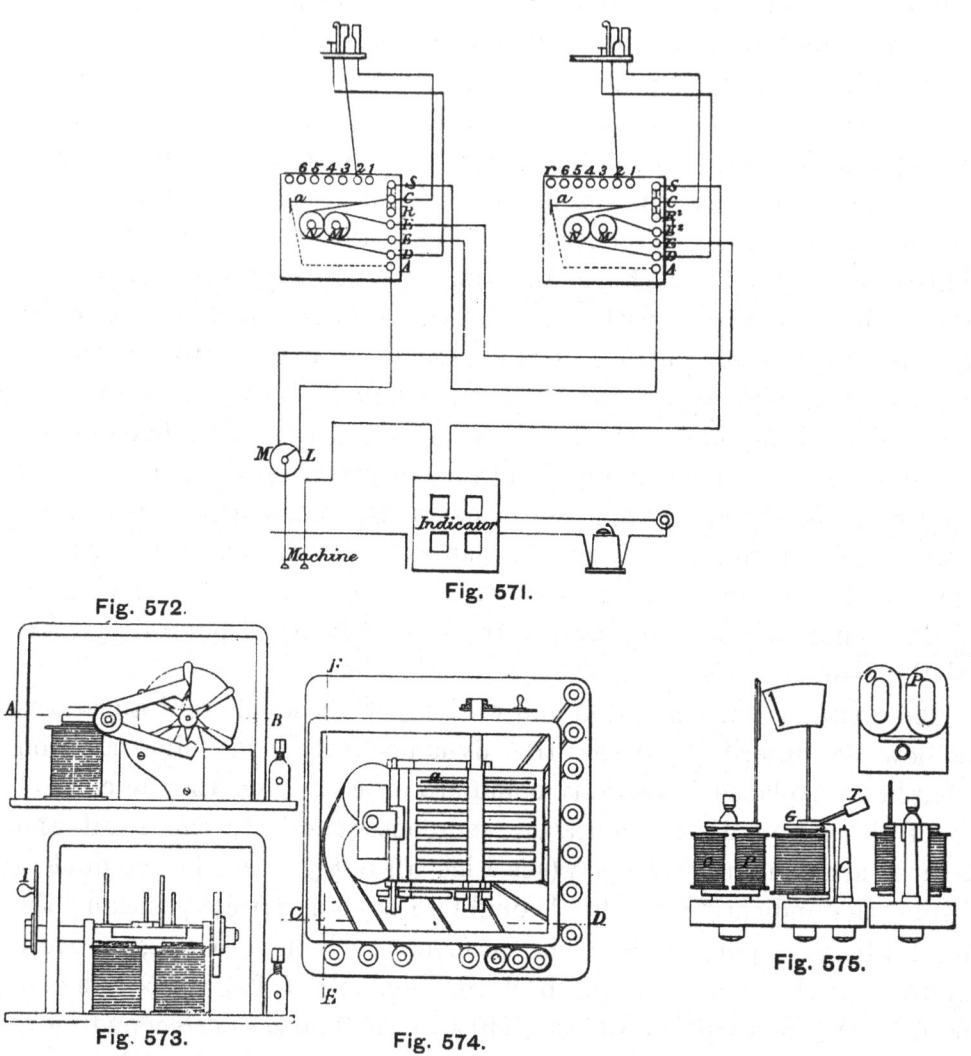

Fig. 571.

Fig. 572.

Fig. 573.   Fig. 574.

Fig. 575.

The current in passing through N magnetises the soft iron core and attracts its armature, which communicates a rotating movement to the shaft and escapement mounted upon it, and consequently, by means of the toothed wheel, rotation through one-seventh of a revolution is imparted to the spindle carrying the arms. By reason of this motion the corresponding arm is lifted out of the mercury, and the arm 2 enters it; but the fixed portion of the clip 1 on the bracket being only in connection with the contact, on account of the position of the expansion strip, if the current ceases to pass by this clip, it also ceases to pass through the electro-magnet, and flows through the second candle on the bracket; the armature is no longer attracted, the escapement falls, and in doing so communicates a second rotating movement to the spindle carrying the arms, and the arm 2 is plunged deeper into the mercury. The same series of operations is performed with each shifting of the candles. As long as the circuit is closed, and, therefore, as long as the candle burns, the current passes through the electro-magnets O and P of the annunciator, attracting the armature G which exposes the blank portion of the indicator plate connected to it. The counterweight T being clear of the column C, the circuit of the small battery is open, and the bell does not work. If, however, through the burning out of a candle, the current ceases to flow through the electro-magnets O and P, the armature G is no longer attracted, that portion of the plate carrying a number, is thrown into view, and the counterweight T falling on the column C, closes the circuit, and sets the bell ringing. The attendant is thus informed that a candle has burnt out on circuit No. 1. He then goes to the two-way commutator on this circuit, and makes contact with the key M, causing the current to flow through the electro-magnet M^1, whence it passes by a special wire through the electro-magnet of the second commutator, and then to the last commutator of the circuit where the contact E^2 is connected to the contact R^2 in order to utilise the ordinary return wire: The current thus passes through all the electro-magnets on the circuit, the armatures of which are all attracted; the succeeding arm is plunged into the mercury, and as soon as the key of the commutator L is pressed, the circuit is relighted. This operation can be performed very rapidly. The essential point in regulating the quantity of mercury in each compartment of the receiver is, that as the arm 1 emerges, arm 2 shall enter its corresponding compartment. When the electro-magnet ceases to attract the armature, the arm 2 should be more deeply immersed, but the arm 3 ought to remain out of contact with the mercury in its respective compartment. If there be too much mercury in the compartment 1, at the moment when the magnet attracts the armature, so that the arm 1 does not emerge and the arm 2 begins to enter, the current would find

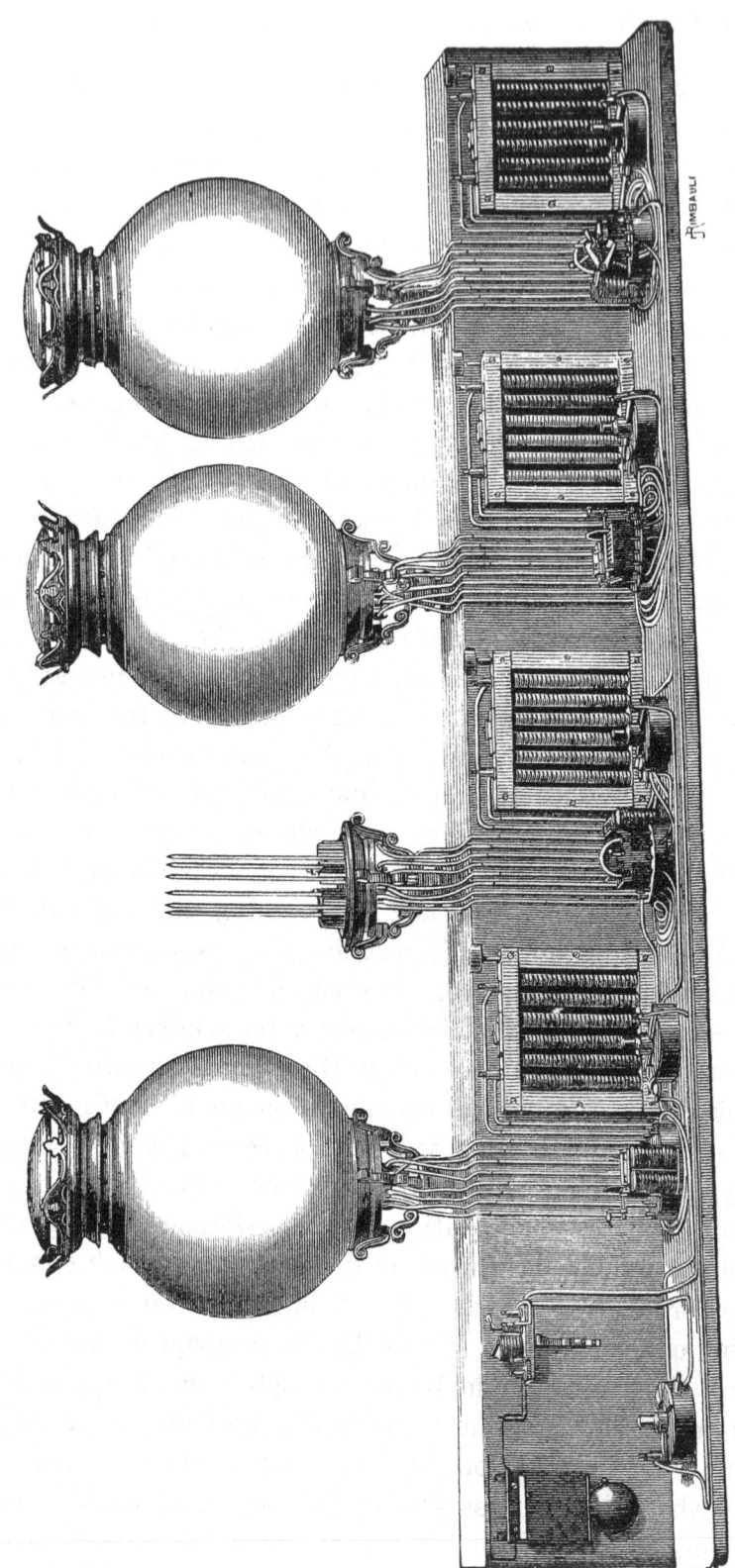

Fig. 576. Jablochkoff Candle with Automatic Bracket.

two ways open to it :—1st, it might pass by the electro-magnet, since the circuit would not have been broken on the fixed part of the clip No. 1 ; or 2nd, it might pass through candle No. 2, but the electro-magnet having a lower resistance than the candle, the current would pass through the magnet, and the candle would not be ignited. The contact R which is in communication with the last compartment of the receiver, is generally employed to throw in a resistance equivalent to that of one candle, and when it is desired to cut out one bracket from the circuit, the 7th point is put into communication with the mercury of the 7th compartment, which corresponds to the required resistance. This operation is affected automatically after the last candle on the bracket is burnt out ; the resistance is made up of coils of galvanised iron wire. Fig. 576 illustrates a complete circuit with automatic brackets.

From the foregoing description it will have been seen that a group, more or less numerous, of Jablochkoff candles can be fixed in brackets of comparatively simple construction. These brackets are supported in a circular frame, the upper part of which is recessed to carry a globe of opal glass which is capped with a coronet and cover plate. A saucer of opal glass rests upon the base of the bracket; it screens the clips and serves to catch the small sparks which are produced when a candle is being lighted. Other lamps are arranged as candelabras on consoles or swing mountings, with the conducting wires enclosed in the supporting arm. An arrangement especially adapted for lighting workshops is shown in Fig. 577. Fig. 578 is a very effective mode of mounting. Here the candle holder is made so as to throw as little shadow as possible, and is fixed to a rod suspended from the ceiling; the globe, which is egg-shaped, is entirely closed below, and the coronet, which forms a finish to its upper part, serves as a balance weight, and rises when the globe is pulled down, to expose the candle holders for the renewal of the carbons. . Fig. 579 shows a holder fitted with candles, as adapted to this arrangement.

Reference may be made here to a special system of contacts for lights placed at a certain height, in such a manner that the apparatus can be drawn down within reach, for cleaning the brackets and renewing the candles. The arrangement for distributing the current consists of two parts; the upper and fixed portion P, Fig. 580, is furnished with a wooden support carrying at the centre a copper rod, corresponding to the return contact of the commutator. The movable portion P^1 is similar in form, and carries at the centre a copper sleeve, into which the fixed tube enters when the candle holder is raised. This copper sleeve is placed into communication with the centre of the bracket by a suitable contact; the outer clips of the bracket are connected with the copper strips $t\ t^1$ by

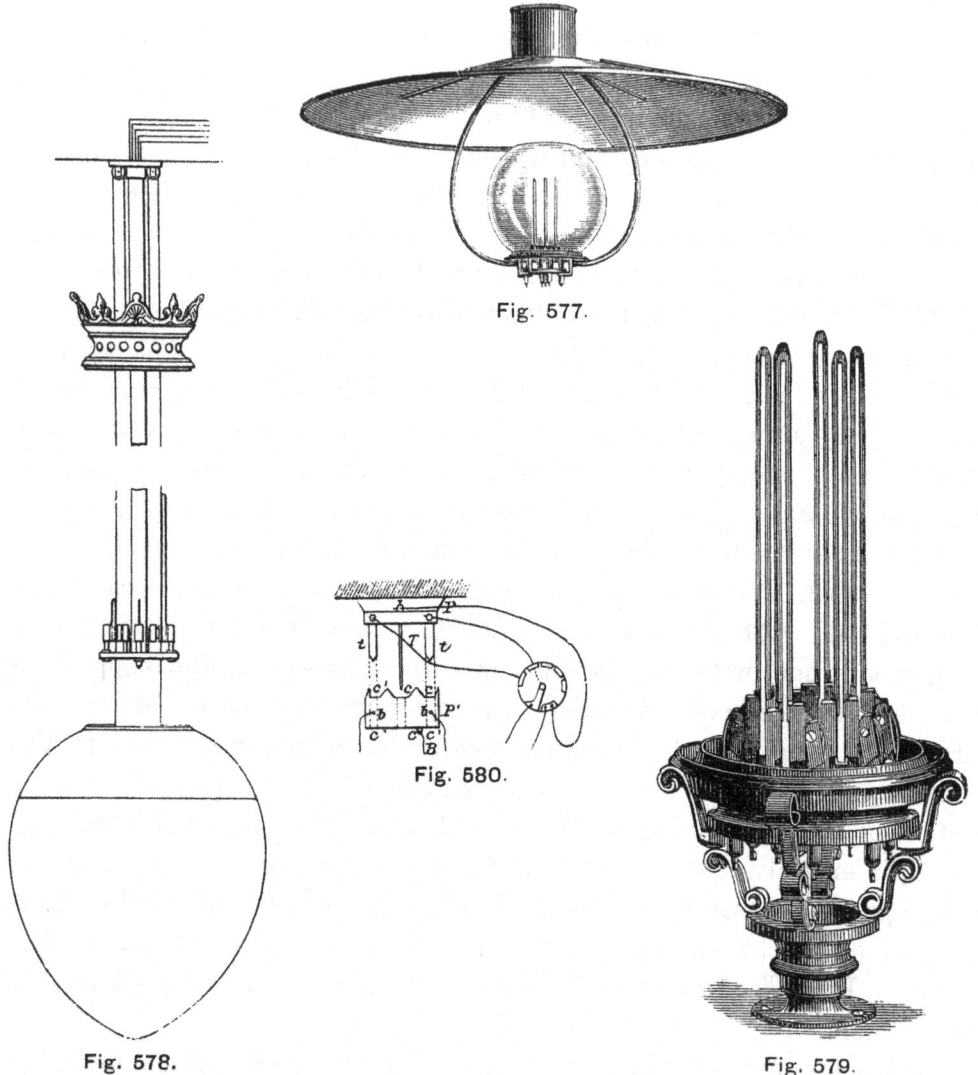

Fig. 577.

Fig. 580.

Fig. 578.

Fig. 579.

means of contacts and the slides $c\ c^1$ placed around the rod P^1, and intended to guide the strips $t\ t^1$. When the apparatus is raised, these strips enter into the slides. The rough sketch annexed explains the working of this very simple apparatus.

The latest and simplest form of holder is that recently devised by

M. Parent, the chief engineer to the Compagnie Générale d'Electricité. The motive power of the apparatus is a spring R composed of two metals expanding unequally, and soldered one to the other. This spring is fixed to one of the clips of the holder by the two screws $v\ v^1$; it is bent round, as shown, the loop pressing against the candle while the end is brought down parallel to the clip. The latter and the various pieces accessory to it, are shown in Figs. 581 to 584, of which the first three are side views, and the last is a horizontal section. As just stated, the spring R is fixed to the clip by the screws $v\ v^1$, the heads of which pass through openings made in the front part of the spring; at the free end of this latter is placed the stud E. When the candle is

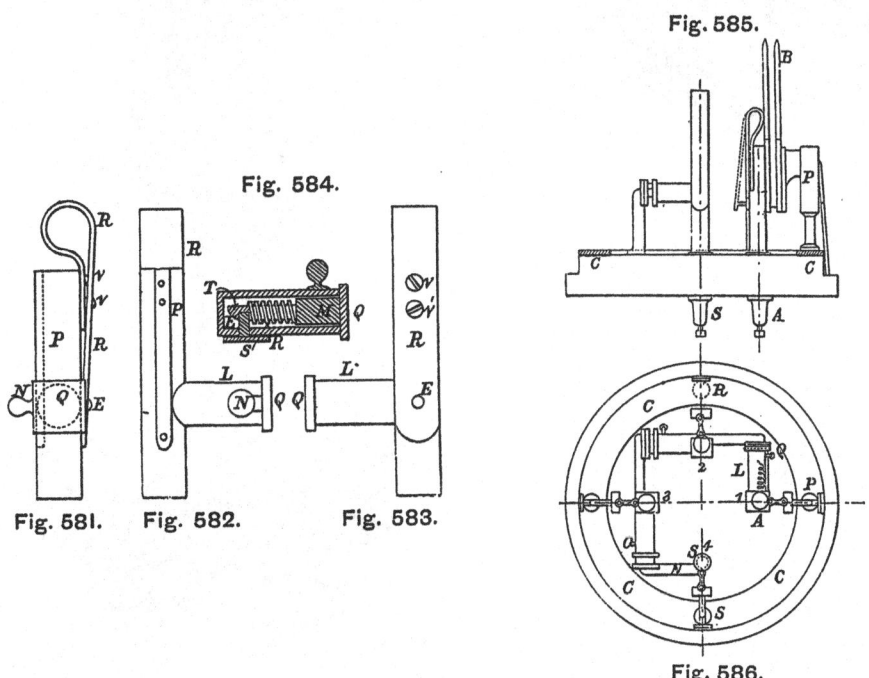

Fig. 585.

Fig. 584.

Fig. 581.   Fig. 582.   Fig. 583.

Fig. 586.

burnt down, the heat acting on the spring causes the latter to open, and the stud E is drawn away from the recess made in the end of the steel pin T that is attached to the head M and the square plate Q; a coiled spring passes over the pin T, as shown. This device is contained within a small tube L, and is fastened at right angles to the holder P. It will be at once seen that as soon as the stud E is withdrawn, the pin T is forced forward by the coiled spring, and contact is made by the square plate Q. The small handle N serves to press back the pin against the coiled spring, until the stud E re-enters the slot and locks it. Figs. 585

and 586 show how this device is employed for a chandelier with a separate commutator. The chandelier carries four candles numbered 1, 2, 3, 4; this number may be doubled, giving 12 to 16 hours of consecutive lighting. In the case illustrated, only three candles, Nos. 1, 2 and 3, are lighted automatically one after the other; No. 4 is a spare candle reserved for emergency. All the outer clips are mounted on the circle

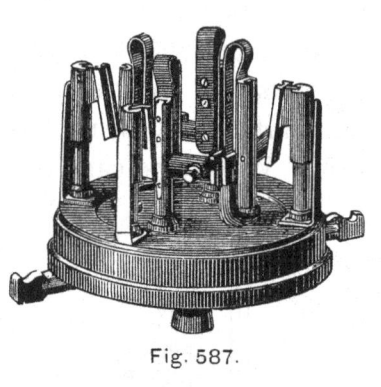

Fig. 587.

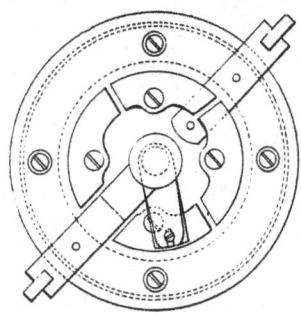

Fig. 588.

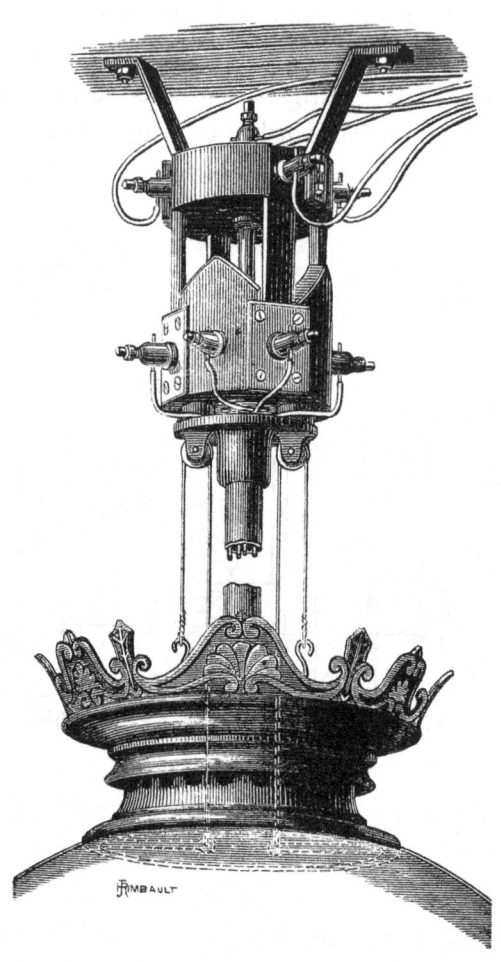

Fig. 589.

C, on which is a terminal R for the return current. The current is led to the first candle by the terminal A; when this is nearly consumed it heats the spring and produces the outward movement of the plate Q as already described. This plate comes into contact with a similar plate carried at the end of an arm fixed to the inner clip of the second candle, to which the current arrives, and finding less resistance passes by the piece L. In the same way it is led to the third and fourth. In the event of any one of the candles, No. 2 for example, becoming extinguished before the spring has been dilated, the attendant, by shifting the commutator key, leads the current by a third wire to the terminal S, which corresponds to the inner clip of the last or reserved candle, and so maintains the continuity of the lighting. The attendant has then to ascertain which of the candles failed to effect the expansion of the spring, and when the reserve candle is burnt out, he restores the key of the commutator to its original position in all the apparatus, and replaces the reserve candle in order to resume the normal working of the system. Figs. 587 and 588 show the arrangement applied to a chandelier having its own commutator. Fig. 589 illustrates the general connection of a balanced chandelier for a group of Jablochkoff candles.

Three other forms of electric candles have to be noticed here: those of Rapieff, Wilde and Jamin.

Rapieff.—Fig. 590 represents the Rapieff electric candle which differs from that devised by M. Jablochkoff in several particulars; in the first place there is no insulating material inserted between the carbon pencils; secondly, the two pencils are not parallel but make with one another a small angle, such that while the resistance of the circuit is diminished by the burning down and consequent shortening of the carbons, it is increased in proportion by the gradually increasing distance between them causing the arc to become longer as the candle becomes shorter. By this means the one is intended to compensate the other in order to obtain a steady light.

The Rapieff candle thus consists of two nearly upright pencils of carbon, the distance of which apart can be regulated by a screw adjustment. The holder of one of the carbons is connected to the armature of an electro-magnet concealed within the stand, and when no current is passing, the upper extremities of the pencils are brought into contact by a spring attached to the armature of the movable carbon. When, however, a current is sent through the apparatus, the armature is attracted and the carbons are separated to the proper distance necessary to produce the electric arc, but the moment that any interruption in the circuit takes place, the armature is released, and the carbons coming. again together the arc and the light are re-established.

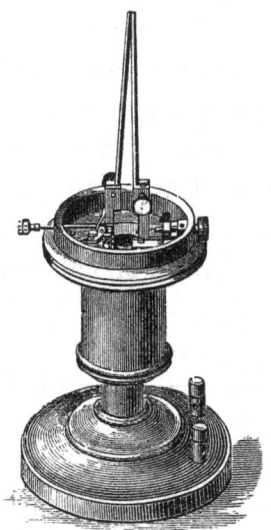

Fig. 590.   Rapieff.

Wilde.—The Wilde candle comprises two carbons, about 22 in. long and 4 mill. in diameter.   These candles are held vertical and parallel to each other in a suitable stand, separated by an air space of 3 or 4 mill. in width.   The holder consists of a tube carrying on top two plates insulated from each other, and with a space between them axial with the tube.   On these plates are placed the terminals for connecting the wires leading from and to the generator.   On each plate is a small vertical standard, enlarged at the top with a projecting block, on the inner face of which a V-shaped grove is formed.   To the outer face of each block is fastened a ⊃-shaped plate spring, one leg being fastened to the block, the other carrying a free running grooved pulley, the axis of the groove corresponding with that of the V groove before mentioned.   The carbons are held between these grooves and the pulleys.   One of the brackets carrying the carbons is somewhat different to the other, being free to have a slight oscillating motion, controlled by the armature of two electro-magnets mounted on the plate carrying the positive terminal.   This con-stitutes the automatic device of the apparatus.   So long as the candle continues to burn, the current passing through the electro retains the armature, and consequently holds the carbon vertical in the articulated holder; but if the candle ceases to burn, the electro becomes inactive, the carbon falls forward against the other carbon, and the light is restored if the circuit is complete; then the electro-magnet is once more excited and the armature with the carbon are separated. A simple mechanical device is introduced in this holder for raising the carbons when they are burnt down.

Jamin.—This system, which at one time seemed to promise highly satisfactory results, does not appear to realise the expectations of its inventor, and has not, in this country at all events, found any successfully prolonged attention. The essential feature of the Jamin system is that the candle is held in the middle of a flat coil or loop composed of about 40 convolutions of insulated wire, forming part of the main circuit. The effect of the powerful currents passing through this coil is to deflect the movable part of the circuit, that is to say the arc. In one of the latest forms, Jamin mounts three pairs of carbons within the coil, and means are provided for establishing the arc between a second pair of carbons when the first are consumed. The device applied to each pair of carbons for starting the arc is much the same as that of Wilde, one of the carbons being movable, and controlled in its motion by the armature of a magnet. The current passes through that pair of carbons which offer the least resistance, somewhat as in the proposal made and experimented upon with the Jablochkoff candle.

XI.

INCANDESCENCE-ARC LAMPS.

THE electric lamps to be considered under this head belong to the category of incandescence lamps burning in air. From the point of view of quantity of light obtained, they occupy a position intermediate between arc lamps, electric candles, and incandescence lamps proper, which have of late been so largely developed, and which are illustrated by the Swan, Edison, Maxim, and Lane-Fox systems. In incandescence lamps burning in air, the production of light results from the passage of the electric current through a rod of carbon of a diameter so small that its extremity becomes heated nearly to whiteness. The idea of obtaining light in this way dates back for a great many years, and in the early days of electric lighting, when the source of power was found in the battery current, numerous variations in its application were proposed, to which we will chronologically refer.

Greener and Staite.—The first lamp of this kind of which we have any record, is one patented by Greener and Staite, the latter of whom appears to have been a most enthusiastic worker in electric illumination. Between the years 1846 and 1849 he took out five patents for lamps, some of which show considerable ingenuity and skill, and had he possessed a cheap and efficient generator of electricity, there is little doubt, that he would have attained very considerable success. In the lamp in question, prisms or cylinders of carbon were enclosed in air-tight transparent vessels and rendered luminous by a current of electricity passing through them. They were divided on the surface into numerous acute points, either by being formed in a suitable mould or by being grooved with a fine saw. Two such prisms or cylinders were placed together, or a hollow cylinder was inserted into two hollow cones of platinum placed base to base, which were connected respectively to the positive and negative poles of the battery or magneto-electric generator which produced the current. In the year 1846, Staite patented five lamps, all more or less of the class we are considering. He appears to have been experimenting towards the production of an arc lamp; indeed in each case his drawings show a small arc between the two electrodes, though it is at the same time evident that a great part of the illumination would be due to incandescence. Fig. 591 shows the simplest of his designs, and one that with slight alterations has reappeared under different names several times in the last few years. The pencil F

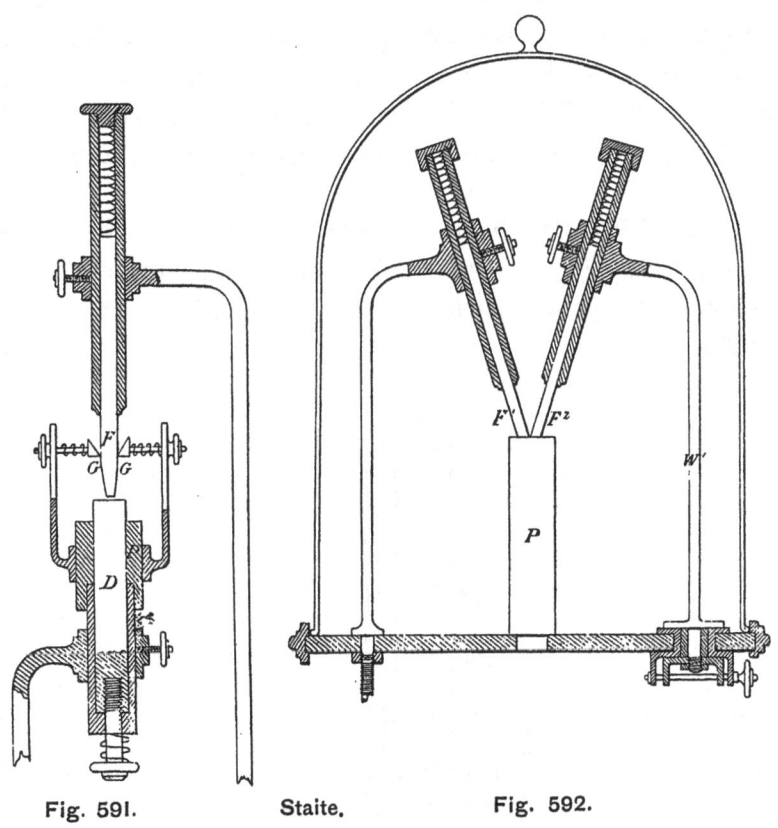

Fig. 591.   Staite.   Fig. 592.

is pushed by a spiral spring towards the carbon cylinder D, and is guided by two insulated platinum prisms G G, which would appear to be intended to act after the manner of Siemens' abutment pole. The lamp is enclosed in an air-tight glass vessel provided with a valve for the escape of gas. The specification drawings also show such a lamp inverted and with the guides G G in electrical connection with the same pole of the generator as the pencil F, so that the current is principally confined to the end of the pencil.

Fig. 592 shows a somewhat different lamp, although of the same type. The two pencils F¹ F² slide through copper cases connected to the generator, and abut at an angle to each other, on the non-conducting cylinder P, which is of compressed pipe-clay, phosphate of lime or other similar refractory material. The screw at the bottom of the case serves to adjust the positions of the electrodes and the length of the arc. The similarity of this arrangement to the Soleil will be at once apparent.

Staite included two other lamps in his specification, one of which had a platinum electrode, but he appears to have abandoned the incandescence method at this time and confined himself to arc lamps, with one exception, in which a strip of iridium, or of iridium alloyed with platinum or copper, was heated to incandescence in a vacuum.

Harrison.—In 1857, Harrison brought out the incandescence-arc lamp illustrated by Fig. 593, consisting of a carbon disc driven by clock-work and having a thin pencil carbon resting vertically on the disc, and in a line with its axis.

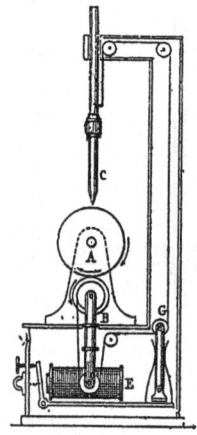

Fig. 593. Harrison.

Shepard.—During the next 10 years the attention of inventors appears to have been directed mainly to arc lamps, but in 1858, E. C. Shepard patented on behalf of some inventor, a method of incandescence in air, and brought out a lamp in which the two electrodes burned in contact. The upper one was loaded and its position was determined and maintained by its point being passed through a guide rather less in diameter than itself. Through this guide, which appears to have corresponded to what is now known as an "abutment pole," the pencil could only be fed as its diameter was reduced under the wasting action of the incandescence. The lower electrode was gently and constantly pressed towards the upper one by a float immersed in mercury. This lamp is illustrated in Fig. 594. A A^1 are two vessels of water designed to keep the apparatus cool; *c* is the upper carbon with its point projecting through a hole in the guide plate *c*, and with its opposite end fitted in a hollow holder filled with mercury, to act both as a weight and a movable

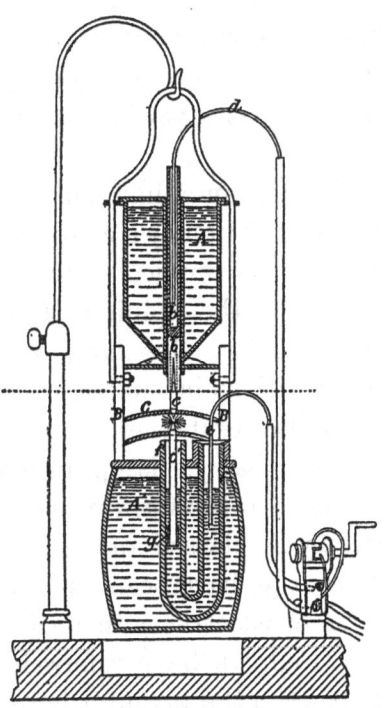

Fig. 594. Shepard.

contact for the leading wire *d*; *c*^{1} is the lower carbon with its float *g* within the vessel of mercury *f*.

In the interval between the two last-named patents, however, a new system of illumination, in which a stream of mercury formed one of the electrodes, appears to have excited considerable interest, and to have monopolised the attention of inventors for two or three years, and quite a number of patents were obtained for devices of this nature; these are all described at the end of this volume. As may be supposed, this system of lighting had no practical results, and was soon abandoned.

After the patent taken by Shepard in 1858, there followed a period of inactivity, but in 1872, the dawn of a new era in electrical research, the problem of carbon incandescence lamps was studied afresh, and especially by several Russian inventors, Lodiguine, Kosloff, Konn and Bouliguine.

Konn.—In that year Konn patented Lodiguine's lamp in England, and described it in his specification as comprising stems of graphite or carbon enclosed in a species of lantern hermetically closed and filled with nitrogen or other gas that does not support combustion. One very neat form of this lamp was provided with a bent or angular conductor.

The Lodiguine lamp imported into France by Mr. Kosloff was an apparatus working *in vacuo*, and does not appear to have had any practical success.

Another form of Bouliguine lamp had only one carbon, which was moved in proportion to its consumption by a counterweight and mechanism of electro-magnets, somewhat too complicated for the efficient working of the apparatus At least this is the opinion of M. Fontaine, the well-known electrical engineer, who had this lamp in his hands for a long time, and who experimented with it in sufficient detail to enable him to form an authoritative and unfavourable opinion. In this class may be mentioned the recent American Sawyer lamp, in which the carbon burns in nitrogen. This, at all events in the form in which it was made known to the public, must apparently be classed among the long list of practical failures, since it does not seem to have been used in practice, and it was not shown at the Exhibition of Electricity in Paris.

In all these lamps to which we have referred, and in others of the same class, the principal object of the inventor is to check combustion, in order to increase the duration of the carbon. In most cases this advantage, important indeed, but to be gained at too great a cost, is realised by a complication of apparatus; moreover, the combustion of carbons in air is accompanied by a more considerable production of light, and from this it results, as will be seen further on, that in incandescence lamps burning in air, the ratio of light produced to power expended, is higher than in the vacuum lamps. In these latter the duration of the carbon pencil or filament is of course indefinitely greater; but it must be remembered that if less carbon is burnt in the lamp, a very much larger quantity is burnt in the firegrate of the boiler producing the steam that supplies the engine driving the electrical generator.

Varley.—The first incandescence lamp burning in the open air, of which we have any record since the revival just referred to, dates from 1876, and is mentioned in a patent taken by Mr. Varley for a dynamo-electric machine. It consisted, as is shown in the annexed diagram, Fig. 594, of a disc of carbon A on which rested a pencil B mounted at the lower end of the lever C. This lamp never appears to have been employed.

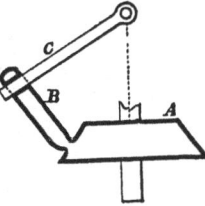

Fig. 595. Varley.

Reynier.—This idea, perhaps, served as the base of later lamps, and, at all events, was revived in a practical form by M. Reynier, who described it on the 13th of May, 1878, in a note presented to the Academy of Sciences at Paris by the Comte Du Moncel. The principle of the Reynier lamp is indicated in the annexed diagrams, Figs. 596 and 597, from

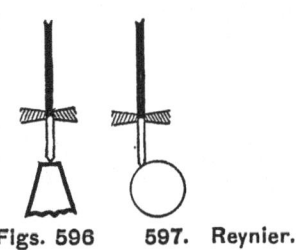

Figs. 596   597. Reynier.

which it will be seen to embody nothing new, the former being—not to speak of earlier arrangements—similar to that of Varley just mentioned, and the second based on the ingenious arrangement of Harrison.

In Fig. 596 it will be seen that the stick of carbon falling by its own weight abuts upon a carbon block. It is maintained by two lateral contacts, just as in Staite's lamp, which only put in the circuit a portion of the carbon, and thus limited the length of incandescence. In the second arrangement, Fig. 597, which is an improvement on the former, the carbon bears eccentrically against the periphery of a little disc of carbon, free to turn around its axis. In the Harrison lamp of 1857 the carbon disc, as we have seen, was driven by a clock train and the vertical carbon was placed in the same line as its axis.

N N 2

In Reynier's it will be noticed that it is the tangential component of the weight of the pencil that produces the movement of the disc. By this arrangement is avoided any accumulation of ash, which is a bad conductor and interrupts the passage of the current, or at least renders it intermittent if care be not taken to remove it as fast as it is produced. Figs. 598 and 599 illustrate complete lamps of this description. This

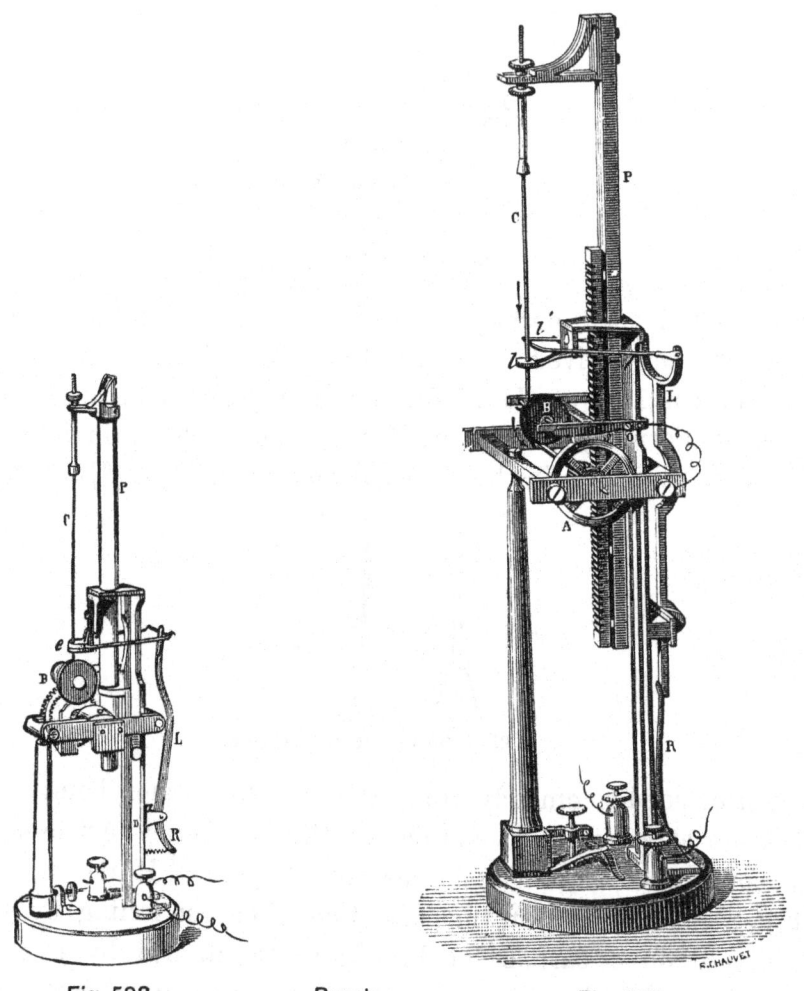

Fig. 598.   Reynier.   Fig. 599.

arrangement has been employed in the construction of a type of lamp lately experimented upon by MM. Sautter and Lemonnier, and which gave the following results: The carbons were 2 millimetres in diameter, and were made by M. Carré. The direct-current Gramme machine that furnished the current was driven with a speed of 920 to

930 revolutions per minute. The conductors between the generator and the lamp were formed of 100 metres of copper wire 3 millimetres in diameter.

The length of the incandescent portion of the carbon was 5 or 6 millimetres.

No. of Lamps in Circuit.	Luminous Intensity, Each Lamp. Carcels.	Carcels. Total.
5	15	75
6	13	78
7	10	70
8	5	50

Under similar conditions a Serrin arc lamp would have given an intensity of 320 Carcels. If a comparison be made between the luminous intensity and the motive force expended, which was about three horse-power, it will be seen that the intensity per horse-power was for the four cases given above, respectively 25, 26, 23 and 16 Carcels. In the case of a Jablochkoff candle it would have been 50 Carcels, and with a Serrin lamp, 106 Carcels. From the foregoing it will be seen that the lighting power of the Reynier lamp is very low, but it may be conceded that this inferiority is compensated for, to a certain extent, by the quality of divisibility that is secured, and an extreme steadiness indispensable for many purposes. We shall see further on how the results obtained by these trials have been modified in subsequent experiments.

Werdermann.—On June 21, 1878, a German inventor, Mr. Werdermann, resident in London, patented a lamp very similar to that of Reynier. The first published description of this apparatus appears in the *Comptes Rendus* of the Academy of Sciences of Paris; it was presented on November 18, 1871, by the Comte Du Moncel, in a note written by the inventor, which is worth reproduction, although as a matter of course the author, and not ourselves, is responsible for the statements it contains. This system, said Mr. Werdermann in his description, is based upon the effect of incandescence of a carbon heated to reddish whiteness. It is so disposed that the electric generator being suitably arranged, it is possible to light simultaneously a certain number of lamps by a simple derivation of the current. It consists essentially of a small carbon pencil, Fig. 600, free to move within a metal tube, which serves for it as a guide, and at the same time as a current commutator. A collar fitted to the lower part connects it by two cords, which pass from the tube by two grooves, and over two pulleys to a counterweight, which tends continually to raise the carbon, and to cause it to press lightly against a large carbon disc about 2 in. in diameter, kept in a fixed position by a vertical support. This support is fastened to a sort of funnel-shaped envelope which receives the ashes from combustion, and facilitates the fitting of a glass globe to the lamp. The upper carbon disc is placed in connection with the negative pole of the electric generator, and the metallic guide of the carbon pencil corresponds to the positive pole in such a way that only that part of the carbon pencil comprised between the metal tube serving as a support, and

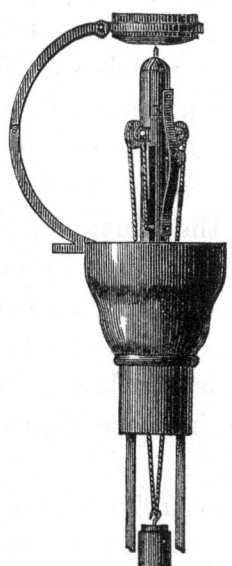

Fig. 600. Werdermann.

the carbon disc, is brought to a state of incandescence; this is a length of about ·75 in. This incandescence is increased by the action of a small voltaic arc, which is produced at the point of contact of the two carbons. The upper block, on account of its larger mass, does not burn, and indeed undergoes no alteration. The action of the counterweight is regulated by means of a spring, furnished with a regulating screw, which presses more or less on that part of the tube guiding the carbon pencil, and so acts as a brake. Experiments made with a Gramme machine of the galvanoplastic type, and driven by a two horse-power portable engine, gave the following results :—

1. When the current was divided between two lamps, the intensity of the light was equal to 360 candles, or about 38 Carcels. This light was white, and seemed quite free of the blue and red rays seen so often with the voltaic arc ; the light, moreover, was absolutely constant.

2. In connecting on the circuit 10 derivations, each corresponding to one lamp (see Fig. 601), 10 luminous sources were obtained, each of about 40 candles. To secure uniform action, a coil of feeble resistance was interposed in each derivation as shown at *a*. Under these conditions the resistance of each lamp was ·392 Ohm.

3. The consumption of the carbons of the small pattern lamp did not exceed 2 in. per hour, and for those of the large size · it reached scarcely 3 in. in the same time. Carbon pencils 1 metre in length could be employed in the lamp ; those used were manufactured by M. Carré. With this system all the lamps could be lighted or extinguished simultaneously or successively, and as their luminous intensity was moderate, transparent globes could be employed instead of those of ground glass.

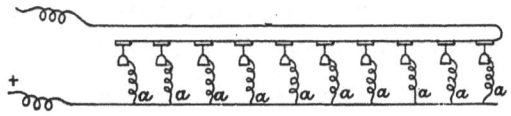

Fig. 601.

So far Mr. Werdermann, and as a consequence of reading this communication to the Academy, M. Reynier protested, in a note addressed to the Academy, claiming the fundamental principle of the Werdermann lamp as his own particular property. This claim of priority was disputed by Mr. Werdermann, who replied to M. Reynier by the following note : " The lamp is not based on the effects of the incandescence of a carbon heated to whiteness. It has for a fundamental principle the production of a voltaic arc infinitely small, and the incandescence of a short length of the electrode is solely the inevitable consequence of the voltaic arc itself." This explanation is clearly applicable indifferently to both systems, and the discussion would probably have terminated in the law courts, if the two inventors had not arrived at the far wiser determination to forget their differences and join together in placing the system claimed by each in the hands of a syndicate that was formed in Paris, and of which the engineer was M. Napoli ;

this gentleman saw the way to adapt from each apparatus its most valuable qualities, and to create the various practical and new types under which the Werdermann-Reynier lamps were made familiar at the Paris Exhibition of Electricity

The various patents belong actually to the Compagnie Générale d'Electricité, which works under one administration, the Jablochkoff candle, the Jamin and the Jaspar lamps, the Werdermann and Reynier systems, and the Maxim incandescence lamp. The numerous experiments made with the Werdermann and Reynier lamps by the Compagnie Générale have been very complete, and are of considerable interest, the more so that they have never yet been published; a summary of the experience obtained will therefore find a fitting place here. The Compagnie Générale owns the patents for the alternating-current Gramme generator, and not possessing any type of direct-current machine, it has been obliged to adapt its various lamps to suit alternating currents. An advantage arising from this system of working is that only one type of generator is necessary, and there are even special cases where one machine can feed at the same time several different classes of lamp. It must be borne in mind that when alternating-current machines are employed for feeding incandescence lamps, they must have the lowest possible internal resistance, to diminish heating, and to transfer it where it will be useful, that is to the carbon filament within the lamp. With this object the alternating-current auto-exciting machines have been made (type No. 2), which usually supply from 16 to 20 Jablochkoff candles, in which the diameter of all the wires has been increased about one-third. The ring is divided into 12 sections, of which six grouped in quantity, form a circuit. able to burn 12 Napoli lamps, or 24 such lamps for the whole ring.

Under these conditions, the speed being 1,450 revolutions per minute, the luminous intensity reached 25 Carcels, and it is probable that the number of lamps could be brought up to 32, but at the expense of the intensity which would be reduced to 15 or 20 Carcels only. With the auto-exciting machine, type No. 1, which is able to feed from 8 to 10 candles, and modified as above described, 16 Napoli lamps of from 15 to 20 Carcels can be worked in circuit, the speed being 1,400 revolutions per minute, and the force absorbed 10 horse-power. The following table shows in a convenient form the data connected with experiments upon the Napoli lamp :—

Electric Generator.	Revolutions.	Number of Centre Lamps.	Intensity.	Power Absorbed.	
				Total.	Per Lamp.
Gramme auto-exciting alternating current. Type 2	1,450	24	Carcels. 25	22	·9
Gramme auto-exciting alternating current. Type 2	1,450	32	15 to 20	22	·65
Gramme auto-exciting alternating current. Type 1	1,400	16	15 to 20	10	·62

The Werdermann-Reynier-Napoli lamps are placed in the circuit, with the positive pole (in the case of a continuous current) connected to the carbon of the first lamp, and the carbon of the second lamp connected with the abutment block of the first lamp. We annex a diagram which clearly represents the arrangement. In this diagram Fig. 602, A is the ring of the auto-exciting Gramme generator divided into

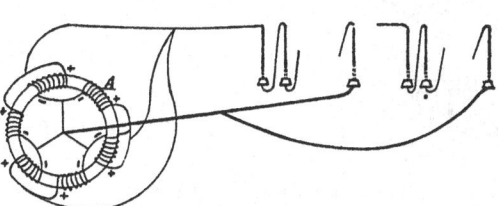

Fig. 602.

two circuits, and having a common return wire. The signs + and — in this sketch do not possess their ordinary values, but indicate the ingress and egress of the wires according to their mode of winding. The carbon pencils employed in the lamps of the latest type are 4 millimetres in diameter; the Reynier lamps are also worked with carbons of 2 or 3 millimetres, but the light is very small, falling to from 4 to 6 Carcels; and, moreover, the frequent breaking of the carbons renders the lighting very unstable. The first Werdermann lamps had much larger carbons, giving a powerful light, but at the expense of a general and dangerous heating of the lamp.

In a circuit of the Werdermann-Reynier-Napoli lamps, the chances of extinction are due to two causes, either the complete consumption of the carbon, or some stoppage in the travel of the carbon pencil, which may produce a voltaic arc, that, from the absence of tension in the generator, cannot be maintained.   The first of these dangers can be easily avoided, by making use of a carbon of sufficient length, and it will be seen further on that all the different patterns are constructed of such a capacity as to secure quite a prolonged period of lighting.   The second danger may be avoided by the use of automatic relays, which cut the extinguished lamp out of circuit.   While upon this part of the subject, we may refer to a system for relighting which forms an integral part of the system of one of the types of lamps devised by M. Napoli, and we annex sketches of another apparatus for the same purpose by M. Reynier; Fig. 603 being a general view, and Fig. 604 a diagram showing the mode

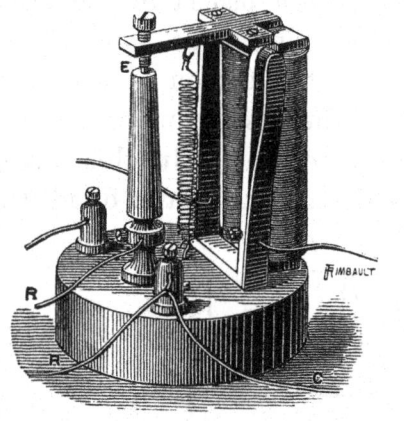

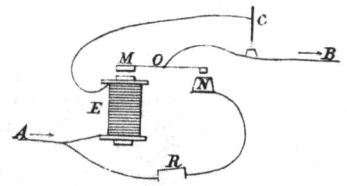

Fig. 603.   Reynier's Relighting Apparatus.   Fig. 604.

of working.   In the circuit A B, feeding the lamp C, an electro-magnet E is interposed.   When the lamp is burning, this magnet attracts the movable key M, which turns around the pivot O.   If the lamp is extinguished, the current ceasing to pass, the electro releases the armature M, and it falls back from the action of the weight, which falls upon the contact N, the lamp C is thrown out of the general circuit, and the resistance it previously exerted is replaced by an auxiliary resistance R, consisting of carbons, wires, or even a second lamp identical with the one extinguished, and which is lighted as the other goes out, and consequently prevents an extinction from reducing the intensity of the light. In this way each lamp in a circuit can be accompanied with an automatic acting duplicate.

M. Reynier has also devised a small auxiliary regulator (noticed on a previous page) to replace the resistance R, and which will burn during the time necessary to replace a new carbon in the lamp that has become extinguished. It has been said above that M. Napoli has largely modified the Werdermann and Reynier lamps, and has created several new types. We will consider those which are actually being employed by the Compagnie Générale d'Electricité. Fig. 605 represents the first of these types. It consists of a plate carrying the two terminals E and F, the first of

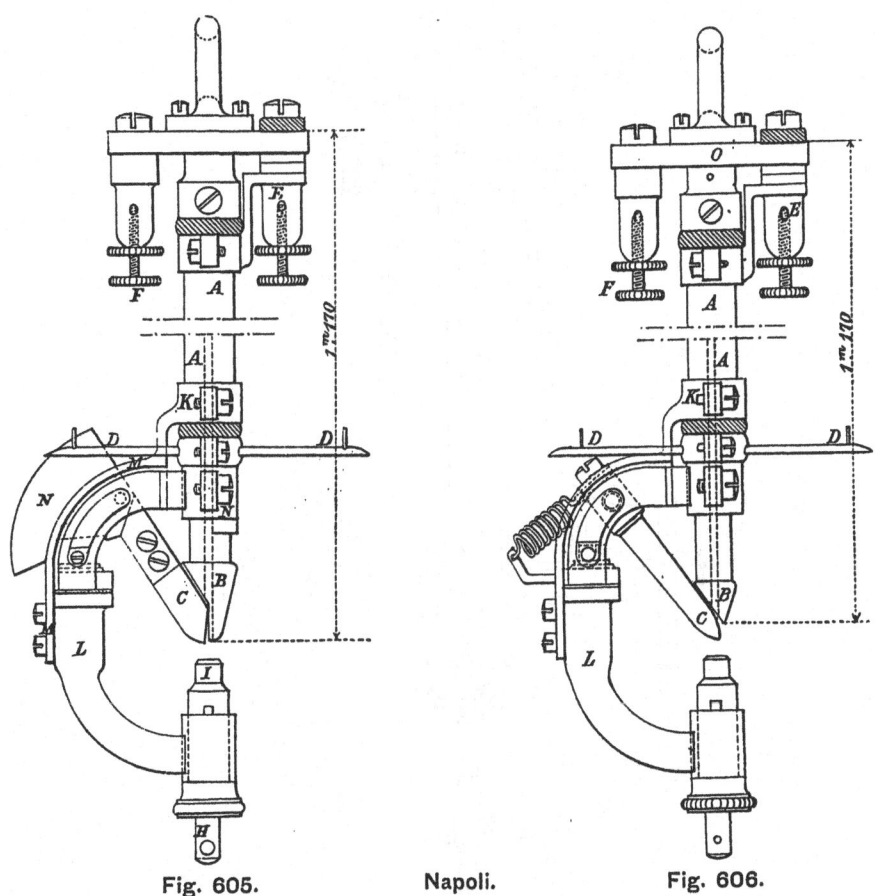

Fig. 605. Napoli. **Fig. 606.**

which, insulated electrically from the plate, is in connection with the tube A, and thence by the piece K, and the square piece M M with the abutment block I made of red copper. The terminal F communicates with a sliding counterweight in the tube A, that rests on the upper end of the carbon to which it passes the current. Sometimes the spring is replaced by a counterweight N. The carbon is brought to incandescence by the flow of the current between the jaw B—C and the abutment piece I; by means of the screw H, the carbon can be set clear of this abutment at will. The apparatus is made of brass, the pieces B, C, and I are of red copper, and they can be easily replaced by others when they are worn. Similar general arrangements will be seen in Fig. 606, which is a variation of Fig. 605, in which the counterweight that produces the locking of the carbon is replaced by a coiled spring; the

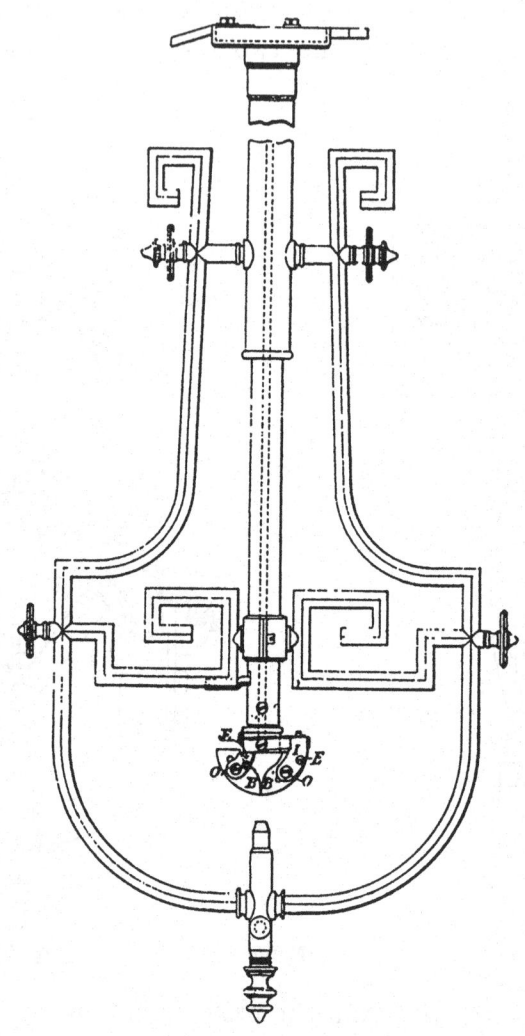

Fig. 607. Napoli.

shape of the movable jaw is also modified. Fig. 607 is another variation of the lamp, in which the current is transmitted to the abutment block by the framework of the lamp. The jaws are two pieces of red copper, B B[1], free to oscillate on the pins O and O[1], and locking the carbon pencil by their weight. The incandescent length can be regulated by approaching or removing the block with the adjusting screw at the bottom of the frame. Fig. 608 shows another variety, the object here being to avoid any danger from extinction. In this, the abutment block I is movable around the axis O, and is supported by a frame which can turn upon the centre L. The position of the jaws B is regulated by the screw V, connected to the mechanism *a, b, c, f,* which when the block I comes in contact with the jaw, bears on the square *d* by the screw *f,* and thus closes the circuit of the lamp. Now the

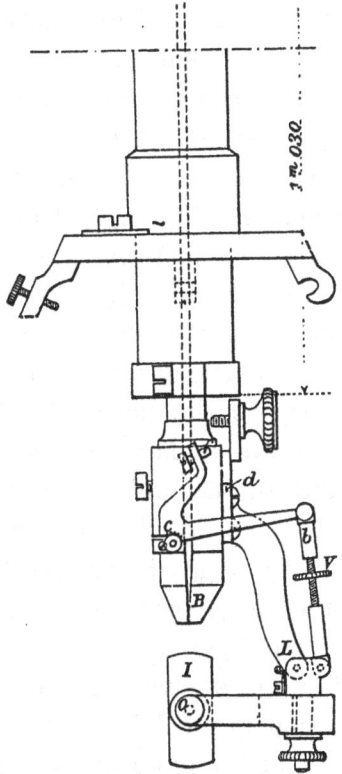

Fig. 608. Napoli.

whole apparatus is so balanced that the abutment block has always a tendency to turn into contact with the jaws; as long as the lamp burns it is prevented from doing so by the weight of the carbon, but when this latter is burnt out the lamp is thrown out of balance, and the movement of the block extinguishes the light.

These various systems of lamps have been largely used by M. Napoli, and adapt themselves well to decorative purposes. It will be remembered that they were employed with a pleasing effect at the Paris Electrical

Exhibition, where they lighted the peristyle of the entrance, the stage of the little theatre, the room in front of the theatre and the salon of the President of the Republic. In these installations the lamps approximated to the earlier forms used by Mr. Werdermann, that is to say, the carbon pencil was raised by a system of counterweights sliding within a cylinder. The pencils were placed within a tube of sufficient length, which formed a tail, cleverly made use of where the lamps were held by statues, placed on brackets, &c. At the Grand Opera, Paris, and in the salon of the Athenæum, where the first experiments were made, lamps of similar form were used.

The light given by the Werdermann-Reynier-Napoli lamp is of great steadiness, except occasionally for a few moments when, in consequence of any abrupt descent of the carbon, there is a rapid eclipse, generally accompanied by a minute detonation. The light is less harsh than that of the voltaic arc, and it has a yellow tint that naturally varies with the degree of incandescence reached by the carbon. This can be regulated by passing the current through a resistance commutator, by means of which very variable intensities can be obtained, an arrangement of much value in certain cases, particularly for the footlights of theatres. However, for this latter purpose the light is not sufficiently diffused, and incandescence lamps—such as Maxim, Swan or Edison—afford a much more satisfactory solution, although at a higher cost. Despite its special qualities, the light given by the Werdermann-Reynier-Napoli lamp has but a restricted use; a great obstacle to the development of the system in workshops and large spaces is the necessity of employing conductors of short length and large diameter, and of which the cost is relatively high. In installations of average importance, the conductor is a rope of 35 strands of copper of 1·14 millimetres diameter, whilst in installations of the Jablochkoff candle or the voltaic arc, the current can be led for a couple of miles by conductors formed of seven strands of 1·14 millimetres.

Ducretet.—Here we may notice some special arrangements of incandescence-arc lamps, which have been proposed, but which have scarcely gone outside laboratory experiment. The first to be named is the lamp schemed by M. Ducretet, presented to the Academy of Sciences at Paris, on December 30th, 1878, in a note, of which the following is an abstract: The principal feature of this lamp consists in the revival of the old idea of using a column of mercury in which carbon pencils are immersed. The difference in density produces a thrust which constantly and regularly pushes the carbon against an abutment in proportion to its consumption, the upper part of the carbons becomes incandescent. The resistance introduced into this circuit remains constant, whatever may be the length of the carbons and their consumption; the portion immersed in the mercury is cut out of the circuit, the length of the part not immersed remains constant. Fig. 609 will explain this arrangement without any further description.

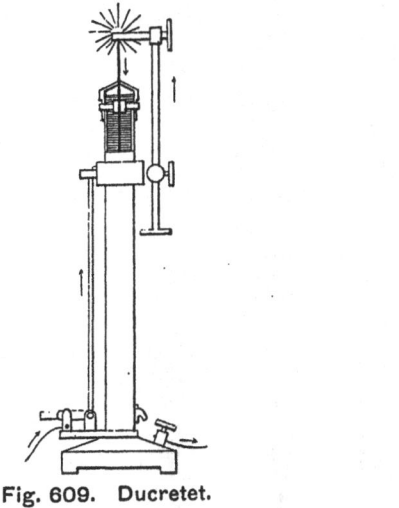

Fig. 609. Ducretet.

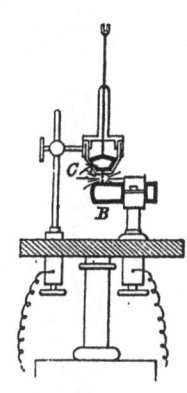

Fig. 610. Clamond.

It should be added that a similar arrangement figures in a patent by M. Reynier, dated November 18th, 1878, and that the inventor deposited a claim of priority with the Academy of Sciences. But this patent not having been made public at the time when M. Ducretet showed his lamp, it was decided that this was another of the very common examples of similar ideas occurring to different minds at the same time; as we have already seen, the idea was then 30 years old. Fig. 610 shows a lamp schemed by M. Clamond, and made by M. Ducretet; the carbon A bears by its weight on the carbon block B, after traversing the little chamber *c* containing mercury, by which the current arrives at the carbon.

Joel.—The Joel incandescence lamp was one of the very numerous systems that was called upon to do useful work at the Paris Exhibition of Electricity, and it must be admitted that, from some cause or other, the results obtained there were not satisfactory; at the more recent Crystal Palace Exhibition, however, there was a marked improvement. One form of the Joel lamp (only the mechanical details in this system are claimed as novel) is illustrated by Figs. 611 to 615. The general character of the lamp is shown both in Figs. 611 and 612, a section and side elevation respectively, to be that of a carbon pencil rising from below into contact with a carbon block held in a bracket attached to the side of the lamp. The light produced, like

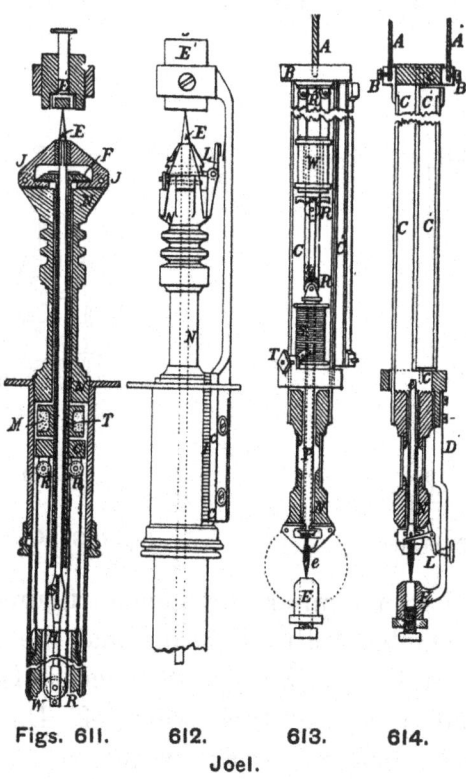

Figs. 611.   612.   613.   614.

Joel.

that in the Werdermann lamp, is partly due to the arc and partly to incandescence. At the top of the body of the lamp N are hinged two clips J J of such a form that when closed as shown in Fig. 611, the faces are parallel and nearly in contact. Passing through the body N, is the tube T terminating above the top of N with a pivotted flange F, and extending below into the base of the lamp. Near the bottom is the collar C to which are attached the two pulleys R R. A cord passes around these two pulleys as shown, both ends being fastened to the annular weight W that slides within the base of the lamp, while the cord also passes round the pulley R which is placed at the bottom of the carbon holder H. The upper part of this holder is formed into a split clip or socket S to receive the end of the lower carbon E. The jaws J K at the top of the tube T are pivotted as shown in Fig. 611, and from the arrangement described it will be seen that the weight W has a two-fold action; it forces the carbon E upwards and at the same time restrains it, and makes a good contact by the pressure it exerts on the jaws F. By varying the length of the inclined part of the jaws, in relation to the horizontal arm, the upward pressure of the carbon can be accurately regulated, and as the flange F is pivotted, any irregularity in the carbon causes the flange to tilt, and throws part of the pressure on to the standard N, thus relieving the carbon. As soon as the lower carbon has burnt out, the clip-holder H, which will then be at the top of the lamp, makes contact with the bell-crank L, and short-circuits the lamp. The upper electrode E^1 is preferably of graphite held in copper clips attached to a bracket on the lamp.

This was the original construction of the Joel lamp. An addition was, however, made subsequently, which is shown in Fig. 611. The tube T is placed within an electro-magnet M coiled with fine wire, fixed in the position indicated, immediately above the collar C, to which the rollers R carrying the annular weight W are attached, and form an armature to the magnet. The action of this coil and armature is such that should the carbon pencil E not be in contact with the upper electrode E^1 the whole of the current will pass through M, which will attract the armature C, and by lifting the weight W will relieve the carbon from the pressure of the jaws K J. The whole lifting action of the weight is then exerted upon the carbon E, which is thus forced into contact; when the current ceases to pass through the electro-magnet M, the armature C falls, and the jaws J are again pressed against the carbon as before. Figs. 613 and 614 show a modification of the Joel lamp in which the fixed carbon block is placed beneath instead of on top. Here the weight W is suspended from the pulleys R^2 and the block B, and the weight acting through the cords W^1 presses down the carbon e, and at the same time through the flanged tube P closes the jaws so as to oblige them to make a good contact with the carbon. The body of this lamp is composed of a split tube, one-half of which, C, is secured to the piece N at the bottom, and to the block B at the top; the other half, C^1, is connected to the bracket D, which carries the bottom electrode E. By

this means each half of the tube is utilised as a conductor. The half C ,
which is hinged to the other, is held close by the button and catch T,
which also makes additional contact. If the body of the lamp be opened,
the piece T must be turned, which makes a contact with the half C of
the tube, short-circuiting the lamp and enabling the carbon to be renewed
with safety. At L, Fig. 614, there is an adjustable short-circuiting

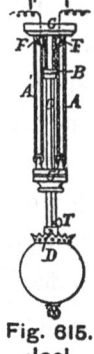

Fig. 615.
Joel.

contact actuated by the magnet S and the armature S¹ when
the former has been excited as described. Or it may be used
to make contact and short-circuit the lamp when the carbon
has been consumed, by the action of the enlarged socket of
the carbon holder. The conductors A and A¹ are attached to
the lamp by the screws B and B¹, which connect it to the
portions of the tube C and C¹. The lamp may be suspended
as shown in Fig. 615, in which the weight G¹ acts as an
ordinary chandelier balance. The copper wires A A¹ are
connected to the ceiling plate G, pass down to the counterweight, and then over the pulleys F F to the top of the lamp. By this
means the contact is always maintained.

Soleil.—As we have seen, almost from the commencement of invention in electric lamps, attention has been devoted to the employment of
refractory materials in connection with the voltaic arc, and their utilisation in increasing the power of the light. One of the earliest of the
constant lamps—that of Staite in 1848—was based on this idea. It
was formed, as we have shown, of two carbons which, fed forward by two
spiral springs, rested on a block of chalk or magnesia, that was brought
to incandescence on the passage of the current. Twenty years later a
French physicist, M. Leroux, professor at the Ecole Polytechnique,
carried out a series of experiments upon the behaviour of refractory
substances in the voltaic arc. This was the period when Tessié du
Motay was working at the oxyhydrogen light, and was exhibiting it
publicly with great success. M. Leroux employed for his investigations
small cylinders of compressed magnesia. He placed one of these
cylinders above and very near the arc, and thus brought it to a state of
incandescence similar to that of the carbons, producing a steady light by
this means. The wear of the magnesia was very trifling; the cylinders
were previously prepared with a silicious solution which rendered them.
extremely hard. At the conclusion of these experiments, M. Leroux

announced his conviction that the principle was well adapted for an electric lamp, but he did not pass from theory to practice, and contented himself with his laboratory experiments. Passing on to 1870, we come to the researches of M. Jablochkoff after refractory materials that might be used for his electric candle. Two or three years later, and following closely the same order of ideas, there were produced two other examples, in which refractory materials were utilised in the arc. In the first of these, the lamps of M. De Bacllehache, the carbons penetrate small bodies of magnesia perforated with conical holes, as in Fig. 616, and are pushed forward by springs. This was an arrangement somewhat similar to the familar candle lamp used in carriages. The refractory pieces here played a double part, to keep the carbons in place and to increase the light by·

Fig. 616.   Fig. 617.

incandescence. The second was the lamp of M. Delaye, to which the inventor gave the name of the solar lamp. Fig. 617 shows the principle of this arrangement. It consists of two parallel carbons, each bearing against blocks of refractory material. These blocks, as well as the ends of the carbons, become incandescent, giving a golden. hue to the light. Both these systems were shown at the Palais de l'Industrie during the Electrical Exhibition of Paris.

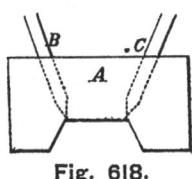

Fig. 618.

The Soleil lamp is based on a similar principle. It is of the same family as the Jablochkoff candle, and belongs to the category of incandescence-arc lamps. It possesses, however, characteristics special to itself, and differs considerably in its latest development from its earlier forms ; it was designed, in 1879, by MM. Clerc and Bureau. Fig. 618 illustrates the principle of its construction. The block A is a parallelo-pipedon of a refractory substance, such as marble, lime, granite, &c., with a cavity on one side shaped like a truncated cone, to the face of which penetrate the carbons B and C, traversing the mass through inclined cylindrical holes, but prevented from slipping through by the contraction of the holes. When the arc passes between the two points it plays on the face of the recess, heats it, and transforms it into a small crater, whence the luminous rays escape in a conical beam directed by the inclined side of the recess. Fig. 619 shows the construction of one of the blocks, of which the pieces A A of white stones are quadrilateral truncated pyramids, having three vertical faces forming the outer sides

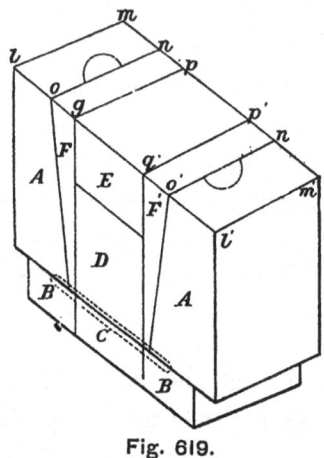

Fig. 619.

of the block and one inclined face ; on this latter a hemispherical groove is cut for the reception of a carbon of similar form (see Fig. 620). Below the pieces A are two thin blocks of granite B, shown separated in Fig. 620 to indicate the position of a small groove cut in their upper face. This

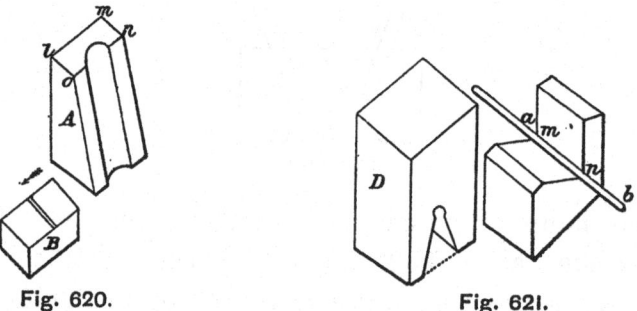

Fig. 620.   Fig. 621.

groove receives a very small carbon pencil, which connects the ends of the carbons and permits lighting to take place. The block D between the granite pieces B, is of white marble, and is surmounted by the block E of white stone, which really only serves as a packing piece, and replaces the more expensive material of which D is composed. This latter is shown in its normal position and also reversed in Fig. 621

in such a manner as to indicate the form of the recess, and the location of the carbon fuze *a b*; the distance M N marks the distance apart of the carbons, and consequently the length of the voltaic arc. The two pieces F F¹ are wedges, and serve to consolidate the various

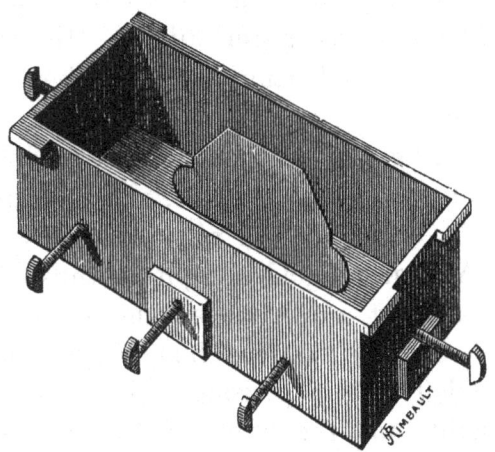

Fig. 622.

pieces of stone, which are then placed in a cast-iron frame, and locked tight by a number of screws (Fig. 622). By reference to Fig. 620 it will be seen that the recess in the marble block is symmetrical for the vertical and even diffusion of the light of the lamp. Other forms of

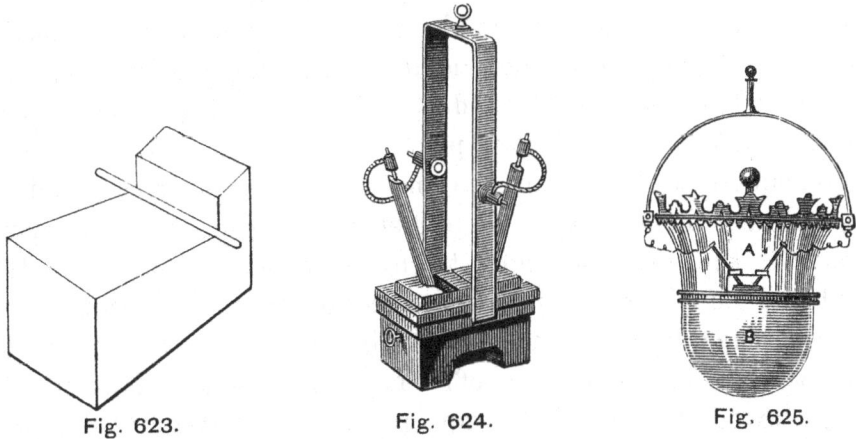

Fig. 623. Fig. 624. Fig. 625.

receivers may be used for the special distribution of the light. Thus in Fig. 623 the block is recessed with one side vertical and the other almost horizontal, in order to throw the light to one side. Fig. 624 shows the lamp complete for lighting, and suspended from a bracket which receives the current. Fig. 625 is a sketch of

the lantern for the Soleil lamp.   In the picture gallery at the Paris
Exhibition there were 10 lanterns of this design, each containing two
Soleil lamps, so arranged that the second ignited on the extinction of the
first.   The part A of the lantern was of copper, B was a bowl of glass
attached to the bottom of A, through which two holes were made to
give passage to the luminous beam diffused throughout the glass.
This arrangement of enclosed lamp reduced the noise produced by
the alternating current.

At the moment of lighting, the current burns the fuse connecting
the two carbons, and establishes the arc, which clinging close to the
marble surface rapidly heats it, and after some few seconds raises it
to incandescence.   A slightly golden light is thus obtained, which if
not an absolute, is at least a relative advantage, placing the light as
it does midway between the familiar hue of gas, and the cold rays
of the arc.   The Soleil system presents two special advantages: the
steadiness of the light, and the fixity of the luminous area.   These
are due to the fact that the light does not proceed direct from the
arc which, as it were, is absorbed into the refractory mass, the
incandescent light of which becomes predominant. and imparts to the
lamp the steadiness which is the especial characteristic of the incandescence systems.   Moreover, if from any cause—the slackening of
a driving belt for example—the intensity of the current be momentarily
reduced, the stored-up heat in the block and the luminosity proceeding
from it, are sufficient to supply the deficiency for a short time, and
if the intensity of the light be reduced, it will be a gradual process,
so that no scintillation will take place.   Steadiness and stability are
features common to lamps on the incandescence principle, and the
Soleil possesses them fully.   The method of grooving the marble
block, allows the rays of light to be diverted in any direction,
and it should be observed in this connection that this lamp cannot
in any case throw light at once in all directions, a point which must
be remembered in considering its efficiency.   It is stated above that the
carbons employed in the Soleil lamp are hard semi-cylindrical pencils,
and generally of so large a diameter as to burn for a considerable period.
M. Desguin, a Belgian engineer, who has made a careful investigation
of the Soleil lamp, says that with carbons of a semi-circular section
and 10 mm. radius, the hourly consumption is 10 mm. per carbon,
or 20 mm. per lamp and per hour.   Usually, however, much larger

carbons are used. In preliminary experiments in lighting the auditorium of the Opera, the engineer of the Soleil Company devised a somewhat different arrangement. Instead of carbon pencils he employed parallelopipedons of carbon, 20 mm. × 20 mm. × 40 mm. (·79 in. × ·79 in. × 1·58 in.). The arrangement is shown in Fig. 626, in which A A represent the blocks of carbon; B a piece of white marble; C C two pieces of granite; the carbons wear, as shown by the dotted lines and in the perspective diagram. At the Opera this lamp was placed in a reversed position, so as to throw the light upwards. It rested simply

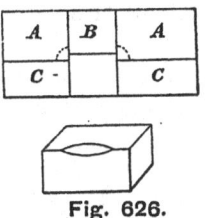

Fig. 626.

on two contracts, and its weight was sufficient to secure the easy and constant passage of the current. In this modification the arc has a greater length than in the ordinary form, and the estimated duration is six hours. The delicate part of the lamp is evidently the refractory block, in which, under the high temperature to which it is subjected in working, the carbonate of lime is gradually transformed into pure lime, and a molecular action is set up, that destroys the cohesion of the material. This action appears most marked when the lamp is subjected to a series of lighting and extinction, which disintegrates the material and reduces its durability. On the other hand, M. Desguin states that he has kept lamps burning 15 consecutive hours without the marble being destroyed, and with compressed magnesia, more favourable conditions of durability can be obtained, but at the same time a white light is produced.

The Soleil lamp has been subjected to a series of trials by a commission consisting of M. E. Bède, past professor of physics at the University of Liège; M. P. Desguin, past professor of physics at the Royal Museum of Industry; M. T. Dumont, telegraph engineer; M. E. Rousseau, professor at the Brussels University, and M. Wauters, doctor of science and controller of the service of gas to the city of Brussels. These trials, which lasted from the 21st of March to the 5th of September, 1881, resulted in a report of which the following is an analysis.

The photometric measurements were made by means of the Bunsen photometer, and the standard was a moderator lamp made by Delèuil, and equivalent to a Carcel burner. The diagrams taken from the 25 horse-power engine with a Richards indicator, were computed with an Amsler planimeter. The electric intensities were measured with an electro-dynamometer, and reduced to Ampères. All the experiments were conducted with an auto-exciting Gramme machine, type No. 1, constructed for four, six or eight Jablochkoff candles. The resistance of the circuits could be varied by means of rheostats of iron wire spirals. The results obtained are embodied in the following table :—

SUMMARY OF TRIALS OF SOLEIL LAMP CARRIED OUT BETWEEN MARCH AND SEPTEMBER, 1881.

Date	Hour.	Revolutions of Gramme Machine.	Total Work, in Horse-Power.	Useless Work, in Horse-Power.	Useful Work, in Horse-Power.	Resistance Introduced in Circuit, in Ohms.	Intensity of Current in a Circuit, Webers (Ampères).	Number of Lamps.	Class of Lamps.	Luminous intensity from Lamp, in Carcels (Maximum).	Consumption in Carcel Lamps (Grammes).	Number of Carcels per Horse-Power.	Number of Horse-Power per Lamp.
1881.									mm. in.				
May 30 ...	...	...	17	8·8	8·2	...	10	3	22= ·87	256	...	93	2·7
June 13 ...	...	1,920	16·5	8·6	7·9	3	10·5	2	30 1·18	280	46·5	70	3·9
,, ...	...	1,920	16·7	8·6	·8·1	3	10·5	3	22 ·87	212	42·2	78	2·7
August 3 ...	8.12 to 8.30	1,959	20·9	7	13·9	...	5·5	12	11 ·39	105	39·3	90	1·16
,, ...	8.33 to 8.45	1,959	18·7	7	11·7	...	5	12	11 ·39	100·5	39·3	103	0·97
,, ...	9.21 to 9.35	1,959	20·9	7	13·9	...	5·8	8	22 ·87	126	33·3	72	1·70
,, ...	9.56	1,959	16·5	9·2	7·3	9·3	5	4	22 ·87	...	45	72	1·80
,, ...	11. 5 to 11.6	1,959	22·8	7	15·8	...	9·6	6	22 ·87	234	45	88	2·6
August 13 ...	7.45	1,986	...	5·5	...	...	...	16	11 ·39	...	...	...	...
,, ...	8. 5 to 8.18	1,986	23·5	5·5	...	...	5·7	16	11 ·39	113	42	100	1·1
,, ...	8.21 to 8.30	1,986	24·3	5·5	18·8	...	5·1	16	11 ·39	89	42	75	1·2
,, ...	8.33 to 9	1,966	23·8	5·5	18·3	...	5	16	11 ·39	70	42	61	1·1
,, ...	9. 3 to 9.22	1,986	23·5	5·5	18	...	4·3	16	11 ·39	60	42	53	1·1
August 20 ...	8. to 8.10	2,060	26·7	7	19·7	...	12	12	11 ·39	140	44·8	85	1·6
,, ...	8.17 to 8.18	2,060	26	7	19	...	11·8	12	11 ·39	126	44·8	79	1·6
,, ...	8.21	2,060	24·9	7	17·9	...	12·3	12	11 ·39	120	44·8	80	1·5
,, ...	8 28 to 8.30	2,060	23·5	7	16·5	...	11	12	11 ·39	85	44·8	61	1·4
,, ...	8 34 to 8.35	2,060	22·5	7	15·5	...	9·05	12	11 ·39	66	44·8	51	1·3
,, ...	9. 8 to 9.13	2,060	18·6	9·2	9·4	...	8·5	8	30 1·18	208	44·8	66	3·1
,, ...	9.21 to 9.22	2,060	18·6	9·2	9·4	4·8	8·13	8	30 1·18	205	44·8	65	3·1
,, ...	9.30 to 9.40	2,060	16·2	9·2	7	9·5	6·9	3	30 1·18	180	44·8	77	2·3
,, ...	9.56 to 9.58	2,060	15·7	9·2	6·5	13·4	6·4	3	30 1·18	165	44·8	76	2·2
,, ...	10.15	2,060	16·4	9·2	7·2	22·9	5·8	3	30 1·18	145	40	60	2·4
,, ...	10.16	2,060	17·7	9·2	8·5	37·2	5·3	3	30 1·18	115	40	40	2·8
,, ...	10.19	2,060	17	9·2	7·8	55·3	4·0	3	30 1·18	98	40	37	2·6
,, ...	10.46	2,060	18	9·2	8·8	...	22·8	2	35 1·38	612	40	139	4·4
,, ...	10.47	2,060	17·3	9·2	8·1	...	22·3	2	35 1·38	562	40	138	4·

From the foregoing, the following conclusions may be derived :—

1. The speed of the Gramme generators was maintained at the high rate of 1,900 to 2,000 revolutions. This was certainly unfavourable to the good maintenance of these machines.

2. The work absorbed was divided into total, useless, and useful work ; the last two columns of the foregoing table are deduced from the useful work only, which it is important to note, since if the calculations

had been based on the total work, the figures would have been greatly modified.

3. The lighting power decreases during a certain time from the commencement of the trial, reaching a normal of from 50 to 60 per cent. of the initial value. At the same time, the intensity of the current and the power absorbed decrease, but at a less ratio.

The luminous efficiency per horse-power at the moment the normal intensity is attained, is, according to the results obtained with the 11 mm. carbons, with which the largest number of trials were conducted:

51 Carcel burners for the 11 mm. lamps to the number of 12.
53 Carcel burners for the 11 mm. lamps to the number of 16.

The individual intensity of the lamps being 66 Carcels in the first case, and 60 in the second. It should be observed that they reached 138 Carcels with the 35 mm. carbons, but the time of experiment was too short to be of use. It must not be forgotten that the luminous zone of each lamp occupies only a portion of the circumference, and that the photometric results recorded are there measured in the direction of maximum intensity. For all these reasons it appears that as regards the power absorbed, the Soleil lamp does not compare favourably with some of the arc lamps.

The system, however, possesses many points of value and interest. It has the great quality of steadiness, and its colour is good, which makes it specially suitable for some installations. The excellent and pleasing illumination of the picture gallery at the Palais de l'Industrie illustrated its value in both these particulars.

As regards economy, two points have to be considered—the cost of power, and the cost of the lamp itself. As regards the former, the experiments have shown that it consumes more power than many arc lamps, while the luminous angle is restricted. The cost of the lamps is made up of the value of the carbons, and the marble blocks. The former consume slowly, as we have seen, so that their cost is insignificant, while the marble blocks, costing about half a franc, will last for about 80 hours. The question then turns on the point whether the cheapness of the lamp compensates for the larger expenditure of power. The friends and inventors of the Soleil lamp believe that it can be made to compensate, while they look to effecting a reduction in the cost of feeding. Time and experience alone can decide the question, experience gained not in the laboratory, but in industrial use. Durability and simplicity it can undoubtedly claim, but for general adoption the question of cost is the most important, and in its present form the evidence in this direction is rather unfavourable. Doubtless, however, the system is susceptible of many useful modifications.

André.—Finally, in connection with this branch of the subject, we may refer to two incandescence-arc lamps, devised by Mr. G. C. André, of Dorking. The object of the arrangement shown in Fig. 627 is to supply a definite quantity of air to the incandescent carbons, while the current is flowing, an automatic ventilation being produced when the current ceases. In the illustration, *a* is a small hole for the ingress of the air, and it may, if desired, be made adjustable in size by any suitable

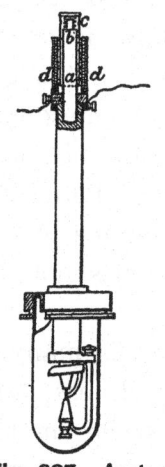

Fig. 627. Andre.

means. The tube *b* for the carbon is provided at the top with a small aperture. As long as the current is passing, the core tube *c* is drawn down by the action of the solenoid *d*, and a leather washer then covers the aperture, preventing escape of the air. When the lamp goes out, and the current ceases, the valve rises and allows the air to escape, thus producing automatic ventilation.

Fig. 628 represents the second form, jointly schemed by Mr. André and Mr. E. Easton. In this lamp the light is produced by the incandescence of a carbon electrode in contact with a copper electrode c^2 as in the Werdermann system, and the novelty consists in feeding a carbon or carbons placed end for end in a holder so as to slide therein, by means of a cord and weight, which, when the negative electrode approaches the positive, are brought into action by means of a regulator or brake controlled by the electric current. The electrode c^2 can slide endwise in the bracket c through the contact spring L, so as to follow the wasting carbon D until it has reached the end of its travel, when a further consumption of carbon will break the circuit, and diminish or extinguish the light. The carbons D are packed end to end in a tube, and are forced up by an arrangement resembling a carriage lamp with a weight replacing

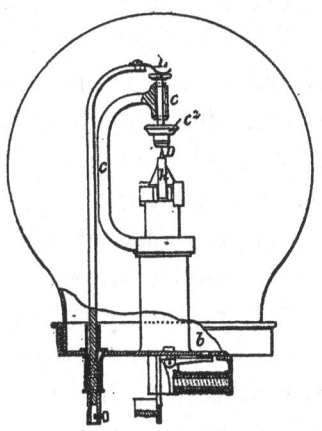

Fig. 628. Andre and Easton.

the usual spring. The weight is controlled by being attached to a cord which passes round a brake pulley held or released by the action of a solenoid (not shown) wound differentially, so that when the current is flowing normally it is out of action, but when the resistance of the lamp is increased, it lifts the brake lever and allows the weight to feed up the carbon; b is a ventilating hole controlled by the apparatus described above. In order to short circuit the lamp when the carbon is exhausted, the upper pole c^2 falls on to the contact point k.

XII.

INCANDESCENCE LAMPS.

THE first person to apply for a patent for an electric lamp in this country was Frederic De Moleyns, of Cheltenham. In the year 1841 he filed a specification of his invention, describing how the light produced by the incandescence of a platinum wire could be increased by a falling stream of charcoal or plumbago particles. This apparatus (Fig. 629) consisted of a glass globe provided with two air-tight

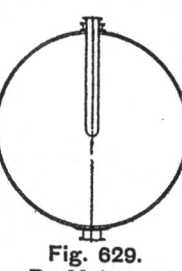

Fig. 629.
De. Moleyns.

brass caps, through one of which the globe could be exhausted. Projecting through the upper part of the globe, and reaching nearly to its centre, was a receiver, furnished at its lower end with a nozzle of such a size as would allow powdered carbon to flow in a regular stream, like the sand of an hour glass. A thick copper conductor passed down the receiver nearly to its orifice, where it was joined to a platinum wire which passed through the nozzle and was wound into a helix one-eighth of an inch in diameter, and one-quarter of an inch long. A similar copper wire, ending in a platinum helix, entered the globe from the opposite side sufficiently far for the two helices to be put in contact. When the copper electrodes were connected to the poles of a battery, and the globe was placed in such a position that the powdered charcoal flowed from its receiver, the metal helices became white hot, as did also the stream of carbon particles that flowed through and around them. The inventor does not appear to have relied upon incandescence solely for the light, as he made provision for moving the helices out of contact with each other, intending, presumably, to maintain the circuit by a number of minute arcs formed between the particles of falling carbon.

In 1845, Augustus King patented a lamp in which the light was produced by the incandescence of carbon in a Torricellian vacuum. In Fig. 630, c is a thin strip of graphite held between two clamps or forceps f and g, which are fixed to a porcelain rod i, the whole being suspended by a platinum wire sealed into the globe. This globe is blown on the upper end of a barometer tube of more than 31 inches in height, filled with mercury in the usual way. The current enters through the wire *d*, flows through the carbon conductor, heating it to incandescence, and passes by the wire *n* to the column o mercury which is included in the circuit. This attempt at incandescence lighting was extremely ingenious and did credit to the real inventor, J. W. Starr, of Cincinnati, who brought his invention to Europe in company with King. The latter, by some means or other, so arranged as to keep Starr in the background, merely stating in his application for Letters Patent, according to the usual method in communications then, that the invention had been communicated to him.

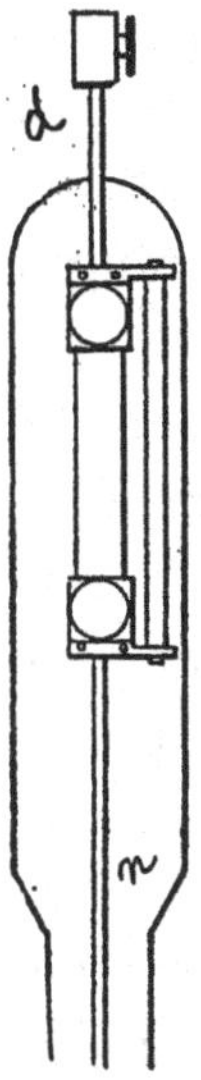

Fig. 630.
King.

In a recent number of *Knowledge*, appears a communication from Mr. Mattieu Williams, which contains some paragraphs relating to this incandescence lamp of Starr. These paragraphs are of sufficient interest to be reproduced *verbatim* :—

"As far back as 1846 I was engaged in making apparatus and experiments for the purpose of turning to practical account 'King's patent electric light,' the actual inventor of which was a young American named Starr, who died in 1847, when about 25 years of age, a victim of overwork and disappointment in his efforts to perfect this invention and a magneto-electric machine, intended to supply the power in accordance with some of the 'latest improvements' of 1881 and 1882.

" I had a share in this venture, and was very enthusiastic until after I had become practically acquainted with the subject. We had no difficulty in obtaining a splendid and perfectly steady light, better than any that are shown at the Crystal Palace.

" We used platinum, and alloys of platinum and iridium, abandoned them, as Edison did more than 30 years later, and then tried a multitude of forms of carbon, including that which constitutes the last 'discovery' of Mr. Edison, viz., burnt cane. Starr tried this on theoretical grounds, because cane being coated with silica, he predicted that by charring it we should obtain a more compact stick or thread as the fusion of the silica would hold the carbon particles together. He finally abandoned this and all the rest in favour of the hard deposit of carbon which lines the inside of gas retorts, some specimens of which we found to be so hard that we required a lapidary's wheel to cut them into the thin sticks.

" Our final wick was a piece of this of square section, and about ⅛ of an inch across each way. It was mounted between two forceps—one holding each end, and thus leaving a clear half-inch between. The forceps were soldered to platinum wires, one of which passed upwards through the top of the barometer tube, expanded into a lamp glass at its upper part. This wire was sealed to the glass as it passed through. The lower wire passed down the middle of the tube.

" The tube was filled with mercury and inverted over a cup of mercury. Being 30 in. long up to the bottom of the expanded portion, or lamp globe, the mercury fell below this and left a Torricellian vacuum there. One pole of the battery or dynamo-machine was connected with the mercury in the cup, and the other with the upper wire. The stick of carbon glowed brilliantly, and with perfect steadiness.

" I subsequently exhibited this apparatus in the Town Hall of Birmingham, and many times at the Midland Institute. The only scientific difficulty connected with this arrangement was that due to a slight volatilisation of the carbon, and its deposition as a brown film upon the lamp glass; but this difficulty is not insuperable."

In a recent number of *La Lumière Electrique* appears a very interesting and somewhat remarkable contribution from the pen of M Ch. De Changy. We do not know whether the writer can substantiate all the statements that he makes, and which give to him and to M. Jobard, of Bruxelles, the credit of being the first to obtain practical results in electric lighting by incandescence. We will let M. De Changy speak for himself, by translating the whole of the article referred to :—

" Electric incandescence lamps increase, multiply, accumulate unceasingly; there is scarcely a day but a new apparatus of this class, is not produced with more or less noise, announced always, be it understood, as the most perfect, or rather the only perfect of those existing. I have thought it would be interesting to expose the origin and the point of departure of this movement at present so considerable. I believe I may, with a better right than anyone, give at least the main outlines, having myself taken, in the early trials, a part, the importance of which will be appreciated by the reader. My labours in this direction reach nearly 40 years back. All authors who occupy themselves with this subject cite them as the oldest, from a theoretical point of view; as to my experiments and my applications of incandescence in a vacuum, their priority over all others is a fact which no one contests.

"I was directed to this research by my respected professor and friend, M. Jobard, of Bruxelles, who had for a considerable time been turning his thoughts in this direction; and he had suggested, in 1838, in the *Courrier Liberal*, the idea that *a small carbon employed as a conductor of a current, in a vacuum,* would give an electric lamp with an intense, fixed and durable light. He advised me to pursue the practical realisation of this theoretical conception. I was an engineer occupied especially in mining work, and the possible application of incandescent lighting to the works connected with my profession struck me, and I entered upon the research that he had indicated to me.

"After a certain theoretical preparation, I began my real experiments in the year 1844. I sought solely to utilise carbon in a vacuum, in the manner indicated by M. Jobard. The striking advantages possessed by this material as regards incandescence—advantages it is unnecessary to enumerate to-day—were known even at that time, and my attempts were

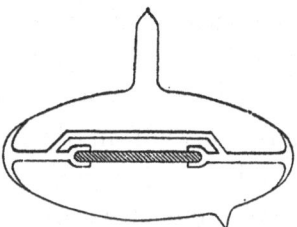

Fig. 631. De Changy.

concentrated on this point. I made use of the only carbon conductor of electricity which we then possessed, that is to say retort carbon. I cut rods as fine as possible of this material, and enclosed them in glass vases from which the air was extracted, placing them in connection with the conductors. I do not dwell on the difficulties I encountered from the commencement in forming the connections and obtaining a hermetic seal; they are common to all lamps of this class; but the nature of the carbon itself of which I made use was accompanied by a special obstacle. This was never homogeneous, and therefore the rod under the influence of the current was always destroyed at one point or another. I attempted to obtain homogeneity by a process that consisted in filling the pores, by saturating the stick of carbon in melted gum or saccharine solution, and in subsequent baking. The result was better, and I made a type of lamp of which I give (Fig. 631) an approximate representation. As an experiment, the success was encouraging. In a practical point of view it was evident that these apparatus could not be seriously utilised. Nevertheless, it was for that time a progress worthy of record; the more so that I was alone upon this path; the curious experiments of Starr and King, to which I shall again refer, and which were presented to the public in 1844, were not made, as has been stated in error, with carbon, but with platinum brought to incandescence in air.* Interesting as this result was for men of science, from a practical point of view it offered little encouragement to me. Moreover, I was at this time called to England as engineer-in-chief of the Weal-Ocean and Weal-Ramoth mines; my researches were therefore naturally suspended.

"Notwithstanding, the idea that had led me to undertake them—that of lighting mines—did not cease to preoccupy me, especially in the function to which I was called. I invented, even at this time, an oil lamp, which was used for a long while under the name of the Victoria safety lamp. This was only, to tell the truth, a sort of *pis-aller*, the electric lamp never having ceased to appear to me the ideal for illumination of this class; therefore, on my return to Bruxelles in 1850, I resumed my researches, at first intermittently, then towards 1855 with great activity. I then directed my efforts in two directions at the same time; without abandoning carbon, I occupied myself in designing a system of lamp in which the incandescent body was platinum; this material is far from having the qualities of carbon, but it is free from some of its faults, and it was reasonable to suppose that it would permit me more rapidly to obtain a system, less perfect perhaps, but sufficiently practical. It is true that the employment of the metal presupposes the solution of various problems, principally two; it is necessary in order to prevent the fusion and the destruction of the incandescent wire even in a vacuum, strictly to limit the intensity of the feeding current, that is to say in a word, to obtain electrical regulation; on the other hand, in order that the system may be practically useful, it was necessary that one should be able to place several lamps in the same circuit, these lights being too feeble to be economically employed if each required a separate circuit. Several attempts had been made in this direction. I have spoken of the experiments of Starr and King, which made so much noise about 1844; they did not last long, the process employed is little known; the death of Starr, and the disappearance of King, abruptly put an end to these

---

* It will be noticed that this statement does not agree either with the Patent Specification, or Mr. Mattieu Williams' remarks quoted above.

trials of which a clear explanation was never completely given. In a patent dated 1848, Staite proposes the employment of iridium rendered incandescent in air; this metal is very refractory; the arrangements indicated by Staite appear reasonable; I do not know if these lamps worked, it is probable that they could be used to a certain point, but Staite was obliged to give a current to each lamp, and, moreover, iridium is very scarce and very costly, so that the system is entirely impracticable. Towards the time when I resumed my studies, Mr. Holmes had undertaken some works in the same direction, but did not meet with any success, his regulation being very insufficient; his lamps melted and were very frequently destroyed.

"Experiments, methodically followed, led me at first to recognise that the metal must receive a particular preparation, it could not be brought at once to incandescence but must be accustomed gradually to the class

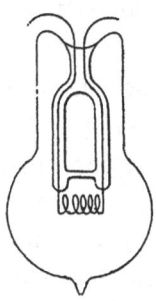

Fig. 632.

of service it had to render; to this effect I maintained it at a moderate red heat and made it slowly rise to the degree at which it should be maintained. Mr. Edison rediscovered these facts about 20 years later. I also discovered that it was advisable not to employ the platinum pure, but very slightly carburated. To this end I subjected it to an operation very similar to the cementation of steel, by heating it in powdered carbon and afterwards drawing it into wire; in this manner I quickly arrived at making lamps which did not destroy, and which gave considerable intensity of light. I give here a diagram of one of these lamps Fig. 632. As may be supposed, I lost no time in proposing the application of which I had always dreamed, that is to say, the lighting of mines; I presented my lamps to Mr. Devaux, engineer-in-chief of mines, who made a singular objection to me, that my apparatus was too safe, that is to say, that it would not notify the presence of fire-damp as the Davy lamp does. I must say that I was not discouraged by this criticism, because it seemed to me that it was more useful completely to avoid the danger than-to allow it to exist on the chance of being informed of it in time; moreover, there are other means of recognising and signalling the presence of fire-damp. I suggested one myself, that I had devised, to Mr. Devaux, but that gentleman's opinion was fixed; none the less I continued my researches, and on the 17th of May, 1858, I obtained a patent for a complete system of regulating and dividing the current for incandescent electric lighting. I will explain soon how the taking of this patent led to the interruption of my work, but I ought first briefly to explain the system that I employed Each lamp was placed on a circuit derived from the general current which traversed besides the lamp, the coil of an electro-magnet; the second derived circuit branching from the first, was formed by the core of this electro-magnet, and its armature. The second circuit was thus only closed if the electro-magnet being active brought its armature in contact with the core; the operation of the system will now be easily understood; in the normal condition the circuit containing the lamp and the wire of the electro-magnet were closed—the opposing spring of the electro-magnet being regulated in such a way that the current suitable for the lamp was not strong enough to overcome it; if the current increased too much the magnetism of the electro-magnet increased with it; the opposing spring was overcome, the armature was brought into contact with the core, and thus closed a derived circuit of low resistance, and diverted the current, grown too intense and dangerous to the apparatus. The amount of electricity absorbed by the lamp being thus limited, it became possible to place several in the same circuit, which thus realised the two objects sought—regulation and division. I have said that in working at incandescence by metals I had not abandoned carbon. During this period as well as afterwards, I attempted, not finding any suitable carbons, to make it in different forms. I passed through dies, plombagine made into paste, so as to form very fine sticks, which I afterwards baked, but I was obliged to introduce agglomerating bodies, such as clay, the plombagine alone not having sufficient consistency. I did not suceed in making pencils so hard as I desired for electrical incandescence; they all had a certain tendency towards softening. In 1859, M. le Comte du Moncel described in his studies on the Ruhmkorf coil, experiments in which he had obtained a very brilliant light by the incandescence of vegetable fibres, such as cork, or of vellum saturated in sulphuric acid and carbonised. The application to lamps in vacuum was indicated, and I tried materials of this kind, but it was necessary to take certain precautions to increase the conductibility and the homogeneity which at first were insufficient. This required a series of experiments which I could not pursue to the end, for reasons I am about to explain. I had perfected at the time certain types sufficiently curious in which the incandescence of platinum was combined with that of carbon that was surrounded by a spiral of the metal ; this type is represented by Fig. 633.

" When I had thus succeeded in the regulation and division with the platinum lamp, I began to consider their applications, and to this effect I showed my investigations to persons competent, and in a position to

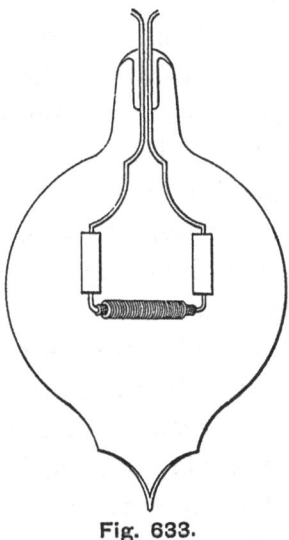

Fig. 633.

adopt my system. Naturally at the head of these was M. Jobard, who had never ceased to follow my work with the most lively interest. Satisfied with the results obtained, he made them the subject of various communications ; in particular he addressed a note to the Academy of Sciences in France, on the 27th February, 1858 ; prior, as will be seen, to my patent. It is unnecessary to reproduce here this complimentary communication. He summarised in it the experiments I had made, and referred to the proposed applications. Of these, I had suggested four principal ones : 1. Lighting of mines ; 2. Submerged lamps for fishing ; 3. Luminous buoys ; 4. Nautical telegraphy obtained by means of coloured tubes containing the incandescent spirals, and combinations of which, effected with a keyboard, formed signals visible from the masthead. The

Academy of Sciences considered it advisable to obtain a special report on the subject, and M. Quételet, perpetual secretary of the Academy of Bruxelles, being requested to undertake this, delegated M. Melsem one of his colleagues. This gentleman took a part in the experiments, followed them closely, and on the 3rd April, 1858, reported to M. Desprez on what he had seen. In this letter he enumerated the results; said that he had seen in one circuit, fed by a Bunsen battery of 12 cells, somewhat modified by me, several lamps which could be lighted simultaneously, in groups or separately, at will, without the normal intensity of each being affected; he reported, therefore, that the regulation and division were effected. The Academy of Science nominated a commission, formed of MM. Becquerel, senior, Desprez, and Babinet. M. Desprez wrote to M. Jobard, to ask of him the detailed description of the apparatus and the processes. M. Jobard replied that it was impossible for him to furnish these details, because their publication would affect the patent which I had applied for. Upon this M. Desprez declared that M. De Changy, wishing to make his invention a source of profit, did not merit the name of *savant*, and the Academy ought to take no further interest in his work. This reply discouraged me profoundly, and shortly afterwards I entirely ceased my researches.

"It may be that the reader will at first feel surprised at the regret with which I abandoned these investigations, the results I have recorded as having obtained, appearing very small; it is certain, indeed, that the experiments I have mentioned would not be striking to an electrician of the present day, but it must be remembered what was the condition of things at that time. Division and regulation—to a limited degree, certainly—that I had realised, were results absolutely new in 1858; these results formed the point of departure to the vast progress obtained since; they showed the possibility of it, and opened (I have the right to say it) the way in which we have advanced since; they had therefore an importance which justified the praises of men of science, and explained the value I attached to them. On the other hand the reader may be surprised that, knowing the value of my researches, I was able to decide on abandoning them, but it must be remembered that the hour of great applications was not come, the powerful sources of electricity to-day so widely known, did not exist; the practical results were certain but not immediate, and discouragement was natural.

"However, I am wrong in saying I entirely abandoned my researches.

A study followed during so many years and with so much ardour cannot be thus abandoned. Only I directed my labour, much less actively, in a different direction. I had already devised a regulator, which there is no occassion for me to describe here, although for the time it was, I think, ingenious and interesting. I occupied myself with incandescence in air, either with carbon pencils falling through an incandescent spiral of platinum wire, or with a carbon pencil falling between two others, or preferably between two discs of carbon forming electrodes.

"Years passed by: everything changed in electricity. At the same time that theory grew clearer every day, inventions increased with wonderful rapidity, especially in mechanical generators of electricity, which within a few years reached the degree of perfection and the great power, that we now see. Then, the propitious hour for the employment of incandescent lighting had arrived, the element wanting for my work had been created, and everyone knows by what general evolution, invention turned almost simultaneously towards creations of this kind.

"It does not fall within my province to remark at what point were utilised my researches which were not unknown, although their author was somewhat forgotten in his silence; it is not for me to show how some have simply taken the fruits of my labour, and re-edited my results, sometimes with inferior practical arrangements. The readers of this journal (*La Lumière Electrique*) have followed the facts during the last few years, they will recognise at once the analogies between recent inventions and the old studies I have described.

"It is quite natural that I should take my part in the present general movement. My friends have urged me, and I have no reason to refuse. The consideration which stopped me in 1858 exists no longer, public opinion having completely changed on that point. To-day it is considered quite natural that everyone should draw from his work what benefit he can, which I think is just; it is considered that far from being detrimental to science, that can only be useful to it, which is reasonable; some of the most illustrious among us have set the example, and the sentiment of M. Desprez would seem to-day very curious. I could only resume my labours at the point where I quitted them 20 years ago, if I found myself ahead of other inventors; my old methods improved, completed and modified on certain points have given me a type of incandescent lamp which I consider satisfactory. I have made several types, indicated in Figs. 634, 635, 636 and 637, varying from 3 up to 50 candle-power. If it should prove of interest, I shall describe all the details of my new processes. For the present I believe it has been of service to recall categorically labours long since old, but the account of which will contribute to throw a light on certain points in the history of science."

In the above interesting and circumstantial narrative, three points strike the reader with some astonishment. The remarkable similarity in processes and forms of lamps devised by M. De Changy and others lately brought forward as new; the foolish narrowmindedness of M. Desprez; and the readiness with which M. De Changy abandoned

Fig. 634.

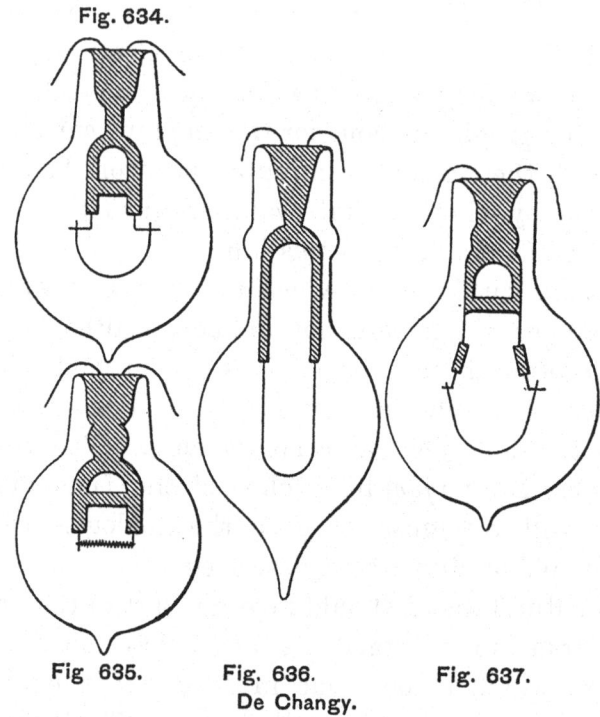

Fig 635.   Fig. 636.   Fig. 637.
De Changy.

a cherished project, brought after many years' work to a successful issue, when charged with the crime of wishing to make money by an invention, the value of which he fully appreciated. It appears strange that it did not occur to him either to abandon his patent, and so regain the favour of M. Desprez, or abandon the patronage of the Academy and make use of his invention. We have reproduced the article because it is of high interest, but we do not endorse the statement it contains; and on the other hand we have examined the specification of the patent obtained in 1858, and find it very obscure, and of too small value, even historically, to reproduce here.

In 1848, W. E. Staite, who seems to have anticipated one-half of the so-called new lamps of the last few years, brought out an incandescence lamp with an iridium bridge. With his usual ingenuity, he not only applied the metal to the purpose he had in view, but he also invented a process of working it. In his specification he said: "I fuse oxide of iridium in a cupel of bone ash under the voltaic arc, by which means it is brought under the action of the most intense heat with which we are acquainted. Having obtained an ingot of the metal, I submit the same to the constant action of heat for the purpose of annealing it, and for this purpose I use one or more large jets from an oxyhydrogen blow pipe. When the ingot is at a white heat, I pass it

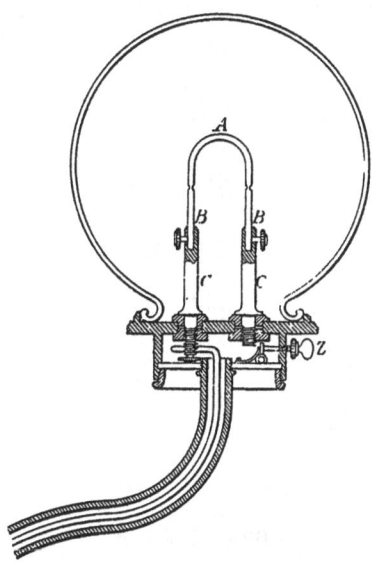

Fig. 638. Staite.

between rollers, or hammer it, repeating the process again and again until the metal is sufficiently annealed to be worked up into the required shape. Sometimes I mix platinum with the iridium, and sometimes copper in small quantities, which facilitates the annealing. Sometimes also I cut it into thin strips from the ingot in the same way that precious stones are cut by the lapidary. For lighting purposes, I use either a thin strip of iridium so prepared and cut off, or a combination of iridium with platinum, or of iridium with copper, or a spiral coil of the iridium as a solid electrode."

Fig. 638 is a sectional view of one of Staite's lamps. A is a strip of iridium fused into two pieces of platinum B B, which are fixed to two

metallic holders C C passing through insulating washers in the lamp
bases. The current enters by the central wire, flows successively through
the conductors C, B and C, and returns through the switch z to the
bracket which is included in the return circuit. As shown in the
illustration, the continuity is broken at z, and the lamp cannot be lighted
until the connection is made by screwing in the knob X.

In 1852, E. C. Shepard, himself a well-known inventor, took out a
patent on behalf of one Floris Nollet, of Brussels, for an apparatus for
lighting by the incandescence of charcoal in a vacuum. From the
arrangement he adopted, it would seem that he did not succeed in
obtaining, or at any rate in maintaining, an efficient degree of exhaustion,
for he made provision for feeding the carbon forward as it wasted.   The

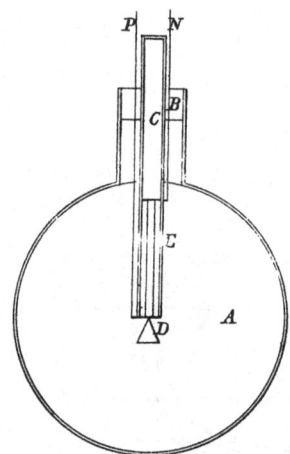

Fig. 639. Shepard.

plan he adopted was to force a little cylinder of charcoal down a tube
open at the bottom, by a lead weight, on to the point of a cone of
"prepared coke."   One conducting wire was attached to the metal
cylinder, and the other to the bracket which carried the cone, and the
whole was enclosed in an exhausted transparent or translucent globe
provided with a stem for insertion in a socket.

The specification is an interesting one, and contains the following
passages relating to "Apparatus for lighting in a vacuum ":—

"When it is desired to light by means of the incandescence of
charcoal, I employ an arrangement of apparatus similar to that which
is represented in Fig. 639. A is a globe of glass, which may, if desired,
be ground or obscured on the inside, in order to disperse and soften the

intensity of the light. The stopper B of this globe is very carefully luted, and two insulated polar wires, P and N, of about 2 millimetres in diameter are passed through it. One of these wires is soldered to the side, and towards the upper end of the copper tube C of about 1 decimetre in length. This tube is intended to hold a little cylinder of charcoal, which is kept constantly pushed down from above by a small cylinder of lead, which is of a weight adjusted to the purpose. The other polar wire comes out under the lower end of this tube and terminates in a little cone D of prepared charcoal or coke, against the top of which the little charcoal cylinder E rests. When the current is established, the cylinder becomes incandescent, and remains in that state a very long time, because the small portions which are consumed are continually replaced by other portions of the cylinder which the metal presser forces on to the top of the cone. A tube of glass, or other suitable substance, also passes through the stopper and admits of its being connected to the pneumatic receiver, in order to create a vacuum in the globe. The tube is afterwards cut and hermetically sealed, in order to prevent the re-entry of the air."

The claim for the foregoing combination is as follows :—

"I claim the apparatus for lighting by the incandescence of prepared charcoal, before described."

This was among the last of the early attempts at lighting by incandescence in a vacuum. Up to this date, and for some four or five years afterwards, inventors struggled manfully to attain illumination by electricity, but one by one they fell out of the race; zinc, their source of energy, could not hold its own against its cheap active rival, coal, and it was not until the latter had been forced into the service of the electricians that they again turned their earnest attention to the old problems. The first inventors to resuscitate the incandescence system were of Russian nationality, among the foremost of whom was M. Lodyguine, a Russian physicist, who in 1873 proposed an incandescence lamp, subsequently improved by MM. Konn and Bouliguine; to M. Lodyguine, the Academy of Sciences, at St. Petersburg, awarded a special prize. His idea was not however practical. It consisted in placing in a dried and exhausted glass chamber one or more carbon pencils, of which the cross section was reduced in the vicinity of the luminous focus, by means of a commutator, by which the second carbon could be placed in circuit when the first was destroyed. A more practical form of lamp based on this order of design is that shown in Fig. **640**, and proposed by Konn; in it a carbon block was connected to two terminals of platinum. Of course the interior of the globe was exhausted of air.

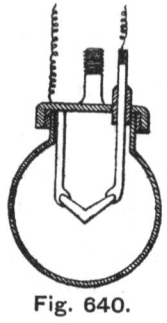

Fig. 640.
Konn.

The standard form of Konn's lamp is however that indicated in Fig. **641**. This was brought out in 1875, and, possessing more ingenuity than practical value, is of interest only historically. It consisted of a copper base A, formed with a socket to receive a glass bell, the lower part being cylindrical; the joint was made air-tight by a gland L and a leather washer. In the base an opening was formed, fitted with a valve K opening outwards, and to this opening an air pump could be connected to exhaust the air from the bell. Two copper rods C C rose

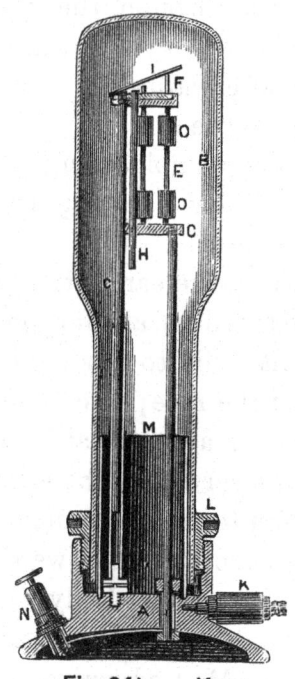

Fig. 641.   Konn.

vertically from the socket within the bell: the one was in electrical connection with the positive terminal, and the other, insulated from the base, communicated with the negative terminal. Both terminals were attached to the base of the lamp. At the top of the negative rod was screwed an insulated plate C, with five small recesses in its upper surface, and a similar plate F was attached to the positive rod in such a position as to be immediately over the plate on the negative rod. Through this second plate five holes were made corresponding in position with the recesses in the lower plate. To one edge of the upper plate a third plate was hinged, and a copper rod H passed through both fixed plates, connecting them together. The carbons employed were small pencils placed between upper and lower blocks of carbon O; beneath the lower blocks was a small copper stud fitting in one of the recesses of the lower fixed plate. On the top of each of the upper blocks of carbon was a copper rod passing through one of the holes in the upper plate. There were five such pencils, carbon blocks and copper rods all alike, except that the copper rods varied in length. At the bottom of the bell was a receiver M to catch the products of combustion from the carbon. The action of the lamp was very simple. As soon as the five carbons were put in place, with the hinged plate resting on the longest copper rod projecting through the hole in the upper plate, the bell was put in its socket, the joint made good, and the air exhausted. The circuit then being completed, the current passed through the positive rod, the hinged plate, and the copper rod with which it was in contact; then bringing the carbon pencil to a state of incandescence, it passed through the negative rod and back to the generator. As soon as one carbon was consumed, the hinged plate fell into contact with the second, and so on, till the whole were consumed; finally it made contact with the rod connecting the two fixed plates, and allowed the current to continue flowing to the other lamps on the same circuit.

Electric lighting by incandescent filaments, which has now reached so high a stage of development, depends in each of the systems successfully introduced, upon the same fundamental principle, the heating to whiteness of a carbon filament, contained in a glass bulb, from which the air has been removed as completely as possible. The distinctions between the different systems lie in the details of manufacture, which, as will be seen from the following descriptions, approximate more or less closely to each other. The systems now chiefly in use are those of Edison, Swan, Maxim and Lane-Fox, all of which are described in detail in the following pages, beside several others which have not yet acquired any practical success.

EDISON.

Before Mr. Edison arrived at his present standard form of incandescence lamp, he had with characteristic enthusiasm, experimented with a vast number of materials which he thought might serve as illuminating media. In 1879 he finished his investigation with metal filaments *in vacuo*, and the results he obtained are far too interesting to be passed over in silence, leading as they did to his subsequent employment of carbon filaments.

It was announced in this country, in the autumn of 1879, that Mr. Edison was, following the steps of old inventors, but with the new experience supplied by Crookes and other modern scientists, experimenting with an alloy of iridium and platinum, and with each metal separately, for the production of the light-giving portion of an incandescence electric lamp. It was then with considerable truth surmised that the former metal was too refractory to work, and the latter substance was too liable to destruction by fusion to render very promising the practical and commercial success on any extended scale of this employment of those metals. Mr. Edison, so far from being blind to these difficulties, set himself to search for fresh and more plentiful sources of supply of metals, and at the same time to prosecute an investigation into the phenomena taking place when they were rendered incandescent under different physical conditions.

Observing that a platinum wire heated to incandescence in a hydrogen flame, gave to the latter a greenish tinge, at the same time losing some of its own weight, Mr. Edison made some careful quantitative measurements to ascertain the nature and extent of the loss sustained by the platinum. He took a bundle of platinum wire four-thousandths of an inch in diameter, and suspended it in a hydrogen flame. Before the experiment the platinum weighed 306 milligrammes (about 4¾ grains), and during its period of incandescence it lost weight at the rate of about 1 milligramme (·015 grain) per hour. Mr. Edison also noticed that a platinum wire within a hydrogen flame produced upon that flame different effects when deriving its temperature of incandescence solely from the flame itself, than when superheated from an external source; thus, a platinum wire glowing with incandescence within a hydrogen flame by which it was heated, imparted to it a pale green colour. When a current of electricity, however, was passed through it so as to heat it above the temperature of the flame, the latter assumed a much deeper tone of green, and it would seem from this, that the depth of colour is in some way connected with the amount of platinum being thrown off per unit of time, in other words, with the rate at which it is losing its weight under the conditions in which it is placed. In one experiment a platinum wire weighing 343 milligrammes (nearly 5·3 grains) was maintained for nine hours in an incandescent state, at the end of which time it was found to have lost no less than 42 milligrammes in weight (·65 grains); and when a spiral ⅛ in. in diameter and ½ in. in length, composed of fine platinum wire, was enclosed within a small glass shade and maintained in a state of incandescence for 20 minutes by the passage through it of an electric current, it was found that the inner surface of the glass shade became slightly coated with metallic platinum, and that after five hours of such treatment the deposit became so thick as to be perfectly opaque to the light emitted by the incandescent spiral. Mr. Edison's explanation of the physical cause of this phenomenon is very interesting. "From this and other experiments," he says, "I became convinced that this effect was due to the wasting action of the air upon the spiral; that the loss of weight, and the colouration of the hydrogen flame, were also due to the wearing away of the surface of the platinum by the attrition produced by the impact of the stream of gases upon the highly incandescent surface, and not to volatilisation as commonly understood." Further experiments supported this theory, for no such wearing away or diminution of weight was produced when the spiral was enclosed in a bulb exhausted by means of a Sprengel pump to such a degree that the discharge spark from an induction coil would not strike across a distance of 1 millimetre. In this experiment the spiral was maintained at the most dazzling incandescence for several hours without the slightest deposit being formed upon the surface of the glass envelope.

It is very interesting in turning over the pages of the history of the progress of science during the last half century, to observe the different aspects with which philosophers at various periods regarded certain facts of science, certain substances, and certain phenomena connected with them, and how in their different views they seemed to arrive at different results; but as the observations of other investigators approaching the subject from other directions came to be recorded, and the great harmony of nature asserted itself in the human mind, things apparently the most opposite, united their forces to support the truth, proving that the conditions were the only agents in opposition, that the different positions assigned to the same phenomena by different observers were the result of a sort of mental parallax, and that Nature could not in these or anything else be divided against herself. Previous investigators had made researches bearing on these experiments of Edison; the names of Döbereiner, of St. Claire Deville, of Troost, of Faraday, and of Graham, will recall a multitude of investigations bearing on the occlusion and dispersion of hydrogen gas by certain metallic substances, more especially platinum, palladium, iridium, and certain other of the tetrad metals, under various physical conditions. It was found by Professor Graham that platinum, whether in the form of wire or of foil, is capable of absorbing when at a low red heat nearly four times its volume of hydrogen (measured cold), that palladium foil condenses more than six hundred times its volume of hydrogen at a temperature below that of boiling water, and that in the spongy state platinum absorbs 1½ times its volume of hydrogen, and palladium nearly one hundred times.* Hydrogen so occluded at comparatively low temperatures, and at ordinary pressures, is driven out again, when the metal which has absorbed it, is subjected to the action of heat and to a diminution of pressure, and Mr. Crookes in some of his beautiful experiments has taken advantage of these phenomena for varying the internal pressure within his exhausted vessels without having to use the Sprengel pump. The facts recorded by Mr. Edison of the diminution of weight in metallic wires when rendered incandescent at ordinary pressures, are enhanced in interest by comparing them with the researches of the pioneers in the science of the subject to whom we have alluded. The modification of the phenomena when atmospheric pressure is removed, and when the same experiments are conducted in high vacua, taken in conjunction with Mr. Edison's explanation of it, adds very great interest to the researches of Mr. Crookes upon radiant matter. The last-named investigator has shown that, by the impact of gaseous molecules upon the vanes of a mill, mechanical motion may be imparted to them, and by the concentration of streams of such molecules upon a small area the most intense heat may be produced; Mr. Edison has shown that these

---

* Philosophical Transactions, 1866.

molecules, infinitesimally minute as they are, are so thoroughly material in nature that their friction on the surface of the heated platinum is sufficient to rub away its substance, transporting its particles to another place in a manner analogous to the action of the sand-blast process of Mr. Tighlman, for the cutting and engraving of stone, glass, and other hard substances.

From these experiments, Mr. Edison was led to undertake a minute examination of the structure of the substance of wires of platinum and other metals before and after being subjected to incandescence. In the course of this examination he found that the effect of incandescence upon the wires experimented upon, was to produce all over their surface innumerable cracks, and when the incandescence was maintained for twenty minutes these fissures became so enlarged as to be visible to the naked eye. When still further continued for several hours the cracks united and the wire fell to pieces, and Mr. Edison pointed out that a somewhat similar phenomenon had been noticed by Professor Draper in platinum plates which had for a long time been subjected to the action of flame. A number of experiments made by Mr. Edison led him to the conclusion that this cracking up of the surface of the metal, is due entirely to the occluded gases imprisoned within its mechanical and physical pores becoming expanded and driven out under the action of heat, these gases first separating the molecules of the metal, then tearing their way through it to the surface, and escaping by the cracks and fissures so produced, a sort of microscopical volcanic action going on within the metal, producing miniature earthquakes· and fissures on its surface out of which the heated gases are emitted. By heating spirals of platinum wire gradually within a chamber exhausted to a pressure of 2 millimetres, by means of a transmitted electric current of periodically increasing strength, the gaseous bodies contained within their substance were gently withdrawn; and, by allowing the metal in the interval, between each ·increase of temperature to cool down, a series of expirations of air from its surface took place, alternating with a closing up and welding again together of the minute fissures through which the gradual heating *in vacuo* had enabled' the gases to escape. The pores of the metal thus became converted into an infinite number of what might be called molecular valves opening outwards whenever subjected to the gaseous pressure from within caused by the action of heat, but closing and welding up again with increased density at each alternate cooling.

By continuing this very simple operation Mr. Edison was able so completely to change the physical character of the metals subjected to the process, that he may almost be said to have made new substances altogether, for he produced metals in a state such as had never been known before, increasing their hardness and density to an extraordinary degree and raising their points of fusion so high that they remained unaffected in temperatures at which most substances would be melted or consumed, and very many would be volatilised. Most persons are aware that when a spiral of platinum wire is heated to a bright incandescence, it softens and loses its elastic or rigid character, but it was reserved for Mr. Edison to discover that when the same spiral is subjected to this alternate gradual heating and cooling in an exhausted chamber it becomes as homogeneous as glass and as rigid as steel, while a fine spiral so treated is, at a temperature at which it is glowing with an incandescence of dazzling brilliancy, as springy and elastic as when cold, and cannot be annealed by any known process.

By the process we have referred to, spirals of platinum wire could be raised by the passage of the electric current to such a high temperature that their incandescence gave out a light of 25 standard candles, while the same coils not treated by the process, gave a maximum light of but four candles at their point of destruction by fusion. One spiral which had been submitted to the vacuum heating process more gradually, and for a longer period, produced a light of 30 candles, and the surface of all the wires so treated was free from roughnesses or cracks of any kind, presenting a brilliantly polished exterior; they were as hard as the steel used for pianoforte wires, and could not be annealed. Interesting and valuable as these experiments were, they did not by any means solve the question of incandescence lighting, and the platinum spirals were afterwards abandoned in favour of the carbon filaments Edison now employs.

In order to indicate clearly the experimental stages through which Mr. Edison's investigations passed, it will be useful to refer in some detail to the patent in which the processes just referred to are described. This patent is dated 17th June, 1879, and is entitled "Improvements in electric lights, and in apparatus for developing electric currents and regulating the action of the same." After referring to the phenomenal changes in the structure of platinum wires, to which we have referred, and having hinted some explanation as to their cause, the specification describes the manner in which the inventor heats the platinum spiral. "A spiral of platinum wire is placed in a glass bulb with its ends passing through and sealed in the glass, and the air exhausted from the bulb by a Sprengel pump until the discharge from a 3-in. induction coil will not pass between two subsidiary wires in the bulb, the ends of which are 4 millimetres apart   The wires of the spiral are then connected to a magneto-electric machine, or battery, whose current can be controlled by the addition of resistances. Sufficient current is allowed to pass through the wire to bring it to about 150 deg. Fahr.; it is allowed to remain at this temperature for 10 or 15 minutes. While thus heated the air or gases in the pores of the metal are expelled by the action of the heat and the expansion of the gases which tend to pass outwards in consequence of the vacuum.   While the air or the gases are passing out of the metal, the mercury pump is kept continuously working.   After the expiration of about 15 minutes the current passing through the metal is to be augmented, so that its temperature will be about 300 deg. Fahr., and it is allowed to remain at this temperature for another 10 or 15 minutes.   If the mercury pump be worked continuously, and the temperature of the spiral raised at intervals of 10 or 15 minutes, until it attains vivid incandescence, and the bulb be then sealed, the metallic wire is then in a state heretofore unknown, for it may have its temperature raised to the most dazzling incandescence, emitting a light of 25 standard candles ; whereas before treatment the same spiral would only emit a light equal to three candles before attaining the melting point."   After describing several metallic alloys and combinations, the specification proceeds, " Carbon sticks may be also freed from air in this manner, and be brought to a temperature that the carbon becomes pasty, and if then allowed to cool, is very homogeneous and hard. Rods or plates made of mixtures of finely-divided. conducting and non-conducting material may thus be freed from air."

Fig. 642 shows one form of lamp schemed at this time by Edison, and embodying the devices described in the patent now referred to.   In this lamp, *a* is the spiral of platinum, carried on an infusible bobbin, and supported within the vacuum tube *b* by the rod *b*[1], of the same material as the bobbin. The vacuum tube *b* is carried by the case *k*, and around the tube there is a glass globe I. Within the case *k* is a flexible metallic aneroid chamber L, that opens into the glass case, so that the air when expanded by the heat, can pass into the aneroid chamber and give motion to the flexible diaphragm and parts connected therewith. When the current circulating round the bobbin a becomes too intense and heats the latter too highly, the air within the glass case is expanded, and causes the diaphragm x to be deflected, so making the pin on the diaphragm

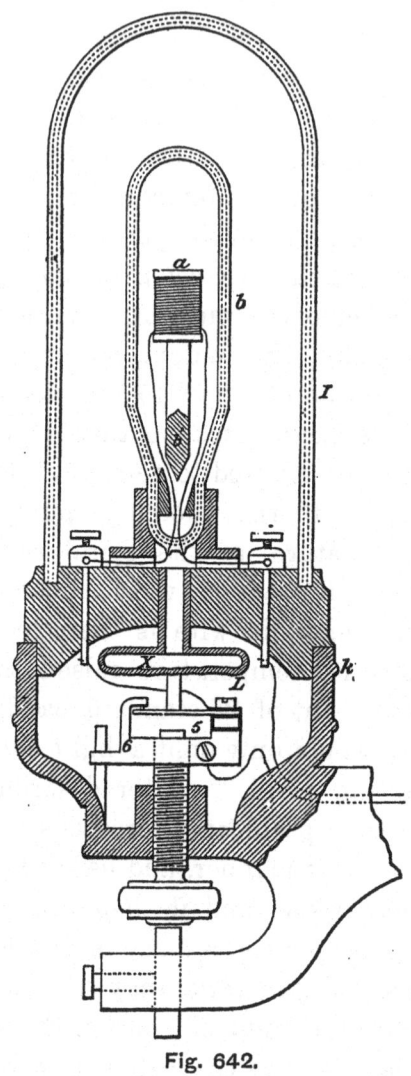

Fig. 642.

press on the spring 5, and separating this spring from the block 6, breaks the circuit to the lamp. The temperature within the globe I is thus immediately reduced, and the parts return to their normal position, the current being restored to the lamp by the contact re-made between the spring and the block 6. All credit must be awarded to Edison for this ingenious arrangement, which appeared full of promise not three years ago, but which to-day is a marvel of complication compared with the present standard incandescent lamp of Edison, Swan, and others.

The claims referring to lamps in this patent of Edison's (17th June 1879) are five in number, as follows:—(1.) The combination with a sealed vacuum chamber made of a glass vessel, of a continuous incandescent conductor wound upon a bobbin of infusible material. (2.) The method described, of preparing metals and metallic conductors for electric lamps or burners, consisting in freeing the metallic conductors of gases in a vacuum, and then hermetically sealing the surrounding air-tight transparent case. (3.) (Sixth claim). The process of moulding sticks of pulverulent material, consisting in filling a mould with the material to be compressed, and then compressing the mass by successive operations by means of a plunger. The seventh claim refers to the moulding apparatus employed. The second patent to be noticed is dated the 10th of November, 1879, the object of the invention being "to produce electric lamps giving light by incandescence, which lamps shall have a high resistance so as to allow of the practical sub-division of the electric light." The intention is stated of using " carbon wire," arranged so as to offer a high resistance with only a small surface for radiation, such carbon wire being of course contained in an exhausted bulb, and the current conducted to it by platina wires sealed into the glass. After referring to the practice which had been followed up to that date (by Konn and others), of using relatively large rods of carbons in closed vessels containing gases that do not combine with the carbon, such rods having a relatively low resistance, Mr. Edison proceeds as follows: " I have reversed this practice; I have discovered that even a cotton thread properly carbonised, and placed in a sealed glass bulb exhausted to one-millionth of an atmosphere, offers from 100 to 500 Ohms resistance to the passage of the current, and that it is absolutely stable at very high temperature; that if the thread be coiled as a spiral, and carbonised, or if any fibrous vegetable substance which will leave a carbon residue after heating in a closed chamber be so coiled, that as much as 2,000 Ohms resistance may be obtained without presenting a radiating surface of more than $\frac{3}{16}$ in.; that if such fibrous material be rubbed with a plastic compound composed of lamp-black and tar, its resistance may be made high or low according to the amount of lamp-black placed upon it." He then proceeds to describe means for making "carbon wire" out of plastic material rolled out and carbonised, or wound into a spiral between a helix of copper wire to be afterwards eaten away by nitric acid. Edison had at that time experimented with carbonised "cotton and linen thread, wood-splints, paper coiled in various ways, also lamp-black, plumbago and carbon in various forms," and so forth. The following important sentences also occur: "By using the carbon wire of such high resistance, I am enabled to use fine platinum wires for leading wires, as they will have a small resistance compared to the light giving body, hence will not heat and crack the sealed vacuum bulb. There are four claims to this patent as follows: (1.) An electric lamp for giving light by incandescence, consisting of a filament of carbon of high resistance, made as described and secured to metallic wires, as set forth. (2.) The combination of a carbon filament within a receiver made entirely of glass, through which the leading wires pass, and from which receiver the air is exhausted. (3.) A coiled carbon filament or strip arranged in such a manner that only a portion of the surface of such carbon conductor shall radiate light. (4.) The method herein described of securing the platina contact wires to the carbon filament and carbonising of the whole in a closed chamber." From the foregoing it will be seen, that some of the most important principles controlling the Edison incandescence lamp are set forth, and it is interesting to note the characteristic way in which he had experimented with a great variety of materials. Cotton thread he had tried, but it did not occur to him to convert the thread before carbonising into a new material by sulphuric acid, as Mr. Swan did a few months afterwards, thus creating an efficient and thoroughly reliable filament.

The next patent (15th December, 1879) refers chiefly to the use of paper for the luminous conductor, and to the means for carbonising the same. In his specification, Mr. Edison explains how he prefers to use the best "Bristol board," from which he punches out narrow elliptical strips, with enlarged ends. A number of these were placed on each other flat-ways, in a mould of wrought iron with strips of tissue paper between each layer, covered with a flat piece of gas coke, and raised gradually to about 600 deg. Fahr. ; even a second carbonising at a white heat succeeded, after which the mould and its contents were allowed to cool gradually. The specification describes the mounting of these strips in the bulb and it comprises the following claims :—" (1.) An electric lamp formed of carbonised paper. (2.) The method herein described of manufacturing carbons for electric lights, consisting in exposing the filament of paper to the action of heat in a mould to drive off the volatile portions and carbonise the paper. (3.) A carbon for electric lights made as a filament with the ends broader for the clamping devices that connect the conductors." The fourth claim refers to the mode of securing the carbon. Fig. 643 shows the form of lamp made by Edison in accordance with this specification; like Swan, and at just about the same time, he obtained promising, but fallacious results, and he was still far from having arrived at the desired end. These paper carbons were **very**

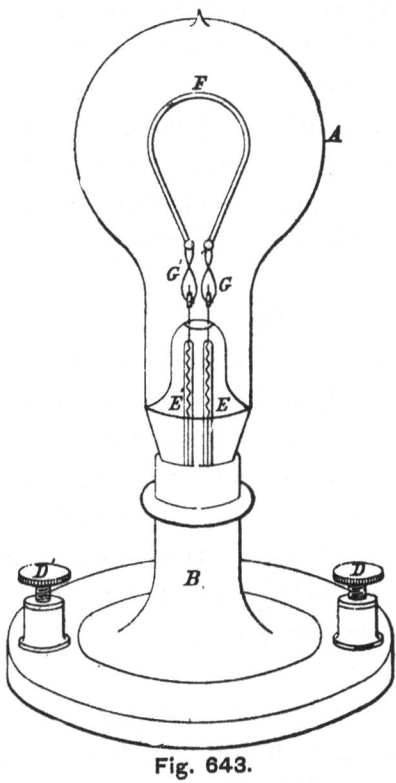

Fig. 643.

unreliable, some were fairly durable, others failed very quickly, their resistances were not uniform. In a word, this device led the inventor a step, and a wide step forward towards ultimate success, and the " electric lamp formed by carbonised paper," as the specification somewhat dubiously defines it, has an historical interest. In the figure, which explains itself, A is the glass bulb, B the wooden stand of the lamp, D D[1] are the circuit wire terminals connected with the platinum wires E E[1] that are fused into the tube forming a part of the lamp, G G[1] are the clamps holding the carbonised paper strip F.

Edison's patent, dated 16th September, 1880, is entitled, " Improvements in electric lamps, and in carbons or incandescing conductors therefore, and in means for and methods of manufacturing the same." The specification of this patent, which is very voluminous, is interesting and important, as it describes in great detail the mode (modified doubtless since that date) of preparing the carbon filaments, and the physical properties which, according to Edison, are necessary in the carbons to produce efficient lighting. This latter point is particularly dwelt upon in the specification, wherein it is pointed out that, prior to the date of the patent (September, 1880), in manufacturing carbons for electric lamps the practice was to make them of as low resistance as possible, a porous carbon having been used which was dipped or soaked in some carbonisable liquid until its pores were filled, and then subjected to re-carbonisation, which process was repeated until the pores of the original carbon were filled. By this process the resistance of the carbon is lessened while its liability to disintegration under high heat is increased.

For incandescent lamps, Mr. Edison states that he discovered that the filament should " have the highest possible resistance in a very small bulk, and be capable of resisting the disintegrating effects of very high heats and the absence of atmospheric pressure, and further that carbons which are purely structural in character alone possess these qualities. By purely structural is meant a carbon wherein the natural structure, cellular or otherwise, of the original material is preserved unaltered, that is, not modified by any treatment which tends to fill up the cells or pores with unstructural carbon, or to increase its density, or to alter its resistance."

The specification proceeds to state that in practice " the incandescence conductor of a lamp should be of about 100 Ohms resistance. While this may be varied within certain limits, the resistance stated is a preferable one, and is a very high resistance compared to the carbons referred to as previously used. It is essential that this high resistance should be had without increase of radiating surface, that is, only the radiating surface necessary to give a certain standard amount of light ·at the proper degree of incandescence should be used."

Filament conductors possessing the qualities here claimed as necessary may be made of such natural fibres or fibrous bodies as are capable of carbonisation, " especially such as are large, filamentary and cellular in character. Of such there are several varieties, of some of which a single fibre or several fibres aggregated together artificially are used, of some a piece composed of numerous elementary fibres are used."

" The preferable single fibres are those of which jute, bast, manilla, hemp, &c., are good types, the more preferable one being a fibrous grass from South America called " monkey bast " fibre, each blade of which is generally round and composed of a great number of elementary fibres held together by a natural cement or resin, which carbonising, locks all the elementary fibres together into a homogeneous filament. These blades vary somewhat in size and are also slightly tapering. It being necessary that the conductors should be of a uniform size, they are reduced to uniformity by passing them through a die that shaves off the superfluous material. It is in this patent that the use of cane or bamboo as a suitable material for carbons is mentioned, and the treatment is thus substantially described :—

" The cane is split into pieces somewhat wider than necessary, and the inner or pith portion removed ; it is then cut into strips, is passed through a shaving tool, in which the knife is fixed, the material being forced against it by a movable block provided with an adjustable stop screw, by which the distance of the block from the knife can be regulated to adjust the thickness to be given to the slip. The slip is thus shaved on both sides until the proper thickness is attained. It is then placed in a clamp made in two halves, of a length equal to the desired length of the slip. In one half a shoulder or off-set is formed at a distance from the edge greater than the desired width, upon which one edge of the fibre rests, which is then clamped between the two halves, and the protruding portion carefully shaved off, which may be done by hand or by a cutting blade moved by machinery. On the opposite side of the clamps is an off-set in one half, at a distance from the edge exactly equal to the width to be given the slip. Upon the same side of the clamps, and at the ends, projections are made of the exact shape and size to be given the broadened ends of the carbons. The slip shaved upon both sides and one edge is transferred to the opening in the clamps, and the extra material shaved or cut off ; it is now of uniform size throughout its body, with enlarged ends (the widening is only on one edge) formed upon it."

Various other devices are described for preparing the filaments, but these need not be referred to here. Detailed descriptions are given of modes of carbonisation, which may be summarised as follows :—A metal (preferably nickel) flask is employed, in the bottom of which is cut a groove of a curved or horseshoe form, which the filament is ultimately to receive; the filament is placed within this groove, and a relatively heavy metal cover is laid upon it in the flask, the upper face of the cover having a similar groove cut in it to receive a second filament, over which a similar cover is placed, and so on till the flask is full, and a considerable number of filaments are ready for carbonisation. The flask so filled is placed in an oven and subjected to an intense heat, produced by gas fuel, and a suitable blast directed upon the flask. In some cases it is stated to be found desirable to maintain an atmosphere of hydrogen, " or some hydro-carbon," within the flask during carbonisation, to prevent excessive oxidation of the filaments. There are no fewer than 44 claims in this specification, of which some of the more important are substantially as follows:—(1.) "An incandescing conductor, formed of one or more carbonised natural fibres," as described. (2.) A carbon clamp for uniting the filament and metallic conductors, formed of a carbonisable material. (3.) The mode of forming the blanks for the carbons, by placing the material between contoured clamps, and cutting away the superfluous material. (4.) "A slip or filament for forming by carbonisation the incandescing conductor of an electric light made of a material composed of fibres laying parallel through the length of the slip or filament." Besides the foregoing there are numerous claims referring to the process of carbonisation, the form of lamps, the use of mercurial column seals and conductors, modes of testing carbons, arrangements of circuits, &c., &c.

The public, through the magnificent exhibits at the Paris Exposition de l'Electricité, the show at the Crystal Palace, and the installation at the Holborn Viaduct, are familiarised with the present form of the Edison lamp. This is illustrated by Figs. 644 and 645, the former being an illustration of the larger, or 16-candle lamp, the resistance of which, when in action, is about 140 Ohms, while Fig. 645 represents the eight-candle or half-lamp, whose resistance is about half that of the larger one. The fittings by which Mr. Edison places his lamps in circuit with the conductors leading from the generators are very simple and ingenious, and are well shown in two forms in the figures. In Fig. 644 the two ends of the carbon filament are connected respectively, by means of platinum wires, to a screw and a cone, each stamped in thin brass, held at the proper distance apart, and insulated from one another by a filling of plaster-of-paris, which surrounds the neck of the envelope and forms a firm and rigid attachment. The socket into which this fitting screws, and which is in permanent connection with the conducting mains, consists of a cone of thin brass, below which is a hollow screw, both being embedded in a cylindrical mass of plaster-of-paris, and forming a matrix, or counterpart, of the fitting on the lamp. The action of screwing a lamp into its socket is to draw the cone on the lamp firmly into its seat, and at the same time to press the threads of the screw against the corresponding threads of its helical socket, thus insuring very perfect metallic contact between the terminals of the lamp and the

Fig. 644.   Fig. 645.

conducting mains. This form, excellent as it is, has certain disadvantages, one of which is that the mechanical action of the contrivance, when being screwed up tight, is to put a tensile strain on the plaster filling, and therefore has a tendency to break the connection by pulling the screw and the cone apart. To meet this objection, Mr. Edison has since introduced the form of fitting shown in Fig. 645, in which the upper cone is removed, and a flat disc, or sole-plate, shown at the bottom of the figure is substituted for it. When a lamp of this form is screwed into its socket, the sole-plate is pressed against a corresponding plate at the bottom of the socket, and the strain on the plaster being one of compression, there is no tendency for it to be pulled apart. Mr. Edison has still further improved the material of his fittings by constructing them of thin copper instead of brass. The latest types which are supplied with these copper fittings, and the pear-shaped glass envelope, are replaced by a very elegant form consisting of a cylinder with hemispherical ends (Fig. 646). By this arrangement, the sides of the envelope are at a more uniform distance from the incandescent filament, and while presenting a much more shapely appearance, they are not so liable to crack under the influence of unequal heating, and there is a smaller loss of light through refraction and internal reflection.

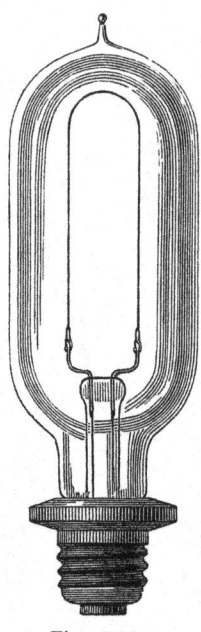

Fig. 646.

A considerable portion of the work of manufacturing the lamps is performed by machinery, by which greater uniformity is insured, but the chain of operations by which the rough materials are converted into a finished incandescence lamp, is identical in its sequence and character with that of hand-made lamps, and will be understood from Fig. 647. The bulbs or flasks are delivered from the glass works in the form shown at No. 1. The first process is to heat the closed end and draw out a hollow cone, shown in No. 2. To this little cone is then fused a short fine tube, 3, by which the envelope is afterwards to be attached to the exhausting apparatus. The bulb then presents the appearance shown at 4. The next process is to prepare the glass stem, by which the carbon filament is held in its place within the globe, and within which the platinum conductors are fixed that are to join it to the connecting fitting. A piece of tube, No. 5, is taken, and its two ends are drawn or tapered off, as shown in No. 6. It is then thickened up in two places (see No. 7), between which it is divided into two separate tubes, shown at 8 and 9; two platinum conducting wires are then passed through the tube

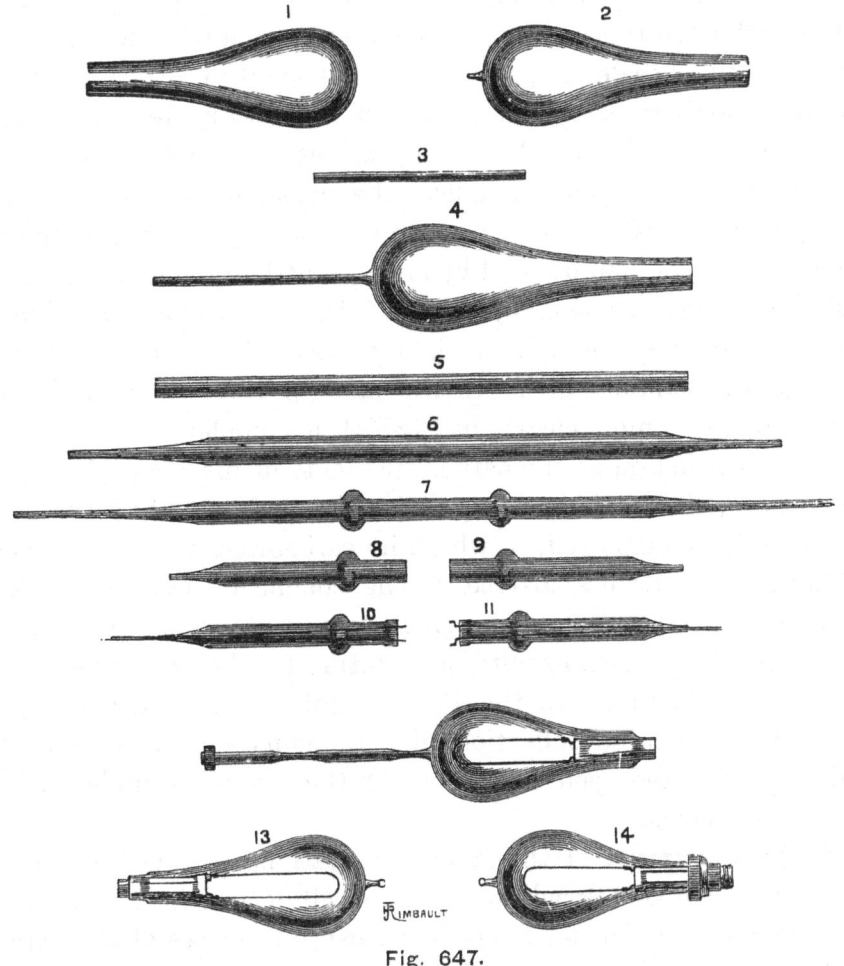

Fig. 647.

(see No. 10), care being taken that they do not touch one another. The large end is then heated nearly to its point of fusion, and is squeezed flat so as to cause the platinum wires to be closely imbedded within its substance. The little cranked pieces, shown in No. 11, are then soldered to the ends of the wires, and to these are attached, by an electro deposit of pure copper, the ends of the carbon loop to be presently described. The glass stem with the carbon filament attached, is then introduced into the neck of the bulb, to the mouth of which its thickened portion is attached by fusion, so as to form with it one piece of glass. The drawn down end is then trimmed off square, as shown in No. 12, which also illustrates the drawing down in two places of the exhausting tube, so as to form two capillary contractions. This tube having been attached by a mercury joint to a Sprengel pump, the bulb is exhausted, and when the required exhaustion is complete, the contraction nearest the pump is sealed off, leaving a short length of tube attached to the bulb by a capillary neck, but exhausted to the same degree of internal residual pressure. A second sealing off at this neck renders the hermetic closing of the lamp complete, the object of the double sealing being to guard against any leaking into the envelope at the moment of detaching it from the pump; and by fusing a little globule of glass upon the sealed off point (see No. 13) all possibility of an opening being left into the outer air, however minute, is removed. Nothing now remains but to attach the terminal fitting to the stem of the bulb, and the lamp (shown in No. 14, or more clearly in Fig. 644) is complete.

In the construction of the carbon filaments in his lamp Mr. Edison, as we have seen, after making a long series of experiments to ascertain the best material to be employed, in which he carbonised a large number of vegetable fibres and tissues, arrived at the conclusion that the fibres of certain kinds of bamboo offered the greatest advantages, both for facility of manipulation and for uniformity of structure for the preparation of the incandescing conductors. In the Paris Exhibition of Electricity, Mr. Edison exhibited a large collection of specimens of fibres and other materials experimented upon, among which there were examples of some 20 varieties of bamboo.

Fig. 648 is illustrative of the general operations involved in the preparation of the bamboo (already referred to), all of which are carried on by boys. Canes of this material are first sawn into pieces of the required length (No. 1) which are from the character of the plant naturally tubular. One of these pieces is then split longitudinally down the middle into two half tubes, 2 and 3, each of which is again split into three narrower strips, shown at 4 and 5; the hard silicious outer covering peculiar to the plants of the bamboo family is then removed, and the straight fibrous portion is shaved down to a uniform thickness; these stages of the process are represented by Nos. 6 to 9, the latter of which is perfectly straight, flat, of uniform thickness throughout, and cut to the required length. The next step in the process consists in pressing one of these flat strips between two metal blocks which are accurately surfaced to one another, so as partly to project throughout the whole of its length; this projecting portion is then cut away, with the exception of a little piece at each end, so as to leave a fine thread of bamboo fibre having enlarged portions at

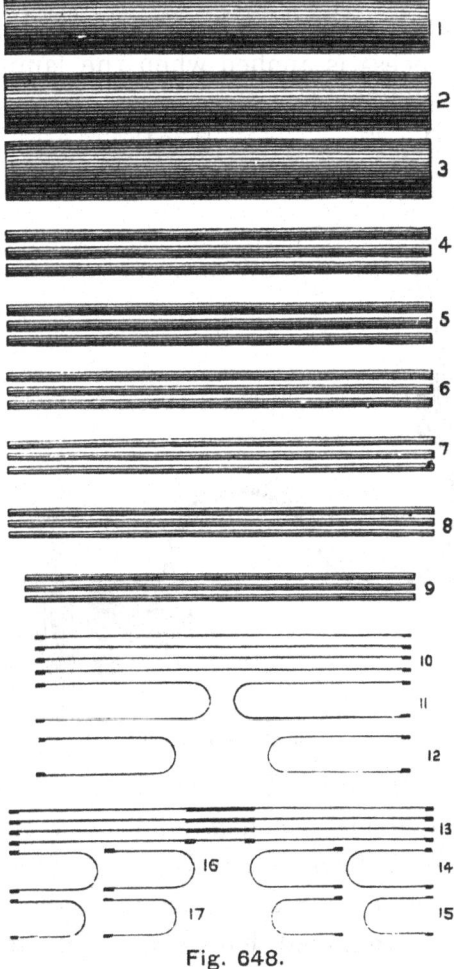

Fig. 648.

either end as shown in the figure at No. 10, or if the intended filaments are for the half-lamps, an enlarged portion is left in the middle as well as at the two ends (see No. 13), so that two filaments may be constructed out of the same bamboo strip. The threads so far prepared are then laid in moulds consisting of nickel plates in which grooves are cut of the forms shown at 11, 12, 14, 15, 16, or 17; a flat nickel plate is then fixed over them, so as to enclose them and prevent the admission of air during the process of carbonisation, which is effected in raising the moulds with their contents to a high temperature by placing them in a muffle and heating them to incandescence. When the moulds are opened, the filaments are in a condition to be electro-plated to their platinum supports, and introduced into the lamp-bulbs, but yet another process is required to confer upon them their perfectly homogeneous and elastic character, as well as their refractory nature at high degrees of incandescence, which are such essential characteristics in an incandescence lamp. This further process is applied when the lamp is attached to the pump, and while exhaustion is going on ; it consists in alternately heating and cooling the filament in the Sprengel vacuum by passing

Fig. 649.

currents of electricity through it of increasing strength until high degrees of incandescence are reached, and between each increase of current allowing it to cool down, exhaustion going on all the time. By this process, which we more fully described on a previous page, not only are the occluded gases eliminated, but as the fibre is subjected to a far more severe test than it can ever be subjected to in working, none but the "fittest" survive, and a healthy generation of lamps is insured.

We now come to refer to the different methods adopted by Mr. Edison for mounting the lamps and applying them to commercial or to domestic use, and the ingenious connections and other details of installation.

Fig. 649 is an illustration of a pivotted wall bracket, carrying an

Edison lamp, which is fitted to it by the screw attachment described in connection with Figs. 644 and 645, and Fig. 650 is a double-jointed bracket lamp which can be turned about and turned on and off like a double-

Fig. 650.

jointed gas bracket. The interesting feature of these brackets is the construction of the joints on which the arms of the brackets are pivotted,

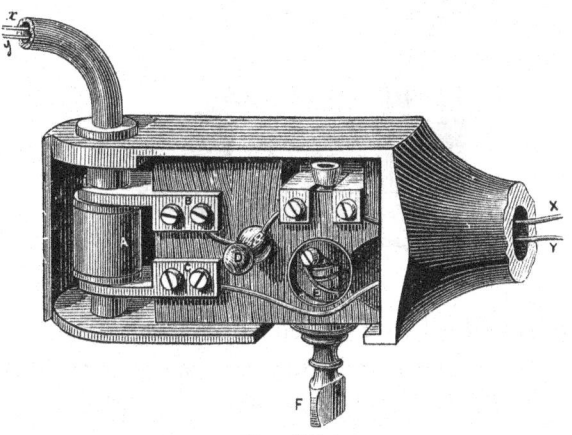

Fig. 651.

and which are so arranged that the current is transmitted to the lamp in all positions of the movable arm.

Fig. 651 is an enlarged view of the joint in Fig. 649, which is identical in construction with that nearest the wall in Fig. 650. Referring to Fig. 651, it will be seen that the two conducting wires X and Y, connected with the main circuit, are attached respectively to two metallic springs B and C, which press against two brass collars insulated from one another, and which are in connection respectively with the insulated wires x and y, that pass through the tubular arm of the bracket, and lead to the lamp. It will be seen that by this arrangement X and Y are

Fig. 652.

always in metallic connection with x and y through the springs and collars, which are always in contact in whatever position the arm may be F is a small quick-acting switch by which the current can be shut off or turned on by an action precisely similar to the turning of a gas-tap. The action of this switch is very imperfectly illustrated in Fig 651, but it is more clearly shown in Fig. 652. By turning the button F the conical plug seen above it at the top of Fig. 651 is, by the action of a screw,

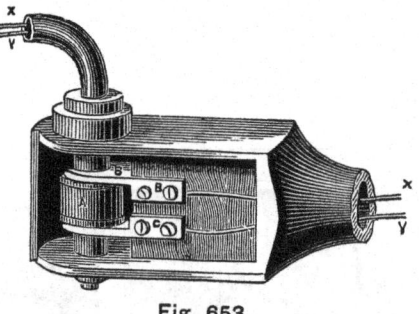

Fig. 653.

drawn into or lifted out of the space separating two plates, thus completing or interrupting an electric circuit passing from one to the other. This plug, as shown in Fig. 652, consists of a split cone with a strong spring keeping the two portions apart, and by this contrivance a very firm contact is insured when the cone is drawn into its seat between the two plates.

Fig. 653 is an enlarged sectional view of the smaller elbow joint of

Fig. 650, and explains itself. It is, in fact, similar in construction to that shown in Fig. 651, but without the switch.

Fig. 654 shows the application of an Edison light to a portable lamp, which is connected with the conducting main by a pair of wires enclosed in a flexible covering, and which can be put into or out of circuit by the turning of a little switch shown in the upper part of the figure.

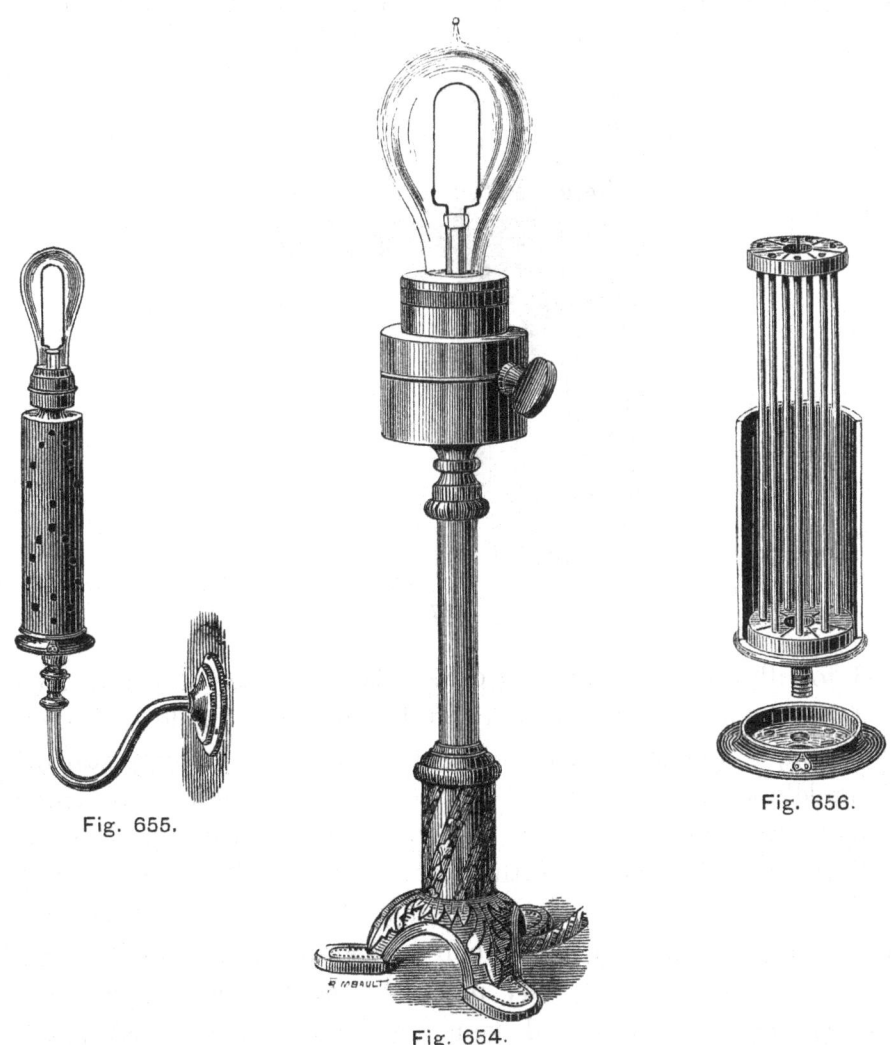

Fig. 655.

Fig. 656.

Fig. 654.

Fig. 655 represents a bracket lamp which is capable of being turned gradually up or down so as to produce varying degrees of illuminating power. Around the vertical supporting stem of this lamp, are fixed a number of vertical rods of carbon, shown in Fig. 656, forming a sort of rheostat or resistance box, the resistance in the circuit of the lamp depending upon the number of rods which are included in it, and this number can be varied at will by turning the circular base of the resistance box through a portion of a circle. The rods are protected from injury by being covered with a cylindrical metal tube, which is perforated with holes to admit of a free circulation of air, and thus to dissipate the heat generated by the resistance offered to the passage of the current.

Fig. 657 is an illustration of a form of Edison lamp, or rather lantern,

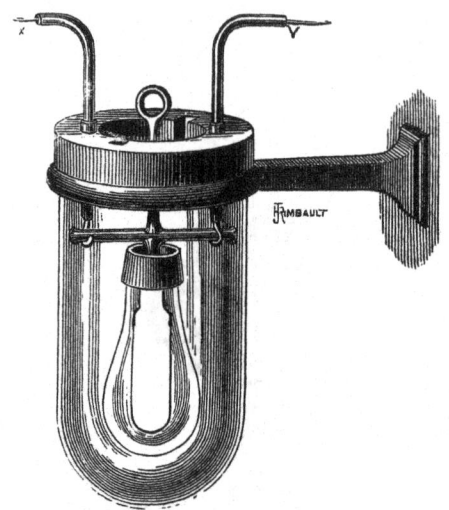

Fig. 657.

designed for illuminating mines or other places where there is danger of fire or of explosion from the ignition of inflammable or explosive vapours. This very simple apparatus consists of an incandescence lamp supported in the middle of a glass vessel of water with which it is entirely surrounded, the light emitted by the lamp passing through the water and through the glass sides of the water vessel. This idea was first suggested by Dr. Tyndall, in the discussion at the Society of Telegraph Engineers on Mr. Swan's paper, when his lamp was shown in London for the first time, and it is very ingenious so far as it goes, but it does not get over the principal point of inconvenience and possible danger involved in the electrical illumination of mines, namely, the difficulty of guarding against injury to the conductors, which might at any time plunge a section of the workings into sudden darkness, or possibly, in a fiery mine cause an explosion by a spark or small arc being formed between two exposed portions of the conducting cables.

We now pass on to consider the conductors adopted under the Edison system for the canalisation of the current, and by which electricity is distributed over a district to be lighted from a central station; we shall then refer to some of the switches and other details connected with the distribution, and lastly to the apparatus devised by Mr. Edison for regulating the strength of the current to the work it is required to do, and the instruments by which the extent to which the current is utilised by subscribers or lost through defective insulation is indicated and recorded.

The conductors may be grouped into three classes, viz., (1.) the principal and submains which carry the great outgoing current beneath the streets and in the subways; (2.) the conducting cables which are employed for derivations, such as house-service leads, supplying the

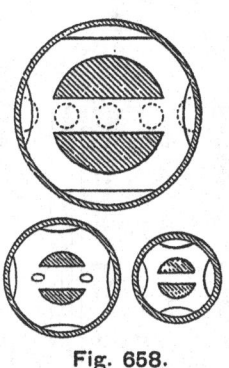

Fig. 658.

current to smaller numbers of lamps than would require the conductors of the first group; and (3.) the single insulated wires of various diameters by which the groups of lamps are placed in circuit with the service conductors and their branches. Those of the first class, or electric mains, alone require special notice, the two other classes differing in no material respect from the electric light cables and wires used under other systems. The mains are, however, of peculiar and interesting construction. The conducting portion consists of two solid copper rods of segmental cross section, clearly shown in Fig. 658, which represents a cross section of three different sizes of these conductors, suitable respectively for large mains or for branch circuits, or for the mains of smaller installations; the two segmental bars are protected from injury and from moisture by being passed through the middle of a wrought-iron pipe, and are held in position in the first place by diaphragms punched out of stout

R R 2

millboard and threaded on to the rods at intervals along the length of the tube; and the whole of the intervening spaces are filled with a bituminous insulating compound which performs the threefold duty of keeping the conductors in the middle of the tube throughout their length, of serving as an insulating material protecting the conductors from damp, and of so completely filling the tube that the latter, with its conductors in place, can be bent to any desired angle without distortion of its sectional cross figure, or displacement of its contents. At both the Paris Exhibition of Electricity and the Electrical Exhibition at the Crystal Palace, specimens of these conductors in various sizes were exhibited; they are made in lengths of from 15 ft. to 25 ft. As copper rods of

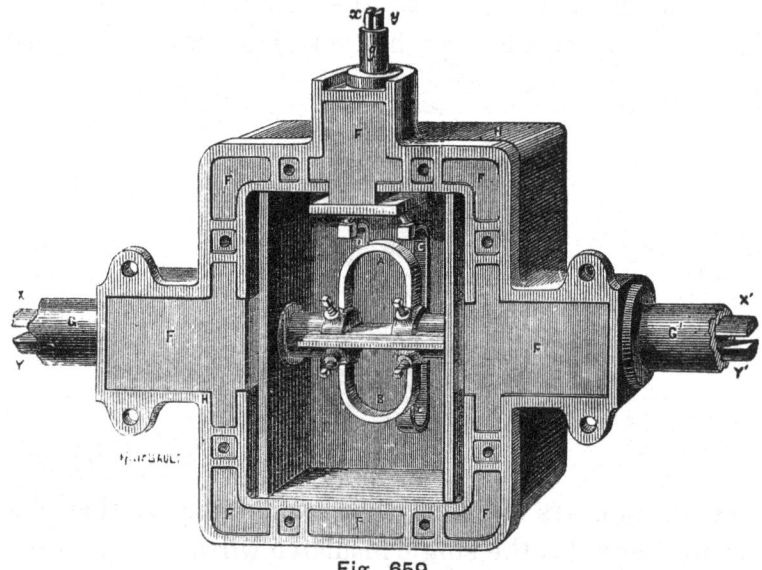

Fig. 659.

such a length must necessarily expand and contract under variations of temperature, it would naturally follow that unless some provision were made to allow for variations of length without distorting the mains, they would in a short time destroy themselves, or at least would soon be injured in both conductivity and insulation. The arrangement for meeting this difficulty is shown in Fig. 659, which is an illustration of one of the main junction boxes with its cover removed. This consists of a box of cast-iron H H with double walls, through the side of which pass the ends of two pairs of conductors X Y and $X^1 Y^1$ to be joined; the connection between them is made by two horeshoe, or U-shaped, strips of copper A B. These strips are of equal sectional area to that of the conductors, and are, on account of their form, capable of closing or opening under the elongation or shortening of the straight conductors. The figure also shows the method of connecting a branch submain $x\,y$ with the main circuit, the conductor x being connected to the loop A by the copper strip D, while y is a metallic connection with B by the bent strip C. Every branch circuit throughout the system is, moreover, protected against any sudden increase in the strength of the current by being provided with a safety guard consisting of a short length of lead wire, through which the whole of the current passing into a branch circuit is transmitted; and as long as this current is of its normal strength, it acts simply as one portion of the conductor, but its cross section is so determined that it is incapable of transmitting a current of a strength

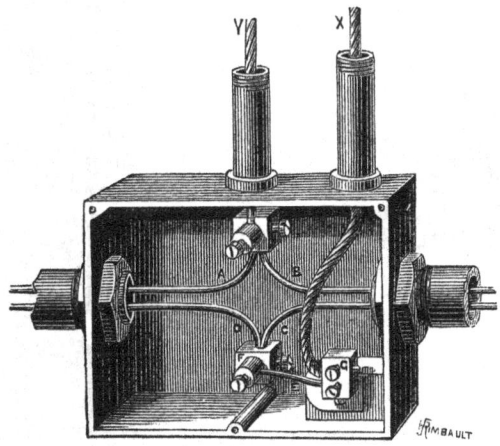

Fig. 660.

above that at which the branch circuit can be worked in safety; the moment this strength is exceeded the lead wire fuses, and the branch to which it is fitted is cut out of the circuit. This safety guard is applied not only at every junction of submains with the main, but at every branch leading from the submains, and at every derived circuit, or group of lamps, branching again within a house; the arrangement is still more clearly shown in Fig. 660, which represents a junction box for connecting the cables for a house service with a submain. In this case the conductors of the two submains are coupled together by the curved copper rods A, B, C and D, and the house cables X and Y enter the box by the tubes shown on the top of the figure being connected respectively with the blocks G and H, the former of which is coupled to C and D through the block F by the short length of lead wire E, which, by fusing, if by any chance the strength of the current rises beyond a certain point, cuts out the branch without affecting the main current passing through the box from right to left. Fig. 661 is an illustration of another form of branch between two medium size mains, the construction and principle of which is sufficiently clear from the sketch.

Fig. 662 shows the arrangement of the branch block and safety guards employed under the Edison system for the smaller branches, such as those within a house leading to one room or to a separate group of lights. This consists of a square block of wood, having on its one face two grooves ploughed into it, capable of allowing the cables to be laid in them, and on the other face are ploughed two smaller grooves, whose

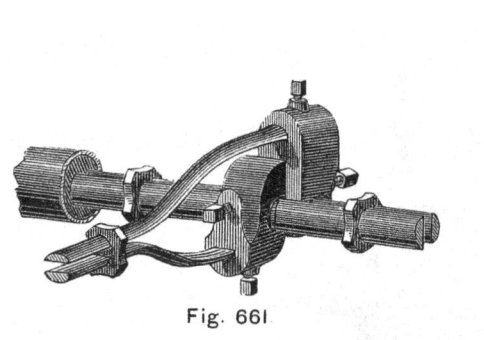

Fig. 661.

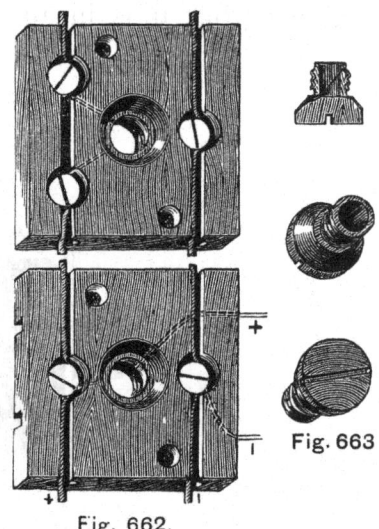

Fig. 662.

Fig. 663.

direction is at right angles to those on the other face. In the centre of the block is imbedded one of the lamp sockets or fittings which have already been described ; and the conical seat and screw connection are so connected to the cables and the branch wires, that if a lamp were screwed into the socket its filament would act as a coupling-piece, placing the branch in circuit with the cable conductors. Instead, however, of a lamp, a little screw safety plug, shown in Fig. 663, is inserted; this plug is in its external fittings identical with the screw and cone fitting of a lamp, but the cone is connected with the screw by a short piece of lead wire within the cavity of the plug, which, when the plug is screwed into its socket, places the branch in circuit with the cables, but fuses the moment that the current approaches a strength dangerous to the safety of the lamps. Mr. Edison has recently introduced a slightly improved form of branch block, shown in Figs. 664 and 665. In this apparatus the new lamp-fitting is introduced, the cone at the larger end of the plug being replaced by a flat bearing plate at the smaller end, which, when the plug is screwed into its place, bears against a flat bearing plate at the bottom of the socket. A plug of this description is shown in Fig. 666, and its flat bearing plate is connected to its screw by a piece of lead wire. The grooves in the block are moreover curved so as to make room for the safety socket in the centre of the block without, however, necessitating the pair of conductors on each side being laid at such a distance apart as was necessary with the earlier form of branch block.

Under nearly all electric light systems, especially those like Mr. Edison's, in which currents of great quantity and of low electromotive

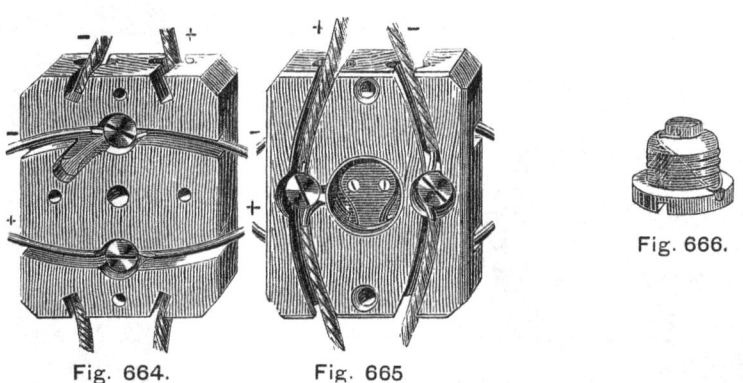

Fig. 664.   Fig. 665   Fig. 666.

force are employed, considerable difficulty is experienced by the burning and consequent destruction of the contact surfaces of the switches, for it must follow that however rapidly a contact may be broken, an electric arc must be momentarily produced, and this means consumption of the metal of which the contact surfaces are constructed, and the amount of the consumption will depend upon the strength of the current, on the area of surface affected by the arc, and on the time during which the current is passing from one contact to the other in the form of a visible arc or spark. In the switches adopted under the Edison system, especially in those used for interrupting strong currents, this difficulty is reduced to a minimum, first, by dividing the spark due to interruption over two, three, or more interruptors at the same moment, and thus reducing the arc at any one interruption to a-half, a-third, &c., according to the number of interruptions produced; second, by increasing thereby the area of surfaces in contact; and, third, by insuring extremely rapid making and breaking of the circuit. Fig. 667 is a view of the back of one of Mr. Edison's three-fold switches, showing the internal arrangements. This switch is so designed that by turning the cross handle M M through a portion of a revolution, the three conical spring plugs E, F and G are simultaneously thrown into or out of their seats, which are cut in the edges of the four plates A, B, C and D. The current entering by the attachment screw H when the switch is " on " passes in succession through D, G, A, E, B, F and C to the other terminal screw K, and as this is accomplished by a cam suddenly setting into action a strong spring the interruption is extremely rapid, and the sparking is distributed

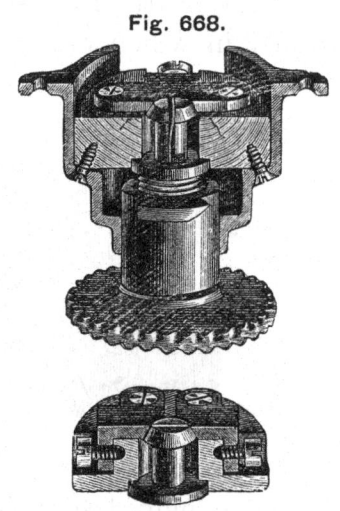

Fig. 668.

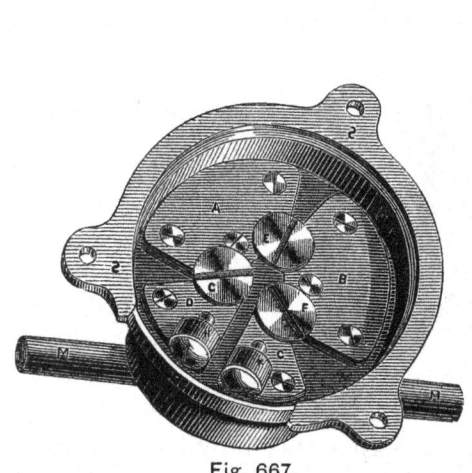
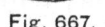

Fig. 667.

Fig. 669.

over no less than six surfaces, so that not only is the spark reduced to a minimum at any one spot, but the area of surface contact is proportionately increased. In the improved switches recently introduced the conical connecting plugs are pushed in from their base instead of being drawn into their seats by their smaller ends; by this means a central rod passing between the fixed plates is avoided, and the cone comes clear away from its seat with no danger of an arc being formed through the rod becoming bent or otherwise eccentric. Figs. 668 and 669 represent a section of a single switch recently sent over from America by Mr. Edison, in which the arrangement of conical plug to which we have just referred is employed. By the rotation of the hand-wheel at the bottom of Fig. 668, a double cam attached to its spindle pushes into its seat or suddenly withdraws, the split spring cone shown in both figures, and thus connects or electrically separates the two plates shown in the upper part of Fig. 669, which represents a section through the switch taken in a plane perpendicular to that of the section Fig. 668.

The strength of the current produced by the generator is regulated to the work it is required to do, and as this is effected by varying the resistance of the exciting circuit of the field magnets so as to proportion the intensity of the magnetic field to the current required for feeding the lamp circuits, the mechanical resistance to the rotation of the armature is varied accordingly, and with it the motive power for driving the machines. Mr. Edison's current regulator is illustrated in Fig. 670. This apparatus consists of a rectangular table supporting beneath it a number of coils of German-silver wire wound upon skeleton bobbins, between which and between the convolution of each helix, there is a free circulation of air so as to keep the whole system cool. All these coils are coupled together in series, the first being connected to one attachment screw of the apparatus, the other attachment screw being connected to the regulating lever, which by being moved over a number of contact-pieces, arranged in a circle on the upper table, and connected severally with the resistance coils underneath, can be made to throw into the circuit of the field magnets of the generator or generators, any number of the resistance coils from one to the full number in the regulating apparatus. In permanent installations the regulating room is provided with photometers by which the photometric value of one lamp in any one circuit can continually be observed, and there is also an electromotive force indicator consisting of a reflecting galvanometer, the resistances of the circuits being so balanced, that when the current is at its normal electromotive force the indicating spot of light rests at the centre of its scale, but moves to the left or right, according as the electromotive force diminishes or is increased.

Having now described the principles and construction of the apparatus employed under this system of electric illumination for the generation, canalisation, regulation and utilisation of the currents, it only remains to refer to the highly interesting and ingenious apparatus devised by Mr. Edison for measuring the amount of current made use of by each " consumer," as well as the total amount passing out of the central station of a district.

The principle of action of the instruments employed depends upon

Fig. 670. Edison Current Regulator.

the fact, that the weight of metal deposited per unit of time in an electro-plating bath by electrolysis is proportioned to the quantity of electricity passing through it in the same period, or, in other words, to the strength of the current, and as the amount of this electro-deposition is equal at all points of a circuit, it follows that by placing an electrolytic cell in any part of a circuit, the amount of electricity transmitted by the cell in a given time can be determined by observing the quantity of metal deposited within the same period.

The apparatus employed in the Edison system in connection with the service of a house is illustrated in Figs. 671 and 672.

Fig. 671.

The house meter consists generally of two electrolytic cells so connected as to form together a shunt to the lighting current, and also so arranged that one cell is capable of transmitting a greater quantity of electricity than the other, the proportion between them, however, remaining the same. The two cells are fitted into a small cast-iron cabinet, as shown in Fig. 671, and below them is a simple automatic arrangement, to be after described, for preventing the temperature of the cells from falling below a certain degree.

In the earlier form of meters devised by Mr. Edison, the electrolyte employed was a solution of copper sulphate, the electrodes being of pure copper, but in the more recent instruments the copper solution has been replaced by a solution of zinc sulphate and the electrodes are of zinc, by which arrangement polarisation currents are reduced to a minimum and more accurate results are insured.

Referring now to Fig. 672, V and V¹ are the two glass cells, each of which is filled with a solution of zinc sulphate, in which is immersed a pair of amalgamated zinc plates. The two cells are precisely alike, but from the fact that the one cell is short-circuited by a shunt having one-fourth the resistance of the shunt by which the other is short-circuited, the quantity of electricity passing through the former

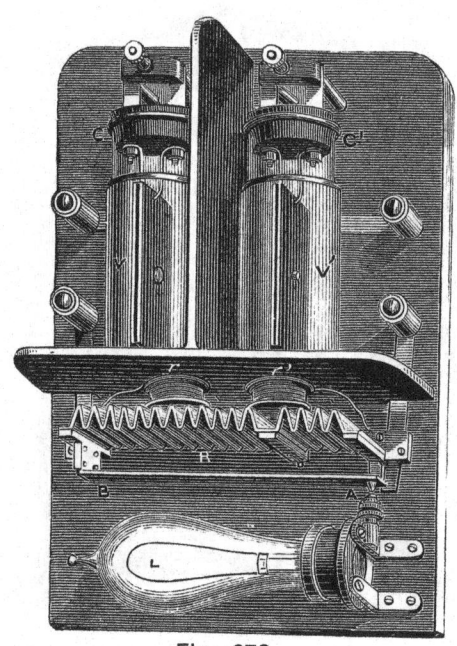

Fig. 672.

per unit of time is four times that which is transmitted by the other, and therefore, within a given period, the quantity of metallic zinc deposited on the receiving plate of the one cell is four times that which is deposited on that of the other. The short-circuiting shunts are shown at R, and consist of a corrugated band of German-silver, of which the 12 corrugations towards the left of the figure constitute the shunt of the cell V, while the remaining four corrugations belong to the right-hand cell V¹. An ordinary Edison incandescence lamp L, is thrown into circuit whenever contact is made at the point A, and this is effected by the bending downwards of the compound expansion bar A B, which is so constructed and adjusted as automatically to throw the lamp into circuit whenever the temperature within the instrument falls below a certain point. The radiant heat from the lamp then raises the temperature of the air within the apparatus until by the upward curling of the compound bar, the lamp is again cut out of circuit; r and r^1 are two compensating resistance coils of insulated copper wire, which are included in the branch circuit of the electrolytic cells, their object being to keep the resistance of the apparatus uniform under variations of temperature, for, as the resistance of the zinc solution within the cells is diminished by a rise of temperature, the quantity of electricity transmitted by it would be greater as the temperature was increased unless this increase were compensated or neutralised. To effect this a coil of copper wire is inserted in the circuit of each cell, and as the electrical resistance of copper increases with a rise of temperature,

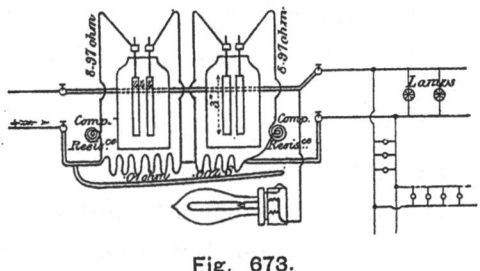

Fig. 673.

while that of the zinc solution diminishes, it is only necessary to make the length of the copper wire such that its resistance rises under an increase of temperature as much as that of its corresponding cell diminishes, in order to compensate the apparatus for variations of temperature. The method by which the various parts of the apparatus are coupled up will be understood by referring to the diagram Fig. 673, which also illustrates the position of a lamp circuit with respect to the meter.

The object of having two cells instead of one is to enable the supplying company to keep a check upon its inspectors, and also to guard against fraud on the part of dishonest consumers, for while the left-hand division of the meter is accessible to the ordinary inspectors, and in some cases to the consumers themselves, the right-hand portion is under the sole control of the company through their superior inspectors, and the readings of the two cells should always be in the same proportion to one another. Again, the object of causing the one cell to deposit slower than the other is that while the ordinary inspectors go their rounds once a month, the upper inspectors visit the meters once a quarter; the one cell is therefore depositing for four weeks between two inspections, while the other is in action for 13 weeks before its plates are weighed again; and, by causing the three months' cell to deposit slower than the one month cell, the amount deposited in each between any two inspections is more nearly equalised.

The zinc plates are shown in detail in Fig. 674; they are 3 in. long, 1 in. wide, and ¼ in. thick, fixed at a distance of ·24 of an inch apart by first placing two little ebonite blocks of that thickness between the plates, the one at the top and the other at the bottom (as shown in the figure), and then holding the whole together by an ebonite screw passing through the middle of the plates; before being put together, however, a coating of chemically pure zinc is deposited by electrolysis upon the surface of each plate so as to reduce local action to a minimum, this operation being performed in a chemically pure solution of zinc sulphate, the plates are afterwards amalgamated with pure mercury.

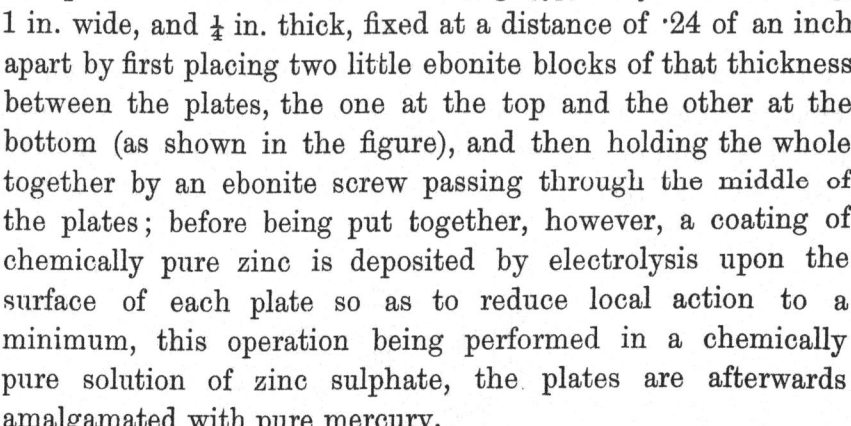

Fig. 674.

When the apparatus is set into action, zinc is dissolved off the one plate of each pair and deposited upon the other, and as the quantity of zinc transferred from the one plate to the other is proportional to the total quantity of electricity which has passed through the cell, and is independent of the time, it is only necessary periodically to weigh the plates in order to ascertain the quantity of electricity which has been utilised, and it is immaterial whether a current of 1 Weber flows for 10 hours, or one of 2 Webers for five hours, or of 5 Webers for two hours, or of 10 Webers for one hour, the transfer and deposition of metal will in all cases be the same.

The consumer is charged so much per Weber of current passed through the meter per hour. The inspectors go round once a month collecting the cells, and substituting fresh ones in their place; each old cell on being removed from the meter is placed in a separate compartment of a basket, its number being carefully noted. The cells so removed are taken to the central station, where the pairs of plates are separated and accurately weighed in a delicate balance, the results being entered in the meter book, of which a specimen page is given below.

The Weber current that has passed through a "25-light" meter since the last observation is calculated by the following formula:—

$$V = \frac{W}{1 \cdot 336}$$

in which V represents the Weber current that has passed per hour, and W the increase in weight of the receiving plate since the last observation, the figure 1·336 being the number of milligrammes of zinc deposited per Weber per hour. Thus supposing that a set of zinc plates before being put into action weigh as follows: those in the monthly cell, 92,800 milligrames and that in the accompanying quarterly cell, 92,600 milligrammes, and those after having been in use for a month in an installation of 20 lights burning about four hours a day they weigh as follows:

		Mill.
Monthly cell	....	95,317
Quarterly cell	....	93,229·2

Subtracting these weights from the first we get a difference in the monthly cell of 2,517 milligrammes, and in the quarterly cell of 629·2 milligrammes, and if all has been working well, the increase in the weight of the quarterly zinc should be about one-fourth that of the monthly plate Applying the above formula to these results we get the figure 1,883, which represent the average number of Webers passed through the apparatus each hour it has been in operation, multiplied by the number of Weber-hours, and it is upon this figure that the consumer is charged. The following is a specimen page of a meter book, showing the general form of the entries and the principle upon which they are made:—

METER NUMBER.	NAME OF CONSUMER.		ADDRESS OF CONSUMER.				

Plates put in.	Weight of Plates.		Plates taken out.	Weight of Plates.		Gain.		Webers Consumed
	Monthly.	Quarterly.		Monthly.	Quarterly.	Monthly.	Quarterly.	
January 20 ..	90,300	90,400	Feb. 20	91,400	..	1,100	..	823·3
February 17 ..								
March 17 ..								
April 14								
May 12								
June 9								

The Edison meters of the form illustrated in Fig. 671 are at present constructed in two sizes, the one for currents of 25 lights, and the other for 50 lights in a circuit. In the former apparatus the monthly cell deposits per Weber 1·336 milligrammes per hour, the deposit in the quarterly cell being about one-fourth of that amount. In the 50-light meter the rate of deposit is 1·377 instead of 1·336. The zinc solution is made as follows :—90 parts by weight of pure zinc sulphate is dissolved in 100 parts by weight of distilled water, and if the solution has been properly made, a hydrometer should show a specific gravity of 1·33 at a temperature of 18 deg. Cent. We may mention here that Mr. Francis Jehl, who has been associated with Mr. Edison throughout the development of his very interesting meters, has recently been conducting a very careful series of experiments upon the specific resistance of various solutions of zinc salts.

The most interesting and elaborate apparatus of this kind designed by Mr. Edison is the automatic meter or electric counter represented in Fig. 675, and which is rather suited to the laboratory of the central station, where it can be in proper and intelligent hands, than to house or factory installation, where it might soon be put out of order. The apparatus consists of two electrolytic cells, each containing a solution of copper sulphate, and in each of them is immersed a hollow cylinder of copper. Within this cylinder is suspended from the end of the beam of a sensitive balance, another copper cylinder ; the two cylinders in each cell constitute the electrodes of an electro-depositing bath, and according as to whether the current traversing the cell is in one direction or the other, so will the metallic copper be transferred from the fixed cylinder to the suspended cylinder and *vice versâ*, the connections being so made that while the apparatus is depositing copper on the inner cylinder of the one cell, metal is being dissolved from the inner cylinder of the other cell ; thus, while the one is, during the passage of the current, becoming heavier, the other is losing weight. It is clear, from this construction, that a time must come when the one will overbalance the other, and the beam will in consequence be deflected ; in doing this the current is automatically reversed, and, traversing the cells in the opposite direction, the process is reversed, copper being dissolved from the heavier suspended cylinder, while the other receives a deposition in the other cell until it overbalances the first. This causes the beam to be depressed, and again reversing the current sets the instrument to its original position ; the process is also thus repeated as long as the current passes. Connected

to the beam by suitable mechanism is an ordinary counter, by which
the number of oscillations of the beam is recorded, and as each
oscillation represents a deposition of copper to a certain weight, and that
weight is a measure of the quantity of electricity flowing through the
instrument, the current is measured by a process almost identical
mechanically with the method by which the quantity of gas utilised
in an establishment, is recorded. The above is a description of the

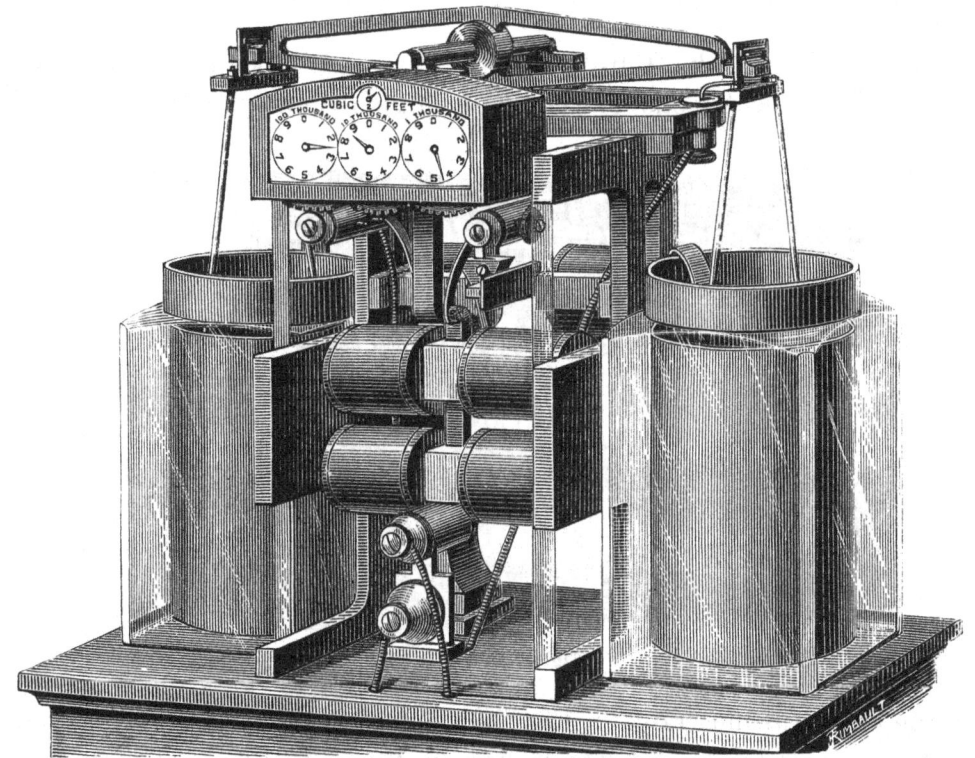

Fig. 675.*

general principle upon which the automatic meter works. It will
be seen, however, on reference to the illustration, that there are certain
electro-magnetic attachments and switches. The principal object of
these accessories is to insure the beam tumbling over to one side or the
other the moment that the quantity of metal to which the instrument
is adjusted, has been transferred from the one electrode to the other,
for as the action is very gradual, it would be quite possible for
the beam to stop at its central position when equilibrium has been

* This figure has already been used on an earlier page of this volume, to illustrate a principle of
electrical measurement.

established between the two balanced electrodes, and contact might be broken on one side without being made on the other; then the action of the instrument would cease, although the current might be utilised throughout the installation. By the electro-magnetic attachment, however, the apparatus can be so adjusted that the very slightest motion of the beam to one side or the other of a certain fixed point makes an electrical contact, thereby throwing one or other of the electro-magnets into circuit, and by the attraction of an armature causing the beam to fall over.

Connected with the Edison system there are several interesting accessory instruments, one of the most original and curious of them

Fig. 676.

being the apparatus illustrated in Fig. 676, which is a resistance " box " devised by Mr. Edison, and which has on the top the usual set of connection plugs and contact pieces; instead, however, of coils of wire forming the resistances, the lower side of the upper table is provided with lamp sockets, the two terminals of each being connected to a corresponding pair of contiguous contact pieces. Into these sockets are screwed as many lamps, and the carbon filaments which constitute the resistances can be thrown into or out of circuit by removing or inserting the connecting pegs in the usual way.

In the instrument shown in the figure there are 10 lamps, and these can be thrown into a circuit in exactly the same way as are the coils of a resistance box. It is not claimed for this instrument that it is an accurate measure of resistance, but it is very useful and convenient for increasing or diminishing the resistance of a circuit by the equivalent of so many lamps inserted either in the circuit or as a shunt, and it may also be used for testing lamps and ascertaining how many may be used in any particular circuit.

Before taking leave of Mr. Edison's work in incandescence lighting, it will be interesting to review in some detail his more recent labours as an inventor, although we are reluctantly compelled to pass over his latest and, it is said, most successful developments, which have not as yet been made public. Some of this work is naturally impractical, and will be useless save as possible stumbling blocks to future inventors, others have found a valuable embodiment in the lamps we have already described, while others still remain to be worked into a practical shape.

Fig. 678.   Fig. 679.

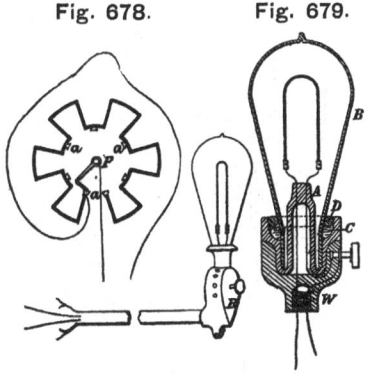

Fig. 677.

The first suggestion we may notice was patented in February, 1881. It relates (1) to incandescence lamps in which the carbon is straight, and is not a bow or curl, as had been previously the case. The bridge is held extended upwards by a vertical glass arm carrying a conductor. (2) In order to prevent heat from the incandescent material being carried down to the point where the conducting wires are fused into the glass, a long interval is interposed between the clamps of the carbon and globe, and the conducting wires are supported by a glass pillar. (3) Lamps to be suspended may be cheaply constructed of a bridge, a sealed glass sphere, and protruding conductors. (4) Figs. 677 and 678 show an arrangement whereby the intensity of a lamp may be modulated at will. The base B contains the resistance ring of heavy carbon, shown in Fig. 678. One main conductor is attached to the stud

s s 2

in the centre, and the current flows through the metal arm F to any of the contact pieces *a a*, to which the arm may be applied by means of its external knob. In the illustration the arm rests on the terminal of the second leading wire and the carbon resistance is out of circuit, but any number of its wings may be included in the path of the current. (5) Several forms of glass supports to prevent the bending over of the fibre are illustrated. (6) The form of the glass stem through which the wires pass is modified to withstand the air pressure better. (7) Fig. 679 illustrates a new method of sealing the lamp ; W is a wooden base, A a glass socket fitted therein, of a form resembling a W in cross section. The globe B is fitted into the socket A by grinding, and the joint is further secured by the mercury C and rubber ring D.

The second patent, also dated February, 1881, refers to the connection of the wires and the carbon filament contained in the

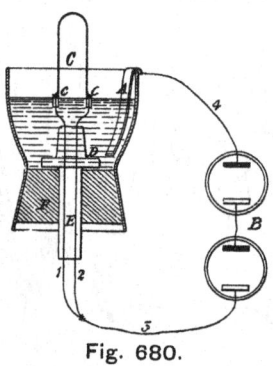

Fig. 680.

hermetically sealed glass globe of an incandescent electric lamp, the object being to dispense with delicate clamps. The invention is substantially for the method of uniting, which consists in first attaching the carbon filament to the wire by mechanical means, and then completing the attachment by plating, the deposited metal securing good electrical connection between the wire and the carbon. The wire can be mechanically attached by flattening and splitting the end, and inserting the carbon in the forked opening, the sides of which are pressed down on the carbon, or the wire may simply have a turn or two round the carbon, or be bound to it by fine wire. The illustration shows an electro-plating cell which is also claimed as part of the invention. In this cell F is a rubber stopper closing the bottom, E is a supporting tube to receive the conducting wires 1 and 2, which are sealed therein, the tube having an enlargement D for sealing into the neck of the enclosing globe of the incandescent lamp. At the upper ends the wires 1 and 2 are secured to the carbon C; the tube E is passed through an aperture in F until the point of union is just covered by the plating solution in G, connections 3 and 4 are then made with a battery B, the plate A forming the anode, and the ends *c c* the cathode of the depositing cells. The parts of the wire not required to be plated are coated with varnish or wax.

Figs. 681 and 682 illustrate a patent dated April 1881, in which several forms of incandescence lamps are described, so arranged that the act of putting them into their sockets connects their terminals to the respective conductors. Fig. 681 is an example of one form of lamp and socket. Two metal bands or rings *a b* are put upon the neck B, one of them, *a*, being formed into a male screw thread; from these bands, wires 3, 4 lead to the clamps *c c* of the bridge. Upon the interior

Fig. 681.

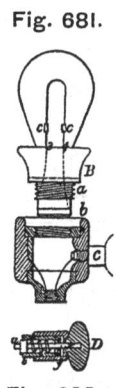

Fig. 682.

of the socket is an internally screwed metal band connected to one main conductor, the other conductor being connected to *c*. Instead of forming one of the bands into a screw thread it may be made concave and be held by spring fingers in the socket. Or the insulated neck may be slightly bevelled and have two concavities, one on either side, in the bottoms of which are metal plates to which lead the wires from the bridge. Spring fingers attached to the socket enter these concavities and make the necessary contacts while they secure the lamp. Each lamp is provided with a circuit controller C, by which it may be turned in or out as desired. In this appliance, which is shown in section in Fig. 682, is a bush which is screwed into the lamp socket, and which carries the spindle *x* ending in the contact point *u*. This spindle is constantly pressed forwards by a spring, its motion being restrained by a pin y, which, as the handle D is rotated, drops into one or other of two cross grooves, one only of which is sufficiently deep to allow the point of the spindle to come with the terminal of the lamp. The invention further relates to inserting a piece of lead wire in lamp circuit, so that in case of an abnormal current the lead melts and destroys the connection. We have already seen what an important part in the Edison system this lead safety wire plays; the present appears to have been the first notification of it. We have also seen that the mode above suggested of connecting the lamps in circuit, has been already rendered obsolete by subsequent improvements of Mr. Edison.

The next patent, dated May 4th, is substantially as follows :—When a number of incandescence lamps are fed by currents from a main kept at a constant potential, and one lamp is taken as a standard, as in the

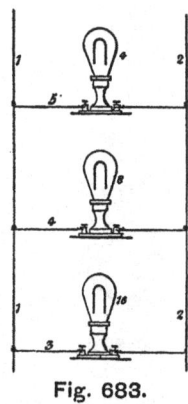

Fig. 683.

Edison system, lamps may be constructed to give a smaller light of any desired amount by diminishing their radiating surfaces and increasing their resistances in definite proportions. Thus, supposing that a lamp giving under the normal potential in the system, 16-candle power, be taken as a standard, then if another lamp with half its radiating surface and twice its resistance be substituted for it, a light of eight-candle power will be obtained; one-fourth the surface and four times the resistance would give four-candle power, and so on. In the drawing 1, 2 are the main conductors with lamps 4, 8, 16 set upon the multiple arcs 3, 4, 5.

Finally we may mention a proposed process for the manufacture of carbon burners, in which flexible carbon sheets or bridges are made by precipitating carbon from a gas or vapour on to sheets of nickel or cobalt heated to a high temperature. The metal is removed from the carbon so precipitated by being dissolved in acid. Blocks or crucibles of carbon are prepared by carbonising paper or pulp between metal dies. The hydrogen that is not removed by the heat is taken up by passing chlorine through the retort. After the bridge is placed in the globe the latter is exhausted to carry off the occluded gases while the current is passing, but as the heat does not extend to the clamps the temperature of such parts is raised by rays projected by a mirror and lens (see Fig. 684) from

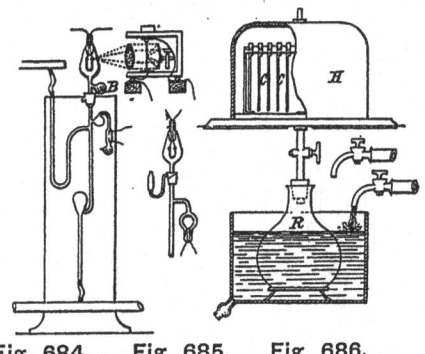

Fig. 684.   Fig. 685.   Fig. 686.

an arc light. After the lamp has been sealed off from the pump a small amount of gas is left which is absorbed by cocoa-nut charcoal contained in the globe B, which is heated to a high temperature, and left upon the stem for 24 hours. In Fig. 685 the bulb charcoal containing is heated by a current of electricity; Fig. 686 shows a method of filling up minute cracks in bridges by the deposition of carbon upon or in them from hydro-carbon vapour. C C are the bridges, R a reservoir of hydrocarbon in a tank of hot water, H a flask heated in any convenient manner.

SWAN.

The first published notice of the Swan incandescence lamp appeared in the issue of the *Photographic Journal* for June, 1880, but Mr. Swan had publicly exhibited a carbon filament lamp, which had given excellent results, 12 months before the above-named date, at the conclusion of a lecture he delivered in Newcastle, Sir William Armstrong having presided at this meeting. It is, therefore, an historical fact that the Swan carbon filament incandescence lamp had been brought to a practical form and was publicly exhibited in the autumn of 1879. As an actual fact of much interest, Mr. Swan had been labouring at this work of incandescence lighting for many years, one of the earliest forms he adopted in those comparatively remote days, having been a horseshoe of carbonised paper, placed beneath a glass bell which was more or less exhausted of air. Figs. 687 and 688 show the somewhat primitive device. The small arch of carbonised paper was about an inch high, and half-an-inch across;

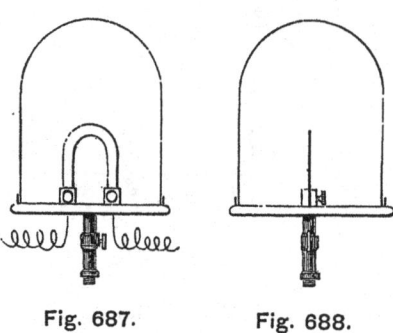

Fig. 687.   Fig. 688.

the lower ends were clamped to small blocks of carbon, and the bell was exhausted as far as possible of air. When an electric current of sufficient strength was passed through this carbon strip it was, owing to its high resistance, brought rapidly to a state of incandescence, but naturally such a device had but a very short duration in service. The filament became hotter on the inner than on the outer edge, and under this unequal influence began to curl over, rapidly bending more and more

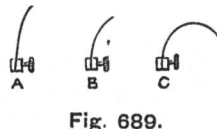

Fig. 689.

until the crown of the filament would touch the base of the lamp, and break up. Fig. 689 illustrates by three stages, A, B, C, this gradual deformation of the carbon.

This, however, was in the very early days of Mr. Swan's experiments, which he appears to have abandoned for a considerable time, resuming them however with great ardour since 1877. In fact, during the whole of 1878-79, his electric light laboratory seems to have been in a constant state of activity. Early in 1879 he realised the fact that to obtain a durability of the carbon filament, it was necessary to maintain it at a high or a higher temperature during the process of exhausting the air from the glass bulb, than it would have subsequently to sustain in actual work. As we have seen, it was just about this time that Mr. Edison was conducting a remarkable series of parallel experiments with platinum and its alloys, and the results he obtained of the changed physical properties of metal wire, raised to incandescence in vacuo, corresponded strikingly to those obtained by Mr. Swan with carbon filaments, treated in a similar way. It was on the 17th June, 1879, that Edison took out his patent in this country for the application of this principle he had discovered, for the manufacture of incandescence electric lamps with prepared platinum, or alloyed platinum luminous loops, but as we have also seen, he, like inventors 20 years before him, quickly abandoned metallic, and availed himself of vegetable filaments. Mr. Swan, on the other hand, had worked with the latter from the beginning, and the evolution of his system, from the first imperfect and rapidly failing horseshoe of carbonised paper, to his permanent metal-like filament of carbonised thread is an interesting one to follow.

Before doing so, it may be advisable to point out in the clearest possible manner the great radical distinction between the Edison lamp and the Swan lamp of the most recent type. Edison insists upon using a "structural carbon," because he says that such carbons alone possess the qualities of the highest possible resistance in a very small bulk, and are capable of resisting the disintegrating effects of intense heat, and the absence of atmospheric pressure. He further says that by structural carbon he means " a carbon wherein the natural structure, cellular or otherwise, or the original material is preserved unaltered, that is, not modified by any treatment which tends to fill up the cells or pores with unstructural carbon, or to increase its density, or alter its resistance." To obtain such carbons, therefore, Mr. Edison is obliged to resort to raw material, such as natural fibre, and now exclusively to bamboo strips, which are subjected to the series of beautiful processes we have already described.

Mr. Swan's object, on the other hand, is to obtain a material suitable for the carbon filament which shall be as far as possible devoid of structure ; he could not, therefore, make use of any vegetable fibre in its natural state. Paper he quickly found was unsuitable even when prepared by his special process, and he ultimately adopted cotton thread, which is susceptible to the parchmentising operation that had enabled him to obtain such promising results with paper prepared in the same manner. Steeping cotton in a solution of sulphuric acid and water, until the tissue is destroyed, produces, when properly washed and dried, a horny homogeneous filament, of very considerable strength. To increase the density and uniformity of the filament thus obtained, it is passed between compressing rollers and flattened, so that a somewhat increased area of incandescent surface is thus obtained. From the great practical success attending both the Edison and the Swan system, is is evident that a " structural " carbon is no more absolutely necessary than is a structureless one, but this is one of the leading points of difference in the two systems. We may now refer to the more important points patented by Mr. Swan in connection with the manufacture of his incandescence lamps, premising that he did not go the Patent Office till years of laboratory work had passed. The first patent to be noticed is that dated 2nd January, 1880, and refers to modes for overcoming defects that had previously existed in electric lamps where a carbon conductor was employed in a more or less exhausted globe; the special defects referred to being the wasting of the carbon, and the obscuration of the glass by the deposition of very minute particles of carbon on its surface. In his specification, Mr. Swan says :—" I have found that if the globe or vessel of glass containing the carbon to be rendered incandescent be exhausted of air to a high degree of exhaustion by means of a Sprengel pump, while the carbon within it is raised (by the passage of a current of electricity) to the same or a higher degree of incandescence as that ultimately required to be produced in using the lamp, and if the lamp be hermetically sealed whilst exhausted of air under the conditions specified, the lamp so prepared is durable, both the faults above-mentioned being eliminated, the incandescent carbons enduring without waste, and the interior surface of the globe keeping free from carbonaceous deposits."

In describing the construction of such a lamp as he then proposed, Mr. Swan says, that he employed " a rod, filament or lamina of carbon of a length and thickness proportioned to the current to be employed in heating it," as, for example, carbons 2 in. long and from $\frac{1}{50}$ to $\frac{1}{100}$ in. in diameter would be suitable. The lamina or filament was held at its ends by platinum clips, that formed a part of the conducting wires which passed through the body of the glass globe, and were sealed in by fusion. The bulb was then submitted to the action of a Sprengel pump and exhausted to the one-thousandth part of an atmosphere, the carbon being raised to incandescence by the passage of the current, and maintained in that condition till the work of exhaustion was complete. It will be seen from the foregoing that Mr. Swan appreciated the general requirements controlling the successful production of an incandescence lamp, and this is clearly set forth in his claim. "I claim the production of an incandescence carbon electric lamp from which the air has been exhausted while the carbon was in a state of incandescence, which is hermetically sealed during exhaustion."

Less than three weeks later, Mr. Swan took another patent, embodying some more of the points which characterise his system. The first part of this patent relates to the prevention of cracking and leaking in the glass bulb, by the use of platinum cups, soldered to the wires and fixed into the glass. Subsequent development in the manipulation of the bulbs rendered this detail unnecessary. The second part of the patent refers to the carbons, and to a process for rendering them more durable. The inventor states that he had found that the cardboard horseshoes, when heated to brightness in the exhausted globe, would curl and break in consequence of the unequal heating of the inner and outer parts, as we have already explained. This he proposed to prevent by using parchment paper, that is bibulous paper soaked in dilute sulphuric acid, and washed and dried. From the paper so prepared strips were cut, bent to the desired form and then carbonised in a close vessel at a white heat. The third part of the patent refers to a means for preventing the escape into the bulb of occluded gases from the platinum conducting wires by coating them with enamel, where they lie within the bulb.

The interesting claims in this specification are those "for forming carbons for incandescent electric lamps in the manner described, in which a strip of paper or card bent to the required form is carbonised by subjecting it to heat in a vessel containing charcoal or other form of carbon in powder." Also the following claim for forming "the carbon for incandescence electric lamps by carbonising vegetable parchment of the required form in the manner described." This patent which describes and claims the use of a new material for the manufacture of the carbons is followed by another in which a similar material in a more convenient form, is described. The use of cotton thread is here claimed, not as cotton thread, but converted into a homogeneous and quite different material by the action upon it of sulphuric acid   After having been thoroughly washed and dried, the filament is "carbonised by any of the well-known means." In the same patent is described a mode of thickening the carbons at the ends by wrapping cotton or other suitable material round the ends of the thread, prior to converting it by the sulphuric acid, which thoroughly united the thickened ends to the thread, after which the whole was carbonised together.   The idea of this thickening up was to facilitate the attachment of the filament to the platinum clips, but in recent improvements in the construction of the lamp devised by Mr. Gimingham, this somewhat clumsy mode of attach-

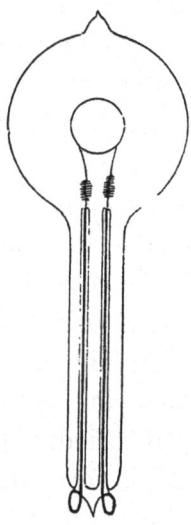

Fig. 690.

ment has been abandoned.   The first claim in this patent, which is dated 27th November, 1880, is for forming the carbon or carbons of an electric lamp from cotton thread converted by the action of sulphuric acid and carbonised.

Fig. 690 shows the form of lamp as made under the process described in the patent, although it had been some time before the date just mentioned brought into the form shown, which was that of lamps exhibited by Mr. Swan in the previous October.

Finally, we may mention a patent dated 2nd December, 1880, which refers to the carbonising of the parchmentised cotton filaments, and more especially to the flattening them previously between rollers, to obtain a more homogeneous material, and to secure " an increased superficial area for lighting."

Since this last date, Mr. Swan has introduced several alterations and improvements (to some of which we shall presently allude), but in these patents which we have analysed, is contained the essence of what is original in his system of incandescence lighting.

We have seen that in his first patent he claims the application to carbon, of a very similar process which Edison had patented in June of the same year, as applied to metals, platinum and its alloys, with more of brilliant promise than subsequent experience confirmed. This was

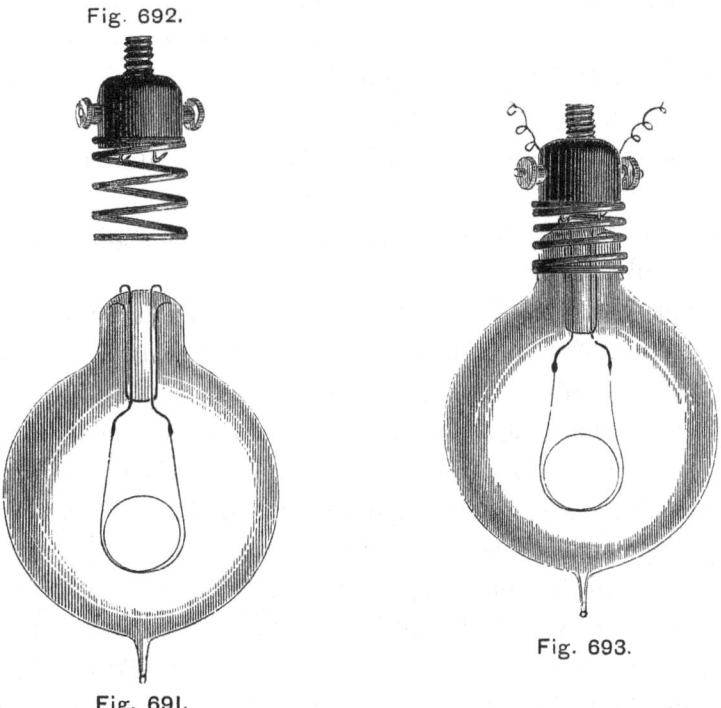

Fig. 692.

Fig. 693.

Fig. 691.

the rendering of the filaments suitable for incandescence lighting, by subjecting them to a high temperature in a vacuum. Though unsuccessful for metal, this process proved eminently successful for carbons. The second point is the material Mr. Swan employs. He does not use cotton thread, but a material produced from cotton thread, in which the fibre and the whole structure, as far as possible, has been destroyed, and nothing but a homogeneous filament is left.

The most recent form of lamp devised is that illustrated by Figs. 691, 692 and 693, which is due to Mr. C. H. Gimingham, of the Swan Electric

Light Company, and which, while being neater and simpler than those of the older types, can be produced more cheaply. As will be seen from the figures, the exterior of the lamp is entirely of glass, and from the short stem at the top project two small platinum loops, the terminals of the platinum wires carrying the carbon filament. The lamp connection is shown in Fig. 692, and is about as simple as can be desired. It is simply an ebonite stud with a screwed plug for attachment to a gas bracket or other stand; on each side of the stud are the binding screws, and in connection with them are the two small platinum hooks projecting from the stud. Attached to this latter is also a light spiral spring. Fig. 693 shows the lamp connected up, and from this figure it will be seen that the loops of the platinum terminals are passed over the platinum hooks on the stud, compressing the spiral spring round the neck of the lamp, and thus making a most efficient fastening in which the electrical contact is perfect, and at the same time there is a total absence of rigidity. It must be admitted that, ingenious and complete as they are, none of Mr. Edison's lamp fittings can compare with this graceful and simple device.

From what has already been said with respect to the degree of attenuity to which it is necessary to exhaust the bulbs of the Swan lamps in order to ensure their permanence and successful working, it will readily be understood that very special means must be adopted, not only to produce vacua of so high a degree, but also to do so with as little loss of time as possible, and in such a manner that the whole process may be under perfect control, and be readily applied and manipulated. It will therefore be of interest in connection with this subject to give a short description of the very beautiful vacuum exhauster of Mr. Gimingham, which is at the same time an elaboration and a simplification of the well-known mercury vacuum pump of Dr. Sprengel, and which in its turn is a practical application of the discovery of Torricelli, by which a vacuum of a high degree of attenuity may be produced by the descent of a column of mercury in a vertical tube closed at the top, and which is so familiarly exemplified in the ordinary mercurial barometer.

Fig. 694 is a general view of the apparatus as seen from the front, while Figs. 695 and 696 represent two of the details drawn to a larger scale. Referring to Fig. 694, A is a reservoir capable of containing a considerable quantity of mercury (say 25 lbs.), and which for convenience of observation

Fig. 694.

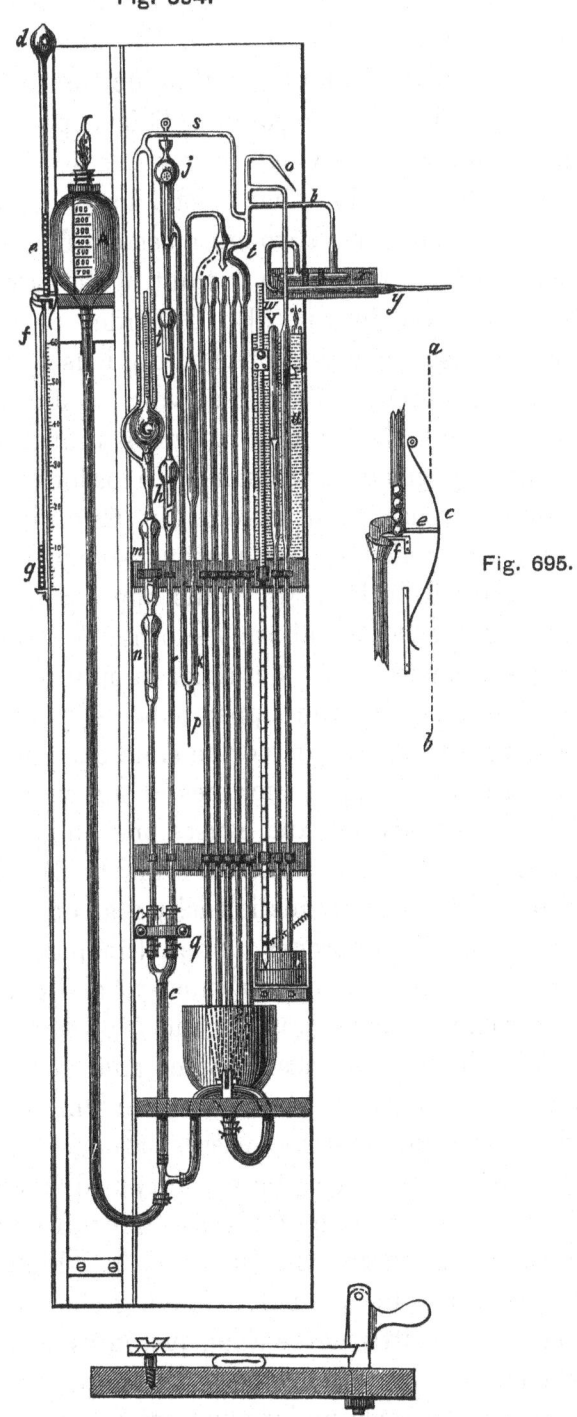

Fig. 695.

Fig. 696. Gimingham's Vacuum Pump.

is divided so that its contents can be read at sight, each division representing 100 cubic centimetres.   To the lower funnel-shaped end of
this reservoir is attached a long caoutchouc tube *t t* communicating with
the rising supply tube of the pump.   This supply reservoir is mounted
upon a sliding support, by which it can be lowered when empty to a
position whose level is below that of the lower bowl or waste reservoir,
with which it communicates through a second flexible tube furnished with
a pinch cock, shown in the figure in front of the bowl, but more clearly in
the detailed side view in Fig. 696.   In order to know the quantity of
mercury that has passed through the pump, it is only necessary to observe
the height of the mercury in the reservoir, as shown by the divisions on
its side, and to count the number of times that the reservoir has been
emptied during any given exhaustion.   In order to record the number of
times the reservoir has been raised, Mr. Gimingham devised the very
simple and ingenious recording apparatus shown to the extreme left of
Fig. 694, and in detail in Fig. 695.   This apparatus consists of two glass
tubes *d e* and *f g*, placed one above the other, but not exactly in the same
vertical axis ; the upper tube contains a supply of lead shot of uniform
size, and the lower end of the upper tube is provided with a releasing
detent so arranged that every time the slide carrying the reservoir passes
a spring, shown at *c*, Fig. 695, which lies in the way of its vertical path *a b*,
the detent is pushed forward and causes one shot to drop into the lower
receiving tube *f g*, against which a scale is fixed, and by which the
number of shot which it contains is easily read off at sight.

The course of the mercury, when the pump is at work, is from
the reservoir A, through the flexible tubes *t t* and *c* and pinch cock *q*, and
from thence through three air traps *h i* and *j*, by which any small bubbles of
air entangled in the mercury are detained and are prevented from passing
into the pump.   From the last air-trap *j*, in which the stream is broken
up into a fine spray, the mercury passes through a syphon tube K to the
distributing nozzle *t*, by which the mercury is divided into five jets, each
of which is projected into one of the five fall tubes which form so special
a feature of the Gimingham exhauster, and by which the high exhaustion
are produced in so rapid a manner.   Each globule of mercury in each of
the five tubes form a little descending piston by which air is drawn from
the upper chamber in which the nozzle *t* is fixed, and which communicates with the vessels under exhaustion, and is discharged into the
receiving bowl below, and moreover, air is also entangled in the body of

the descending mercury and carried down with it, and to such an extent that it is doubtful which of the two actions contribute most to the process of exhaustion.

One of the most interesting features of Mr. Gimingham's exhausters, is the method by which the degree of exhaustion is measured, and as this is accomplished with accuracy to small fractions of one-millionth of an atmosphere, it will be interesting here to describe the principles upon which the measuring apparatus depends. For what in so perfect an exhausting apparatus must be regarded as the low degrees of exhaustion, that is to say, exhaustion between ordinary atmospheric pressure and that represented by the drawing up, so to speak, of a mercurial column to within about one millimetre of the barometric height, the barometer gauge shown at U and V, Fig. 694 is employed. This consists of two mercurial barometers, one of which, V, is of the ordinary construction of a good barometer, being closed at the top in the ordinary way and dips into a cistern at its lower end, and side by side with this is a second barometer tube, U, whose upper end is pneumatically connected with the vacuum side of the pump, and in order to measure such vacua as can be indicated by such means, all that is necessary is to compare the reading of the gauge U with that of the barometer V. In order that the readings may accurately measure the heights of the mercury in the tubes from the surface of the mercury in the cistern, Mr. Gimingham devised an extremely ingenious electrical means of ensuring that the zero of the scale shall be exactly at the surface of the mercury in the cisterns. For this purpose, the scale is divided on a glass rod W, which is capable of sliding through a short range in a vertical direction. The lower extremity of this rod, which is the zero of the scale, is furnished with a platinum point, which by a wire fused into the rod is connected with one pole of a voltaic cell, while the mercury in the cistern is similarly connected to the other pole, a small galvanometer (shown in the figure to the right of the letter V) being included in the circuit. Before taking a reading, it is only necessary to lower the scale until the galvanometer is deflected, which thus indicates the moment that the platinum point makes contact with the mercury, or in other words, that the zero of the scale is equally at the surface of the mercury in the cistern. The difference between the readings of the two barometers is then determined by the aid of a sliding index and vernier, and although great accuracy is attained in determining

T T

the difference between the readings, it is a measure of the vacuum only in the low degrees of exhaustion, and is perfectly useless as an indicator of exhaustion at which the readings in the two tubes appear to be alike.

For the higher degrees of exhaustion, the vacuum gauge of Professor Herbert McLeod is employed, and with this instrument, which has been further improved by Mr. Gimingham, exhaustion up to one ten-millionth of an atmosphere may be determined with accuracy, and with careful observation, measurements to one hundred-millionth of an atmosphere have been taken. In Fig. 694, the McLeod gauge is shown at G; the principle of its action is to measure the degree of exhaustion within a glass bulb of known capacity (and which is in communication with the vessel to be exhausted), determining the amount of residual gas within it by compressing its contents into a smaller space, such as a capillary tube divided into fractions of millimetres. This is effected by filling the bulb with mercury and driving its gaseous contents into a divided capillary tube communicating with its upper part, in which tube its amount is determined by the space it occupies, and as the value of each division of the capillary tube is known in terms of the capacity of the bulb, the quantity of residual gas in the tube is a measure of the degree of exhaustion within the bulb, and therefore in whatever vessels are being exhausted. The mercury is introduced into the gauge G by the pinch cock *r*, and air is prevented from being carried into it by the interposition of two air traps *m* and *n*.

There are in this beautiful apparatus many points of great interest to the physicist, but which can hardly be regarded as coming within the scope of the present work. For instance, there are, attached to the pump, tubes such as that shown at *x* for chemically drying the residual gas within the lamps or other vessels to be exhausted, and which are attached to the pump at *y*, and in some instances tubes are provided which contain substances to absorb certain gases or metallic vapours which interfere with the perfection of the vacuum or cause impurities to exist within the residual gas which can never be entirely removed.

We have described in considerable detail on a previous page Mr. Edison's installation at the Crystal Palace, in order to give an idea of the mode in which his lamps are arranged and connected. We cannot do better therefore than describe here a large, and a yet more interesting, installation of the Swan lamp at the Savoy Theatre, London. This theatre is lighted by no less than 1,158 Swan lights of the form we have just described, and which has been adopted by the Swan Company as their most improved pattern. Of these 1,158 electric lights, there are 114 in the auditorium, attached in groups of three, and supported on very elegant threefold brackets projecting from the different tiers and balconies, each lamp being enclosed within a ground, or opaloid shade, by which arrangement a most soft and pleasant light is produced.

Fig. 697 is a view of one of these bracket lamp holders, which have been designed and constructed by Messrs. Faraday and Son, of Berners

Fig. 697.

Street, London. Two hundred and twenty lamps are employed for the illumination of the numerous dressing-rooms, corridors, and passages belonging to the theatre, while no less than 824 Swan lamps are employed for the lighting of the stage.

The stage lights are distributed as follows :—

6	rows of	100	lamps each	above the stage	 600
1	,,	60	,,	,, ,,	 60
4	,,	14	,,	fixed upright	 56
2	,,	18	,,	,,	 36
5	,,	10	,,	ground lights	 50
2	,,	11	,,	,,	 22

824

In addition to the above-mentioned lights within the theatre, there are eight pilot lights within the engine-room, which serve the purpose of illuminating the machinery; and as they are in the same circuit with

T T 2

some of the lights in the theatre, they indicate to the engineer in charge of the machines, by the changing of their illuminating power, when the lights on the stage are turned up or down.

The lamps are worked in parallel circuit in six groups, five of which comprise 200 lamps each, and the fifth embraces 166 lamps. The current of each group is produced by one of Messrs. Siemens Brothers' W_1 alternate-current machines, illustrated on a previous page by Fig. 272, the field magnets of which are excited by a separate dynamo-electric machine of the Siemens type, known as D_7, in general form similar to that shown in Fig. 273.

The six alternating generators are driven at a speed of 70 revolutions per minute, and the six exciting machines at 1,150 revolutions, by three steam engines—a portable 20-horse engine by Garrett, a 12 horse-power portable by Marshall, and a 20-horse semi-portable engine by Robey—but the horse-power actually utilised, as measured by a Von Hefner-Alteneck dynamometer, is between 120 and 130 horse-power. We must not, however, omit to state that in addition to the six pairs of machines for working the 1,166 incandescence lamps, there is also a Siemens dynamo machine for producing a powerful arc electric light suspended outside the theatre, and that the power to drive this machine is included in the above-mentioned horse-power employed.

The most interesting feature, however, from a scientific point of view, of this installation, is the method by which the lights in all parts of the establishment are under control, for any of the series can in an instant be turned up to their full power or gradually lowered to a dull red heat, as easily as if they were gas lamps, by the simple turning of a small handle. There are six of these regulating handles—corresponding to the number of the machines and circuits—arranged side by side against the wall of a little room, or rather closet, on the left of the stage, and each of these handles actuates a six-way switch which, by throwing into its corresponding magnet circuit greater or less resistance (increasing or decreasing it in six stages), the strength of the current passing through the lamps is lessened or increased by as many grades. The special interest of this part of the installation, however, is the fact that the turning down of the lights is accompanied by a corresponding saving of motive power in the engine, for the variable resistance which is controlled by the regulators is not thrown into the external or lamp circuit of the alternate-current machines, but into the circuit by which their field magnets are excited. When a series of lights is lowered, increased resistance is thrown into the circuit of the dynamo machine, which is exciting the magnets of the alternate-current generator corresponding to that particular series of lights; the intensity of the magnetic field of the latter machine is thereby reduced, and consequently the currents induced from that field and transmitted to the lamp circuit are diminished in strength; but by the weakening of the magnetic field, the mechanical resistance to rotation is correspondingly reduced, and therefore less power is required to drive the machine. The resistances thrown into the circuits are of two sorts; the switches, to which we have just referred, transmit the exciting current into long spirals of iron wire supported on a frame and having a free circulation of air around them by which the heat generated by the current is rapidly dissipated, and the switches of the other two circuits operate in a similar way upon resistances composed of zig-zag bands of hoop iron similarly arranged. It has often been argued by those whose interest it is to oppose the introduction of electric illumination, that lighting by electricity is accompanied by two sources of danger—the one, that of causing fire through improper contacts or the overheating of conductors, and the other, the giving of dangerous shocks to persons who incautiously handle the wires. These objections to electric lighting have just that substratum of truth in them which makes it necessary to explain what they mean. It cannot be denied that fires have been caused by badly laid or badly constructed electric light conductors, and it is also well known that several fatal accidents have occured from shocks received from some of the higher electromotive force machines. We venture, however, to affirm without the slightest fear of intelligent contradiction, that there cannot exist the very smallest fear of fire occurring in an installation of incandescence electric lighting, if the conductors are properly constructed, and put up by a person who understands his business; and the same remark applies with equal force to the question of the danger of electric shocks. We would also venture to say that accidents arising from either of the above causes in an incandescence installation is altogether inexcusable; in fact, there is no more excuse for an electrician to fix an unsafe conductor, then there is for a gasfitter to lay a leaky or otherwise defective gas-pipe, which would be attended with still greater certainty of disaster. It is not too much to say that, with but very few exceptions, if any, all the accidents which have occurred, either of fire or of serious shocks in electric light installations, may be traced to badly constructed or improperly fixed conducting wires. With the splendidly constructed cables of Messrs. Siemens, accidents of this description are practically impossible, and it should be remembered that it is part of the Swan system, as it is of that of Mr. Edison, to make use of little fusible safety shunts at various places in the circuits, so that if from any cause there occurs any liability for the conductors to become overheated the current is instantly interrupted; this is, however, not

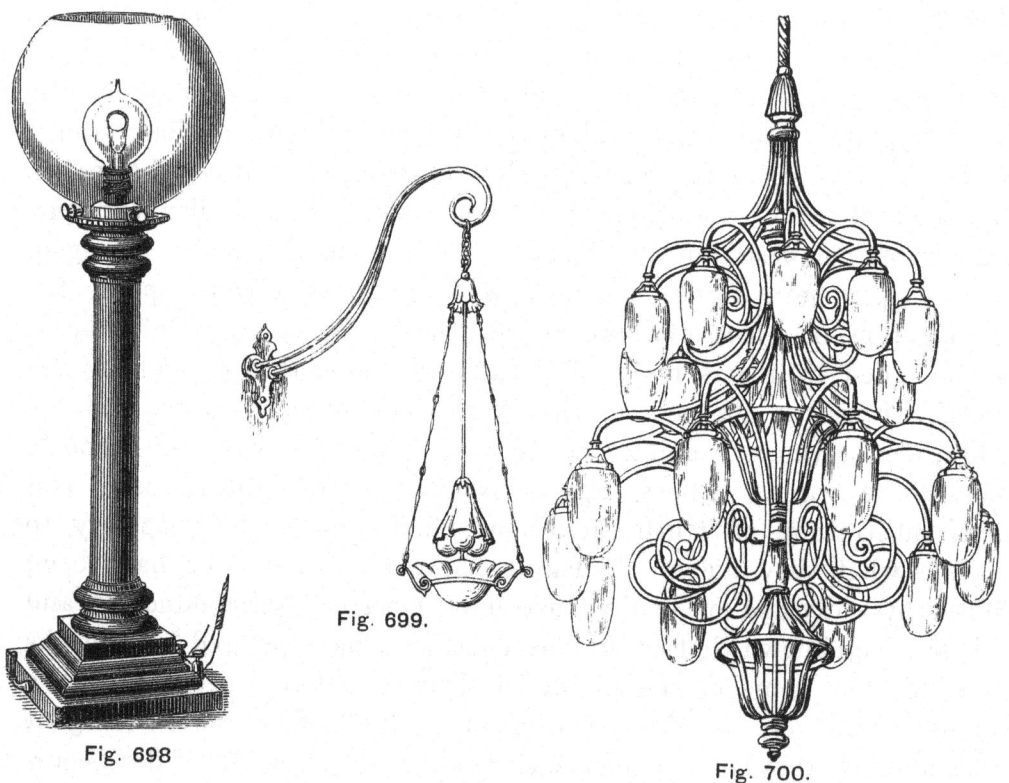

Fig. 699.

Fig. 698

Fig. 700.

intended so much to guard against a danger which is next to impossible to occur in practical working, but to protect the lamps themselves from destruction from too powerful a current being transmitted through them.

In an artistic and scenic point of view nothing could be more completely successful than this lighting of the Savoy Theatre; the illumination is brilliant without being dazzling, and while being slightly whiter than gas, the accusation of ghastliness, so often urged against the light of the electric arc, can in no way be applied. In addition to this the light is absolutely steady, and it is now possible for the first time in the history of the modern theatre to sit for a whole evening and enjoy a dramatic performance in a cool and pure atmosphere.

Figs. 698, 699 and 700, show three adaptations of the Swan system. Fig. 698 is an ordinary table lamp; Fig 699 is a bracket light, suitable for house lighting; and Fig. 700 represents a chandelier in which the glass bulbs are enclosed in globes somewhat similar to those used in the Savoy Theatre.

LANE-FOX.

The English patents taken out by Mr. St. George Lane-Fox in 1878, show that he was an early inventor in incandescence lighting. In that year he not only described an incandescent lamp with a platinum-iridium wick or " burner," and specified a complete system of distributing the currents by aid of electric mains and branch wires with current meters and regulators, but he also added secondary batteries to the circuit of the current at suitable points to act as electric reservoirs and keep the electro-

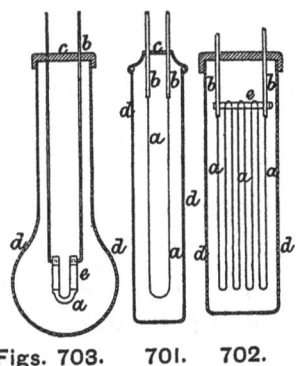

Figs. 703. 701. 702.

motive force throughout the system as uniform as possible. Mr. Fox's first lamps are illustrated in Figs. 701 and 702, and consist of glass envelopes $d\ d$ enclosing loops of platinum-iridium wire, either in air or a passive gas-like nitrogen. In the figures, $a\ a$ is the wire suspended from two conducting electrodes $b\ b$ which pass through a cover c, which may be hermetically sealed to the envelope. In Fig. 702 the wire a is looped several times round a bar of glass or other non-conductor e, so as to give a greater illuminating surface. Fig. 703 represents another form of lamp described by him in 1878, and one which may be regarded as an intermediate form between his first lamps and the carbon filament lamp which he has now adopted. In this form the incandescent wick *a* is made of a refractory material, such as asbestos impregnated with carbon, and it is held by metal clips *e* connected to the electrodes *b b*, which pass through the cover *c* of the glass envelope *d d*, which is filled with nitrogen gas.

Passing from these early and tentative forms of lamp, Mr. Lane-Fox adopted a carbonised filament in place of the wire and refractory arch or bridge shown in Figs. 701, 702 and 703, and contained in an exhausted bulb.

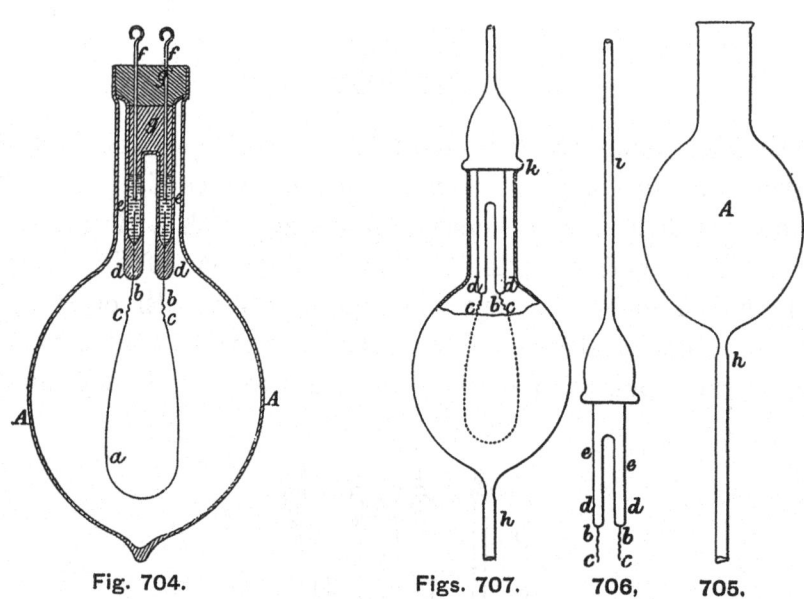

Fig. 704.   Figs. 707.   706,   705.

One of the latest forms of his lamp is shown in Fig. 704, and the method of its construction will be gathered from the accompanying details in Figs. 705, 706 and 707. In Fig. 704, *a* is a loop of carbonised fibre connected to the platinum wires *b b* by small spirals *c c* at their ends. The connection is made more perfect by a serving of Indian ink round the joint. The upper parts of the electrodes *b b* are fused into solid pieces of glass *d d* forming the bottom of glass tubes *e e* containing mercury, into which the ends of the wires dip, and thus make contact with the external electrodes or terminals of the lamp *f f*. The tubes *e e* are closed at their upper ends by a layer of marine glue *g*, and over that a cap or luting of plaster-of-paris g^1. Fig. 705 shows the glass envelope or flask A as it is first blown, with a hole at the bottom leading into the tube *h*. Fig. 706

represents the device inserted into the mouth of the flask to support the electrodes. It is a hollow tube of glass of peculiar shape, and bifurcated at the lower extremity as shown at *e e*, into which the wires *b b* are fused. After insertion, the bulge in the upper part and the mouth of the flask are fused together at *k* (Fig. 707) by the blow-pipe, and the upper piece is then severed. The mercury is then poured into the tubes *e e*, the stout copper wire terminals *f f* are inserted, and the marine glue and plaster luting cap the whole.

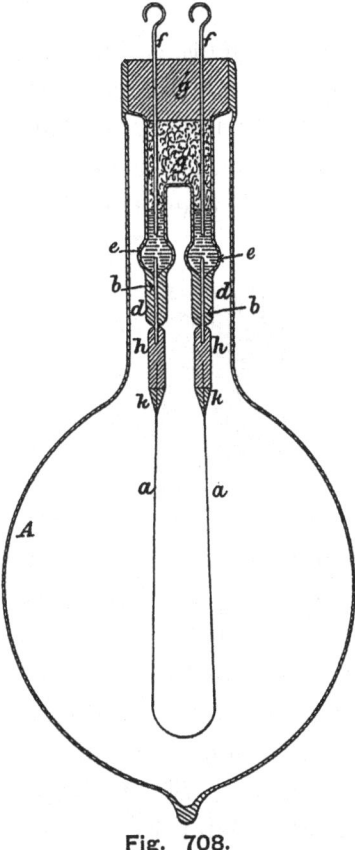

Fig. 708.

It now remains to produce the requisite vacuum within the lamp, and this is effected through the tube *h*, which is afterwards fused in the blow-pipe, and hermetically sealed up. During the operation, which we shall describe presently, the filament is kept incandescent by passing the current through it so as to make it give off any occluded gases. The body of the flask or globe may be of any kind of clear or coloured glass to subdue or tinge the light, but the branches into which the platinum wires are fused must be of lead glass, as without the lead the platinum and glass will not readily adhere, and the glass is liable to crack through unequal expansion.

Another pattern of the improved lamp of Mr. Lane-Fox, that shown at the Crystal Palace Electric Exhibition with beautiful effect in the Tropical Department and the Alhambra Courts, is illustrated in Fig. 708. Here the carbon filament $a\ a$ is joined to the platinum electrodes $b\ b$ by small ferrules of carbon $h\ h$, made of carbon cylinders, through which a fine hole is drilled to admit the ends of the wires. A cap or luting of Indian ink $k\ k$ is then added to taper off the joints between the ferrules and the filament. The electrodes $b\ b$ are fused into lead glass tubes $d\ d$ as before, and the connection between them and the outer electrodes or terminal wires $f\ f$ is made by the mercury at $e\ e$. Instead of employing marine glue, however, it is replaced by an elastic packing of cotton wool g, and over this is the cap or cork g^1 of plaster-of-paris.

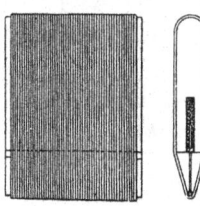

Fig. 709.

The filament employed in these lamps is made from grass fibres, preferably that known as French whisk or bass broom, and used in making certain kinds of carpet brushes. The fibre is first cleaned by boiling in a strong solution of caustic soda or potash, and the outer skin scraped off. The soda or potash is then boiled out of it, and a number of fibres are stretched round a mould or shape of plumbago, as shown in side and end view in Fig. 709; they are then baked in a plumbago crucible at a white heat. After being baked in this manner the fibres are further carbonised by depositing carbon upon them from a rich hydro-carbon gas, such as benzole. For this purpose they are suspended in large globes filled with benzole or coal gas, and then heated to incandescence by the current. The white-hot filament decomposes the gas, and carbon is deposited on its surface, especially at the thinner parts where the temperature is highest. In this way the fibre is covered with a hard skin of carbon, which brings it to the required resistance and renders it more uniform throughout, a point of some importance as affecting the durability of the filaments. Instead of employing the electric current in this way, Mr. Lane-Fox also carbonises his filaments by raising the benzole receptacle to a white-heat in a furnace.

After being carbonised in this manner, the fibres are classed according to their thickness, and are ready for mounting in the lamps. Slight differences of thickness occasion great differences in resistance, and the thicker specimens are reserved for lamps of 30 to 60 candle-power, whilst the smaller ones are kept for lamps of 10 to 20 candle-power. The

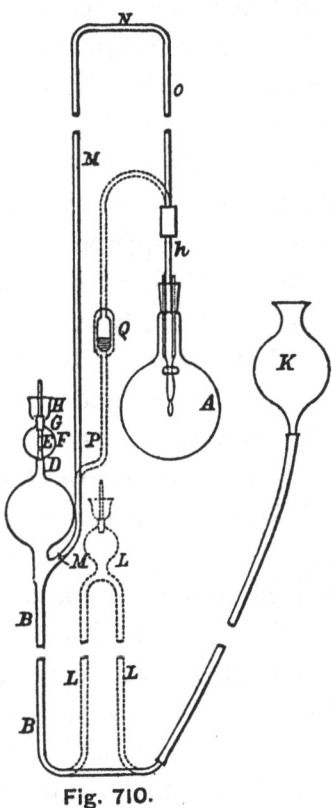

Fig. 710.

classification is facilitated by the use of a small galvanometer mirror reflecting a ray of lamplight on a vertical scale graduated in candle-power. The thickness of the fibre is ingeniously caused to alter the angle of this mirror and deflect the beam of light up or down the scale.

The exhaustion of the bulbs is effected by a very simple and ingenious mercurial air pump invented by Mr. Lane-Fox. As illustrated in Fig. 710, it consists of a vertical glass tube B, 30 in. high or more, and terminated at its upper end in a bulb C having a ground neck D, which at certain times receives the ground lower end of a glass rod E. This neck opens into another bulb F above, through the neck G of which the glass rod E also passes. The part of the rod which passes through the neck G is covered with india-rubber, so that the rod, while capable of being forced up and down in opening or closing the neck D of the lower bulb, shall all the while preserve a tight joint in the neck of the upper bulb. Above the upper bulb is a cup H. The lower end of the glass tube B B is connected to a strong flexible rubber pipe I fitted at its other end with a glass vessel K; and an air trap L L is advantageously interposed between the pipe and tube, but this is not essential. To the glass tube B just below the bulb C is connected another vertical tube M about 40 in. long above the level of the cup, the upper end of which tube is connected by a bend N to a tube O hermetically joined to the tube *h*, which communicates with the interior of the globe A of the lamp through an india-rubber stopper.

This constitutes the exhausting apparatus, and it operates in the following manner: The open vessel K is partially filled with mercury, and raised by hand until it becomes nearly empty, the mercury rising in the two bulbs C and F, and filling them up to the cup H. The neck G of the upper bulb F is then closed by the rod E, the neck D of the lower bulb being left open. The open vessel K is then lowered about 36 in. in order that the mercury may sink well below the point where the tube M communicates with the tube B. The consequent fall of the mercury in the bulbs produces a vacuum, which will be filled by the air in the lamp escaping through the exhausting tube *h o*, N M. The open vessel K is again raised and the neck G of the upper bulb opened to liberate the air. In continuing the operation the two bulbs C and F are refilled with mercury, the neck G is once more closed, and the vessel again lowered so as to exhaust the air as before. When this process has been carried on until the exhaustion is nearly complete, the pumping is then modified in this way: The open vessel K is raised and lowered several times while the rod E is out, so as to make the mercury rise and fall in the bulbs, which should now be slightly warmed, in order to evaporate any moisture on the interior. The mercury is not allowed to fall below the point of communication with the exhausting tube M while the rod is out, otherwise the mercury will be forced up the said tube into the lamp. Having got rid of all traces of aqueous or other vapour, the rod E is again inserted (while the open vessel is raised) into the neck G of the upper bulb, leaving the neck D of the lower bulb open. The open vessel is further raised or lowered several times, so that the mercury rises and falls in the bulbs, and in this manner all traces of air from the surface of the tubes or bulbs will collect in the upper bulb F. The open vessel K is then placed at such a height that the mercury fills the lower bulb and is just above its neck D. The rod E is then forced down, closing the neck D, and it is then lowered until the mercury is below the point of communication with the exhausting tube M, again raised, and so on. The upper bulb F, above the neck of the lower renders the vacuum more perfect; but by using very pure sulphuric acid so as to wet the surface of the glass bulb C, and the ground joint at the neck D, this upper bulb may be discarded, provided the cup is closed in at the top with just sufficient opening to allow the rod to move freely up and down.

When the lamp has been thoroughly exhausted in this manner, a current is sent through the filament, rendering it incandescent, and the pumping action is continued as rapidly as possible. This process is continued from time to time for two or three days, so as to draw off the occluded gases from the heated carbon. The fine exhaust tube h (Fig. 705) is then sealed up and broken off, leaving the lamp complete.

Sometimes Mr. Lane-Fox dispenses with the long tubes M Q, and connects the tube B with the tube h of the lamp by means of a pipe P shown in dotted lines on Fig. 710. In this tube there is a valve Q opening downwards, and when the mercury in the pipe P falls, there is a free passage between the lamp and the bulb C through the valve Q, but when the mercury again rises as high in the valve Q the valve closes. It should be added that Mr. Lane-Fox also employs sulphuric acid in place of mercury in the pump, but with a somewhat modified apparatus.

The plan of Mr. Lane-Fox for distributing the currents from a central station to the incandescent lamps is illustrated in Fig. 711 where A A are a pair of generators driven by a steam engine as shown. One pole of each generator is connected to the prime conductor or " electric main " C, and the other poles are connected to " earth " by the earth plates E or water pipes. Branch conductors $C^1 C^1$ or submains start from the main, and other branch conductors $C^{11} C^{11}$ lead to the lamps and through the lamps l l to the return main. The lamps are all in multiple arc, and a constant electromotive force of about 100 Volts is kept up in the circuit. To insure this result an automatic governor F, for controlling the generators is included in a circuit between the main and " earth." The regulator consists of an electro-magnet, through which a shunted portion of the current passes, and attracts a lever armature in front of its poles, as shown in Fig. 712. This lever plays

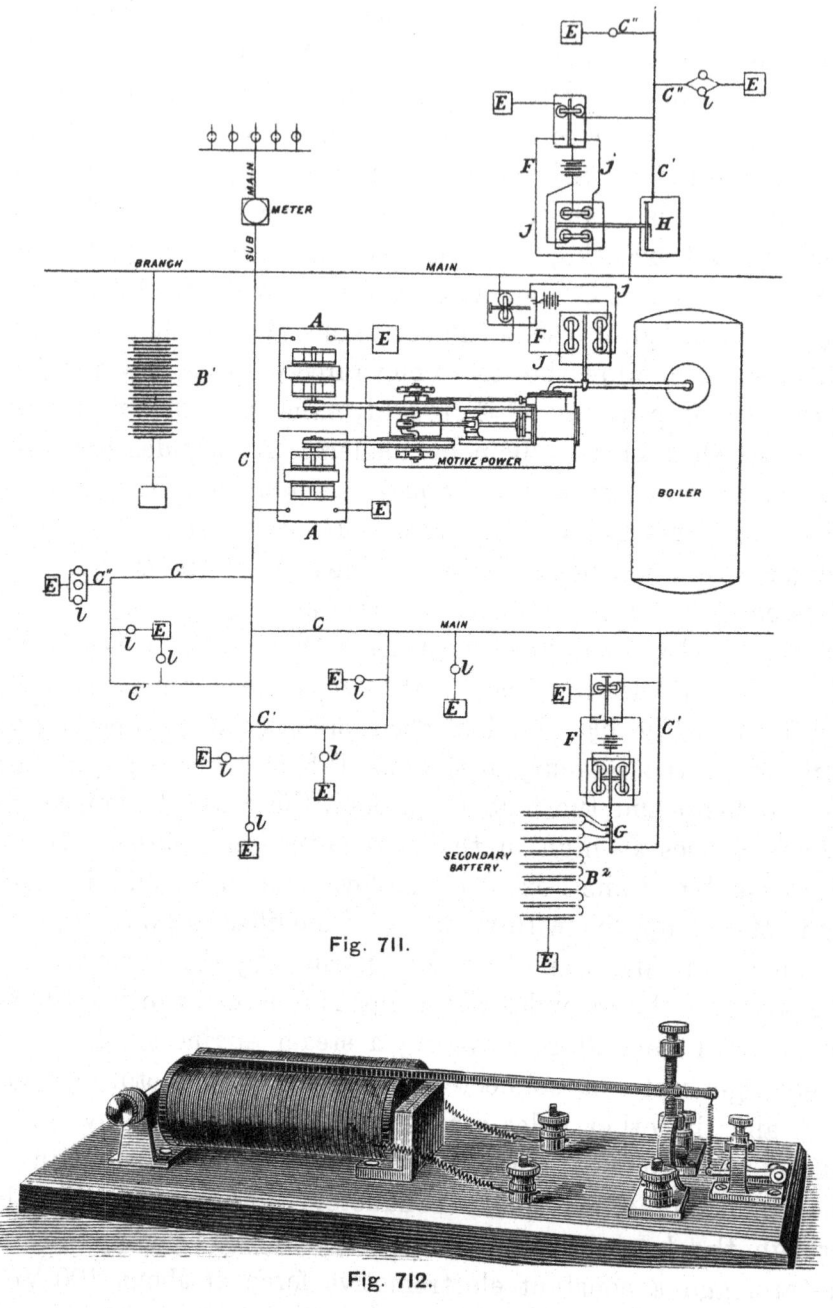

Fig. 711.

Fig. 712.

between two adjustable contact pins resting midway when the current is of the proper intensity, but striking the upper pin if the current becomes too strong, and the lower pin if it becomes too weak. By this means a local circuit is closed, and more or less resistance is inserted in the circuit of the field magnets of the generator, so as to bring the lighting current to its normal value, and the lever of the regulator to its mid position. This resistance is inserted by the automatic device shown in Fig. 713, where the middle pair of electro-magnets are caused by the local current to rotate a vertical axis, carrying at its upper end a small toothed pinion, which can gear with either of the two like wheels, enclosing it between their toothed edges, as shown. If it gear with the right-hand wheel, the horizontal spindle carrying these two wheels will be rotated in one direction, and if it gear with the left-hand wheel the

Fig. 713.

same spindle will be rotated in the other direction. Hence the second vertical axis seen on the left of the figure will be rotated in one direction or the other, and as it carries at its lower end a sliding contact arm which moves over the studs of a circular resistance box, partly visible in the figure, it throws in or takes out resistance from the field magnets. The direction in which the horizontal spindle shall turn is determined by two double electro-magnets, only parts of which are seen. According as the lever in Fig. 712 touches the upper or lower contact, one or other of these electro-magnets attracts an armature placed between their opposed poles, and brings the pinion-headed axis into gearing with one or other of the two wheels on the horizontal spindle.

The secondary batteries B¹ and B² are used to store up any surplus current and return it to the circuits when there is any falling off in those below the normal current. At B¹ they are arranged in "series," and at B² they are so connected up by means of a commutator C, that the battery in discharging will automatically adapt the potential of its discharge to the needs of the mains as determined by the regulator E.

The meters employed by Mr. Lane-Fox for measuring the current used in the lamps are of three kinds, one, in which a derived current from a supply conductor is passed through an electro-magnet and made, by attracting an armature, to open and close more or less a conical valve regulating the flow of air through a species of windmill counter. The other and preferable kind is illustrated in Fig. 714, where *a* is a similar

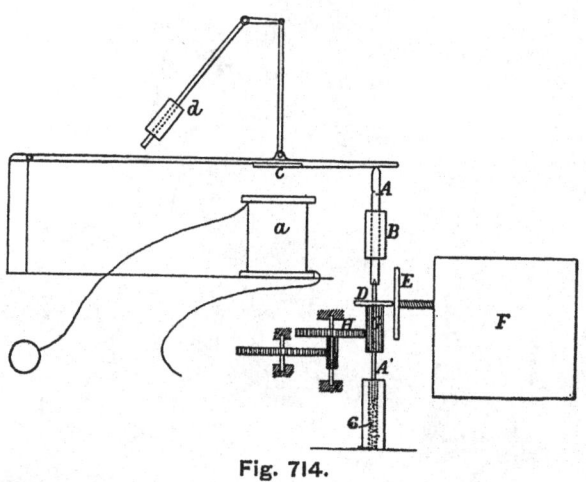

Fig. 714.

electro-magnet in a derived circuit from the main, and *c* is the lever armature on its poles. The lever bears at its outer end on a spindle A, which is free to move vertically in a guide B, and forms a pivot for a spindle A¹, which is supported by a spring C to keep the spindle A in contact with the lever armature. On the spindle A¹ is a disc D in frictional contact with another disc E, which is driven by clockwork F, and pressed against D by a spring. On the same spindle is a long pinion G gearing with a toothed wheel H and actuating a counter. When no current passes in the electro-magnet *a* the disc D is exactly in the centre of E, and therefore does not rotate, but when the armature is attracted by a current the lever bearing on A displaces D from the centre of E and D begins to rotate with a speed proportional to its displacement, that is approximately proportional to the current strength. Thus the counter will indicate the quantity of current used.

The meter chiefly used by Mr. Lane-Fox is, however, that shown in Fig. 715. It is an integrating meter, and consists of a double-poled electro-magnet wound with stout wire and having a hinged armature of soft iron inclined over it, as shown on the right of the figure. This

Fig. 715.

armature is supported in its inclined position over the poles by a spiral spring hung from a striding support. The armature is branched at its extremity, and the fork is attached to a vertical stem seen on the left of the figure. This stem gears with a mechanical counter or indicator at its lower end as shown, and on its upper part carries a small horizontal disc which rolls on the rounded surface of a piece of boxwood, which is rotated round a vertical axis by means of a small oscillating electro-motor worked by a shunted portion of the main current. This piece of boxwood is of a semi-oval form, narrow at the top, so that when the horizontal disc touches at the upper parts it revolves less quickly than when it touches at the lower and thicker parts where the moving circumference of the boxwood is greater. And as the horizontal disc derives its rotation by frictional contact from the rotation of the boxwood surface, it follows that the position of the disc on the boxwood surface affects its rapidity of rotation, and consequently the indications of the counter. This position is altered by the strength of current in the main circuit passing through the doubled-poled electro-magnets. The attraction of this magnet pulls the armature down against the force of the spring, and with it the horizontal disc carried by the vertical stem. Further, the surface of the boxwood is turned to a curve, found by experiment to give the proper speed of indicator for the corresponding strength of current. The electro-motor, which gives a continuous rotation to the boxwood barrel, consists of two electro-magnets on a shunt circuit from the main current, and these by proper interrupting spring contacts are magnetised alternately, and keep a spring balance lever in oscillation. This lever in turn works a ratchet wheel, and rotates a shaft carrying the boxwood barrel. Clockwork could also be applied for this purpose, but the electro-motor requires no winding up.

The Lane-Fox system is now worked by the Anglo-American Brush Electric Light Corporation, and was recently exhibited at the Crystal Palace, where the beautiful chandeliers designed by Mr. E. R. Johnson, for the Alhambra Courts, produced a very fine effect. These Courts were lit by Lane-Fox lamps fed from a Sellon-Volckmar secondary battery charged by a Brush machine, and the light was conveniently graduated by switching on a greater or less number of cells of the battery.

MAXIM.

We have already described the Maxim generator and regulator of the electric current, and have now to refer to the incandescence lamps devised by the same inventor. In this form of lamp, Mr. Maxim makes use of a strip of carbonised paper, which is rendered durable by covering its surface with a hard deposit of carbon; this is effected by heating the filament to incandescence within a closed chamber filled with the vapour of gasoline. As the current raises the temperature of the filament to a white heat, the gasoline is decomposed, and the carbon becomes attached to the filament. The degree of incandescence of a filament varying with the resistance to the current, it follows that the thinner and more imperfect parts of the paper strips, become more highly heated than the remainder, and the deposit of carbon on those parts is

Fig. 716.

more active. In this way it is intended to obtain a perfectly homogeneous filament, which shall be durable, and at the same time one in which the light shall be uniform. The Maxim lamp appears to fulfil both these requirements to a high degree. Fig. 716 illustrates a common form of this lamp, intended to yield a light of from 20 to 60 candles.

For electric lighting installations consisting of a number of lamps upon the incandescence principle, Mr. Maxim arranges his lamps as so many bridges or shunts between two main conductors, the ends of which furthest from the generator, are insulated from one another. This arrangement will be understood by reference to the diagram, Fig. 717, in which M represents the generator, and P P and N N the two main leading wires or cables connected respectively to the positive and negative terminals of the machine; the lamps are represented by the circles marked L, L, L, L, &c., being connected to the main wires by the fine conductors $p\,n$, $p\,n$, $p\,n$, &c. If another group of lights in one direction has to be illuminated by the same current, such for instance as a branch street or a number of lights in a public building, a submain is laid connected with the principal main in the manner shown on the diagram, where $P^1\,P^1$, $N^1\,N^1$ represent the two wires of the submain, two of the lamps on which are shown at $L^1\,L^1$, they being connected to it by the wires $p^1\,n^1$ in a manner precisely similar to those in the principal circuit.

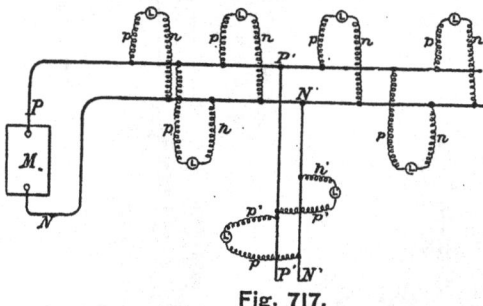

Fig. 717.

With respect to the illuminating power of the lamps as well as to the horse-power absorbed by them when giving out a light of a given photometric value, we must refer to a very interesting series of measurements made with the Maxim lamps by Mr. Henry Morton, the eminent President of the Stevens' Institute. Mr. Morton found that the resistance of the carbon in one lamp which he measured was 20·4 Ohms when cold, but that when raised to a degree of incandescence at which it emitted the light of 50 candles, its resistance was reduced to 8·3 Ohms. To produce this light a current of 4·07 Ampères was necessary, thus indicating an expenditure of energy of about 5,850 foot-pounds per minute. This represents a total illuminating power of about 275 candles per horse-power of energy of current expended, and allowing for a difference between the theoretical and the actual energy given off by the steam engine and dynamo machine, Mr. Morton calculated that when the lamps are giving a light of 50 candles, the average efficiency of the installation may be represented by a light of from 115 to 140 candles per horse-power of engine absorbed.

Figs. 718 to 723 illustrate a recent modification of the Maxim lamp embodied in a patent dated July, 1881, and which describes several methods of manufacturing incandescing conductors. The first consists in cutting the blanks from fibrous material, carbonising them, bending the carbonised strips into the desired shape, and then electrically heating them to give them a permanent set. In the second mode the strips are cut from previously carbonised sheets, and then bent and heated. The third process is for carbonising and electrically heating in a carbonaceous vapour, narrow sheets, and cutting the filaments from them. In

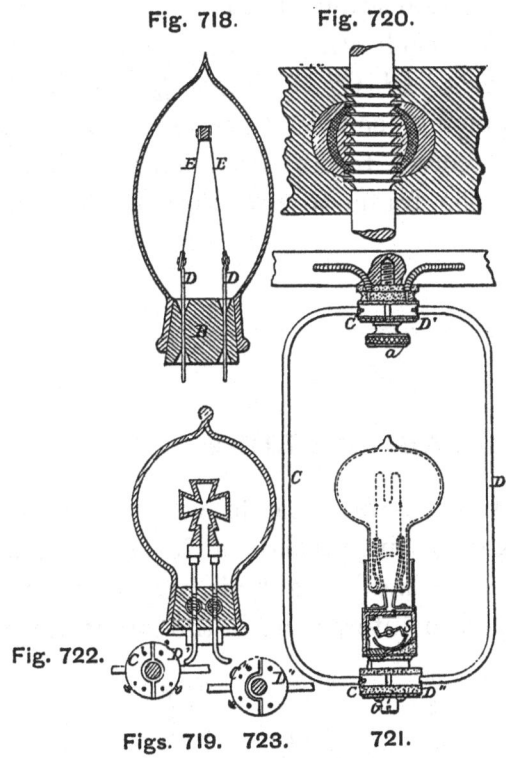

Fig. 718.   Fig. 720.

Fig. 722.

Figs. 719.   723.   721.

Fig. 718, B is a stopper moulded with two indentations on each side. By means of a drill these indentations are extended through the glass, and through the holes thus obtained slightly tapering wires D D are drawn, covered with powdered gum copal. The whole is then heated, and the wires drawn further into the holes. E E are two conductors united at the upper end to a block of the same or other material. Another method of inserting the wires D D is to make them in the form of taper steel plugs, and grind them into their places. The lamp can be rotated to bring the ends of the bridges successively in contact with the line wires.

Another method of sealing the conductors into the glass stopper is shown in Figs. 719 and 720. Each conductor has grooves or threads formed around part of it (Fig. 720), and on such portion a small quantity of vitreous cement or a composition of potash, silica and the oxides of iron and copper, is caused to adhere while hot, about which when hard a second layer of the material, but containing a smaller proportion of metal, is formed, and then over this a third layer of nearly pure glass. The wires with their adhering mass of cement are then laid in a mould and a glass stopper is formed about them, which is ground into the open neck of the lamp globe. Fig. 721 is an elevation of a lamp and bracket. The bars C and D are insulated from each other, and are attached at their ends to circular segmental plates $C^1 D^1$, $C^{11} D^{11}$ (Figs. 722 and 723). These are bound together by clamping screws between washers of insulating material. The screw G serves to fix the bracket to its support; the other screw G^1 clamps the frame to the lamp holder; S is a circuit breaker.

JABLOCHKOFF.

M. Jablochkoff's inventions for incandescence lamps date back from 1876, the patent covering it, bearing the date of November 30th of that year, about the same time that he had brought his candle to a practical form. The arrangement consisted in interposing in the circuit an induction coil

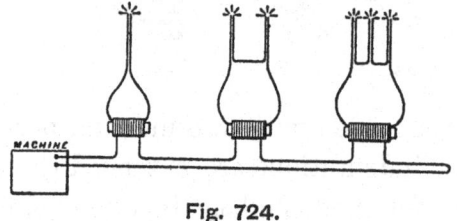

Fig. 724.

developing an induced current in a second coil. This second current was utilised according to its tension to feed one or several electric candles, and each coil could thus be assimilated to a distinct supply. The accompanying diagram (Fig. 724) will explain the arrangement in which the coils, being of small dimensions, the carbons of the candles required also to be very small, and they burnt out rapidly. M. Jablochkoff then substituted wires for the carbons with the result of obtaining a very reduced light. Now, the spark from the coils is able to raise to incandescence, refractory materials placed in its path, their conductibility being increased when they are raised to a red heat; and in placing between the electrodes of the coil a refractory strip, a luminous band can be obtained of a considerable length. It was in this way that M. Jablochkoff arrived at his kaolin incandescence lamp, which is illustrated in Fig. 725. The lamp is carried on a wooden base *a, b, c,* the upper side of the disc *a* being formed with a groove for the reception of an opal or other diffusion globe. The plate *d e* is of porcelain held in brass clips *f, g*. The rods

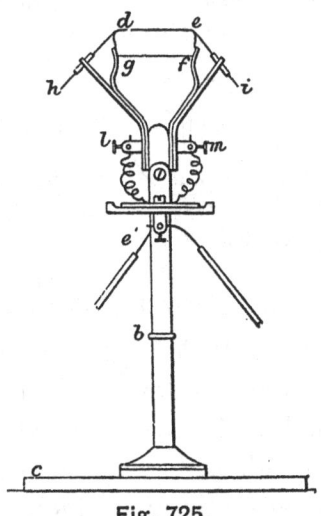

Fig. 725.

h, d, e, i, are of stout steel wire in connection with the contacts, *l, m,* and by the contact *e*[1] and a second corresponding contact at the back, with the induction coil. As soon as the spark is produced it is led on to the porcelain plate by means of a carbon pencil set in an insulated handle, the upper part of the plate becomes heated, and very quickly incandescent. The same result can be attained in coating the upper part of the plate with a mixture of gum and carbon, similar to that used for the fuse of the Jablochkoff candle. The light thus obtained is quite steady, the intensity from two to three Carcels (between 20 and 30 candles), the colour of the light is yellow, and the length of the incandescent strip is about 1½in. The porcelain wears away very slowly and can be renewed at a nominal cost; the amount of power absorbed, however, is relatively considerable.

Amongst the numerous devices that have recently been suggested in relation to incandescence lighting, we may select a few for notice here, without, however, offering any criticism upon the merits of the various designs.

Gatehouse.—Mr. T. E. Gatehouse has proposed a lamp in which the current passing through the bridge is regulated differentially. As is well known, the resistance of metals increases, and the resistance of carbon decreases as the temperature rises. It follows, then, that if a platinum bridge be short-circuited by a carbon filament, a greater portion of the current will be deflected through the shunt as the temperature approaches to an excessive height. Fig. 726 shows the arrangement diagrammatically, where P is a platinum wire, C a carbon shunt adjust-

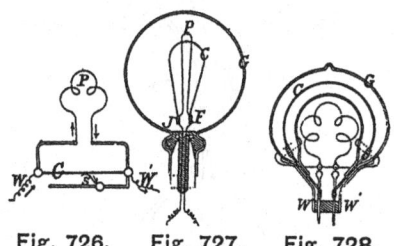

Fig. 726.   Fig. 727.   Fig. 728.

able by the sliding contact S; W W¹ are the leads. Fig. 727 shows an actual lamp, where the same letters indicate similar parts. The connections J F are bound with iron wire which combines with any oxygen that leaks into the globe. The conducting wires are passed through a mass of Chatterton's mixture, which slightly melts and forms a seal. Fig. 728 represents a lamp in which a carbon filament C is alone placed inside a crescent-shaped exhausted globe G, the platinum incandescent filament being placed outside.

André.—By this arrangement, in order to form the filament, a vegetable fibre, such as rattan, is immersed in a mixture of one part nitric and four parts of sulphuric acids, and, after washing, it is steeped in a solution of nitro-cellulose, in ether, or other solvent. It is allowed to remain there until it assumes a semi-dissolved state. The material thus prepared, when dry, is ready to be formed into the shapes required. It is then carbonised in the ordinary way and treated in ether, tar, boiled linseed, or other drying oil, sugar syrup, solution of starch, dextrine, and the like, to fill up the pores. The filaments are then calcined to carbonise the absorbed substances, or the pores may be filled up by heating the carbon in a vessel of hydrocarbon gas. Or instead of this, the filament may be steeped in drying oil, and the oil allowed to dry before calcining; burnt linseed oil may be used mixed with lamp-black and rubbed into fibre. The ends of the bridge may be inserted into sockets at the ends of the conducting wires, and be fixed by dissolved nitro-cellulose, wood pulp, or burnt boiled oil carbonised by the current. This is shown in Fig. 729, where *b* are carbon blocks or metallic sockets attached to the conducting wires *c c*, *a* is the filament, and *d d* are wires connected to each of the blocks, which are united by a wire *e* to complete the circuit during the carbonisation of the contacts. Another

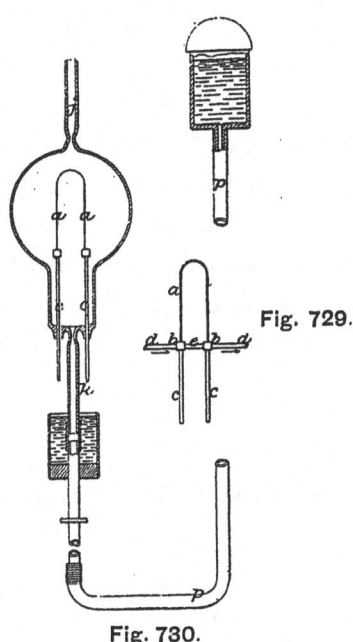

Fig. 729.

Fig. 730.

method is to couple the filament to the wires by aid of a clip of mica or steatite. Fig. 730 shows the exhausting apparatus. At each end of the bulb there is fixed a contracted tube, and to the lower of these there is coupled a vessel of mercury by means of a flexible tube *p*. When the vessel is raised to the position shown, the globe is filled with mercury, and the greater part of the air expelled through the tube *j*, which is then filled with a soft stopping. The vessel of mercury is then lowered more than 30 in. to create a barometrical vacuum in the globe. This liberates all the occluded gases, and when the vessel is again raised these gases are forced out of the tube *j*, which is unstopped and then sealed; on the vessel being again lowered the tube *k* is sealed.

Gimingham.—This very ingenious device for connecting the filaments to the terminal has been devised by Mr. C. H. Gimingham, whose name we have had occasion to mention already in the course of this work, and is most successfully used in the manufacture of the Swan lamp. The ends of the terminals (composed of platinum or other suitable metal) are flattened, and the flattened ends are formed into a tube by drawing through a wire plate or otherwise; the carbon filament is mounted directly in these tubes.

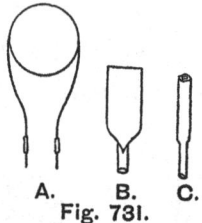

A.   B.   C.
Fig. 731.

Fig. 731, B represents a terminal wire with its end flattened out into a thin plate, which is thereafter bent by drawing through a wire plate, or by any other convenient means, into a tubular form as shown at C. Both terminals of the lamp are thus formed, and the ends of the carbon are placed in the tubular parts, as shown at A, and by the combined spring of the carbon and the metal tubular parts the said carbon is held firmly in place in the terminals.

Faure.—This invention refers (1) to the holders for the carbons; (2) to means whereby the carbon may be removed without the destruction of

Fig. 733.

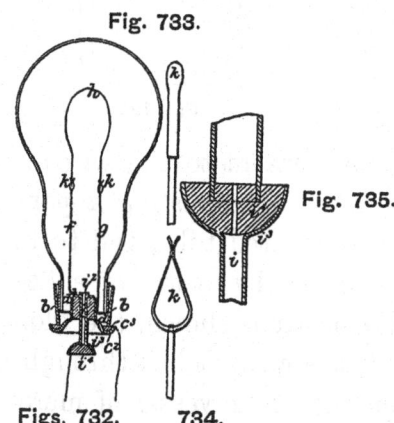

Fig. 735.

Figs. 732.   734.

the body of the lamp; (3) to improved means of exhausting the lamp; and (4) to the production of the carbon itself. The globe has a tubular stem to which a metallic collar *b* is attached by fusing or by electro-plating the stem and fixing the collar to it by solder or by fusible metal that will expand in cooling. The carbon holder consists of a metal tube *c* closed by a plug *a* of glass, or other non-conductor. The tube *c* is splayed out into a trough *c²* to contain a metal *c³* which melts at a low temperature for receiving the outer edge of the socket *b* so as to make an air-tight joint. The two conductors *f g*, upon which is fixed the filament *k*, are arranged as follows: The one electrode is formed by a piece of wire *g* soldered to the tube *c* and the other holder is formed by a small tube passed through the plug *d* and through the plate *i²* to which the wire *f* is fastened. The filament is secured by two spring clips *k* of metal which are made by bending the same into an elongated loop and then flattening it. Figs. 733 and 734 are two views of such a clip. To effect the final sealing the metal *i⁴* in the cup *i³* at the end of the tube *i¹* is perforated with a very fine hole (Fig. 735.) After exhaustion the fusible metal is heated and the hole thereby closed. The filament is manufactured by drawing or cutting graphite into small strips, which are heated and bent into loops or other forms while hot.

Société Force et Lumière.—Figs. 737 and 738 show modifications of incandescence lamps belonging to the Société La Force et la Lumière, of

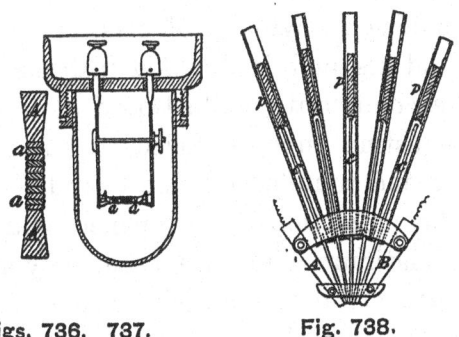

Figs. 736. 737.   Fig. 738.

Brussels. To obtain sufficient resistance in this incandescent lamp without the use of an attenuated conductor, the bridge is divided in several places, the parts lying in contact with one another, the "augmentation of the resistance being directly in proportion to the number of divisions." In Fig. 736, *a a* is the bridge of carbon, irridium, or other suitable materials held between platinum terminals A A, and divided into sections as shown. Fig. 737 shows such a bridge set horizontally in an exhausted globe between two spring arms. Fig. 738 is an

example of an incandescence lamp to burn in the open air; *c c c* are
carbon rods impelled towards a common point by the weights *p p*; A B
are the two terminals of the lamp between which the current passes,
making its way transversely across the carbon rods and heating them to
the necessary degree.

Jameson.—In this arrangement, a series of relays of carbon filaments are sealed up in incandescence lamps, so that when one fails
it may be removed, and a fresh one be substituted without access to the
interior of the lamps. The filaments are mounted on an endless band
passing over rollers, and arranged to bring each into position successively.

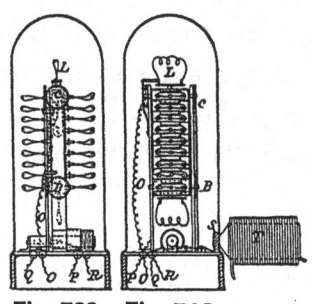

Fig. 739.   Fig. 740.

Crookes.—By this device a cylinder of glass is blown into the shape
shown, A, Fig. 741, and its end is doubled inwards by a two-pronged
tool B, till it takes the configuration illustrated in elevation and plan C.
At this stage an exhausting tube is fused to the globe, the neck is
divided, and the upper end rounded off and closed, D. The two hollow
points at the top of the projection B are opened, and conducting wires
passed into them. If platinum wires are used, they are sealed directly
to the glass, but if other metal is employed the wires are coated with
white enamel or arsenic glass, and passed through short cylinders of glass
into which the enamel is melted; such a conductor and cylinder are
shown. The wire is then passed through the open point of the glass
projection until the glass cylinder on the wire rests against the end of
the glass projection, and the two are then sealed together in the blow-
pipe either with or without the interposition of arsenic glass or enamel.
The parts now have the appearance shown in E. After the carbon
filament has been connected to the wires the lamp neck can be joined (F),
and the exhaustion be effected. The difficulty of making a good con-

nection between conducting wires other than platinum and the glass can be avoided by using a compound wire with a core of copper or silver and a platinum sheathing. Fig. 742 shows an apparatus for adjusting the resistances of the bridges. A is a glass vessel, C a plate of vulcanised fibre, D D¹ glass-covered wires, E a mercury seal, G the exhausting pipe,

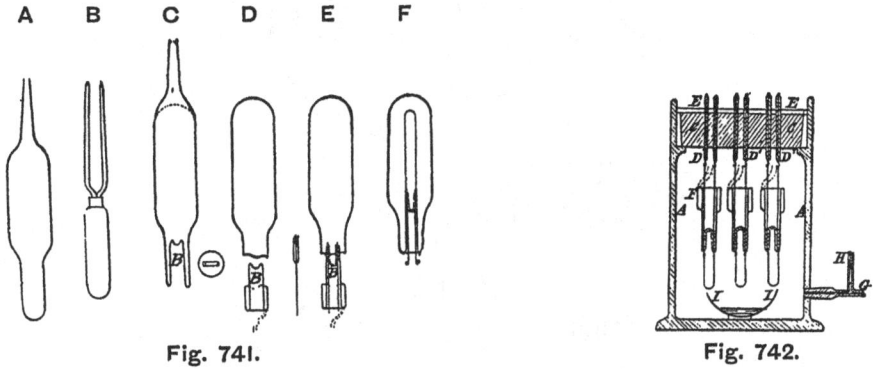

Fig. 741.   **Fig. 742.**

H tube leading to a vessel containing chloroform or other substance. The chloroform may also be placed in the cup I. The resistance of each bridge is measured, and, if it be too great, a current is sent through it until a sufficient quantity of carbon is deposited on it to augment the conductivity.

Hussey and Dodd.—Figs. 743 and 744 show this arrangement, which consists of a translucent globe with prolonged ends, through which extend holders B B¹ for carbons C, one of which can be heated to incandescence at a time. The holders B are all connected to the metal band D, and the holders B¹ to insulated plates E on a commutator, the upper one of which, for the time being, is in contact with spring G, which forms one of the lamp terminals. The current enters the lamp at D, traverses one carbon and leaves at E. When that carbon breaks or becomes inefficient, the lamp can be partially rotated around its longitudinal axis, and a second carbon be brought into circuit. The arrangement includes an electro-magnetic appliance for effecting the rotation automatically on the failure of the current. The current can be regulated by resistance coils upon a spool O actuated by the handle Q. Outside the coils are metal bars R R¹ R², &c. (Fig. 745), which are connected at short distances apart to the flanges of the spool O, and are thereby insulated from each other. They are severally coupled to the wires of the respective coils. One wire of the circuit connects to a loose sleeve O¹, on a metallic stem O², extending from the spool, so that the wire will not be turned when the spool turns. The bar R is in electrical com-

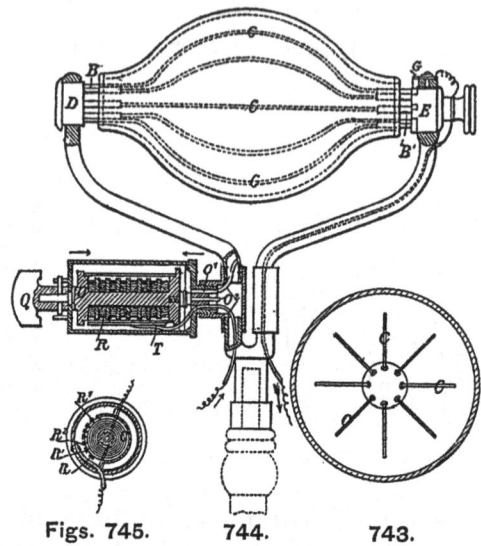

Figs. 745.   744.   743.

munication with the stem O². The leading wire goes to the fixed contact spring T, which bears on the bars R¹, R², &c., when the spool is rotated. When T rests on R the coils are out of circuit, when it rests on R¹ the first coil is in circuit, and so on.

Nichols.—Mr. J. V. Nichols, of Brooklyn, proposed an incandescence lamp, shown by Figs. 746 and 747. For this the conducting wires are

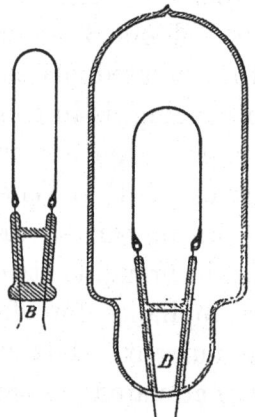

Fig. 746.   Fig. 747.

first embedded in a small circular disc of vitreous cement, such as a composition of potash, silica, and oxide of iron and copper; this disc is sealed directly into the opening of the main globe by a blowpipe, and the exterior surfaces of the two parts are made smooth and even. By this arrangement copper conducting wires may be used. In the illustration, B is the disc of cement with extensions upwards around the conductors. The globe is made in the form of a cylinder with hemispherical ends; the ends of the carbon bridge have holes in them, through which the conductors are passed and clamped over.

Riverton.—This invention, which was patented in November, 1881, relates to the utilisation of electricity for lighting and heating. The first part is illustrated in Fig. 748, in which there is shown a conductor having its two main portions connected by a series of transverse filaments, which form so many passages for the current. The ends of the main parts of the filaments are held between split clamps of metal, which obtain an elastic hold. In a modified lamp the bars a^1 are arranged vertically, the conductor when in the blank resembling a gridiron. Secondly, a number of independent filaments are included

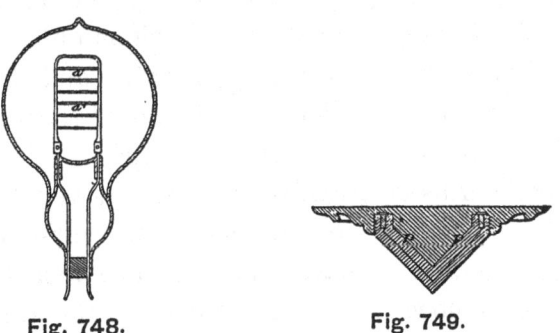

Fig. 748.   Fig. 749.

in one globe, and are furnished with contact pieces, so that one or more may be placed in circuit at one time. These contact pieces are operated by hand. Thirdly, a number of platino-iridium wires are arranged in an exhausted globe upon a form by which they are bent into a balloon-like, fan-like or other shape. Fourthly, an incandescing bridge is made of a number of thin laminæ laid together to the end, that a rupture may not extend across the whole of them. Fifthly, the light-emitting carbon is encased within a hollow piece of glass and surmounted by a reflector p (Fig. 749). The specification illustrates several modifications of this idea. Sixthly, the invention relates to apparatus wherein the electric current is caused to develop heat.   Many methods of applying this are illustrated, employing terra-cotta, water and air for the diffusing media, and controlled, as to temperature, by hand or by the expansion of fluids or metals.

THE MAGNETO-ELECTRIC GENERATOR OF M. WORMS DE ROMILLY.

THE early history of magneto and dynamo-electric generators would be incomplete without a detailed notice of the patent of M. Worms de Romilly, of Paris, which is one of the most suggestive of all the early inventions, as will be seen from the following translation of the specification deposited at the French Patent Office.   The illustrations which accompany the description are made from machines actually constructed by M. de Romilly.   These illustrations unfortunately arrived too late for publication in their chronological order, and the notice of this generator has therefore to find a place at the end of the volume.

The date of the patent is 3rd March, 1866; it is granted to M. Henri Louis Felix Worms de Romilly, and is entitled "the collection, development and application of induced currents produced from 'magnetism in rotation,' or 'magnetism in movement.'"   The text of the specification is as follows : "When we cause a metallic plate to pass before the pole of a magnet, perpendicularly to the axis of this magnet, whether the movement is reciprocal, or whether, having the form of a disc, it revolves without interruption before the poles, there are set up in this plate; induced currents perpendicular to the direction of movement, and of the same sign, before and after the passage in front of the pole. This phenomenon, discovered by Arago, explained afterwards by Faraday, and studied in their details by Poisson, by Nobili and Mateucci, have received the name of 'magnetism of rotation,' or 'magnetism of movement.'   It is the collection and utilisation of induced currents, which are produced when a metal plate passes before the pole of a magnet, perpendicular to its axis—currents which are set up perpendicular to the direction of the passage of the current—that is to say, the collection and utilisation of those currents due to what is named 'magnetism by movement,' or 'magnetism of rotation,' which I propose to reserve by the present patent. Induced currents, which up to the present time have been utilised by employing artificial or natural magnets, have always been produced by the successive reciprocations of a bar of soft iron carrying one or several wires coiled in a direction perpendicular to its axis, and passed in front of the inductor magnet in such a way as to be magnetised or demagnetised rapidly (as in the Clarke and Alliance machines), or they have been produced by the passage, and the successive and rapid interruptions of a current from a voltaic battery, in an inductor wire, coiled round a bar of soft iron, and itself covered with a coil, in which is set up the induced current (as in induction coils or Rumkhorff machines). Here it is necessary to pass before the pole of a magnet, or between the two poles of a magnet, a piece of metal of high conductivity—a plate or a copper wire, for example—and to collect the current thus obtained. Experiments made up to the present have been conducted with revolving discs, or with plates passing and re-passing before the poles. The currents thus produced are very weak, which has rendered it impossible to collect them so as to employ them as ordinary induced currents. What renders the current so feeble is, in the first place, the continuity of the plate or disc, which allows it to be dispersed with extreme facility, so that only a small part can be taken off to demonstrate its existence.

"To obviate this inconvenience, it would suffice to pass before the pole a succession of wires, separated from each other, and each extremity of which would communicate with a fixed conductor that would come in contact successively at the moment of their passage before the pole. The current would thus be produced successively on the passage of each separate wire before the inductor pole, without any part being lost, as is the case with a continuous plate. This current, very feeble when the wire passes before a single pole, is greatly increased when it passes between two opposite poles.

"I now assume that it is possible to connect each of these separate wires to the adjacent wire, in such a way that the current obtained in one may be in the same direction, and, as it were, a continuation of that which is obtained in the other; in this manner a kind of battery would be obtained, each wire of which would represent an element. To arrive at this, we cause to revolve between two poles of the same name—both north, or both south—a plate of soft iron covered with well insulated wire, and coiled in such a manner that the flattened spirals in this iron core are presented perpendicularly to an imaginary line joining the two poles, and perpendicular also to the line indicating the direction of movement of the plate. The soft iron will thus assume a polarity contrary to that of two similar poles. The passage of the plate covered with wire will give rise, on the one side, to a current procured by the wire travelling before a north pole, for example, and a south pole is produced in the plate itself by the influence of the fixed pole, while on the opposite side of the plate there is set up in the coil of wire, a contrary current produced by the passage in the same direction before a south pole, resulting from the influence on the soft iron plate, and the other fixed north pole. The semi-spirals on one face of the plate partake of the same movement as the semi-spirals of the other face, in a contrary position, referred to a fixed pole, and the pole produced by induction.

"This is precisely what is required in order that the currents which have to flow around the core may form at each semi-coil, at each coil, and throughout the series of coils, a continuation of each other. If, then, we suppose a long plate of soft iron thus surrounded, and passing between two similar poles on the two extremities of the coil, being connected by a conductor to a galvanometer, this latter will indicate the presence of a continuous current as long as the plate moves in the same direction, but which will be interrupted when the movement is arrested, and which will take the contrary sign when the plate travels in the opposite direction. The sign of the current can also be changed by the introversion of the poles, by the direction of the movement, and also by the direction of winding the coils to the right or left. Further, this wire coiled singly, can be covered by a second wire wound parallel to it, and brought back over itself, so as to surround the soft iron core several times. Each thickness of the coil superposed will add its current to that of the preceding coil. In this manner, instead of a feeble current, we shall have one, the energy of which will depend on the power of the magnets, on their proximity, on the purity of the soft iron core, on the number of superposed coils, and upon the velocity of movement.

"Now to consider an arrangement which will make this process practical, and by which one can collect the induction currents of which we have been speaking. Instead of a soft iron plate, we will take a plate curved in such a way as to form a cylinder, of which the diameter is much greater than the height. We will connect this cylinder by radial arms converging from one side to a rigid shaft passing through the centre of the cylinder; this shaft will rest in bearings in which it may revolve rapidly. The cylinder, entirely open within, has thus only one of its sides partially occupied by the radial arms, that serve to fix it to the central shaft. This arrangement allows of winding longitudinally on the cylinder, a copper wire covered with an insulating material. We will first assume only half the circumference of the cylinder to be thus covered, then we will place in close proximity to one side of the cylinder, two magnets, each formed of a bar of steel magnetised; the similar poles of these two magnets—the north, for example—are opposite to one another, and allow the iron cylinder, one half of which is surrounded with insulated wire, to pass between them; in this way the wire receives an opposite polarity. We will now do the same thing with the side of the cylinder diametrically opposite, in such a way that on this, as on the other, a magnet passes within the cylinder (and parallel to it), having its south pole in close proximity, while a second similarly magnetised bar is placed outside, in such a way that the cylinder passes between two south poles. Now we will connect each end of the wire coiled on the half-circumference to two terminals placed near the shaft, and insulated from each other; and we will connect, by a conductor, these two terminals with a galvanometer. Then, by means of a pulley, fast on the shaft of the apparatus, we will impart a movement of rotation to the cylinder. The coil successively revolves perpendicularly between the two poles, north on one side, and south on the other. The soft iron core will assume a polarity contrary to that of the magnets, in the portion comprised between the poles, and subjected to their influence. From this there results, as in the plate to which reference has been made, similar currents, as long as the half-circumference on which the wire is coiled, passes between two poles of the same name; but so soon as it passes between magnets of the opposite polarity, the current will change its sign. Thus one half-turn will give a current of one sign and the complimentary half-turn will give a current of the opposite sign. If we continue the coiling around the cylinder in the same direction—to the right, for example—in such a way as to surround the whole cylinder, we shall not perceive any current, since, whilst half the circumference passed before one pole, the other half-circumference would pass before the opposite pole, giving a current the inverse of the former. To avoid this, it is necessary to utilise the effect of the two poles, which is easily done by coiling the wire of one half-circumference in a direction contrary to that of the other half.   Thus, if the former be coiled to the right, the latter must be coiled to the left; in this manner, the effects of the pole being reversed with each half-revolution, at the moment when the half-circumference to the right passes between the north poles, the other half, coiled to the left, passes between the south poles, and they will give simultaneously a current of the same sign, the opposite directions of winding serving to destroy the opposition of the poles.   The first half-circumference having passed before the north poles, will present itself before the south poles, passing between them, while the second half-circumference passes before the north poles. The current in the first half-circumference will change its sign in passing before the south poles, at the same time that a similar change will be effected for the second half-circumference before the north poles.   At each half-revolution, the current in the two semi-circumferences with the wire coiled in opposite directions, will change its signs simultaneously. Thus for each revolution there will be two opposite currents which will follow one after the other through the whole of the wire coiled around the soft iron core.   If, instead of simple insulated terminals we place a commutator on the shaft of the cylinder, we shall perceive currents always in the same direction, and which will produce the effect of continuous currents.

"In place of a single coil of wires, several thicknesses of wire may be superposed, and instead of a pair of magnets on each side we may place several pairs in such a way as to surround a half-circumference inside and outside with similar poles, and the whole of the other half-circumference within and without with poles of an opposite sign. (See Figs. 750 and 751.)   Collectively these series of magnets form two concentric cylinders between one end of which the soft iron ring covered with its coils revolves.   The electromotive force of the current increases with the number and length of coils, with the speed of rotation, and the number and power of the magnets.   Each complete turn or half turn of wire is attached by the two ends, to a brass terminal placed on a wooden bar, and fastened to the radial arms of the cylinder.   With these terminals we connect the various superposed wires, either in such a way as to form only one large cable, or end to end, so as to be able to vary at will the electromotive force of the apparatus.

"But it must be clearly understood that such an apparatus is only one of a thousand possible arrangements that may be indefinitely varied, and by which the principle protected by this patent may be put in practice.

"For example, the soft iron ring may be given the form of a torus

Fig. 750.

A Driving pulley.   **B** Commutator.   **D** Copper ring, connecting circles of magnets P_N P_S

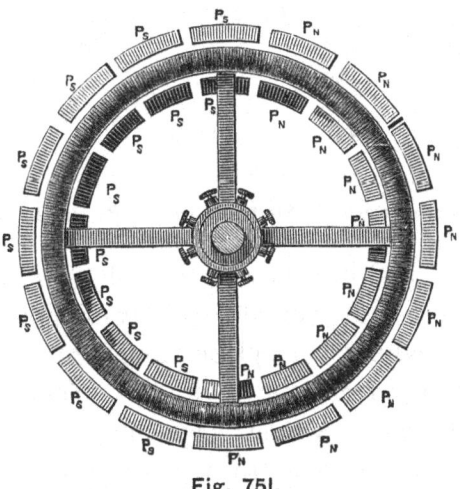

Fig. 751.

flattened perpendicularly to the axis of rotation, and having its wire coiled radially, so as to form a sort of disc pierced with a large central hole, in order to allow the radial coiling of the wire, the ring being mounted rigidly on a shaft passing through its axis. Magnets would be placed radially around the disc on each side, with similar poles opposite each other, as in Fig. 752.

" Or, the soft iron core may be formed as a cylinder with flat parallel ends, having a steel shaft running through its axis, and fixed to it either by means of radial arms or by blocks of wood driven into the cylinder at intervals; this shaft rests in bearings, and a pulley is arranged to drive it; I surround this cylinder parallel to its axis with an insulated copper wire, following always the same direction and passing from one side to the other by a turn diametrically across the ends of the cylinder. On the first coil thus made I superpose (always winding in the same

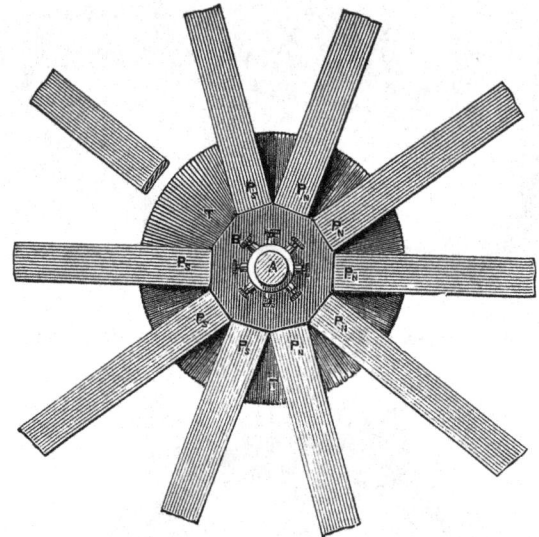

Fig. 752.

A. Driving shaft.   B Commutator.   T Flattened ring. P_N P_S North and South poles.

direction) a second turn, and in this way several coils in succession, and I join the ends of the wire thus wound to a galvanometer. Diametrically on each side of the cylinder, I place magnets, each having their opposite pole turned towards the cylinder. The magnet on one side, turns its north pole, whilst opposite, the magnet on the other side, turns its south pole to the cylinder. These magnets may each of them extend over one-half the surface of the cylinder. If now we cause this cylinder to revolve, we shall set up, at each half-turn, a current of an opposite direction, but simultaneously of the same sign in the coils. An analysis of the manner in which each element of the wire is presented to the poles, is too similar to the preceding explanation to require repetition; a commutator placed on the axis of the apparatus allows the currents to be collected so as to be practically continuous. Each turn and each half-turn on the cylinder can be made by a separate wire, the two ends of which, each going to a terminal, can be joined to the other wires in such a way as to give to the current the required electromotive force. Practically the envelopment of the cylinder cannot be effected by wires placed one after the other, and coming in succession across the diameter of the two bases. It is necessary that they should clear the shaft, and then, superposed on each other, they would form around the shaft a thickness equal to the sum of the collective thickness of wire in all the coils. This thickness would be doubled with the second coil of wire, tripled with the third, and so on. To avoid this inconvenience, I divide the circular bases of the cylinder into six or eight equal parts, and I wind the wires in such a way as to form in each division a parallel band like a ribbon, along each side of the

Fig. 753.

cylinder, and passing diametrically across the ends. These bands would be necessarily opened in the middle to make room for the shaft, and then brought together again. Each band of parallel wires is laid close to the succeeding one upon the cylinder, and they are superposed on each other in passing across the ends, so that the entire surface of the cylinder will thus be covered, as shown in Fig. 753. In this way we shall have at each end a height of superposed wires which will only be the sum of the thickness of one wire multiplied by the number of superposed bands.

"The detailed description of two apparatus, the indication of other forms infinitely varied, which can be devised, the general explanation of the principle to which I refer—that is to say, an induced current set up, when a metal plate is pressed near a pole, with a movement perpendicular to the axis of the magnet, a current perpendicular to the direction of movement, and to the axis of the magnet; this description and this explanation, appear to me to define and limit sufficiently the present patent, by which I reserve the putting into practice and the utilisation of this induced current, arising from 'magnetism of rotation.' The application embraces all the uses to which the current can be applied, such as chemical decomposition, electro-plating, the electric light, telegraphy, magnetisation, &c. Practice alone can specify to what science the current can be applied most easily and most economically."

END OF VOLUME I.

INDEX.

— · • · —

684

686

690

692

APPENDIX.

APPENDIX

APPENDIX.

—•••—

NOTE.

—•—

IN the following pages will be found Abstracts of English Patent Specifications, bearing dates earlier than the year 1873, in so far as they relate to the following subjects :—The Mechanical Generatio of Electricity; the Production and Application of Light by Electric Currents; the Manufacture of Carbon Electrodes; the Manufacture, Insulation, Sheathing, Laying, Supporting and Coupling of Electric Conductors; the Manufacture of Insulators and Insulating Compounds; Electro-Magnets; Current Regulators and Measurers.

Great care has been taken to include all specifications bearing upon the subject. The total number of specifications filed in the Patent Office is, however, far too great to allow of a separate examination being made of each to ascertain if, perchance, it contain anything relating to electric lighting, consequently reliance has had to be placed upon official indexes, which, as regards the earlier years, will probably be superseded by new editions hereafter. Hence the specifications examined have been only those which according to the indexes would appear to have some reference to the subject. However, it is confidently believed that nothing of any importance has been omitted.

In order to complete the list of machines for the mechanical generation of electricity, it has been necessary to include many electric motors, as it is obvious that, although the fact was not recognised by the inventors, these in numerous cases were reversible, and if they had been driven from an external source of power they would have generated currents. At the same time, it was necessary to draw a sharply defined line in respect of such apparatus. A galvanometer needle oscillating in its coils will generate currents in the convolutions, but it is evident that to include a galvanometer among generators of electricity would be absurd, and would lead only to confusion. All electric motors, therefore, have been included, in which rotary motion was produced by tangentia action between magnets, while all those iu which the motion was produced by the reciprocation of magnets or armatures, have been excluded.

The subject of conductors and insulators has been found to be a very heavy one. The interest taken by the public in the success of the first Atlantic Cable induced a crop of inventions by persons of little technical knowledge, whose productions, in the main, were either old or worthless. Combinations of tar, pitch, india-rubber, gutta-percha, resin and other bitumens and gums, were patented by dozens, and it has been found necessary to keep the notices of such inventions within very moderate limits. Processes bearing any resemblance to those at present in vogue have been treated in more detail, and allusions to the use of substances that have recently found a useful application in electricity, such as colophane and ozokerit, have been accentuated. The laying of subterranean conductors has been in all cases fully treated, as this is a matter of great interest in view of the certain extension of electric lighting in our streets. Oceanic conductors, on the other hand, have received less attention, and schemes for carrying them on buoys submerged one or two hundred yards below the surface, or for suspending them from balloons in mid air have been merely noted, or in some cases omitted entirely.

It is to be feared that in some instances the abstracts will appear unsatisfactory to the reader, and that either the very point that he desires to learn will be wanting, or he will find a difficulty in conceiving that the apparatus could effect the results which the inventor has set forth as his object. Unfortunately, this is a defect that cannot be avoided. A slight study of many of the specifications, filed in the early days of electrical science, would show that the inventors were totally ignorant of the laws governing the phenomena that they were striving to turn to account, and that their speculations were guided more by their imagination than by knowledge. It would be a herculean task to write a critical account of all the inventions that have been patented in relation to (say) electric motors, pointing out how far they were founded on correct principles and assigning to each its proper theory of action. Fortunately this is not needful for the purposes which this appendix is intended to fulfil. The law does not require that an inventor should have a clear or a correct conception of the scientific principles underlying his invention, but it does require that the invention should be novel. It is hoped that these abstracts, aided by the illustrations, will prove helpful to intending patentees in forming a decision on this point at least. As an aid to the reader, each paragraph has a heading in large type, indicating the general nature of the subject of which it treats. It must, however, be borne in mind that such prefixes refer only to the leading features of the invention, and that the greater must always be understood to include the less. Thus, under the designation of "Electro-motor" descriptions of an armature, a commutator and a field magnet may perhaps be found, while an allusion to reflectors may be included under "Electric Light." In the index, however, which has been compiled with great care, the same conditions do not obtain: therein, down to 1872, reference is made to the several details. It must be remembered that from its nature an abstract cannot be as full as the original from which it is drawn, and inventors are therefore cautioned, in cases where they feel any doubt, to refer to the specifications themselves. Printed copies of these can be consulted free of charge at the Great Seal Patent Office, Chancery Lane, London, and in the public libraries of most large towns. In cases where they are still in print, they may be obtained by applying, either personally or by letter, enclosing amount of price and postage, to Mr. H. Reader Lack, at 38, Cursitor Street, Chancery Lane, London. When specifications are out of print, manuscript copies and tracings of the drawings may be obtained, but at an increased cost.

Abstracts marked with an * relate to applications not proceeded with. The number of views given in the specification drawings is stated in each case after the title, where the invention is illustrated. When the inventions are patented as communications, the names of the inventors are given in italics, wherever this has been possible. Many of the earlier specifications, however, bear only the patent agent's name, or that of the person to whom the invention has been communicated, and in such cases it has been impossible to record those of the true inventors. Thus it will be noticed that the name of Shepard appears frequently, but it must not be assumed that he was the inventor; on the contrary, he acted as agent to a variety of persons. It has been found impossible in the present volume to include abstracts of such patents as we are dealing with later than 1872. The remainder, which are, perhaps, as numerous, and of course of far more importance, will appear in a subsequent volume. In the meantime, it is hoped that the classified lists which follow the abstracts will be found of some value. W. L. W.

ABSTRACTS OF PATENTS

RELATING TO

ELECTRIC ILLUMINATION.

COMPILED BY W. LLOYD WISE.

1837.

7,386.—M. Berry, London. *E. Williams, New York, U.S.A., on behalf of T. Davenport, Brandon, Ver., U.S.A.* **Obtaining Motive Power.** 7d. (5 figs.) June 6.

ELECTRO-MOTOR.—A vertical spindle carries four bar electro-magnets arranged radially in a horizontal plane. The pole faces of these magnets rotate as closely as possible within the inner peripheries of two semi-circular permanent magnets set with opposing poles, and forming a ring encircling the movable magnets. The coils of the electro-magnets are connected together in pairs, and the ends of their conductors are led down to insulated plates on a revolving commutator carried on the spindle, by which they are alternately put in contact with other plates connected to the poles of a battery. By this arrangement, the polarity of the electro-magnets is reversed each time they pass the poles of the fixed magnets. The magnetic ring may be formed of more than two magnets, and these may be either permanent or electro-magnets.

7,390. — W. F. Cooke, Hastings, and **C. Wheatstone,** London. **Electric Telegraphs.** 2s. 9d. (27 figs.) June. 12.

CONDUCTORS.—Conducting wires are laid in resinous cement in channels in wood rails, lined and covered with a lid. Several wires may be laid in one channel and insulated from each other by wrappings of coarse thread and by varnish. The troughs may also be of metal, and be formed in two parts, or have an opening along one side for the entrance of the conductors.

1838.

7,614.—W. F. Cooke, Hastings. **Electric Telegraphs, &c.** 2s. 4d. (32 figs.) April 18.

CONDUCTORS.—One part of the invention deals with laying conductors within solid lead or iron pipes, and with methods of connecting the pipes and bringing out the conductors where required.

7,729.—L. C. Callett, New York. (*A communication.*) **Propelling Vessels, Carriages, &c.** 7d. (2 figs.) July 11.

COMMUTATOR.—The current from a battery is distributed to the coils of a motor by means of two plates of silver pressed on to a revolving cylinder of wood, each plate extending half-way round. The plates are on opposite sides of the cylinder, but are displaced sideways with relation to each other, so that the conducting springs, which bear upon them rest alternately on the metal and the wood.

1840.

8,345.—C. Wheatstone, London, and **W. F. Cooke,** Slough. **Electric Telegraphs.** 2s. 11d. (20 figs.) January 21.

Magneto-Electric Generator.—One part of this invention relates to a magneto-electric machine "which may be of ordinary construction." As shown in the drawings, it consists of a compound horseshoe magnet, set vertically, with a horseshoe electro-magnet rotating upon a horizontal spindle in front of it. The spindle carries a commutator with two insulated plates connected to the two ends of the coil, and rubbed, as they revolve, by two springs which are the terminals of the external circuit. The commu-tator delivers the successive electrical impulses to the springs in one continuous direction.

8,644.—H. Pinkus, London. **Applying Motive Power to the Impulsion of Machinery, &c.** 6s. 8d. (23 figs.) September 24.

Electric Lighting.—Among many other things, this invention alludes to electrically producing a light, glow, sparks, or heated surfaces. The currents are distributed by mains or circuits, and the light may be enclosed in hermetically sealed glass globes containing rarefied or compressed gases, and situated in reflectors.

1841.

8,783.—J. Barwise and **A. Bain,** London. **Electric Clocks.** 2s. 8d. (12 figs.) Jan. 11.

Conductors.—When a number of clocks are worked in parallel circuits, all the wires that supply them are twisted together, the positive and negative groups of wires being insulated from each other by hemp.

8,937.—W. Petrie, Croydon. **Obtaining Motive Power.** 7d. (12 figs.) April 27.

Electro-Motor.—The patentee claims the application of the "deflective action" which exists between electric currents and magnets for the purpose of obtaining a moving power. A horizontal shaft carries an armature M, consisting of a large

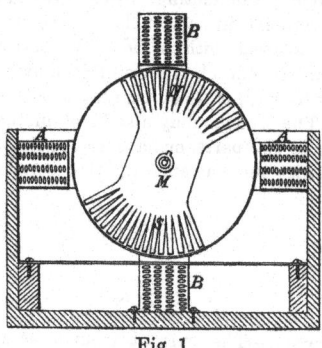

Fig. 1.

number of bar or electro-magnets so arranged that the north and south poles occupy respectively about 90° of the circle. This armature rotates within a case wound longitudinally with two coils, at right angles to each other, and traversed by a current from a battery. The action of the coils upon the armature is similar to that of a galvanometer coil upon the needle, tending to cause its polar surface to move tangentially to the wires. A commutator upon the axis directs the current during the first quarter revolution into the first coil, during the second into the second coil, during the third into the first coil again, but in the reverse direction, and in the fourth quarter revolution into the second coil, also in the reverse direction. The specification contains a detailed description of the commutator, and also an account of two appliances

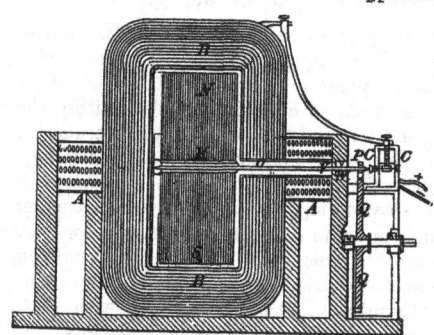

Fig. 2.

for connecting the two coils, immediately before the current is cut off from one of them, to lessen the spark of the extra current. The brushes "may be divided longitudinally from within a short distance of the end which touches the current-changer's cylinder, into several strips, so that each may be certain of being in contact with the cylinder by reason of its own individual elasticity."

9,022.—C. Wheatstone, London. **Producing, Regulating and Applying Electric Currents.** 2s. 2d. (30 figs.) July 7.

Magneto-Electric Generator.—Consists in the employment of several machines, such as that already described with reference to Specification No. 8,345 of 1840, with their armatures arranged on the same spindle. In the drawings a composite machine is shown with five sets of permanent magnets and five sets of rotating bobbins, so arranged that the current in any one coil commences before the currents in the other coils have ceased. All the positive and negative commutator brushes are respectively connected to the same leading wires, so that all the coils are coupled together in parallel circuit.

ELECTRO-MOTOR.—The field electro-magnets are placed in the circumference of one circle and the moving magnets in the circumference of another circle, which is not concentric with the first, and which does not revolve about any fixed point whatever situated in its own plane; in other words, the axes of the two circles never coincide, but are either parallel to each other or inclined at some angle. When the engine is in action the electro-magnets are excited successively under the control of a commutator, and, in consequence of the attraction, the circle which carries the armature rolls upon the other circle, and different part of the circumferences of the two circles are brought successively into near proximity. The armature may be a ring of soft iron, or a number of separate iron plates, or it may carry electro-magnets. Two forms of engine are illustrated; in the first a number of electro-magnets are placed radially around the inner circumference of a hollow cylinder, within which there rolls a circular armature carried on a crank pin at the end of the driving shaft. In the second the polar faces of a number of rectangularly-shaped electro-magnets form the frustum of a very flat cone, over which there rolls a disc; the axis of this disc describes by its motion a pointed cone on the same axial line as the flat fixed one. One end of the axis is carried in a ball and socket joint, and the other engages the crank pin of the engine.

Another improvement consists in a mode of producing rotatory motion by the mutual actions of permanent magnets, or of electro-magnets, upon electro-magnets or upon electric coils without magnetic cores. In each case the essential feature is that the magnetic or electro-magnetic bars are arranged as radii of a circle in proximity to electro-magnets or to electric coils, which never extend over the centre of motion of the magnetic bars. The specification shows a machine with three horseshoe magnets attached to a rotatory disc, and set with their branches projected radially. On a fixed disc, immediately behind the other, there are six straight bar electro-magnets arranged symmetrically and tangentially to the circle, and so coiled that, when connected in series, they present poles of like name to each other. On the passage of a current from a battery, the attraction or repulsion of the electro-magnets moves the permanent magnets around their common axis until each of

their poles arrives at the centre of a polar field of opposite name to itself; at that instant the current is momentarily stopped by the commutator, and then reversed to propel the movable magnets towards the next polar field and so on. Several such discs may be fixed parallel to each other on a common axis.

ELECTRO-MAGNETS.—Insulated copper wire or ribbon is coiled in concentric grooves in the face of a soft iron disc. The grooves communicate with each other by notches cut in the intervening rings.

9,053.—F. De Moleyns, Cheltenham. Production or Development of Electricity and its application to Illuminating and Motive Purposes. 7d. (2 figs.) August 21.

ELECTRIC LIGHT.—A glass globe, "similar to that in use for showing the electric light in vacuo," has two copper electrodes inserted into it from opposite sides, each ending in a coil of fine platinum wire, and one being further furnished with a piece of spongy platinum. Within the exhausted globe there is also a tube or phial, filled with pulverised charcoal, and ending in a nozzle like the tube of an hour-glass; from this nozzle the charcoal is continuously delivered in a stream through the two platinum coils. On the current being turned on, "one of the fine platinum coils becomes white hot, the charcoal which falls through and around both coils is ignited, and a light pure and intense is thus obtained, which is kept up by the continually changing points of contact of the coils with the charcoal powder."

ELECTRO-MOTOR.—An electro-magnetic engine is described, in which series of electro-magnets are arranged at regular distances in the rim and spokes of a wheel, parallel to its axis. These magnets project at either side of the wheel, which rotates between frames upon each of which are electro-magnets corresponding with the number in the wheel. The poles of the fixed magnets are opposed to the poles of the moving magnets, and, during the revolution of the wheel, attraction is periodically converted into repulsion by a change in the polarities of the fixed magnets, effected by a commutator worked by the wheel. This machine is not illustrated.

ELECTRO-MAGNETS.—A strip of sheet iron, with insulated wires upon it, is wound upon a cylindrical rod, so that upon its withdrawal the sheet iron and wires form a compact electro-magnet, each wire having a surface of iron on both sides. The central portion may be filled up with an iron rod.

9,204.—T. Wright and A. Bain, London. Application of Electric Currents to Railway Signalling, Telegraphs, &c. 2s. 5d. (14 figs.) December 21.

The inventors claim laying down conducting

c

wires in asphalte, pitch, or other cement in a trench.

9,431.—J. S. Woolrich, Birmingham. **Coating Metals.** 10d. (8 figs.) August 1.

MAGNETO-ELECTRIC GENERATOR. — Before the poles of a powerful permanent compound magnet there rotates an armature, consisting of two coils and ·iron cores united to form a horse-shoe, and mounted on a horizontal spindle. The currents generated in the coils are transferred by a commutator of the usual construction to two springs, and thence to the leading wires. The inventor claims "the method of coating with metal the surface of articles, formed of metal or metallic alloys, by means of magnetic apparatus in combination with metallic solutions."

1842.

9,465. — W. F. Cooke, London. **Electric Telegraphs.** 2s. 3d. (40 figs.) September 8.

CONDUCTORS.—The invention relates : (1.) To various methods of stretching, suspending and nsulating overhead wires. (2.) To apparatus for straining conductors. (3.) To insulators. (4.) To conductors, for use on railway trains, composed of two pairs of twisted copper wires, woven into a web or threadwork, of which the conductors appear to form the selvage ends.

1843.

9,745.—A. Bain, London. **Electric Timepieces and Telegraphs.** 3s. 5d. (35 figs.) May 27.

CONDUCTORS. — The wires are to be laid in asphalte and embedded in grooves in railway sleepers or in wood pavement.

9,786.—O. W. Barratt, Birmingham. **Gilding, Plating and Coating Metallic Surfaces.** 4d. June 15.

ELECTRIC GENERATOR.—Relates to obtaining electricity, continuously, from the magnet by an apparatus in which a number of magnets are fixed in wood at equal distances, in an upright position, with their poles connected by iron wire. Leading wires are attached to the north pole of first magnet, and south pole of the last magnet, to carry off the current which is stated to be produced without rotation or the expenditure of power.

9,982.—J. Schottlaender, London. **Electro-depositions of Metals on Felted and other Fabrics.** 7d. (6 figs.) December 8.

MAGNETO - ELECTRIC GENERATOR. — What the inventor calls a "magnetic battery" consists of two circular arrangements of horseshoe, or other magnets, placed one within the other. The magnets in the outer circle, which are apparently permanent magnets, have their legs parallel to each other, and to the central axis. Each branch is wound with a coil of insulated wire. The inner circle magnets are similarly arranged to the outer circle, but carry no bobbins. "When the poles of the inner circle of magnets stand opposite to the contrary poles of the other magnets no electricity passes, but, on moving the inner circle of magnets and bringing similar poles in opposition, an electric current passes through the wire coil. By communicating a rapid motion to the inner circle of magnets, a rapidly intermitting flow of electricity passes off from one wire to another, its intensity being proportional to the number and size of the magnets, the quantity of covered wire and the velocity" of revolution. The illustration in the specification does not show the connections or the method of receiving or gathering the currents.

1845.

10,548.—T. Wright, Thames Ditton, Surrey. **Electric Light.** 5d. (2 figs.) March 10.

ARC LAMP.—The light is produced between the edges or peripheries of carbon or plumbago discs, steadily rotated by clockwork, so as to continually present fresh surfaces in the path of the current. The illustration shows a frame of wood carrying five carbon discs in the same vertical plane. Each disc is held between metal plates. The axes of the first, third, and fifth discs are set in fixed bearings, while those of the second and fourth are in sliding blocks. On passing a current of electricity through the whole series of discs, while they are rotating, a brilliant light is produced at the edges of contigous discs. As soon as the points of contact are sufficiently heated, the discs B² and B⁴ are drawn away from B¹, B³, and B⁵ by screws C¹, C², "when a brilliant and permanent

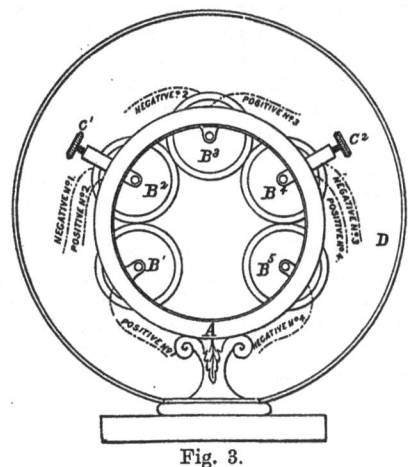

Fig. 3.

light will continue to be evolved at the several proximate portions of the discs, so long as the current passes and the discs continue to revolve." The light may be surrounded with a ground glass globe D, or by a suitable arrangement of reflectors.

10,655.—C. Wheatstone, London, and W. F. Cooke, Blackheath. Electric Telegraphs, &c. 3s. 2d. (37 figs.) May 6.

ELECTRO-MAGNETIC GENERATOR.—The twelfth part of the specification refers to Patent No. 8,345, of 1840, and provides for the substitution for the permanent magnet forming part of the magneto-electric machine there described, of an electro-magnet excited by a battery.—"Such battery being so used for keeping up voltaic magnetism in such machine, a far greater effect is produced than such battery could produce if used in electric telegraphs without the intervention of such machine."—The thirteenth part consists in enclosing conductors in lead tubes. The separate wires, when more than one is used, are wrapped with worsted thread and varished with shellac. They are made up into a bundle with white lead and starch, and enclosed within a tube of lead which may either be formed with sheet lead and soldered along the joint, or may be moulded around the bundle of wires by hydraulic pressure in the well-known manner of manufacturing lead tubes from semi-molten metal.

10,799.—W. Young and A. McNair, Paisley. Apparatus for Manufacturing Electric Conductors. 1s. 5d. (24 figs.) August 4.

CONDUCTORS.—The conductors are covered with thread to insulate them, and, after being coated with asphalte, pitch, wax, resin, or other suitable non-conductor capable of being liquified by heat,

are enclosed in tubes of lead, or other soft metal, to preserve their insulation and protect them from injury. The process and apparatus for applying the soft covering and the metal tube is very fully described and illustrated in the specification. In one arrangement the conductor, or conductors, are drawn through a vessel containing bituminous or resinous matters, and leave it by a tube ending in a nozzle which retains or strokes back all the superfluous dielectric. This tube, which is capable of endwise motion, passes through the core of the cylinder which contains the lead, or metal, to form the external coating, and abuts, or nearly so, against a die plate attached to the front of a hollow ram that is forced into the said cylinder by a hydraulic press. The hole in the die plate is of such a size as to allow a tubular stream of metal to flow between it and the nozzle of the movable pipe to enclose the conductor, which is drawn forward through the vessel, the movable pipe, the nozzle, the die plate and the hollow ram at the same speed as the lead pipe is forced from the cylinder. The metal is poured into the cylinder in a fluid state for mechanical reasons, but it is forced out by preference at a temperature varying from 250 to 400 deg. Fah., although it may be operated upon cold. In a modified machine the same or similar parts are somewhat differently arranged. It is proposed to use the lead covering as a return conductor.

10,919.—E. A. King, London. (*A communication.*) Electric Lights. 7d. (2 figs.) November 4.

INCANDESCENCE LAMPS.—In the first form of lamp described in this specification, the bridge

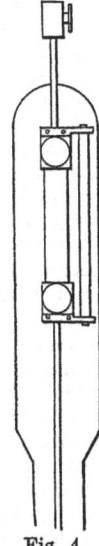

Fig. 4.

consists of a strip of platinum reduced by a gold beater, or rolled between plates of copper, as thin as possible. It is held between clamps and heated by the passage of the current. In another form (see illustration), the bridge is a piece of carbon in the shape of a small pencil or thin plate, and is situated in a globe blown at the upper end of a mercurial barometer tube. Two or more lights may be arranged in series by suitably increasing the number of armatures, if the generator be an electro-magnetic engine, or the number of cells if it be a battery. When an intermittent light for use in lighthouses, or for other purposes, is required, it may be obtained by breaking the circuit at intervals by clockwork.

10,838.—A. Bain, Edinburgh. Electric Clocks and Telegraphs, &c. 7s. (42 figs.) Sept. 25.

CONDUCTORS. — This specification describes a method of carrying conductors between asphalte insulators in a trough fixed upon posts, the whole forming a railing or fence. Overhead wires are placed on the metal covers that protect the insulators from wet.

10,939.—J. Brett, London. Machinery for Printing Communications made by Electric Telegraphs. 1s. 5½d. (32 figs.) Nov. 13.

CONDUCTORS.—Submarine cables are formed of two or three wires coated with copal or other varnish, bound with waxed or sere cloth and then plaited, in the manner of a riding whip, with waxed or tarred threads. Around this a plaited cable is formed saturated with bitumen. The manner of laying is described.

11,010.—J. Church, Colchester. Manufacture of Coke and Construction of Coke Ovens. 9d. (8 figs.) December 20.

CARBONS.—" When it is desired to obtain a coke more than usually free from sulphur and other metallic admixtures," the coal is subjected at the completion of the process of carbonisation to the action of a current of electricity. This purified coke is several times alluded to in later specifications as a material for the manufacture of carbon electrodes.

1846.

11,076.—W. Greener, Birmingham, and W. E. Staite, London. Electric Light. 4d. Feb. 7.

INCANDESCENCE LAMP.—Consists in effecting the illumination of streets, squares, and public places by means of solid prisms or cylinders of carbon enclosed in air-tight transparent vessels, and rendered luminous by currents of electricity. The carbon is previously freed from impurities and divided on the surface into numerous acute points. It may be replaced by platinum or other infusible metal, or by hollow cylinders of carbon, either plain or acuminated, partially inserted within and in perfect contact with hollow cones of platinum.

CARBONS.---To obtain pure carbon, the inventors take a quantity of charcoal or of lamp black, or of powdered coke purified by electricity (see No. 11,010), and digest the same in dilute nitro-muriatic or other acid. They then wash it and mould it under a screw press into prisms, cylinders, &c., with numerous acute points on the surface. By opposing such acuminated surfaces to one another, the current passes from one to the other by many points. The same purpose may be effected by means of a multitude of thin strips of charcoal, separated from each other by platinum foil and bound together by platinum wire. When a combination of hollow cylinders of carbon with hollow cones of platinum is employed, the leading wires are connected to the cones, which are placed with their bases facing each other to receive the ends of the hollow cylinder of carbon. Such a bridge can either be stationary or rotate on its axis.

11,188.—E. A. King, London. *(A communication from abroad.)* Magneto-Electric Machines. 6d. (7 figs.) April 30.

MAGNET COILS.—The improvements consist, firstly, in constructing the coils on the armatures of magneto-electric machines from a ribbon of rolled copper instead of copper wire. The strip is carried round a rectangular or triangular bar with its edge towards the bar. As it is too wide to be bent edgewise it is turned over on itself at each corner, somewhat in the manner that notes and letters are occasionally folded. This arrangement causes the corners of the strip to be double the thickness of the other portions, and therefore it is sometimes previously grooved at such places so as to form, when bent, a series of half lap joints. In the formation of such a coil on a cylindrical core the following plan is adopted. A circular plate of copper has a hole cut in its centre the size of the intended bar, and a broad flat ring is thus obtained. This ring is cut open on one side and one of the edges thus produced is slightly bent above the other. A second ring is prepared similar to the first, and its raised edge is soldered to the edge of the first ring, and in this way two convolutions of the coil are formed, which is completed by the addition of a sufficient number of rings.

MAGNETO-ELECTRIC GENERATOR.—All the currents produced in magneto-electric machines, having more than one armature, are collected by a combination of springs pressing on separate segments of the commutator. Each current may be used separately, or the whole may be collected into two currents. Fig. 5 shows a front view of such an arrangement. *d d d*

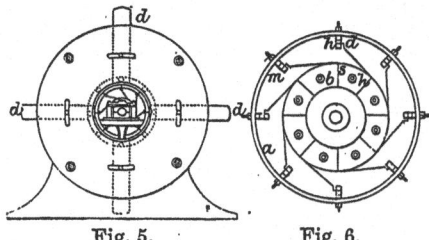

Fig. 5.   Fig. 6.

are magnets supported in a circular frame with their poles directed towards the centre. The planes of the magnets are parallel to the axis of the machine, upon which is fixed a wheel made of two parallel circular brass plates. The armatures are on the periphery of the wheel, and, when it is put in rotation, they are made to pass between the poles of the magnets. The commutator *a*, which is shown to a larger scale in Fig. 6, consists of a brass ring divided into as many insulated segments as there are armatures. Upon each segment there s made to press a spring *d*, connected to an insulated bolt *h*. The terminations of the armature coils are connected at one end to the armature segments, and at the other end to a brass connection ring fixed on a cylinder of wood at the opposite extremity of the axis. This ring is also provided with a collecting spring. Eight separate circuits may be led from the eight springs to the single spring at the other end, or they may be combined to form one, two, or four circuits.

INDUCTION CURRENTS.—To prevent the formation of induction currents in the wheel which carries the armatures, it is divided by a saw gate from the centre to the periphery.

POLE-PIECES.—It is stated that by fixing bent pole-pieces to each magnet a double magnetisation of the armature cores is effected. The connection in such cases requires to have double the number of segments described above.

11,428.—H. Mapple, Hendon. Transmitting Electricity, &c. 1s. 4d. (10 figs.) Oct. 27.

CONDUCTORS. — According to this invention, electric conductors are covered with cotton, and, after having been thickly coated with tar, pitch, or other insulating material, are threaded through a lead pipe or pipes. The pipes are protected by an external wrapping of coir rope or suitable fibrous material, and are drawn successively through a bath of hot pitch and a trough of sand, after which they are laid underground in cast-iron tubes. To thread the wires the pipes have longitudinal slits cut into them at intervals, and a long needle carrying a cord is passed into the end of the pipe and out at the first slit; it is then inserted again into the pipe and out at the second

slit, and so on, until it reaches the farther extremity. The wires are then drawn through and the slits closed and soldered. It is necessary to employ a pipe larger in internal diameter than the bundle of wires in order to admit of their easy introduction, and to avoid waste of material; the pipe containing the wires and the melted dielectric is passed through grooved rollers, or drawn through die plates, until it is sufficiently diminished in diameter to tightly embrace the wires.

ELECTRO-MAGNETS.—The end of a soft iron core stands within a thin hollow reel wound with insulated wire, and is drawn inwards on the passage of a current.

11,449.—W. E. Staite, Peckham, Surrey. Lighting by means of Electricity. 9d. (7 figs.) November 12.

ARC LAMPS.—Five different arrangements of lamps are described in this specification. In each case the lamp is enclosed within a glass globe with a small valve opening outwards by which means much of the oxygen becomes expelled by the heat, and the remainder converted more or less into carbonic acid. All openings necessary for the introduction of carbons, &c., are carefully covered to prevent the entrance or circulation of air. In the first form of lamp the upper carbon, which is relatively very thick and short, is carried by a curved copper arm, while the lower one is forced upwards through a tube by a piston which forms the termination of a rack. The feed is effected by a pinion and a train of clockwork set to advance the carbon at a definite speed equal to its rate of consumption. Two triangular prisms of platinum, each with its apex pointing inwards, are attached to the free ends of two vertical springs in such a way as to abut against and serve as guides to the lower electrode, which projects some little distance above the mouth of the tube. To start the lamp the lower carbon with its tube and train of wheels is raised by hand, against the action of a spring, until the carbons touch, and it is then allowed to drop a regulated distance to form the arc. A second lamp is shown in Fig. 8. The two carbons F F are forced against the cylinder P of compressed pipe-clay, phosphate of lime, or other refractory and non-conducting substance, by springs. The holder W¹ is moved to the left till the pencils touch, and is then brought back the length of the arc which plays between the two rods. In a third form of lamp this arrangement is modified and inverted. The two pencils are pressed upwards through two passages bored in a block of pipe-clay, and abut against a second block, through the centre of which a third pencil descends, resting on the lower block between the two lower pencils. The arc is formed by moving the lower block from side to side. In Fig. 7 the upper carbon F is pressed down on the lower one D by the spring above it. The exact use of the insulated platinum "guides" G G is not described, but they appear to be for the purpose of forming a double abutment through which the pencil will be fed as it wastes. The fifth lamp has for its lower electrode a copper cylinder enclosing a hollow cylinder of pipe-clay, the interior of

three, or more wires, and a corresponding number of electrodes to each battery. For signalling purposes the various electrodes may be covered with differently coloured glasses.

CARBONS. — Equal quantities of coal and "Church's patent coke" are mixed and reduced to powder. The result is heated and compressed

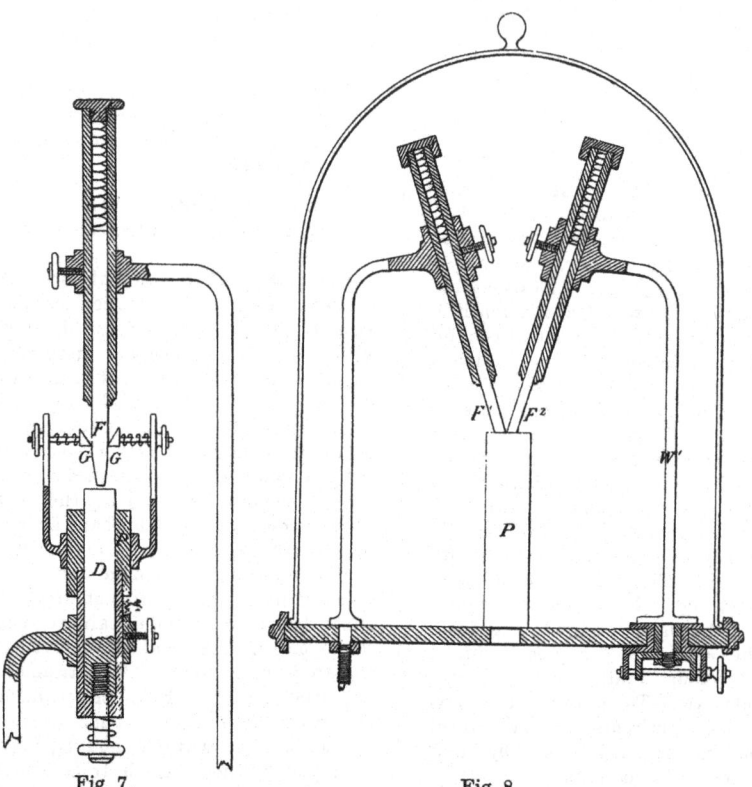

Fig. 7.

Fig. 8.

which is slightly coned in a downward direction. Both cylinders are surmounted by a thick ring of platinum with a countersunk hole in it. The upper electrode is a solid cylinder of pipe-clay surrounded by a cylinder of carbon, the whole being of such a diameter that when pressed downwards by a coiled spring it will just pass through the bevelled ring of the lower electrode without touching it, but be stopped immediately below by coming in contact with the interior of the clay cylinder until its diameter is lessened by the burning away of the carbon under the action of the current. "The contact between the electrodes is produced and broken off laterally by means of an arrangement similar to that employed" in the first lamp. The method of action of this lamp is not very evident.

FLASHING LIGHTS. — These are obtained by making and breaking the circuit by means ordinarily employed in telegraphs. The rapidity of the flashes may be increased by employing two,

in moulds, after which, it is soaked in a solution of sugar, and raised to a white heat in a furnace.

11,481.—M. Poole, London. (*A communication.*) **Constructing and Working Electric Telegraphs, &c.** 3s. 6d. (41 figs.) December 14.

MAGNETO-ELECTRIC GENERATOR.—In magneto-electric generators, for use with step by step telegraphs, the armature coils, instead of being rotated, receive a slight reciprocating motion to and from the permanent magnets, the cores coming into actual contact with the steel bars at one end of the stroke. The coils are made up of three concentric portions, the end of each portion being brought out to a terminal, so that a part or the whole of the wire may be included in the circuit by moving a finger into contact with one or other of the terminals. The successive portions of each coil are of different diameters, the finest being nearest the centre, so that as they are farther removed from the magnetic influence they offer less resistance to the passage of the current.

CONDUCTORS.—Relates to various methods of suspending and insulating overhead wires, and comprises (1) a straining apparatus; (2) a protective system to prevent injury from broken wires; (3) the employment of rolled instead of drawn wire, and also of lead and zinc wire enclosed in tubes.

INSULATORS.—The insulators are of earthenware, and are effective to prevent the entrance of rain blown laterally, as well as descending vertically.

ELECTRO-MAGNETS.—The excited metal may be nickel. Two forms of magnets are described in the specification, and also a keeper curved in a form analogous with the magnetic curves, and arranged to enter partly within the coils. Coils working within one another may replace magnets.

1847.

11,576. — **A. Brett** and **G. Little**, London. **Electric Telegraphs, Time Keepers, &c.** 7s. 9d. (104 figs.) February 11.

CONDUCTORS.—Relates to means of stretching, nsulating and supporting overhead wires.

1,634.—**W. H. Hatcher**, London. **Electric Telegraphs and Timekeepers.** 3s. 2d. (23 figs.) March 23.

MAGNETO-ELECTRIC GENERATORS.—The coils are moved towards and from the magnets by a hand lever, and are not rotated.

11,751.—**A. Symons**, London. **Preventing Accidents on Railways, &c.** 2s. 4d. (29 figs.) June 15.

CONDUCTORS.—Wires or rods are embedded in a groove in each carriage. The coupling is effected by elastic lengths and bracelet clasps.

11,783.—**W. E. Staite**, London. **Lighting by Electricity.** 2s. 9d. (10 figs.) July 3.

ARC LAMPS.—The invention is for improvements on that described in Specification No. 11,449, of 1846, and is for the purpose of imparting to one of the electrodes such motions that the light may be preserved from going out, be kept more uniform, and "be renewed by the action of the apparatus itself whenever it has been put out, providing always that the current has been maintained."

The upper electrode is carried in a stationary holder, the regulation of the arc resulting from the movement of the lower electrode. The illustration shows a vertical section through the lower part of the lamp. *d* is the lower carbon holder adapted to receive pencils of various diameters; it is mounted on a rod O, which is screwed for a part of its length, and works through a nut *e* attached to the frame D. At its lower end the rod O carries a cross-piece *s*, which fits across the tube P, taking into slots in each side of it, so that it can slide up and down in it. The tube P, which takes a bearing round the nut *e* at the top, and stands on a silver-plated footstep at the bottom, is caused

to revolve by a worm wheel S driven by a worm on a horizontal shaft. The end of this

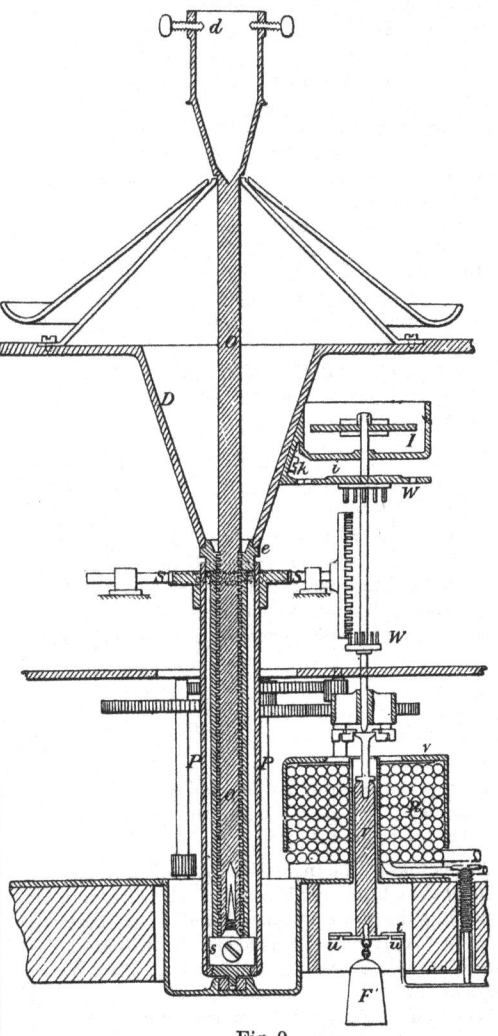

Fig. 9.

shaft carries a crown wheel so situated that it can

be driven by either of the unequal-sized pinions W W, or be out of gear with both of them, according to the endwise position of the spindle upon which they are fixed. The spindle is subject to a constant force, tending to rotate it in one direction, by clockwork actuated by a spring or weight, and is raised or lowered by the iron core *r* moving in the coil R. In its mid-position the shaft is prevented from rotating by the arm *i* coming in contact with the stop *k*, and when raised or lowered beyond this stop, its revolution is controlled by the centrifugal brake I. The solenoid R, which is provided with a soft iron cap *v*, forms part of the electrical circuit, and exercises varying upward attraction upon its core, dependent upon the resistance of the arc and the condition of the battery. The lamp is regulated by the employment of graduated weights, such as F¹, and, in order to prevent needless reversals of the apparatus, it is provided with a loose weight *t* which is dropped on to the adjustable shelf *u* when the core falls below its mid-position and is taken up when it rises above it. From the system of feeding adopted in this lamp it follows that the lower carbon is rotated at the same time as it is raised or lowered.

CARBON PENCILS.—Equal parts of coal and of "Church's patent coke," (see Specification No. 11,010), are reduced to powder and intimately mixed. The result is consolidated in moulds by heat and pressure, baked repeatedly and plunged in melted sugar. When cold the pencils are placed in a charcoal fire and raised to an intense white heat. Gas carbon may be treated in the same way.

11,894.—P. A. J. Dujardin, Lille, France. Electro-Magnetic Telegraphs. 1s. 2d. (12 figs.) October 7.

MAGNETO-ELECTRIC GENERATOR.—This invention comprises among other things, three forms of magneto-electric generators, in each of which both the permanent magnets and the coils are stationary. In the first generator a coiled bobbin

envelops each branch of a compound magnet, whose armature, a soft iron bar, receives a reciprocating motion towards and from the poles by a hand lever. In the second the coils are wound on soft iron pole-pieces in contact with and extending outwards in a line with the branches of the magnet. The third generator is like the first, but its armature is fixed on a spindle and rotates in front and just clear of the poles. Provision is made for combining several such generators to form a battery.

11,926.—G. Petrie, London. Electric Telegraphs. 4s. 10d. (84 figs.) October 26.

INSULATORS.—Overhead wires are suspended by being attached to metallic rings. These rings are insulated by being placed over alternate series of waterproof and insulating covers. An additional ring is in metallic contact with the earth, and each ring is provided with metallic points, which are capable of being conveniently adjusted so as to conduct atmospheric electricity from the wires to the earth.

11,974.—W. Reid, London. Protecting Telegraphic Wires and Communicating Intelligence. 1s. 7d. (26 figs.) November 23.

CONDUCTORS.—The wires are strained through insulators of stoneware or gutta-percha fixed in troughs or grooved sleepers of wood. The vacant space is filled with marine glue, gutta-percha, asphalte, Stockholm tar, or other resinous substance. A cover plate is added, a joint being made with sheet gutta-percha between it and the trough. The wires may also be insulated with cotton and laid in a trough or groove without the resinous material, or they may be separately encased in lead pipes. Troughs of stone, concrete or earthenware may also be used. Conductors on railway carriages are carried in tubes and sustained by disc insulators. They are coupled by short tubes and universal joints.

INSULATORS.—Several forms of insulators are described.

1848.

12,039.—H. Highton, Rugby, and E. Highton, London. Electric Telegraphs. 8s. 3d. (117 figs.) January 25.

CONDUCTORS.—The insulation of overhead wires is effected by threading beads upon the wires and imbedding the beads in holes bored in the posts or arms. A straight or curved slot is cut from the bore of the bead to the outside to facilitate its application.

INSULATORS.—Enamelled iron is used in place of porcelain. When a covering of cotton and varnish is applied to the wires, it is protected by an

external metal tube at the points liable to be damaged by friction.

12,054.—J. Brett, London. Electric Printing and other Telegraphs. 2s. 5½d. (63 figs.) February 8.

INSULATING CONDUCTORS.—The wires are coated with a varnish of caoutchouc and india-rubber dissolved in benzole. In some cases benzole, alliole or toluole are used "as solvent for and to combine." List of cotton, or linen cloth saturated with boiled oil or wax, asphalte, &c., may be employed as an outer covering.

MAGNETS.—Combining permanent magnets with electro-magnets for the purpose of increasing the action of the former.

12,079.—F. Wishaw, Hampstead, Middlesex. Manufacture of Pipes. 10d. (16 figs.) March 8.

INSULATING CONDUCTORS. — One object of the inventor is to provide straight or curved pipes, channels or ducts, suitable for the passage of electric conductors in a state of insulation. (1) These ducts are interstices surrounded by solid earthenware or pottery, and constitute a cluster of channels within one external pipe. They are made, when straight, in short lengths, in a suitable machine consisting of a cylinder with a piston and a screw, whereby the clay is forced through a die, and are connected by socket joints united by cement. Curved pipes are made in moulds. A register mark on each length forms a guide for laying them in line. (2) The die or "dod" is made longer than those previously in use to ensure the clay being pressed firmly together. (3) Clusters of separate glass or earthenware pipes are combined together by cementing the separate pipes into clusters of collars. Pipes of ordinary construction are employed in combination with internal plugs or diaphragms consisting of a cluster of pipes of short length, such pipes and plugs being held together by external cemented collars. The plugs, which may be of earthenware, glass, gutta-percha, &c., act as insulators to the wires strained through them. (4) A dove-tailed shaped groove is formed in each spigot and socket, and a mixture of asphalte and gutta-percha, or other suitable cement, is poured in to make an air-tight joint. The illustrations show the moulding machine and dies, various sections of pipes and different arrangements of joints.

12,136.—W. H. Barlow, Derby, and T. Forster, Streatham, Surrey. Electric Telegraphs. 2s. 2d. (11 figs.) April 27.

CONDUCTORS. — The conductors are enclosed between strips or fillets of gutta-percha, or gutta-percha mixed with other substances, particularly cowrie, or New Zealand gum, and flowers of sulphur, by aid of a machine which forms the subject of the invention. The sheets of gutta-percha compound are passed between steam-heated rollers to soften them, and are pressed on the wire. When more than one wire is included between the sheets, the rollers have cutting rings and grooves to divide the fillet into separate strips.

12,212—W. E. Staite, London. (*Partly a communication from abroad.*) Galvanic Batteries, Magnets, Application of Electricity to Lighting and Signalising, &c. 4s. 4d. (30 figs.) July 12.

ARC LAMPS.—The tenth part of this invention relates to an arc lamp, somewhat similar to that described in Specification No. 11,783, and consists principally in an improved mode of giving motion to the lower electrode. The upper carbon is fixed in place by three keys or wedges. The lower one is carried in a socket terminating in a rack in gear with a pinion. Around the pinion shaft is a cord connected to a weight that counterbalances the electrode and its holder. Upon the same shaft is a spur wheel, acting as a ratchet wheel, and designed to drive the pinion shaft and consequently the carbon holder, in one direction or the other, accordingly as the forward or backward end of a double-ended pawl is in gear with it. This pawl is carried by an arm to which a tendency to constant reciprocation is given by a train of clockwork. A counterbalanced lever is connected to the core of a solenoid, weighted as shown in the illustration to the previous patent, and by its movement it controls the position of the double-ended pawl. In the mid-position, that is when the lamp is burning steadily, the counterbalanced lever arrests the motion of the reciprocating arm. In action the clockwork, acting through the reciprocating arm, ratchet wheel, pinion and rack raises the lower electrode into contact with the upper one; as soon as the circuit is established, the solenoid draws its core upwards and reverses the pawl, whereupon the pinion is rotated in the opposite direction until the arc is drawn to the right length, at which point, if the parts are properly adjusted, the core will fall and stop the clockwork.

DISC ELECTRODE.—In place of the upper vertical pencil, a circular electrode may be employed, fixed on an axis, to which a slow rotary motion is imparted by the moving power of the lamp. A steel scraper impinges on the periphery of this disc to keep the edge free from lumps. The cross section of the disc resembles two cones placed base to base.

INCANDESCENCE LAMPS.—Iridium is the material employed as the conductor in the bridges of these lamps. To prepare it, the oxide of the metal is fused in a cupel of bone ash under the voltaic arc. The ingot obtained is annealed in an oxyhydrogen jet, and at a white heat is worked under hammers and between rolls. Platinum and copper are sometimes mixed with iridium to facilitate the annealing. In certain cases thin strips of metal are cut from the ingot "in the same way as precious stones are cut by the lapidary." To prevent loss of heat and light by radiation, the metallic holders are passed through a block of glass or other non-conducting material. In the illustration, A is a bridge of iridium, fixed to two pieces of platinum B B, supported on two copper holders, C C. The circuit can be broken and the lamp extinguished by unscrewing the rod Z. The specification also contains a drawing showing three such bridges enclosed in one globe.

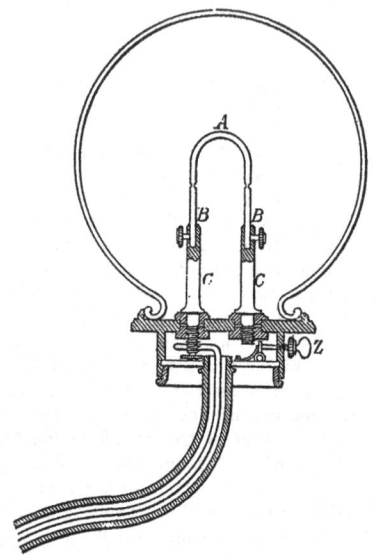

Fig. 10.

INTERMITTENT LAMP.—The invention further consists of certain arrangements for producing a regularly intermittent light, "especially suitable for lighthouses." A lamp for this purpose has no clockwork gear, the moving power being derived from a solenoid traversed by the current, and wound with a greater number of convolutions at the ends than in the middle. The stem of the lower carbon holder descends through the axis of the solenoid; it is rather more than counterbalanced by a weight attached to a string passing over a pulley, and is also connected by a train of wheels to a fly to regulate the speed of its movements When the circuit is completed the coil draws the carbons apart, and the light is evolved until the electrodes become too far separated, when the current suddenly ceases and the light is extinguished; the weight then draws the lower carbon up until it again touches the upper one, when the same series of operations is repeated. Various methods are described of varying the lengths of the flashes and intervals by different arrangements of gearing and flies.

ARC LAMP.—The apparatus above described, may be modified to produce a steady light by the size of the solenoid being increased, and the weight being adjusted to exactly balance its action when the normal current is passing. To compensate for the decreasing weight of the carbon, as it consumes, the counterweight in rising picks up a chain, equal in weight, per unit of length, to the carbon rod employed. By another modification the core is made to shunt a portion of the current from

the coil when it moves an excessive distance in one direction, and to take up additional weights as it rises.

Fig. 11 shows a part of a lamp so arranged that the distance through which the carbon holder can be moved, and consequently the length of the carbon pencil, is independent of the length of the solenoid. B^{11} is a rod of cruciform section, forming a continuation of the carbon holder, and furnished with ratchet teeth on one side. B^1 is a hollow soft iron cylinder, acting as the core of the solenoid and connected to B^{11} by the spring catch g. When the current diminishes the counterweight (not shown) raises the rod B^{11} and core B^1 until the wedge piece H forces the catch g out of the ratchet teeth, when the rod rises independently. As soon as the arc is reduced the increased current causes the

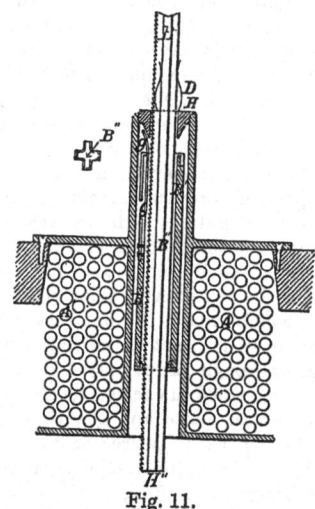

Fig. 11.

core to descend and the catch to re-engage with the ratchet teeth. This occurs periodically so long as the lamp is burning. The coil may have its power augmented by exterior casings of soft iron of various forms. It may also be replaced by an electro-magnet with obliquely formed poles, in which case the core must be suitably modified to act as the armature.

REFLECTORS.—Portions of the reflectors are placed inside the glass shade and their continuation is effected by a separate piece outside. Their form is described by the revolution of a hyperbola, about its focal co-ordinate, which is coincident with the axis of the electrodes.

INTENSITY COIL.—It is stated that the intensity of a current may be increased by causing it to traverse a coil of insulated wire wound in an iron case.

CARBON HOLDERS.—When the electrodes are required to be longer than is convenient for their strength they are fed through tubes, and in such case they may be formed in several portions slightly connected at their ends. The upper part of each tube is furnished with a cap provided with a number of spring fingers which embrace the pencil and conduct the current to it. As it would be difficult to draw built-up carbon rods backwards through the springs, the whole tube, which is nearly balanced, is made capable of endwise motion for a distance rather longer than the span of the arc.

CARBON PENCILS.—Purified plumbago powder, lamp-black, charcoal powder " of sundry kinds of wood," graphite in powder and also in grains are mixed with brown sugar, and the compound is melted with heat. The product is passed into perforated moulds lined with plaster-of-paris or paper, and slowly dried and baked. After removal from the moulds the rods are packed in sand in a crucible, and exposed to a white heat. External wrappings of metal foil, such as tin, may be employed to increase the strength and conducting power of the pencils.

12,219.—A. E. Le Molt, London. Constructing Electric or Galvanic Piles for obtaining Electric Lights. 6d. (5 figs.) July 20.

ARC LAMP.—The electrodes are discs of carbon so arranged that by their revolutions near each other in the same plane, or in planes at an angle to each other, they constantly present fresh surfaces, and when they have completed a revolution, they approach each other to maintain a constant length of arc. The discs are rotated at a uniform speed by clockwork acting through chains and chain wheels. Each disc is carried by an arm centred on the chain wheel shaft, and the two are impelled towards each other by a spring, the distance between them being regulated by a spiral cam.

CARBONS.—Gas retort carbon is used purified by immersion, first in nitro-muriatic acid, and afterwards in fluoric acid. Carbon battery plates have their ends electrotyped with copper or other metal to enable them to be soldered to the strip connecting them to the next cell.

12,236.—W. T. Henley and D. G. Forster, Clerkenwell. Telegraphs and Timekeepers. 1s. 3d. (18 figs.) August 10.

MAGNETO-ELECTRIC GENERATORS.—According to this invention, " two distinct currents may be derived from the same magnet, and the reversed current can be made of equal intensity with the primary induced current." In one arrangement two coils of wire, wrapped on soft iron pieces, are placed between the poles of a permanent magnet and a soft iron armature. Each coil is separately centred on a spring lever that can be operated by hand to move the coil away from the pole of the magnet. In another arrangement the coil is mounted on an arm on a rock shaft, and lies

between the two poles, the direction of the current depending on the direction of motion. A third generator has its coils mounted on an axis between the poles of the magnet; each pole has two branches, and the coils " are so placed that when in motion one pole of the magnet is not released from its opposition to the armature until the other just touches it. By this means the two opposite currents have an equality of power."

INSULATOR.—Gutta-percha mixed with highly pulverised glass or sand is used as an insulator.

12,262.—J. L. Ricardo, London. Electric Telegraphs, &c. 5d. (5 figs.) September 4.

CONDUCTORS.—The specification drawings show six conductors laid parallel to each other in the same plane, and embedded in gutta-percha, or in a mixture of gutta-percha, gum cowrie and milk of sulphur. The wires are fed between two fillets of gutta-percha that have been drawn over steam-heated surfaces, and the whole is passed between a pair of heated grooved rollers. The rings between the grooves do not cut the fillets, but merely press them more closely together between the wires. The edges are closed by a solid bead being formed upon them not containing any wire.

INSULATOR.—An insulator for overhead wires is also described.

12,276—F. Allman, London. Producing Light by Electricity. 2s. 1d. (16 figs.) September 28.

ARC LAMPS.—The striking of the arc and its subsequent regulation is effected according to this invention by power derived from the current alone. The means employed are (1) coils and permanent magnets; (2) electro and permanent magnets; (3) electro-magnets or solenoids; (4) metallic bars expanding and contracting in accordance with their temperature; (5) gasometers raised by the pressure of gas resulting from the electrolytic decomposition of water. In the first example shown in the specification, the upper carbon holder is suspended from the end of beam turning upon gudgeons. This beam, which is a compound magnet, is so hung as to be partly within a long flat coil traversed by the current, under the action of which it is deflected from its horizontal position, and caused to assume an inclined attitude; by these means the carbons are drawn apart and the arc is regulated. The adjustment of the lamp is effected by means of balance weights.

In a second example the coil is dispensed with, and in place of it two electro-magnets are employed, fixed respectively under and over the two ends of the permanent magnet and so wrapped as to offer opposing poles to it. A third arrangement shows a beam of non-magnetic metal carrying an electro-magnet at each end opposed to a fixed electro-magnet. Solenoids may be substituted for the electro-magnets. In a fourth apparatus, the upper electrode is hung from the end of a relatively long arm that is constantly impelled upwards by a spring, and is held down by a tie-rod which also acts as a conductor for the current. As the tie-rod elongates under the heat produced by the passage of the current, the arm rises and separates the carbons. The electrical continuity of the tie-rod is broken by the insertion of a "cock," consisting apparently of a taper plug, working in a socket, and having a flat filed on one side. It is stated that by turning the cock so as to bring a greater or less portion of the plug to bear on an insulated conductor inserted in the socket the quantity of the current can be regulated. Instead of the tie-rod above mentioned, a compound bar of steel and brass may be used to raise the arm. When the electrolytic action of the current is made use of as the means of regulation, a bell containing two platinum plates is suspended in a vessel of acidulated water, and is connected by a lever to the upper carbon holder. As the gas forms, the bell rises, separating the carbons; the gas gradually escapes through an adjustable orifice.

12,295.—S. Hjorth, London. Applications of Electro-Magnetism as Motive Power. 1s. 2d. (20 figs.) October 26.

MAGNETO-ELECTRIC GENERATOR.—To generate electricity for the purpose of increasing the adhesion of locomotives to the rails, a number of electro-magnetic coils are placed between the arms of the wheels within the rim. Attached to the framing there are arranged several compound permanent magnets set astride of the rim, so that the coils pass between their poles when the wheel is rotating. The electricity produced may be made available for, among other things, "moving the engine itself by means of a number of electro-magnets fixed to the wheels of the engine, and so revolving by the motion of the train and brought within the action of fixed permanent magnets, the electricity produced being conducted to the electro-magnetic engine as an auxiliary power."

ELECTRO-MAGNETIC MOTOR.—A shaft carries a number of arms, each of which is a magnet or an electro-magnet. As the shaft revolves these arms pass through U-shaped cages of magnetic bars. The bars are of various lengths, so that as the arm is emerging from the cage it passes through a long polar field. The magnetic bars are so placed that "one set of magnets commences to exercise its power before the current passing round through the other is broken, in order that a constant flow of electricity may be maintained." The direction of the current is controlled by a commutator.

1849.

12,482.—C. T. Pearce, London. Obtaining Light by Electric Agency. 1s. 10d. (9 figs.) February 16.

ARC LAMP.—The lower carbon is a pencil of cylindrical, triangular or other section; when other than cylindrical, one of its projecting corners is presented to an inclined circular electrode which is steadily rotated by clockwork. The pencil is held and guided in a split sheath, up which it is constantly impelled. The sheath ends in a ring, in which are inserted three platinum wires, tipped with iridium, which bear on the pencil near its extremity, and conduct the current to it. The circular electrode revolves in contact with two scrapers, and has a slide and a screw for the manual adjustment of the arc. Two or three circular electrodes may be similarly arranged around one triangular pencil.

REFLECTORS.—Made by electro-precipitation on a mould or matrix.

GLOBE.—The lamp is covered with a globe furnished with a syphon tube opening into a vessel of liquid.

ARC LAMP.—Four carbons, forming a cross, are impelled towards a common centre (the arc) by weights or springs. The two horizontal carbons are "non-conducting," and two vertical carbons "conducting."

ABUTMENT POLE.—The upper carbon holder is a tube of metal and pipe-clay, and terminates in a conical nozzle, which may be of pipe-clay, phosphate of lime, or other infusible substance. This cone prevents the rod of carbon from being urged forward too rapidly. Two slots admit conducting wires as in the previous lamp. "As the lower carbon becomes wasted by the action of the current, weights urge it upwards, until checked by the lateral bars of carbon which press on it near its upper end." The specification does not make the action of this part of the arrangement very clear.

CARBON PENCILS. — Pieces of carbon can be "grafted" together by a dowel inserted into an axial hole drilled in each, or into a groove cut in the side. Two or more pieces of conducting and non-conducting carbon may be combined by grooving or "union of surfaces" to form an electrode in which the surface presented to the action of the current shall be of conducting matter, while the rest is of non-conducting matter.

ARC.—An arm, impelled in contrary directions by a spring and an electro-magnet, lays a piece of carbon across the interval, and momentarily bridges the arc, when the light is extinguished.

FLASHING LIGHTS.—A rotating cylinder with alternately conducting and non-conducting segments breaks and makes the circuit at determined intervals.

SWITCH.—A similar arrangement may be used to throw two or more batteries successively in and out of circuit.

REGULATOR.—The regulation of an arc lamp may be effected by the repulsion between an electro-magnet and a bar-magnet, electro-magnet or coil. The electro-magnet is fixed and the opposing magnet is suspended on a spindle over it in the manner of a compass needle. When the current passes the suspended magnet is deflected, and by means of a toothed wheel moves one of the electrodes to strike the arc.

12,772.—W. E. Staite and W. Petrie, London. Electric and Galvanic Instruments and Apparatus, and their Application to Lighting and Motive Purposes. 1s. 10d. (15 figs.) September 20.

IRIDIUM.—Articles are manufactured from this metal by fusing grains of it together by the disruptive discharge from a galvanic battery. Iridium powder is laid on a copper bed and an iridium electrode, attached to a battery of 70 elements, is moved over it in lines, in such a way as to form an electric arc that fuses the surface together. The partially formed ingot is then turned over, and more grains are added before the under side is treated. The solid ingot is afterwards held to be finished in a hand vice connected to one pole of the battery, and is then further worked with the iridium electrode.

RHEOSTAT.—The current from a battery is maintained constant by being made to traverse more or less of a resistance coil. The controlling agent is a solenoid, which raises or lowers a weighted core according to the strength of the current. The resistance coil may be varied in several ways —it may be partly immersed in mercury, or may be moved with relation to fine contact pins.

ARC LAMP.—A chain and barrel are used " in place of a rack and pinion for supporting and imparting motion to the electrode shaft. The shaft or holder is a hollow slotted cylinder ; the chain is fixed to the bottom, passes up the middle and out through the slot to the barrel." On another part of the said barrel is hung a counterpoise weight and "compensation chain." The end of the latter always rests on the bottom of the lamp case, and its weight is, inch for inch, equal to that of the electrode, the connecting chain, and the chain which suspends the balance weight all taken together.

In an electric lamp movement in which the electrodes " cannot be damaged by the surplus clockwork power," the carbon holder is a screw longitudinally key grooved. It is raised or lowered by the rotation of a crown wheel driven either way by a two ended click controlled by the soft iron core of a solenoid. The wheel has a spring support, which gives way and allows it to descend

and encounter a stop as soon as the carbons come into light contact. The solenoid core has also a spring tip to prevent the click being jammed hard into the teeth, and an arm to stop the clockwork when it is in mid-position. A weighted lever serves to adjust the regulation of the lamp.

12,899.—I. L. Pulvermacher, London. Galvanic Batteries, Electric Telegraphs, &c. 4s. 3d. (113 figs.) December 15.

CARBON PLATES AND VESSELS. —" Plastic graphite" for vessels, diaphragms or battery plates is prepared from carbonaceous graphite by washing, heating and pulverising it, and afterwards mixing it with a bituminous substance and heating the whole to the temperature of boiling oil. The articles, when moulded, are placed in a tight case and subjected to the action of high pressure super-heated steam. The addition of sal-ammoniac renders the product more porous, and dippings in coal tar make it a better conductor.

ELECTRO-MAGNETIC ENGINES.—The illustrations, Figs. 12 and 13, show respectively an elevation

Fig. 12.

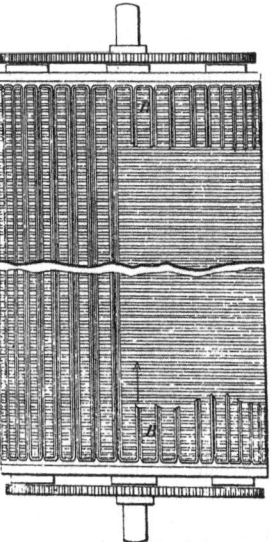

Fig. 13.

and a plan of one form of the engines. The general arrangement resembles a long pinion, rotating within a hollow internally toothed cylinder, all the teeth being wound lengthwise with insulated wire. The diameters of the two parts are such that the points of the pinion teeth just clear the points of the cylinder teeth. The moving part or armature B B, and the fixed part or field magnets A A, are each formed of thin plates of soft iron, each plate separated from the adjoining one by a layer of non-magnetic material. One method of making the armature and magnets is to cut the rings out of tin plate and to unite them into a solid mass by the action of heat. The insulated wire is wound round the teeth in such a way that all the currents in a given groove flow in the same direction, and that adjoining teeth have alternate polarities. The polarity of the teeth of the armature remains constant, while that of the field magnets is periodically reversed by a commutator. In another arrangement, Fig. 14, there

raised and lowered or rotated in the liquids another consists of a pair of lunette-shaped plates of platinated silver or graphite, over which rollers, connected to the coils of the magnets, revolve. As the width of the lunette-shaped plates gradually diminishes to a point, it is sought by their use to steadily decrease the current from its maximum, until, at the moment of reversal, it has almost fallen to zero ; by this device the maximum effect of the current is produced on the electro-magnets at the proper time. Other commutators are also described acting on the same principle ; they are of various forms, such as a gradually deepening groove filled with spongy platinum, or a divided ring in connection with the battery poles, and several continuous circular rings in connection with the electro-magnets ; or mercury and metal springs instead of spongy platinum and rollers, or rings coiled with German-silver wire and traversed by springs ; vertical mercury tubes whose openings are in connection

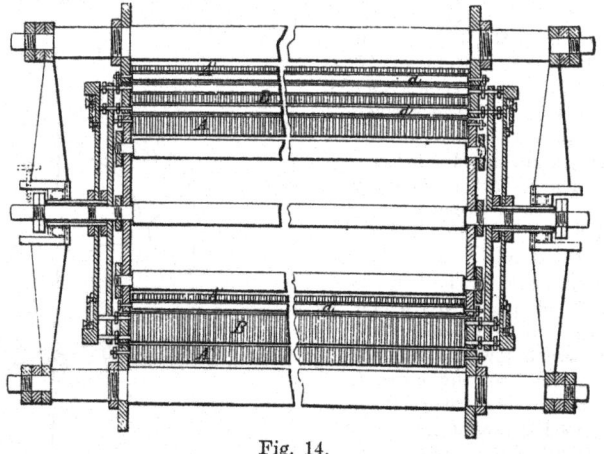

Fig. 14.

are two sets of field magnets A A[1] placed concentrically with their teeth opposed, leaving an annular space between them in which the armature B rotates. This armature is a hollow cylinder with teeth both on its external and internal surfaces. The conductors are insulated rods *d d*, strained by nuts between two end discs. Both machines have elaborate appliances for setting the armature concentric with the field magnets.

COMMUTATORS.—"In order that the changes in the poles of the electro-magnets may be gradual and be thus effected without sparks, and that the secondary currents accompanying such changes may subserve, instead of counteracting, the production of power by the machine, and further in order to the avoidance altogether of secondary currents from the movements of the electro-magnets," various forms of commutators are described. In one arrangement the batteries are

with pins and wires over which rollers traverse, the tubes being in two series, one half circle of which is in connection with one battery pole, the other with the remaining battery pole, and opposing resistance to the current in proportion to the amount of mercury included in the circuit.

CURRENT GOVERNOR. — A pair of centrifugal governors regulate the amount of surface of the battery plates exposed to the exciting fluid, and also control a mercurial resistance.

A MAGNETO-ELECTRIC ENGINE. — This engine comprises a revolving battery, a regulator, a system of electro-magnets, a commutator and a governor. The interposition of the commutator between the electro-magnets and the battery, changes the poles of the magnets, and "also serves to render the secondary electric currents available in communicating motion or power to the machine, but to effect this latter object the commutator must have some lead in advance of the magnets,

The effect produced by the lead is that the secondary current, which is produced by the primary current following, augments the power of the preceding current, and acts in unison with, instead of adversely, to it, which it would certainly do if there were no lead given to the pole changer."

MAGNETO-ELECTRIC GENERATOR.—This is a modification of the machine shown in the illustrations, the electro-magnets of the inner cylinder B B being replaced by permanent magnets made of steel rings. The currents developed from the different rows of electro-magnets are in opposite directions, and therefore they require to be reduced to a continuous current by means of a commutator.

COMMUTATOR.—C^1 C^2, Fig. 15, are two indented rings fixed to the end of the electro-magnet cylinder. The conducting wires of one polarity are connected to the ring C^1 and those of the other polarity to C^2. To each side of the

Fig. 15.

machine there is affixed a collecting roller; these are so disposed, that when one roller is on C^1 the other is on C^2.

FLAT MAGNETIC MACHINE.—This is constructed on the same principle as the cylindrical machine, and is suitable for therapeutic purposes.

1850.

12,959.—E. Highton, London. Electric Telegraphs. 3s. 6d. (67 figs.) February 7.

CONDUCTORS.—Relates, among other things, to (1) enclosing insulated conductors in masonry or cement instead of metallic tubes; (2) to encasing conductors in lead or other pliable metallic tubes, by compressing the tube on to the wires by rollers, or by doubling a fillet of sheet metal around them; (3) to enamelling gutta-percha coverings by solvents and friction; (4) to straining and suspending overhead wires; (5) to constructing telegraph posts.

12,991.—W. Brown, Airdrie, Lanark, and W. Williams, St. Dennis, Cornwall. Electric and Magnetic Apparatus for Communicating Intelligence. 2s. 10d. (40 figs.) March 7.

ELECTRO-MOTOR.—A tubular or hollow cylindrical magnet, with its poles running longitudinally on either side, is fixed on an axis between two horseshoe electro-magnets, whose poles are each provided with a continuation having the form of a section of a cylinder. These sections are connected by brass strips and together form a complete cylinder, within which the tubular magnet rotates. The manufacture of the tubular magnet is described and the process whereby it may be caused to have two, four, or other number of poles parallel to its axis—other forms of magnets may be employed, and for the electro-magnets there may be substituted coils wound with their convolutions parallel to the axis of the tubular magnet, and with their axes crossing in the common centre. A commutator intended to adapt the machine for use as a telegraph indicator is described.

ELECTRIC GENERATOR.—Currents "are obtained by induction from electro-magnets formed by a coil of insulated wire upon iron or nickel charged with voltaic electricity."

BRAIDING MACHINE. — The invention further includes a braiding machine for covering gutta-percha coated wire, which is afterwards sheathed with coir and pitch.

COUPLING CONDUCTORS.—A screw connection is described.

13,062. — E. W. Siemens, Berlin. Electric Telegraphs. 3s. 6d. (37 figs.) April 27.

CONDUCTORS. The tenth and following portions of this specification relate to laying conductors underground by an instrument similar to a mole plough, and also to depositing them under water. Insulating compounds of gutta-percha and sulphur, mechanism for treating them, and a machine for applying them to wire, are also described, as well as a method of protecting wires by sheet lead wound helically around them. A process of testing the insulation of the conductors and a device to prevent them from becoming charged, in the manner of a Leyden vial, are also included in the invention.

ELECTRO-MAGNETS.—The cores are iron tubes split longitudinally and provided with large polar masses.

13,128.—A. V. Newton, London. (*A communication.*) Production of Gases for Lighting, Heating, and Motive-Power Purposes. 2s. 3d. (13 figs.) June 12.

MAGNETO-ELECTRIC GENERATOR.—To increase the decomposing power of the electric current, the wire of the helices, which are of the usual form, is replaced by copper tubing filled with water or other "electrical absorbent"; the cores also may consist of tubing filled with water. The generator in other respects does not present any features of novelty.

GOVERNOR.—The current is regulated by an electro-magnet, by which, when it becomes too "intense," the circuit is broken and connected to earth.

13,302.—E. C. Shepard, London. (*F. Nollet, Brussels.*) Electro-Magnetic Apparatus for obtaining Motive-Power, Light, and Heat. 2s. 1d. (12 figs.) October 24.

MAGNETO-ELECTRIC GENERATOR.—The armature comprises eight helices arranged in two groups of four each, symmetrically situated around a common horizontal axis. The field magnets are four horizontal compound permanent magnets, disposed in two pairs with opposing poles. The general arrangement is such that four times in each revolution of the armature, which rotates between the poles of the magnets, the outer end of each of the helices is opposite a pole of one of the field magnets. Two ivory discs with inlaid copper segments enable the currents to be commuted into óne direction. By these means two pairs of " primitive currents" constant in one direction are obtained; they "are caused to communicate and pass into a system of spirals," or are otherwise treated or utilised.

SYSTEM OF SPIRALS.—The spiral is composed of a hollow central cylindrical axis of magnetised metal, terminating in two discs of dry wood. Two series of insulated wires, exactly the same length as the distance between the discs, are arranged on a horizontal plane like the warp of a web. The wires thus arranged are wound on the central axis and parallel to it, so as to form a cylinder. Several spirals are connected together in such a way that all the spirals of odd numbers are in communication and are traversed by the primitive current, and all the spirals of even numbers are individually traversed by a current of induction. The specification gives an illustration of the arrangement, but even with its aid the construction and method of action are not made clear.

OSCILLATING GENERATOR. — The eight helices mentioned above are fixed at the end of a "pendulum," having a vertical oscillatory movement between the poles of horizontal magnets.

ROLLING GENERATORS.—The helices are mounted on tramway carriages and are set with their axes horizontally and at right angles to the track. A to and fro motion between fixed magnets is imparted to the carriages by cams or eccentrics. The currents pass through the wheels and rails, the latter being made to act as a commutator by being laid in short insulated lengths, the alternate rails on either side being connected together by diagonal conductors.

INDUCTION COIL.—If two insulated helices be mounted end for end upon one hollow axis, and "the primitive current be made to pass into the upper helix, a current of contrary direction is produced by induction in the lower helix, while at the same time the soft iron centre is magnetised,

which in its turn reacts on the lower helix in a manner corresponding with the action of the upper helix. There is thus obtained in the lower helix a very powerful current by double induction, and the primitive current, far from being weakened, is increased by the reaction of the electromagnet, so that it can be made to pass into a numerous series of double helices before closing the circuit of the primitive current."

INCANDESCENCE LAMP.—A glass globe, which may be ground or otherwise obscured, has two insulated wires passed through the stopper. One of these wires is soldered to a copper tube which holds a cylinder of charcoal pushed down from above by a small cylinder of lead. The other polar wire comes out under the end of this tube, and terminates in a little cone of prepared charcoal or coke, which, when the current is established, becomes incandescent, the small portions that are consumed being continually replaced by a portion of the cylinder which the lead weight forces on to the top of the cone. When constructed, the globe is exhausted and sealed.

COVERING CONDUCTORS AND MAKING HELICES.—The invention includes a machine for these purposes.

13,336.—E. Clark and H. Mapple, London. Electric Telegraphs, &c. 11d. (8 figs.) November 12.

INSULATORS.—Relates in part to applying metal hoods or shields to earthenware insulators to prevent the deposition of dew.

13,363.—G. Shepherd and C. Button, London. Electric Telegraphs. 2s. (30 figs.) November 23.

CONDUCTORS.—The coated conductor is wrapped in fibrous material impregnated with one of several mixtures, of which the formulæ are given in the specification. These consist mainly of gutta-percha, tar, pitch. creasote, red and white lead, and boiled oil. The cable, after being sheathed, is further protected by being fixed within angular spaces formed by the links of chains; or the line may be entwined or woven through the links of a chain, or attached thereto by means of extra links or double links; or it may be attached to or intertwined with a wire rope. The shore end of the cable is encased in iron pipes or tubes fixed on the foreshore. Also tubes connected together by bell and socket joints can be used in place of the chains mentioned above to enclose the cable.

13,427.—G. E. Dering, Lockleys, Herts. Electric Telegraphs. 10d. (13 figs.) December 27.

INSULATORS.—According to the tenth head of this specification, line wire insulators are made with an inverted second bell of insulating material within the first.

1851.

3,497.—F. M. A. Dumont, Paris. **Electric Telegraphs. 2s. 3d. (5 figs.)** February 7.

CONDUCTORS.—Submarine conductors are suspended by rods from floating buoys placed at suitable intervals.

13,536.—W. Millward, Birmingham. **Electro-Magnetic and Magneto-Electric Apparatus. 6d. (5 figs.)** February 28.

MAGNETO-ELECTRIC GENERATOR.—A brass wheel, rotating in a horizontal plane, carries an even number of electro-magnetic coils, which pass between the poles of half their number of permanent compound magnets, arranged radially and similarly in vertical planes around the wheel. A commutator acting on "the whole of the magnets at the same instant" directs the currents in the same direction. The generator is intended specially to produce currents for the excitation of a large electro-magnet used for magnetising bars of steel. "The power of such an electro-magnet depends upon the strength of the permanent magnets used in the machine; * * * it will be therefore evident that by having two sets of the permanent magnets and charging them in such a machine that their supporting power may be increased by continual passes or charges from the electro-magnet thus produced."

Referring to the illustrations, *c c c c* represent eight bars, which may be of steel or of cast or soft iron. When soft iron is used, bars of steel permanently magnetised are used in connection with

Fig. 16.

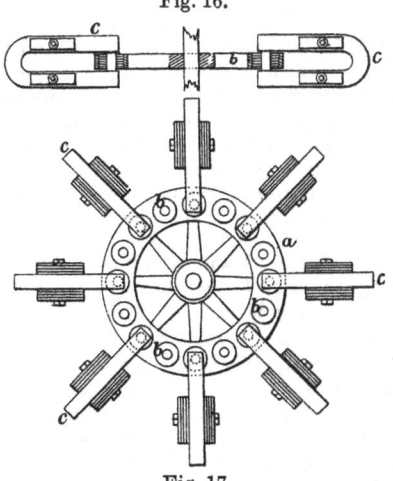

Fig. 17.

them as shown. These bars are placed so as to be out of the influence of the armatures *b*, which revolve either between the poles or in face of them. In another form of generator the magnets and coils are both stationary, and a combined commutator and soft iron keeper rotates between them.

13,613.—J. J. Greenhough, Washington, U.S.A. (*A communication.*) **Obtaining and Applying Motive Power. 1s. 9d. (29 figs.)** May 3.

ELECTRO-DYNAMIC AXIAL ENGINES.—The invention is to render available for mechanical purposes the axial forces reciprocally existing between a helix, traversed by an electric current, and a magnet in the line of its axis. The helices employed are composed of wire of square, rectangular, hexagonal, or other section, insulated on two or more sides; wedge pieces or "risers," in metallic contact with the coil, are inserted at the end of each wrapping to preserve the symmetry of the ends of the bobbin. The cross section, and consequently the conducting power, of the wire of the helices is greater in the external than in the interior convolutions. Each helix is composed of a number of separate coils placed end to end in electrical connection, a certain number only of adjacent ones being traversed by the current at a time. By means of a commutator fresh coils are regularly taken into circuit at one end, while an equal number of coils are cut out of circuit at the other end, the effect being that a zone of magnetic influence, denominated "the helix of actual operation" is rapidly carried along the whole length of the coils, the remaining portions of the helix being inert. If a soft hollow bar be made to traverse the interior of such a series of helices, and be at the same time caused to actuate the commutator, it will receive a steady and continuous impulse in one direction, until it emerges from the coils. In some cases a helix is wound on the bar to increase its electro-dynamic effect. The specification illustrates several forms of pumping engines and reciprocating, oscillating and rotary motors. The figure herewith shows the latter. C C are a number of

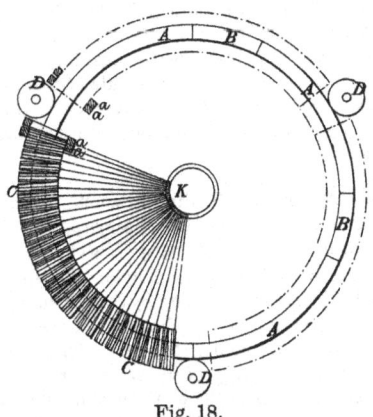

Fig. 18.

coils, forming a circular helix; all the coils are connected together and also to the commutator K. A A are soft iron bars united by brass distance pieces B B, the whole forming a continuous ring, supported by the three rollers D D. The coils and the commutator are stationary, while the ring revolves and drives, in any convenient manner, contact pieces around the commutator. The coils *a a* are made specially large to compensate for the adjoining empty spaces.

COMMUTATOR.—The segments can be made of square or circular form, and be arranged to be turned round occasionally to present clean edges. Or they may be provided with edges of refractory metal that can be renewed. The "communicators" or brushes are similarly constructed.

CARBON PLATES.—Sheet iron is placed in gas retorts to receive a deposit of graphite.

13,660.—J. Chatterton, Birmingham. Protecting Telegraphic Wires. 6d. (1 fig.) June 12.

CONDUCTORS.—A stationary naked wire is fixed to the mandrel of a lead pipe-making machine, and is carried through a water trough to a length of gutta-percha covered wire. The pipe passes over the wire, and, when cooled, continues its course until it has enveloped the covered wire, after which it is rolled to effect complete contact. The specification explains the method of jointing the lengths.

13,698.—Earl of Dundonald, London. Sewers, Drains, Waterways, Pipes, &c. 10d. (18 figs.) July 22.

CONDUCTORS. — Wires, after passing through melted bitumen of a flexible description, are drawn through a die, and then through water or water and clay, or water and lime, to deprive the surface of its adhesive qualities. The coating will be more uniform if the wire be previously covered with filamentous matter and passed through naptha or fluid bitumen; or the filamentous matter may be previously saturated with melted bitumen. When it is desired to lay down several wires, a thick covering of flexible bitumen is put upon a rope, and the wires are laid in longitudinal grooves made in the surface. A sheet of plastic bitumen is bent transversely round the rope and wires and united longitudinally.

13,755. — C. Watt, London. Decomposing Saline and other Substances. 8d. (6 figs.) September 25.

CARBON ELECTRODES. — Lamp-black or tar is coked and pulverised, mixed with sugar and rammed into a mould. The mould is gradually raised to a red heat, and the carbon, while warm, placed in a solution of sugar or in hot tar, subjected to the action of an air pump while still in the fluid, and afterwards heated red hot with powdered charcoal or coke in a covered crucible. This is repeated until the required density is attained, after which it is subjected to most intense heat for several days.

1852.

13,938. — E. Highton, London. Electric Telegraphs. 1s. 11d. (7 figs.) January 29.

CONDUCTORS.—The wires are suspended from arms inclined to the posts at an angle of 45 degrees. They are strained by hydraulic machinery, or by being caused to assume a zig-zag form. For their better insulation they are varnished for 2 feet at each side of the point of support, and are held by a covering of india-rubber or gutta-percha, and metal tags clamped to the arm.

14,021.—W. and A. Smith, London. Electric and Electro-Magnetic Telegraphic Apparatus. 1s. 7d. (24 figs.) March 8.

CONDUCTORS. — The continuity of the line is broken three times at each post by three insulators, one on the post and two at the ends of two short pieces of wire attached to the first. The circuit is completed by a strip of sheet copper bent to clear the post, &c. The invention includes machines for manufacturing submarine cables on board the ship from which they are submerged.

14,057.—M. Poole, London. Covering Telegraph Wires. 1s. 3d. (4 figs.) April 6.

CONDUCTORS.—The features of novelty are (1st) Employing flexible varnish, pitch or other bitumen alone, or combined with india-rubber or gutta-percha, on wire before coating it with insulating materials; (2nd) Applying coatings or successive coatings of gutta-percha or india-rubber, or of both (a machine is described for the purpose); (3rd) Constructing apparatus for applying two or more coatings simultaneously to the wire.

14,146.—A. Bain, Hammersmith. Electric Telegraphs and Timekeepers. 2s. 3d. (18 figs.) May 29.

MAGNETO-ELECTRIC GENERATOR.—Six electro-magnets are arranged vertically in a circle, in the centre of which is a spindle driven by clockwork and carrying two permanent magnets set back to back in a vertical plane. Two poles of like name pass respectively over the electro-magnets and two under them.

14,166.—W. Reid and T. W. B. Brett, London. Electric Telegraphs. 1s. 6d. (28 figs.) June 12.

CONDUCTORS.—Insulated wires are laid in pipes with a trough-like body and a cap or cover of wrought or cast iron, wood, slate, or earthenware. The specification illustrates various sections of pipes, and forms of joints, bends, tees, and testing boxes. Wires intended to traverse a body of water are enclosed in a "vertebral chain," consisting of a series of short hollow portions, united somewhat similarly to the portions of the back bone, so as to leave a free passage throughout, while at the same time the whole is flexible and each part capable of turning freely.

14,190.—T. Allan, Edinburgh. **Producing and Applying Electricity.** 11d. (12 figs.) June 24.

MAGNETO-ELECTRIC GENERATOR. — Permanent horseshoe magnets are placed in one and the same plane with their poles arranged alternately and touching an imaginary circle. In the centre of the circle, and perpendicular to the plane of the magnets, is a spindle, which has as many radiating soft iron arms as there are magnets. Each arm carries a coil of insulated wire, the coils being connected in series, or parallel, as desired. Two or more such sets of magnets and radiating arms can be arranged one above another, one spindle serving for all.

COMMUTATOR.—The alternate currents generated can be turned in one direction by a wooden lever, mounted on an axis, and provided with insulated plates which make contact with fixed pins; or by two reciprocating arms oscillating between two contact pieces.

14,197.—E. C. Shepard, Westminster. (*A communication.*) **Electro-Magnetic Apparatus for the Production of Motive Power, Light and Heat.** 1s. 10d. (11 figs.) July 6.

MAGNETO-ELECTRIC GENERATOR.—A wooden frame carries a horizontal shaft B on which are a number of non-magnetic wheels E E. Around each wheel there are fixed electro-helices C, each wound between wood flanges on a soft iron tube, and retained in position by insulated brass clips. The field magnets D D are arranged in circles with their poles between the electro-helices as shown. There are twice as many helices on a wheel as there are fixed poles on a circle of field magnets, and at a given moment one half the helices are traversed by a current in one direction, and one half by a current in the other. p^1, n^1, p^2, n^2, are four brass conducting rings on each wheel for the collection of the currents from the respective positive and negative ends of the two sets of bobbins. Four strong wires $f f$ severally connect these rings to the four rings $g g$ of the commutator, in each of which are 16 alternately conducting and non-conducting segments. Each ring passes under two springs so adjusted that as one is leaving a non-conducting segment the other is leaving a conducting segment; also the springs on one ring are

in the middle of a pair of segments as the springs on the corresponding ring are making their change. The commutators can be moved on the shaft, so as to give them a little "lead" as regards the helices, the amount depending on the speed of rotation. In a modified generator the field magnets are shown as rotated 90 degrees and set

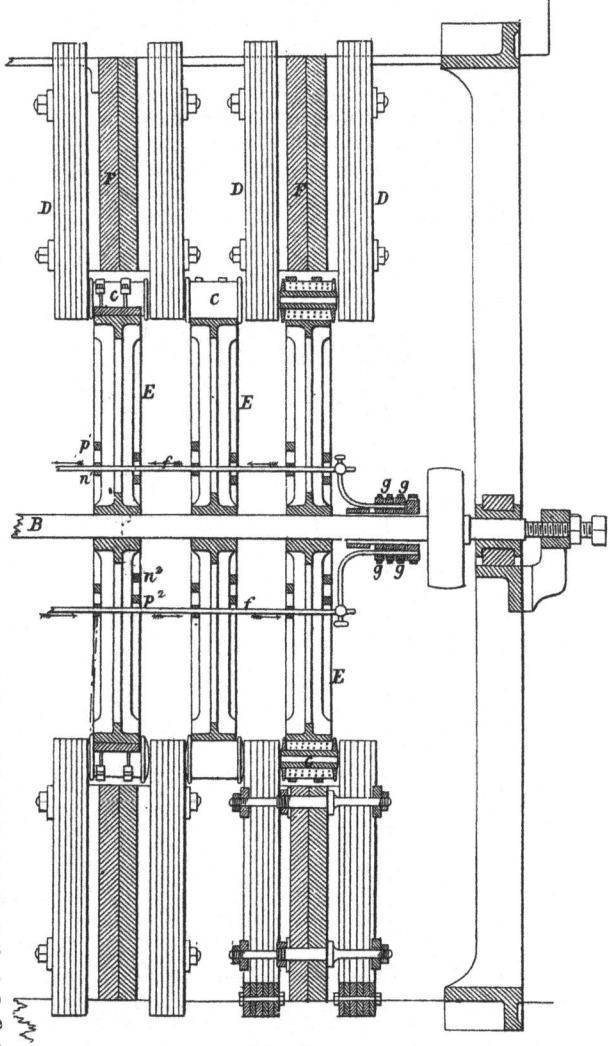

Fig. 19.

astride of the wheels. The arrangement may also be reversed and the permanent magnets attached to the wheels while the helices are stationary. In each machine either straight or horseshoe magnets may be used.

14,198.—M. J. Roberts, Gerrard's Cross, Bucks. **Obtaining Light, Motion, &c., by the Agency of Electricity.** 1s. 7d. (10 figs.) July 6.

Arc Lamp.—One electrode is fixed; the other is maintained at the striking distance, and is from time to time allowed to fall, and is then raised the length of the arc by clockwork or by the agency of the current. In one form of lamp the upper carbon is pushed by a weight through a tube with a platinum-lined spring mouth-piece. Two clips or tongs pass through slots in the tube and engage the carbon. When the current fails, the pencil slides through the tube and clips, until it rests upon the lower carbon and completes the circuit. Upon this an electro-magnet attracts its armature, and by a system of rods and levers binds the clips on to the pencil, and at the same time raises both the tube and the clips, and holds them until the current again fails. In another lamp (see Fig. 20.) the tube E, guided at two points, is

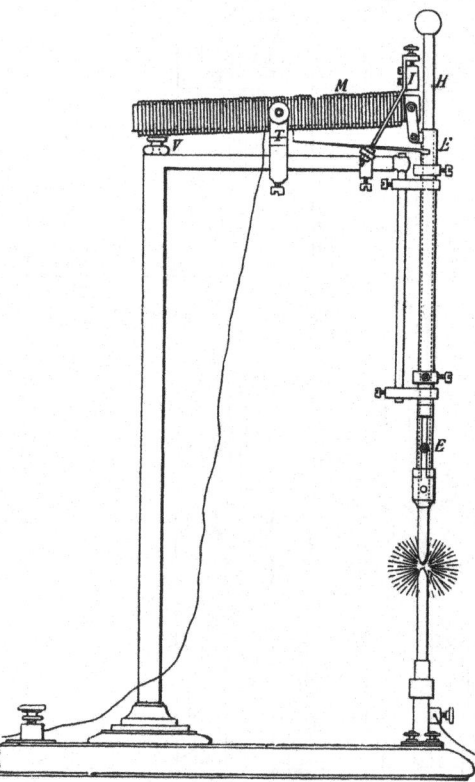

Fig. 20.

connected by a link to the electro-magnet M. H is an iron weight, faced with brass, fixed to or placed upon the carbon. When the current begins to circulate, the magnet attracts the weight H, and rotating on the studs or trunnions T, rises towards the fixed armature I, until stopped by the set screw V. The friction of the tube is sufficient to raise the carbon, and maintain it so long as the magnetic force acts upon the weight H. When the current fails, the magnet rotates in the opposite direction, the weight H is

released, and the carbons come together, when the same cycle of operations is again repeated. It is not stated that either of the above lamps can be adjusted to feed the carbon before the failure of the arc.

Multiple Electrodes. — If several sets of carbons be combined in one lamp and connected to one or more sources of electricity, "a fair average of light" will be obtained.

Intermittent Lamp. — In clockwork lamps suitable for lighthouses, two cams, upon a constantly revolving shaft, act at intervals, the one to raise and then depress the lower carbon, and the other to release and then retain the upper carbon in its holder. The number and duration of the periods of extinction depends on the contour and speed of rotation of the cams.

Incandescence Lamps.—In an exhausted glass globe, a piece of graphite, "as thin as conveniently can be made," is supported between two clips connected to the poles of a battery.

Electrodes.—The brilliancy of the light will be much increased if lime, or other alkaline earth, be interposed between the electrodes, as in the manner of the Drummond lamp, or by adding lime to the carbons during manufacture.

Magneto-Electric Generator.—A ring of alternate sections of magnetised and non-magnetic metals is supported on india-rubber-covered rollers and rotated through helices. (See Specification No. 13,613 of 1851.)

Electro-Motor.—Several wheels, similar to Barlow's rotating wheel, are mounted on an axis and work in separate troughs of mercury. Radial currents of electricity through the wheels cause their rotation. In another motor, a wheel with small electro-magnets around its periphery rotates so that the electro-magnets pass close to the poles of a fixed permanent or electro-magnet. The novelty consists in making these small electro-magnets neutral or feebly repulsive of the fixed magnet immediately they come opposite its pole.

Spark Preventer.—Two apparatus are described and shown in the specification for diminishing the spark which occurs on breaking an electric circuit. In one the electricity, before the rupture of the primary circuit, is made to flow through a shorter circuit by means of a spring attached to the keeper of an electro-magnet in the primary circuit. In the other the primary circuit is never broken, but excites a magnet which completes a shorter circuit, through which the greater part of the current is diverted.

14,330.—E. H. Jackson, London. Obtaining Light and Motive Power. 1s. 2d. (15 figs.) October 21.

Arc Lamps.—The upper carbon is fixed; the lower holder carries a rack in gear with a pinion on the frame. This pinion also acts as a ratchet wheel and gears with a pawl, the whole arrangement serving to prevent the lower carbon from descending. The movable electrode is carried by a bar electro-magnet suspended axially in a second hollow magnet, and so counterbalanced by a spring or weight that it has a tendency to rise and put the carbons in contact. The current traverses first the coil of the hollow magnet, then that of the bar magnet, and finally the carbons; its action is to draw one magnet into the other, in opposition to the counterweight.

GOVERNOR, METER AND CURRENT INDICATOR.— The current passes between two terminals, through a bath of dilute sulphuric acid. One terminal is adjustable to regulate the distance traversed by the current through the liquid. A needle, "acting in connection with the above, measures the amount of electricity used."

ELECTRODE.—Platinum or mercury placed in a recess in a piece of carbon constitutes a "non-consuming electrode."

14,331.—E. B. Bright, Liverpool, and C. T. Bright, Manchester. Electric Telegraphs. 2s. 2d. (38 figs.) October 21.

CONDUCTORS. — Relates to various ways of suspending and insulating overhead wires by marine glue and gutta-percha. Subterranean wires are laid with marine glue in longitudinal grooves in wood. When wires are protected by a helically wound riband of iron a flexible washer, apparently a strip of cloth, is interposed between each overlap and the turn preceding it. Submarine wires have a considerable amount of insulating material between them and are helically wound as above.

MAGNETO-ELECTRIC GENERATOR.—Two electro-magnetic coils can be vibrated before a fixed permanent magnet by a handle. The invention consists in a form of commutator by which for every movement of the handle only one current of magneto-electricity is passed from the machine, so that either positive or negative currents may be used separately.

ELECTRO-MAGNETS.—The core is divided into sections by non-conducting rings or collars, and each section is wound full before the succeeding one is commenced.

14,343.—C. Liddell, Westminster. (*A communication.*) Electric Telegraphs. 6d. (4 figs.) November 11.

INSULATORS.—A hollow iron upright is stopped with a plug of wood, in which is fixed a bar carrying an enamelled iron hood, provided with legs for the reception of an overhead wire.

14,346. W. Petrie, Woolwich. Obtaining and Applying Electric Currents. 1s. 4d. (24 figs.) November 13.

MAGNETO-ELECTRIC GENERATOR.—Two pairs of permanent magnets are arranged so that their poles correspond with the corners of two opposite and parallel squares. A coil of insulated wire stands between the poles, and is attached or not, as the case may be, to a soft iron core whose ends are branched to reach across the diagonals of the squares. The revolution of the core induces alternate currents in the coil. In another machine a brass spindle has one or more iron rods let into it longitudinally, and is divided into sections by iron collars or rings. Insulated wire is wound between the collars and the various coils are suitably connected together. Each collar rotates between the poles of one or more field magnets, the poles being arranged alternately. There should be one permanent magnet at each collar for every rod let into the spindle. The currents will be alternate in direction, but may be changed by a commutator. It is not necessary that the coils should rotate they may be carried in a loose tube.

ALL SPECIFICATIONS AFTER THIS DATE WERE FILED UNDER THE
PATENT LAW AMENDMENT ACT.

1852.

17.—C. H. Newton and G. L. Fuller, London. Protecting Electric Telegraph Wires. 4½d. (7 figs.) October 1.

CONDUCTORS.—Troughs of wood, slate, &c., covered with lids, receive the wires. The lengths of trough and lid break joint and are secured by hoops and wedges.

28.—M. Poole, London. (*Goodyear, America.*) Coating Metals and other Substances with India-rubber Compositions. 2½d. October 1.

CONDUCTORS AND INSULATORS.—The articles are coated with india-rubber mixed with sulphur, and, in some cases, with pitch or coal tar deprived of water, and are finally vulcanised. The material is laid on in the form of overlapping fillets, or is forced from a die like gutta-percha. Insulators are coated inside and out.

*55.—G. Mumby, London. Envelopes. 2½d. October 1.

GENERATOR.—Electricity is used in the feeding process and is generated by the usual apparatus, consisting of a cylinder, rubber and conductor. The cylinder is made to revolve by a band attached to the pulley on the mainshaft.

212.—T. Slater and J. J. W. Watson, London. Applying Electricity to Illuminating Purposes. 11½d. (14 figs.) October 4.

Arc Lamp.—The carbon holder is free to slide downwards except when gripped by levers attached to the armature of an electro-magnet. Referring to the illustration, C C is an electro-magnet carried by a bracket on the column B. The armatures H H are attached to "brass levers H¹ H¹, which are fitted with eyes, through which the carbon holder D passes freely when induction ceases, but when the current passes through the magnet the outer extremities of the compound levers H¹ H¹ are raised, and their eyes simultaneously grasp the rod, which is thus held at a proper elevation; but when the current is broken the levers assume a horizontal position and allow the rod to drop down."

In a modified lamp the electro-magnet is situated in the base, and its armature is connected

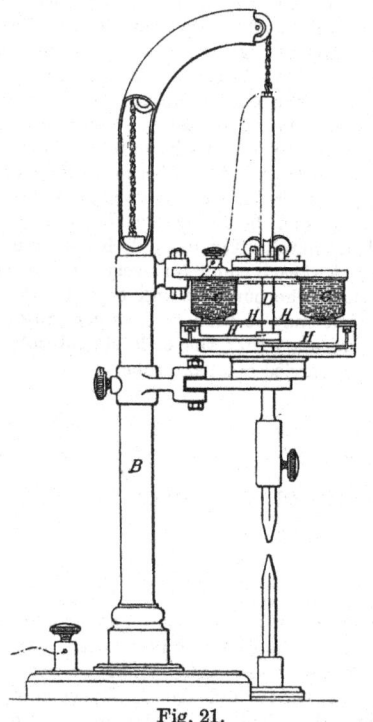

Fig. 21.

to a rod passing up the hollow pillar to a gripping lever, shown in plan in Fig 22. The spring G keeps

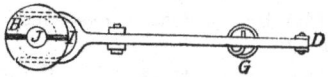

Fig. 22.

the end D elevated, when the magnet is not in action. The forked end of the lever embraces the collar H, which consists of two semi-conical pieces hinged together at I. The pieces grasp the carbon holder J, when the fork moves upwards, and carry it with them the length of their stroke. The striking distance can be regulated by a screw adjustment. The lower electrode is carried on a screw that passes through an internally threaded ratchet wheel operated by a pawl on the lever armature of an electro-magnet. The pawl makes a stroke every time the armature leaves the magnet. The screw may also be operated by hand or by a cord. A rack and vertically acting pawl may also be employed. The invention also relates to "the use of two or more electrodes in the same or different planes, so as to produce more than one point of illumination in the same lamp at the same time." This is not described further.

Electrodes.—Several sets of electrodes may be combined in one lamp, or a number of small electrodes may form the positive pole to one large electrode.

Electrical Governor and Indicator.—An electro-magnet draws down an armature in opposition to a spring, and thereby actuates an indicator.

Alarum.—An electro-magnet holds the detent of a mechanical gong in opposition to a spring. When the current weakens or fails the bell rings.

Current Regulator.—Iron wires of different thicknesses are embedded in pipe-clay, or alumina, or asbestos. The current enters all the wires, and leaves by a movable contact piece that can be connected to any one of them.

Switch.—A suitable switch serves to bring a fresh battery into circuit and to cut out an exhausted one. The poles of the lamp may be reversed by a slide with metallic ends moving over plates connected to the battery in proper succession.

Electro-Magnets.—An insulated band of copper is bent backwards and forwards and plates of soft iron are laid between the folds.

Carbon Pencils.—70 parts of powdered coke, boiled in acid and afterwards neutralised, are mixed with 20 parts of red beech charcoal, 5 parts of gas pitch, 4 parts of finely-divided Newcastle caking coal, and 1 part china clay. The mass is carefully heated in a screw mould and the lump sawn into pencils. Electrodes may be prepared from beech or box-wood charcoal; they are sawn to the required size, and successively steeped in caustic lime, dried, raised to a white heat with powdered charcoal in a closed crucible, cooled, soaked in alum solution, dried, brought again to a white heat, plunged in treacle, and "finally once more ignited." For powerful currents soluble glass is substituted for the alum solution to prevent "excessive convection."

277.—Earl of Dundonald, London. Coating and Insulating Wires. 5½d. (2 figs.) October 6.

CONDUCTORS.—Consists "in various particulars whereby ordinary concrete bituminous substances are rendered applicable to guide and transmit the galvanic or electric fluid through subterranean excavations as effectually as the well-proved bitumen of Trinidad, or that of the British North American provinces" (see No. 13,698 Old Law). The substances employed are gums, resins, oil of petroleum, distilled tar, bitumen and caoutchouc. The claims refer (1) to the adaptation and manipulation of such gums and resins; (2) to producing an insulating covering by influence of heat applied to the wire; (3) to preserving gums and resins from being decomposed or destroyed by combining them with antiseptic oils and mineral substances. Single wires, or groups of wires, so insulated and preserved, may be covered with pipes or laid between sheets of bituminous coating combined with filamentous, metallic, textile or fibrous materials.

459.—C. W. and J. J. Harrison, Richmond, Surrey. Protecting Insulated Telegraph Wires. 5½d. (2 figs.) October 20.

CONDUCTORS. — Gutta-percha covered wire is painted with a mixture of caoutchouc, naptha, creasote and shellac, and is further protected by helically-wound overlapping strips of caoutchouc similarly painted. Wires so prepared may be laid in earthenware pipes or channels, or have a lead tube drawn closely over them. (See No. 13,660.) For submarine purposes the core is served with flannel or horse-hair cord, and sheathed with galvanised wires, whose section resembles a triangle with the point removed. The wires when laid together form a cylinder, lying side by side like the stones in a circular culvert.

481.—J. Fowler, Bristol. Laying Wires of Electric Telegraphs. 6½d. (3 figs.) October 21.

CONDUCTORS.—Describes a plough for laying conductors, or conductors and tubes, underground in the manner in which draining tiles are laid.

*497.—L. N. Legras and W. L. Gilpin, London. Generating Electricity. 2½d. October 23.

ELECTRIC GENERATOR.—Refers to "a battery partly magnetic and partly voltaic, consisting of a cylinder or other suitable piece of iron, or compounds of iron or metal, used with a solution composed of any common salts and water or other suitable liquid."

*511.—J. Hunter, London. Electric Telegraphs and Apparatus Connected Therewith. 5½d. October 25.

INSULATOR.—Comprises an iron standard, a porcelain or glass cone, a cast-iron cap or cup, a porcelain rest, a catch, and a "pierced clay cup."

CONDUCTOR.—A new insulating composition is alluded to.

561.—J. G. Wilson, London. Communicating Signals for Railway and other Purposes. 5½d. (4 figs.) October 29.

ELECTRIC LIGHT.—The signals may be "illuminated by the electric light," and the lighting rendered "simultaneous and continuous with the action of the signal."

* 566.—L. N. Le Gras and W. L. Gilpin, London. Transmitting Electric Currents. October 29.

CONDUCTORS.—Non-conducting pipes containing water, with or without ropes to prevent rending when used for submarine purposes, replace the ordinary covered wires.

595.—J. J. W. Watson and T. Slater, London. Galvanic Batteries, Illuminating and Heating by Electricity, and Obtaining Chemical Products. 4½d.

ELECTRIC LIGHT.—To make the carbon electrodes waste equally, and "to cause the convection of the particles from one pole to the other to be performed with a certain lateral divergence, causing great dissemination of the light," two batteries are employed, and the positive pole of one and the negative of the other are connected to one electrode, while the remaining poles are connected to the other electrode.

INCANDESCENCE.—Fine platinum or iron wires, coated with pipe-clay or plaster-of-paris, or other non-conductors, are again encompassed with metallic rods, and, when traversed by a current, constitute a source of heat for culinary and other operations.

613.—M. Bianchi, London. (*A. Carrosio, Genoa.*) Electro-Magnetic Apparatus for Generating Gas. November 2.

Consists "in new arrangements and combinations of electro-magnetic apparatus for the decomposition of water."

Provisional protection refused.

652.—J. H. Young, London. Weaving or Manufacturing Fabrics. 1s. 8½d. (10 figs.) November 5.

It is proposed "to form the shuttle race of soft iron points, having insulated copper wound round part of their length; then, if an arm carries with it the two poles of a galvanic battery, such section of points would become magnetic in succession, and draw the shuttle round the circle."

680.—W. T. Henley, London. Electric Telegraphs. 9½d. (21 figs.) November 9.

INSULATORS.—Relates to insulators in the form of sheaves, and to non-conducting shackles to break the continuity of the wires at stations.

CONDUCTORS.—Describes several apparatus for straining overhead conductors. Underground and submarine wires are coated with india-rubber, wrapped with tarred yarn and sheathed with galvanised wire. Troughs and their lids for underground conductors are made, by preference, of rolled zinc or galvanised iron.

740. — Earl of Dundonald. Apparatus for Laying Telegraphic Wires in the Earth. 4½d. (2 figs.) November 13.

CONDUCTORS.—Relates to an apparatus similar to a drain plough for laying conductors underground.

750.—J. Mirand, Paris and London. Electric Apparatus for Transmitting Intelligence. 11½d. (15 figs.) November 15.

ELECTRO-MOTOR. — A fixed horseshoe electro-magnet stands opposite a similar magnet upon a spindle, which also carries a commutator.

COUPLING CONDUCTORS. — The communicating wires of railway carriages are connected by hooks that can be drawn out of telescopic cases or holders, against the tension of springs.

778.—H. V. Physick, London. Electric Telegraphic Apparatus. 5½d. (8 figs.) November 17.

CONDUCTORS.—Calico or other fibrous material, impregnated with insulating medium, is wrapped helically around a conductor by means of a triblet or die. Strips of metal may be laid on by the same apparatus, and the overlapping edges subsequently soldered. The sheathing may be divided in places to prevent it forming a continuous conductor. The die is a taper hole through a block of metal, which for convenience of manufacture is made in six pieces. The wire is rotated on its axis as it is drawn through the die, and the calico or sheet metal is fed in at an angle to the wire and laid in position by a "tongue."

827.—J. Kilner, Dewsbury. Insulating the Wires of Electric Telegraphs. 9½d. (16 figs.) November 23.

INSULATORS.—Relates to tools and moulds for making glass insulators.

834.—O. Watt, London. Obtaining Currents of Electricity. 5½d. (3 figs.) November 23.

ELECTRO-MAGNETIC GENERATOR.—The generator comprises two horseshoe electro-magnets, excited by a battery. The one acting as the armature has both a rotating and a reciprocating motion before the poles of the other. This is effected by the shaft, on which it is carried, being moved endwise by a cam or eccentric at each revolution.

*842.—A. Brackenbury, London. Electrifying Machine. 2½d. November 24.

ELECTRIC GENERATOR.—"What is claimed as new is the production of electricity by drawing gutta-percha between cushions made of silk or other suitable material, so as the cushions when pressed towards each other by springs or any other means would rub the surfaces of a sheet or riband made of gutta-percha sufficiently hard so as to produce or excite electricity when the sheet or riband of gutta-percha is moved backward and forward between them." A machine for the purpose is described.

*849. — A. J. L. H. Tourteau, Comte de Septeuil, Paris. Electro-Magnetic Engines and Batteries. 4½d. (14 figs.) November 24.

ELECTRO-MAGNETIC MOTOR. — The illustration shows a side view of the motor. Electro-magnets, whose cores are built up of thin plates, arranged as in Nos. 1 and 2, or as in Nos. 3 and 4

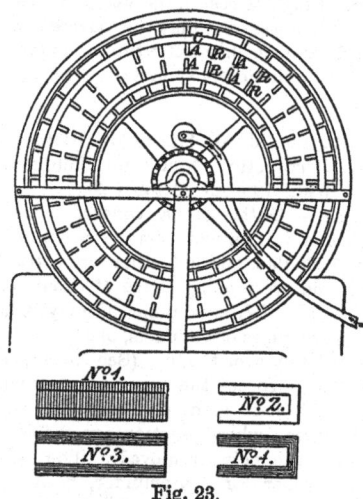

Fig. 23.

are fixed to the outer periphery of the rotating wheel and to the inner periphery of the fixed case. The cores only of the magnets are shown in the illustrations. The coils are wound in the usual way. A commutator reverses the battery connections at suitable intervals.

1,046.—W. H. F. Talbot, Lacock Abbey, Wilts. Obtaining Motive Power. 6½d. (5 figs.) December 13.

ELECTRO-MAGNETIC MOTOR. — "A heavy iron cylinder is made to roll upon a long, but narrow, metallic table or plate, close beneath which a row of horseshoe electro-magnets is placed. The magnets stand vertically with their poles uppermost, so that as the cylinder rolls along the plates it unites the two poles of each magnet consecutively. If the magnets are placed close enough together, and on a level with each other, their summits will form a sufficiently firm surface for the purpose required. The machine is so contrived that the cylinder is always attracted forwards till it reaches the end of the row of magnets; the action is then reversed and it returns in the opposite direction till it reaches the other end, and so on. The whole distance traversed by the moving cylinder makes one stroke of the engine." The motion of the cylinder sets in action the commutator, "which magnetises the magnets which are before the cylinder and unmagnetises those that are behind it." The commutator is a reciprocating slide, and it is designed so as to excite the magnets in advance of the cylinder whichever way it runs.

1,134.—J. F. Kingston, Carrol County, Maryland, U.S.A. Obtaining Motive Power by Electro-Magnets. 1s. 0½d. (14 figs.) December 22.

ELECTRO-MAGNETS.—Plates of iron or nickel are bent into U-form, with their branches very close together, and are packed parallel in a frame by means of brass bolts and a cradle. Insulated bundles or strands of wire are wound round and between these plates, so that the current passes in the same direction from end to end on one side of each plate, and in the opposite direction on the other side of each plate.

ELECTRO-MAGNETIC MOTOR. — Two rotatory engines are arranged in such manner that the electric current, when acting on the magnets of one engine, shall be cut off from the magnets of the other engine. A number of compound magnets, such as described above, are applied to a rotary wheel moving freely on an axis; on another wheel of the engine, affixed to the axis, there are an equal number of keepers; when the magnets are excited their keepers are attracted and motion is given to the axis, and to the other parts, in such manner that the keepers of the corresponding rotatory engine on a different axis, are removed a distance from their magnets equal to the distance through which the other keepers are attracted. Spur gearing, connecting the two engines, communicates with the frames of the electro-magnets, and with their axes by means of a ratchet wheel and click motion. A commutator directs the current alternately to each engine.

1853.

5.—J. J. W. Watson and W. Prosser, London. Manufacturing Steel and Carburising Iron. 3½d. January 1.

ELECTRODE.—Electricity is employed to facilitate the conversion of iron into steel, and is applied to the metal by means of electrodes "of graphite or any form of carbon."

96.—J. W. Wilkins, Hampstead. Electric Telegraphs. 1s. 0½d. (10 figs.) January 13.

INSULATORS.—Around a core of any material there are placed one, two, three or more concentric tubes with a space intervening between each, so as to form chambers round the core, up and down the sides of which the electricity must travel to connect the point of suspension of the wire with the exterior. The wire is suspended from an insulated hook, and the insulator is covered by a non-conducting film. Variously modified examples are given.

*100.—J. H. Vríes, London. Obtaining Motive Power. 2½d. January 14.

ELECTRIC GENERATOR. Electricity is generated "by the friction of two metals to which motion is given by the engine."

*119.—C. Binks, North Woolwich. Electric Light. 4½d. January 18.

ELECTRODES.—One or both electrodes of an arc lamp may be formed of several rods placed close together without actual contact. The rods may have one common connection with the battery, or each may constitute an independent conductor. It is preferred to apply this arrangement to the electrode that consumes most rapidly only. Charcoal obtained by subjecting lignite to destructive distillation is "admirably adapted" for electrodes. Tar, pitch, bitumen, asphaltum, resin, or lamp black, painted on bars, rods, or wires, and then dried and heated, leave a residue of solid or compact carbon. The bars and their covering together form electrodes of low resistance. The same result may be attained by inserting metal rods into "masses of charcoal," or by attaching slips of charcoal to plates or bars.

ARC LAMP.—The electrode is carried by a floating stem immersed in liquid (water, mercury, oil &c.), so balanced that the stem rises exactly as the carbon consumes.

ELECTRODE AND LAMP.—The electrodes are placed and act concentrically, as for example, a long cylinder within a hollow cylinder or ring, or truncated cone of charcoal. The arc is formed between the ends of the rod and of the cylinder. The rod is made to move progressively through the cylinder by any convenient means, and may be of any section. Several such sets of electrodes may be combined in one lamp.

By another method, one "slip of charcoal" is placed close and parallel to a second, and the long one is made to slide gradually past the other

f

in a parallel direction, so that the side of the long slip and end of the short slip form the striking points. Or the long slip may be placed between two others and both its sides utilised.

According to another arrangement, mercury from a reservoir, connected to the positive battery pole, flows or drops into a receiver, or on to charcoal or metal connected to the negative pole, in such quantities that it is partially or fully vaporised. The vapour is collected and condensed.

ARC LAMP. — The electrodes are caused to alternately mutually approach (short of actual contact) and recede, crossing the point of maximum light, by a movement imparted to one or to both, or to all of them, by any mechanical agency. The movement is obtained "by the vibration of wires," or of a metallic fork carrying a carbon on each branch; by an eccentric wheel, or by the revolution of two discs on a common axis, revolving in contrary directions and carrying carbon arms or radii, acting as the electrodes and crossing each other scissorwise; by the rapid revolution of two carbon wheels with serrated edges, placed edge to edge, or by rings placed concentrically "with the light emanating positions serrated or made angular." The motive power may be from clockwork or magneto-electric action, or otherwise. An apparently uniform emission of light is obtained by bringing the electrodes into actual contact and separating them with great rapidity. A rotatory movement may be combined with the former to the end that the several independent flashes may be blended together into an apparently uniform ring of light.

*149.—E. Edwards, Birmingham. Constructing and Attaching Knobs, Handles, and other Articles of Glass, Earthenware, &c. 2½d. January 21.

INSULATORS —The attaching screw is of the same material and in one piece with the article itself.

185.—W. T. Henley, London. Covering, Laying and Uniting Wires and Ropes for Telegraphic Purposes. 11½d. (4 figs.) January 25.

CONDUCTORS.—Relates to sheathing conductors in various ways, especially with a short lay "as harp strings are covered," and describes machinery for the purpose. Lengths of submarine cables may be united by iron clamps or screw joints. Land wires may be laid by a kind of trenching plough, and furnished with testing boxes at intervals.

231.—R. A. Brooman, London. (*A communication.*) Diving Bells and Apparatus to be used in connection therewith. 8½d. (3 figs.) January 29.

CONDUCTORS.—The invention is for apparatus for laying pipes or telegraph wire in a trench below the sea. It is described at length and illustrated.

*303.—D. L. Price, Beaufort, Brecknock. Signalling by Electricity for Railway Trains and Railways. 4½d. (4 figs.) February 4.

COUPLING CONDUCTORS.—Various means are described of connecting the conductors of successive carriages, such as metallic tapes coiled in boxes and provided with springs, or plates with nibs and holes, or helical springs united by spring snaps or union joints, or a rod with two collars and a spring clasping piece. The ordinary buffers, or auxiliary buffers may be utilised for the same purpose.

338. — T. Allan, Westminster. Protecting Telegraph Wires. 4½d. (4 figs.) February 8.

CONDUCTORS.—Consists in protecting insulated wires by means of iron wires so formed into a rope, around a central iron wire core, as to leave spaces between the helical twists for the reception of the insulated wires, with or without smaller wires or bands of metal exterior thereto. Another mode is to form the longitudinal wires into links a yard long, or thereabouts, which are to be linked to a pair of semi-circular cross-pieces of iron, between which a hempen rope containing the insulated conductors is made to pass.

*405.—J. Day, London. Apparatus for Holding and Protecting Insulated Telegraphic Conductors. 2½d. February 16.

CONDUCTOR. — The insulated wire is enclosed between two semi-circular troughs, held together by rings or wire.

570.—J. J. W. Watson, London. Producing Light 5½d. (1 fig.) March 7.

ELECTRIC LIGHT.—The specification mentions the combination of an incandescent platinum coil, or any other form of electric light, with an oxyhydrogen light.

634.—W. E. Staite, Manchester. Producing and Applying Electricity and Obtaining and Treating Chemical Products Resulting from Electrolytic Action. 9½d. (5 figs.) March 14.

ARC LAMP.—Within the tube A, filled with liquid, is a stem b, terminating at its lower end in a float b b, and at its upper end entering the carbon holder C. "The weighting of the float is so regulated that its tendency to rise shall cause the lower electrode to be very slowly pushed up and only as fast as it is consumed." D is a helix wound on a metal reel, and provided with a hollow soft iron core capable of free motion up and down within it. "When the lamp is not in action 'the core' is elevated by a spiral pring, or other suitable contrivance, placed beneath it." The power of the coil to draw down the lower electrode varies with the current passing through it. Instead of the ratchet and click *f*, it is preferred to attach " to the electro-magnet at the top a friction clip, which as the

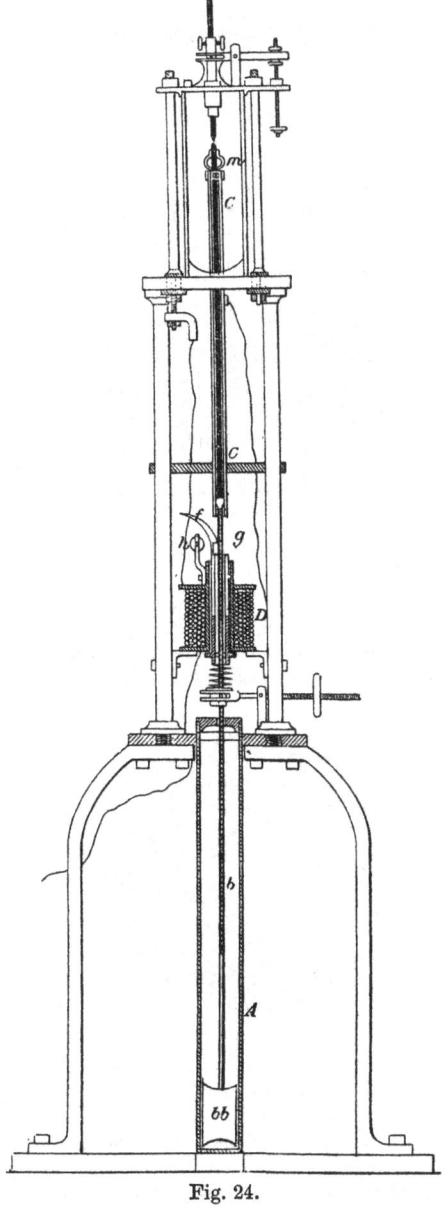

Fig. 24.

electro-magnet " (core?) " is pulled down by its attraction towards the centre of the coil, clasps the rod *b*, and holds it firmly to the magnet, so that it cannot ascend until such diminution in the power of the magnet be produced as shall

allow it to rise to a position at which the clip is detached from the ratchet." *m* is a loose fitting guide of flexible springs, and is capable of end-wise motion equal to the length of the arc. The light is enclosed within a glass cylinder with a ventilating top plate. The action of the lamp is not made very clear in the specification, but the reader is referred to previous specifications, Nos. 11,783 and 12,212 O. L.

Judging from the drawing and the materials of the three specifications, it would seem that the lower electrode is constantly impelled upwards by the buoyancy of the float, just as in most modern lamps the motion of the upper electrode is derived from gravity. The stem *b* passes through the hollow core of the solenoid D, which acts as a guide to it, and would quickly move through the whole of its stroke or travel, were not its teeth (which are just visible in the drawing) engaged by the click *f* from time to time. When the arc becomes too short, the increase of current in the solenoid draws down the core and brings the curved tail of the pawl *f* against the roller *h*, by which it is forced into the rack, stopping its upward progression and slightly drawing it downwards. When the arc has again become unduly lengthened, the core rises and releases the stem.

In a second lamp the carbons are horizontal. The movable one is carried in a slotted tube and is impelled forwards by a driver, to which is attached a cord, passing over a fixed pulley on the frame, and supporting a weight which gradually descends in a tube of liquid. A second cord is attached at one end to the weight and at the other to an electro-magnetic arrangement similar to that in the figure.

CARBON ELECTRODES.—The plates or rods are soaked or boiled in oil previous to the last baking. Carbon battery plates are drilled with holes and metal connections, preferably lead and tin, are cast on to them.

CURRENT INDICATOR.—When the current falls a little below the normal, the core of a solenoid rises and releases a small clockwork alarum. With still less current it releases a large alarum, and on the breaking of the circuit both alarums are set in operation. This can be used in connection with the indicator described in the inventor's previous specification.

*700.—J. H. Johnson, London. Smelting Iron and other Ores. 2½d. March 22.

SMELTING METALS.—Consists in smelting metallic ores, particularly iron ores, which have been previously mixed with charcoal, by the application of the electric arc. This is effected by dropping the ore or metal between the poles of two large electrodes connected to a battery. The ore is melted as it passes through the arc, and the molten metal falls into a receiver where it is kept in fusion by a furnace. By another arrangement the two electrodes are placed at a slight angle, and the higher of the two is made hollow and filled with ore to be reduced. This ore is gradually pushed forward, as it melts, by a piston and screw. The electrodes are also advanced as they consume.

881.—R. J. Kaye and J. O. Openshaw, Bury. Obtaining Power by Electro-Magnetism. 7½d. (5 figs.) April 12.

COMMUTATORS.—Various forms of commutators for electro-magnetic motors are described, such as springs moving over an inlaid disc, spring armatures attracted by permanent magnets, revolving armatures with raised non-conducting portions, and small levers pressed by the teeth of a ratchet wheel on a shaft.

CURRENT GOVERNOR.—An ordinary centrifugal governor raises and depresses wires of different lengths dipping into mercury. The wires are connected to independent sets of cells and come out of the liquid in succession.

906.—J. W. Duncan, London. Combining Gutta-Percha with other materials and applying the same. 11½d. (11 figs.) April 14.

INSULATING MATERIALS.—The compounds are made from gutta-percha, caoutchouc, Canada balsam, Venice turpentine, styrax, shellac, pitch, gum ammoniacum, gluten, &c., in various proportions. Machinery is described for working and heating them, and for covering conductors and uniting them with fabric and sheet metal to form cables.

1,023.—W. Reid, London, Apparatus for Testing the Insulation of Electric Wires. 2½d. April 27.

CONDUCTORS.—Insulated conducting wires to be tested are placed in a vessel. The air is exhausted and then water is admitted under pressure, after which the tests are made.

*1,281.—W. Bauer, Munich, Bavaria. Vessels to be used chiefly under water, and apparatus for manipulating same, &c. May 25.

CONDUCTORS.—These vessels are applicable for laying submarine cables.

1,312.—W. Smith, London. Making and Laying Submarine and other Telegraph Cables, and Making Ropes and Cables generally. 11½d. (8 figs.) May 28.

CONDUCTORS.—Describes machinery for serving and sheathing conductors, in which the reels rotate on their axes, but are not carried round in a circle as is usual. The machinery may be placed on board ship and the cable delivered directly overboard.

1,562.—A. E. L. Bellford, London. Magneto-Electric Machines. (*A communication from C. Carpenter, New York.*) 10½d. (7 figs.) June 28.

MAGNETO-ELECTRIC GENERATOR.—" Several currents are developed at the same time whose changes of direction take place at successive intervals so that only one is broken at a time." Fig. 25 is a vertical section through the machine,

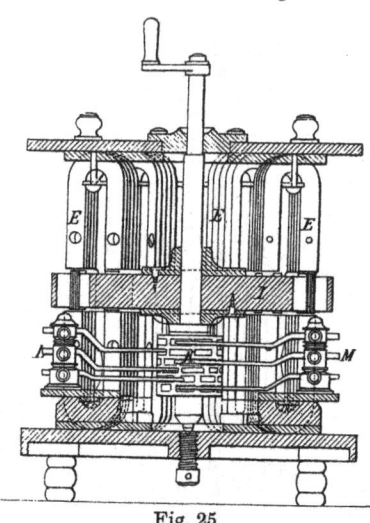

Fig. 25.

except as regards the commutator, and Fig. 26 a horizontal section through the rotating wheel, and electro-magnetic coils. E E are U-shaped permanent magnets arranged symmetrically and

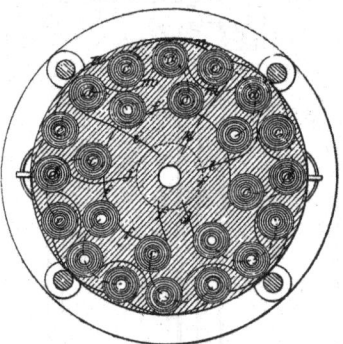

Fig. 26.

radially in a circle, with north and south poles occurring alternately and poles of like name diametrically opposite. The number of magnets in a circle should be a multiple of four, so as to always bring two north and two south poles opposite each other. The distance between the poles of the two circles of magnets " should be about one-third greater than the distance across the end of the pole to obtain the maximum power." In the wheel I there are two circles of coils wound on soft iron cores. There are as many coils in the inner circle as there are permanent magnets in a circle, and in the outer circle twice as many. The coils in the outer circle are at equal distances apart, and the centres of no two coils, of which one is in the outer and the other in the inner circle, can be at the same time opposite the centres of the poles of the magnets. The coils of the inner circle constitute one series (c), and those of the outer circle two distinct series (d and e). The three series c, d and e, form three circuits. The illustration shows the method of connection to produce currents of high potential. The outside end of each coil is connected to the outside end of the next coil of the series, and the inside end to the next inside end, and so on, so as to leave but two terminal wires. f^1, f^1, represent the connections between the coils c c, and f and g are the terminals; m m are the connections between d d, and i and j are the terminals; n n are the connections between e e, and k l are the terminals. If a current of low potential be desired, the inner ends of one half of the coils of each series, and the outer ends of the other half, should be connected to form one pole, and the remaining ends connected to form the other pole. The commutator is a cylinder, K, of non-conducting material, inlaid with segments of copper. The commutator comprises three minor independent commutators, one for each series of coils. Each minor commutator is composed of two parallel rows of segments, the number in each row being equal to the number of magnets in a circle. There is a small break between every two segments consecutively, and all the evenly numbered segments of one row are connected to the odd numbers of the other row, and *vice versâ*. The collecting springs, mounted on the terminal posts L, M, bear directly on the divisions between the segments when the centres of the coils are opposite the poles of the magnets. The breaks between the segments of the different series on the commutator are arranged "spirally," and are caused to pass the points of the springs in the same order of succession as the coils, with which they are respectively connected, pass the poles of the magnets; hence the breaks must follow in regular succession and at the proper intervals, two currents being always flowing between the segments and springs to the posts L, M, which unite them and form one positive and one negative pole. Several circles of magnets and coils may be combined in one machine.

1,587.—E. C. Shepard, London. (*A communication.*) Magneto-Electric Apparatus suitable for Producing Motive Power, Heat and Light. 1s. 10½d. (8 figs.) July 1.

MAGNETO-ELECTRIC GENERATOR.—This is substantially the same apparatus as that described and illustrated in Specification No. 14,197, the point of novelty being the mode of coupling up or combining the metallic circuits of the several series of coils. In the example given in the specification there are, on each rotating wheel, 16 coils arranged in four sets of four, all the coils of each set being always similarly situated as regards the magnetic fields through which they are at any moment moving. The armature carries four pairs of insulated copper rings, each running in contact with a collecting spring. All the inner ends of one set of coils are connected to one ring, and all the outer ends to the next, and so on. "By this mode of combination it will be seen that the metallic circuit in the pairs of contact springs will be uninterrupted and there will be no making and breaking of contact."

1,779.—W. T. Henley, London. Protecting Wires of Electric Telegraphs. 1s. 8½d. (22 figs.) July 30.

CONDUCTORS.—Relates (1) to the construction of troughs and pipes for protecting telegraph wires, and the connecting pieces used therewith or cast thereon. (2) To various methods of constructing the lids or covers of the said troughs or pipes, in such manner that they may be fixed to the troughs or pipes without the aid of bolts or screws, and, where desired, so that they cannot be removed at any intermediate point without cutting or breaking; also to a method of ribbing the lids. (3) To a mode of securing the lids of testing boxes.

1,806.—P. Armand Le Comte de Fontaine Moreau, London. (*A communication.*) Apparatus for Regulating Electric Lights. 8½d. (2 figs.) August 2.

ARC LAMP.—The carbons are horizontal and the lamp can be arranged to stand on a table or to be suspended. In the latter case it casts no shadow. The electrodes are forced towards each other by springs, their motion being restrained by cords wound in opposite directions round a grooved pulley, which is rotated by clockwork when the fly is released from engagement with the armature of an electro-magnet on the decrease of the current. A second pair of electrodes is brought into action as soon as the first pair is consumed. The lamp is illustrated and described at length in the specification, but the method of its action is not made very clear.

1,889.—T. Allan, London. Improvements in Electric Conductors and in the Mode of Insulation. 3½d. (1 fig.) August 12.

CONDUCTORS.—In submarine and subterranean conductors iron is substituted for copper. Strands of two different metals may be also employed as conductors. The dielectric is a compound of caoutchouc, sulphur, and coal tar or other similar substance, baked hard (see No. 14,299, Old Law). A method is described of simultaneously applying several coats of insulating material to a number of wires.

1,909.—G. E. Dering, Lockleys, Herts.   Insulating **Electric Telegraph Wires.**   4½d.   August 15.

CONDUCTORS.—If the line wire and return wire form the sides of a parallelogram, whose width is one-tenth to one-twentieth of its length, they need not be insulated, as the intervening earth resistance will be greater than the metallic resistance. Wires of two different metals may be used to generate the required currents.

* 2,307. – W. Wilkinson, Nottingham.   **Protecting Telegraph Wires.**   2½d.   October 8.

CONDUCTORS.—Insulated wires are laid in longitudinal corrugations in a "piece of metal" with two plain or corrugated sides or flaps. The flaps are pressed down, and the whole forms a band that can be coiled.

2,340.—N. Callan, Maynooth, Ireland.   **Coating Iron with certain Metallic Alloys.**   2½d.   October 12.

CONDUCTORS.—Describes a method of coating iron with an alloy of lead and tin.

*2,345.—H. and D. M. Mapple, London.   **Electric Telegraph and Alarum.**   2½d.   October 12.

MAGNETS.—In order to enable the permanent magnets of electric telegraphs to retain their power they are "tipt" with iron points, which points are carried into the electric coils.

2,371.—J. Farrel, Stangate, Surrey.   **Insulating Telegraph Wires.**   2½d.   October 14.

INSULATORS.—Consists in applying a composition of tar and powdered charcoal in place of, or in addition to, other insulating materials.

2,372.—F. W. Cadogan, London.   **Locomotive Electric Telegraph.**   1s. 1½d.   (3 figs.)   October 14.

CONDUCTORS.—Describes and illustrates a field telegraph apparatus with means for laying and coiling conductors and for throwing them over rivers, &c.

*2,457.—J. B. Verdun, Paris.   **Globes.**   6½d.   (2 figs.)   October 25.

ELECTRIC LIGHT.—Consists in a new mode of constructing globes representing the earth or the heavens, and also uranographic machines, georamas, and cosmoramas. An electric or other light may be set in the centre.

2,498.—J. W. Wilkins, London.   **Obtaining Power by Electro-Magnetism.**   4½d.   (3 figs.)   October 28.

ELECTRO-MAGNETS. — The illustration clearly shows the form and arrangement of the magnets.

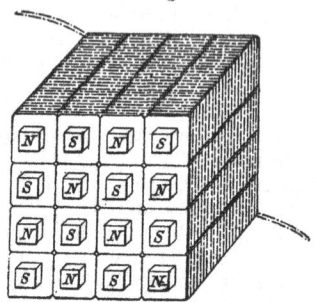

Fig. 27.

The poles are alternated so there is no antagonism in the currents flowing in neighbouring conductors.

*2,677.—J. Gall, Edinburgh.   **Electro-Magnetic Engines.**   2½d.   November 18.

ELECTRO-MOTORS.—Consists in adapting and applying to such engines, magnets and magnetic cylinders, and also in their combination with internal and external drums or cylinders for the passage of electricity. "The magnetic cylinders form the terminations of the poles of double magnets, and are mounted on axles at each end ; and the internal and external electric cylinders are also mounted in such a manner as to be capable of revolving with their cylindrical surfaces in close proximity with the magnetic cylinder. All the cylinders are connected together only with small portions of communication in such manner that the currents of electricity" may flow in opposite directions along the two cylinders. The electric cylinders are covered with tinfoil connected to collecting rings on the axis. This apparatus appears to be based on the principle that a current will rotate round a magnetic pole.

2,764.—J. S. Rousselot, Nimes, France.   **Electro-Magnetic Apparatus for Obtaining Motive Power and Operating Brakes of Railway Carriages, &c.**   7½d.   (8 figs.)   November 26.

MAGNETO-ELECTRIC GENERATOR.—The apparatus consists of a series of permanent horseshoe magnets, fixed to the framing of a locomotive in such positions that they will act in connection with coiled armatures of soft iron mounted on the running wheels.

2,846.— W. T. Henley, London.   **Electric Telegraphs.**   4s. 5½d.   (60 figs.)

MAGNETO-ELECTRIC GENERATORS.—Fig. 28 shows a generator in which a straight bar carrying two coils rotates between curved polar extensions on the branches of a permanent magnet. In Figs. 29 and 30 several bars are combined to form a star and rotate with their coils within a number of fixed poles. The invention includes several modifications of these two arrangements. In one modification they appear to have a reciprocating rotary motion only. "The currents are transmitted in either direction

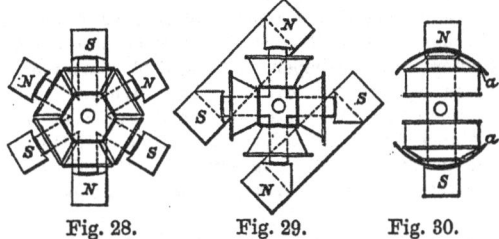

Fig. 28. Fig. 29. Fig. 30.

according to the position in which the handle may be placed, either to the right or left, and when moving back to the vertical position the 'back current' is interrupted or cut off" either by a circular slide in conjunction with springs and studs, or by a permanent magnet fixed transversely on the shaft, in connection with iron rods in the circuit, working on centres, and metallic and spring connections.

In another generator the armature is detached from the permanent magnets, and moved a short distance by a lever and cam or roller.

INSULATOR.—The invention includes a novel form of insulator for overhead lines.

2,956.—J. Latimer Clark, London. Insulating Wires of Electric Telegraphs. 2½d. December 20.

CONDUCTORS.—To prevent return or induced currents in long gutta-percha or india-rubber coated conductors laid in earth or water, they are further enclosed in cheap insulating material of resinous or bituminous character, to increase the distance of the water from the wire.

3,007.—R. Green, Brettell Lane, Staffordshire. Insulators for the Wires of Electric Telegraphs. 2½d. December 28.

INSULATORS.—The whole insulator, including the screw, is made in one piece of the same material.

3,031.—H. V. Physick, London. Electric Telegraphs. 5½d. (6 figs.) December 31.

INSULATOR.—Consists of an inverted cup with a central projection. Apertures are made through the two sides of the cup and the projection for the passage of the wire, which is thus surrounded by a dry zone of earthenware or glass. The wire rests on a sharp edge in the centre of the projection.

1854.

52.—E. Tyer, London. Electrical Signalling Apparatus. 3s. (42 figs.) January 10.

MAGNETO-ELECTRIC GENERATOR.—The current necessary for working certain railway signals is produced by a generator, the coils of which are moved, in opposition to a spring, by the weight of a passing train.

INSULATORS.— Spikes, provided with eyes and covered with non-conducting material, serve as insulators. Their shanks stand within a recess cut in the post.

63.—J. J. W. Watson, London. Electrical Signalling Apparatus. 7d. (2 figs.) January 11.

ELECTRIC LIGHT.—The signalling apparatus described may be employed in conjunction with the electric light. A magneto-electric generator is also shown driven from the axle of a train.

108.—E. Highton, London. Method of Suspending the Wires of Electric Telegraphs. 7d. (6 figs.) January 17.

CONDUCTORS.—Overhead wires are so arranged with relation to each other that when they break they can fall clear. The lower wires have more dip than the upper ones,

117.—C. S. Cahill, Greenwich. Electric Telegraphs. 3d. January 18.

CONDUCTORS.—The wires are insulated by a substance "mostly of paper." They are sheathed with iron strips, and "a continuous vacuum" extends "from end to end." In laying submarine cables spring stoppers are used.

235.—C. Erckmann, Paris. Wires for Electric Telegraphs. 3d. January 31.

INSULATORS.—Glass, paper, cardboard, wood, or bone are the materials to be used.

CONDUCTORS.—Subterranean wires are coated with resinous or poisonous substances to defend them from insects.

459.—C. W. Siemens, London. Electric Telegraphs. (*Partly a communication.*) 11d. (14 figs.) February 25.

INSULATORS.—The line wire is supported by a wrought-iron hook, whose stalk is cemented, preferably by sulphur, into an inverted insulating cup, covered by a cast-iron bell. The bell is secured to the post or wall by screws. A large insulator to be used at stretching points is also described.

COUPLING CONDUCTORS.—In signalling apparatus

on trains the coupling chains are utilised as means for connecting the conductors.

*713.—H. A. Archereau, Paris. Galvanic Batteries. March 28.

CARBON ELECTRODES.—Consists in obtaining dense pieces of blacklead or graphite by compressing blacklead powder.

731.—J. Sandys, London. Electric Telegraph Instruments. 1s. 3d. (4 figs.) March 30.

ELECTRO-MAGNET.—The core is a bundle of soft iron wires.

*759.—P. A. F. Bobœuf, London. Application of Electricity and Aerostation to Military Strategy and Pyrotechny. 3d. April 3.

CONDUCTORS.—Ropes attached to balloons enclose wires for the transmission of signals to the ground.

*779.—W. Gilpin, London. Electric Telegraphs. 3d. April 5.

INSULATING MATERIAL.—A new plastic material " composed of certain proportions of gutta-percha, pitch or tar, resin and oil."

CONDUCTORS.—Naked or covered wires are protected by coatings of fibrous material steeped in a hot solution of oil, resin, and tar, after which they are laid together in the form of a rope with more tar and resin, and are pressed while in a plastic state into a solid body. The rope is laid in a trough, or in a trench, with pitch, resin and sand. It may also be sheathed with wire and afterwards galvanised with zinc or alloy.

1,088.—G. E. Dering, Lockleys, Herts. Obtaining Motive Power by means of Electro-Magnetism. 4d. May 16.

COMMUTATORS.—To prevent the effects of the spark, the parts exposed to its action are surrounded by a hydro-carbon or other non-oxygenated matter, or the air is excluded from them by extraction.

CURRENT REGULATORS.—A tapering bar of iron or steel is immersed to a greater or less depth in a vessel of mercury small end downwards.

*1,187.—C. J. Pownall, London. Communicating on Railway Trains. 3d. May 29.

COUPLING CONDUCTORS.—The coupling chains of railway carriages are also caused to act as couplings for the electric conductors.

1,225.—E. O. W. Whitehouse, Brighton. Electric Telegraphs. 4s. 7d. (72 figs.) June 2.

CONDUCTORS.—The " residual electricity or earth current " of overhead wires may be got rid of by inserting at various points along the line lengths of wire, " at which an accurately adjusted amount of conduction to earth can be established and regulated at pleasure." Another mode is to " short circuit the current in order to divert it from the instrument."

1,242.—J. B. Lindsay, Dundee. Electric Telegraphs. 6d. (1 fig.) June 5.

CONDUCTORS.—In transmitting electric signals through or across bodies of water, metallic conductors may be dispensed with, if the terminal wires of both the battery and instrument be connected to the water or moist earth at a distance from each other. It is preferred that the going and returning currents should be separated by a space greater than their length.

1,357.—H. V. Physick, London. Electric Telegraphs and Connected Apparatus. 7d. (5 figs.) June 21.

CONDUCTORS.—The cores of submarine cables are made of " more wires than one, plaited or twisted together." When several cores are combined in one cable they have distinctive coverings to facilitate their recognition. Cotton is substituted for hemp for the heart worming, sewing and taping of telegraph ropes.

INSULATORS.—Various methods of attaching the wire to the insulator and the insulator to the post are described.

1,412.—A. Smith, London. Telegraph Conductors. 11d. June 27.

CONDUCTORS.—The object of this invention is to form strands for telegraph and other cables, without putting individual twist into the wires or yarns forming the strand.

1,606.—N. Callan, Maynooth. Coating Metals with an Alloy of Lead and Tin. 3d. July 21.

CONDUCTORS.—The invention relates to coating metals with alloys of lead, tin, zinc, or antimony, or of some of them.

1,665.—R. Johnson, Manchester. Coating and Insulating Wire. 3d. July 28.

CONDUCTORS.—Wires may be defended from oxidation, and also be insulated by mixtures of solutions of gutta-percha, caoutchouc, tar, pitch, asphaltum, resins or wax in naptha, benzine, or benzole, &c. The wire is drawn through the mixture, and the solvent driven off by heat.

1,714.—C. W. Harrison, Richmond, Surrey. Galvanic and Electrotype Apparatus, Electro-Magnets, &c. 11d. (15 figs.)

ELECTRO-MAGNETS.—The cores are formed of thin plates of iron bent, or doubled lengthwise with rectangular poles and a convex or rounded face. The coils are bundles of wires or ribbons insulated with strips of gutta-percha applied longitudinally. By the use of square or rectangular wire or ribbon "greater power is gained from the more abundant intersections of the direct lines of magnetic force which this peculiar form affords."

ELECTRO-MAGNETIC MOTOR.—The illustrations show, in horizontal and vertical section respectively, an electro-magnetic engine, with field magnets of the kind above described. The

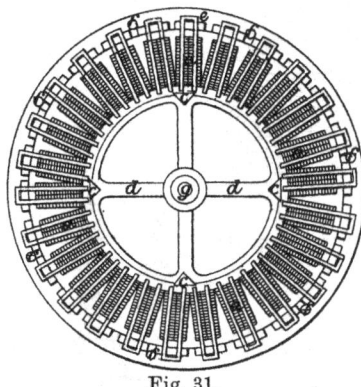

Fig. 31.

armature, as drawn, carries soft iron keepers, but the specification provides that "in place of keepers electro-magnets may be mounted on the axis to operate with the fixed electro-magnets."

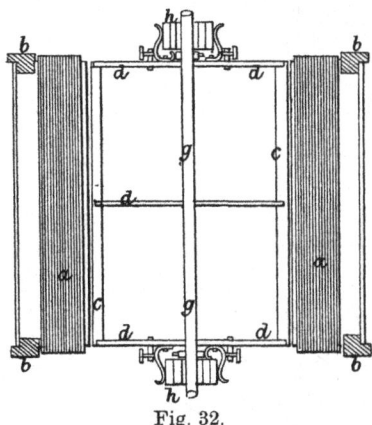

Fig. 32.

a a are long thin electro-magnets fixed in a circular framing consisting of two rings *b b* connected by bolts; *g* is the central axis with three wheels *d d*, upon which are mounted the keepers *c c*; *h h* are commutators.

*2,198.—Soren Hjorth, Copenhagen. Magneto-Electric Battery. 6d. (6 figs.) October 14.

DYNAMO-ELECTRIC GENERATOR.—This specification and the succeeding one, No. 2,199, appear to have originally formed one document, and to have been afterwards hurriedly divided, one sheet of drawings being annexed to this and the remaining three sheets to the other. In consulting the

specifications considerable difficulty will be experienced in following the references to the drawings unless this partition be borne in mind.

"The main feature of this battery consists in applying one, two, or several permanent magnets A of cast-iron, and shaped as shown, in connection with an equal number or more electro-magnets B, shaped as indicated in Fig. 35, in such a manner that the currents induced in the coils of the revolving armatures are allowed to pass round the electro-magnets, consequently the more the electro-magnets are excited in this manner, the more will the armatures C be excited, and more electricity, of course, induced in the respective coilings; and while a mutual and accelerating force is thus produced in this manner between the electro-magnets and the armatures, an additional or secondary current is at the same time induced in the coiling of the electro-magnets by the motion of the armatures, the said current flowing in the same direction as that of the primary current after having passed the commutator. The direction of the current induced in the coils of the armatures will of course be reversed according to the change of the respective polarities, and the commutator D is therefore applied for the purpose of causing the same to flow constantly in one direction. This commutator consists, as shown in Fig. 34, of four or more segments of copper, each opposite pair of which is fixed respectively to one or two smaller rings of different breadth, applied inside the said segments in such a manner that while two opposite segments are fastened to the broad ring, placed between two narrow rings, the other opposite segments are fastened to the latter rings, in order to form communication between the same, while the said commutator is insulated from the shaft; each opposite pair of segments is also insulated from the other pair. The conductor E, extending from the coils of the armature, is fixed to the segment F, communicating, by means of the said ring, with F[1], and the other conductor G to the segment H, communicating with H[1]. The current induced in the coils of the armatures, and flowing through the conductor E, extending from the coils of the armature and fixed to the segment F[1], communicates by means of the said ring with F[1], and the other conductor G, to the segment H, communicates with H[1]. The current induced in the coils of the armatures, and flowing through the conductor E to the segment F, during one stage of polarity, will, during the succeeding stage, flow towards the segment H, through the conductors G, and so on; the armatures having in the meantime made one quarter of a revolution, the conductor I, between the commutator and the coils of electro-magnets, will now, after the next change of polarity having taken place, be in contact with the segment F, while the other conductor will be in contact with H, and the current, having previously flowed towards F, but now flowing towards H, will thus pass in the same direction as it did previously, and of course constantly in one direction.

of attraction of the armatures, while sharp points of different lengths are exposed to the points of separation, and the edges extending from the said points are chamfered off towards the points of

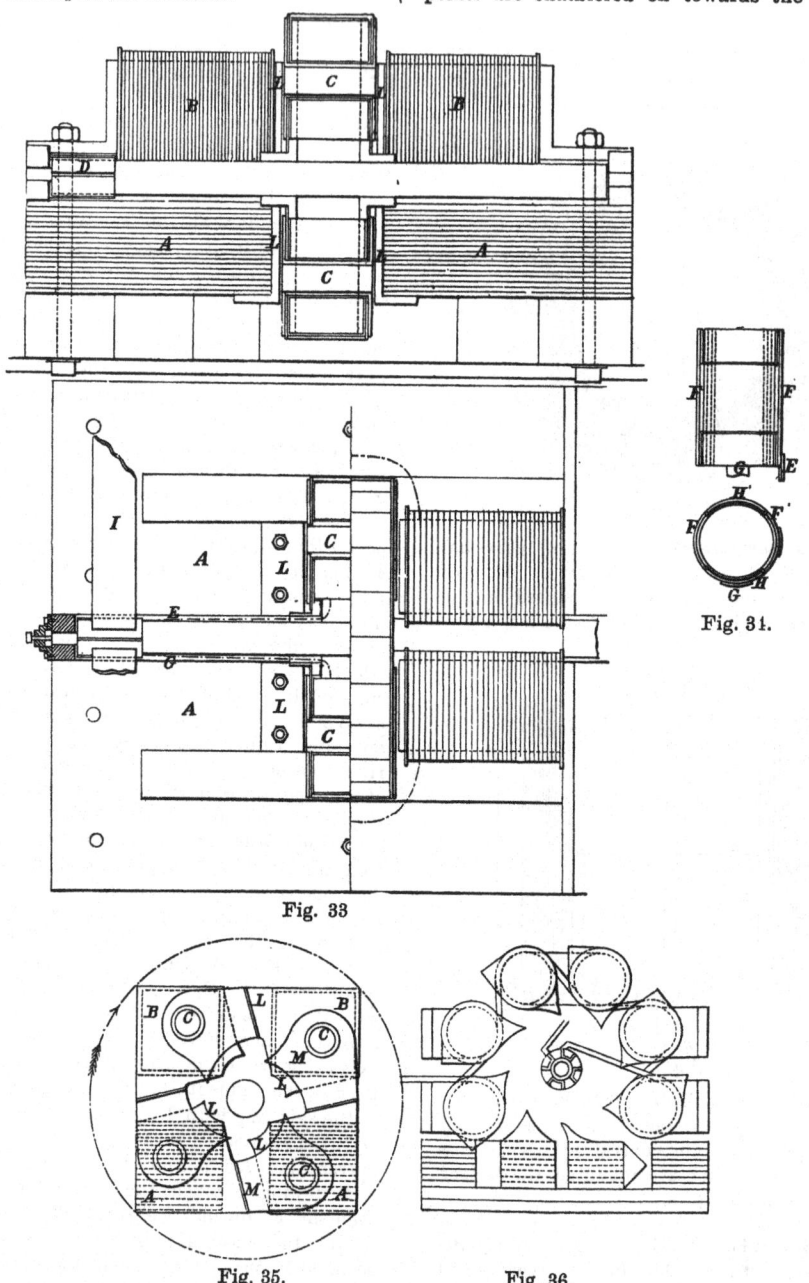

Fig. 31.

Fig. 33

Fig. 35.   Fig. 36.

Figs. 35 and 36 represent iron plates or false poles L applied on the ends of the permanent and electro-magnets, shaped in such a manner as that a large surface with sharp edges is exposed to the points

attraction, the latter having the same length. Corresponding iron plates M are applied on the ends of the armatures and shaped in such manner that the points of attraction expose a large surface with sharp edges to the false poles of the magnets, being at the same time as distant from the centre as possible, while the points of separation are near to the centre and pointed; and the edges of the plates from the said points towards the points of attraction are chamfered off, all with a view to avoid reactionary currents, and at the same time to facilitate the motion of the armature. The permanent magnets A and electro-magnets B are shaped in such a manner, as shown in the figures, that the very ends of the poles face the revolving armatures, while this shape of the magnets also allows the arrangement of the poles on the one side of the revolving armatures to be all north, and those on the other side to be all south. While steel magnets also may be applied instead of cast-iron magnets, the permanent magnets may be coiled like the electro-magnets, which also will serve to make them more permanent."

*2,199.—Soren Hjorth, Copenhagen. Electro-Magnetic Machine. 11d. (4 figs.)

ELECTRO-MAGNETIC MOTORS. — The first is a reciprocating engine containing the following parts : (1.) Several hollow stationary electro-magnets composed of frusta of slightly tapering cones, placed alternately base to base and point to point. They have "double poles, placed in opposite directions, in order to serve for double strokes. This part of the engine may be termed the cylinder." (2.) "A hollow electro-magnet moving within the other and put together in such a way that its single parts terminate in an iron ring. This may be termed the piston in which the magnetism is collected." (3.) A slide communicating with the respective coils of the cylinder for regulating the current round the same. (4.) A commutator for changing the direction of the current round the piston at each end of the stroke. "As one part of the piston enters the cylinder the other part of the same passes out of it, and induces thereby a secondary current in the coiling round the cylinder flowing in the same direction as the primary or battery current, adding thereby to the power produced by the latter."

The specification also describes a rotary motor which " consists of several hollow electro-magnets placed inside a large ring of iron in such a manner that their poles radiate towards the centre of the ring. On the poles of the said electro-magnets are applied plates of wrought iron or false poles, which are curved so as to form a circle concentric with the outer ring. On the shaft, applied in the centre of the engine, are fixed two brass wheels in order to serve as carriages for revolving keepers. The perimeter of these keepers is shaped into several segments, corresponding to the curvature of the false poles of the said electro-magnets. These segments are, as well as the said false poles, grooved in such a manner, as shown in the section, that while the concentric part of the said keepers rolls on the top of the electro-magnets, the segments are attracted by the latter, so that the grooved parts of the first pass between the grooved parts of the false poles of the latter, without however the iron touching or coming in contact with the same. A ring of brass is applied on the top of the false poles of the magnets in order to prevent the keepers coming in contact with the iron in the said poles, and part of one of the said brass rings is formed into a circular rack, into which work corresponding teeth on the circular edge of the keepers, with a view to regulate their revolving motion. The end of the wires forming the coiling are, at one side of the engine, connected to a large brass or copper ring, serving as conductor, while on the opposite side they are connected to as many insulated blocks of copper as the number of electro-magnets, in order that the electro-magnets may be charged in succession as the revolving keepers roll on the top of the false poles. Three conductors I with slides are, for this purpose, fixed to a ring of copper applied round the shaft so as to move with the same, but insulated from it. The said ring serves for conducting the electric fluid to the conductors, and to the coiling of the electro-magnets by means of the blocks. Between the latter, and in contact with the same, are to be applied long and slender wires in order to prevent the current being broken. A circular metallic spring is applied in order to keep the armatures towards the false poles when not attracted by the same."

The dynamo-electric generator, Specification No. 2,198, is well adapted to supply the current for both these motors.

2,243.—T. Allan, Westminster. Electro-Magnetic Apparatus. 10d. (11 figs.) October 20.

COMMUTATOR.—Is for improvements on Specification, No. 14,190 (old law), particularly as regards the commutator, which is formed of concentric rings mounted on the face of a wooden disc. A radial arm on the engine shaft, during its rotation, completes the various circuits. By using a "double sectional zone," the engine can be reversed on shifting the arm accordingly. The zones or rings may be placed on the periphery of the wooden disc.

2,457.—R. Knight, London. Magnetic Apparatus. 6d. (3 figs.)

ELECTRIC GENERATOR.—Two bars of iron in the same straight line are fixed in the magnetic meridian. The two bars do not touch each other, but are placed at such a distance apart as to admit of a piece of soft iron, with coils of wire thereon, revolving between them, and " thus indicating on a galvanometer the properties of the iron." In place of the coils being situated on the piece of soft iron, they may be placed on the two bars, and the piece of soft iron alone revolved in the presence of the ends of the bars.

MAGNETO-ELECTRIC GENERATOR.—The coils are wound on bobbins (see illustration), and placed

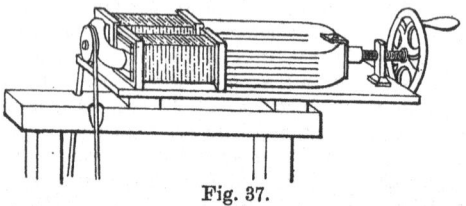

Fig. 37.

on the branches of the permanent magnets. The armature is a piece of soft iron, rotating before the poles of the magnets.

2,555. — C. F. Varley, London. Dynamic Electricity. 7d. (8 figs.) December 5.

CONDENSER.—Sheets of tin foil, alternated with sheets of oiled silk, and alternately connected with the battery poles, serve to store up electricity when the battery is at rest. The arrangement "is very serviceable for electric illumination."

2,708.—J. H. Johnson, London. (*T. O. Avery, New York, U.S.A.*) Electro - Magnetic Engines. 9d. (7 figs.) December 21.

ELECTRO-MAGNETIC MOTOR. — The framework carries four electro-magnets B B¹, two arranged vertically and two horizontally. Upon the bend

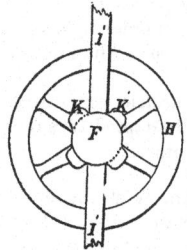

Fig. 38.

of each magnet is a coil D excited by a battery; I I' are "magnetic bars" upon the axis F. K K' (Fig. 38) are commutator rings designed to regulate

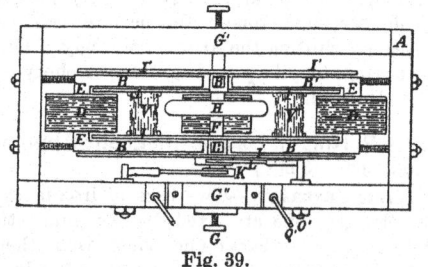

Fig. 39.

the flow of current to the coils, so as to insure the continuous rotation of the magnetic bars and axis. "If found desirable a coil or helix V may be interposed between the inner bars of the

revolving magnetic levers, the currents so produced being auxiliary to the other currents."

2,710.—F. M. Baudouin, Paris. Electric Telegraphs. 1s. 6d. (32 figs.) December 22.

CONDUCTORS. — Relates to (1) a method of insulating wires by means of fatty varnishes and bitumens to be used as a coating, with or without alternate layers of fibrous material; (2) a method of insulating wires by means of bitumen "used as cement or puddings," solidified by cooling, or worked into multi-tubular blocks or kennel-like pieces in which the wires are inserted but not fixed, or else moulded into blocks or courses in which the wires are embedded; (3) apparatus for stretching and straightening wires; (4) the use of stops or guides for holding the wires apart till the bitumen has solidified, the said stops being then withdrawn; also of guides, diaphragms or supports of bitumen or other non-conductor to be left in the course of bitumen; (5) the method of construction of testing boxes or pits.

2,759.—G. E. Dering, Lockleys, Herts. Obtaining Motive Power by the use of Electric Currents. 4d. December 30.

ELECTRO-MOTOR.—The "invention consists in applying that law, which was discovered by Œrsted, that a magnet, when free to move in the neighbourhood of a wire or other conductor through which a current of electricity is passing, will have a tendency to place itself at right angles to the conductor." The magnet is supported by an axis, passed through it about half-way between its extremities, and moves within a coil, like the needle of a galvanometer, the wire of the coil being disposed "so as to circumvent the magnet at all parts of its revolution." The continuous rotation of the magnet is effected by employing suitable arrangements for letting on and cutting off the flow of electricity into the different parts of the coil. Several magnets may be applied to one axis, each with a separate coil. "Magnets of various shapes may be employed, and they may be attached in many different positions to the axis. If iron or unmagnetised steel be used it becomes a powerful magnet through the influence of the current upon it. In some cases there may be coiled" around the bar a separate conducting wire, through which currents of electricity are made to pass continuously. The coils are of wire of rectangular section, laid into a rope and held together by a fibrous covering. The commutator is of the ordinary kind; a short circuit is made between the ends of the coil just previously to the connection with the source of electricity being broken. To avoid the retarding influence of induced currents generated in the coils by the motion of the magnets within them, plates of non-magnetic metal are interposed between the wire and the magnet. The specification is not illustrated.

1855.

83.—F. V. Guyard, Gravelines, France. Electric Telegraphs. 10d. (12 figs.) January 12.

CONDUCTORS.—A series of double posts between the railway tracks, carry wires which lie between two vertical copper rollers on the locomotive. The wire hangs in curves from post to post, and the length of the rollers is greater than the versed sine of the curve. Short rollers on elastic supports may also be employed, and rollers covered with metallic brushes.

148.—P. A. Le Comte de Fontaine Moreau. (*A communication.*) **Obtaining Electromotive Power.** 1s. 3d. (11 figs.) January 19, 1881.

MAGNETO-ELECTRIC MOTOR. — This invention relates to engines, in which the coils are of the kind described and illustrated with reference to Specification No. 13,613 (Old Law). A ring made up of alternate segments of magnetic and non-magnetic metal, rotates through or within coils arranged in a circle. The coils are divided into as many groups as there are magnetic bars in the ring, part only of each group being included in the electrical circuit at a time. A commutator, actuated by the revolving ring, distributes the current, adding active coils to one end of each group and dropping them off at the other end at a speed equal to the speed of rotation of the ring. Fig. 42 is a diagrammatic view of such ring, coils and commutator. The first part of the invention consists of means for carrying the ring, and transmitting the power generated by its rotation. Rollers inserted between the coils support the ring, and cords, bands and wheels of various descriptions, both toothed and frictional, serve to communicate the motion to the outside. Secondly.—The commutator is arranged that the active portion of each group of coils is so situated with respect to the magnetic bar within it as to give the maximum of attraction. Thirdly.—The contact springs never leave one plate of the commutator before attaining its successor, so that the circuit is never broken. Fourthly.—"The currents of induction which have a tendency to form on the cessation of a primary current are sent in the same direction as the latter. To this end each period of the conductor is closed in itself as soon as the latter opens to the primary current."

COMMUTATOR.—Figs. 40 and 41 show in part elevation and in section a commutator for distributing the current under these conditions to a motor comprising three concentric rings of coils. The wire, after traversing the solenoids M, M^1, M^2.(Fig. 41) goes to the contiguous coils M^1 M$^1_1$ M$^2_1$, and so on, the triple coils of the figure corresponding to the simple coils of Fig. 42. In the junction wires between the triple coils are attached respectively two wires, which are two by two in communication with the plates of the commutator. The plates, equal in number to the triple coils, are arranged on the insides of two wooden rings, and are distributed in the order of the respective coils, but alternately on one and the other. "Thus, while the groups M, M^1, M^2 communicate with the clamp screw and plate b b^1 on the upper ring, the clamp screw and plate b^2 b^3_1 on the ring U^1, communicate with the third group M^2, M$^1_1$, M$^2_1$, and so on." Two rollers P, P^1, communicating with each pole of the battery, revolve against the plates of the commutator. Each roller comprises two cylinders u z or u^1 z^1, of which the former is two-thirds the width of the plates. The parts z z and z^1 z^1 are respectively electrically connected and are insulated from the parts u, u^1. From the oblique form of the plates the rollers touch three or four

Fig. 40.

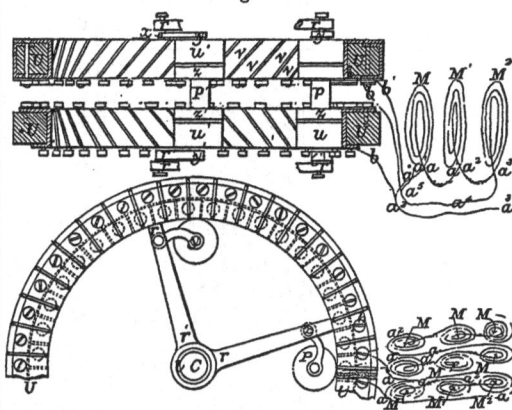

Fig. 41.

simultaneously, and the plate on each ring is also always in communication through the rollers z z and z^1 z^1 with the corresponding plate of the other ring. The action of the commutator is as follows: the roller P, which is the trailing roller, takes part in the successive breaking of the current of the different coils. "It will be understood that at the moment that the extremities in communication with the pole of the battery pass successively and alternately to the plates v, v^1, the intermediate rollers z, z^1, are found exactly at one point tangentially to the last plate v, which is in contact with the posterior extremity of the excluded coil; and that the other extremity touches the plate v on the other ring, which communicates with the anterior extremity." This closes the circuit of the last coil and allows the secondary current to circulate in it. The rollers P^1, P can be adjusted to vary the number of coils included between them, and their lead with respect to the magnetic ring. A regulator for effecting the adjustment of the rollers to the requirement of the work is described, but not very clearly.

MAGNETO-ELECTRIC MOTOR.—Fig. 42 is an electro-motor according to this invention. M, M¹, M², are three rings of coils, each of which includes a hollow cylinder, two parts iron and one part brass, capable of rotation within the coil. Between the coils are four other cylinders, G, G¹, G², G³, suspended from an upper cast-iron circular frame R upon the vertical shaft C. Each of the sets of coils is divided in three places to admit conical pinions, I, I¹, I², to support the magnet rings and to transmit their movement to the exterior. Upon the pinion shaft Q, which runs on anti-friction wheels T, is a pinion S, gearing with a toothed ring in the cylinder G³. The distributing apparatus is similar in principle to that already explained, but considerably more complicated.

The illustration shows one example in which a number of arms, each terminating in a ring P, are fixed in a boss D upon a rotating shaft. Clusters

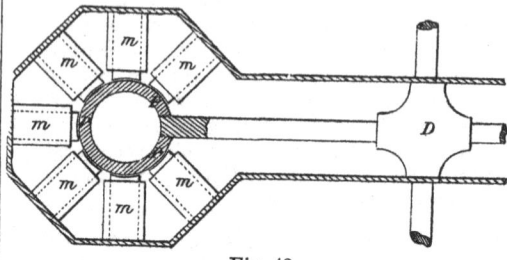

Fig. 43.

of electro-magnets m m are arranged at intervals around the inside of the case, the number of clusters being one more, or one less, than the number of spokes. The magnets are alternately,

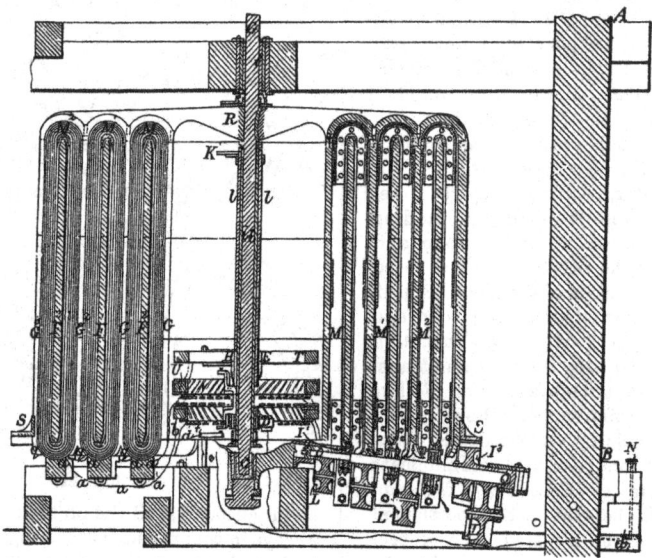

Fig. 42.

191.—J. H. Johnson, London. *G. Bonelli, Turin, Italy.* **Electric Telegraphs.** 8d. (8 figs.) January 25.

CONDUCTORS.—For railway and other purposes a conducting bar, carried on non-conducting supports, lies between the rails and in contact with a "slider" on the locomotive. For ordinary purposes the bar may be coated with tar and provided with a cover.

261.—T. Allan, Westminster. **Obtaining and Transmitting Power.** 1s. 4d. (8 figs.) February 3.

ELECTRO-MAGNETIC MOTOR.—Several motors are described in which motion is produced by the attraction of electro-magnets on soft iron radial bars, arranged in a boss like the spokes of a wheel.

connected and disconnected to the battery in the usual way.

739.—H. Chapman, London. **Apparatus for the Production of Electric Light.** 1s. 5d. (9 figs.) April 3.

ARC LAMP.—The figure herewith illustrates the invention, of which four modifications are described in the specification. Within the sheath D is a smaller tube divided longitudinally into two parts pressed together by springs. This tube carries the upper electrode G, which is fitted into a socket H and can slide through the split tube. I is a tubular weight connected by a pin to H, and suspended from a chain passing over pulleys and down to the barrel K. The sheath D may be square instead of cylindrical, one side being made to open like a door. On the axis of the barrel K is fixed a brake wheel L, embraced by a brake shoe fixed on a lever, which carries an armature N, situated over the electromagnet O. When the normal current is passing through the coils of the magnet the brake wheel is held fast, but when the arc lengthens the adjustable spring *d* overcomes the attractive force and allows the brake wheel to revolve and the

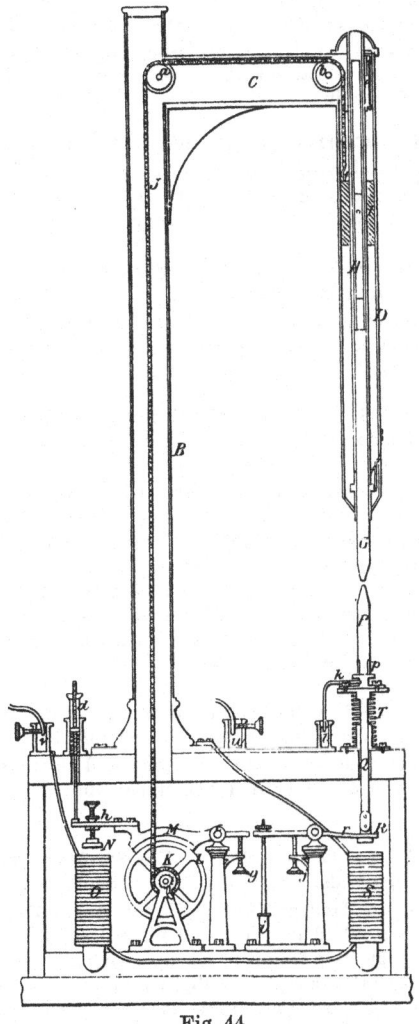

Fig. 44.

chain to be paid out, whereupon the upper electrode descends by its gravity. The lower carbon P is fixed in a socket on the rod Q, which is connected to the armature of the magnet S. The spring T serves to balance the weight of the electrode and holder, and the spring *i* to oppose the magnetic force. As soon as the circuit is completed the magnet S attracts its armature and establishes the arc. Electrical connection is made to the lower holder by the wire *k* and mercury

cup *l*. By a slight modification of the gearing the carbons may be arranged horizontally. In this case the movable carbon holder has a rack cut upon it and gears with a pinion on the brake shaft, which derives its motive power from a weight attached by a cord or chain wound round it.

The feeding of the carbon and the striking of the arc may both take place on the same electrode, and the other electrode be fixed. In such case the chain passes round a guide pulley situated directly below the hollow pillar B. This pulley rides on a lever which is depressed by a magnet as soon as the circuit is completed. The chain is paid out to feed the carbon, as shown in the illustration. In a focussing lamp the same arrangement obtains, but the lower carbon holder carries a rack geared to the brake shaft. If the gearing be fixed to a swing frame, centred on the brake shaft, this frame may be moved to strike the arc instead of the guide pulley mentioned above.

806.—Soren Hjorth, Copenhagen. Magneto-Electric Battery. 6d. (4 figs.)

In this specification, an auto-exciting dynamo-electric generator is described and claimed. The specification and illustrations will be found in the earlier part of this volume.

807.—S. Hjorth, Copenhagen. Electro-Magnetic Machine. 6d. (1 fig.) April 11.

Electro-Motor.—A series of hollow electro-magnets, having their poles in opposite directions, are combined with a series of annular projecting surfaces within them in metallic connection with the poles of such electro-magnets, and arranged so as to act in succession upon a movable compound electro-magnet, and thereby produce a stroke of length sufficient for practical purposes. In action, " the extreme electro-magnets being respectively charged, so that one is neutral while the other has a dissimilar polarity, the moving compound electro-magnet is attracted to the next series of electro-magnets charged, in like manner, while the preceding are discharged, and so on in succession, the return stroke being performed in the same way." The plates of the sliding commutator are connected together by metallic pieces of high resistance to mitigate the spark. In such an arrangement the secondary current induced in the coil of the neutral electro-magnet passes in the same direction as the battery current, thus assisting to increase the power.

808.—S. Hjorth, Copenhagen. Electro-Magnetic Engine. 5d. (2 figs.) April 11.

Electro-Motor.—This engine appears to be very similar to the one described and illustrated in Specification No. 2,199 of 1854.

875.—J. H. Johnson, London. (*A communication.*) Manufacture and Application of India-rubber, &c. 4d. April 19.

CONDUCTORS.—Treats of coating and covering metallic and other surfaces with compounds of india-rubber, gutta-percha and sulphur.

998.—J. Lacassagne and R. Thiers, Lyons, France. Electrometric Apparatus. 10d. (2 figs.) May 4.

ELECTRIC METER AND REGULATOR.—The apparatus is for rendering electric currents regular and constant, whatever may be the variation in the battery employed, and also for regulating and indicating the amount of electricity expended upon any kind of work. The current passes from one platinum plate to another through a liquid conductor, and the depth to which the plates are immersed determines the resistance through the liquid. The plates are attached to a bell which receives the gases from the decomposed liquid and is raised as the gases accumulate. The current also traverses an electro-magnet whose armature, which is drawn upwards by a weighted lever, controls a valve by which the gases can escape from the bell. When the current is strong the valve is maintained shut, and the accumulating gases raise the bell and the plates until the magnet no longer attracts the armature, whereupon the valve allows part or all of the gases to escape, and the bell descends. An adjustable tap can be set to let out the gases at about the proper speed to avoid oscillations of the armature.

*1,050.—J. W. Lewis, Manchester. Lightning Conductors. 3d. May 10.

CONDUCTORS.—A number of strands are worked into a flat rope.

1,091.—R. S. Newall, Gateshead. Apparatus for Laying Down Submarine Electric Telegraph Wires. 8d. (3 figs.) May 14.

CONDUCTORS. — Relates to methods of coiling submarine cable in tanks on board ships, &c.

1,199. — C. W. Harrison, Woolwich. Manufacture of Metallic Ropes, Cables, &c. 7d. (9 figs.) May 26.

CONDUCTORS.—Electric and other cables, either solid or hollow, are formed from a number of angularly shaped plates or strips, placed side by side upon a core and parallel therewith. " Each strip in its transverse section forms the sector of a circle, the angles of two of its opposite sides being either radial lines from a centre or central core, or forming a tangent thereto." Cables may also be constructed from ribbon metal instead of wire. A novel apparatus for and a method of manufacture of such cables is fully described.

1,268.—P. A. Godefroy, London. Treatment of Gutta-Percha. 3d. June 4.

INSULATING MATERIAL.—The ground shells of the coco-nut tree are mixed with gutta-percha.

1,318.—C. F. Varley, London. Electric Telegraphs. 2s. 4d. (41 figs.) June 9.

ELECTRO-MAGNET.—The coil is wound on a reel of soft iron, over each end of which a hollow casing or cap, also of soft iron, is fitted. The wire coil is thus almost completely encased in soft iron and the magnetic polarity is developed at the inner end of the casings, which for that purpose do not quite meet, but leave the central portion of the coil exposed.

PERMANENT MAGNETS.—Water from an elevated cistern is forced over a red-hot steel bar placed within an electric helix provided with an iron exterior casing.

1,646.—C. Deschamps and C. Vilcoq, Paris. Free Diving Boat. 6d. (7 figs.) July 20.

ELECTRIC LIGHT.—The boat carries an electric light in a glass chamber at its upper part.

1,649.—P. A. Le Comte de Fontaine Moreau. Voltaic Batteries. 10d. July 21.

CARBON ELECTRODES.—The upper edge of the plate has a thin coating of copper electro deposited upon it.

1,848.—S. Statham and Willoughby Smith, London. Electric Telegraph Cables or Cores. 6d. (10 figs.) August 15.

CONDUCTORS.—One or more wires are laid helically round a core of fibrous material coated with an insulating substance, or round a core of gutta-percha, with or without a conducting wire in the centre, and are insulated externally by successive coatings of gutta-percha. Wires may also be woven into a cloth or braid, and laid on a core in a similar manner.

* 2,039.—P. A. Balestrini, Brescia, Lombardy. Electric Telegraphs. 3d. September 8.

CONDUCTORS.—The wires are enclosed in alternate coverings of yarn, india-rubber and marine glue.

2,043.—E. Grenet, Paris. Electro-Magnetic Apparatus. 2s. 2d. (17 figs.) September 8.

ELECTRO-MAGNETIC MOTOR.—Upon a revolving shaft is a cylinder with radial plates, resembling teeth, projecting from it. Each tooth is divided into several parts placed slightly in advance of one another, so that they together form an approximately helicoidal line. Each part of each tooth is surrounded by a coil of wire, which is at the proper moment traversed by a current of electricity. Around the movable cylinder is a hollow fixed cylinder with internally projecting teeth, also surrounded by longitudinal wire coils. A current changer and a contact breaker effect the simultaneous magnetisation of the fixed and revolving parts of the apparatus by two different currents. The commutators for the two sets of magnets are separate, and the battery current is divided between the two. The specification describes the distributing apparatus at length, but it does not appear to present any features of interest.

2,089.—L. D. B. Gordon, Westminster. (*E. W. Siemens, Berlin.*) **Electric Telegraphs.** 3d. September 15.

Conductors.—Two or more insulated wires are embedded in the middle of a mass of gutta-percha, or one insulated wire is covered with iron or copper wires, which are surrounded by fibrous materials and then with gutta-percha or other insulating substance.

2,336.—S. Statham, London. Electric Telegraph Cables. 3d. October 18.

Conductors.—Wires insulated in gutta-percha are served with yarn, or yarn and wires, previously coated with marine glue.

2,613.—F. Puls, London. Obtaining Electric Light and Heat. 5d. (4 figs.) November 20.

Electric Light.—The current from a battery renders incandescent a wire of platinum, palladium or iridium, while the gases generated in the same battery burn in contact with the said wire.

2,862.—D. L. Price, Beaufort, Brecknock. Electric Telegraphs. 1s. 5d. 15 figs. December 18.

Conductors.—To make connections between conductors on railway carriages, a coiled spring, that has a tendency to recoil into a case, is fixed to the end of each carriage. A suitable appliance is attached to each spring to unite it to the corresponding spring on the next carriage. When two or more wires are used, one spring, which does not form part of the circuit, is used to tighten them all.

2,895.—E. Tyer, London. Electric Telegraphs. 7d. (9 figs.) December 21.

Magneto-Electric Generator.—The invention includes a generator in which the coils are placed on the axle or wheel of a railway vehicle and the magnets on the frame. The coils may also be independent of the axle and be driven by a band.

1856.

216.—S. Statham, London. Electric Telegraphs. 3d. January 26.

Conductors.—An extensible conductor is made of a hollow strand or cord of metal wires laid helically, or of wires braided into a hollow tube with or without a core. Instead of a tubular conductor a flat metal band, made up loosely of wires and threads of india-rubber or gutta-percha, or their compounds, may be used.

*335.—J. Woodman, Manchester. Telegraph Insulator. 3d. February 8.

Insulator.—The stalk of the hook or eye is fastened into a conical-shaped non-conductor; this non-conductor with the hook or eye is passed through the bell at the back, and by its conical shape forms a wedge, whereby the more strain that comes upon it the tighter it will fit, and thus become waterproof.

573.—F. H. Holmes, London. Magneto-Electric Machine. 10d. (6 figs.)

Magneto-Electric Generator.—Fig. 45 is a front elevation, and Fig. 46 a part side elevation of a machine with six wheels of helices. A A are the magnets set radially and fixed to the main framing, and B is one of the wheels of helices. The poles of the magnets occur at regular intervals, and the helices are so arranged, on any one wheel, that all their centres are always in

Fig. 45.

similar positions with respect to the corresponding poles of the magnets at every part of the revolution. When a machine has several wheels of helices the number should always be an even one, and the first, third, &c., wheels should be placed on the axle, relatively to the second, fourth, &c., in such angular positions that their helices have a little lead in the rotation. The method of connecting them is shown in the diagram. D is the commutator made in two parts, in each of which there are two rings with projections on one side, the projections of one ring fitting into the

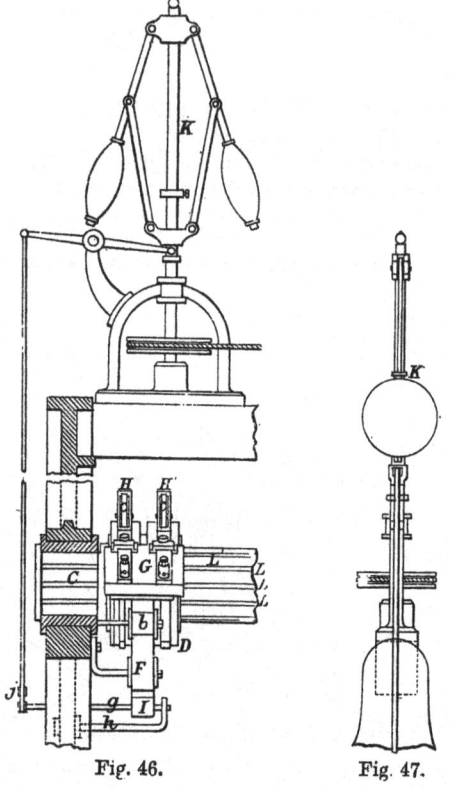

Fig. 46.   Fig. 47.

spaces of the other ring. F is a movable segmental frame for supporting the insulated conductors; it is carried on three rollers and can be moved by the centrifugal governor K acting through the tooth *g* and suitable connection *j*. L L are the wires for conveying the currents from the helices to the commutator. The conductors or collectors are so adjusted on the commutator D, that when the centres of the helices are opposite a point distant about one-third from the side on the breadth of the pole of a magnet in the direction in which the machine turns, the rollers of the conductors shall be on the divisions of the commutators. It is stated that when the machine runs at low velocities there

will be no spark, but as the speed increases the spark will become destructive. To obviate this, the point of contact of the conductors with the commutators is varied as the speed increases or diminishes by the governor K.

The currents produced may be employed for the electric light; in the manufacture of iron and steel for the purpose of obtaining a better product and in less than ordinary time; in the separation of metals from their ores; and for separating the sulphur in the process of making coke.

876.—R. S. Newall, Gateshead. Electric Telegraphs. 6d. (5 figs.) April 12.

INSULATOR.—The insulator is made in two or more separate pieces, so that the part which holds the wire may be easily removed from the bell or other support, which is fixed to the post. The part which holds the wire is formed with a screw to fix the wire after it has been strained.

1,723.—M. Vergnes, New York. Electro-Dynamic Engine. 10d. (6 figs.) July 21.

ELECTRO-MAGNETIC MOTOR.—The apparatus resembles two galvanometers with their magnets upon the same spindle and their coils parallel to each other. Two flat hollow coils are arranged in parallel vertical planes, and a horizontal shaft is passed through them. An electro-magnet is fixed on the shaft within each coil, in such way that it can revolve freely, the two magnets being set at right angles to each other. Thin plates of iron similar in shape to the electro-magnets are fixed on the shaft, so that they revolve close to the outer sides of the coils. The machine is put in motion by two separate batteries, one for the coils and the other for the magnets. The commutators are so arranged that whatever may be the position of the electro-magnets, " their polarisation will remain the same." The current is cut off from the electro-magnets and coils at suitable intervals, to ensure their regular and continuous rotation.

*1,868.—J. Woodman, Manchester. Telegraph Insulator. 5d. August 8.

INSULATOR.—This invention is the same as No. 335 of 1856.

*2,107.—C. W. Siemens, London. (*A communication.*) Electric Telegraphs and Apparatus. 3d.

MAGNETO-ELECTRIC MACHINE.—A cylindrical bar of iron is taken, and two broad and deep longitudinal grooves are cut in the opposite sides. A quantity of insulated or coated wire is wound in a longitudinal direction in these grooves. The wire may be protected by brass covers, which complete the original cylindrical form of the bar. This bar is mounted upon a vertical axis, and revolves between the poles of a series of permanent magnets, placed one above another, and which may consist of steel bars fixed at one end to a flat iron bar, and having grooves at their other end to correspond with the form of the aforesaid cylindrical bar. The bar is made to revolve, and at each revolution two successive electrical currents are produced in its coil in one direction, and then two in the contrary direction. These currents may be sent in one direction by aid of a commutator.

2,124.—P. A. Balestrini. Brescia, Italy. **Electric Telegraphs.** 7d. (11 figs.) September 11.

CONDUCTORS.—Insulated wires· are enclosed in fibrous cords rendered waterproof by a mixture of india-rubber and pitch. The protected wires are laid together parallel or helically on a core of fibre, or of iron wire coiled with fibre, and are served with waterproofed hempen cords, and sometimes with wires. Parachutes are attached to submarine cables at intervals to retard their descent.

2,456.—J. Lacassagne and R. Thiers, Lyons. **Electric Lamps.** 7d. (3 figs.) October 20.

DIFFERENTIAL ARC LAMP. — The feeding is effected by the passage of mercury from a reservoir into a receiving cylinder underneath. The tendency of the mercury to rise in this cylinder lifts a piston or float supporting the lower electrode, which approaches the upper electrode at a speed dependent upon the flow of the mercury from the reservoir into the cylinder. Referring to Fig. 48, a is the reservoir from which the mercury descends into the vessel b, to raise the float and stem x; d is a flexible tube leading to the regulator e, in which there are two vertical ducts, communicating at their upper extremities and closed by a common valve f, of india-rubber. The mercury rises up one duct and, when the valve permits it, passes down the other and through the flexible tube d to the cylinder b; $g\ g$ are bobbins or coils traversed by the main or principal current; $h\ h$ is a coil traversed by a derived current producing a force acting in opposition to the first coils $g\ g$; i is the soft iron armature " which shuts the valve by the action of the current producing the light. It is attracted in the opposite direction by a spring or by a second electro-magnet receiving magnetic action from the derived current, the intensity of which is determined by the resistance offered to it by a coil interposed for the purpose, consequently the passage of the mercury from the reservoir to the cylinder tends to establish itself under the action of the derived current which opens the valve, and to be interrupted under the action of the principal current, which closes it; the resistance opposed to the derived current is invariable. The resistance afforded to the principal current increases by the widening of the distance or space between the two electrodes, and diminishes by their

coming near to each other. If the distance between them increase, the principal current, meeting a greater resistance, diminishes its intensity, which also diminishes the magnetic power of the electro-magnet due from the principal current which closed the valve; the

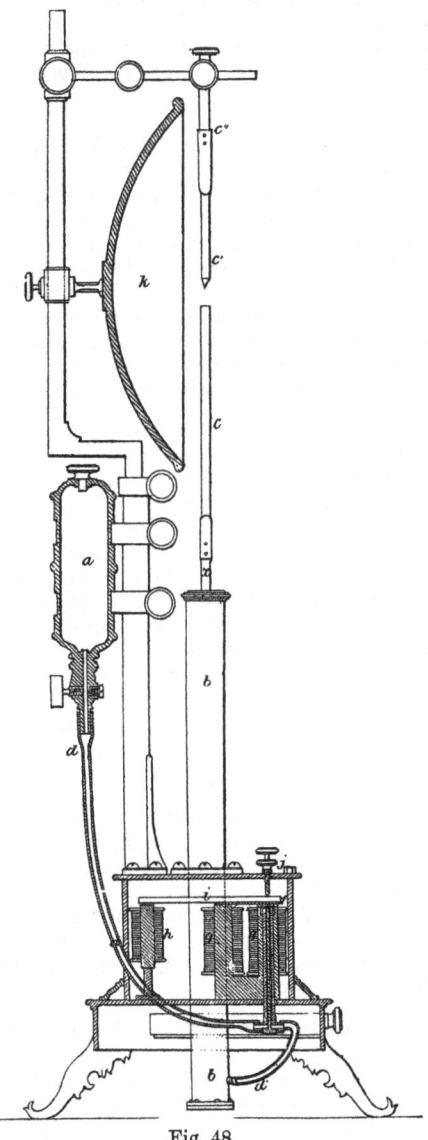

Fig. 48.

intensity of the derived current (the resistance to which is invariable) increases in the same ratio, and tends by its action on the second electro-magnet to open the valve. If the electrodes approach each other the contrary effect is produced. The principal current increases by the diminution of resistance through the electrodes, while the derived current decreases in the same proportion and tends to close the valve to prevent further passage of the mercury. It is then evident that after having determined the resistance of the derived current by the length of the wire of the resisting bobbin or coil, there will be established between the two currents a kind of equilibrium, which will maintain a passage for the exact quantity of mercury that should necessarily flow from the reservoir to the electrode cylinder to lift and regulate the space between the electrodes."

*2,483.—C. W. Harrison, Woolwich. Electric Conductors. 3d. October 23.

CONDUCTORS.—Alternate strips of gutta-percha and cloth are wound around the wire, and the two are united by heat. Several such conductors may be associated in one cable. One conducting medium may likewise be insulated by winding another helically around it, one or both being covered with a fibrous material. For hot climates insulated wires are enclosed in lead. The interstices of cables are filled with tar and cement.

2,547.—J. T. Way, London. Electric Lights. 3d. October 29.

ELECTRIC LAMP.—Mercury is caused to flow through an orifice, or orifices, on to points of steel, platinum or charcoal. The mercury is in connection with one of the poles of the battery, and the points are in connection with the other poles of the battery, and they are so arranged that the distance between them and the orifices, from the mercury escapes, can be adjusted so as to bring the points to the level at which the

streams of mercury break into drops. In place of using points of steel, &c., for the lower electrode, a regulated surface of mercury may be employed if desired, and the apparatus may be surrounded by a glass to prevent the escape of mercurial fumes; means may also be provided for raising the mercury from the lower receiver, into which it falls from the orifices, to the upper receiver or cistern which supplies the jets.

*2,747.—C. F. J. Fonrobert, Berlin. Electric Telegraphs. 3d. November 20.

CONDUCTORS.—Wires are coated with gutta-percha and coal tar.

2,769.—W. T. Henley, London. Electric Telegraphs. 1s. 3d. (8 figs.)

MAGNETO-ELECTRIC MACHINE.—Two coils upon soft iron cores are mounted parallel to and partly between the legs of a horseshoe magnet. Two curved pieces of iron are attached to a spindle in such a way that as the spindle revolves the curved pieces alternately connect each of the cores to the N and S poles of the magnet, and so cause reversions of polarity in the same, and currents of electricity in their coils.

CONDUCTORS.—Gutta-percha covered wires are coated successively with tarred yarn, tarred tape, iron wire and hempen strands.

2,931.—J. Latimer Clark, London. (*Partly a communication.*) Electric Telegraphs. 9d. (5 figs.) November 29.

INSULATORS.—Relates to the form of the insulator, the stalk and socket, and the method of securing the wire.

1857.

67.—E. J. Hughes, Manchester. (*A communication.*) Manufacture and Application of Elastic and Adhesive Compounds. 4d. January 8.

INSULATING COMPOUND.—From substances containing fibrine, cellulose, gelatine, chondrine, colline, caseine, resins, soap, tar, asphalte, pitch, oils and tannin.

129.—G. Bedson, Manchester. Coating and Insulating Wire. 3d. January 15.

INSULATING COMPOUND.—Vegetable or mineral pitch, tar oil, caoutchouc, gutta-percha, naptha and shellac.

211.—P. A. Balestrini, Brescia, Italy. Electric Telegraphs. 1s. 1d. (14 figs.) January 23.

CONDUCTORS.—The shore end of a submarine cable is passed through a series of cylindrical blocks of wood placed at short distances apart, round each of which a bar of steel is wound helically, the blocks of wood being held together by

chains. The joints of the several parts of the cable are enclosed in boxes and run in with cement. A long thin copper wire is wound in long helices around the centre core of submarine cables to carry off the induced electricity from the insulating material.

232.—E. Highton, London. Electric Telegraphs. 10d. (1 fig.) January 26.

CONDUCTORS.—A current from the zinc pole of the battery is caused to pass through the cable at all times when it is not in use, to preserve it from decay.

484.—D. L. Price, Beaufort, Brecknock. Signalling Apparatus. 7d. (13 figs.) February 19.

CONDUCTORS.—Conductors on railway carriages are connected by "suitable metal catches or fastenings. For electric bells they comprise one insulated and one naked wire twisted together."

588.—C. W. Harrison, Woolwich. Electric Light. 1s. 11d. (21 figs.) February 28.

ELECTRODES.—These are formed by placing pieces of metal, or other material, in gas retorts, in order that they may receive a deposit of carbon. If hollow electrodes are required, a current of coal gas is made to pass through a chamber containing metal tubes heated to redness, which thereby receive a coating of carbon. By another process, spongy or powdered metals, either with or without carbon or plumbago, are compressed and moulded into suitable form.

ELECTRIC LAMPS.—A cylindrical carbon rod is capable of rotating on, as well as moving along, a horizontal axis. This forms the positive electrode. The negative electrode is a rod or pencil, which rests on the surface of the cylinder, and is connected with the armature of an electromagnet in such a way that, when the current is passing, a suitable distance is maintained between the electrodes, but on breaking the circuit, contact between them is renewed. Clockwork is employed to impart motion to the cylindrical electrode. In order to regulate the distance between the carbons, the electro-magnet is made adjustable with respect to its armature. One form of lamp illustrated shows a cylindrical electrode mounted on a hollow axis which traverses a screwed shaft, resting in supports attached to the base of the lamp. A clockwork train conveys motion to the screwed shaft by means of a chain running on a pulley fixed to one of its ends. The negative electrode is held vertically in a pendent support, and passes freely through a platinum lined nipple Its upper end is hooked to a cord which is led over pulleys, and down the inside of the vertical support. The other end of the cord is attached to a vertical rod. An electro-magnet contained within the base of the lamp is provided with an armature, fixed to the short arm of a bell crank lever, the long arm of which terminates in a forked end. This fork carries a pair of spring nippers which, at the moment the armature is attracted by the electro-magnet, take hold of the vertical rod fixed to the end of the cord, and carry it down, thereby separating the carbons; at the same time a bent rod, also attached to the armature, actuates an arm carrying a pin, that drops between the teeth of a wheel in the clockwork train, and thus arrests the motion of the cylindrical electrode. A spring is provided to withdraw the armature when the attraction of the electro-magnet ceases. The direction of the current is from the positive terminal of the lamp, through the support, to the screwed axis and cylindrical electrode, thence through the negative electrode and its support to the electro-magnet coils, from which it passes to the negative terminal. Another form of lamp consists in an arrangement for producing the light between the edge of a constantly rotated disc and a pointed rod or pencil. The invention also comprises a flashing lamp. In this lamp the sparks are given out by electrodes, to which a vibratory motion is imparted by suitable means. Mercury by preference is employed for the positive electrode. Another part of the invention relates to the use of an induction coil in connection with a primary current. The primary current is made to give motion to the electrodes, and the induced current to produce the light. An increased light is obtained by combining the induction currents with magneto-electric currents. The inventor applies the light to the production of signals and for indicating the depth of water to vessels.

680.—J. A. Cumine, Belfast, and O. Hunter, Islandreagh, Antrim. Electro-Magnetic Engines. 6d. (1 fig.) March 9.

ELECTRO-MAGNETIC MOTOR.—Electro-magnets are fixed radially on a stationary circular frame with their poles pointing inwards towards the centre. Within the circle of their poles there rotates a wheel carrying soft iron armatures or keepers of such a length that they just clear the above-named poles. The keepers are separated from each other by a smaller angular distance than the electro-magnets, so that some are near to the poles when others are comparatively remote. The electro-magnets are divided into sets, all those of one set being excited at a time, and being separated from each other by intervening magnets of the other sets. In the motor illustrated in the specification every sixth magnet appears to belong to the same set. When the battery is connected to the engine one set of armatures or keepers is attracted, and brings the next set within the attractive influence of the next set of electro-magnets, to which the current is transferred by the action of the commutator just before the armatures are in a radial line with the excited electro-magnets. The commutator consists of wheels with alternate conducting and non-conducting segments and distributing rollers in the usual way, and is made in two portions, one for motion in the forward direction and one for the backward direction.

869.—H. B. Girard, Paris. Insulating and Stretching Telegraph Wires. 7d. (10 figs.) March 28.

CONDUCTORS.—Telegraph wires are insulated by the application of two coatings, the first consisting of graphite or plumbago, sifted and mixed with glue or size, and the second composed of linseed oil, sulphur, gutta-percha, orcansson and tar. The materials may be applied after the wires are in position. Underground wires are insulated by threading glass beads upon them.

933.—F. M. Baudouin, Paris. Electric Conductors. 1s. 6d. (6 figs.) April 3.

CONDUCTORS.—The invention relates to the insulation of of conductors by the application of insulated coatings or coverings, consisting of tapes or ribbons of textile material, either previously impregnated or not with insulating matters, applied alternately with layers of insulating bodies. The tissue may consist of paper, cotton, gummed taffetas or other woven fabrics. Bastennes bitumen is the insulating material preferred, but other bodies may be used. Two machines are described for applying the various coatings. Wires covered according to this invention may be combined in the form of flat cables or straps, either by attaching them to flat impermeable bands, or by weaving them with hempen threads, which may previously be rendered impermeable to electricity. These bands may likewise be covered with thicknesses of tissue. Subterranean cables are simply coated with hot bitumen and deposited underground or buried in a long block of bituminous mastic, with which sand or dry earth is incorporated. To protect underground conductors from the effects of lighting gas, the bituminous covering is surrounded with metal or with paper treated with sulphate of copper or with a silicate and an insoluble soap. In some cases the block in which the wires are embedded is covered with potters' clay or plaster.

947.—E. Testelin, Ghent, Belgium. Electro-Magnetic Motor. 10d. (4 figs.) April 4.

ELECTRO-MAGNETS.—This specification begins with a long disquisition upon magnets, which apparently does not contain anything of value. An electro-magnet is illustrated in which the branches are wound with coils of insulated ribbon copper.

*1,033.—J. B. Pascal, Lyons. Electric Lamps. 3d. April 13.

ELECTRIC LAMPS.—Two electric currents are employed, a main current to produce the light, and a secondary current to regulate the electrodes. The upper electrode is fixed, and the lower one supported on a piston or float resting in a cylinder containing mercury. This cylinder communicates with a vessel of mercury at its own level and connected at its upper part with another vessel containing acidulated water, in which the poles of the battery supplying the secondary current terminate. The main current producing the light passes through the coil of an electro-magnet and attracts an armature, whereby the secondary current is interrupted. A spring re-establishes the circuit as soon as the electro-magnet ceases to act. In this way, so long as the main current maintains its full strength, the secondary current is suspended, but should the strength of the main current diminish, the armature is released, and the secondary current decomposes the acidulated water, thereby establishing a pressure within the

vessel, which is communicated to the mercury in the cylinder, raising its level and bringing the electrodes nearer together at the same time breaking the secondary circuit.

*1,097.—G. Davies, London. (*M. S. and H. E. Goddier, Paris.*) Laying Telegraph Wires. 3d. April 18.

CONDUCTORS.—The invention consists, firstly, in placing telegraph wires either in grooves formed in wooden rails, or flutes formed in sheet metal, and then filling the grooves or flutes with liquid bitumen or other water-resisting medium. The rails are laid in the ground and filled in with bitumen. Secondly, a space is left between the ends of adjoining rails to allow of them being folded up into bundles for transport, and of being unfolded when they reach their destination and immediately laid in the trenches.

1,258.—J. T. Way, London. Obtaining Light by Electricity. 8d. (3 figs.) May 4.

ELECTRIC LAMPS.—This invention relates to lamps in which flowing mercury electrodes are employed, and consists in improvements on Patent No. 2,547 of 1856. Two jets, made of pipe-clay, or a mixture of pipe-clay with silica or phosphate of lime, are so placed that the streams of mercury issuing from them meet each other at a point where one or both of them falls into drops. When cups are used, as described in No. 2,547 of 1856, they are made of similar materials. The glasses which enclose the light are placed sufficiently close to it to become heated, and thus prevent the mercury adhering to their surfaces and obscuring them. Mica, in some cases, is substituted for glass. Economy in the battery is obtained by rapidly making and breaking the electric circuit. A small electro-motor effects this, and at the same time operates a pump, employed to return the mercury to the cistern from which it has issued. To regulate the height of the mercury column above the jet the mercury is made to fall from an upper cistern through a delicate stop-cock into a graduated glass tube, or the jet is connected to a cistern in which a constant level is maintained by a float, arranged to regulate the descent of the material from the upper cistern. For signalling purposes a make and break key is included in the circuit, and if a flashing light is required this key is made self-acting by clockwork or other means. When employed in lighthouses the light is surrounded with the usual glass lenses. The lamp illustrated consists of a short, horizontal cast-iron cylinder, with glass ends, supported by a tube, connected to a cistern which forms the base of the lamp. This cistern is intended to receive the condensed mercury. Tubes pass through the cylinder sides, and convey the mercury from suitably placed separate cisterns to the jets, which are directed, one vertically downwards, and the other horizontally. The former is provided with an up and down adjustment in a vertical plane, whilst the latter can be advanced and withdrawn, as well as moved across the axis of the vertical jet. "By these arrangements the streams of mercury can be made to strike each other at any required distance from the orifices of the jets." A modification of the lamp, for use when it is desired to obtain an all round illumination, has a cylindrical glass closed by metal plates at top and bottom, the supply pipes being introduced obliquely through the top one, so as not to interfere with the light.

MAGNETO-ELECTRIC MOTOR.—In connection with the electric lamp, already described, a magneto-electric motor is employed, which consists of an electro-magnet mounted on a vertical axis between the poles of a permanent magnet. The ends of the electro-magnet coils dip into two mercury cups, or are in contact with two strips of metal, which act as a commutator, and by continually reversing the current produce rotation of the electro-magnet on its axis. In the present invention this motion is used to effect a rapid make and break in the lamp circuit, and at the same time to drive the mercury pump already mentioned.

1,350.—R. S. Newall, Gateshead. Manufacture of Wire Strands for Electrical Purposes. 3d. May 13.

CONDUCTORS.—To obtain the maximum conducting area with a minimum of surface, and consequently of electrostatic capacity, wire strands are drawn through a die or rollers to "efface their irregularities."

1,412.—C. W. Harrison, Woolwich. Obtaining Light by Electricity. 4d. May 20.

ELECTRODES. — The following substances are employed:—Potassium, sodium, zinc, tin, lead, bismuth, and alloys of these metals; the oxides of potassium and sodium, chlorides of sodium and aluminium, fluorides of lead, sulphides of copper, the mixed carbonates of potash and soda, and the fused nitrate and chlorate of potash. These are all employed in a fluid state, in combination with a second electrode, either fluid or solid.

ELECTRIC LAMPS. — One of the substances, already enumerated above, is placed in a reservoir, so constructed that its contents can be heated by convenient means, in order to liquify the said substance. A stream of the melted electrode is then caused to flow through an orifice in the reservoir, the volume of the stream being proportioned to the strength of the electric current, and this again having relation to the amount of light required and the nature of the electrode used. In a suitable position, in or near the line of the stream, is placed a pencil or disc of carbon, or other substance, or a reservoir of

metal, upon which the escaping stream will flow. The fluid electrode is connected to one pole of the battery, and the carbon, or second electrode, with the other pole. Those substances which admit of being liquified by the passage of an electric current through them, are preferably formed into pencils and enclosed in a clay or glass tube having an orifice through which the point of electrode projects, so as to make contact with the second electrode. By preference, two carbon pencils are used for the negative electrode in combination with one positive stream electrode, such pencils being arranged one on each side of the stream, or the negative electrode may consist of a reservoir of fluid in conjunction with a positive stream electrode. The light is protected by a glass tube, in the ends of which are apertures "by which the heated vapours may be directed around the flowing electrode," and, if desired, a pipe may be attached to the upper end of the glass tube to carry away the vapours to be condensed in a suitable receiver. When a reservoir containing a fluid positive electrode is employed, the surface of such electrode is covered with a conical float, having a small aperture in the apex, by which evaporation of the material is confined to a small area. The negative electrode is a carbon pencil, placed in a tube which works vertically in a frame, and is "controlled by an electro-magnet and keeper, or by a helix of the conducting wire." The point of the carbon pencil makes contact with the fluid electrode, but as soon as a battery is connected to the lamp negative electrode is raised and the light produced. Consumption of the fluid electrode is compensated for by the sinking of the float, and as the carbon electrode diminishes it advances constantly toward the mouth of the float, and thus preserves a constant degree of separation between the electrodes.

*1,609.—J. H. Tuck, London. Application of Light for Facilitating Operations under Water. 3d. June 9.

ELECTRIC LIGHT.—The diver carries a lamp connected by insulated wires to a battery on shore or on board ship.

1,687.—W. B. de Blaquiere, London. Connecting the Ends of Submarine Electric Telegraph Cables. 9d. June 16.

CONDUCTORS.—To connect the ends of submarine cables a tube of metal is used with a part at each end made movable on a hinge or joint. The portion of the tube intermediate of its two ends is filled with gutta-percha, having therein a wire or wires suitable for connecting the two cables. The ends of the two parts of the cables are brought near together, and coupled to the wire in the tube, gutta-percha in a soft state being pressed around the joints. The movable part is then pressed down to grasp the cable, which is roughened to give a firmer hold. The two parts of the tube are held together by a nut. The tube may likewise be made in two parts united by nuts.

1,754.—J. S. Rousselot, Paris. Obtaining and Applying Motive Power. 9d. (13 figs.) June 23.

MAGNETO-ELECTRIC GENERATOR.—Sixteen horse-shoe magnets are fixed vertically in a circle. The north and south poles of each magnet are in the same radial line. Above the plane of the polar faces there rotates a horizontal disc carrying sixteen U-shaped electro-magnets similarly arranged to the permanent magnets, but with their poles pointing downwards. The central shaft carries a commutator to direct the reverse currents into a continuous current. The negative end of one coil and the positive end of the succeeding coil are connected to the same commutator plate all the way round, thus placing all the coils in parallel circuits.

*1,767.—J. Church, London. Manufacture of Artificial Fuel. 3d. June 24, 1881.

CARBONS.—Sifted breeze and tar are subjected to " streams of electricity " while in a retort.

1,998.—F. H. Holmes, Blackwall. Magneto-Electric and Electro-Magnetic Engines. 6d. (4 figs.) July 20.

MAGNETO-ELECTRIC GENERATOR.—The poles of the magnets are arranged in a circle and at equal distances apart, and the helices are disposed in a similar manner, their number being always a multiple of the number of poles of the magnets. A separate and distinct commutator is provided for each series of helices, that is, supposing the number of helices to be double the number of the poles there will be two commutators, and they will be so arranged that the collecting rollers will pass over the divisions in each commutator at the instant when the whole of the helices, in communication with that commutator, are at the dead point of the current, or on the point of reversal. As there are twice as many helices as poles, it follows that when one series are on the dead point, the other series will be in the condition of maximum electric force. By connecting the wires from the two commutators, either parallel or in series, a "compound current" will be the result, which has no dead point, and varies only from the maximum of current of half the number of helices to the mean of the current of the whole of the helices.

HELICES.—These are wrapped on sheet metal bobbins with removable cores that screw into one of the end flanges.

COMMUTATORS.—These are formed of two rings each with projections on one side, which engage like the claws of a claw coupling. They

are each insulated from, but connected to, the same shaft. The divisions between them are not filled with solid non-conducting material, as is usual, but constitute air spaces of sufficient width not to become choked with dust and dirt. To prevent these spaces forming hollows, into which the collecting rollers might descend, they are arranged at an angle to the axis, or they have a curved or sinuous configuration. See also Specification No. 2,628 of 1857.

*2,297. — E. Grenet and A. Vavin, Paris. Electro-Magnetic Machine. 3d. September 1.

ELECTRO-MAGNETIC MOTOR.—The machine comprises a set of upright electro-magnets, the upper faces of which are plain and pass through a table or plate of bronze, which is flush with them and perfectly level and smooth. A corresponding set of magnets with curved polar extremities rock or roll upon the table, and their motion is communicated by connecting rods to a crank shaft. The battery current is transferred from one magnet to another by a commutator consisting of plungers alternately inserted into receivers and withdrawn therefrom ; water, or other non-conducting liquid, is placed in the receivers above the mercury for the purpose of extinguishing the spark which results from the rupture of the galvanic current.

2,341.—B. Sharpe, Hanwell. Electric Telegraph Cables, &c. 10d. (11 figs.) September 8.

CONDUCTORS.—The exterior protecting wires of submarine cables are, according to this invention, laid parallel to the core and fastened into their places by helical coils of fine wire, either wound around them or arranged like the weft in tubular weaving. Instead of the longitudinal wires strips of metal may be used either bound on or soldered along the joint or joints. A metallic covering may also be produced by electro-plating.

2,449.—J. Absterdam, Massachusetts, U.S.A. Electric Telegraph Cables. 10d. (5 figs.) September 21.

CONDUCTORS. — Submarine cables are made elastic lengthwise by giving them bends, flexure or corrugations by passing them through indented rolls.

*2,467.—J. De la Haye and M. Bloom, Salford. Laying Submarine Telegraph Cables. 3d. September 23.

CONDUCTORS.—Submarine cables are rendered temporarily buoyant by calico, &c., being glued to them. On immersion the glue dissolves, liberating the calico, and the cable gradually sinks.

2,628.—F. H. Holmes, Blackwall. Magneto-Electric Machines. 11d. (6 figs.) October 14.

MAGNETO-ELECTRIC GENERATOR.—This machine

is driven direct from an engine fixed beside it at a speed of from 85 to 90 revolutions a minute. Upon the central axis are two wheels H (see figs. 49, 50 and 51) each with a brass trough-like rim in which are arranged the cores K of a series of helices L. The ends of the cores K are rounded off, and the rim J has a slightly projecting fillet

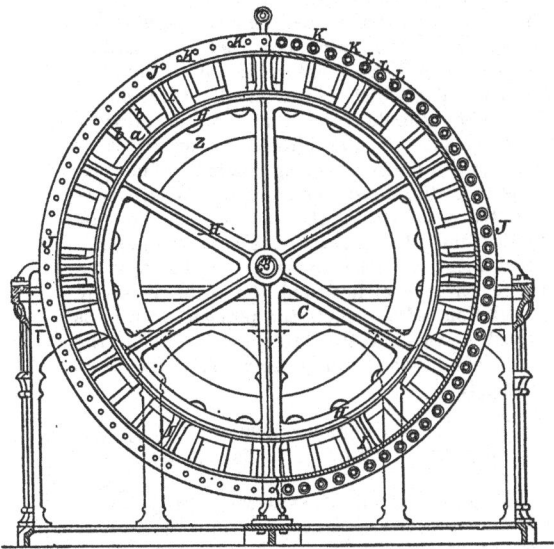

Fig. 49.

on each side to prevent accidental contact with the magnets *b b*. The ends of the helices are connected together in the usual way and the terminal wires are carried through the axis to the commutators N N, which are made in accordance with Patent No. 1,998 of 1857. There are three

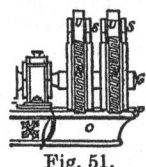

Fig. 51.

series of stationary magnets in the machine, with 20 magnets in each series. The number of helices is twice as great as the number of magnetic poles (*i.e.*, 80), and the number of commutators corresponds to the same multiple.

2,707.—J. Mackintosh, London. Constructing and Laying Telegraphic Cables. 4d. October 24.

CONDUCTORS.—The central strand of a cable is surrounded by gutta-percha, over which a bat or fleece from a carding engine is laid and firmly pressed. An outer coating of gutta-percha and iron-filings is added.

2,841.—J. T. Way, London. Obtaining Light by Electricity. 9d. (6 figs.) November 10.

ELECTRIC LAMP.—One part of the invention consists in so arranging the charcoal, or other conducting substance forming the solid electrode, that it shall be constantly renewed and presented to the stream of the flowing electrode at a uniform distance from the orifice of the jet from which the latter issues. This may be effected in two ways.

Fig. 50.

Two horizontal carbon rods may be pressed towards each other by helical springs, or by fluid pressure acting upon pistons in cylinders, and jets of mercury may be directed so as to break upon their point of junction. Instead of two rods, one rod may be pressed against a piece of lime or other infusible material, the flowing electrode being directed against the point of contact. "A surface of charcoal, or other conducting material forming one electrode may be presented at a constant distance from the orifice from which the flowing electrode issues, by causing a cylinder or rod of such conductor, upon which the flowing electrode is thrown, to move forward along its axis; the required motion being given by suitable self-acting machinery."

"A rod of charcoal, forming one of the electrodes of a lamp, may be projected by a spring, or otherwise, through a ring of iridium, so that as the end is burned a fresh portion is pushed forwards in a similar manner to that in which the candle in a Palmer's candle lamp is forced forwards as its end is consumed." *a* is an iron tube communicating at its upper end with a reservoir of mercury, and carrying at its lower end a jet of fire-clay *b* cemented into it. It is so held that its position can be adjusted in every direction. *l* is the lower electrode which is pressed constantly upwards by the spiral spring *m*. Its higher end is prevented from rising above its proper position by a cap of fire-clay m^1 carrying a ring of iridium of smaller internal diameter than the charcoal rod. This ring is shown separately in Fig. 53, and it will be seen that it is provided with four projections which are embedded in the fire-clay cap before it is burnt. An alternate method of manufacture is to embed in the clay a

Fig. 52.

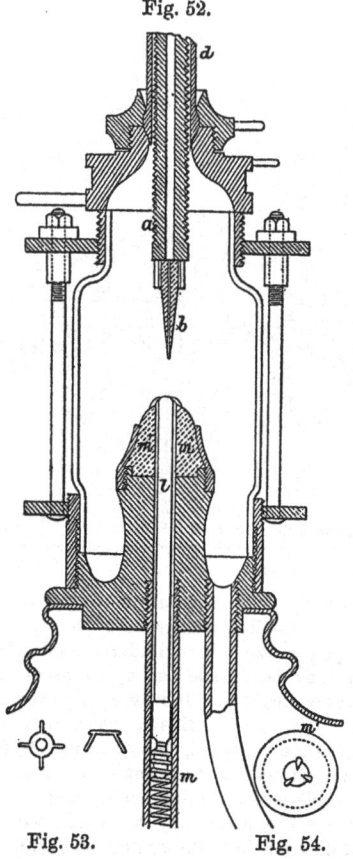

Fig. 53.   Fig. 54.

series of separate points projecting inwards as in Fig. 54. The mercury may be heated before using and the breaking point of the jet be regulated by altering the head of the fluid, which, when once set, may be made constant by an arrangement acting on the principle of the bird fountain.

2,866.—J. Mackintosh, London. Preparing Telegraphic Wire, &c. 4d. November 14.

CONDUCTORS.—The outer surface of gutta-percha covered wire is subjected to the action of sulphuric acid, and then to water, or to the action of chloride of sulphur and a solvent to render it able to withstand tropical heat.

2,868.—M. Henry, London. (*A communication.*) Electric and Galvanic Conductors, &c. 7d. (4 figs.) November 14.

CONDUCTORS.—The new conductor consists of wire insulated by a coating of gutta-percha or india rubber, and firmly enclosed in a tube of lead or ductile metal. The apparatus by which the lead sheathing is applied consists of an inverted hydraulic cylinder arranged vertically, with its ram entering a cylinder filled with fluid lead. At the bottom of the lower cylinder is a die whose diameter equals the external diameter of the lead pipe to be produced. The gutta-percha covered wire enters the die through a vertical pipe, which passes axially through the hydraulic cylinder and ram, and is provided with means for the circulation of water to prevent the dielectric being melted. This pipe ends in a die, the diameter of the covered wire, situated just within the opening of the first named die. In operation the lower cylinder is filled with fluid lead, the wire is drawn through the central tube dies, and the water is set in circulation, after which the plunger or ram is forced down and the lead pressed out in the form of a tube tightly embracing the core within it.

2,987.—E. C. Shepard, London. Magneto-Electric Machines. 1s. 5d. (5 figs.) December 2.

MAGNETO-ELECTRIC GENERATORS.—As shown in the drawings, the apparatus is composed of four permanent magnets, placed in a vertical plane with their poles symmetrically arranged in a circle. Several such circles, each comprising four magnets, may be united in one machine. Between every two circles is mounted a revolving wheel, having a series of coils of insulated wire fixed on its periphery, so that when the wheel rotates the coils pass between the "sides of the ends of the magnets" composing the circles, for the purpose of generating induced electric currents. To this end the magnets are so arranged that adjacent poles in either direction are of opposite natures. Several methods of connecting the coils to the collecting rings are described, accordingly as they are to be arranged in series of one, two or four. The commutator is of the ordinary description, with plates which receive alternate positive, and negative impulses from the collecting rings. A modified commutator "for the purpose of a continued current for the electric light" appears to consist simply of two rings, to which the alternate ends of the coils or groups of coils are connected.

3,020.—W. T. Henley, London. Ropes and Cables for Telegraphic and other Purposes, &c. 1s. 4d. (9 figs.) December 5.

CONDUCTORS.—According to this invention, submarine cables are made from strands of wire and vegetable fibre, having a central thread or core of hemp or other similar material. The strands are then laid round another and larger core into a rope. The invention includes machinery the manufacture of ropes and cables.

3,101.—E. Highton, London. Electric Telegraphs. 5d. (2 figs.) December 17.

CONDUCTORS. — To protect underground wires they are laid in cast-iron boxes, triangular in cross section and set with the apex upwards to glance off the point of a pickaxe or other tool in repairing the roads.

3,164. — B. Burleigh and F. L. Danche`l,

London. Manufacture of Articles from Plastic and other Materials. 4d. December 24.

ELECTRODES.—Relates to the manufacture of, among other things, cells, cylinders, plates and electrodes of solidified carbon. Carbonaceous matter may alone be taken, or it may be rendered resinous, gummy, oleaginous, saccharine, glutinous or plastic by the admixture of moist bituminous, or other suitable cementing medium. It is forced into suitable moulds by percussive pressure and baked or burned in a closed vessel.

1858.

18.—G. E. Dering, Lockleys, Herts. Electric Telegraphs, Insulated Cables, &c. 6d. January 5.

CONDUCTORS.—To insure the adherence of the gutta-percha, the strand is first covered with varnish, secondly wetted with a solvent, and finally heated before the application of gutta-percha.

The specification also describes a telegraph cable, the peculiarity of which consists in the combination, in one and the same cable, of two or more separate conductors formed of iron or steel wire, constituting also the chief source of strength.

252.—J. Chatterton, London. Electric Telegraph Wires, &c. 3d. February 10.

CONDUCTORS.—Wires or strands, that are about to be covered with gutta-percha, are passed through a vessel fitted with gauges and containing an insulating compound of gutta-percha, rendered fluid by solvents, and the application of heat. By this process complete adhesion between the wire and the covering is secured. The compound employed consists of gutta-percha three parts, Stockholm tar one part, and resin one part. When the core receives more than one coating of gutta-percha a layer of the fluid is added before each succeeding coat. The copper conducting wire is silver-plated to prevent oxidation. In long conductors, in which the signals are sent in one direction only, the cross section is gradually diminished " so that as the electric currents become weaker they will have suitably sized conductors."

277.—J. C. H. Sievier, London. Submarine Electric Conductors. 4d. February 13.

CONDUCTORS.—To aid the insulation of copper conductors they are coated with metals of high resistance before the gutta-percha is applied. Such metals are bismuth, tin, iron, lead, brass, antimony, zinc, nickel or German silver, or alloys of those metals. They may be used singly, or one over the other, the outer one being by preference lead. The same metals may also be

applied alternately with gutta-percha in constructing a cable, and a sheathing of lead be placed over the fibrous materials to preserve the whole.

*282.—E. Hunt, London. Voltaic Batteries, &c. 3d. February 15.

ELECTRODES.—Carbon points for the electric light are manufactured from the residuum from the distillation of tar or pitch, which, after being reduced to an impalpable powder and mixed with tar or other hydro-carbon, is moulded to the required form. The pencils so produced are alternately raised to a red heat and immersed in hydro-carbon until the required density is obtained.

293.—H. Wilde, Manchester. Connecting the Ends of Lightning Conductors and Submarine Telegraph Cables. 9d. (7 figs.) February 16.

CONDUCTORS.—In connecting the ends of submarine cables a small metal cylinder is put over the end of each of the cables, the holes in the cylinders being the same diameter as the outside of the cable. The outside wires of each cable are then bent over the outside of each cylinder, which is grooved for their reception. Two other metal cylinders are then driven over the wires and smaller cylinders to prevent the wires being drawn out by the weight of the cable. The conductors are joined, and the space formed by the difference of the diameters of the cable and the outsides of the cylinders, filled with gutta-percha; the whole of the coupling is then enclosed in a metal tube, screwed internally at each end, and a cap, slipped over the end of each cable before the joint is made, is then screwed into the tube. The outsides of these caps are made conical, and screwed to about three-quarters of their length, two longitudinal slots being cut through each cap as far as they are screwed. A nut is also screwed over each cap to press it to the cable.

333.—F. M. Baudouin, Paris. Electric Telegraph Cables. 5d. February 20.

CONDUCTORS.—The essential feature of this invention is that the same wire serves for the passage of the current, and to resist the tensile and other strains on the cable. It is made in the form of a twisted strand and is insulated in the usual way. Above the gutta-percha a covering of fibrous material is placed to defend it against the effects of friction, and, when the cable is to be laid in shallow water, a sheathing of iron wires may be added, or, under special circumstances, alternate coverings of hemp and wire may be used until sufficient strength is attained. Multiple conductors are made by uniting several, as above described, in the same common envelope, or by interposing insulated copper wires in the strands of the core of the cable. Aluminium and its alloys may be used as the core of the cable, and the said core may be covered like the bass strings of a musical instrument, so that if the central wire should break there will still be some metallic contact. It is proposed to add the sheathing to the core as it is delivered from the ship to the sea.

393.—M. Henry, London. *Grenet and Vavin, Paris.* Electro-Magnetic Motors. 10d. (5 figs.) February 27.

ELECTRO-MAGNETIC MOTOR. — Electro-magnets with straight or curved polar faces roll on similar magnets.

COMMUTATOR.—The currents are distributed by a metal rod free to move in a vessel containing mercury and a bad liquid conductor. The current passes when the rod is in the mercury, and the spark is suppressed when the current is broken by the entry of the rod into the bad conductor. A "counter current" is employed to destroy the residual magnetism of the electro-magnets "at the exact moment at which the main current ceases to magnetise."

444.—J. N. Hearder, Plymouth. Submarine Telegraph Cables. 4d. March 5.

CONDUCTORS.—To lessen the statical charge which accumulates at the two sides of the insulating sheath, after the manner of a Leyden jar, the inventor first covers the conductor with cotton, silk, wool, hair, flax, or other fibrous material, in one or more layers, and then applies the insulating compound. Or he begins with the insulating material and adds alternate coats of it and the fibrous material. According to a third method the fibre lies next the wire and is followed by alternate coverings of fibre and gutta-percha. When two or more conductors are embodied in one cable the same plan is followed in the construction of their common envelope.

*486.—J. F. Gee, Wrexham, Earthenware

Pipes or Tubes Suitable for Protecting Underground Conductors. 8d. March 10, 1881.

LAYING CONDUCTORS.—The invention consists in the use of lock, semi-circular, rabited joints in connecting pipes or tubes. The pipes may be made in circular or semi-circular portions and the wires may be placed in eyelet holes or guiders.

782.—W. Rowett, Liverpool. Electric Telegraph Cables or Ropes. 4d. April 10, 1881.

CONDUCTORS.—Submarine cables, according to this invention, have a "semi-floating quality" imparted to them by being largely constructed of Indian-grass fibre, New Zealand hemp, Manilla hemp, cotton wool, coir fibre, coco-nut fibre and similar materials. They are preserved from decay by being steeped in a solution of some or all of the following materials: turpentine, resin, paint, tar, oil, naptha, water, copperas, bitumen, alum, bichloride of mercury, gambier, india-rubber, shellac, copper soap and brimstone.

883.—J. Chatterton, London. Combining and Coating Metal Conductors for Electric Telegraphs. 3d. April 22.

CONDUCTORS.—When combining two or more insulated conductors that have been coated after the manner described in Specification No. 252 of 1858, or otherwise, they are laid together and passed through a vessel containing heated insulating compound, and placed between the laying machine and the gutta-percha covering machine. When the conductors are laid round a central core it is passed through insulating compound before it enters the laying machine. In like manner an insulated wire conductor may be laid round with a series of strands of fibrous material by laying machinery, and then be coated with gutta-percha, or a compound containing gutta-percha, at one operation.

1,090. — J. Macintosh, London. Insulating Telegraph Wires. 4d. May 14, 1881.

INSULATING CONDUCTORS.—A yarn, made from gun-cotton, is wrapped round the wire which is then subjected to the action of a fluid, that acts as a solvent to the gun cotton so as to cause it to adhere to the wire. Sometimes the gun cotton is first dissolved and india-rubber and other substances added to it. Over this coating, or in place of it, a layer of india-rubber or india-rubber compound is applied, and is afterwards treated with sulphuric acid or with chloride of sulphur. The rubber is applied in a plastic state by grooved rollers which draw it from a hopper, and feed it forward in a cylinder compressed round the wire. Several successive coats may be thus laid on the conductor.

1,099.—C. W. Harrison, Woolwich. Obtaining Light by Electricity. 10d. (6 figs.) May 17.

This invention relates to the use of mercury or other fluid or semi-fluid body, as one or more of the electrodes in obtaining light by electricity, and comprises several forms of lamps. In the first the electrodes are two streams of mercury. The upper one falls from a small nozzle and breaks into drops just as it reaches a vertical tube of lamp-black and silica up which mercury constantly rises and overflows. The arrangement, or a number of them, may be enclosed in a lantern similar to that described in Specification No. 558 of 1857. The flow of fluid is regulated by a tap controlled by an electro-magnet. In another lamp the waste or escaped mercury is collected in a vessel, and serves to raise a float upon which a carbon electrode is carried to receive the flowing electrode. By this arrangement the carbon rod is raised at the same speed as the head of mercury decreases. The reservoirs for holding the fluid electrodes are formed of lamp black and other carbon and silica or china clay, rollerstone, &c. The condensation of vapours on the glass is prevented by means of a stream of water flowing over the glass, or by partly or wholly filling the glass or case with water, alcohol, bi-sulphide of carbon or other liquid, and in causing it to circulate around the light or to pass away through the apparatus by means of a pump. The rise of vapours from the waste fluid electrodes is prevented by a stratum of water in the waste receiver. It is stated that the best results are obtained by connecting the upper electrode to the negative battery pole.

1,241.—C. Wheatstone, Hammersmith. Electro-Magnetic Telegraphs, &c. 4s. 6d. (41 figs.) June 2.

ELECTRIC MOTOR.—Two magnetic needles, or bars with their poles in opposite directions, are placed near each other on opposite sides of, and parallel to, the axis round which they move ; two straight electro-magnets are placed one on each side of the axis of motion of the needles and parallel thereto. When a current is transmitted through the coils of these electro-magnets in such manner that the adjacent poles of each shall have at the same time opposite polarities, the dissimilar poles of the two needles are simultaneously acted upon, so that each of their poles is attracted by one pole of one electro-magnet, and repelled by one pole of the other electro-magnet, the attracting and repelling actions all conspiring to move the axis in the same direction. Instead of being straight, the magnetic needles or bars may be curved inwards towards each other, so that their weight shall be thrown nearer their axis of motion. By this construction the force of the electro-magnets is applied near the axis of motion of the magnetic bars, instead of at

a considerable distance from it, as in the case when electro-magnets are applied to the poles of a straight magnetised bar moving on an axis to which it is perpendicular. This motor, which is intended to be actuated by alternating currents, admits of various modifications: (1) The magnetic needles or bars may be inclined to each other at a small angle crossing the axis of motion, their opposite poles being at opposite sides of the axis. (2) A single magnetic needle, balanced or unbalanced, may be employed in conjunction with two electro-magnets, or a single electro-magnet may be used in conjunction with two magnetic needles, the electro-magnet being placed at one side of the needles only. (3) Instead of the electro-magnets being parallel to the magnetic needles or bars, they may be curved, so that their two poles shall act, one attracting, the other repelling, on the poles of the magnetic needles between them. Two curved electro-magnets arranged in this manner, each to act on one of the points of each of the magnetic needles, will operate as efficiently as straight electro-magnets applied in the manner described above. (4) Several such pairs of balanced magnetic needles may be fixed on the same axis, the poles of those adjacent being of opposite kinds, so that the planes in which they lie shall be at equal angular distances from each other. As many straight electro-magnetic bars as there are needles are placed parallel to the axis, and in the angles between them, and the wires are so disposed that when the current in either direction passes through the coils, the joint attracting and repelling actions drive the needles to move round the axis in the same directions. In such an arrangement there must be an even number of magnetic needles and electro-magnets.

MAGNETO-ELECTRIC GENERATOR.—According to the first modification, instead of the armature revolving before the poles of a magnet, it is caused alternately to approach and recede from them by means of a cam fixed to a wheel or by other suitable means. If a single magnet is employed a contrivance for alternately inverting the direction of the currents must be had recourse to for telegraphic purposes, but this is unnecessary when the magnet oscillates between the poles of two similar magnets so disposed that the currents shall pass through the circuit in opposite directions as the armature moves towards one or other magnet. The drawings show two horizontal magnets lying, with their similar poles opposed to each other, on either side of an armature suspended from a vertical pendulum which is caused to oscillate by a wheel with a waved cam path cut in its face.

According to a second modification neither the magnet nor the coil of wire is put in motion, but only the soft iron core which is enveloped by the coil. A coil of insulated wire is fixed, and within it is fitted a cylindrical core of soft iron capable of rotating on its axis without carrying the coil with it. The ends of this soft iron core have lateral projections also of soft iron; these are of equal length, and at equal angular distances from each other, and extend over the ends of the coil, so as to be brought into close proximity to the poles of a magnet, situated so that one of its poles is near one end of coil and its other pole near the other end of the coil. On causing the core to rotate once, as many double currents are produced in the coil as there are prolongations on each end of the soft iron cylinder. Instead of a single magnet several may be arranged with their poles in two parallel circles and placed at distances from each other corresponding with the distances of the prolongations; by this means the strength of the currents is considerably augmented. If the poles of the magnets contained in the same circle are similar, the successive currents are alternate in direction and unequal in intensity; but if the poles are alternately dissimilar two currents in one direction are followed by two currents in the opposite direction available for the purpose required; in the latter case the number of the magnets must be even and the angular distances of the prolongations must be double the angular distances of the magnetic poles, so that they shall not be at the same time opposite adjacent magnets. Instead of the stationary coil being wound concentrically with the central axis a separate coil may be placed on each pole of the magnets.

A third modification consists in combining several magneto-electric generators, and gearing them together in such way that the armatures have always the same relation to the magnetic fields, and are all traversed by the same current. Electro-magnets may be substituted for permanent magnets in any of the constructions described above.

1,268.—C. Hancock, London. Manufacture of Electric Telegraph Cables. 3d. June 5.

CONDUCTORS.—The invention consists in covering submarine cables with hair, such as horsehair.

*1,379.—R. S. Newall, Gateshead. Cords, Ropes or Cables. 3d. June 18.

CONDUCTORS.—To form a submarine cable, yarns are laid into a strand in such a manner as to retain the twist given to them in the process of spinning; then a number of such strands are laid round a core without altering their twist.

*1,385.—J. Bradshaw, Bolton-le-Moors. Apparatus for Obtaining and Applying Motive Power. 3d. June 19.

MAGNETO-ELECTRIC MOTOR.—Upon a vertical shaft a wheel is secured in a horizontal position,

such wheel being provided with a number of permanent magnets placed at intervals, side by side, and radiating from the centre of the wheel, and "the two poles of each magnet being placed alternate to that next in succession throughout the entire circle." Around the periphery, and slightly above the circle of permanent magnets, are arranged a number of electro-magnets, their cores being connected in pairs, and their coils joined up in series "forming a complete electro-magnetic circuit." The poles of the electro-magnets "project over the outer ends of the radial permanent magnets, and according to the alternating direction of the electric current, or the polarisation of their extremities, both repel and attract the poles of the permanent magnets alternately," and thus give rotation to the wheel and its shaft. The direction of the electric current is reversed by a suitable commutator. The power of the motor "may be greatly increased by fixing the permanent magnets in a vertical position upon the wheel," "and employing two circles of electro-magnetic coils placed upon platforms above and below, so as to act upon both ends of the magnets simultaneously; the magnets may also be further employed in rows or tiers upon the same shaft to increase the driving power."

*1,483.—C. F. Vasserot, London. *L. B. S. Charrier de Sainneville, La Freta, Lyons.* Electric Conductor. 3d. July 2.

CONDUCTOR.—A fibrous cord is surrounded by a wire of conducting metal.

1,491.—J. Latimer Clark, London. Electric Telegraph Cables or Ropes. 6d. (12 figs.) July 2.

CONDUCTORS.—In order that one insulated wire may be distinguished from another, various numbers of grooves or ribs are formed longitudinally on the coatings.

1,605.—C. De Bergue, London. Submarine Electric Telegraph Cables, &c. 7d. (7 figs.) July 16.

CONDUCTORS.—To make a cable that shall be only slightly heavier than water, and shall have no tendency to kink, a gutta-percha covered core is surrounded with fibrous or metal cords running longitudinally, and is then served with two layers of twine. The invention includes a machine for making such a cable.

1,687.—P. A. Godefroy, London. Cleansing Gutta-percha, &c. 10d. (9 figs.) July 26.

INSULATING MATERIAL.—Gutta-percha is first cut into slices and immersed in hot water, after which it is compressed by hydraulic pressure into a trough. When thoroughly cold the mass is granulated by a rasping machine, and the particles agitated in water to separate the dirt, after which it is ready for the masticator and subsequent operations.

CONDUCTORS.—The wire is warmed before it enters the covering machine.

*1,708.—W. Buckingham, C. Humfrey, and L. R. Sykes, London. Telegraph Cables. 3d. July 28.

CONDUCTORS.—The ropes which enclose the wire in submarine cables are made of fibrous materials, saturated with gutta-percha mixed with powdered glass.

1,740.—C. De Bergue, London. Submarine Telegraphic Cables, &c. 7d. (9 figs.) July 31.

CONDUCTORS.—This invention is for improvements on Patent No. 1,605 of 1858, and consists, firstly, in making cables of the kind described in the former specification with one layer of serving instead of two. Secondly, the longitudinal cords are made of two or more strands twisted together, so that the projecting surfaces become embedded in the gutta-percha. Further, the invention comprehends cementing the longitudinal cords together, and to each other, by Jeffrey's marine glue, saturating and coating them with gutta-percha; also saturating the binding twine with gutta-percha, and consolidating it with the longitudinal cords with heat or otherwise; also creosoting the materials, or some of them, employed in the manufacture of electric cables.

*1,742.—W. H. Crispin, London. Electric Telegraph Cables. 3d. August. 2.

CONDUCTORS.—The use of metallic sheathing is abandoned, and the conducting wire is covered with fibrous material steeped in gutta-percha, and surrounded with hemp strands or plaiting.

1,752.—H. Greaves, Westminster. Construction of Streets, Roads and Ways, &c. 10d. (5 figs.) August 2.

LAYING CONDUCTORS. — In combination with overhead street railways and also with cast-iron footways the inventor makes channels for the reception of electric conductors.

*1,773.—C. M. Archer, London. Electric Telegraph Cables and Wires. 3d. August 4.

CONDUCTORS. — The wire sheathing is wound round the inner part with a very short lay like a helical spring.

1,797.—J. Walker, London. Manufacture of Electric Telegraph Cables. 6d. (5 figs.) August 7.

CONDUCTORS. — To construct a telegraph cable according to this invention three copper conductors are separately covered with sheet caoutchouc, having powdered glass on the surface next the wire. One or more sheets of caoutchouc,

either alone or stiffened with pulverised charcoal or fine glass, are added, and then the three wires are included in one gutta-percha envelope. A tube of wire is formed round the gutta-percha and if necessary helical wire sheathing is applied above this to resist longitudinal extension. A tube of india-rubber or gutta-percha may be formed over them all for a protection.

1,811. — Willoughby Smith, London. Compound for Insulating Electric Telegraph Wires, &c. 3d. August 9.

INSULATING COMPOUND. — Gutta-percha, three parts; Stockholm tar, one part; and resin, one part.

*1,828.—J. G. Appold, London. Manufacture of Wire Ropes or Cables. 3d. August 11.

CONDUCTORS.—Two sets of sheathing wires are applied, twisted and laid in opposite directions, or the wires are braided on to the cable.

1,848.—C. L. Light, London. Electric Telegraph Ropes or Cables. 5d. (1 fig.) August 13.

CONDUCTORS.—The cable is composed of the following parts, beginning from the centre: (1) a gutta-percha cord; (2) a copper wire wound helically on the gutta-percha; (3) a layer of insulating material; (4) a casing of tarred hemp; (5) a layer of wire covered with tarred hemp; (6) a layer of wire laid in the opposite direction to the former.

*1,889.—M. F. J. Delfosse, London. Electro-Magnetic Machines. 3d. August 19.

MAGNETO-ELECTRIC GENERATOR.—The object of the invention is improvements: (1) "in the apportioning of the number of coils composing each disc or wheel to the magnets between which they pass;" (2) in the "conducting rings;" (3) in the employment of "unsplit annular rings;" (4) in the means of coupling the coils; (5) in the use of split coils for the purpose of increasing the energy of the currents. No details are given.

1,924.—J. Macintosh, London. Insulating Telegraphic Wires or Conductors, &c. 4d. August 24.

CONDUCTORS.—The insulating compounds dealt with under this invention are india-rubber and shellac, and gutta-percha and shellac, and also gutta-percha and fibre. A machine consisting of grooved and beaded rollers is described for applying these compounds to conductors. A second machine for applying yarns and gutta-percha simultaneously to a wire is also included in the specification.

1,965.—J. Latimer Clark, F. Braithwaite and G. E. Preece, London. Telegraph Cables. 6d. (3 figs.) August 30.

CONDUCTORS.—Refers to Patent No. 2,956 of 1853, and relates, firstly, to a more effectual method of applying the coating there described. Insulated wires are wrapped with a coating of fibrous material, which has been highly dessiccated, and then saturated with a hot mixture of turpentine, resin, pitch, tar, asphalte, or similar substances, such saturation being completed by passing the materials through or placing them in air-tight vessels, from which air is exhausted before the admission of the non-conducting mixture, and into which the air or the mixture can afterwards be injected under pressure. The insulated wires are surrounded with this perfectly saturated mixture of fibrous and non-conducting material, which may either be laid on in a mass, or preferably in the form of strands coiled around the wires either with or without an admixture of iron wires. The cable thus constructed is again saturated with the mixture by placing it in or passing it through an air-tight vessel, and submitting it as before to the influence of a vacuum and subsequent pressure. The wires thus become surrounded with a perfectly non-inductive material, and are at the same time preserved from decay and injury. Secondly, cables are preserved from decay by surrounding the iron wires with a compound of pitch, tar, asphalte, or other material retained upon and around the cable by strands or flat bands of fibrous material saturated with preservative substance, and coiled around the cable so as to envelop it completely, or to form a worming in the interstices between the wires. Thirdly, cables are preserved by intermingling zinc with the wires, to keep them in an electro-negative condition. Fourthly, iron and other conducting wires are joined by right and left hand threads and couplings, or by placing the two ends in a ferrule, and flattening the points by pressure or hammering. Fifthly, the interior conductor of cables is made in segments, which, when laid together, form a circle in cross section.

1,994.—J. Bleakley, Accrington. Apparatus for Communicating Railway Signals. 7d. (2 figs.) September. 2.

CONDUCTORS.—The wires of adjoining carriages are connected together by means of a small hook at each end, or by twisting the ends around each other.

2,137.—A. F. Jaloureau, Paris. Manufacture of Pipes for Water, Gas, &c. 7d. (18 figs.) September 23.

LAYING CONDUCTORS.—This specification describes the manufacture of pipes by rolling up in a cylindrical form thin sheets of paper, cloth, wire gauze, metal or other suitable material into pipes, and cementing them by bituminous or caoutchouc mastic. The process may be applied to rendering pipes of glass, wood, &c., impermeable; pipes may be jointed by means of cemented metal collars,

and they may have branch pipes attached to holes bored therein by means of india-rubber collars.

2,180.—C. W. Siemens, London. (*Partly E. W. Siemens, Berlin.*) Electric Telegraphs. 3d. September 30.

CONDUCTORS.—A machine, consisting of beaded and grooved rollers for laying semi-tubes of insulating material on to wire, forms part of this invention.

*2,188.—J. W. Wilkins, and J. B. Dunn, London. Construction of Electric Telegraph Cables. 3d. October 1.

CONDUCTORS.—The interior of an electric telegraph cable is made of a helix constructed from a band of copper. The helix is formed by binding the strip on a cylindrical mandrel. On this helix a coating of gutta-percha is applied, so as to produce a cylinder on which are placed a series of narrow bands of copper with their edges apart. The bands are wound on the cylinder with a very long lay. Over them gutta-percha is applied, and above it more spirals if desired. A serving of hemp completes the cable.

*2,189.—Sir E. Belcher, London. Manufacture of Telegraphic Cables. 3d. Oct. 1.

CONDUCTORS.—The interior conductor of a cable is first covered with beads of glass, porcelain, bitumen, resin or other non-conducting material, and is then passed through a fluid cement of resin, over which a tube of gutta-percha is formed.

2,192.—J. Rogers, London. Submarine Electric Telegraph Cables. 5d. (4 figs.) October. 2.

CONDUCTORS.—Around the wire, either first covered with gutta-percha or not, there is laid longitudinally a number of plaited bands of fibre; the bands are secured by binding them round with twine, and the whole is covered with plaited yarns. Before their application, the bands and yarns are saturated with a mixture of resin, tallow, linseed oil and patent driers. Several conductors may be included in a cable formed according to this invention.

*2,208.—C. E. Oldershaw, Aldershot. Electric Telegraph Cables. 3d. October. 5.

CONDUCTORS.—The conducting wire is insulated with gutta-percha, or other suitable dielectric; round this a layer of buoyant substance, such as cork, is applied, and over the whole two wrappings of fibrous material, laid in opposite directions.

2,225.—C. Baylis, London. Underground Chambers for Gas and Water Pipes and Telegraph Wires. 10d. (3 figs.) October 6.

LAYING CONDUCTORS.—The object is effected by placing the pipes and wires in underground chambers, composed of lengths of cast iron, or other tubing joined together at their ends, so as to form a continuous chamber or gallery in which openings are made at the top, either continuously or at intervals. These openings may be fitted with sliding plates or shutters, made with tongues and grooves to fit into one another. In some cases two parallel chambers may be constructed, and the gas and water pipes placed in one, while the telegraph conductors are laid in the other.

*2,239.—R. Searle, Woodford Wells, Essex. Insulating, Preserving, and Laying Submarine and other Telegraphic Wires or Cables. 3d. October 8.

CONDUCTORS.—The invention consists, firstly, in the application of vulcanised india-rubber, wood, cork, or cork wood, for the purpose of insulating and protecting submarine and other conductors. Secondly, to the application of the above materials for reducing the specific gravity of cables. Thirdly, to the application of tubes of metallic wire to the protection of electric conductors.

*2,245. — J. T. Smith, London. Electric Cables. 3d. October 8.

CONDUCTORS.—A corrugated metallic tube acts as the conductor for the current, "such corrugation being either in unconnected indentations directly transverse to the axis of the tube, or in helical convolutions of one or more helices. The metallic tube, which may be of copper or of any suitable metal or alloy, is covered with gutta-percha or other suitable insulating material, and may be shielded with strands of hemp," or in other well-known way. In some cases a wire or wires may be included within the tube.

*2,250.—J. Tatlock, Chester. Electric Telegraphs. 3d. October 9.

CONDUCTORS.—"To obviate the influence produced by the action of the induced current," two insulated wires, one of which acts as the return wire, are included in one gutta-percha envelope.

2,251.—L. Hope, London. (*A communication.*) Electric Telegraph Cables. 3d. October 9.

CONDUCTORS.—The conductors are first coated with india-rubber, or gutta-percha mixed with sulphur, and are then vulcanised. Afterwards they receive other coatings of gutta-percha and are finished in the usual way. The india-rubber or gutta-percha is combined with the sulphur in a masticator and applied by means of an ordinary covering machine. To prevent the action of the sulphur on the copper wire the latter may be tinned.

2,270.—L. Wray, London. Compounds for Insulating Electric Telegraph Wires. 4d. October 12.

INSULATING COMPOUNDS.—The compounds are made from the following three substances variously combined. First, vegetable caoutchouc, comprising all the elastic gums, mineral caoutchouc (bitumen) or gutta-percha; second, powdered flint, flour of glass, or other suitable silicious or aluminous matter, such as kaolin; third, lac, or shellac, or other resinous substance. The compound having been prepared by masticating, grinding and mixing, is rolled or cut into thin sheets, and the conductor is coated by wrapping slips of it helically around it until it attains the desired thickness. Each coating is consolidated by being passed between grooved rollers or through dies. Or the wire is covered with thin sheets of the compound and passed over a steam table or through grooved rollers. Or the compound is laid on in a plastic state by a covering machine of the usual construction. When found desirable, the wire is coated with a suitable gummy, gum-elastic, resinous, gum-resinous, bituminous or silicious solution. Wires and also hemp strands may be rendered waterproof by an adhesive product from the tree known as *Diospyros embrooptens.*

2,271.—T. C. Shaw, and F. H. Cooper, Hanley. Obtaining Motive Power. 6d. (2 figs.) October 12.

ELECTRIC GENERATOR.—In combination with an engine driven by condensed air or other gas, an electric generator is employed to produce discharges within such air or gas, in order to increase its expansive force. The apparatus consists of a hydro-electric generator, used to excite the primary helix of an induction coil, from which the required discharges are obtained.

2,321.—C. West, London. Insulating and Covering Wire. 4d. October 18.

COVERING CONDUCTORS.—Ribbons of india-rubber are wound helically on the conductor in such manner that each round overlaps about one-half of the previous round. Under the reel, from which the india-rubber is paid out, is a flannel-covered roller running in "mineral ether," or other solvent, for the purpose of moistening the under side of the strip, so as to cause it to adhere readily. Another process consists in submitting the covered conductor to the action of moist heat by plunging it into hot water. For the purpose of preserving one or more conductors when placed in the interior of a cable, a covering of hemp is used, saturated with a composition consisting of some one or more of the metallic carbonates or oxides mixed with some of the fixed oils, either alone or in conjunction with a volatile oil, or saturated with a solution of Peacock and Buchan's composition for ships' bottoms. If the cable is to be constructed with a plurality of wires the conductors are laid together, and surrounded with yarns of hemp saturated with one of the above compositions, and the whole is bound together by a plaiting of yarns. The outside yarns may have a coating of shellac.

2,326.—A. W. Drayson, Plumstead, and C. R. Binney, Woolwich. Submarine Telegraphic Cables. 6d. (7 figs.) October 18.

Conductors.—The wire is enclosed in a tube of india-rubber or gutta-percha much larger than itself. To provide additional elasticity the conductor may be wound helically or corrugated in parts.

2,363.—R. Waller, London. Obtaining Motive Power. 7d. (5 figs.) October 22.

Electro-Magnetic Motor.—In one form of this motor a cylindrical armature, or keeper, revolves within a circle, formed by the poles of a series of electro-magnets fixed radially to the inner surface of a ring. The armature is somewhat less in diameter than the circle of magnet poles, and is placed eccentric thereto. The magnets are excited in succession by a suitable commutator actuated by mitre gear from a shaft which, in turn, receives motion by means of a series of links and levers from the armature. In another arrangement two series of electro-magnets are fixed to two slightly conical plates, and between them is a soft iron disc which acts as a keeper; when the magnets are successively excited the disc oscillates upon a ball and socket joint, and gives motion to the shaft to which it is attached. In a further modification the magnets are arranged in the form of a segment or quadrant, so as to produce a reciprocating motion.

2,368.—E. C. Shepard, Westminster. (*A communication.*) Electric Lamps. 8d. (2 figs.) October 23.

Incandescence-Arc Lamp.—The " improvements consist in certain means of controlling the positions of the electrodes by which they are kept properly in contact with each other, as they are reduced by oxidation and by the disengagement of particles, when the current of electricity used is of large volume and low intensity, or preserving a proper and uniform distance of separation when the current used is of high intensity." In the accompanying drawing, A A¹ represent reservoirs containing water to prevent undue heating of the electrodes. The upper reservoir has a vertical opening through its centre to receive the sliding tube *b*, the lower end of which holds the upper electrode *c*. The tube is filled with mercury to fulfil the double purpose of forming a connection with the conducting wire *d*, and of weighting the carbon so as to keep its point within a seat in a metallic bridge plate C. This seat consists of an orifice, or circular opening, of a size a trifle less than the body of the carbon pencil. It will allow the point of the electrode to protrude, but will not permit it to pass through, except so fast as it is

reduced. In the lower reservoir there is secured an iron tube *f*, bent into the shape of an inverted syphon. It is partly filled with mercury which

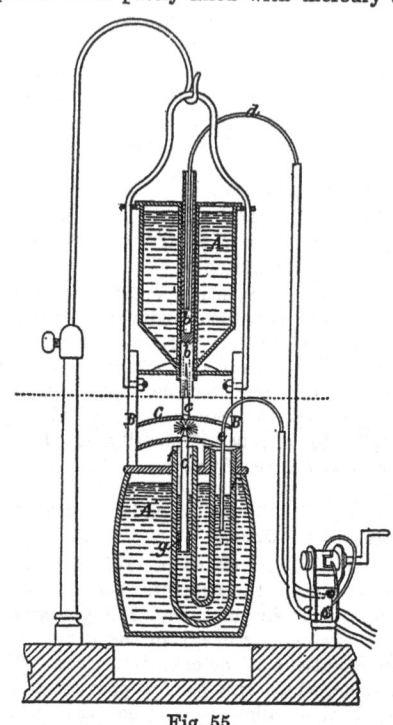

Fig. 55.

buoys up the carbon, and presses it gently and steadily against the point of the upper carbon. The lower bridge merely acts as a guide, and allows the pencil to pass freely through it. A pole changer constantly reverses the current in order to keep the electrodes pointed. When the light is fed with alternating currents this is not required.

*2,394.—L. Wray, London. Substitute for Gutta-Percha, Caoutchouc, &c. 3d. October 27.

Insulating Material.—The material is the juice of the " susu poko " or " milk tree," and is obtained in the islands of the Malayan and Eastern Archipelago.

2,411.—W. Hall and A. Wells, Erith. Electric Telegraph Cables. 1s. 10d. (9 figs.) October 28.

Conductors.—The insulating process is commenced by covering the wire with cotton yarn; over this strips of india-rubber are wound, cut from the bottle without mastication or other treatment, and above all, a helical thread of vulcanised rubber covered with cotton. The three operations are effected simultaneously by one machine, and sometimes a fourth layer of helical strips is added at the same time, Insulated wires may be protected by a covering of gutta-percha surrounded with hemp, braided or laid on helically. When gutta-percha is not used, a layer of hemp saturated in tar is substituted for it. An outer covering of wires is laid on by a hollow tube braiding machine, whereby one set of wires is laid longitudinally without any twist, while another set forms the braiding and encloses the longitudinal wires. The specification includes descriptions of a helical covering machine and a hollow tube braiding machine.

2,419.—G. Zanni, Barnet. Arranging Magneto-Electric Machines for Medical and Other Purposes. 9d. (3 figs.) October 29.

MAGNETO-ELECTRIC GENERATOR.—The invention consists in combining magneto-electric machines with maintaining springs or weights and wheel-work. When the springs are liberated the armature of the machine rotates, and so produces a succession of currents.

2,434.—E. Maynard, Brooklyn, U.S.A. Submarine Telegraph Cables. 6d. (4 figs.) November 1.

CONDUCTORS.—The invention consists in the use of parallel cords of flax, hemp, or other non-conducting fibrous material in connection with one or more twisted metallic conductors, whereby the strain in laying the cable is taken by the parallel cords and the wire is free from tension. To accomplish this object a conductor, consisting of one or more wires, is wound around a strong cord of fibrous material. It is then surrounded with parallel strands which are bound to it by a serving. Two or more surrounding layers may be thus formed, and, if a second conductor is to be used, it is wound around the outside of the aforesaid layers of non-conducting fibrous material and then similar outer layers are applied outside these fibrous layers. Iron or other wires may be applied either parallel or twisted. A saturation of wax is also to be used.

2,601.—Sir C. T. Bright, Harrow. Insulators, &c. 6d. (9 figs.) November 18.

INSULATORS.—This consists in combining with the insulator a protecting cover of papier-maché, wood, leather, iron, or other suitable substance, and also in attaching insulators to their stems by means of an elastic medium, such, for instance, as a compound of gutta-percha and pitch; or the stem may be served with tarred yarn upon which the insulator is screwed. The protecting covers may be adapted to the shackle insulators described in Specification No. 14,331 (Old Law).

2,616.—W. Hancock, London. Manufacture of Electric Telegraph Wires and Cables. 3d. November 19.

CONDUCTORS.—This invention is described under three heads, viz. :—1st. The use of an insulating material composed of gutta-percha, india-rubber, or both, combined with shellac, resin, or Venice turpentine. 2nd. "Preparing" the conducting wire by passing it through a cement composed of bitumen and Venice turpentine, with or without the addition of a resinous substance. It is preferred to apply this cement "simultaneously with the laying on of the coating of gutta-percha." 3rd. Applying the above cement to the insulated core prior to serving it with other coverings.

2,626.—J. H. Johnson, London. (*H. J. Rogers, Baltimore, U.S.A.*) Electric Telegraph Cables or Conductors. 5d. (2 figs.) November 19.

CONDUCTORS.—The insulating core is covered with a braiding of fibrous material, and iron sheathing wires are dispensed with.

*2,670.—J. H. Johnson, London. (*A communication.*) Electromotive Apparatus. 3d. November 24.

MAGNETO-ELECTRIC MOTOR:— "This invention relates to a peculiar construction and arrangement of machinery or apparatus for obtaining motive power by the aid of electricity, and consists in the employment of a number of permanent magnets, so disposed that no two similar poles shall face each other. Between these magnets are fitted, to play freely, a number of soft iron electro-magnets of the ordinary well-known construction, which will alternately be attracted and repulsed by the permanent magnets on reversing their polarity. The electro-magnets may be made to transmit their motions by any convenient arrangement of mechanism, when a simple electro-magnetic motion will be obtained."

"In place of having movable electro-magnets and fixed permanent magnets, the former may be stationary, and the latter movable, and the electro-magnets may be employed only for polarising the permanent magnets."

"*It is also proposed to employ the electro-magnet in obtaining induced electricity, which supplies wholly or partially the electricity necessary for polarising the electro-magnets, which electricity would otherwise be required to be obtained from batteries or other known sources.*"

2,714.—C. Hancock, London. Manufacture of Electric Telegraph Wires and Cables. 4d. November 29.

CONDUCTORS.—Two strips of sheet india-rubber, having their inner surfaces coated with india-rubber solution, are made to enclose the conducting wire, their solutioned surfaces being pressed together by grooved rollers or other means. Several coverings may be applied in this way, in succession, care being taken to "break the joints."

Another mode of insulating conductors consists in applying to them a helical wrapping of cloth (preferably woollen), which has been waterproofed with india-rubber. When two or more such coverings are used, it is preferred to wind them successively in opposite directions, and to employ the strips doubled.

Several insulated wires may be combined to form a cable according to either of the above plans.

* 2,756.—J. Rogers, London. Manufacture of Ropes, Cables, Cords and Lines. 3d. December 2.

CONDUCTORS.—Fibrous yarns are laid longitudinally around the insulated conductor or conductors, and bound together with a serving of wire, thread or other suitable material. The covering is then saturated with a composition consisting of "boiled oil, resin, patent dryers, tallow and pitch, tar or dissolved india-rubber, or some of them."

2,773.—L. W. Fletcher, Bleaklow, Lancashire. Electric Telegraph Cables. 7d. (6 figs.) December 4.

CONDUCTORS.—The object of this invention is to reduce the weight of cables, and to prevent injury to the conducting wires by a longitudinal stress. To attain this, the inventor surrounds the exterior of the gutta-percha covered core with iron wires, placed longitudinally and interwoven with tarred cords. A similar covering may be applied to the wires, previously to the insulating coating.

Another part of the invention consists in reversing the lay of the wires, at short intervals, so as to allow of extension under a longitudinal stress.

*2,835.—A. Barclay, Kilmarnock, N.B. Electro-Magnetic Telegraph Ropes or Conductors. 3d. December 10.

CONDUCTORS.—The component wires of cables "are twisted after the fashion of a common rope," and separated from each other "by mechanically separative matter only, or by an insulating medium." In conductors made in this way "the strain is borne equally by all the wires." The same result may be attained by plaiting the wires. An electro-deposit of metal may be employed to preserve the conducting wires.

2,836.—A. Barclay, Kilmarnock, N.B. Obtaining Electromotive Power. 11d. (9 figs.) December 10.

ELECTRO-MOTOR.—The power is generated by the alternate attraction and repulsion of electro-magnets and permanent or electro-magnets. In one arrangement a number of fixed electro-magnets are arranged radially like the spokes of a wheel. In the neutral portion of each magnet an eye is formed, the metal being thinned and spread out, so that when three such magnets are threaded on a shaft the eyes unite and form a

hub in such a way that the six arms stand in the same plane. Several such wheels (say three) are placed at intervals along a horizontal shaft, and are fixed in a frame. Between these wheels other wheels of permanent or electro-magnets, fixed to the shaft, rotate, and are subject to the influence of the field magnets. The current is brought from the battery to two rollers which run upon a commutator of usual construction, the office of which is to reverse the current in each of the stationary magnets six times in each revolution. A second arrangement resembles a galvanometer with a six-armed magnet rotating within a flat box, wound with three coils 120 degrees apart. In each case the motor is placed in a closed case from which the air can be exhausted.

2,891.—W. Clark, London. (*F. M. Baudouin, Paris.*) Submarine Electric Cables. 5d. December 16.

CONDUCTORS.—The conducting wire is laid around a core of suitable elastic material, in two superimposed contiguous helices, the second helix falling into "the furrows of the first." By this means a conductor is obtained which is capable of extension without "direct continuity of the metallic mass" being interrupted. The sheathing wire strands may be also provided with an extensible core.

2,937.—A. Barclay, Kilmarnock. Obtaining and Applying Electricity and Magnetism to Motive Power and Telegraphic Purposes. 8d. (8 figs.) December 23.

ELECTRO-MOTORS.—This is substantially the same motor as that described in Specification No. 2,836 of 1858, adapted to move the needle of a telegraphic instrument.

2,997.—J. W. Duncan, London. Electric Telegraph Cables, &c. 1s. 1d. (17 figs.) December 30.

CONDUCTORS.—The central wire of the conductor is passed through an amalgam of mercury and tin, which is applied by means of a "volute cone piece," similar to the arrangement described in Specification No. 906 of 1853. Metallic ribbons are then laid on "convolutely or concentrically in layers superimposed on each other." Stranded conductors may be treated with the above amalgam and united by being passed through a die, or in place of such amalgam a cement, preferably that described in the above-mentioned specification, may be used. Other modes of uniting conducting wires are stated. Compound conductors are made by inserting a rod of copper or silver in concentrically arranged cylinders of other metals, and drawing the whole into wire. These metals may also be combined by forming them into tapes, which are then wound one on the other. Compound metallic woven tapes may be employed for electrical conductors.

Tapes saturated in linseed oil or india-rubber solution and gutta-percha, treated according to Patent No. 13,788 Old Law, and compounded with linseed oil, are used as insulating media. An outer covering of strong canvas, sewn together with a lip joint is applied as an outer protection to conductors.

In covering machines the cylinders are arranged in a horizontal line, whereby one or more coatings of insulating material may be simultaneously applied.

The water which has penetrated into "invalid" cables is displaced by Canada balsam applied with a force pump.

A protecting sheathing for cables consists of corrugated hoop iron.

1859.

*31.—L. J. Higham, London. Means for Obtaining Submarine Electrical Conduction. 3d. January 4.

CONDUCTORS.—To prevent displacement of the insulating coating by straining of the cable, the inventor proposes to enclose the insulated conductor in a jointed case, which is to form a permanent shield thereto. This protecting case is formed of "base plates or troughs" linked together. Over these plates "semi-tubular" covers are placed "for securing the insulated medium within the jointed casing."

*56.—A. Barclay, Kilmarnock, North Britain. Electric and Magnetic Telegraphs. 3d. January 7.

CONDUCTORS.—The cable is connected to an iron strip "or collection of metallic surfaces" "in any fluid matter," such connection being made by means of a series of wires, in accordance with the principle explained in Provisional Specification No. 2,937 of 1858, the object being to secure a "reservoir," "distributor and conductor." Another part of the invention relates to the use of sheathing wires as conductors, and also to uninsulated conductors which are mechanically protected.

84.—D. E. Hughes, New York. Insulating Electrical Conducting Wires. 4d. Jan. 11.

CONDUCTORS.—The object of this invention is to make a conductor, which shall not lose its insulation, even after the gutta-percha or other covering becomes impaired. To attain this end a soft non-conducting substance is introduced between the conducting wires and the gutta-percha, which, when the gutta-percha coating is pierced, oozes out and fills up the fissure. For this purpose it is preferred to use resin dissolved in oil or turpentine, or a solution of india-rubber in a hydrocarbon. In some cases it is preferable to employ a substance which shall be decomposed on coming into contact with water, and deposit a varnish at the point of rupture. The conducting wires may receive a coating of gutta-percha, india-rubber or fibrous material before the application of the restorative medium.

87.—C. W. Siemens, London. Constructing Supports for Telegraphic Wires, Joining Pipes, &c. 1s. 6d. (22 figs.) January 11.

INSULATORS.—An inverted cast-iron bell incloses "a wide-mouthed cup or tube of porcelain or other vitreous material" cemented in place with a mixture of sulphur and oxide of iron. The mouths of the cups are some distance apart, and the inner cup has no shoulder at the point where it enters the iron case. A stalk is cemented into a cylindrical cavity in the porcelain cup, and is made U-shaped, with a horizontal extension which is screwed and provided with a nut. The line wire lies in a notch formed by a projection on the cast-iron cup, and is secured by a ring driven down on it, or the projection may be hooked and a little on one side, "so that the wire may be angled into its place and then drawn tight." At the stretching places the wires are secured by placing them in split conical tubes, which are then driven into a hole or between the two projections on the cast-iron cup.

*119.—O. Rowland, London. Laying Electric Telegraph Wires in Streets. 3d. January 13.

CONDUCTORS.—Iron troughs, "furnished with covers suitably constructed to form part of the pavement of a street," are laid end to end by the side of the kerb-stone and fitted together with sockets. The insulated wires are placed in the troughs and the covers dropped into position. By sliding the covers sideways, a short distance, they are locked in place by hooks, with which they are provided, passing under projections formed on the troughs.

181.—J. Latimer Clark and J. Muirhead, London. Electric Telegraphs, &c. 8d. (7 figs.) January 20.

INSULATORS.—Brackets for supporting insulators are cast hollow or tubular, malleable cast iron being used for the purpose.

CONDUCTORS.—Gutta-percha or india-rubber coated wires are protected from the action of the weather by a serving of spun glass, saturated with a suitable oily or resinous material. It is preferred to paint the whole over with tar.

*263.—A. Barclay, Kilmarnock, North Britain. Obtaining, Distributing, or Applying Electricity and Magnetism. 3d. January 28.

CONDUCTORS.—Rails of railways, lines of pipes, railways, and other continuous conducting lines are employed for the transmission of electric currents.

*329.—A. Barclay, Kilmarnock, North Britain. Manufacture of Electro-Magnetic Cables, &c. 3d. February 4.

CONDUCTORS.—The protecting wires or strands of cables are covered individually, or in groups, with hemp, and passed through a bath of melted tar or suitable substance, and afterwards through a reservoir of same. This process is repeated as often as is necessary, and the strands are then compressed between rollers to give the required consolidation to the coating. Finally the whole cable is treated in the same way. It is preferred to wind the successive servings of hemp in opposite directions to prevent them opening. The conducting wires may be separated from each other and twisted in the rope, in order to give elasticity.

423.—G. Bedson, Manchester. Joining Wire for Telegraphic and Other Purposes. 3d. February 15.

CONDUCTORS.—This relates to joints in the conductor. The ends of the wires are placed in a short tube and pressure is applied "so as to cause the wires and the tubes to be bound together." The whole may then be jointed by soldering, "galvanising" or otherwise.

The wires may be joined by this process, omitting the compression above mentioned.

434.—W. H. Horstmann, New York. Telegraphic Cables, &c. 10d. (13 figs.) February 16.

CONDUCTORS.—The surface of the conducting wires is brightened and then coated with shellac, over which in succession are applied cloth or fibrous material, an elastic insulating compound, preferably a mixture of resin and tar, a second covering of fibrous material, and an iron wire sheathing. This last is coated with "a resinous compound in a melted state," and the whole served with a helical wrapping of fibrous cord. The protecting wires must in all cases be straight and parallel. Another part of this invention consists in constructing the cable while it is being laid.

512.—C. W. Siemens, London. (*Partly W. Siemens.*) Electric Telegraphs, &c. 2s. 8d. (27 figs.) February 25.

MAGNETO-ELECTRIC GENERATOR.—A coil of wire is wound longitudinally in grooves on a bar of soft iron, the whole forming a Siemens arma-

ture. (See Specification No. 2,107 of 1859.) It is placed free to revolve, or oscillate, upon its axis between the poles of a compound permanent horseshoe magnet. By another method the coil is placed in trough-like polar extensions of the magnet and a bar oscillates within the coil. Another generator consists of an armature movable towards one pole of a permanent magnet by vibrating on an axis in connection with the other pole of the magnet. By moving the armature upon its axis it approaches or recedes from the other pole of the permanent magnet, and alternating currents are thereby produced in a coil of wire wound upon the armature, or within which the armature oscillates while the coil itself is fixed. A second permanent magnet may be applied with its poles in proximity to the similar poles of the first magnet with the advantage of increased power. Another arrangement consists of a pair, or several pairs, of horseshoe magnets, each pair having its opposite poles connected by a piece of soft iron, and the oscillating opposite poles free to act by induction upon the oscillating armature; the opposite poles that are connected by the soft iron piece belong to two separate magnets.

INSULATORS.—The insulator described in Specification No. 87 of 1859, which consists of an inverted cast-iron bell enclosing a vitreous cup, has a screw thread at the lower end of the iron stalk; a bent iron plate, with a hole in its upper part, is slipped on to this stalk, and afterwards a nut. When the nut is screwed up, the conducting wire is gripped between the lower part of the bent plate and the end of the stalk. The iron bell has a rib which connects it to a flange; a ring passed over the post and flange secures the insulator in place. In another arrangement the insulator stalk is bent into a hook at the lower end and a saw-cut is made in the bend for the reception of a cotter; the wire rests in the hook and is secured by driving in the cotter.

*607.—W. Clark, London. (*A. H. S. Treve, Paris.*) Submarine Telegraph Cables. 3d. March 8.

CONDUCTORS.—To obviate the static charge in working an insulated conductor, the inventor proposes to wind a copper wire helically around it, through which an electric current is made to flow to earth. This "also conduces to the intensity of the current which would be in power nearly doubled."

653.—W. Clark, London. (*V. L. M. Serrin, Paris.*) Apparatus for Regulating Electric Lamps or Lights. 10d. (4 figs.) March 15.

ARC LAMP.—The illustrations are respectively vertical and horizontal sections of a focussing lamp. The two carbon holders are sustained by two cords, each of which passes round a drum fixed to the wheel *k*. One of these drums is double the diameter of the other, to compensate for the difference of consumption of the two electrodes. The wheel *k* forms part of a train of wheels, of which the last member is provided with ratchet teeth, and revolves, or is stationary, according as the pawl *m* is out of, or in gear, with it. The arc is

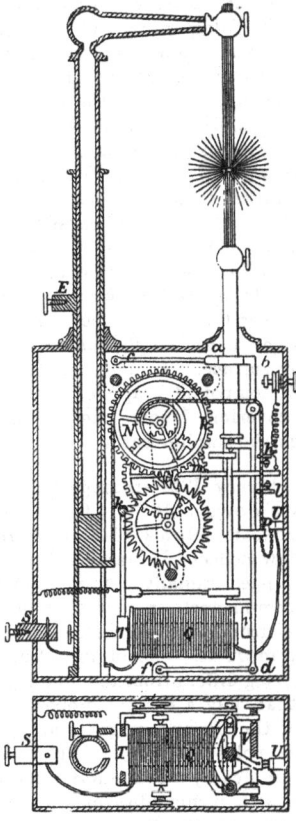

Fig. 56.

struck and the feed regulated by an oscillating system "which forms the particular feature of this improved regulator." Unfortunately this system is not very clearly shown in the drawings. It is composed of two horizontal bars, *a e*, *f d*, and one vertical bar *b d*, pivotted together and to the frame. The vertical bar carries an armature V, which is attracted downwards by the electro-magnet Q when the current is flowing. The cord which supports the lower carbon holder passes round a pulley N on the vertical bar, and, when the latter descends, the cord, pulley and carbons all fall with it, thus striking the arc. The pawl *m* is carried on the same system, and can oscillate between the fixed stops *h* and *l*. As soon as it touches the upper one it is raised out of gear with the ratchet wheel, and allows it, and the train of wheel work, to revolve, paying out the two cords

and permitting the electrodes to approach each other. T is a second armature opposite the other end of the magnet. It is suspended from an arm and is connected by a rod to an arrangement of mechanism on the lower holder, whereby, when the armature is attracted, the electrode is slightly rotated to break it off from the upper one in case the two have become welded together. The course of the current is from the binding screw S, round the magnet coil, through the chains U P, and the two electrodes to the terminal E. The specification includes a simpler form of lamp in which there is no torsional arrangement, and where the ratchet wheel is replaced by a fly that is alternately engaged and released by a projection on the oscillating system. A rack and pinion is substituted for one of the drums and cords or chains.

*687.—J. Molesworth, Rochdale. (*F. N. Gisborne and F. O. J. Smith, Boston, U.S.A.*) **Electric Telegraphs.** 3d. March 18.

Conductors.—This relates to the use of insulated conductors in connection with an "entire metallic conducting circuit."

716.—W. Warne, J. A. Fanshawe, J. A. Jaques, and T. Galpin, Tottenham. **Compound for Insulating Electric Conductors.** 3d. March 21.

Conductors.—An insulating compound is prepared by combining india-rubber or gutta-percha, or a mixture of the two, with bituminous substances, to which may be added animal or vegetable oils and earthy matters.

Conductors coated with this compound are served with a wrapping of fibrous material prepared as follows:—

A suitable woven fabric is saturated with a bituminous, resinous or other substance, so as to form an adhesive surface, on which is dusted pulverised glass or flint; the fabric so coated is then passed between rollers in order to force the silicious particles into the material, and thus produce a vitreous non-conducting fabric.

*755.—C. Cowper, London. (*R. J. Bingham, Paris.*) **Telegraphic Cables.** 3d. March 25.

Conductors.—A copper wire, "thicker than those usually employed" for cables, is first covered with a suitable insulating coating and then with a mixture composed of caoutchouc, gum lac, tar, and powdered cork in such proportions that the specific gravity of the cable shall be slightly above that of water.

The composition may be applied at the time of paying out the cable. For shore ends the cork is replaced by emery or similar hard substance.

785.—R. Searle, Woodford Wells, Essex. **Electric Telegraphs.** 3d. March 29.

Conductors,—The metallic conductor is supported on a foundation of bamboo, rattan canes, or similar substances. The inventor prefers to form channels in these canes to receive the conducting wires and to cover the whole with caoutchouc or vulcanised india-rubber.

The conductor may be applied helically and may consist of a plain, grooved or corrugated wire or ribbon.

*837.—C. F. Kirkman, London. Protecting Telegraph Wires. 3d. April 4.

CONDUCTORS.—A covering of "network, plaiting or braiding of coco-nut fibre" is applied over the insulated conductor, "with or without the exterior service of small wires."

863.—J. Rogers and E. J. Tweed, London. Coating Electric Telegraph Wires. 6d. (4 figs.) April 6.

CONDUCTORS.—Longitudinal bands or yarns of fibrous material are applied so as to surround the insulated conductor; they may be secured by binding with yarn over which a coating of wires previously served with hemp, is braided; or in place of the longitudinal braiding, or in addition thereto, one or more series of wires, previously served with hemp, are laid around the insulated conductor, and then covered with a braiding of hemp or other fibrous material.

881.—W. Hooper, Mitcham. Insulating and Protecting Telegraphic Conductors. 4d. April 8.

CONDUCTORS.—A double insulating covering is applied to the conductor, consisting first of a coat of india-rubber and secondly a coat of vulcanisable compound.

The two coats are separated, the one from the other, by a layer of thin metal, or other material, in order to limit the action of the sulphur to the exterior coat.

A second part of the invention relates to the production of a vulcanised india-rubber coating, on conductors which have been covered with gutta-percha.

942.—W. Sinnock, Brompton, Kent. Electric Telegraph Cables, &c. 7d. (2 figs.) April 14.

CONDUCTORS.—This invention consists in applying one or more layers of fibrous material, saturated with a mixture of india-rubber, gutta-percha, and bitumen, to the insulated or uninsulated conductor; these are arranged helically, by preference, and around them is placed an outer casing of fibrous materials "woven as warp and weft circularly round the core." The inventor claims also the machinery for conducting these processes.

943.—A. McDougall, Manchester. Coating Metallic Surfaces. 8d. April 14.

CONDUCTORS.—Part of this invention relates to coating electrical conducting wires with a "mixture of the products of coal-tar or natural asphaltum, in combination with sulphur and beeswax." This is effected by passing the heated wire through a bath of the melted compound.

951.—H. A. Silver, London Insulating Wire for Electric Telegraphs. 3d. April 15.

CONDUCTORS.—The insulating process is effected as follows: One or more coatings of india-rubber are wrapped helically around the conductor in opposite directions, with their edges overlapping, and one helix breaking joint with that previously applied. The whole is then subjected to steam for a few minutes to cement the surfaces together. This last process constitutes the peculiarity of the first part of the invention. Heated air may be used in place of steam.

Another part of the invention consists in passing the insulated conductor, as above described, through a bath of melted sulphur in order to vulcanise its outer surface. In some cases a coating of vulcanisable compound is applied over the first india-rubber covering, before passing the conductor through the sulphur bath.

1,039.—H. C. Hurry, Wolverhampton. Obtaining Motive Power. 10d. (4 figs.) April 25.

ELECTRO-MAGNETIC MOTOR.—A drum rotates within a hollow concentric frame. Around the drum are arranged radially a number of small horseshoe electro-magnets in four tiers or rings, and upon the drum there are also fixed four rings of similar magnets. The number of magnets in a rotating ring differs from the number in a fixed ring so that only a certain proportion (say one-twelfth) of the moving magnets are exactly opposite fixed poles at a time. The battery current is distributed by a commutator consisting of a fixed ring with insulated segments traversed by rollers connected to the moving magnets. The magnets are excited in sets, the current in all those that have just passed the centre of a pole being reversed together.

*1,152.—C. Frost, London. Electric Telegraph Cables. 3d. May 7.

CONDUCTORS.—The insulated core is surrounded with rings of metal alternating with rings of cork or other elastic material; a covering of hemp is then put over the rings. The object of the invention is to produce a flexible cable, the conductor of which is "effectually protected from damage." The rings also "prevent the covering of hemp from pressing on the conductor."

1,194.—W. Warne, J. A. Fanshawe, J. A. Jaques, and T. Galpin, Tottenham. Insulation of Telegraphic or Electric Conductors. 11d. (6 figs.) May 12.

CONDUCTORS.—A compound for insulating conducting wires consists of caoutchouc, gutta-percha, coal tar, pitch, plumbago, oxide of·lead, and·silicate of magnesia; or when no gutta-percha is employed the plumbago and oxide of lead are also omitted. The material thus produced is applied to the wires in a plastic state through suitable dies, and the insulated conductor is then enclosed in an outer covering of fibrous material prepared in accordance with the process described in Specification No. 716 of 1859.

1,213.—J. Chatterton, London. Covering Wires and other Metal Conductors for Telegraphic Purposes. 3d. May 14.

CONDUCTORS.—The object of this invention is to afford protection to insulated wires during and subsequent to the processes of manufacture. This is effected by applying over the gutta-percha covering a helical wrapping of india-rubber or india-rubber cloth, preferably vulcanised, over which the external yarns and wires are laid in the usual manner. An adhesive cement is sometimes interposed between the insulated conductor and its elastic covering, by preference that described in Specification No. 1,811 of 1858.

*1,285.—B. F. Greenough, Boston, U.S.A Hydro-Electric Conductor. 3d. May 24.

CONDUCTORS.—A metal sheathed gutta-percha or india-rubber tube contains pure water, through which the current is passed " in an analogous manner to that employed in metal conductors."

1,289. — R. A Glass, London. Submarine Electric Telegraph Cables. 3d. May 25.

CONDUCTORS.—This invention consists in insulating conducting wires by covering them, first with a layer of gutta-percha, and then with a layer of one or more of the compounds described in Specification No. 2,270 of 1858.

1,325.—A. Smith, London. Machinery for the Manufacture of Lines, Ropes and Cables. 1s. 4d. (16 figs.) May 30.

CONDUCTORS.—These are improvements on those described in Specification No. 14,021 (Old Law), and relate partly to cable-making machinery. The core is protected by passing it through a hollow tube, which rotates with the reel frame but does not rotate on its own axis, and, therefore, does not in any way affect the twist or condition of the fibres of the core passing through it. Another part of this invention consists in various modifications of the machinery described in the specification above referred to.

1,647.—W. E. Newton, London. (*G. W. Beardslee, Flushing, N.Y., U.S.A.*) Magneto-Electric Machines. 11d. (5 figs.)

MAGNETO-ELECTRIC GENERATOR.—A vertical shaft carries three sets of horizontal permanent magnets arranged in three stages. Each set is built up of thin circular plates cut into a star-like form resembling the spokes of a wheel. Between the upper and middle magnets, and also between the middle and lower ones, are fixed vertical helices with soft iron cores, carried by adjustable brackets projecting from two rings concentric with the main spindle. When the axis is rotated the poles of the permanent magnets pass over and under the cores of the helices, and induce currents in their coils. The ends of the fixed helices are connected to stationary rings, from which they go to a commutator that changes the alternate impulses into a direct current. This resembles a polarised relay. A steel tongue plays between two electro-magnets traversed by the alternate currents, and is attracted or repelled according to the direction of the currents. It moves a vibrating beam which is covered with contact pieces that dip in and out of mercury cups connected to the main conductors.

*1,749.—C. W. Smith, Evans, Erie, New York, U.S.A. Electric Telegraphs. 3d. July 28.

CONDUCTORS.—It is proposed to surround the conducting wires with a lapping of thread and varnish, over which are applied in succession a gutta-percha coating, a serving of hemp, and a second coating of gutta-percha. The whole is finally enclosed in an external covering of small iron wires.

*1,767.—G. Gurney, Woodleigh, Cornwall. Electric Telegraph Conductors. 3d. July 30.

CONDUCTORS.—Instead of a wire or stranded conductor the inventor employs " a flat ligature or ribbon of copper." This is coated with " a vitreous covering or varnish," over which is placed india-rubber or gutta-percha. To support this ligature, flat steel ribbons are fixed on either side. Iron ribbons may be used where less strength is required. The whole may be bound together with thin wire or string and several conductors may be united in one band.

1,812.—W. R. Drake, London. (*E. Schneider, St. Petersburgh.*) Deep Sea Telegraph Cables. 8d. (3 figs.) August 6.

CONDUCTORS.—The conducting wire is wound helically about a central linen thread, and then served with a coating of " black oakum," over which is a covering of gutta-percha, and a second serving of black oakum. The sheathing consists of iron wires applied with a long twist, and held together by helically wound wires or cords.

1,905.—W. T. Henley, London. Machinery for the Manufacture of Ropes and Cables, &c. 1s. 2d. (10 figs.) August 19.

CONDUCTORS.—This relates to " a system of

l

rollers consisting of three or more, used for compressing a cable when made, so as to prevent the wires riding, and to give an even, uniform appearance."

*2,079.—F. N. Gisborne and L. S. Magnus, London. Telegraph Cables. 3d. Sept. 12.

CONDUCTORS.—Longitudinal fibrous, metallic or other strands "ranged round an insulated conductor" are held together by two layers of wire applied helically in opposite directions. Such longitudinal envelopes are covered "with a preservative compound of india-rubber, shellac, resin and vegetable wax, with pitch, so as to give the cable a smooth and regular exterior surface."

In some cases the cable, constructed as above, is covered with a "network of wire, hair or fibrous material."

2,135.—L. Engler and E. F. Krauss, Paris. Insulators for Electric Wires. 8d. (4 figs.) September 19.

INSULATORS.—The principle of this invention consists in the use of a "bearing of glass or crystal" to carry the line wire, such bearing being supported in a suitable way by a hook or ring formed at the end of an iron bracket.

2,170.—T. B. Daft, London. Coating Metal Conductors for Electric Telegraphs. 3d. September 24.

CONDUCTORS.—A brass or brass-coated conducting wire is used; it receives an insulating coating of vulcanisable india-rubber compound, by preference, and then several layers of yarn previously saturated with india-rubber compound. A "finishing coating" of compound is applied, and the whole is cured in the usual way. In some cases a flattened conductor is employed. The inventor claims that by the above means a good adhesion is obtained between the conductor and its insulating coating.

2,269.—J. Macintosh, London. Coating Metallic Conductors for Electric Telegraphs. 3d. October 5.

CONDUCTORS.—"Lamp-black, naphthaline, and gutta-percha, are combined into a plastic compound" and employed for insulating conducting wires. It is preferred to apply the above material over a coating of gutta-percha. In some cases a compound of india-rubber, or india-rubber and gutta-percha combined with lamp-black, is used for the same purpose.

2,402.— P. A. Godefroy, London. Submarine Cables. 5d. (8 figs.) October 20.

CONDUCTORS.—The wires receive an insulating coating of gutta-percha, india-rubber, or the compound described in Specification No. 1,268 of 1855; over this is applied in a longitudinal

direction a complete covering of fine steel or iron wires, the machine described in Specification No. 1,687 of 1858 being employed for the purpose. An external covering of "any suitable material of a light texture" completes the process.

2,433.—H. S. Rosser, London. Electric Telegraph Cables, &c. 7d. (3 figs.) October 25.

CONDUCTORS.—This relates to the manufacture of cables composed of "two or more wires partially insulated." Each wire is wound with a very long twist around a small cord, and the whole is then laid into a strand of hemp or fibrous material saturated with a preservative substance, such as tar. It is preferred to coat the wires with platinum. The cable may have a central core of insulated wire, around which two or more cotton-covered wires are laid, and the whole may be worked into a hempen cable as before.

*2,440 —H. C. Hurry, Worcester. Electromotive Engines. 3d. October 25.

This is practically the same engine as that described in Specification No. 1,039 of 1859. It is preferred that the polar faces of the magnets shall not be tangential to the peripheries of the drum and frame.

*2,454.—I. Zacheroni, Liverpool. Electric Telegraph Cables, &c. 3d. October 27.

CONDUCTORS.—The insulated conducting wires are laid longitudinally around a central core of vegetable fibre, and over them is applied a serving of fibrous yarns; the whole is then coated with gutta-percha. "The vegetable fibres may be covered with beeswax."

2,503.—C. W. Siemens, London. (*Partly by W. Siemens, Berlin.*) Insulating Electric Telegraph Conductors, &c. 1s. 6d. (16 figs.) November 3.

CONDUCTORS.—The wire is insulated by a covering of india-rubber applied in continuous strips, the edges of which are cut by circular shears and united by the pressure of grooved rollers.

The machines for carrying out this process are illustrated and described. The application of a moderate heat facilitates the adherence of the newly cut surfaces of the india-rubber.

INSULATORS.—A wrought-iron stalk is partly encased in a tube of vulcanite, or "hornified" india-rubber and then inserted in an iron protecting hood. The lower end of the stalk is screwed and receives a small nut and bent plate, on which the conducting wire is laid. By screwing up the nut the wire is gripped between the bent plate and the end of the insulator stalk. The cup or hood is fixed to the post, or the hood may carry the wire, and the stalk be the means of support. A stretching insulator is also described and illustrated,

2,521.— D. Joy, Manchester (*C. De Bergue, Barcelona, Spain*). **Telegraphic Cables.** 3d. November 5.

CONDUCTORS.—This refers to machinery for the manufacture of cables, in accordance with the method described in Specification No. 1,605 of 1858, by which a gutta-percha insulated conductor is enclosed by longitudinal strands, and the whole laid over with a binding of " strands, wires or other covering."

2,546.—J. Hamer, Manchester. **Material for Insulating Electric Wires and Lining or Covering Various Articles.** 5d. November 9.

CONDUCTORS.—The object of this invention is to utilise silk waste, one of its applications being for the insulation of conducting wires. " Silk noils and winders' or weavers' waste" are spun into a coarse yarn, according to the process described in Specification No. 3,145 of 1857 (not included in these abstracts), and such yarn is " wound round the wire in close successive coils " in the ordinary way, or it may be braided upon the wire by the machinery according to Patent No. 12,991 of 1850.

2,552.—W. Clark, London. (*E. D. Rosencrantz, New York*). **Electric Telegraph Wires or Conductors.** 6d. (4 figs.) November 9.

CONDUCTORS.—This invention consists in the employment of a compound conducting wire, consisting, preferably, of a silver central wire with an outside copper casing. The two metals are united by heat previous to being drawn into wire, the union being effected by pouring melted silver into a hollow heated copper ingot. In drawing a wire the fibres or laminæ become arranged so as to " overlap in conical sections," and advantage is taken of this peculiarity to send the current through the wire in the same direction as it passed through the draw plate, whereby the current is " kept more in the centre of the wire."

2,559.—G. Seymour, London. **Insulating and Protecting Electric Telegraph Conductors.** 3d. November 10.

CONDUCTORS.—Animal hairs or fibres are employed for insulating and protecting conducting wires, but such hair or fibre should not " be used in the first coating, though it may be applied therein." It is preferred to use the fibre in the form of narrow strips of felt, which are wound helically around the insulated conductor.

2,685.—E. Tomey, Perth, N.B. **Apparatus for Insulating Telegraphic Wires or Conductors.** 7d. (10 figs.) November 28.

INSULATORS.—These are made of glass, by preference. Each insulator consists of a short standard, formed with a collar and screwed end, for attaching it to the post or support. The other end terminates in a " thick flat head " for carrying the wire. The head has a hole through the

centre and a transverse, and slightly oblique, slot by which the wire is introduced. The insulator may be placed vertically or horizontally, the slot being made accordingly.

2,710.—H. De Matthys, Antwerp. **Electric Telegraph Cables.** 5d. (3 figs.) November 30.

CONDUCTORS.—A gutta-percha coated copper wire is surrounded with iron wires, helically arranged, over which is applied a second covering of gutta-percha, another set of iron wires and a third covering of gutta-percha ; the whole then receives one or more layers of " tarred ropes," laid longitudinally and secured by a helical serving of the same material.

2,757. — F. Coignet, Paris. **Manufacture of Beton.** 5d. December 6.

CONDUCTORS.—This relates to the manufacture of an artificial stone which may be used for " continuous protective coverings " for electrical wires.

Lime is slacked by grinding with a small quantity of water, and then baked earth, cement, sand, or puzzolanas are added, according to the qualities required in the resulting material.

2,759.—J. Shaw, Manchester. **Insulating and Laying Down Telegraphic Wires.** 11d. (36 figs.) December 6.

CONDUCTORS.—Naked wires are supported and kept apart in continuous watertight troughs, by suspending them between suitable rests, props or hooks, arranged at intervals in such troughs. The rests are formed in one piece with the trough, which by preference is made of glazed earthenware in 3-feet lengths. Socket joints are used, and a channel is formed along the top of each side of the trough in which lids with inverted side flanges fit. The whole is made watertight by running in a cement composed of coal tar and gas pitch.

2,809.—J. Chatterton and W. Smith, London. **Insulating Telegraphic Conductors.** 4d. December 10.

CONDUCTORS.—The insulated core is placed in a chamber, from which the air is exhausted ; slightly warmed Stockholm tar is then introduced, and a pressure applied, by which means the tar is made to penetrate the minute pores existing in the insulating coverings, and thereby to improve the insulation. Thin sheet gutta-percha is subjected to the same process before use. To obtain perfect union between successive coverings of gutta-percha, the conductor is drawn through a flame as it passes to receive an additional covering. To cool and harden gutta-percha coverings they are passed through a bath of cold Stockholm tar.

2,857.—C. Hancock, London. **Insulating Telegraph Conductors, &c.** 3d. December 15.

CONDUCTORS.—The insulating material is vulcanised india-rubber, used in the form of a tube through which the conducting wire is passed; the tubes are retained in place by binding or braiding with cords, tapes or other suitable material. In place of tubes, fillets or ribbons of vulcanised india-rubber may be used. A cement, preferably consisting of a compound of masticated india-rubber dissolved in oil, gold size and coal tar naphtha, may be employed, to give an adhesive property to the india-rubber surface.

2,956.—L. S. Magnus, London, and W. Sinnock, Brompton, Kent. Preparing Yarn, Twine, &c., for Submarine Telegraph Cables. 3d. December 28.

CONDUCTORS.—This consists in saturating yarn, or strands of fibrous material, "with a composition of india-rubber, gutta-percha, vegetable or other wax, resin, pitch, with or without other matters." Such yarns are then fit to be employed in the manufacture of electrical cables.

1860.

*47.—W. Hooper, Mitcham. Reworking Compounds of India-rubber and Sulphur, and Insulating Telegraphic Wires. 3d. Jan. 6.

CONDUCTORS.—Insulating compounds of india-rubber are reworked by grinding them to a pulp with naptha, and adding raw india-rubber and sulphur to the mixture. The vulcanising process is conducted in the usual way, after the compound is applied to the wires.

To protect copper conductors from the injurious action of sulphur, the inventor coats them with copper or other metal, applied in a suitable manner with, or without, the interposition of a layer of paper, tannin, shellac or other material. In some cases the wire is coated with vulcanite before the india-rubber covering is applied.

Part of this invention relates to joints in electrical conductors which are insulated with vulcanised india-rubber, and consists in the use of a wrapping of pure rubber, over which is placed a quantity of vulcanisable compound, and then a sheet of vulcanised india-rubber.

*84.—W. Sinnock, Brompton, Kent. Insulated Wires for Submarine Telegraph Cables. 3d. January 12.

CONDUCTORS.—The object of this invention is to maintain uniformity of tension in the yarns composing cabled conductors. This is effected by placing the "forming tube" at some 10 to 20 feet from the "make" tube, thus giving sufficient space for "assimilation and uniformity of tension" before the yarns enter the "lay."

*93.—F. N. Gisborne, Sydenham. Insulating Material for Telegraphic Purposes. 3d. January 13.

INSULATING MATERIAL.—An insulating compound is made of 3 parts bituminous pitch, 1 part "coal silicate" or earthy substance, 1½ parts gutta-percha, ½ part sulphur — 6 parts in all.

109.—J. Chatterton and W. Smith, London. Coating Telegraphic Wires. 3d. January 14.

CONDUCTORS.—In order to improve the pliability and durability of conductors insulated with india-rubber or gutta-percha, they are immersed in a heated insulating liquid which fills the pores of, but has no solvent action on, the insulating coverings.

122.—J. H. Johnson, London. (*J. M. Bachelder, Cambridge, Massachussetts, U.S.A.*) Insulating Submarine Telegraph Cables. 3d. Jan. 17.

CONDUCTORS.—This invention consists in the use of an insulating compound, for electrical conductors, made of "pulverised silex, glass or other non-conducting material, mixed with india-rubber and sulphur, and subsequently vulcanised," or of a compound of pure india-rubber and powdered silex.

206.—C. F. Varley, London. Electric Telegraphs, &c. 1s. (21 figs.) January 27.

CONDUCTORS.—Two or more wires or strands, each separately insulated, are compounded as follows:—If three are used then 1 and 2 are joined, a little further on 2 and 3 are joined, and further on still 1 and 3 are joined, and so on "alternately throughout the cable." All the junctions are carefully insulated.

INSULATORS. — For overground lines metal insulators are used, covered with vulcanised caoutchouc "by Daft's patent process, dated 1860, No. 227" (not included in these abstracts).

STATICAL GENERATOR.—An insulated plate is placed "near or between two other suitable conductors; these being charged statically induce in the former the contrary state," the inner plate being momentarily put to earth. This plate then passes between two other plates to which it delivers its charge; when contact has ceased it is put to earth and thus acquires by induction a charge of the opposite kind to its former one; it then passes back to the first pair of plates and the same actions are repeated, "thus by continuing the process the charge rapidly augments to the required amount."

A generator constructed on this principle is illustrated in the sheet of drawings, and consists of a rotating glass or vulcanite disc having on its periphery a number of wooden carriers which revolve between two cheeks or inductors, made by preference of polished wood, and supported on sheets of glass. Opposite inductors are connected by bars, with which the revolving carriers make contact as they pass. Two earth connections with terminal knobs also touch the carriers at such time as they are between the inductors. "Conductor balls" project, one from each pair of cheeks, and represent the terminals from which the electricity may be taken when the generator is in action.

*217.—J. Wilkes, T. Wilkes, and G. Wilkes, Birmingham. Wire for Electric Telegraphs, &c. 3d. January 28.
CONDUCTORS.—A compound wire for use as an electrical conductor, in which the chances of a flaw extending across the whole of its section are eliminated, is made by enclosing a number of metallic wires within a tube of the same material, and afterwards rolling or drawing down the tube and wires until the whole have acquired the requisite diameter.

261.—A. Stoddart, Tours, France. (*J. S. Davison, New Jersey, U.S.A.*) Submarine Telegraphic Cables. 5d. (2 figs.) February 1.
CONDUCTORS.—"According to this invention the conductor is made capable of greater extension than the exterior of the cable." A bundle of comparatively short straight wires are bound together by a wire lapped helically around them, and each of the short wires is attached, at one end, to one of the convolutions of the binding wire. A conductor constructed in this manner will elongate by the coils of the helices opening and the interior straight wires sliding one on the other.

519.—C. W. Siemens, London. Electric Telegraphic Cables and Conductors. 1s. 11d. (15 figs) February 25.
CONDUCTORS.—This relates to conductors insulated with "Wray's mixture" (see Specification No. 2,270 of 1858). A helically stranded conductor is passed through a bath of molten bituminous or resinous cement, and then through a heated conical die in order to compress and unite together the several wires. After passing the conductor through a straightening machine it is enclosed in successive tubular coverings of Wray's mixture applied in strips by means of a machine. Each additional covering has its joints at right angles to the last. Yarns, saturated with a resinous or bituminous cement, are then applied at a uniform tension and with a long twist—the twist of each successive layer being in the opposite direction to that preceding. The whole is then encased in an outer protective sheating, consisting of a flat metallic ribbon wound helically round the cable. The inventor claims, in addition to the process above described, the several machines for conducting the same.

590.—W. Bauer, Munich, Bavaria. Apparatus for Diving, and for Raising and Lowering Heavy Bodies, &c. 2s. 11d. (32 figs.) March 3.
ELECTRIC LIGHT.—This invention relates chiefly to diving chambers, and includes the use of such as light-ships. During stormy weather "balloons" or buoys, are employed, carrying an electric light as a substitute for the usual light displayed from a tower on the diving chamber. The electric current may be produced in the balloon itself or conveyed thereto from the principal structure.

*764.—H. V. Physick, London. Electric Telegraphs. 6d. (2 figs.) March 24.
INSULATORS.—A bell-shaped iron protecting hood carries the line wire and rests on the insulator, "but is not attached thereto by cement of any kind." The objects attained by this arrangement are safety of the insulator during transit, facility for cleaning, and protection from moisture. A screw in the hood bearing against a collar on the insulator, prevents the former rising. Fog is excluded by a specially shaped mouth to the bell and insulator.

*856—C. V. Walker, Redhill, Surrey. Troughs for Receiving Electric Telegraph Wires. 3d. April 3.
CONDUCTORS.—This invention relates to troughs for receiving electrical conductors. Each length of wood employed for this purpose has cut from it longitudinally a piece of triangular section. The conductors are placed in the groove so formed, and covered by the triangular piece which has been cut out therefrom.

912.—C. Newbold, Nottingham. Manufacturing Vessels, &c., from Bitumenised Paper. 1s. 6d. (16 figs.) April 12.
CONDUCTORS.—Semi-circular troughs or half-pipes for containing electrical conductors are made as follows: A sheet of bitumenised paper, or other material, is coiled around a core of suitable diameter, which is then withdrawn and the tube is collapsed to a semi-circular form, its edges and ends being shaped, as required, by pressure.

*994.—H. A. Silver and J. Barwick, Silvertown, Essex. Moulding Battery Cells and Insulators. 3d. April 20.
INSULATORS.—These are constructed "of a combination of metal and hardened rubber or other gum," their union being effected by moulding under pressure by the following process: India-rubber, gutta-percha or other similar gums, together with sulphur, are exposed to a temperature of 400° Fahr. in a suitable mould, and then the mould and its contents are cooled suddenly by plunging them in cold water.

1,146.—J. Reid, London. Electric Telegraph Conductors. 3d. May 9.

CONDUCTORS.—The wires which are to form a stranded conductor are first passed through a mixture of " the ordinary gums used in coating telegraph wires," or other suitable compound, kept in a heated state, and are then twisted together " so that the whole becomes one solid mass," impermeable to water. The same object may be attained by passing the wires through a bath, containing a mixture of tin and lead in a molten state, adhesion of the metal being assisted by a suitable flue.

1,164.—J. Grantham, W. Sinnock, and L. S. Magnus, London. Wire Rope. 3d. May 11.

CONDUCTORS.—This relates to machinery for applying the protecting wires to cables, in which the usual bobbins are dispensed with. A rope walk is necessary. The wires are contained in a long hollow shaft, and before reaching the lay plate, pass through a separating plate, so as to form a circle about the core, which is fed through a tube. Motion is imparted to the shaft containing the wire by suitable gearing, and the usual draw-off is provided. The wires are guided while passing through the shaft by long steel blades inserted therein at intervals in its length. The wires are joined by welding under the heat of a blow-pipe flame.

1,169.—W. E. Newton, London. (*B. H. Wright, Rome, U.S.A.*) Electric Conductors for Telegraphic Purposes, &c. 1s. 2d. 5 figs. May 11.

CONDUCTORS.—Around a hempen core is wound two separate layers of copper wire, applied helically with the coils in close contact. Over this a covering of insulating material is applied, and then " tarred cordage of about three-sixteenths of an inch in diameter is longitudinally disposed along the conductor," and secured by a covering of twine, which is to be wound helically round the same. This mode of construction produces a strong, light and flexible conductor, in which kinks will not be formed.

1,178.—J. Chatterton and W. Smith, London. Electric Telegraph Conductors. 3d. May 12.

CONDUCTORS.—In forming a stranded conductor the central wire is first caused to pass through, and be coated with, a composition, preferably consisting of a mixture of gutta-percha, Stockholm tar and resin, so that when the outer wires are laid around it they may embed themselves in such coating whilst it is in a plastic state. To protect gutta-percha or other insulating material, when used for coating conductors, from the effect of heat, beads of wood are threaded upon the conductors, and over these beads a tube of lead is applied.

1,209.—C. M. Guillemin, Paris. Submarine Electric Telegraphs. 7d. (2 figs.) May 16.

CONDUCTORS.—The object of this invention is to eliminate the influence of induction in subaqueous conductors. Around the conducting wire is a coating of gutta-percha, and then a conducting envelope composed of a coiled wire or ribbon; over this is another insulating coating of gutta-percha, and lastly the usual exterior protecting envelope.

1,275—R. H. Collyer, London. Tubes, &c. 5d. May 28.

CONDUCTORS. — Troughs or channels, for containing electrical conductors, are made from fibrous material, with which a bituminous composition has been incorporated. The material is forced through a suitably shaped die by hydraulic pressure.

1,329. — R. H. Collyer, London. Telegraph Cables, &c. 4d. May 29.

CONDUCTORS. — An insulating composition for coating electrical conductors is made by mixing bitumen, shellac and oil in a heated fluid state and adding thereto the short hair of animals, the shorts of wool or other fibrous material, and also a suitable quantity of calcareous matter. This composition is used either alone or in combination with the bitumenised fibre described in Specifications Nos. 564 and 804 of 1860 (not included in these abstracts), and is applied to the metallic conductors by means of a die and tapering tubes through which it is forced in a heated state.

*1,515.—W. Morris and H. Mapple, London. Electric Clocks and Telegraphs. 3d. June 22.

INSULATORS.— The supporters of insulators are made in one piece " grasping the post, so as to carry several insulators with few bolts or borings." The metal shanks of earthenware insulators are embedded in the insulators at the time of making.

*1,531. — R. Jobson, Dudley. Moulding Earthenware or Porcelain. 3d. June 23.

INSULATORS.—When insulators are moulded by the pressure of a plunger upon plastic material contained in a die, there is a tendency for the plunger to carry some of the material back with it, to obviate which the inventor makes the plunger conical, with a loose split ring expanded and forced upon it. On withdrawing the plunger the ring is detained by suitable means, and then springs clear of the insulator. A slight modification of this principle adapts the plunger to produce insulators having a double bell.

1,546.—W. Hooper, Mitcham. Reworking Compounds of India-rubber, &c.; Insulating Wires. 4d. June 25.

CONDUCTORS.—Insulating compounds of india-rubber and sulphur are reworked by a process described in Provisional Specification No. 47 of 1860.

To prevent the injurious action of sulphur on the conducting wires they are coated with a varnish consisting preferably of oxide of zinc and shellac dissolved in methylated spirit, a lapping of cotton yarn being previously applied to the wire. The insulation of the wires is completed by the application of, first, a coating of pure india-rubber, and, secondly, a coating of vulcanisable compound. It is preferred to do this by wrapping a narrow fillet of india-rubber helically around the conductor in such a way that the edges overlap each other, the fillet being kept in tension during the process. The outer coating is cured by the application of heat. Joints are made by first hard soldering the conductor and varnishing the metal surface as above described; successive coatings of pure india-rubber, vulcanisable compound, and vulcanised sheet respectively are then applied, and the joint subjected to heat in a metal case to effect union of the parts.

1,560.—J. Macintosh, London. Insulating and Impervious Compounds. 3d. June 27.

CONDUCTORS.—This relates to an insulating compound which is made by mixing gutta-percha or india-rubber with paraffin or stearic acid, the object of the invention being to obtain a material free from minute pores. Conductors are rendered more impervious by surrounding them with yarns saturated with paraffin or stearic acid. A coating of india-rubber is put over the yarns "so as to render them sufficiently adhesive to be put on the cable by grooved rollers or other suitable means." An outer coating of india-rubber compound is then applied and cured as described in Specification No. 2,866 of 1857, which renders the conductor "not liable to be injured by tropical heat."

1,561.—J. C. Evans, Greenwich. Rolling Metals; Covering Electric Cables; Wire and other Ropes. 7d. (7 figs.)

CONDUCTORS.—This invention consists in a peculiar arrangement of three or more rolls, the axes of which are placed at an angle with the substance to be operated upon, less than a right angle, but preferably at a very small angle, so as to be nearly parallel to such substance. This apparatus is adapted amongst other things to the "lapping" of submarine electric cables.

1,603.—R. N. Reid, London. Insulators for Electric Telegraphs. 3d. July 3.

INSULATORS.—A thickness of india-rubber is applied around the stem where it is inserted into the insulator and the junction rendered complete by filling in with a suitable cement, disposed either between the stem and the india-rubber, or between the india-rubber and the insulator.

The object in either case is to prevent fracture of the insulator from the unequal expansion of the metal and porcelain, or other substance, of which it is composed.

*1,626.—S. Krotkoff, London. Electric Light Reflectors. 3d. July 5.

REFLECTORS.—The objects of this invention are: 1st, to diffuse the electric light equally on the entire radius of illumination; 2nd, to render the luminous rays supportable to the naked eye; 3rd, to "unite an undetermined number of the lights, which separately and collectively would participate in these advantages."

These objects are attained by the use of "one or more reflectors acting circularly in open spaces" or "obliquely" for streets, houses and covered spaces.

*1,765.—A. V. Newton, London. (*G. Caselli, Florence.*) Regulating the Force of Electric Currents. 3d. July 30.

RESISTANCES.—"Rheostatic" fabrics are made as follows:—A ribbon is woven, of which the warps are cotton threads and the weft two threads, one of cotton and the other of thin iron wire. The wire is thus insulated throughout its whole length. This insulation may, however, be rendered more perfect by steeping the fabric in a solution of gum lac or in liquid tar. Another process consists in winding the wire in zig-zag form around pins placed in two parallel rows on a board, and varnishing it with gum lac, after which it is enclosed between two sheets of paper. These fabrics are folded up and disposed in wooden boxes with suitable connections for making them available as resistances.

*1,918—J. S. Gisborne, Birkenhead. Telegraphic Insulators. 3d. August 8.

INSULATORS.—These consist of a "dome-shaped cup with a hollow projection downwards from the inside, and two or more studs or projections on the top, with one or more cross holes or apertures in the same." The hollow projection downwards receives the shank, and is preferably coated internally with india-rubber or other non-conducting material. The wire passes between the studs on the top and is secured by a cotter which is placed in the cross holes and depresses the wire into a hollow formed in the top of the dome.

1,978.—P. A. Godefroy, London. Insulating and Laying Down Telegraphic Wires. 4d. August 15.

CONDUCTORS. — Wire is wrapped with unbleached cotton, and suitable lengths of it are enclosed in metal or earthenware pipes. An insulating mixture, consisting of coal pitch and wood tar, is then poured in, after which the tubes are heated to insure complete penetration of the compound. The wires should be stretched during the process, and balls of wood may be threaded upon them to maintain them in position. The pipes are laid in trenches, and the wire ends joined in the usual way, or by enclosing them in a tube and applying pressure; the pipe ends are then surrounded by two semi-circular pieces of iron, and insulating material is poured in through an aperture formed for the purpose. The ends of the joint are caulked with white lead and hemp.

When "flexible tubular insulation" is required lead pipes are employed.

2,051.—J. Wilkes, T. Wilkes, and G. Wilkes, Birmingham. Wire for Electric Telegraphs, &c. 6d. (6 figs.) August 25.

CONDUCTORS.—This relates to compound conducting wires, made in accordance with the method described in Provisional Specification 217 of 1860.

*2,056.—J. Chatterton and W. Smith, London. Telegraph Cables. 3d. August 25.

CONDUCTORS.—In order to avoid damaging the insulating material of cables by heat, in the process of applying the protective coverings of wire or yarn, such coverings, instead of being saturated with hot tar or other matters, are coated with a composition of oxides or carbonates of lead with oil or varnish. It is preferred to bind an uncoated tape exterior to the saturated coatings in order that the cable may be coiled immediately after the process.

2,116.—C. W. Harrison, Plumstead. Electric Telegraphs. 10d. (7 figs.) September 1.

CONDUCTORS.—Stranded conductors are passed through grooved rollers, by which the wires are pressed together and made to "fill up the space that is lost in using ordinary round wires."

The sheathing wires of cables are protected by lapping each separately with a fibrous material, saturated with a composition consisting of a mixture of 1 lb. of gutta-percha or india-rubber dissolved in 4 gallons of tar oil, with 24 lbs. of resin dissolved in 12 gallons of bone or Dippel's oil.

2,145.—M. Vergnes, New York. Magnetic or Electric Helices. 7d. (7 figs.) September 5.

HELICES.—This invention "consists in making a helix of a number of independent wires, each of which has only length sufficient to extend from the outside of the helix to the inside and back again, and is so wound or coiled as not to cross itself, and in so connecting or joining the several wires that the cross section of the helix shall be increased as the distance from the battery or length of the wire is increased, whereby a greatly increased power is obtained with a comparatively small quantity of wire."

A number of "double lays" or spiral coils are combined to form a helix, their ends being so

connected that the electrical resistance is diminished towards the centre of the length of the helical circuit.

In forming elongated helices grooved metallic plates are used as guides for shaping the wire. The "effective power" of these helices is increased by the use of two iron plates fixed one on each side of the helix.

2,249.—S. Barnwell and A. Rollason, Coventry. Combining Pyroxylene With Other Substances for Producing Various Compounds. 5d. September 15.

CONDUCTORS. — An insulating compound, for coating electrical conductors, is made by "dissolving pyroxylene in any of its known solvents, and adding thereto oils, animal, vegetable, or mineral. If desired, gums, resins, or oxidisable salts may be added to the compound.

*2,355.—G. H. Birkbeck, London. (*G. Perrin, Paris.*) Electro-Magnetic Apparatus. 3d. September 28.

ELECTRO-MAGNETS.—The purpose of this invention is to obtain "an indefinite length of current, preserving the same attractive force whatever may be the distance which is given to the armature."

A series of cylindrical iron bars, connected by slotted links of copper are placed in the interior of an electro-magnetic helix, rods of insulating material being employed as guides. This arrangement is "applicable to electro-magnets of every form, and may have two arms, straight or of other form, and a suitable armature or armatures adapted thereto." By this plan a "current" "as long as desired" is obtained, "the power of attraction increasing with the length of the cylinder."

2,457.—G. Bonelli, Milan. Electric Conductors, &c. 1s. 2d. (8 figs.) October 9.

CONDUCTORS.—A number of copper wires, either bare or covered with insulating material, are made to form part of the warp of a fabric. The compound conductor thus produced is coated with gutta-percha or other insulating substance, and then rolled up into the form of a cable, so that the wires lie longitudinally. A binding of twine and a coating of pitch, or tar, are afterwards applied in succession.

2,462.—C. Wheatstone, Hammersmith. Electro-Magnetic Telegraphs, &c. 2s. 5d. (20 figs.)

MAGNETO-ELECTRIC GENERATOR.—Two soft iron or hard steel cores with their coils are fixed to each pole of a permanent magnet, so that the centres of their unfixed ends are equidistant from each other in the circumference of a circle. An axis passing through the centre of this circle carries a soft iron armature, the breadth of which is a little greater than the distance between two adjacent cores. The wires of the coils are so connected together that the currents produced simultaneously in each coil, two by the approach and two by the recession of the armature to and from the cores, all concur to produce a current in the same direction in the communicating wire, and during one revolution of the armature four instantaneous currents are produced alternately in opposite directions. Several magnets of this description may be combined to form a compound magneto-electric generator. The magnets are placed so that the axes carrying the armatures are vertical, and arranged at equal distances in a circle; each of these axes is furnished with a pinion, and all the pinions gear into a wheel fixed on an axis. The positions of the armatures may be so varied that the currents produced by all the magnets may be simultaneous, or they may be successive. The impulses may be augmented by short circuiting the coil "the instant after the shock is thrown into the telegraph circuit, without breaking the continuity of the latter." The specification illustrates several forms of this generator specially applicable for telegraph purposes and for ringing bells.

CONDUCTORS.—Cables are supported over towns by links from wires strained above them, or by a stout iron or steel wire within them. A cylinder of vulcanised india-rubber, interposed between the support and the insulating apparatus of the wires, prevents the communication of sound and vibratory motions caused by the wind.

*2,546.—M. Wesolowski, Cincinnati, U.S.A. (*Partly a communication by J. N. Reithoffer, Vienna.*) Obtaining Light. 3d. October 18.

STATIC ELECTRICAL GENERATOR.—A disc of vulcanite is rubbed on both sides by a cloth pad charged with an amalgam consisting of zinc, tin and mercury in suitable proportions. The spark obtained by the use of this generator is employed to ignite a wick saturated with an inflammable fluid, the flame so obtained being available for lighting lamps or tapers.

2,548.—W. Andrews, London. Insulators. 7d. (5 figs.) October 18.

INSULATORS.—These are of the inverted type. The insulator is cylindrical and of two diameters, the large upper portion containing the head of the suspension hook, which is cemented in place by a filling of marine glue. A screwed cup closes the upper end. The insulator thus constructed is fixed in a vertical hole made through the cross arm of the post, and secured in place with an insulating cement. India-rubber packings are inserted under the shoulder of the insulator and under the head of the shank. The cross arm is hollowed along the length of its under surface, and coated with a waterproof paint.

Modifications of this arrangement are described and illustrated.

*2,587.—W. H. Walenn, London. Magneto-Electric and Electro-Magnetic Machines. 3d. October 24.

MAGNETO-ELECTRIC GENERATOR.—"The invention consists, first, in revolving the coils or the permanent magnets about their own axes or about axes concentric to the cylinder on which they are placed. Second, in never changing the magnetic polarity of the cores of the coils, therefore in enabling the current always to traverse the coils in the same direction." Six machines embodying the principle are briefly described. In the first an electro-dynamic coil is mounted on an axis which passes through the centre of its soft iron core. This coil is rotated between the opposite poles of a permanent or electro-magnet or magnets, and the current is led away from the two ends of the coil. A number of such coils may be mounted side by side like bobbins in a spinning machine. In the second the field magnets embrace the coil and induce opposite magnetic polarities at its two sides. A third machine has magnets mounted radially on a shaft rotating within a fixed coil.

"ELECTRO-MAGNETIC ENGINES.—*All the hereinbefore mentioned methods are adapted for use in electro-magnetic engines, by simply transmitting a suitable electric or galvanic current through the coils, and causing them to produce mechanical motion as a result, instead of using the mechanical motion of the coils or magnets, and causing them to produce electricity as a result. In all the above methods electro-magnets may be used instead of permanent magnets.*" This is a provisional specification only, and is of considerable interest, as containing a distinctly recorded appreciation of the reversibility of the electric current for the transmission of power.

*2,661.—T. G. Ghislin, London and South Africa. Preparing and Applying Substitutes for Horn, Bone, &c. 3d. October 31.

CONDUCTORS.—This relates to the use of certain "fibrous bulbs, viz., Josephina berglilly, &c. (the amarillydeæ family) and a fibrous aquatic known by botanists as the pincus serratus, or grass tree," for insulating electrical conductors. The fibre of these plants is employed either separately or mixed with other fibres, and is applied in conjunction with "a solution composed of zopissa, marine glue, Trinidad pitch, Egyptian asphalte, and Judean gum."

2,674.—W. E. Newton, London. (*Messrs. Claes, Vandenhest and Co., Menin, Belgium*). Insulating Electric Conductors. 4d. Nov. 1.

CONDUCTORS.—The wire is insulated by covering it with a silk, linen, or cotton fabric, "attached together by caoutchouc" or similar gum, which is then vulcanised. The stranded conductor is protected from the action of the sulphur by tinning, gilding or otherwise, and in order to prevent the formation of air bubbles it is passed through "a solution of caoutchouc in naptha, mixed with calcined magnesia or French chalk, sulphur, and such other materials as are usually employed in vulcanising," contained, when necessary, in an exhausted chamber. The fabric is coated on both sides with a vulcanisable layer of caoutchouc solution, to which on one side a thickness of vulcanisable sheet is applied. The whole is then wound helically around the conductor, "care being taken that half the width of the cloth always covers the half already wound on." Several successive coatings of solution and cloth are given to the conductor in this way till the desired thickness is obtained.

*2,678. — R. Murray, Newcastle-upon-Tyne. Manufacture of Telegraph Cables or Ropes. 3d. November 1.
CONDUCTORS.—This invention consists in employing the cable laying or making machinery to actuate a make and break apparatus, by which a succession of electrical currents are sent through the conductor to a galvanometer, the object being to obtain tests as to its condition during the process of manufacture.

2,770.—F. Walton, Manchester. Insulating Telegraphic Conductors. 3d. November 12.
CONDUCTORS.—For the purpose of insulating electrical conductors, the inventor employs a material obtained from linseed and other drying oils, as described in Specification No. 209 of 1860. This material is mixed, by preference, with shellac or other gum, and applied by pressure when it is in a sufficiently plastic condition.

2,862.—R. Jobson, Dudley. Moulding Earthenware or Porcelain. 10d. (8 figs.) Nov. 22.
INSULATORS.—The principle of this invention has already been described in Provisional Specification No. 1,531 of 1860. In addition to the details then given the following process is set forth :—In the manufacture of insulators, in some cases, the plunger is made in parts, concentric or not with each other, and capable of independent motion up and down, so that they may be driven into the mould "in succession up to a certain extent, and then driven simultaneously, so as to finish the interior" of the insulator.

The plunger may be made separate from the ram of the fly press, so that it "may receive a succession of blows without being raised out of the mould."

The concentric parts of such plunger are sometimes composed of two or more pieces.

*2,871.—E. Keirby, Halifax, Yorks. Insulating Telegraphic Wires and Cables. 3d. November 23.
CONDUCTORS.—"This invention consists in covering such wires and cables with yarn of hemp, tow, flax or other fibrous substance saturated in a mixture of native asphaltum or coal-tar pitch and gas tar, or other oily substance by which they are insulated and preserved from decay."

2,957.—W. P. Piggott, London. Generating Electric Currents, &c. 7d. (3 figs.) December 3.
CONDUCTORS.—Two or more copper wires are insulated by embedding them in "well tarred hemp," over which a sheathing of iron wires is applied.

2,980.—C. S. Duncan, London. Electric Telegraph Cables or Ropes. 3d. December 4.
CONDUCTORS.—The central wire is enveloped in a covering consisting of "strips of cane (bamboo or rattan)" bound round helically "with or without unspun silk or other suitable cane or wood, the same being well tanned with a preparation of oak bark, terra japonica, mammore or sea-squill tan, or any other known tanning agent. The whole is then covered with a thin coat of gutta-percha, india-rubber, or marine glue." Subsequent coverings of yarn, hemp or wood shavings and strands of wire or hempen rope are lastly applied.

The object is to attain "great strength and flexibility," together with lightness and good insulation.

*2,991.—R. A. Glass, Greenwich. Preserving Electric Telegraph Cables and Wires. 3d. December 6.
CONDUCTORS.—This invention consists in enclosing conductors in air-tight boxes, and in the use of jacketted boxes, between the sides of which cooling mixtures are circulated. In some cases the box is filled with water and a cover applied. The object is to preserve the conductors prior to their immersion under water.

3,047.—A. F. Jaloureau, Paris. Holding, Protecting and Insulating Telegraph Wires. 7d. (22 figs.) December 12.
CONDUCTORS.—The wires are arranged in a circle parallel to each other, and enclosed in pipes, preferably those described in Specification No. 2,137 of 1858. Bitumen in a hot state is then introduced, glass ferrules being employed to maintain the wires in position. A socket is formed in the insulating material at the ends of the pipe lengths, in which a plug of non-conducting material is inserted when it is desired to make a joint. This plug has grooves in which the wires are placed after their ends have been united. An outer ferrule encloses the joint. The wires may be covered with india-rubber, and the india-rubber coated with an electro-deposit of copper.

*3,103.—F. Silas, London. Aerostatic Signal Apparatus. 3d. December 18.

ELECTRIC LIGHT.—For signalling and other purposes a powerful electric light is adapted to an apparatus called a "semasphere," which consists of a balloon, steadied by kites or parachutes, and secured by ropes. The ropes may serve as conductors.

3,138.—J. Chatterton and W. Smith, London. Electric Telegraph Cables. 4d. Dec. 21.

CONDUCTORS.—This invention consists :—firstly, in keeping the yarns of fibrous material, used for serving insulated conductors, saturated with a conducting liquid, preferably tar water or tan liquor, for the purpose of applying an electrical test of the insulation; and secondly, in coating the external protecting wires with a mixture of tar thickened with wood or cork dust, or mineral oxides, such mixture being used in a cold state. Over this is wound a wrapping of strong fabric impregnated with a preservative compound such as marine glue.

*3,142.—J. H. Johnson, London. (*H. Worms*, Paris.) Magneto-Electric Machines. 3d. December 21.

MAGNETO-ELECTRIC GENERATOR.—The helices are grouped in fours or in pairs, and their opposite poles are connected to a common conductor which communicates with the commutator. "The commutator consists of a number of divided rings, to each half of which is connected a wire or conductor leading to a corresponding circle of helices." On these rings spring "electrodes" with anti-friction rollers are caused to run. The cores of the helices are made of tubular sheet metal, split down one side in order to obviate the development of induced currents. The wire "is wound in the form of a flat band or ribbon of several wires. By this means a current of greater quantity in proportion to the number of the wires in the band is obtained, and of increased tension in proportion to the fineness of such wires, which is of great advantage in many cases."

To collect the currents without the aid of a commutator, one end of the wire of the helices is connected to an insulated conducting rod contained within the main shaft of the generator. The outer end of this conductor is in contact with a screw, also insulated from the framing which supports it, and from this screw the current is led off. The other end of the coil wire is simply connected to the surface of the main shaft, and a binding screw on the plummer block serves for the second terminal of the generator.

1861.

*47.—H. Hirsch, Berlin. Insulating Telegraphic Wires. 3d. January 8.

CONDUCTORS.—This consists in insulating conducting wires by covering them with spun glass; when it is required to fill up the interstices betwen the glass fibres, "resinous, tarry or other suitable material" is employed.

175.—J. Chatterton and W. Smith, London. Telegraphic Cables. 4d. January 22.

CONDUCTORS.—In order that the gutta-percha insulating coatings shall not be damaged by heat, in the process of applying hot pitch, resin and other matters, with and after the application of the fibrous and wire coverings, the inventors employ, in place of such heated substances, a combination of oxides or carbonates of lead or other preparation of that or other metal, with tar, oil, or varnish, together with the usual external fibrous envelope. This last is saturated or coated with the composition and applied helically or longitudinally over the gutta-percha. It is preferred to bind an uncoated tape exterior to the coatings of saturated fibre.

*260.—S. Moulton, Bradford, Wilts. Telegraphic Cables. 3d. January 31.

CONDUCTORS.—The conducting wire is formed into a helix and embedded "in a strand or cord of india-rubber" which is afterwards vulcanised. A protecting covering of hemp, wire or other material may be applied, provided such covering does not affect the "elasticity" of the cable. It is stated that conductors of this construction "will readily accommodate themselves to any ordinary strain."

271.—J. J. De Arrieta, London. Application of Chapapote for Preservative and Other Purposes. 5d. February 1.

CONDUCTORS.—"A certain natural matter called in Cuba chapapote," is employed for "insulating and protecting" electrical conducting wires.

INSULATORS.—Chapapote is combined with paper or fibrous substances to produce a material which may be formed into insulators.

322.—J. Branscombe, London. Telegraphic Cables. 3d. February 8.

CONDUCTORS.—The insulated wire is coated with leather, which serves as a bed for the protecting wires, and at the same time, when moist, being a sufficiently good conductor, allows any defect in the insulating material to be detected by tests. The leather is joined into a continuous strip and applied preferably by "an ordinary taping machine,"

*329.—D. Ker, Plymouth. Submarine Telegraphic Cables. 3d. February 9.

CONDUCTORS.—A single wire or stranded conductor is coated with a composition consisting of gutta-percha combined with a small proportion of vegetable or other wax. The wire thus insulated is enclosed in a metal tube, which is painted over with a suitable waterproof composition. A protective covering of "whalebone, sliced cane or tough wood" is then laid round the cable, over which is applied a coating of marine glue or other composition, and a helical wrapping of sheet metal. Three or more of these protective coverings may be used, if desired, and the external one can be secured by a serving of tarred string. A second part of the invention consists in adapting conductors to chain cables, preferably constructed for the purpose, with long links having holes through their ends through which the conductor is passed. The object of this arrangement is to afford protection from abrasion.

*353.—A. Parkes, Birmingham. Electric Telegraph Conductors. 3d. February 12.

CONDUCTORS. — Copper conducting wires are constructed as follows:—Tubes are placed one within the other, and the centre is filled with a solid copper rod. The whole is then extended by drawing or rolling. A compound conductor of silver and copper may be built up in this way, by using silver and copper tubes alternately, or silver tubes within an outer tube of copper. The silver may be introduced into the copper tubes by electrotyping.

449.—J. Reeves, New York, U.S.A. Electro-Magnetic Engines. 1s. 3d. (4 figs.) Feb. 22.

ELECTRO-MAGNETIC MOTOR.—A circular metallic toothed ring, the outer surface of which is composed of alternate surfaces of iron and brass, is supported on a series of friction rollers, and rotates through a number of hollow helices or coils of insulated wire wound upon tubes or barrels, fixed to a stationary frame. When a current of electricity is passed through any coil, the iron portion of the ring will be attracted, and drawn into the tube. By suitably making and breaking the communication between the coils and the battery, a continuous motion of the ring will be obtained, which can be transmitted by its teeth to external mechanism. The commutator consists of a number of levers, carrying contact pieces, and alternately raised and depressed by cams.

*466.—W. O. Brooke, London. Apparatus for Suspending Electric Telegraph Wires. 3d. February 23.

INSULATORS. — An inverted cast-iron cup, of rectangular form, has a recess formed on the top, in which a wire may be fixed with solder. The sides of the cup are turned up at their lower ends into a gutter-like form. A smaller inverted cup of vulcanite is fixed to the interior of the cast-iron cup, and secured by running melted sulphur in the space between the two. The whole is supported on a wrought-iron stalk, which is cemented in place with melted sulphur.

*508.—M. Henry, London. (G. F. Tournachon, Paris.) Photography. 3d. February 27.

ELECTRIC LIGHT.—This invention relates to the employment of the electric light to produce photographic images.

549.—H. Hirsch, Berlin and London. Insulating Telegraph Wires. 3d. March 4.

CONDUCTORS.—Spun glass is applied to the conducting wire "in thin loose fibres" so as to completely cover its surface. This coating is then enveloped in a helical wrapping of thin metal strips, and the whole is passed through a "fluid metallic bath of high temperature," by which means the entire surface of the metal is "covered with a metallic base." When it is desired to employ several separate conductors, strips of copper are wound with a long twist around the glass core, so as to leave a "wide serpentine space" between them. Another layer of spun glass may be applied, and then a metallic envelope as above described. Gutta-percha or india-rubber may be used in combination with these processes, but in all cases the outer covering "must be metallic."

569.—H. A. Silver and H. Griffin, Silvertown, Essex. Manufacture of Insulators, and other Articles of India-Rubber, &c. 6d. (2 figs) March 6.

INSULATORS.—These are formed of "two or more sheets or surfaces of india-rubber, or india-rubber compound," with a layer of cloth "interposed for the purpose of supporting" them "during the process of curing or manufacture." Hard india-rubber insulators are formed in moulds made of French chalk.

579.—T. W. Evans, Paris. Telegaphic Cables. 4d. March 7.

CONDUCTORS — Seven pure copper wires are placed parallel, and in close contact with each other, and are then passed through a draw plate, which reduces their volume and compels them "to assume the form of longitudinal sections of a prolonged cylinder." This is preferably electro-gilded. Three insulating coats are employed: "an interior one of pure caoutchouc, a second of gutta-percha, and a third of caoutchouc." The last may be vulcanised, and it is stated that during the process, the interior coating, which is to be put on in a stretched state, expands under the action of the heat applied, and forces out any air that may be enclosed; at the same time, the gutta-percha being softened, effectually cements the two india-rubber coverings together.

710.—W. Andrews, London.   Insulators for Telegraphic Wires.   10d. (10 figs.) March 21.

INSULATORS.—The essential part of this invention is "the interposing a gutter or ditch fitted with a cover between the wire and the stem, rod or other instrument which carries the insulator," the object being to "secure a dry sheltered chamber between the wire and the arm, or bracket and post." The inventor also claims the use of an insulated bolt or stem.

734.—W. T. Henley, London.   Electric Telegraphs, &c.   2s. 2d. (46 figs.)   March 23.

MAGNETO-ELECTRIC GENERATOR.—In Specification No. 2,769 of 1856, a machine was described in which a fixed armature or temporary magnet was employed, arranged between the poles of a permanent magnet, induction being produced and reversed in the temporary magnet by curved soft iron pieces attached to a movable disc. According to the present invention the temporary magnet is fixed at right angles to the permanent one, instead of parallel thereto, and an oscillating motion is given to the disc and soft iron pieces by means of gearing or otherwise.

POSTS, INSULATORS, &c.—The posts are made from four large sheets of metal rivetted together at the longitudinal edges and fixed to a cast-iron foot. In the insulators, the bell and the sharp edges, as made in the ordinary shackle insulator, are combined.

748.—J. Morgan, A. T. Jay, E. Edwards and J. Tilston, London.   Ropes or Cables.   7d. (2 figs.)   March 25.

CONDUCTORS.—A conducting ribbon of copper is coiled helically, corrugated, or bent so as to allow of longitudinal extension without fracture, and is insulated by one or more coatings of pure india-rubber, or other elastic material. The insulated conductor is introduced into a wire helix, which receives a covering of vegetable fibre, wires, or metallic ribbon, applied helically in the opposite direction.

753.—J. Chatterton and W. Smith, London.   Submarine Telegraph Cables.   3d. March 25.

CONDUCTORS.—"The object of this invention is to prevent oxidation and decay" of the wire sheathing of electrical cables. This is carried into effect by separately coating the protecting wires with an adhesive material, such as that described in Specification No. 1,811 of 1858, on which are laid yarns of fibrous material, by preference applied longitudinally. The method described in Specification No. 883 of 1858 is followed in preparing the sheathing wires. In some cases the yarns are saturated with an adhesive material.

800.—R. Searle, Woodford Wells, Essex.   Telegraph Cables and Wires, &c.   3d.   April 1.

CONDUCTORS.—Improvements on Patent No. 785 of 1859. Corrugated, grooved, or other than round wires are insulated by threading thereon beads, tubes, or discs of glass or other insulating substance, which are "secured into one continuous flexible line by an outer covering of plaited material." Washers of india-rubber, if necessary, are placed between the glass beads. The first set of beads may be covered with a layer of india-rubber or gutta-percha, over which is threaded a second series of beads. An outer covering of "fibrous material made up with bitumen" is lastly applied.

*853.—T. G. Ghislin, London.   Preparing and Applying Vegetable Productions.   3d. April 6.

CONDUCTORS.—The fibres of "juncus serratus (an aquatic) and certain members of the armyllideæ" are applied, with or without admixture of other materials, for insulating and protecting electrical conducting wires. Specification No. 1,094 of 1857 refers to the same plants, but the above is one of several new applications.

858.—H. Wilde, Manchester.   Electro-Magnetic Telegraphs, &c.   1s. 6d. (12 figs.)

MAGNETO-ELECTRIC GENERATOR.—Two straight bars of soft iron, having lateral projections at each end, are screwed on to the opposite sides of a bundle of rectilinear permanent magnets, all the lateral projections being on one side and enveloped with coils of insulated copper wire. Two of these bundles of magnets are laid parallel and arranged with their dissimilar poles and lateral projections opposite each other. Two magnetic or soft iron bars are fixed by their centres at right angles to an arbor, and made to vibrate between the polar surfaces of the coiled iron armatures, by hand or power, for the purpose of generating the electric current, the coils being coupled together so as to deliver the currents simultaneously in one direction. A number of magnets with their coiled armatures may be ranged side by side, the polar surfaces in each row being in the same line, and the requisite number of oscillating iron bars being fixed in the arbor.

924.—T. Miller, Fossaway, Perth, N.B.   Preparing and Applying India-Rubber, &c., for Insulating Telegraph Wires.   4s. 3d. (27 figs.)   April 15.

CONDUCTORS.—This invention relates, first, to the machinery for preparing india-rubber for insulating purposes, and secondly, to improved machinery for laying india-rubber and its compounds, and fibrous and other material, on the wire. The object of the first part of the invention is to cut the india-rubber strips with a knife, so arranged that the marks or corrugations made by such knife shall run in a longitudinal direction.

Wire covering machinery is described at some length, and illustrated in detail in the drawings.

981.—J. B. J. Noirot, Paris.   India-Rubber Pipes.   6d. (4 figs.)   April 20.

CONDUCTORS.—A machine for producing continuous india-rubber tubes, applicable also to the insulation of electrical conducting wires, consists of "two flatting cylinders turning towards one another, and compelling the material to pass through a gauge plate with central core." In coating conductors, the wires are "drawn through the internal part of the mandrel the lower portion of which has been previously cut out."

*1,113.—O. Rowland, London.   Electric Telegraphs.   3d.   May 3.

CONDUCTORS.—In place of galvanising overhead iron or steel conducting wires, the inventor coats them, by preference, with a mixture of lead, antimony and tin.   Wires similarly coated are employed for sheathing cabled conductors.

A method of supporting overhead wires, consists in hanging them to a stout wire, strained between points in long spans.

1,322.—E. H. C. Monckton, London.   Obtaining and Applying Magnetic Motive Power, &c.   4d.   May 25.

ELECTRIC MOTOR.—Motive power is obtained "through the agency of positive or negative electric currents, however derived or applied, by causing the said electric currents to be diverged at certain points by the introduction of suitable mediums at such times as the electric currents are made to pass near each other, the attraction or repulsion of the electric currents being thereby momentarily diverged or sufficiently neutralised to admit of other similarly applied, electric currents exerting or exercising their forces in like manner." "A suitable arrangement" removes the said media (which may consist of "pure soft iron" or other "para-magnetic" substance) when they have performed their office. The provisional specification mentions a means of obtaining continued rotary motion by having two magnets "so placed as to rotate past the centres of each other's attraction by means of their own momentum acquired by mutual attraction."

MAGNETS.—Fused steel is cast in moulds, hardened and magnetised.

In rendering hardened steel uniformly and "in a compound manner" magnetic, a ring of the same is placed vertically before the poles of a powerful horizontal magnet "in which position either the ring or magnet is caused to revolve till the former is sufficiently charged." It should then be withdrawn in a straight line whilst still revolving, "the same relative positions" being preserved.

The poles of magnets are tapered "to concentrate their magnetism" and such magnets are applied in the construction of motors.

The provisional specification mentions plating magnets with "gold, silver or other suitable substance to protect them from the atmosphere."

1,329. — C. S. Duncan, London.   Electric Cables or Ropes.   4d.   May 27.

CONDUCTORS.—This relates to improvements in the process of construction of electrical conductors, described in Specification No. 2,980 of 1860.

The insulated conductor is enveloped in casing of cane applied either helically or horizontally.   In some cases a covering of tanned fibrous material is laid on before the cane is applied.   Joints are made by a suitable splice, united with marine glue.   A helical binding of cord or wire is used to retain the cane in position, and to obviate "kinking."

*1,341.—E. H. C. Monckton, London.   Obtaining and Applying Magnetic Motive Power.   3d.   May 29.

ELECTRO-MAGNETIC MOTOR. — This invention relates "to cutting off or diverging magnetic attraction at certain points for the purpose of obtaining motive power, and consists in the application of galvanic or electric currents conveyed by wire into contact with magnets." Also "in the direct application to the poles of magnets" "of pieces of iron or other similar paramagnetic mediums."

To obtain a rotary motion, magnets are "placed on the periphery of a wheel or attached to a frame so as to attract, and be attracted in their turn, by other magnets in a fixed position, these magnets" "having their attraction cut off or diverged by the above applications in such manner as to admit of other magnets being successively brought into their sphere of attraction, whereby a rotary" motion is produced. The magnets are protected by electro-plating.

*1,406.—H. G. B. Roeber, Silvertown, Essex.   Insulators, &c.   3d.   June 4.

INSULATORS.—Two or more insulating materials are used in the manufacture of insulators, such as india-rubber, gutta-percha, shellac or enamel, in combination with porcelain or glass, such materials being preferred as are "differently affected by changes in the atmosphere, in order that the effect of one material may counteract that of the other," and the insulation remain perfect in all weathers.

CONDUCTORS.—Conducting wires are coated with "india-rubber or its compounds, in combination with sulphurets, sulphates and oxides of metals," such materials alternating with the india-rubber coverings. The india-rubber coatings are applied helically by means of revolving bobbins, or longitudinally by means of grooved rollers.

*1,484.—C. F. Varley, London. Electric Telegraphs. 3d. June 10.

INSULATORS.—The inventor states:—" I use very small insulators, reducing the diameter of the insulating portions as much as possible, and giving the insulator always two chances. By using a steel, iron or other suitable pin entirely covered with vulcanised caoutchouc" or other similar material, " if one end be imperfect the other prevents the electricity from escaping. These pins, I again generally partially cover" with caps of wood, glass or other material. This invention applies to erect as well as inverted insulators.

1,567.—W. E. Newton, London ; (G. Beardslee, Long Island, U.S.A.) Electro-Magnetic Engines. 11d. (4 figs.)

COMMUTATOR.—Two rings, each with teeth projecting from one side, are placed on a shaft with the teeth engaging like the claws of a claw coupling. Insulating material is placed between the teeth, and each collector is made wide enough to bridge across the insulating material between two teeth and connect both rings at each change of current. The two collectors are separated by a distance equal to the width of a tooth and an insulating lamina.

MAGNETO-ELECTRIC GENERATOR.—The invention consists in connecting the cores of the helices by a ring of soft iron instead of fixing them on non-magnetic materials.

,800.—Sir W. O. Brooke, London. Suspending and Insulating Electric Telegraph Wires. 9d. (9 figs.) July 17.

INSULATORS.—This relates to insulators, substantially similar in construction to those described in Provisional Specification No. 466 of 1861.

1,806.—C. West, London. Insulating and Covering Wires. 3d. July 18.

CONDUCTORS.—In insulated wires manufactured under Patent No. 2,321 it was found in some cases that " the india-rubber in contact with he wire became soft and treacly." To obviate this, the inventor proposes to use only the best quality of South American rubber, " not masticated, but in its native state." In order " to consolidate it after it has been placed on the wire " it is subjected to a " process of desiccation." Solvents are applied to the rubber during the covering process, by which the successive coats are made to unite.

1,907.—J. Rylands, T. G. Rylands, and P. Rylands, Warrington. Joining Wire for Telegraphic Conductors, &c. 5d. July 31.

CONDUCTORS. — Joints in overhead wires are made in the following manner :—The ends of the wires are twisted together in the usual way, that is, by winding the end of each wire helically about the other, and the space between the two twisted portions is then wrapped with wire, after which the whole is heated and welded into a solid mass tapering from the centre. The wire wrapping may be dispensed with, and the joint welded and shaped as before, or a block of metal may be cast on the joint. The object in all cases is to obtain a joint tapering from the centre " so that it may pass freely through the insulators without catching."

1,994.—H. Wilde, Manchester. Electro-Magnetic Telegraphs, &c. 2s. 6d. (26 figs.)

MAGNETO-ELECTRIC GENERATOR.—Several forms of generator are included in this specification. In one the permanent magnets are bent in the form of a hoop with the ends at a short distance from each other. Upon each of the poles, formed on the interior surface of the hoop, a cylindrical iron core wound with insulated wire is placed at right angles to the chord of an arc, the centre of which is the opening of the ring. The other extremities of the iron cores point towards the centre, and before them a small piece of soft iron is made to revolve on an axis at right angles to that of the hoop, for the purpose of generating the current. Two of these hoops with their cores may be combined, and the extra currents made to succeed each other as required. The permanent magnets may also be made from strips of spring steel within a wooden ring. In a second generator a compound permanent magnet with radial arms, like the spokes of a wheel, has a soft iron core, wound with a coil, attached to each pole perpendicularly to the magnet. A disc of soft iron with sections cut out similar to those of the magnet is made to revolve on an axis concentric with that of the permanent magnet. The passage of its arms over the cores generates current in the coils. Any number of compound magnets and discs may be arranged around one central shaft. A third form of generator has two bundles of permanent rectilinear magnets, laid parallel with dissimilar poles opposite each other. Upon each of the poles is a cylindrical iron core wound with insulated wire, so placed as to form a prolongation of their extremities. Between the two magnets there revolves an axis with a disc cut out, as already explained, at each end, and placed before the polar extremities of the two cores. The currents will circulate in the same, or alternate directions, according to the relative positions of the two discs. In place of the disc, rectangular pieces of iron may be employed.

2,292. — F. Barnett, London. Automatic Electric Signals for Railways. 8d. (2 figs.) September 14.

ELECTRIC LIGHT.- In railway tunnels a series of buttons are arranged in such a way that, when depressed by the wheels of a passing train, an electrical circuit shall be completed and an electric lamp thereby lighted. The buttons should be placed at intervals of 200 yards, the train always lighting the lamp 200 yards behind, as a signal to any approaching train.

2,303.—J. Reeves, New York, U.S.A. Electro-Magnetic Engines. 10d. (3 figs.) Sept. 16.

ELECTRO-MAGNETIC MOTOR.—This machine is an improvement on that described in Specification No. 449 of 1861. The commutator is a ring with alternate conducting and non-conducting segments, the non-conducting portions being composed of sand, emery or other sharp substance, to clean the ends of the contact springs. These springs are mounted on a slotted ring which can be moved to alter the positions of the springs, and thus control or reverse the engine.

2,310.—R. A. Brooman. London. (*C. Pougnaire, Marseilles, and J. S. Bourcy, Paris.*) Stretching, Supporting, and Uniting Telegraph Wires. 1s. 1d. (18 figs.) Sept. 16.

CONDUCTORS.—The inventor claims the following improvements in apparatus employed in erecting overhead wires:—1st. Forming the rollers in stretching winches with a slot extending to the axis, and forming the axis hollow from the point where the slot terminates. By this means a third wire, for uniting the two wire ends at a joint, may be dispensed with. 2nd. Pincers with grooved jaws, and legs provided with links, arranged so that the jaws close on the wire with the pull of the winch. 3rd. A ratchet key for the winch. 4th. Certain apparatus for supporting wires, described and illustrated. 5th. A tool for joining the ends of electrical wires.

2,464.—W. T. Henley, London. Magnetic and Electric Telegraph Apparatus. 2s. 9d. (18 figs.)

MAGNETO-ELECTRIC GENERATOR.—The specification refers to previous Patents 2,769 of 1856, and 734 of 1861, which relate to machines in which the armature has its polarity alternately reversed by the inductive influence of the permanent magnet without the necessity of moving either the temporary magnet or the permanent one. The latter magnet is entirely separated from the former until brought into magnetic connection by the intervention of moving pieces of soft iron; these are so arranged that they cause both ends of the armature to be brought alternately into connection with each pole of the permanent magnet, thus causing electric currents of exactly equal power, moving alternately in opposite directions, to be induced in the coils of wire surrounding the soft iron. Several forms of generator, according to this principle are illustrated. In some the coils lie between the poles of the magnets, and in others they are placed above them. Wheels with soft iron projections around their rims serve

to connect the magnetic poles to the soft iron cores.

2,656.—I. L. Pulvermacher, London. Galvanic and Magneto-Electric Apparatus. 11d. (10 figs.) October 22.

MAGNETO-ELECTRIC GENERATOR.—A U-shaped electro magnet rotates before the poles of a permanent magnet in the usual way. To facilitate the starting of the instrument the two sets of poles are at a considerable distance apart when at rest, but as the speed increases the centrifugal power of two weights produces a longitudinal motion of the soft iron magnet along its axis, until it just touches the fixed poles as it revolves. The drawing shows a coil on each branch of the steel magnet, but the use of them is not indicated in the description.

2,661.—T. Morris, R. Weare, and E. H. C. Monckton, London. Magnets, Induction Coils, &c. 4d. October 24.

MAGNETS.—Permanent magnets are made of cast iron which is decarburised, shaped, recarburised, hardened, and magnetised in succession.

HELICES.—"Coils for armatures of magneto-electric machines, where great intensity is required," are insulated by introducing gutta-percha between the successive layers of wire; "the thickness of the gutta-percha must keep pace with the intensity of the electric current as the coil proceeds in thickness." Helices of "graduated wire," which becomes thicker as it proceeds outwards, are employed in order to enable them to "discharge freely the increased current of electricity." Plates of metal may be substituted for wires. Collodion, gun cotton, and combinations of shellac, gutta-percha, india-rubber and their solutions, may be used for insulating the helices, as well as ammoniacal solutions of horn and other animal substances, and metallic oxides.

CONDUCTORS.—May be insulated by the above processes.

*2,682.—F. Barnett. Electric Danger Signals for Railways, &c. 5d. (1 fig.) October 26.

ELECTRIC LIGHT.—The inventor proposes to erect electric lamps on tall posts for the purpose of signalling on railways. The light would be obscured on one side, so as to only be seen by an approaching train.

*2,755.—T. Walker, London. Cables or Chains for Telegraphic and Other Purposes. 3d. November 2.

CONDUCTORS.—Metal pipes are joined together "so as to form a continuous chain of any desired length," and insulated wires are passed through. The pipes may be coated with a vitreous enamel. Machinery for making joints is described.

2,997. — H. Wilde, Manchester. Magneto-Electric Telegraphs. 1s. 7d. (6 figs.)

This invention is for improvements on a previous one, No. 1,994 of 1861.

MAGNETO-ELECTRIC GENERATOR.—A number of rectilinear permanent magnets, with soft iron cores placed on their extremities and enveloped with coils of insulated wire, are fixed upon a frame in the circumference of a circle and at right angles to its plane. Several such circles may be fixed concentric with each other. Through the centre of the circle, or circles, a shaft with a number of radial bars of soft iron is made to revolve before the polar extremities of the permanent magnets and stationary coils, the wires of which are so coupled up that the electric currents generated in each are propagated simultaneously through the same conducting wire.

In a modified form of generator the rectilinear magnets are replaced by horseshoe magnets, set with their flat sides either parallel to or at right angles to the plane of rotation of the soft bars. In some cases the soft bars are arranged parallel to the axis, and the magnets are radial to it. The momentary current may be used for telegraphing through uninsulated submarine cables.

3,015.—E. Tyer. London. Electric Telegraphs. 3s. 11d. (39 figs.)

MAGNETO-ELECTRIC GENERATOR. — Two nearly semi-circular magnets are mounted on a framework so as to form a ring with two breaks or intervals. This magnetic ring is capable of turning freely round a fixed shaft as a centre, and it has four horns of soft iron fixed to the poles of the magnets, so as to project radially from the wheel. The horns, as the wheel revolves, pass over the ends of the iron cores of a series of coils, arranged in a circle and mounted on an iron ring, which couples them altogether; as the horns pass in succession over the poles of the coils alternate induced currents are developed. In a second generator, a stationary magnet, preferably of horseshoe form has three coils placed in a row near to its poles, one of them being between the poles. A plate or disc of iron, with radiating spokes, is mounted on an axis, so that when it revolves the spokes pass over, and in close proximity with, the poles of the permanent magnet and the coils without actually touching

them. Several such magnets and coils may be arranged round a shaft, upon which is a soft iron wheel with radiating spokes. A simple form of generator has a notched iron ring, which is mounted in permanent contact with magnets, and coils are placed between openings cut in the ring. Two strips of soft iron mounted on an axle can be moved by a hand lever to bring each of the coils into magnetic connection, first with one pole and then with the adjacent pole induced in the said iron ring.

3,078.—C. F. Varley, London. Electric Telegraphs. 11d. (15 figs.) December 9.

INSULATORS.—These improvements consist in reducing the diameter of insulators as much as possible, and in " constructing them of two or more insulating portions, so that if one fail to insulate, the other part arrests the escaping electricity." Several arrangements to effect this object are described, amongst which the following are claimed :—The various forms of caps illustrated, " trumpet " shaped lips to the insulator, a special form of pin, the use of double earthenware, porcelain or glass cups, ebonite in conjunction with stoneware, an earthenware pin, projecting ears on porcelain cups for securing the wires without the necessity of binding, varnishing porcelain insulators, a pin entirely covered with insulating material, caps of a "fusible resinous compound," and " the insertion into aerial circuits at the points of support of lengths of wire partly covered with vulcanite."

3,109. — J. Potter, Leeds. Connecting Telegraph and other Wires. 5d. (6 figs.) December 11.

CONDUCTORS.—This relates to joints in conductors. The wires are introduced into conical ferrules, and rivet heads are formed on their ends. The ferrules are united by screwing them together.

*3,170.—W. Dicey, Waltham Abbey, Essex. Submarine Telegraphic Cables. 3d. December 18.

CONDUCTORS.—Strips of metal are laid upon the insulated core with a long twist, and with the edge of each overlapping the one adjoining. These metal strips are secured by a helical binding of similar strips, wire or yarn, applied in the opposite direction to the first.

1862.

59.—C. W. Siemens, London. (*Dr. W. Siemens, Berlin.*) Insulating and Protecting Wires, &c. 3s. 4d. (31 figs.) January 9.

CONDUCTORS.—The conductor is formed of a

wire or a strand of two or more wires ; or sometimes six wires are wound into the helical recesses in a rope formed of seven wires laid together. For the purposes of insulation the conductor is heated and led through a vessel of Chatterton's compound. On emerging it is passed through water, and then covered with india-rubber or its compounds, as described in Specification No. 2,503 of 1859. Joints are made by overlapping the uncovered parts of the wire by india-rubber strips. By means of a steam-heated dye the india-rubber is compressed to the proper form. Various machines for effecting these operations are described.

ELECTRO-MAGNETS.—A mass of iron is added to that part not directly under the influence of the coil to exalt the magnetism at the poles.

MAGNETO-ELECTRIC GENERATOR.—A cylindrical bar, with two deep longitudinal grooves, containing a coil of insulated wire, rotates in one direction, between the poles of a series of permanent steel magnets.

188.—T. Morris and R. Weare, Birmingham, and E. H. C. Monckton, Fineshade, Northampton. Producing Electric Signals. 4d.

ELECTRIC LIGHT.—This invention relates to the use of electric light for the purposes of signalling and communicating intelligence. The light is produced in vacuum tubes.

194.—C. West, London. Insulating and Covering Wires, &c. 4d. January 25.

CONDUCTORS.—This relates, in the main, to the method of insulating wires described in Provisional Specification No. 1,806 of 1861. When the same process is employed for covering gutta-percha coated wires, a solution of gutta-percha and india-rubber is used as a cementing medium between the two coatings. For the outer coating of any insulated conductor a solution of india-rubber, in combination with metallic oxides or carbonates, is applied either singly or mixed with oil or resinous and bituminous substances.

236.—J. B. Harby, Leytonstone, Essex. Preserving Electric Telegraph Cables and Wires. 4d. January 29.

CONDUCTORS.—Yarn employed in covering conductors is impregnated with a compound, consisting of solution of caoutchouc, resin, powdered chalk and arsenic, in suitable proportions. The composition is applied hot.

*261. — J. Hargreaves, London. Pipes or Tubes. 4d. January 30.

CONDUCTORS.—Pipes for containing electrical wires are made from a bituminous mastic containing all or any of the following ingredients: bitumen, cement, chalk, clay, flint, lime, sand, slag and slate, together with fibrous materials, gutta-percha, caoutchouc, gums and oils. The pipe-making machinery is described.

359.—R. Johnson, Manchester. Welded Wires for Telegraphic and Other Purposes. 6d. (3 figs.) February 11

CONDUCTORS.—A helical metallic coil is placed over the welded joint of a conducting wire, and secured thereto by galvanising. Should the wires separate, the coil allows of extension, and at the same time preserves continuity of the conductor.

*397.—A. J. Dodson, London. Composition for Coating Ships' Bottoms, &c. 4d. February 14.

CONDUCTORS. — Pulverised slate is combined with pitch to form a material suitable for insulating electrical conductors. Tar is sometimes added.

*404.—J. H. Johnson, London. (*J. H. Koosen, Dresden.*) Electro-Magnetic Timekeepers. (4d.) February 15.

ELECTRO-MAGNETIC MOTOR. — This is so constructed that "the accelerating force acts nearly uniformly in all phases of its rotation, and that no intervals of suspension of the acting power can interfere in any part of the revolution."

Three vertical electro-magnets are disposed so that "their six poles will form points equidistant from each other in a horizontal circle, the whole of the poles being in the same horizontal plane." Alternate poles have opposite polarity throughout the series. Immediately above, and very close to the poles of the electro-magnets are two soft iron bars with cylindrical ends, which bars are secured to a vertical spindle at right angles to each other, "so that their four extremities shall form a rectangle in a perfectly horizontal plane." A commutator is employed for distributing the current, consisting of a disc of wood fixed concentrically on the spindle, having its upper surface inlaid with 12 silver-gilt plates. The coils of the electro-magnets are connected to these plates in the order "required by the relative positions of the magnetic poles" and armatures. The other ends of the coils "are connected with each other and with one pole" of the battery. From the other pole of the battery the current passes into the spindle, whence it enters a contact brush of fine gold wires, which slides over the commutator plates, thereby conducting the current to the electro-magnets in succession, and effecting the required rotation. The governor employed to maintain a constant speed consists of "a steel rod pivotted at its centre of gravity," upon an axis supported in bearings on a metal ring, secured to and encircling the spindle of the motor. Adjustable balls are placed on the end of the axis. When the speed of the motor becomes excessive, the rod tends to assume a horizontal position, and by its movement overcomes the force of a spring, and thereby effects an interruption of the electrical circuit, by which means the speed of the motor is slackened.

413 —J. Chatterton and W. Smith, London. Telegraph Cables. 4d. February 15.

CONDUCTORS.—The iron or steel protecting wires of cables are coated with lead, applied by means of the "ordinary hydraulic lead pipe making machine," the object being to render them incorrodible. The wire is by preference tinned, or coated with an adhesive compound, before it enters the lead covering machine. Strands, preferably of three wires, are coated with gutta-percha and the usual serving" is dispensed with, "the gutta-percha covering acting as a cushion between the core and metal protecting wires." In place of a gutta-percha covering, a coating of marine glue, or other compound, mixed with 10 per cent. of fibrous material, is used, over which is applied one or more helical servings of tape.

*458.—Lord A. S. Churchill, London. Electric Telegraphs. 4d. February 20.

CONDUCTORS.—The return wire takes a different route to the outgoing wire, by which arrangement it is possible to dispense with insulation. For subaqueous conductors a thin coating of insulating material is applied, and conducting wires of larger sections than usual are employed.

*468.—S. Smith, London. Electro-Magnetic Engines. 2d. February 21.

ELECTRO-MAGNETIC MOTOR.—An endless chain with links, alternately of magnetic and non-magnetic material, is stretched over two equal wheels or pulleys. The parallel portions of the chain pass through hollow coils, which are at proper intervals excited by the current from a battery, directed by a commutator of any suitable construction.

594.—G. F. Guy, Bury St. Edmunds. Electro-Magnetic Motive Power Engines. 1s. 2d. (10 figs.) March 4.

ELECTRO-MAGNETIC MOTOR. — The invention consists in the employment of one or more series of electro-magnets arranged in circles, and made to act successively upon a ring of soft iron, which is thus caused to roll over their faces or poles. This ring is connected to the main shaft by a loose joint that permits of its gyrating, and still carrying the shaft round with it. The polar faces of the magnets form a ring which is a frustum of a flat cone. The surface of the soft iron rolling ring is turned to a cone of a greater angle, and is set upon the shaft so that the apex of its coned face would coincide with the apex of the coned face of the circle of magnets if produced at their centre. "Therefore, as the circumference of any part of this ring is larger than the path on the poles or faces of the magnets over which it rolls in proportion to the greater or less angular difference of the two cones, it follows that at each roll it will make a proportionate advance in the direction of its rotation, such advance being imparted to the main shaft of the machine." Any convenient commutator for directing the

current into the several magnets may be employed, the one shown in the drawings comprising rollers running over a cylinder, the opposite halves of which are connected to the two battery poles.

*665.—A. J. Russell, Edinburgh. (*F. Bensa and W. C. I. Anstruther, Florence, Italy.*) Electric Conductors for Submarine Telegraphs. 4d. March 12.

CONDUCTORS.—The conducting wires are twisted helically around the core, each wire being separately insulated by a covering of silk, over which a coating of gutta-percha or india-rubber is used, and then a serving of hemp saturated with tar or grease. An external sheathing of hempen cords is lastly applied.

708.—A. J. Paterson, Edinburgh. Electric Telegraph Cables. 6d. (2 figs.) Mar. 14.

CONDUCTORS.—Conducting wires are protected from strain and injury by enclosing them in a flexible or jointed tube.

*776.—R. M. Roberts, London. Obtaining and Applying Motive Power. 2d. March 20.

MAGNETO-ELECTRIC MOTOR.—After an unintelligible description of a perpetual motion machine, the inventor adds "instead of weights electric or magnetic force may be applied to the wheel."

*1,207.—F. Barnett, Paris. Signals for Railways, &c. 6d. (2 figs.) April 25.

ELECTRIC LIGHT.—For night signals on railways it is proposed to employ tall posts surmounted by electric lamps, which show their light in the direction of the approaching train only. A reflector and coloured glasses may be used if desired.

*1,255.—J. Cliff, London. Insulators. 4d. April 29.

INSULATORS.—This invention consists in constructing a compound insulator as follows:—The outer part has a notch to receive the conducting wire; the other part is a shaft, "carrying near the bottom thereof a disc, the diameter of which is nearly equal to that of the inside of the mouth of the outer part." The bottom of the shaft has a hole in which the insulator shank is inserted, and the head of the shaft is lodged in a recess formed in outer part, and is secured therein with a suitable cement. A coating of shellac is applied to the "upper angle or head inside the outer part," and also to the mouth of the before-mentioned recess.

1,304.—A. V. Newton, London. (*R. Cornelius, Philadelphia.*) Electrical Apparatus for Lighting Gas. 1s. 4d. (20 figs.) May 2.

STATIC ELECTRICAL GENERATOR.—This is a modification of the ordinary electro-phorous, and in one form consists of a metal disc, covered with lamb's skin and fitted with a non-conducting handle. "One half of the lamb's skin is overlaid with silk. The apparatus is completed by a disc of metal fitted with a disc of hard rubber and with a metal handle, the attaching metal pin of which passes through the hard rubber disc."

1,398.—F. J. Bolton, London. Telegraphing. 2s. 4d. (13 figs.) May 9.
ELECTRIC LIGHT.—For signalling purposes the inventor employs an electric light in combination with "lenses, reflectors or other well known optical arrangements."

1,482.—R. Laming, Kilburn. Electric Telegraphs. 6d. May 16.
CONDUCTORS.—The conductors of subaqueous cables are surrounded by "insulators of dried organic materials in a fibrous condition saturated with insulating cements." This invention also consists in the application and use of tubes of lead and india-rubber, and of other impervious coverings for keeping dry the above insulators, "that depend for their efficiency upon an artificial desiccation."

1,516.—T. Morris and **R. Weare**, Birmingham, and **E. H. C. Monckton**, Fineshade, Northampton. Obtaining and Applying Heat by Electricity. 4d. May 19.
ELECTRIC LIGHT.—The invention consists in obtaining light by passing electricity through vacuum tubes or other suitable vessels. The currents are generated by means of the batteries and induction coils mentioned in Specification No. 2,661 of 1861. The vacuum tubes are first exhausted and a gas, such as carbonic acid, carburetted hydrogen, phosphuretted hydrogen, nitrogen, hydrogen or oxygen is injected, or vapours, such as those of sulphuric, nitric, hydrochloric and other ethers. The gas or ether is again exhausted till a sufficient vacuum is obtained. Leyden jars may be included in the circuit "by which means the illuminating power is much increased." Reflectors of various constructions may be combined with the light, or silver may be deposited on the back of the vacuum glasses in lieu of having separate reflectors. The glasses may also be dished and covered with foil to form a parabolic or other reflector. Also lenses and ground glass covers may be used in conjunction with such tubes. Heat may be obtained from currents by the aid of coils of flattened platinum wire. It is stated that currents may be produced by generators driven from the axles of railway trains.

*1,540.—C. W. Siemens, London. (*Partly by Dr. W. Siemens, Berlin.*) Electric Telegraph Apparatus. 4d. May 22.
MAGNETO-ELECTRIC GENERATOR.—The inventor states that the electric currents, required for working the apparatus to which this invention relates, may be generated by "a magneto-electric inductor" similar to that described in Provisional Specification No. 2,107 of 1856.

1,924.—E. De la Bastida, London. (*A. Cohen and C. Vaillant, Harbourg-upon-Elbe.*) Manufacturing India-Rubber Articles. 4d. July 2.
CONDUCTORS.—The insulating coating consists of a wrapping of vulcanisable india-rubber. The curing is effected by embedding the wire and covering in grooves formed in plates, which are put under a press and heated. If great lengths are required in one unbroken piece, the ends of the several portions are left unvulcanised, and may then be spliced, after which the joint is vulcanised in a short mould.

2,070.—E. Bazin, Angers, France. Electric Railway Carriage Signal. 1s. 6d. (11 figs.) July 21.
CONDUCTORS.—In arranging an electrical circuit, in a railway train, the inventor proposes to "galvanise" the coupling chains, and employ them as conductors between carriage and carriage.

*2,131.—P. S. Devlan, Manchester. Telegraphic Cables. 4d. July 28.
CONDUCTORS.—The insulating material consists of paper or other fibrous pulp, combined with caoutchouc or gutta-percha and resin. This is applied to the conducting wires whilst it is in a soft condition, and when dry, a covering of "fine strong cord" is plaited over it. Several alternate coatings of composition and cord may be applied in succession. The object of the invention is to secure "flexibility, durability, tenacity and strength."

*2,238.—H. Fenton and **W. Stubbs**, Liverpool. Telegraph Wires. 4d. August 11.
CONDUCTORS.—This invention consists in covering iron wires with a thin coating of copper, and then with a coating of "asphalte, black varnish, or other like suitable non-conducting substance." "It is preferred to introduce into the asphalte a short fibrous substance to ensure its tenacity." By treating iron wires in this way it is unnecessary to galvanize them.

2,305.—J. H. Johnson, London. (*J. H. Koosen, Dresden.*) Electro-Magnetic Timekeeper. 10d. (5 figs.) August 15.
ELECTRO-MAGNETIC MOTOR. — This motor is substantially similar to that described in Provisional Specification No. 404 of 1862.

2,307. — H. Garside, Manchester. Engraving on Cylindrical and Other Surfaces. 1s. 10d. (7 figs.) August 16.
MAGNETO-ELECTRIC GENERATOR.—Part of this invention relates to the use of "magneto-electric"

currents for actuating engraving machinery. In one of the drawings a generator is shown, consisting apparently of two bobbins revolving on a horizontal shaft in front of the poles of a compound permanent horseshoe magnet.

2,432.—Sir W. O. Brooke, London. Submarine Telegraphic Cables. 4d. September 2.

CONDUCTORS.—The object of this invention is to construct a subaqueous cable, so that its specific gravity may be "regulated with precision," and at the same time its core may be protected from mechanical injury.

To attain this end, it is proposed to thread upon the core, wooden tubes, by preference of American beech, impregnated with drying oils or melted paraffin under pressure. "The core is enclosed in hemp or silk lines rendered waterproof, and preserved by any suitable preparations and coated with tape." Flexible and elastic washers are interposed between the wooden tubes, and to regulate the specific gravity, tubes of iron, hard wood, or marble, may be used at intervals.

2,461.—J. Snider, jun., Pennsylvania, U.S.A. Applying Graphite for Preserving Fabrics and Materials. 4d. September 6.

CONDUCTOR. — The "graphite composition," described in Specification No. 3,024 of 1861 (not included in these abstracts), is applied as a preservative to electrical wires, and when dry, one or more coatings of the "graphite paint," described in the same specification, are added.

2,488.—F. Hands and H. Holland, Birmingham. Imitation Jet. 4d. September 10.

INSULATORS.—A material suitable for the manufacture of insulators is made as follows :—A solution of india-rubber in varnish is effected by means of heat, and to this is added suitable proportions of mucilage and black pigment. The whole is incorporated in a heated state and allowed to cool, after which the insulators are moulded from it and stoved to harden them.

2,580. — H. R. Fanshawe, London. Fishing. 8d. (5 figs.) September 20.

ELECTRIC LIGHT.—For the purpose of alluring fish, the inventor employs a submerged light, which may be "derived from electricity." The light is enclosed in a suitable lantern, to which a supply of air is conveyed by flexible tubing. The globe of the lantern is protected by a wire netting, or a suitable metallic frame.

*2,619.—A. Potter, Birmingham. Electro-Magnetic Engines. 4d. September 25.

ELECTRO-MOTOR.—"The engine consists of a fixed circle or band of soft iron bars screwed straight across the inside edges of two or more stout wheels, the widths of these bars corresponding with the thickness of the coils of the electro-

magnets to be used; the distance between each bar is to be double that of the width of the bars themselves; the number of bars to be used in the circumference of the wheels may vary according to the measurement of the circumference, which measurement must be equally divided by the thickness of the coils of the electro-magnets." These bars form "armatures" for electro-magnets arranged radially upon a rotating shaft in the centre of the circle. The commutator consists of insulated metallic plates studded with platinum points. Platinised rollers with spring bearings are connected to the two poles of a battery, and, as they come in contact with the studs, direct the current into the appropriate coils. In order to reverse the engine, a second commutator is attached to the shaft, or the rollers are shifted by a revolving gear, so as to reverse the direction of the electro-magnets. A reverse current is occasionally sent through the coils to demagnetise the cores.

*2,734.—G. Baguley, Hanley, Stafford, and H. Greener, Sunderland. Insulator for Telegraph Wires. (4d.) October 10.

INSULATORS.—The insulator is formed with a longitudinal slot, into which the wire is introduced through a zig-zag opening. "By this arrangement the tendency of the wire (through the continued action of the wind upon it) to escape from its seat is completely checked," and "wire bindings and other attachments" are dispensed with.

2,772.—E. H. C. Monckton, London. Induction Coils, &c. 10d. October 14.

This specification wanders on for 19 pages, describing a great variety of apparatus, without a single illustration. In many places it is almost unintelligible, and it is seldom that the reader can be sure that he has acquired an accurate conception of the mechanism or appliances that the inventor is endeavouring to explain.

ATTRACTING MEDIUM.—It commences with the description of an "attracting medium," which is a metal plate of any shape and configuration wrapped with projecting pegs which are covered with coils of insulated wire to convert them into electro-magnets when desired. Such "media" may be used as the field magnets of various forms of magneto-electric motors.

As regards magneto-electric generators, the patentee says: "for large permanent magnets I substitute smaller electro-magnets charged by a battery or otherwise, so as to render them magnetic, and then cause the revolutions to take place either in the same way as in the common magneto-electric machines, or else I cause the magnets themselves to revolve between armatures I thereby gain not only a stronger current of electricity but I obtain the galvanic current in a new form, modified by the magneto-electric apparatus."

INSULATORS.—The parts of electric apparatus are insulated by asbestos, either as paper or cloth, or powdered and mixed with varnish.

ELECTRO-MAGNETS.—Sheets of iron and copper are rolled up together, the two being separated by silk. The iron acts as the core and the copper as the coil.

ELECTRIC TRANSMISSION OF POWER.—Windmills and water-wheels are employed to drive generators and the current is led by conductors to electro-motors situated at a distance.

The specification then goes on as follows :— " Reversing the American invention of a ring composed of magnetic and diamagnetic metals drawn through a magnetised helix, I construct solid coils of insulated copper wire or sheet . . . by winding it around the periphery of a wheel of a diamagnetic material . . . The periphery of this wheel is made to revolve " (within ?) " a slotted hollow ring composed alternately of iron and brass segments, made so that part of its section may be removed to admit the periphery of the coil within it, or the reverse of this arrangement may be effected by placing the solid coils on a frame, and causing a wheel with a hollow periphery composed of magnetic and diamagnetic metals to embrace them." Another motor is made with " attracting media " that contain coils only without iron cores. After this there follows the description of several more motors, but in the absence of drawings it is impossible to define their construction or operation.

2,815.—J. Fuller, London. Treating India-Rubber on Wires for Insulating. 4d. October 18.

CONDUCTORS.—India-rubber, after it has been placed on a wire to insulate the same, is subjected to the action of a solvent, so as to render its surface adhesive. A covering of powdered sulphur is then applied, after which the rubber is exposed to heat to " cure " it.

Before the curing process, another layer of india-rubber may be placed over the sulphured coating.

*2,845.—H. Wilde, Manchester. Electro-Magnetic Telegraphs. 4d. October 22.

MAGNETO-ELECTRIC GENERATOR. — Relates to combining the machine described in Specification 1,994 of 1861 with mechanism to adapt it for telegraphic purposes.

*2,986.—J. E. F. Ludeke, Marke, Hanover. Magneto-Electric Apparatus for Obtaining and Applying Motive Power. 4d. November 4.

MAGNETIC MOTOR.—" A series of permanent magnets are placed horizontally on a standard and within a drum, which rests on an axis near the centre of the magnets; each side of the periphery of the drum is supplied with a row or circle of ' rocking needles,' each needle working in a slot or opening in the periphery of the drum."

As the needles pass in front of the north and south poles of the magnets they " become magnetised, and diverge, the one series to the north and the other to the south," " and a continuous rotary motion " is effected.

" A modification consists of a rotating horizontal drum, through the centre of which passes a fixed axis supporting a series of permanent magnets. " As in the last instance, four or more rows of rocking needles are set in bearings from end to end of the cylinder." " The apparatus must be placed so that the needles shall rock towards the north and south poles."

3,240. — H. Wilde, Manchester. Electro-Magnetic Telegraphs. 8d. (4 figs.) December 3.

MAGNETO-ELECTRIC GENERATOR.—Relates to combining the machines described in Specifications 858 and 1,994 of 1861, with mechanism to adapt them for telegraphic purposes.

CONDUCTORS.—Overhead wires are made by twisting several fine copper wires round a core of steel. To prevent the singing of the wires, they are connected to their supports by thongs of leather, hemp, &c., in such way that the thong is maintained at the same tension as the line.

3,331.—C. Hancock and S. W. Silver, London. Compounds for Electric Insulation, &c. 4d. December 12.

CONDUCTORS.—An insulating compound, which may be applied to electrical conducting wires through a die, consists of a combination of gum ballata with some or any of the following materials : — caoutchouc (pure or vulcanised), colootha, shellac, paraffin, resin, pitch, tar, marine glue, gum copal, gum mastic, or gutta-percha. These compounds may be subjected to vulcanisation by the usual processes.

In insulating conducting wires india-rubber, in one or more layers, is first applied, and the several laps are consolidated by subjecting them to the action of steam, after which a coating of ballata or its compounds is added. In some cases the plastic compound may be first applied, and over it the coating of india-rubber.

*3,434.—F. N. Gisborne, London. Indicating the Speed of Ships. 4d. December 24.

MAGNETO-ELECTRIC GENERATOR.—A speed indicator for ships is actuated by electric currents, " excited by a soft iron armature revolving before electro-magnets attached to a permanent magnet, and motion is given to the armature by the action of the water upon screw fans attached to a shaft."

1863.

31.—E. B. Keeling, London. Lighting Halls, Theatres, &c. 4d. January 5.

ELECTRIC LIGHT.—The main feature of this invention consists in the employment of plain or tinted screens, in conjunction with the electric light, by which means the light is diffused and deep shadows are avoided.

41.—W. E. Newton, London. (*G. W. Beardslee, College Point, New York, U.S.A.*) **Magneto-Electric Telegraphs.** 1s. 6d. (4 figs.) January 6.

MAGNETO-ELECTRIC GENERATOR.—This generator consists of a "wheel of radial magnets," constructed on the plan described in Specification No. 1,647 of 1859. There are six magnets with alternate N and S poles, and above them is arranged a series of bobbins, there being as many bobbins as there are arms to the wheel of magnets. Rotation is imparted to the magnets by a handle and suitable multiplying gear. The wires of the several bobbins are connected with each other "in the usual manner."

101.—J. B. Fenby, Worcester. Printing the Score of Music. 1s. 2d. (13 figs.) Jan. 13.

ELECTRO-MAGNETIC MOTOR.—The inventor proposes to drive the wheel-work of a music-printing machine by an electro-magnetic motor.

*104.—W. Platts and J. Bailey, Manchester. Telegraphic Cables. 4d. January 13.

CONDUCTORS.—A method of protecting electrical conducting wires consists in enclosing such wires (presumably first insulated) in a metallic casing constructed in short lengths, which are united by ball and socket joints. The lengths are made in halves longitudinally and secured at their edges with bolts, or otherwise.

*160.—Sir W. O. Brooke, Paris. Insulators. 4d. January 19.

INSULATORS.—For the ebonite cell mentioned in Specification No. 1,800 of 1861, the inventor substitutes "a cell of hard and unvarnished papier maché of the same shape as the ebonite cell, but thicker, so as to fit with strong friction into the iron cap and to receive the iron pin also with strong friction." The several parts are united by lever pressure, the external cap being warmed so as to contract on cooling round the papier maché cell. The latter is saturated with paraffin. Plates of talc, or layers of asbestos fibre are also employed in the same way as the papier maché, either alone or in conjunction with it, or with some cement such as sulphur or melted resin.

*163.—W. H. Harrison, Haverfordwest. Coating and Protecting Wire, &c., from Oxidation. 4d. January 19.

CONDUCTORS.—The following process is adopted in order to protect the sheathing wires of electrical cables from oxidation:—The wire is passed through a bath consisting of a molten alloy of tin and zinc, with a small proportion of lead, after which it is drawn, rolled or burnished, and then treated in a second bath of alloy, composed almost entirely of lead, "tempered in its degree of hardness" by the addition of bismuth. The drawing or other mechanical processes are then repeated.

464.—C. W. Siemens, London. Insulating and Supporting Telegraph Wires. 1s. (11 figs.) February 20.

CONDUCTORS.—The juice extracted from india-rubber or gutta-percha trees is used as an insulating material. The wire is first covered with a layer of fibrous material, and then passed through a bath of the juice, from which it is slowly raised and drawn up through an air flue of considerable height, kept at a temperature of from 100° to 300° Fahr. whereby the coating becomes consolidated.

The above-mentioned juices may be used to saturate the hempen coverings of electrical cables. The core, after it has been spun round with hemp, is passed through a bath of the juice, and then receives a second layer of hemp applied in the reverse direction to the previous one.

The covering thus formed may be enveloped in a "metallic flexible sheathing," as described in Patent No. 59 of 1862.

INSULATORS.—It is preferred to use an insulator consisting of an inverted bell of non-conducting material, cemented into a cast-iron hood. An enamelled stem of iron or steel is cemented into the bell and supports the wire in a hook at its lower end. The cast-iron hood has an arm terminating in a socket, which is slipped over the post. In place of a socket a curved flange may be formed on the arm, and the attachment made by means of a horseshoe strap passing over it and through an iron saddle on the opposite side of the post. Two such insulators may be formed on the same bracket.

505.—W. Hooper, Mitcham. Insulating Wires, &c. 4d. February 24.

CONDUCTORS.—India-rubber sheets for insulating purposes are cut from a block of the material, which is caused to rotate in front of a knife, and the sheets are divided into narrow strips by a series of cutting edges fixed at suitable distances apart. These strips are usually applied to the wire by lapping them round the same. Vulcanisable india-rubber compounds may be cut into strips and used as above described, and the curing may be effected after the wires are coated.

Sheets or strips of india-rubber are varnished

with a mixture of shellac, water, and strong liquid ammonia, to keep them from adhering together. The strips are submitted to a uniform temperature of 60° Fahr. before they are applied to the wire.

In making joints in india-rubber or gutta-percha, two steam heated chambers are used, which, when brought together, form a tube in which the joint is laid. Joints are sometimes immersed in a tar and sulphur or other bath. To cause the vulcanised india-rubber to unite, the inventor coats its surface with india-rubber solution and ignites the same " to produce tackiness." Over the surface so prepared, a coating of sulphur compound is laid, bound with cloth and heated.

The carry-off motion of the lapping machine is regulated by the use of two cones, upon which the driving belt can be shifted as required.

516.—H. Wilde, Manchester. Electro-Magnetic Telegraphs. 10d. (9 figs.)

MAGNETO-ELECTRIC GENERATOR.—The cylindrical soft iron armature *k* (Fig. 57) is grooved at each

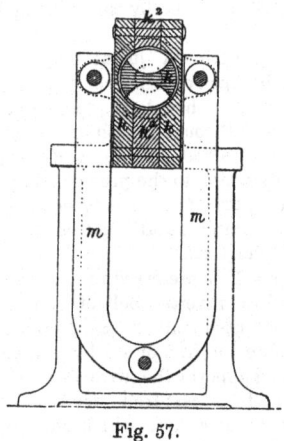

Fig. 57.

side to receive a coil of insulated wire wound

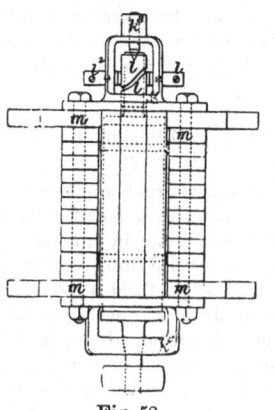

Fig. 58.

parallel to its axis. At one end it carries a commutator formed of two cylinders of hardened steel insulated from each other and divided by a diagonal cut so that the springs may run easily over the junction. The armature revolves in a cylinder formed of two outer pieces of iron k^1, and an inner piece of brass k^2. The alternate currents from the coils are produced with too great rapidity for use in telegraph lines, so they are turned in one direction by the commutator and then are reversed by a more slowly revolving commutator, before they are sent to the step-by-step instruments.

963.—R. Knight, Dunkirk. Treating and Preparing Telegraphic Wires. 6d. (2 figs.) April 17.

CONDUCTORS.—Iron, copper or other wires are covered with a coating of pure metallic tin, which is afterwards rendered homogeneous by compression between rollers. The object is to obtain a non-oxidisable and protective coating.

1,550.—C. Peterson, Newport, Isle of Wight. Compound for Manufacturing Pipes or Tubes, Caulking, &c. 4d. June 20.

CONDUCTORS.—Vegetable fibre is boiled in a solution of caustic alkali, and rendered into a pulp, to which is added an equal weight of saponified tar, and about 10 per cent. of alum or sulphuric acid, or other "powerful chemical decomposing agent."

The resulting composition is employed for coating electrical conducting wires.

1,620.—W. Andrews, London. Insulating Telegraph Wires. 8d. (4 figs.) June 29.

INSULATORS. — In addition to the ordinary insulators, the arms and brackets to which such insulators are attached are kept free from moisture by the use of sheds or covers, open at the bottom. The sheds are made of glazed earthenware, closed at the top, two sides and outer ends, and such sides and ends descend below the under surfaces of the arms. The sheds do not come in contact with those parts of the arms which they envelop. When erect insulators are employed their stems are insulated, and pass through the upper parts of the inverted sheds.

*1,635.—W. Snell, London. (*J. Wiese, Paris.*) Waterproof Material. 4d. July 1.

CONDUCTORS.—A material possessing the qualities of " resistance to moisture and changes of temperature, elasticity, durability, cheapness, and facility of renovation," and suitable for use in the manufacture of submerged cables, is composed of gutta-percha, new or waste caoutchouc, or any other kind of elastic gummy matter, mixed with hair, wool, or other fibrous matter. The inventor calls the material " elastic fibrine " (gorme indéschirable) and does not refer to it as an insulator.

1,668.—H. A. Bonneville, Paris. (*G. E. M. Gérard, Paris.*) Telegraph Wires, &c. 4d. July 4.

CONDUCTORS.—" The wires, whether naked or already protected by cotton, are covered with one or more coats of collodion," which forms an efficient insulating envelope. Oleic acid, raw or boiled castor oil or other greasy matter may be added to the collodion to render it more "supple." A coating of india-rubber, or india-rubber combined with shellac, may be placed over the collodion if desired, and it is preferred to apply such coating by means of two fluted rollers. The inventor states that the collodion prevents contact between the india-rubber and the wire, and consequently prevents the deterioration of the former.

1,733.—E. D. Chattaway, London. Railway Signals. 4d. July 11.

CONDUCTORS.—The inventor encloses insulated wires in a cord, for use in railway trains.

1,751.—P. C. A. Jodocius, Dunkerque. (*J. d'Atouquia de Franca-Netto, St. Petersburg.*) Fishing. 1s. 6d. July 13.

ELECTRIC LIGHT.—The electric light is employed as a lure for fish, the rays being passed through white or coloured glass, the different colours to be chosen according to the phosphoric state of the water, or the kind of fishing undertaken. A battery is mentioned as the source of current, and it is preferred to produce the light in vacuo.

*1,931.—W. Storer and J. Hancock, Nottingham. Electromotive Engines. 4d. August 5.

ELECTRO-MAGNETIC MOTOR.—" A ring of magnets" is attached to a suitable frame, and within them is fixed a circular rack with internal teeth. Upon a central shaft, to which rotary motion is to be communicated, four or more arms are fixed, each of which carries a wheel with six armatures, " more or less free to revolve upon an axis," and upon this axis is also a toothed pinion gearing into the circular rack. " Upon the current being applied to the magnets, in proper order, the armatures are attracted, revolve on their axes, and carry round the arms and shaft to which they are connected." In some cases the arms carrying the armatures are kept stationary and the circular rack is driven round.

*2,007.—A. E. Brae, Leeds. Conducting Electric Currents for Signalling in Railway Trains. 4d. August 14.

CONDUCTORS.—A helical wire is enclosed within an india-rubber tube. In using this conductor for coupling railway carriages electrically, the joint is made by means of a pin dipping into a mercury cup.

2,192.—J. Rowell, Aberdeen. Fences, Gate Posts, &c. 1s. 4d. (45 figs.) September 5.

CONDUCTORS.—Improvement on Patent No. 2,516 of 1862.

This relates to a " diagonal system of straining and strengthening sets of wires or wire cords."

Insulation is effected at their crossing points by means of eyes "constructed of such form and material as will prevent their contact and more completely bind them together."

2,373.—L. H. Norris, London. (*L. Holl, Paris.*) Manufacture of India-Rubber and Gutta-Percha Compounds. 4d. September 26.

CONDUCTORS.—An insulating material suitable for coating electrical conducting wires is made by combining vulcanised india-rubber waste with gutta-percha and tar. A small proportion of sulphur is added, by preference, before the curing process, but the above compound may be subjected to heat, without the use of such additional sulphur.

2,386.—F. G. Mulholland, London. Manufacturing Submarine Telegraph Cables. 6d. September 29.

CONDUCTORS.—Pure copper wire of No. 14 B. W. G. is employed for the conductor, and is " prepared for receiving the insulating mediums" by passing it through a solution of shellac in naptha, to which a small proportion of muriatic acid is added. Allowance for expansion and contraction is made by overlapping the ends of the wires at the joints, and " taking one or more round turns with each fag end over the standing parts of the conductors."

The insulation is effected by passing the conducting wire, thus prepared, through a " semi-fluid admixture" of the solution above mentioned with caoutchouc dissolved in benzine. A lapping of solutioned india-rubber strips is then applied, and the conductor is afterwards passed through a die " arranged to allow under any desirable degree of pressure a permanent and positive covering," at least three-sixths of an inch thick. Phosphorus is "invariably" used to prevent decomposition of the caoutchouc, and vulcanisation is effected by the "well-known cold process."

Any number of prepared wires may be strained and embedded in a composition consisting of vulcanised caoutchouc, resin, coal tar, shoddy, and phosphorus, applied in vacuo by means of grooved rollers.

The conductor thus constructed is passed through a protective composition, consisting of shellac and caoutchouc solutions, combined with red lead and phosphorus.

A sheathing of hemp, or preferably " green hide," is applied by means of braiding machinery.

2,387.—S. Mendel, Manchester. Weaving Fabrics for Covering Telegraph Wires. 4d. September 29.

CONDUCTORS.—This invention relates to a method of producing strips of cloth without selvages, to be used for insulating purposes.

*2,413.—J. E. F. Ludeke, London, and M. Fischer, Paris. Obtaining Motive Power. 4d. October 2.

MAGNETIC-MOTOR.—" On the circumference of a circular frame are placed a series of magnets, some of which on the upper part incline inwards, and the others outwards, but the lower parts of the said magnets are in the same line, some with the north pole to the left, and the others in the opposite direction. Inside of the frame a wheel, provided with armatures, revolves, which armatures, as they come in contact with the magnets, close the current, whilst pieces of iron placed in a direction opposite to the south or north pole of the magnets, and at some distance from them, have the effect of putting in contact and breaking the current." A vertical rod moves round with the armature wheel, " and carries with it one or more magnets." This rod carries a cross-piece from which descends another vertical arm, also carrying one or more magnets. " These magnets, as they move round with the wheel, serve to renovate those fixed to the frame; the outer ones renovating from north to south, and the inner ones from south to north, or *vice versâ.*"

2,893—J. G. Jennings and M. L. J. Lavater, London. Manufacture of Tubes, Rings, and Cords of India-Rubber. 4d. November 18.

CONDUCTORS.—India-rubber strips are cut from a block of the material by a knife set at an angle of 45°, whereby a bevel is given to the edges of such strips, rendering them suitable for lapping helically around conducting wires.

3,006.—H. Wilde, Manchester. Electric Telegraphs. 1s. 6d. (11 figs.) December 1.

CONDUCTORS. — A series of iron wires are varnished and placed in the interior of iron pipes similar to those used for gas or water. The wires are supported and separated from one another, and from the sides of the pipes by means of perforated earthenware cylinders. The ends of the wires are connected by inserting them in a short metal tube and soldering. The joints in the pipes are made watertight by filling the annular space between socket and spigot with an india-rubber ring. To prevent the pipes being flooded they are laid on an incline and drained by means of syphons. Junction boxes are provided where necessary.

MAGNETO-ELECTRIC GENERATOR.—This invention further consists in employing the magneto-electric generator described in Specification No. 516 of 1863, for signalling through uninsulated submarine cables.

Electro-magnets may be employed in place of permanent magnets in the above generator, and they may be excited " by means of a voltaic battery," or preferably by the small magneto-electric generator described in the above-mentioned specification. These generators may be used for producing the electric light.

3,012.—J. G. Redman and G. Martin, Brompton, Kent. Compositions for Coating Ships or Vessels, &c. 4d. December 1.

CONDUCTORS.—A preservative composition for coating electrical cables is made as follows :—Oxidised brass or protoxide of copper is ground in water, dried, and mixed with oxide of zinc or alumina and oxide of lead. The compound is then boiled in pitch, tar, or resin, or a suitable combination of these ingredients, until " a quick drying metallic varnish is formed." The varnish may be thinned by adding mineral naptha or " other cheap spirits."

*3,151.—J. A. Bailey, Brooklyn, New York, and J. J. Speed, Gorham, Maine, U.S.A. Insulating Telegraphic Wires. 4d. Dec. 14.

CONDUCTORS.—The main feature of this invention consists in covering the wire conductor with fibrous cords, which are then compressed by suitable mechanism to a solid mass " so hard and compact as to resist the action of moisture upon it." The covering should first be saturated with a non-conducting substance. Single wires thus covered may be made into a cable, and such cable be wound and compressed in the same manner.

3,252. — F. Walton, Chiswick. Telegraphic Cables. 6d. (5 figs.) December 23.

CONDUCTORS.—The insulating material, preferably consisting of oxidised oil, with or without the admixture of shellac, resin, and powdered silica, is applied to the wire by means of a series of dies, arranged in a circle around a central hole through which the wire is fed. The dies are of such a shape that the issuing strips of insulating material form a complete cylinder about the conducting wire, and are consolidated thereon by " pressing rollers." Over this are applied linen tapes, coated with a composition of oxidised oil, Kaurie gum, powdered silica, sugar of lead, and coal naptha. If several conductors are laid together, the interstices should be filled with composition " so as to make of them one solid coil."

The inventor employs protecting wires " hollowed out in two opposite sides, and conveniently flat on the other sides," which are applied helically, with the flat sides in contact with the core, and between the wires is laid a tarred yarn. The cable is served with strands of hemp or wire as it leaves the rollers of the sheathing machine. The addition of silica to the insulating composition prevents the attacks of boring animals.

1864.

169.— F. J. Ritchie, Edinburgh. Applying Magneto-Electricity for Propelling Clocks. 1s. (6 figs.) January 21.

MAGNETO-ELECTRIC GENERATOR.—Four coils are arranged on soft iron polar extensions of a permanent magnet and a keeper is rotated over them, or a disc wheel carrying small bars of iron is rotated before the poles of a permanent magnet.

*637.— F. H. Needham, London. Electric Telegraph Cables. 4d. March 12.

CONDUCTORS.—The inventor proposes to insulate wires " by means of an oxidised surface or coating which will form a non-conducting covering." In submarine wires there should be added " a deposit of saline earthy substances similar to those found in the sea." The oxidised surface may be obtained by immersion in dilute acids, exposure to moist air, or by other means. " A communicated leaden surface may be applied to the copper," or an oxide of lead, and the sea water " would then increase such oxidation." Submarine cables should be strengthened with fibrous material.

833.—W. E. Newton, London. (*A. Holtzman, The Hague.*) Electric or Telegraphic Conductors. 8d. (12 figs.) April 2.

CONDUCTORS.—The wires are enclosed in an iron tube of suitable diameter, care being taken to keep them from touching each other or the sides of the tube, and the latter is then filled with liquid pitch " forced in with moderate pressure." The lengths of tube are connected by butt joints, made watertight by means of short tubes of gutta-percha, the whole being secured by a metal clamping band.

1,072.—T. G. Ghislin, London. Treating and Applying Seaweed. 4d. April 28.

INSULATORS.—Seaweed, prepared in accordance with Patent No. 2,035 of 1862, is combined with gums, gum resins, including gutta-percha and india-rubber, resins, bituminous, fibrous, and fatty substances, and the silicates of soda and potash, chalk, talc, metallic oxides, gelatine, farina, alum, tungstic acid, powdered charcoal, and other analogous substances, and the resulting compound, called by the inventor " algœite," is applied to the construction of insulators.

1,126.— W. T. Henley, London. Telegraph Wires and Cables. 3s. (17 figs.) May 4.

CONDUCTORS.—" When the line is not of very great length " the inventor prefers to use " homogeneous metal " in place of copper for the conducting wires. Alternate coatings of insulating composition and fibrous material are applied to the wires by suitable machinery at one operation.

After the application of the outer bituminous coating, the cable is compressed, first, by a split die or tongs, and subsequently by grooved rollers so constructed as to obviate a " fin " being formed on each side of the cable.

Several methods of welding the iron sheathing wires are described and illustrated.

1,310. — J. H. Brown, London. Treating Animal Substances for Manufacturing Size, Pulp, &c. 6d. May 26.

CONDUCTORS.—An insulating material is made from " the refuse of animal hides, known as scrolls or glue pieces," or from the skins of animals, generally, which are " convertible into fibres."

The substances are treated in succession with alkalies, sulphuric acid, and a " bleaching liquor " " until the fibrous effects are produced."

The fibre is then mixed with india-rubber, gutta-percha, or other like material, and applied to the wires in a plastic state.

1,386.—W. Clark, London. (*J. H. Cazal, Paris.*) Electro-Magnetic and Magneto-Electric Apparatus. 1s. 2d. (5 figs.)

ELECTRO-MAGNETS.—The magnet is very short, in comparison with its diameter, and has two iron heads or flanges which extend over the coils. The edges of the flanges are notched at intervals to form, as it is stated, magnetic poles.

MAGNETO-ELECTRIC GENERATOR AND MOTOR.— Such a coil placed upon an axis and alternately magnetised and demagnetised would form a motor if provided with stationary keepers. If it were rotated by power and magnets were substituted from the keepers it would generate currents.

*2,047.—T. P. Tregaskis, Perran-ax-Worthal, Cornwall. Magnetic Engine. 4d. August 18.

MAGNETIC MOTOR.—A series of magnets, or pieces of soft iron, are placed around the periphery of a wheel in such a manner that half their ends or poles face the centre of the wheel and the other half face in an opposite direction. They work in radial guides and are supplied with catches to stop them at any point in the line of their motion. The axis of the wheel is horizontal. A stationary magnet, the face of which is curved eccentrically to the wheel, is fixed so that the upper part of the curved portion " may touch (if necessary) the outer face or faces of one or more of the sliding magnets," " at some part of the wheel, which is uppermost," and the lower part may be at " some convenient distance " from the faces of the magnets " in some lower part of the wheel." The poles of the sliding magnets glide over the inner face of the stationary magnet and the catches retain the sliding magnets in position until they are sufficiently low down to be released.

They are then attracted to the centre by a fixed magnet of suitable form and held there also by catches. Springs may be substituted for this inner magnet. By the first arrangement "a continuous eccentric motion" is obtained, and the second arrangement (viz., the springs) produces a "semi-eccentric motion." "A continual overbalancing of the sliding magnets" resulting. The power produced is proportional to the size and weight of the sliding magnets.

2,158.—A. M. J. Count De Molin, Paris. Electro-Magnetic Engine. 8d. (2 figs.) September 2.

ELECTRO-MAGNETIC MOTOR.— The main features of this invention are (1) the very small distance between the armatures and their respective electro-magnets before the latter are magnetised, and (2) the rolling motion of the parts carrying the armatures. A series of vertical electro-magnets are disposed upon a bed plate in concentric circles, and at equal distances apart, their poles also being in radial lines. A shaft, with its lower end terminating in a steel cone, rests on the top of a pillar in the centre of the bed plate, and carries the rings which support the armatures. The surfaces of these armatures, presented to the electro-magnets, are arranged so as to form part of a very obtuse cone, in order that when any one armature is attracted to its respective electro-magnets, the armature and electro-magnets radially opposite to it are only separated by a short distance. The top of the shaft already mentioned revolves in the hollow part of a crank arm, so as to form a ball-and-socket joint. The crank is fixed to the lower end of a vertical shaft, which rotates in bearings, and is provided with a fly wheel and pulleys. A commutator is employed to direct the current to the electro-magnets in succession, and consists of a ring attached to, but insulated from the crank shaft. Against this ring bears a spring in connection with one pole of the battery, and the ring at the same time carries a curved arm, the end of which passes, in turn, over a series of brass plates fixed to a board, carried by the frame of the motor. Each of these plates is in connection with one end of the coils of its respective set of electro-magnets, the other end of such coils being joined to a common conductor attached to the second pole of the battery. By this arrangement of parts, rotation is imparted to the vertical shaft as the electro-magnets are successively excited by the battery current.

2,341.—A. V. Newton, London. (*D. H. Southworth, New York ; B. Lorillard and C. Ferris, White Plains, New York, U.S.A.* Telegraphic Cables. 8d. (11 figs.) September 23.

CONDUCTORS.—" The principal feature of this invention consists in a mode of applying an insulating piece of gutta-percha, india-rubber," or other material, "whereby the said piece is made to serve both as a central core for the separation of the several wires, and an envelope for enclosing the same," and insulating them from each other.

The insulating material is forced through a die "having radial grooves corresponding in number with the metal strands intended to be insulated."

" These grooves serve to form fins or flanges which radiate from a central core, and in the recesses formed by the flanges the metal strands are laid."

A suitable train of machinery performs the operations of forming the cable. A hollow rotating mandrel with tapered entrance turns over the fins or flanges of the core and makes them envelop the wires ; the core then passes through two hollow spindles in succession at which it receives, first, a serving of wire or cord, and then a serving of cord and tape, applied by means of rotary fliers.

The cable thus prepared is passed through a bath of waterproof cement, then cooled in water, and finally it receives an external serving of wire or cord.

*2,486.—C. H. Collette, London. *T. Faucheux, Paris.* Magneto-Electric Machines. 4d. October 10.

MAGNETO-ELECTRIC GENERATOR.—The machine works in an atmosphere artificially raised to a temperature of 70 degrees.

*2,536.—L. J. Crossley, Halifax, Yorks. Supporting and Insulating Telegraph Wires. 1s. (6 figs.) October 14.

INSULATORS.—This invention relates to peculiar forms of insulators constructed of earthenware or other non-conducting material. In some cases the supporting arm and insulator are made in one piece, intended to be bolted to the posts. Iron may be embedded in such insulators to increase their strength. An insulator having a spiral channel leading to the groove, in which the wire rests, is illustrated. "Shackles" and other insulators of somewhat fantastic form are shown in section in the drawings.

2,675.—A. Parkes, Birmingham. Preparing Compounds of Gun Cotton and Other Substances, &c. 4d. October 28.

CONDUCTORS. — An insulating material for electrical conducting wires is prepared from gun cotton by treating it with a solvent, obtained by distilling wood naptha with chloride of calcium, and adding to the gun cotton so prepared, half its weight of castor oil. The compound can be applied with dies in the same manner as gutta-percha.

2,681.—L. P. G. Bellet and C. M. P. De Rouvre, Versailles. Motive Power. 10d. (4 figs.) October 29.

ELECTRO-MAGNETIC MOTOR.—This invention is applied to locomotion, and consists in actuating a pair of driving wheels by electro-magnetism. Each driving wheel carries 20 electro-magnets whose cores form the spokes, and the poles of which penetrate the tyre and are flush with its surface. The wheels are set on an axle, so that the electro-magnets of one wheel alternate with those of the other. The axle carries a commutator " similar to those employed in apparatus for natural philosophy," composed of two ebonite discs on the circumference of which are alternately set 20 ivory and 20 platinum plates. To each of the latter is fixed one wire from each of the electro-magnet coils, the other ends of such coils being in connection with a platinum disc carried by the axle. Two springs, in metallic connection with each other, and also with one pole of the battery, bear against the two commutator discs respectively. A third spring bears against the platinum disc, and conveys thereto the current from the other battery pole. Midway between the rails on which the motor runs are two metallic wires in communication with a battery. Movable rods with grooved rollers depend from the motor and serve to establish a metallic connection between these wires and the commutator springs. The current is so distributed that the electro-magnets are excited in turn as they approach the rail, and by their attraction thereon cause the wheels to rotate and the motor to advance.

3,092.— C. Hancock and S. W. Silver, London. **Electric Insulators.** 4d. December 14.

CONDUCTORS.—The wire to be insulated is passed through a vessel containing milk of ballata, or of caoutchouc, or a mixture of the two. The coating is allowed to dry and the process is repeated till any desired thickness is obtained. The wire may also be insulated by the same materials in " the sheet state," by means of a taping machine. The wire may be covered with cotton or other suitable material, and then coated with the insulating milk as above.

" The insulation may be further protected " by vulcanising.

The compounded milk of ballata and caoutchouc may be masticated, and applied to wires by means of the ordinary covering machine.

1865.

*22.—W. Clark, London. (*C. F. Carlier, Paris.*) **Electro-Magnets.** 4d. January 4.

ELECTRO-MAGNETS.—Uninsulated wire is used in winding the bobbins, but it is essential that each layer of wire " be separated from that superposed by means of a paper covering of sufficient length to envelop the extreme coils." The first layer of wire should be insulated from the core of the bobbin, if such core be of metal.

It is stated that electro-magnets thus constructed are more powerful than those in which insulated wire is used.

*98.— J. Fuller, Silvertown, Essex. **Covering for Telegraphic Cables.** 4d. January 12.

CONDUCTORS.—Aerial cables are protected from injury by sheathing them with metal, applied helically in the form of wire or tape.

To obviate the difficulty of making joints in wires coated with vulcanite, the inventor proposes to place over such coating, a covering of ordinary rubber or gutta-percha, " which can be acted on by solvents or heat," and the ends can thus be " easily and perfectly joined."

269. — R. A. Brooman, London. (*J. A. E. Laloubère, Paris.*) **Railway Rail Enclosing a Telegraph Wire.** 4d. January 31.

CONDUCTORS.—Lengths of tubular rail are filled with a bituminous composition, leaving a passage down the centre about one-third of an inch in diameter, to receive the conducting wire. The rails are laid end to end at a suitable gauge for use with rolling stock. Joints are made by forming discs on the end of the wire lengths, which make contact when the rail lengths are brought together, or by a ring slipped over the wire ends, or lastly by twisting one wire helically about the end of the other.

The ingredients of two bituminous compositions are given.

*362.—W. A. Marshall, London. (*J. Erckmann, Paris.*) **Using Asbestos as an Insulating Material.** 4d. February 9.

CONDUCTORS.—The wire is enveloped in a covering of asbestos, and then passed through a bath of resinous material or other insulating composition, or it may be protected in any other manner.

The main feature of the invention, however, is the employment of asbestos for insulating purposes.

*461. — T. P. Tregaskis, Perran-ar-Worthal, Cornwall. **Obtaining Motion by Magnets.** 4d. February 18.

MAGNETIC MOTOR.—This provisional specification is substantially identical with No. 2,047 of 1864.

619.—C. F. Varley, Beckenham, Kent. **Giving Electric Signals for the Protection of Property.** 8d. (4 figs.) March 6.

INSULATORS.—" A bar of ebonite with a hole

bored across each end to carry the wire through," is used as a "terminal insulator," the bar being of sufficient strength to bear the stress of the wire upon it.

Intermediate insulators are constructed of "a straight piece of solid ebonite," affixed to the pole, to which the conducting wires are attached; or an ebonite hook or ring may hold the wires.

*847.—A. I. L. Gordon, London. Signalling in Railway Trains and Preventing Burglary. 4d. March 25.

ELECTRIC LIGHT. — For the purpose of communicating in trains the inventor proposes to place on the roof of each carriage "a small electric machine provided with a burner enclosed in a glass frame; this burner may be either a charcoal point or any other combustible substance which can be ignited by electricity."

For the prevention and detection of burglary, a contact button is arranged so as to be actuated by the door or window, which when opened is thus made to close the ciruit to an electric lamp placed in a conspicuous position outside the building.

*975.—J. S. Watson and A. Horwood, London. Transmitting Signals and Alarms. 4d. April 6.

CONDUCTORS.—These consist of "one, two or more metal tubes placed one within the other, the interior of such second metal tube being insulated," preferably with cotton or silk. This "second metal tube" may contain one or more wire or wires, also insulated. The tubes as well as the wires are employed as conductors.

1,031.—W. E. Newton, London. (*J. L. Arman, Paris.*) Submarine Telegraph Cables. 8d. (5 figs.) April 11.

CONDUCTORS.—Insulated wires are wound helically around a hempen or other light, but un-absorbent, core, and the whole is covered with Indian hemp, to protect it and "give it sufficient floating power."

1,230.—C. W. Siemens, London. Regulating Machinery in Motion. 1s. 6d. (13 figs.) May 2.

ELECTRO-MAGNETIC MOTOR.—This motor consists of a fixed horseshoe electro-magnet between the arms of which, and parallel thereto, is a spindle, carrying an iron bar, in close proximity to the poles of the electro-magnet.

"When currents of electricity are passed at proper intervals" through the coils of the electro-magnet, these intervals being regulated by means of a contact lever, acted on by a cam on the spindle, rotation of the iron bar and spindle is obtained.

1,368.—T. Faucheux, London. (*H. Brandon,*

Paris.) Rotary Magneto-Electric Machines. 8d. (1 fig.)

COMMUTATOR.—The current is conducted from the generator to an insulated collar on its axis. The upper part of the collar runs in contact with a metal step, and the lower part in an oil box. There is no provision for the return current. The invention is applicable to any generator.

*1,376.—S. A. Varley, London. Insulators. 6d. (1 fig.) May 18.

INSULATORS.—The earthenware cups which form what is known as "double V insulators," are made with bands at the top and bottom of the upper portion of the cups. "The effect of this is to counteract any irregularity in the shape, and to insure the full advantage of the length of the fitting."

The cement preferred for uniting the separate portions of the insulator, consists of asphaltum, shellac and india-rubber or gutta-percha, to which is sometimes added fine sand, or other suitable powder.

*1,412.—H. Wilde, Manchester. Producing and Applying Electricity. 4d. May 23.

ELECTRO-MAGNETIC GENERATOR.—This invention refers to improvements on that described in Specification No. 3,006 of 1863. "A short circuit is made between the coils of the small magneto-electric machine and the coils of the electro-magnetic machine whenever the commutator, on the axis of the armature of the small magneto-electric machine, is at the dead point. The large armature of the electro-magnetic machine is enveloped with coils made of sheet copper the whole width of the armature, instead of being bound with wire as in the drawing of the specification, and the coils are insulated from one another by means of a sheet of cardboard or other insulating substance." The current from the machine may be used to produce the heat necessary for the working of metals, by connecting the terminals to two insulated rolls between which the bar or plate to be heated passes. The machines described in Specifications Nos. 516 and 3,006 of 1863 may be applied to the prevention of the fouling of ships' bottoms.

1,543.—A. I. L. Gordon, London. Signalling on Railways, &c. 1s. 4d. (22 figs.) June 5.

CONDUCTORS.—Three insulated wires are stranded together, and joints are made by fastenings so "arranged that one movement connects all the wires and it is impossible to connect them improperly." These conductors are for use for communicating in railway trains.

ELECTRIC LIGHT.—The electric light is applied to the prevention of burglaries in the way described in Provisional Specification No. 847 of 1865.

1,544.—J. Kennedy, London. Submerging Telegraphic Cables. 8d. (6 figs.) June 5.

CONDUCTORS.—A copper wire is "surrounded by a coating of insulating material and its specific gravity reduced so that it will only just sink in water."

This reduction of the specific gravity may be effected by the use of floats of vulcanised india-rubber filled with cork, or by enveloping the cable in a tubular covering, and filling the space, between such covering and the cable, with cork dust, or by other suitable means.

1,962.—F. A. Abel, Woolwich. Compounds for Waterproofing and Insulating. 4d. July 29.

CONDUCTORS.—An insulating material is made by combining india-rubber or gutta-percha with beeswax, paraffin, and like substances or mixtures thereof. The inventor states, "instead of mixing, masticating, or incorporating the india-rubber and gutta-percha with the paraffin or beeswax in definite proportions, I simply place the india-rubber or gutta-percha in a bath of the paraffin or beeswax heated by preference to a temperature of " from 120° to 150° or from 220° to 250° Fahr., accordingly as gutta-percha or india-rubber is used respectively. The wires may be coated with these compounds by forming the latter into ribbons and winding them helically around it, while in an adhesive state.

1,979.—A. V. Newton, London. (*J. Kidder, New York.*) Obtaining Induced Currents of Electricity. 6d. (3 figs.) July 31.

ELECTRO-MAGNETIC GENERATOR.—An induction coil is applied at the middle of the length of a soft iron core, so arranged relatively to a revolving permanent magnet that the poles of that magnet will revolve in proximity to the ends of the core, and thereby magnetise the core.

MAGNETO-ELECTRIC GENERATOR.—The induction coil may be placed around the central or neutral portion of a permanent magnet to obtain an induced current from such magnet.

2,025.—F. G. Mulholland, London. Submarine Telegraph Cables. 4d. August 4.

CONDUCTORS.—Improvements on the processes set forth in Specification No. 2,386 of 1863.

1st. Two or more chemically pure copper wires, ranging from 12 to 20 B. W. G., are loosely inter-twined "in their entire length," so as to admit of elongation under stress. These wires are coated with insulating compositions, and subsequently covered with "repeated layers of india-rubber immersed in composition."

2nd. Conductors are insulated with the composition described below, served with hemp, and sheathed with iron wires. The complete cable is passed through rollers under pressure, to insure

homogeneity and uniformity, and is afterwards splashed through a trough of heated composition.

3rd. The composition above referred to consists of asphalte, plumbago, caoutchouc, naptha, sulphur, silicic acid, phosphoric acid, and bisulphide of carbon.

The inventor claims also the "deoxidation of the surface of the conducting wires" before insulation, and "the combination of any number of insulated conductors to form a cable."

2,088.—H. R. Guy, London. Submarine Telegraph Cables. 4d. August 11.

CONDUCTORS.—The object of this invention is "to diminished the specific gravity of submarine cables." To effect this, a coating of granulated cork, in combination with india-rubber, is applied over the insulated conductor. Specification No. 1,837 of 1861 (not included in these abstracts) describes the manufacture of the above compound, which is commonly known as "kamptulicon."

The cork may also be combined with gutta-percha.

*2,155.—F. Jenkin, London. Machinery for Manufacturing Telegraph Cables. 4d. August 21.

CONDUCTORS.—This invention consists in adding to the cable-making machine a second drawing off drum, which is caused to rotate slightly faster than the first one, so as to put a proof stress on the cable.

*2,161.—C. Marsden, London. Electric Cables and Wires. 4d. August 22.

CONDUCTORS.—Each wire, after having received one coating of insulating material, is surrounded with pipes of glass or glazed earthenware about an inch long placed at intervals of about two inches. The pipes are preferably made in halves longitudinally. A coating of insulating material is applied externally to the pipes.

This invention also consists in embedding any number of wires, prepared as above or otherwise, in gutta-percha or india-rubber to form a core, and encasing such core in a metal tubing made in halves longitudinally.

The tubing should be in lengths not exceeding six inches.

2,209.—S. T. Jones, London. Submarine Telegraph Cables. 4d. August 28.

CONDUCTORS. — Cork, preferably in sheets, is applied to submarine cables, by means of an adhesive cement, the object being to render such cables more buoyant.

2,213.—W. P. Piggott, London. Electric Telegraph Cables. 8d. (4 figs.) August 28.

CONDUCTORS. — Improvements on Patent No. 2,957 of 1860.

"Several wires partially insulated from each

other" are used. One or more of these wires "must be of opposite electrical denomination to the rest, and the several wires are so arranged that no wire circuit is at any time established within the cable itself."

Hemp or other fibrous substance is employed to separate the wires.

The several wires, thus covered, are laid together and saturated with "a liquid paste consisting of deliquescent salts," in order to "impart thereto the affinity of absorption." The whole cable is then served with a tape coated with insulating material, or a composition consisting of red lead, asphalte, resin and sulphur, boiled in linseed oil and laid on hot, the cable being afterwards drawn through a silicious powder to "prevent its sticking when coiled."

2,217.— R. Laming, Kilburn. Electrical Telegraphy. 4d. August 29.

CONDUCTORS.—In order to eliminate the effect of induction in cables, the inventor proposes to surround a gutta-percha insulated conductor with "a cylinder of small metallic wires" laid on with a slight twist in a bed of cement, over which is to be placed another coating of gutta-percha. Both insulating coatings are to be applied in several layers, the layers being cemented together with a mixture of Stockholm tar, thickened with gutta-percha.

The cable thus prepared may be served with hemp and sheathed with iron wires. Steel may be employed as the material of both the inner and outer conductors.

*2,326.—S. Inkpen, London. Covering, Paying, and Hauling Telegraph Cables. 4d. September 11.

CONDUCTORS.—The object of this invention is to cover electrical cables so as to avoid strain on the insulated conductor. This is effected by laying the insulated core between one or more flat longitudinal continuous layers of stout canvas, which are cemented firmly with india-rubber, and stitched together, so as to form flat longitudinal projecting webs on each side of a cable so covered.

2,332. — J. Macintosh, London. Insulating Telegraphic Conductors. 4d. September 12.

CONDUCTORS.—A stranded conductor is passed through revolving dies so as to compress it into a cylindrical form. It is then bound with fillets of sheet gutta-percha, or other insulating material, and cemented externally with a coating of collodion as described in Specification No. 1,090 of 1858.

Wires may be tinned and caused to unite by passing them through heated grooved rollers.

The conductor, thus formed, is covered with fillets of insulating material, as described in

Specification No. 1,560 of 1860, stretched and put on longitudinally.

A semi-vulcanised outer covering of india-rubber may be applied, or the compound of india-rubber and carbon described in Specification No. 2,269 of 1859.

Another part of the invention relates to improvements in the covering machinery described in Specification No. 1,924 of 1858, and consists in insulating conducting wires with "numerous thin coatings in one operation" by means of two grooved rollers.

*2,341. — J. O. C. Phillips, Birmingham. Submarine Telegraph Cables. 4d. September 13.

CONDUCTORS.—Cables as ordinarily constructed are surrounded with tubes or chambers, containing air or a "gaseous mixture," in order to increase their buoyancy.

These chambers may be attached in any convenient manner, and should preferably be made in compartments.

*2,416.—W. Boggett, London. Manufacturing Electric Wires. 4d. September 22.

CONDUCTORS. — This invention relates to the manufacture of a compound conductor consisting of an iron core, to which is applied an outer sheath of copper, by means of a draw plate. The object attained, according to the "view" of the inventor, is economy, in accordance with the showing of eminent electricians (including Faraday), that "the power of conduction depends wholly on the amount of surface," and not on the section of the wire.

2,509.—J. A. Mee, Manchester. Telegraphic Cables. 6d. (4 figs.) September 30.

CONDUCTORS.—These improvements consist in covering the insulated conductor with a tube, composed of a number of wires of segmental form, applied helically, over which the usual servings and helical metallic wire sheathing are placed, the twist of the former being laid in the opposite direction to that of the tube segments.

*2,530.—H. A. Bonneville, Paris. (*C. E. L. De Nozan, Paris.*) Submarine Telegraph Cables. 4d. October 3.

CONDUCTORS.—The insulated conductor is surrounded with a tube or envelope of tin or other suitable metal, before applying the hemp and iron sheathing. Between the gutta-percha and the metal tube a layer of plaited asbestos is placed for the protection of the insulating covering, in case it is necessary to solder the metal tube or envelope.

*2,605.—F. T. Hubert, Deptford. Submarine Telegraph Cables. 4d. October 10.

CONDUCTORS.— Two or more conductors are

embedded in only one covering, by which the strength of a cable is "multiplied."

The specific gravity of cables is regulated as follows:—A pressing roller shapes the gutta-percha. The thickness and width of the gutta-percha is determined by "cutting knives applied at each end of the roller."

A punching press perforates the gutta-percha so as to form air chambers.

The "external protective covering is also of gutta-percha boiled twice in water and manipulated by the pressing roller before named." A powder is then dusted on, which is prepared as follows:— Phosphorus is dissolved in hot water, and prussiate or azotate of potassium, or deutoxide of lead, and a solution of starch, is added. This preparation, after having been triturated in hot water, and immersed in lime water, is dried and reduced to a powder.

2,733.—A. Parkes, Birmingham. Electric Telegraph Conductors. 4d. October 21.

CONDUCTORS.—These consist of tubular conductors of copper of suitable section strengthened by internal cores of steel, iron, or aluminium bronze.

It is preferred to insulate such conductors with the compound known as parkesine.

Silver solder is used in making joints, in order to render their conducting power equal to that of copper.

2,762.—H. Wilde, Manchester. Electric Telegraphs. 10d. (6 figs.)

MAGNETO-ELECTRIC GENERATOR.—This invention relates to improvements on the generator described and illustrated in Specification No. 516 of

the machine is used to excite the coils of an electro-magnetic generator the springs are connected to the ends of such coils. Twice in each revolution the armature is at its dead point, that is when the division across the commutator is horizontal (Fig. 59), and then the two springs of each pair touch both semi-circles and short-circuit them. Ordinarily, the armature coils of the first machine and the magnet coils of the second form one complete circuit that may be represented by the figure 0, but in the position shown they form two contiguous circuits like the figure 8. If one pole of the combined generator be put to earth, currents for lighthouse illumination may be sent through insulated or uninsulated cables.

*2,941.—A Wells and W. Hall, London. Submarine Electric Cables. 4d. November 15.

CONDUCTORS.—The object of this invention is to relieve the insulated conductor of tensile strain. This is effected by combining one or more conductors with steel or iron ropes, covered with hemp, and saturated with tarry material. The conductor and ropes are secured together with a binding of hemp applied in the process of paying out the cable.

2,948.—J. De La Haye, Manchester. Laying Submarine Electric Cables. 10d. (13 figs.) November 16.

CONDUCTORS.—This invention consists:—

1st. In gilding the copper conducting wire.

2nd. In giving the conductor an undulating form by means of fluted rollers, so as to increase its length about 10 per cent.

3rd. In sheathing an insulated conductor with a plaited metallic covering.

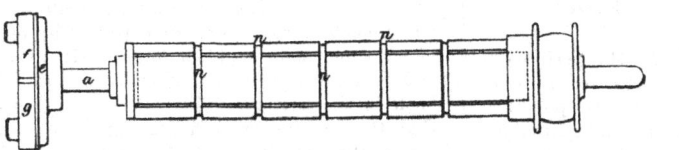

Fig. 59.   Fig. 60.

1863, and also forming part of the combination claimed in Specification No. 3,006 of 1863. The general form of the combined generator is shown in connection with Specification No. 842 of 1867. The armature is by preference of cast iron with a slot extending through it for three-quarters of its length for the prevention of induced currents. It is wound longitudinally with ribbon sheet copper and sheet gutta-percha (Fig. 60), and is surrounded with strong brass bands let into grooves *n n* to prevent the coils flying out of place. The two ends of the coil are led through the spindle to the two insulated semi-circular plates *f g* of the commutator. Two springs of unequal length are placed on each side of the commutator, and each pair is connected to one of the leading wires. When

4th. In giving electrical cables a triangular form in cross section, in order to obviate "chafing" where they are exposed to currents of water.

3,121.—J. Prest, H. Harrison, and B. Roeber, Bedford Leigh, Lancashire. Insulators. 8d. (3 figs.) December 5.

INSULATORS.—Horizontal surfaces are made use of in designing insulators, so as to prevent moisture forming streams, and thereby impairing the insulation of the conductor.

Any two suitable insulating materials which have different specific heats, as for instance glass and shellac, are combined "so as to form one surface." On insulators thus constructed moisture only collects in drops, between which is a comparatively dry surface.

The insulators illustrated have concentric grooves, in which is placed the material having a "different specific heat," the ridges between the grooves have horizontal surfaces.

*3,180.—W. Boggett, London. Telegraphic Conductors. 4d. December 9.

CONDUCTORS.—Conducting wires are enveloped in a metallic sheathing, consisting of a flattened wire folded lengthwise around the conductor by means of a die. Various forms of stranded or compound conductors are specified. The use of amalgamation of the outer covering, in order to diminish induction, is mentioned.

*3,192.—T. Berrens, Paris. Manufacturing and Submerging Telegraphic Cables. 4d. December 11.

CONDUCTORS.—Three or four steel wires are coated with Chatterton's compound, and tightly bound together with hemp. An insulating layer of gutta-percha is then applied, over which is laid copper conductors prepared as follows:—A strand of copper wires is bound round with yarn impregnated with "an elastic and drying composition" containing tar, resin, beeswax, and tallow, and then receives an insulating coating of gutta-percha.

These conductors are laid around the central core of steel wires, and the interstices are filled with strands of hemp impregnated with a preservative material. The whole is then served with hemp and passed between grooved rollers to give it a uniform diameter.

*3,260.—C. L. W. Reade, Parkhurst, Isle of Wight. Obtaining Motive Power. 4d. December 16.

ELECTRO-MAGNETIC MOTOR.—To the periphery of a wheel are applied a number of hinged armatures in the form of arcs of corresponding curvature. The wheel is set on a shaft running in bearings, and a series of electro-magnets are arranged "around and outside the wheel and armatures, whereby the armatures will successively be attracted during the revolution of the wheel, and thus create a continuous motive power."

3,282.—A. V. Newton, London. (*L. C. Stewart, Mamaroneck, New York, U.S.A.*) Electro-Magnetic Engines. 1s. (5 figs.) December 19.

ELECTRO-MAGNETIC MOTOR. — This invention relates (1st) to a radial arrangement of two or more series of electro-magnets in concentric circles with their poles very close to one another, and their cores so shaped and situated, and the wire so wound upon them that "the force generated by the electricity which passes through the enveloping wire acts not only upon the enveloped core, rendering it magnetic, but also with opposite polarity upon the adjoining core on either side thereof."

(2nd) To a mode of reversing the polarity of the magnets without interrupting the circuit. The motor shown in the drawings consists of two vertical concentric cylinders, the outer one fixed and the inner one secured to a vertical shaft running in suitable bearings. The outer periphery of the inner cylinder carries a series of radiating electro-magnets whose poles come very close to the poles of a similar series of electro-magnets secured to the inner surface of the outer cylinder. There must be an equal number of electro-magnets in each series, their cores by preference being made rectangular in cross section, and the insulated wire being so wound upon them that their poles are alternately north and south. Upon the vertical shaft, and below the electro-magnets, are fixed two metallic rings having upon their peripheries alternate depressions and projections, their number corresponding with the number of radial electro-magnets on the inner cylinder, and the depressions of the upper ring being situated over the projections of the lower ring. To each ring one terminal of the inner series of electro-magnet coils is attached. Against the circumferences of these rings two insulated metallic rollers are pressed by means of helical springs, each roller being of such a width as to bear upon both rings as they rotate. One terminal of the coils of the outer series of electro-magnets is attached to one pole of the battery, whilst the other terminal is in connection with one of the rollers. The second roller is joined to the other pole of the battery. By this arrangement of connections the current traverses the coils of the outer electro-magnet, and then passes by means of one roller and ring to the inner series of coils, and thence through the other ring and roller back to the battery. The rollers are so placed that they are never both bearing on the same ring at the same time, and therefore as the rings rotate there is a change of polarity in both series of electro-magnets. By shifting the rollers the distance of one section of the rings the motor is reversed. The projecting sections of the rings are made slightly longer than the depressions in order that the roller may come in contact with one ring before it leaves the other, thus short circuiting the coils at the moment when the magnets of the inner cylinder come into the same radial lines with those of the outer cylinder.

3,299. — W. Boggett, London. Telegraphic Conductors. 8d. (11 figs.) December 21.

CONDUCTORS.—This relates to compound conductors already described in Provisional Specification No. 3,181 of 1865.

*3,347.—H. A. Silver, Silvertown, Essex. Electric Conductors. 4d. December 27.

CONDUCTORS.—The conductor is first coated with pure india-rubber, over which is applied a coating of vulcanisable compound. The whole is cured in a mould made in halves.

3,357.—C. F. Varley, Beckenham, Kent. Telegraphic Cables or Conductors. 4d. Dec. 29.

CONDUCTORS.—The metallic conductor is made partly of platinum so as to ensure continuity, even if all the copper portion is " eaten away by elec-trolytic action." The invention may be carried out by inserting a platinum rod into a hollow cylinder of copper and drawing the whole down into wire, or by making one wire of a strand of platinum and the rest of copper.

The stranded conductor is preferably made with a more rapid lay than usual. Platinum is used in place of the usual binding of copper wires in making joints. Cables have their conductors made tapering towards the centre of their length, at which point the best quality of gutta-percha is to be used.

1866.

*96.—W. A. Rudling, London. Alarm Apparatus. 4d. January 11.

LAYING CONDUCTORS.—In order to admit of additional wires being laid in pipes under the street, the pipes terminate at distances of 100 yards, or more, in boxes or chambers of cubical form, in each of which there are two spindles at right angles to the length of the tube. Each spindle has on it two sheaves or pulleys: one is fixed so that a wire rope or chain, which passes over it to the pulley in the next box, shall be in the centre line of the top of the tube. A portable windlass is connected with the other pulley, which is toothed to receive a chain from the windlass. All that is necessary in placing an additional wire is to remove two of the manhole doors or covers, and connect the insulated wire by means of a clamp or swivel with the endless chain passing over the pulleys in the tube, when by turning the windlass the wire will be carried on to that manhole where the windlass is stationed.

*195.—T. Hutton, Derby. Submarine Telegraph Cables, &c. 4d. January 20.

CONDUCTORS. — The central copper wire is covered with three coats: first, india-rubber or gutta-percha; second, cloth; and third, india-rubber or gutta-percha again. To render it partially buoyant, it is surrounded with a jacket of cork, or india-rubber inflated bags are attached to it at intervals by spring clips.

*222.—F. Wibratte, Paris. Tubes for Submarine Telegraph Wires. 6d. (1 fig.) January 23.

PROTECTING CONDUCTORS.—A gutta-percha coated wire is led through short lengths of tube articulated together by plates and links to form a chain.

*237.—S. M. Martin, S. A. Varley and F. H. Varley, London. Electric Telegraph Apparatus. 4d. January 24.

ELECTRIC MOTOR.—A series of electro-magnets are fixed in a circle on an iron plate; these magnets are at equal distances from each other, and are so arranged that the opposite ones have opposite polarities. Over the poles of the electro-magnets a permanent or induced magnet is mounted on an axis terminating in pivots; the lower pivot rests on a spring which can be adjusted to the requisite degree of strength; on the upper pivot a light indicator is placed. When a current is sent through one pair of electro-magnets the mounted magnet immediately flies round and, arranging itself over the electro-magnet, is attracted down and its motion arrested. On the current being cut off, the spring, on which the pivot of the mounted magnet rests, raises it above the poles of the electro-magnet, leaving it free to move in any direction. To assist the action of this spring, the electric current after passing through the electro-magnets and previously to its entering the earth, is made to pass through another electro-magnet, over which, but insulated from it a series of convolutions are wrapped; the ends of this secondary wire are connected to a series of coils wrapped on the poles of the electro-magnets, but insulated from them. On arresting the current in the electro-magnets a momentary current is developed in the secondary wire, and this current passes through the coils of the electro-magnets. The secondary current passes round the electro-magnets in the opposite direction to the previous current, and momentarily magnetises them in the opposite direction to destroy the residual magnetism and liberate the mounted magnet. In some cases circular coils are substituted for the electro-magnets, or the magnets are constructed with iron cores that do not come up to the surface of the coils.

*311.—W. Darlow, Tottenham. Electro-Magnetic Engines. 4d. January 31.

ELECTRO-MAGNETIC MOTOR.—A series of helices are mounted upon a disc, which rotates upon a central axis. Upon one or both sides of this disc a " magnet ring " is supported on friction rollers so as to revolve in the opposite direction to that in which the disc and axis is intended to rotate. The helices are arranged, by preference, in connection with a galvanic battery, so that the circuit is completed through each helix " to magnetise the successive sections of iron forming the ring (or partial ring) of magnets, as each helix arrives on the descending side of the disc; by this means preponderance is given to that side of the disc by the additional weight of the magnets on the descending side, whilst they are retained to the disc by magnetic attraction. During the retention of the magnets within the helices the ring is caused to rotate with the disc and axis, but the moment the electric circuit is broken the ring of magnets is caused to rotate in the opposite direction to the disc, the electric circuit having been previously completed through the next descending helix, and so on." Continuous rotary motion is thus imparted to the disc and axis. A suitable commutator is employed to distribute the current, and several " magnet rings" may be used if desired.

394.—H. E. F. De Briou, London. Composition for Preserving Metals, Wood, &c. 4d. February 8.

INSULATING COMPOUND.—66 lbs. of vulcanised rubber, 20 lbs. of vegetable pitch, 10 lbs. of shellac, and 10 lbs. of resin are mixed with heat and then dissolved in bisulphide of carbon. The compound is useful for insulating conductors, and preserving the hemp wrappings and wire sheeting.

*466.—H. E. Baron de Gablenz and H. Mahler, Berlin. Telegraphic Cable. 4d. Feb. 14.

CONDUCTORS.—The cable " is composed of a central part or core of hemp, and the copper wire fluid conductor is rolled up in a spiral line around the said core, the copper wire is previously lined with an insulating silk thread. The cable is then strengthened inside from place to place by iron wire bundles, and a similar wire is placed in the centre of the hemp core. This central wire can also be used for the transmission of dispatches by means of the induction fluid which is produced."

480.—D. Nicoll, London. Electric Telegraph Conductors. 4l February 15.

LAYING CONDUCTORS.—Rectangular troughs of earthenware, wood, iron or other suitable material are taken and partly filled with asphalte, bitumen, paraffin, gutta-percha, caoutchouc or other insulating material, either alone or mixed with dried chalk. Short lengths of wire are then laid in the troughs with their extremities projecting beyond its ends. Two troughs are placed face to face to form a tube, and, when these tubes are laid in a line, the neighbouring ends of the wires may be twisted together, or one may be made into a spiral and the other be placed within it. Before this is completed, however, a collar is placed on each end of the tube, and the remaining space is filled up with molten asphaltum. Instead of troughs the wires may be laid in grooves in cast blocks of asphate, and the blocks may be built up of alternate layers of wires and asphalte. In carrying out the former part of the invention, a sufficient number of wires are cut to the required length and held taut in the tube by a winding screw, while the asphalte is being poured in; or a number of wires are coiled round a drum after the manner of warp threads, and are drawn through perforated guide plates into the troughs.

*726.—J. Baker, London. Magnetic Engines. 4d. March 9.

ELECTRO-MAGNETIC MOTOR. — Electro-magnets are fixed transversely at intervals upon a rotating wheel, and a fixed electro-magnet is placed with the pole as near as possible to the poles of the revolving electro-magnets without touching them. The helices are so wound that, when the current is circulating, similar poles are opposed in the fixed and movable electro-magnets. There may be as many fixed electro-magnets as may be thought necessary. At a short distance from each fixed electro-magnet " a powerful permanent magnet is placed, so that when the electric current is passing through the rotating electro-magnets, opposite poles will be as near as possible in contact without actually touching. Each rotating electro-magnet is provided with a small arm, which grazes over an insulated metal channel at the instant the said magnet is exactly opposite the fixed electro-magnet, and leaves the channel when it is exactly opposite the permanent magnet; to these channels the electric wires are attached," and by these means the rotating electro-magnets are excited as they come exactly opposite the fixed electro-magnets. The action of the motor is as follows:—The moment the current passes, " the fixed electro-magnet will attract in succession the soft iron cores until they are opposite, but at that moment the current will pass through each successive helix, magnetise the iron core, oppose similar poles to each other on the electro-magnets and opposite poles on the rotating electro and fixed permanent magnets, and as a consequence the former pair will repel each other, and the latter attract each other until exactly opposite, at that moment the current will be broken in the rotating electro-magnet, and the reactionary current will for a moment reverse the poles of the latter and give it an impulse from the permanent fixed magnet, and so on in succession."

931.—W. Read, Coventry. (*W. H. Read, San Francisco.*) Electro-Magnetic Power Engines. 10d. (5 figs.) March 31.

ELECTRO-MAGNETIC MOTOR.—Continuous rotary motion is imparted to a spindle carrying a number of permanent or electro-magnets, by the successive attractions and repulsions of a series of fixed electro-magnets. A commutator is provided to effect the distribution of the current. Fig. 61 shows a vertical section of this motor. R is a disc, mounted on a vertical spindle l, running in bearings. T, V are two of four permanent compound magnets attached radially to the disc. g is one of a series of electro-magnets fixed to the base of the motor at equal distances apart, and arranged so that their poles $e f$ "command one

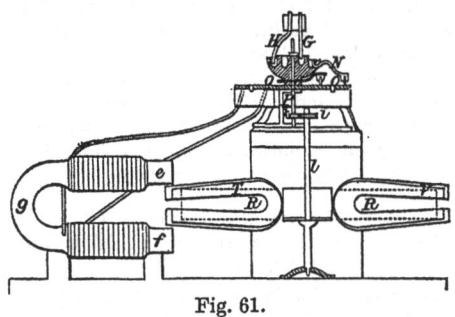

Fig. 61.

quarter of the circle surrounding the rotating magnets." Rotation of the disc R results from the successive attractions and repulsions of the series of electro-magnets upon the permanent magnets, each permanent magnet in turn coming under the action of electro-magnets. The commutator by which this is effected is shown in 'Fig. 61 in section. Fig. 62 is a plan of

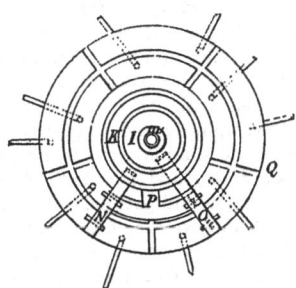

Fig. 62.

the same to a larger scale. The spindle m is driven from the shaft l by means of the spur gear, which is so proportioned that the spindle m makes one revolution while the shaft l makes a quarter of a revolution, that is, while one of the permanent magnets passes one set of the electro-magnets. Q is an annular conducting plate divided into five equal segments by narrow radial insulating strips. Each segment is again divided concentrically as shown, and the whole ten divisions by preference are "inlaid with platinum." The two divisions of each radial segment form the terminals of one of the electro-magnet coils respectively. The spindle m carries an insulated cup with two concentric channels K I filled with mercury into which dip the ends of the battery wires H G. The current reaches the commutator plate Q by means of two brass

springs O N, having platinum edged rollers at their lower ends. P is a third spring attached to the cup, but insulated therefrom, serving to connect the two inner divisions of the same compartments of the plate Q, on which the springs O N rest.

As the springs O N P pass in turn from segment to segment, the electro-magnets are endued with alternately opposite polarities and cause, by their successive attractions and repulsions on the permanent magnets, rotation of the disc R in the manner above described. A switch is used to reverse or cut off the battery current.

947.—C. F. Carlier, Paris. Galvanic Batteries. 8d. (5 figs.) April 3.

ELECTRO-MAGNETIC MOTOR. — This invention in part relates to an electro-magnetic motor in which the residual or "remanent" magnetism of the electro-magnets is neutralised by the use of a second and independent coil wound in the opposite direction to the primary helix. Through this second coil, at each interruption of the circuit, a current is caused to pass in a direction opposite to that of the current which has passed through the primary wire, and of such a strength as to neutralise the residual magnetism without counteracting the rotation of the motor. A motor is illustrated in which two electro-magnets are attached to a frame, with their axes in the same horizontal straight line. Between their poles is an iron disc which rotates in a vertical plane on a shaft running in bearings, and has on its periphery an unequal number of teeth. As the disc rotates, these teeth pass within a very short distance of the poles of the electro-magnets. The shaft also carries a notched commutator wheel, for directing the current alternately to each electro-magnet. The electro-magnets are constructed as already described with two coils, and the connections are made as follows :—Calling the electro-magnets A and B, one end of the primary coil of A is connected to the battery and the other end to the secondary coil of B. The other end of this secondary coil goes to a spring which bears against the commutator wheel. On the other hand the connections are similarly made as regards the primary coil of B and the secondary coil of A. "Great lightness" of construction is obtained, the iron supporting frame serving also to add to the mass of the cores of the electro-magnets.

1,226.—G. Davies, London. (*D. Brooks, Philadelphia, U.S.A.*) Insulators. 6d. (4 figs.) May 1.

INSULATORS.—Consists principally of three parts, an outer hollow cylindrical casing of cast iron, having a projection whereby it is attached to the telegraph post, a hollow glass cylinder of smaller diameter and shorter than the casing, and a double hook or wire holder, the stem of which is smaller in diameter than the interior of the glass cylinder, and is indented at the sides, so that when fixed it cannot be withdrawn from the glass. This stem is cemented in its place by melted sulphur, and the same material is interposed between the glass and the cast-iron casing. The parts over which the leakage would take place are covered with paraffin, varnished with resin and beeswax.

*1,503.—W. E. Newton, London. (*N. Bloch, Paris.*) Connecting Metallic Wires. 4d. May 27.

CONDUCTORS.—Two half-round pieces of a double tapering form are employed; these are hollow inside for the reception of the wire, and are screwed at their ends to fit screw nuts. One of the pieces is provided at its middle with an oblong slot for the reception of the turned up ends of the wires. To effect the junction of the wires, their ends are bent up at right angles, and are passed through the oblong slot in one of the taper pieces; the other piece is then applied to the opposite side of the wires and is secured in position by the nuts.

1,521.—J. H. Johnson, London. (*J. J. E. Lenoir, Paris.*) Transmitting Telegraphic Despatches. 1s. 6d. (8 figs.) May 31.

ELECTRIC-MOTOR. — The invention includes a simple form of electro-motor in which the rotation of a vertical shaft carrying radial soft iron arms is effected by an electro-magnet round which the circuit is alternately made and broken. To avoid the spark at the commutator, "a bottle is employed, which is filled with water and a little quantity of carbonate of soda or other salt, or with an acid; the cork or stopper which closes the bottle is provided with two binding screws, to which are connected the wires proceeding from the apparatus, and which carry platinum wires immersed in the solution; the extremity of each of these wires, insulated by a cotton thread, is bent or curved, and it is between these hooks that the spark is produced." "This apparatus, to which is given the name of condenser, prevents the burning of the wires near the distributing disc, as well as the formation of 'coom,' resulting from the heating of the lubricating oil which would impede the passage of the current."

1,651.—A. Miroude, Paris. Lighting Buoys. 8d. (1 fig.) June 20.

The buoy is lighted by a Geissler tube, and a battery and induction coil.

1,718.—J. Baker, London. Thermo-Electric Magnetic Batteries and Engines. 4d. June 27.

ELECTRO-MAGNETIC MOTOR.—"Insulated electro-magnets" are arranged radially on the circumference of a rotating wheel, with the planes of

their poles parallel to a plane passing through its axis. An exterior set of fixed electro-magnets have their poles very close to those of the revolving electro magnets, and are so placed that contrary poles are opposed. A wheel commutator of the usual type distributes the current so that it passes at the same moment through one stationary and one revolving magnet, contact being made as the moving magnet is approaching the fixed magnet, and broken when it is opposite to it. The number and position of the fixed magnets is such, that half will always be excited. Secondary coils may conveniently be placed on each electro-magnet, and the induced currents obtained from them applied for the purpose of magnetising fixed electro-magnets arranged so as to act on the wheel and thereby increase its velocity.

1,749.—H. A. Bonneville, London. (*P. F. L. Pérémé. Paris.*) Submarine Telegraph Cables. 8d. (6 figs.) July 2.

CONDUCTORS.—The inner copper conductor is surrounded by copper-plated steel wires, and the two are included within a coating of insulating material, over this comes the sheathing protected by a second envelope of fibrous material or gutta-percha, and last of all steel ribbon is laid on helically.

1,751.—H. A. Bonneville, London. (*P. F. L. Pérémé, Paris.*) Preparing Telegraph Wires or Cables. 8d. (3 figs.) July 2.

CONDUCTORS.—The so-called cables are flat bands with conductors running longitudinally of them, about three-quarters of an inch apart. The wires are wrapped with "bandages" of textile material arranged spirally and impregnated with tar taken from the distillations of bituminous schists very rich in oil and paraffin. Thus prepared, these wires are laid parallel, and are sewn in any number in an envelope of textile material, which is enclosed in a second covering of felt, done over with schist oil charged with paraffin. The cable is wrapped in tarred paper, and laid in a trench in the ground; it is applicable "as a conductor of electricity to manufactories where electro-metallurgy and electromotive machines are employed."

*1,825. — C. W. Farmer, W. E. Partridge, Birmingham; B. J. P. Webb, Belfast. Collectors for Wires, Ropes or Cables. 4d. July 11.

COUPLING CONDUCTORS.—The invention consists in laying the ends of the wires together and encompassing them with a metal clip, with or without an inner bearing plate. A modified clip for flat conductors is also described.

*1,843.—R. Jobson, Dudley. Holding Telegraph Wires. 4d. July 13.

INSULATORS.—Relates to the application to insulators of eccentric or nipping devices to grip the wire. There may be two devices on each insulator, or the devices on succeeding insulators may be set in opposite directions, so that while one prevents the wire slipping in one direction the other holds it in the opposite direction.

1,867.—C. Varley, Kentish Town, and S. A. Varley, Holloway, London. Electric Telegraph Apparatus. 1s. 18 figs. July 17.

GALVANOMETER.—A number of hollow segmental shaped coils with their convolutions held together by means of cement, are packed in a hollow cylindrical case, and are traversed by the current to be measured in such directions that their upper faces have alternate north and south polarities. The index consists of soft iron arms, equal to half the number of the coils, radiating horizontally from a vertical spindle free to rotate in a socket mounted in the end of a powerful magnet. By induction, the spindle and arms are magnetised and are at the same time attracted strongly downwards towards the magnet. As this involves considerable friction on the pivot, a second magnet of opposite name is placed at the upper end of the spindle, and a second set of radiating arms arranged to come between or break joint with the first. To insure both sets of arms, which stand in the same horizontal plane, having their proper polarity, they are cranked near their inner ends and carried upwards and downwards to their bosses, which are respectively situated near the opposing inducing magnets. When an electric current is sent through the wires the arms arrange themselves over the centres of the coils. Although this instrument is described in the specification as a galvanometer it would seem to be more properly termed a galvanoscope, as no means are provided for obtaining graduated readings.

*1,878.—J. P. Gillard, Paris. Electricity. 4d. July 19.

GENERATOR.—Relates to a mechanical arrangement called an " electro-polyphore, in which a quick revolving motion is imparted to an arbor of iron or other suitable metal provided in an isolated manner with a series of pairs of metal blades, to be carried in contact with a series of metal conductors fixed in the inner periphery of a drum, in the centre of which revolves the above-mentioned arbor. The metal blades are connected to a galvanic battery, or magneto-electric machine, or other suitable generator, in such manner that one blade of each pair forms the positive and the other blade the negative pole of the circuits; the conductors also form pairs, one conductor of each pair being set in contact with the positive blade and the other with the negative blade of a pair of blades; besides one conductor of each pair is connected to one wire and the other conductor of the pair to another

wire, so as to enable each pair of wires to form an electrical circuit, or the current may be caused to pass through an induction coil or bobbin. A revolving motion at high speed being imparted to the arbor, an electric circuit will be formed at each time each pair of blades comes in contact with one of the pairs of conductors, and as this will take place many times in a second the apparatus will cause the electricity to be attracted in an efficacious manner from the galvanic battery or other generator."

*1,989.—W. A. Marshall, London. Insulating and Protecting Submarine and other Telegraph Wires. 4d. August 2.

CONDUCTORS.—A leaden tube is formed, and while in course of formation is made to enclose one or more insulated wires without injury to the insulating medium by the application of paraffin wax, or other suitable insulating substance, reduced to a liquid state, and introduced in and around the covered wire while the same is being enclosed by the metal, so as to form a moist heat and prevent the substance around the wire being burned or injured by the high temperature of the metal.

2,052.—W. R. Lake, London. (*A. J. B. De Morat, Philadelphia.*) Telegraphic Cable. 6d. (3 figs.) August 9.

CONDUCTORS.—The invention consists in combining a number of separate tubular, concentric, conductors to form a cable through which a number of different despatches or telegrams may be sent at the same time. The core of the cable is formed of one or more wires. Around these a strip of copper is wound helically, and over the joints of this helix another strip is laid. The whole is then thinly, but compactly, covered with india-rubber, gutta-percha, or other insulating material. This process is continued according to the required number of conducting cylinders. The whole is then covered with jute and protected in the usual manner.

*2,329.—J. H. Johnson, London. (*J. M. Batchelder, Cambridge, Massachusetts, U.S.A.*) Electric Telegraph Conductors. 4d. September 10.

CONDUCTORS.—The essential feature of this invention consists in the manufacture of electric telegraph conductors by braiding or interlacing, in lieu of twisting together, any desired number of wires made either of the same metal or of two or more dissimilar metals. Two of these braided conductors may be combined in one cable.

*2,542.—C. E. Spagnoletti, London. Signalling on Railway Trains. 4d. October 3.

JOINING CONDUCTORS.—The wires are joined between each carriage by means of iron bars, pieces of metal cables, or connecting rods, sufficiently strong to supply the place of the present coupling chains, so that the act of coupling up the carriages makes the electrical connections, the bars being brazed, galvanised, or plated at the points of contact. For additional means of contact, side springs may be used at each joint. The connections can also be made by means of a half ball of brass so fitted to the end of each buffer with springs that it will retreat under pressure into the body.

2,747.—Sir E. F. Piers, Manchester. Working Railway Signals. 10d. (9 figs.) October 24.
ELECTRO-MAGNET.—The apparatus includes an electro-magnet composed of concentric iron tubes, each coiled with insulated wire.

***2,836.—O. Rowland, London. Galvanic Batteries, &c. 4d. November 1.**
CONDUCTORS.—To test the insulation of covered wires they are placed in a bath and caused to form one element of a galvanic couple. If the insulation is defective a current will be set up.

***2,870.—T. Walker, London. Electric Telegraph Cables, &c. 4d. November 6.**
CONDUCTORS.—Naked conductors are employed for the return current, or an insulated and an uninsulated conductor are incorporated in the same cable.

2,880.—C. E. Spagnoletti, London. Signalling in Railway Trains. 4d. November 6.
JOINING CONDUCTORS.— The wires are joined between each carriage by means of iron bars, pieces of metal cable, or connecting rods sufficiently strong to supply the place of the present coupling chains, so that the act of coupling up the carriages makes the electrical connections, the bars being bossed, brazed, galvanized or plated at the points of contact. For additional means of contact, side springs may be used at each joint. Also the cable ends may have hooks and eyes, the hook being so constructed as to form either the hook or eye. According to another method a half ball of brass may be fitted to the end of each buffer and supported on springs, an insulated wire being run through the buffer rod.

3,038.—J. Latimer Clark, London. Electric Telegraphs. 1s. 2d. (19 figs.) November 19.
INSULATORS.—The insulator has a bracket with a cast-iron base. The iron or steel arm is cast into the base. The wire is attached to the insulator by a cast-iron cap carrying a snug.

***3,039.—J. Baker, London. Magnetic Engines. 4d. November 19.**
ELECTRO-MAGNETIC MOTOR.—Three or more sets of straight electro-magnets are arranged on the circumference of a wheel parallel to its axis, and upon the poles of each set a soft iron armature is fixed. A system of fixed permanent or electro-magnets is placed on either side of the wheel, also parallel to its axis, and has its poles on the outside provided with a soft iron armature, "which unites the opposite poles of both sets." One set of the fixed magnets has different polarity to the other, and their armatures are brought close to the armatures of the wheel magnets, but without touching them. The commutator, by which the polarity of the rotating magnets is reversed, consists of four small metal rings, two of which encircle the axle of the wheel, one on each side, but are insulated from it. The peripheries of these two rings are divided into the same number of spaces as there are sets of rotating magnets, and have the wires of each set attached to them. Opposite each of these rings is a similar ring divided to correspond with the number of fixed magnets. All the spaces on the rings are insulated from each other, and each space of the wheel ring carries a metal roller, which traverses the fixed ring opposite to it. The electric current is thus made to pass in at a space of a fixed ring and out at the opposite space of the opposite fixed ring, passing in its turn through a set of electro-magnets on the wheel. In the next space the current is reversed as soon as the armatures of the fixed and revolving magnets coincide with each other. Similar magnet poles are attached to each armature, and the magnet coils are so wound that attraction takes place between approaching sets, and repulsion between receding sets. There is no interruption of the battery current. Secondary coils may be placed on the revolving magnets, and the currents induced in them utilised for exciting the fixed magnets. If permanent magnets are used, they are made by applying to the poles of a powerful electro-magnet a steel cylinder, which has previously been compressed and brought to a white heat, and allowing it to cool in that position. The cylinder may also be encircled with a helix, through which an electric current is passing during the process of cooling.

3,124.—W. Clark, London. (*J. H. Delaunay, Paris.*) Time and Distance Indicator. 1s. 8d. (13 figs.) November 27.
GENERATOR.—A Clark magneto-electric machine is used in combination with counting and recording mechanism.

***3,192.—W. A. Marshall, London. Insulating Telegraph Wires. 4d. December 4.**
CONDUCTORS.—The apparatus employed to effect the insulation and protection of wires that are to be enclosed in a leaden sheath, is provided with a reservoir adapted to hold paraffin wax in a fluid state. The wires, which have first been coated with a non-conducting fibrous material are passed through the reservoir, and out at the lower end by an orifice sufficiently large to admit of their passing freely and taking with them a sufficient quantity of wax. About this point the semi-

fluid metal is admitted to flow under pressure around the wires and their coatings. In order to prevent immediate contact of the metal and the coating of the wires, the outlet from the reservoir is provided with a tubular triblet or mandrel to keep the wires and their coating in correct relation, whilst the annular space formed between the outside of the triblet and inside of another surrounding tube, serves as a gauge for the size of the metal tube. The tubular triblet is continued until the metal tube is sufficiently cool to come in contact with the paraffin coating to the wire. The apparatus is applicable when other insulants than paraffin wax are employed.

*3,209.—H. Wilde, Manchester. Electro-Magnetic and Magneto-Electric Machines. 4d. December 6.

ELECTRO-MAGNETIC GENERATOR.—Two wrought-iron discs, placed on the same axial line, are each provided on the inner face with a circle of electro-magnets, arranged parallel to the common axis of the discs, and at equal angular distances apart. The coils are connected in such a way that alternate as well as opposite poles differ in polarity. Between the circles of electro-magnets, and concentrically therewith, a wheel is made to rotate, and on its periphery at equal distances apart are fixed a series of helices, with iron cores, equal in number to the electro-magnets on one of the discs, and placed at the same distance from, and parallel to, the axis of rotation. "When the compound electro-magnets are excited by means of the electricity from a magneto-electric machine or other source of electricity" and rotation is imparted to the wheel, alternating currents are induced within the helices and are collected "in the usual manner from the bearings of the shaft which carries the wheel." A second part of this invention relates to commutators for turning the currents in one direction. A cylindrical commutator of the form shown and described in Specification No. 516 of 1863, or of any other form which inverts the current twice only during one revolution, is employed, and is driven by a toothed wheel and pinion from the axis of the generator, the gear being so proportioned that the direction of the current is reversed as many times as the magnetism of the cores of the helices is inverted. Instead of the commutator being cylindrical it may consist of several flat metallic surfaces insulated from and sliding over each other.

3,281. — C. C. Adley, Dublin. Telegraph Standards and Insulators. 10d. (12 figs.) December 13.

INSULATOR.—A cup-shaped insulator is formed with two concentric lips, care being taken that the edge of the inner lip does not dip below the outer one. The confined space between the two lips will arrest the splashed rain and secure a dry inner surface for the outer lip.

*3,287.—A. W. Hosking, Manchester. Signalling the Collection of Railway Tickets, &c. 4d. December 14.

JOINING CONDUCTORS.—The current is carried along the carriages by an insulated conductor, the connections between the carriages being formed by the coupling chains, or by the gas couplings when such exist.

*3,351.—J. Baker, London. Thermo-Electric and Magnetic Apparatus. 4d. December 20.

ELECTRO-MAGNETIC MOTOR. — A rotating axis carries a disc composed of an odd number of electro-magnetic sections connected together by a non-magnetic material. Each end of each coil is connected to a spring, bearing against a commutator. Facing each side of the disc, and very near thereto, are arranged an even number of electro or permanent magnets with alternately opposite poles. The commutator consists of a disc formed of insulated conducting portions in connection, "some with one pole and some with the other pole of the battery," so that when the springs attached to the wires of the rotating disc pass over these conducting portions the electro-magnets of the disc are alternately reversed in polarity. Sparking is avoided by arranging the commutator so that contact on one portion commences before it is broken with another.

Permanent magnets are made by compressing the steel in the direction of the intended magnetic axis, after which it is heated and allowed to cool within a coil, the ends of which are applied to the poles of electro-magnets, so arranged that the intended magnetic axis lies parallel to the magnetic meridian.

*3,394.—C. and S. A. Varley, London. Generating Electricity. 4d. December 24.

DYNAMO-ELECTRIC GENERATOR.—"Our invention consists in an improved method of developing electricity either by mechanical force alone, or by mechanical force in combination with chemical action.

"We construct our apparatus as follows:—We wrap soft iron bars with insulated wire in a similar way to an ordinary electro-magnet; these bars may be U-shaped, and become electro-magnets when the apparatus is in use; we also construct iron bobbins of such a length that they will pass just freely between the poles of the electro-magnets and wrap them with insulated wire.

"In constructing the apparatus we prefer generally to use two electro-magnets and two bobbins; the bobbins are mounted on an axle and revolve between the poles of the electro-magnets and are so arranged on the axle that when one of the bobbins is between the poles of one of the electro-magnets the other bobbin is between the poles of the other and *vice versâ.* On the axle carrying the bobbins there is a commutator; the ends of the insulated wire surrounding the bobbins are connected to this commutator and through it to the insulated wire of the electro-magnets, forming the whole into one electric circuit. Before using the apparatus we generally send an electric current through the electro-magnets; the object of this is to secure a small amount of permanent magnetism in the direction we wish in the soft iron cores of the electro-magnets. On revolving the axle the bobbins become slightly magnetised in their passage between the poles of the permanent magnets and as the electro-magnets are so arranged that the north pole of one magnet is opposite the south pole of the other, the bobbins are magnetised first in one direction and then in the other, generating weak currents (corresponding to the direction in which the bobbins are magnetised) in the insulated wire surrounding them. The commutator changes the connections of the electro-magnets at the same time as the magnetism of the bobbins is reversed, and consequently the electric currents developed in the bobbins always flow in one direction through the insulated wire of the electro-magnets. The effect of the current passing through the electro-magnets is to increase their magnetism, and to magnetise in a higher degree the bobbins when passing between the poles of the electro-magnets; more powerful currents are now developed in the insulated wire surrounding the bobbins, which in their turn induce a higher degree of magnetism in the electro-magnets; in this way the magnetism developed in the electro-

magnets and the bobbins act and react on one another, causing the circulation of increasing quantities of electricity. In constructing this apparatus to prolong the contact between the bobbins and the poles of the electro-magnets we generally arm the poles of the electro-magnets or the bobbins with horns of soft iron. We call this apparatus a magnetic multiplier. In some cases we use only one iron bobbin, and make it oscillate between the poles of two electro-magnets; this we do by mounting the bobbin on a lever to which motion is communicated by means of an eccentric or crank, or by any other suitable way."

MAGNETO-ELECTRIC GENERATOR.—Currents for ringing bells and for telegraph purposes are produced in the following way:—A number of bar magnets are taken and made into a compound horseshoe magnet by arranging two sets of them one over another. The north poles of one set are connected to the south poles of the other set by plates of iron placed between the magnets. A bobbin of iron, wrapped with insulated wire, is mounted on a lever, and works between the poles of the compound magnet, developing a current as it moves. To prolong the duration of the current a second compound magnet is arranged with its north pole opposite the south pole of the first, and the iron bobbin passes between the poles of both magnets. " The magnetic multiplier may be used in conjunction with a galvanic battery, the effect of which is to bring it into action more quickly."

When using the magnetic multiplier the circuit is sometimes completed with a large electro-magnet, which may be placed near the machine; " this acts as a sort of reservoir or fly wheel to the machine, making the currents more constant."

1867.

212.—J. H. Johnson, London. (*J. M. Batchelder, Cambridge, Massachusetts, U.S.A.*) Electric **Telegraph Conductors and Cables.** 6d. (3 figs.) January 26.

METALLIC CONDUCTORS.—These are formed by the braiding of two or more dissimilar metallic wires, one series of wires serving to increase the conductivity and the other the strength of the conductor. Or an iron or steel wire is coated with copper to increase its conducting power. Either of these conductors can be made into a submarine cable by suitably insulating and protecting them.

261.—C. W. Siemens, London. (*Partly a communication from abroad by Dr. Werner Siemens, of Berlin.*) Producing Electric Lights at Sea, &c. 2s. 2d. (18 figs.) January 31.

DYNAMO-ELECTRIC GENERATORS.—The first part of this invention consists in obtaining powerful electric currents without the aid of batteries or permanent magnets. A soft iron armature, surrounded by a coil of insulated wire (preferably the armature described in Provisional Specification No. 2,107 of 1856), is caused to rotate in front of the poles of an electro-magnet in a direction opposite to that in which it would move consequent to an electric current passing through the electro-magnets, thereby developing within the armature coil electric currents, which by means of a commutator are made available for lighting or other purposes.

A magnetic impulse must at first be imparted to the electro-magnetic arrangement " either by the momentary insertion of a galvanic battery, however small, into the system, or by a touch from a permanent magnet, or by dipping the iron bars of the apparatus in a direction parallel with the earth's magnetic axis." For reproducing a current after the rotation has been arrested the residuary magnetism in the iron alone suffices to commence the inductive action. Fig. 63 is a side view of a generator constructed on these principles. C C¹ are two of the four field magnets, A A¹ their pole-pieces, between which the armature D (Fig. 64) is made to rotate, G G G are the armature bearings, H the driving pulley; the commutator is shown at the extreme right of the figure, where F F¹ represents the armature spindle cut in half longitudinally with insulating material inserted between the two portions; B¹ is one of the collecting brushes. The piece B, which connects the poles of the field magnets, carries a plate of insulating material N. To this are fixed four

electrical resistances, consisting of two coils, both connected to one battery, but one including the resistance to be measured. The two coils are made to simultaneously advance and recede from a galvanometer needle to produce a balance of effect thereon, their position is then made to indicate the unknown resistance.

*632.—George Davies, London. (*D. Brooks, Philadelphia, Pennsylvania, U.S.A.*) Insulators. 4d. March 7.

INSULATORS.—Paraffin is applied either as a coating to their inner surfaces or to impregnate the substance of the insulators. Paraffined paper is inserted between the insulator and its stem and cover.

694.—Donald Nicoll, Kilburn. Electric Telegraph Conductors. 8d. (1 fig.) March 11.

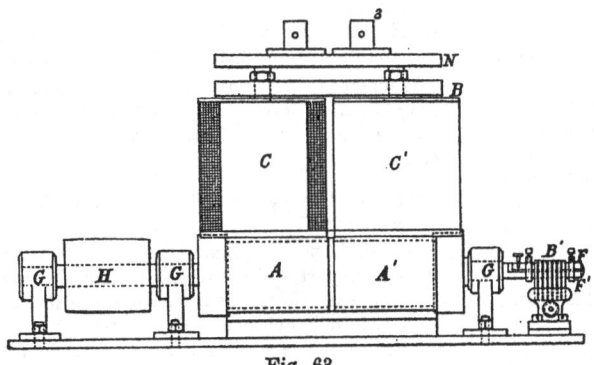

Fig. 63.

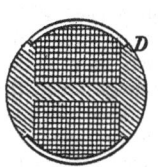

Fig. 64.

mercury cups (two only are shown) to which one end of each of the field magnet coils is connected. By means of these mercury cups the field magnets may be coupled in series or parallel, according as it is required to produce intensity or quantity currents. In both cases the current from the armature is made to pass through the field magnet coils, and thence to the outer circuit. Fig. 64 shows a cross section of the armature referred to above. Another form of generator consists of a circularly curved steel magnet, which is made to revolve through a number of coils arranged in a circle. The four coils nearest to the N poles of the magnet are always in metallic connection with rings forming terminals to the outer circuit, the contacts being made by springs and bell-cranks, actuated by the revolving magnet.

ELECTRIC LAMP.—Buoys are lighted by means of currents generated on land, as above described, and transmitted through a submarine cable. The light results from a series of sparks produced by a contact-breaker actuated by the current itself. Two forms of this contact-breaker are described and illustrated.

RESISTANCE MEASURER.—As an accessory to the above, an apparatus is employed for measuring

CONDUCTORS.—These are served with a plaited or twisted covering of hemp or other fibrous substance which is then impregnated with the insulating substance.

842.—H. Wilde, Manchester. Electro-Magnetic and Magneto-Electric Induction Machines. 1s. 6d. (9 figs.)

This specification describes and illustrates fully the well-known historical type of the Wilde generator. For illustrations and detailed particulars the reader is referred to the body of this volume.

951.—J. J. McComb, Liverpool. (*D. McComb, Memphis, Tennessee, U.S.A.*) Coupling Telegraph and Other Wires. 6d. (4 figs.) March 30.

CONDUCTORS.—This relates to a method of coupling conductors by means of hooks and eyes formed on their ends.

1,015.—J. M. Kilner, Chester. Apparatus for Towing, Laying Submarine Cables, &c. 1s. 6d. (9 figs.) April 4.

CONDUCTORS.—An inclined tube passes through the vessel at or near the centre of its length, through which the cable is allowed to run, its speed being checked by a suitable brake.

1,022.—T. B. Marshall, London. Insulating and Laying Telegraph Wires. 8d. (8 figs.) April 5.

CONDUCTORS.—This invention relates to a method of placing bare wires in channels or tubes and insulating them when in position, by the application of a suitable substance in a plastic or fluid condition. The operation is conducted in successive stages by alternately adding lines of wires and layers of insulating material, until the desired number of conductors is enclosed.

*1,210.—J. H. Johnson, London. (*J. N. M. Van Malderen, Paris.*) Magneto-Electric Machines. 4d. April 26.

MAGNETO-ELECTRIC GENERATOR.—Long parallel coils, extending the full width of the machine and wound longitudinally on cores of a T section, revolve in front of the extremities of permanent magnets, at uniform distances therefrom.

1,307.—Leon Delperdauge, Schaerbeck, near Brussels. Laying and Protecting Telegraph Wires. 8d. (8 figs.) May 4.

CONDUCTORS.—The wires are laid in cast-iron pipes which have a slot extending their whole length, through which the wires are introduced. The slot is then closed with a suitable cement, which can be supported by a strip of T iron, inserted in the pipe, and held in position by keys or other means.

The pipe lengths are connected by india-rubber straps, which overlap their ends.

1,308.—J. H. Johnson, London. (*A. Berlioz, Paris.*) Magneto-Electric Signal Lights. 10d. (11 figs.) May 4.

MAGNETO-ELECTRIC SIGNAL APPARATUS.—A copper cylinder encloses the electric lamp, and is capable of rotating in a horizontal or vertical plane on suitable centres. The front of the cylinder carries a plano-convex lens, and the back a concave reflector. For these a parabolic reflector may be substituted.

*1,503—E. H. C. Monckton, London. Steam Vessels, &c. 10d. May 21.

ELECTRIC GENERATOR.—The armature consists of a " large wheel with two brass sides," which carry a number of wrought iron bars, "each of sufficient bulk to form an armature large enough to close the circuit." These bars are surrounded by coils of insulated copper wire or foil. The armature revolves between the poles of a permanent magnet, and the currents generated in the armature coils are sent through a coil surrounding the permanent magnet in order to increase its power, after which they are "drawn off for the purposes of lighting."

This specification is not illustrated, and the description of the generator is somewhat obscure.

ELECTRODES.—Are made of a mixture of alumina or clay with gas coke, or of clay with plumbago, or a mixture of these is made, moulded and baked.

1,611.—M. A. F. Mennons, London. (*Sören Hjorth, Copenhagen.*) Magneto-Electric Batteries. 1s. 8d. (9 figs.) May 30.

MAGNETO-ELECTRIC GENERATOR.—Bar magnets are arranged cylindrically about the same axial line without touching each other, and between their poles armatures or coils are caused to revolve. These coils are provided with false poles, the dimensions of which correspond with those of a certain number of magnets of similar polarity. An advantage is obtained when the number of magnets and armatures is equal, and when more than one series of armatures is used they should be arranged " step-ways." The specification is not very clear.

1,720.—J. C. Fuller, London. Telegraphic Insulators and Caps. 8d. (3 figs.) June 11.

INSULATORS.—The cover or cap is perforated in order to allow the wind to come in contact with the insulator. The insulator itself may also be perforated.

1,755.—C. Varley and S. A. Varley, London. Electric Telegraphs. 1s. (8 figs.) June 15.

DYNAMO-ELECTRIC GENERATOR.—This invention consists in part of improvements in the method of generating electricity by mechanical force, as described in Provisional Specification No. 3,394 of 1866. Bobbins with soft iron cores and curved pole-pieces are arranged in pairs on opposite sides of an axle, and revolve between the poles of electro-magnets so placed that the bobbins and electro-magnets have always similar relative positions. Before using the apparatus an electric current is sent through the electro-magnets to give them a small amount of permanent magnetism, and this magnetism is afterwards increased by the mutual action of the revolving and fixed electro-magnets, the currents of electricity produced in the revolving bobbins being sent in one direction through the coils of the electro-magnets by means of a commutator. When four electro-magnets are used, two commutators and four bobbins are employed. The electro-magnets are wrapped with a double series of convolutions of insulated wire, one set being connected to one commutator and the other set to the other commutator.

MAGNETO-ELECTRIC GENERATOR.—A compound bar magnet is constructed with soft iron pole-pieces so arranged that the pole extensions of an insulated wire bobbin in contact therewith, can be moved from one pole to the other by a lever, thus developing currents of electricity.

1,770.—M. Gray, London. Telegraphic Conductors. 6d. (1 fig.) June 17.

TELEGRAPHIC CONDUCTORS.—A protective and preservative covering is given to the insulated conductor by enclosing it with strands of hemp or other yarn, which have been saturated with bitumen and gutta-percha or other suitable compound.

1,771.—M. Gray and L. Gibson, London. Examining Telegraphic Communicators. 10d. (3 figs.) June 17.
CONDUCTORS.—This relates to a method of starting and stopping the drums on which the conductor is wound, during the process of examination.

1,772.—M. Gray, London. Telegraphic Conductors. 2s. 4d. (17 figs.) June 17.
CONDUCTORS.—When several wires are to be twisted together previous to their being coated with gutta-percha or other material, they are first passed through a chamber containing a hot solution of gutta-percha, bitumen and tar in suitable proportions, which adheres to them and fills up the interstices between them. This invention also relates to a method of cooling the conductor by passing it through troughs of water, after it has been coated with its insulating covering, and to a chamber in which it is dried after such process.

2,016.—W. S. Andrews, London. Telegraphic Apparatus. 2s. 2d. (17 figs.) July 9.
INSULATORS.—The bolts of the insulators are covered with a "luting" of cotton saturated in paraffin.

2,147.—W. Thomson, Glasgow. Receiving or Recording Instruments for Electric Telegraphs. 10d. (7 figs.) July 23.
ELECTRIC GENERATOR.—This is a generator for the production of static electricity, and consists of a vulcanite disc, having on its rim a number of metallic sectors or "carriers." These carriers are lightly touched at opposite ends of a diameter by two fixed tangent springs, one of which is connected to earth and the other to an insulated metallic receiver. The point of contact of the earth spring is exposed to the influence of an electrified "inductor." On rotating the disc rapidly, each carrier, on leaving the earth contact, takes with it an induced charge, which it gives up to the receiver. An auxiliary generator, termed a replenisher, and constructed on similar principles, is employed to maintain the electrical charge on the before-mentioned inductor.

2,192.—G. Davies, London. (*D. Brooks, Philadelphia, Pennsylvania, U.S.A.*) Insulators. 6d. (3 figs.) July 29.
INSULATORS.—This invention has reference: (1) to immersing the insulators whilst hot in paraffin; (2) the use of blown glass as the material of the insulators; (3) the construction of an insulator consisting of a blown glass vessel embedded in a hood or casing, and carrying a hook of the usual form, both casing and vessel receiving a coating of paraffin; (4) to making the hook approach the outer casing so as to give an easy path to lightning and thus prevent damage to the insulator.

2,221.—F. H. Holmes, Paris. Apparatus for Producing Electric Light. 4d. August 1.
DYNAMO ELECTRIC GENERATOR.—From 16 to 40 armatures are fixed round and parallel to a shaft, and a corresponding number of electro-magnets are attached to a frame which surrounds the armatures, or the shaft may carry the electro-magnets and the frame the armatures. A small portion of the wire on each armature is used to excite the electro-magnets, a current from a galvanic battery in the first instance being sent through the coils of the electro-magnets to magnetise them. The armature bobbins, or the electro-magnets may each consist of two or more separate coils which can be brought into action as required, according to the amount of light to be produced.

2,241.—T. Allan, London. Submarine Telegraph Cables. 4d.. August 2.
CONDUCTORS.—These are formed of iron or steel and copper wires, so proportioned as to give the maximum conductivity consistent with strength.
Joints are made by inserting the ends of the conductor in a steel tube, and soldering them.

2,307.—F. H. Holmes, Paris. Apparatus for Producing Electric Lights. 10d. (13 figs) August 10.
DYNAMO-ELECTRIC GENERATOR.—From 16 to 40 electro-magnets are secured to the inner surface of a brass drum G, each electro-magnet E having

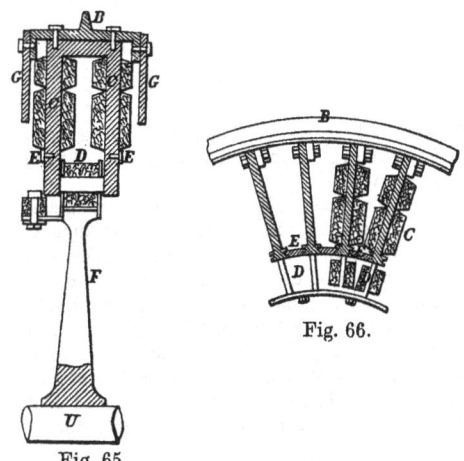

Fig. 66.

Fig. 65.

two or more separate coils of insulated wire (Fig. 65). A wheel F carries on its periphery the armature coils D D, the soft iron cores of which revolve close to the poles of the electro-magnets. The coils are D connected to rings on the shaft, from which the current is led off by brushes of soft copper wire. The current from the coils is partly used to produce the light, and partly to excite the electro-magnets, after having been made continuous by a suitable commutator, which is so constructed as to send the current through part or all of the coils of the electro-magnets as desired. Fig. 66 is a view of a portion of another form of generator, and shows in section the electro-magnets and armature coils. In this arrangement the electro-magnets are secured to a cast-iron drum, and consist of flat plates of wrought or cast iron with the poles enlarged so that the breadth of the poles shall be equal to their distance apart. Plates of metal (iron and lead excepted) or hard wood are inserted between the poles to keep them in position. The coils of the armature are compound, a part of the current produced being used to excite the electro-magnets. In other respects this form of generator resembles that last described.

2,480.—D. Nicoll, Kilburn. Electric Telegraph Conductors, &c. 4d. August 31.
Conductors.—A traction engine performs the operations of opening a trench, and depositing therein a suitable material for the protection of the conductor, which is previously prepared according to Patents 480 of 1866, and 694 of 1867. At junctions the wire ends are twisted together, and the joints completed by pressure between metallic matrices.

2,942.—A. F. Jaloureau, Paris, and C. L. Lardy, Neufchatel. Telegraphic Cables, &c. 1s. (9 figs.) October 19.
Conductors.—The copper conductor is first covered with bituminously prepared twine, and the insulation completed by successive coverings of bitumen and bituminised paper.

*3,046.—J. T. Carter, Sydenham. Manufacturing Circular Webbing or Gasket. 4d. October 29.
Conductors.—This relates to the manufacture of a circular webbing suitable for covering telegraph cables.

3,108.—W. R. Lake, London. (*A. G. Day, Seymour, Connecticut, U.S.A.*) Artificial Rubber Compound. 4d. November 4.
Insulating Material.—Relates to a method of manufacturing an insulating compound for use as a substitute for india-rubber.

*3,390.—M. F. Maury, London. Protecting and Paying Out Electric Cables. 4d. November 29.
Conductors.—The cables are formed into coils of convenient length, and packed in tanks or hogsheads. These are filled with water and sealed.

*3,542.—E. R. Sintzenich, London. (*D. Reed, New York.*) Preparing Varnish, Cement, &c. 4d. December 13.
Insulating Material.—Gutta-percha, india-rubber and other gums are treated with benzine or benzole and alcohol, in order to separate the barky, resinous and foreign matters. The purified product is used for coating telegraphic conductors.

3,609.—L. M. Becker, London. Arranging Telegraph Wires. 1s. 6d. (16 figs.) December 19.
Conductors.—The inventor claims a method of carrying overhead wires on brackets, and also of maintaining a uniform tension on such wires.

1868.

130.—L. M. Becker, London. Laying and Supporting Telegraph Wires. 1s. (8 figs.) January 15.
Laying Telegraph Wires.—Blocks of glass, clay or other suitable material, of triangular section are connected in continuous lengths, and have a number of holes or tubes made through them in a longitudinal direction, through which the wires are passed by attaching to each a piston fitting the tubes and exhausting or compressing the air at one or the other end of the line of blocks. The tubes can be exhausted and hermetically sealed after the wires are introduced.

315.—S. M. Martin, Pinner, and S. A. Varley, London. Signalling on Railway Trains. 1s. 6d. (20 figs.) January 29.
Conductor.—This consists of a core of wire rope or crinoline steel, served with hemp or other suitable material.
Magneto-Electric Generator. — Two sets of straight magnets are mounted parallel to each other and attached by their ends to an iron plate, their free ends being provided with soft iron pole-pieces or "horns," between which a soft iron toothed wheel and axle is capable of revolving. The pole-pieces have teeth, and the toothed wheel is so constructed that when one tooth is opposite the soft iron "horn" of one set of magnets, the space between the teeth is opposite the other soft iron "horn."
The axle carries a coil of insulated wire in which electric currents are generated when the toothed wheel rotates. Fig. 67 is a top view and

Fig. 68 a side view of the generator. s s^1 are the "horns," q the toothed wheel, r the axle, and t

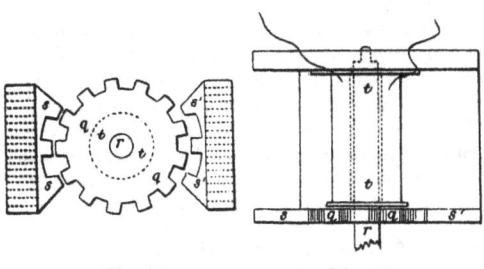

Fig. 68. Fig. 67.

the coil. The compound magnets are shown by the parallel dotted lines in Fig. 68.

535.—W. Perkins, London, and G. G. Tandy, Penge, Surrey. **Insulating Electric Conductors, &c.** 4d. February 18.

INSULATING MATERIAL.—The inventor claims the production of an insulating material by the combination of india-rubber, gutta-percha or other vulcanisable substances with anthracene, napthalene (coal tar products) and sulphur.

647.—A. V. Newton, London. (*L. C. Stuart, Mamaroneck, New York, U.S.A.*) **Electro-Magnetic Apparatus.** 1s. (6 figs.) Feb. 26.

ELECTRO-MAGNETIC MOTOR. — A horizontally rotating shaft carries a series of rotary electro-magnets, so arranged that their poles pass in close proximity to a series of fixed electro-magnets. The rotary electro-magnets are placed on the shaft radially and are angularly displaced with respect to each other, so that when the poles of one set have passed the poles of the fixed electro-magnets the poles of another set are approaching them. A battery is used to excite both series of electro-magnets, and a commutator is provided by means of which the current is transmitted to that set of rotary magnets which is approaching the fixed electro-magnets, the current being cut off at the proper time and transmitted to the next approaching set, and so on in succession. The commutator consists of two wheels, each of which is made up of three distinct plates, insulated from each other, and of such a shape that, when placed together, the periphery of the wheel, so formed, is divided into six sections, every section being insulated from that one adjacent to it, but in metallic connection with that one opposite to it. From each of these plates an insulated wire passes to one pair of the rotary magnets, and the return wires are joined to a series of insulated collars on the shaft. Against these collars rest springs, which conduct the current to the fixed electro-magnets, the return wires from which are all joined to two contact springs, bearing on the two commutator wheels, and in connection with one pole of the

battery. Two other similar springs are placed behind these, also bearing on the commutator wheels, and convey thereto the current from the other pole of the battery. The relative positions of these two pairs of springs determine when any series of magnets shall be demagnetised, the pair in advance acting as a cut-off, and at the same time conveying the induced current from either series of magnets, which may have been just cut off, into that series which at that instant is already under the full action of the battery. This arrangement of parts also effects a short circuit for the magnet coils as they arrive at their dead centres, by reason of the two sets of contact springs, already mentioned, bearing at that moment on the same section of the commutator wheel. These sections are separated by an oblique joint, for the purpose of avoiding a break in the circuit during the transfer of the current from one set of magnets to the next set. Fig. 69 shows a cross section of the motor, A A is

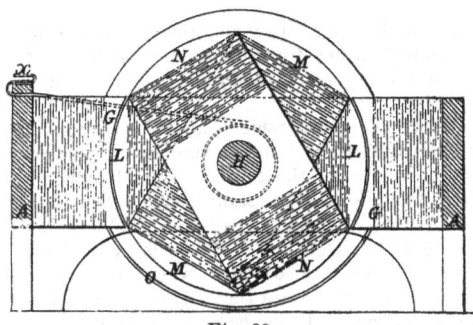

Fig. 69.

the frame, G G the fixed electro-magnets, L M N the rotary electro-magnets, and H the axle.

*668.—W. M. Bullivant, London. **Composition for Ships' Bottoms, &c.** 4d. February 27.

CONDUCTORS. — A preservative composition, suitable for applying to submarine cables, consists of red lead, beef suet, wax, pitch, brimstone, arsenic, Russian tallow, resin and blue butter in suitable proportions.

691.—H. B. Wilder, Sulham, Berks. **Telegraphic Apparatus.** 1s. (9 figs.) Feb. 29.

MAGNETO-ELECTRIC GENERATOR.—A permanent compound horizontal horseshoe magnet carries on each pole a vertical electro-magnet. A circular cast or wrought iron plate has its periphery divided into a number of equal teeth and spaces, and is so placed with respect to the magnets, that when it is rotated the teeth and spaces alternately pass over the poles of the electro-magnets. By this means positive and negative currents are generated in the coils of the electro-magnets, the number of such currents at each revolution corresponding with the number of teeth in the rotating plate.

773.—I. L. Pulvermacher, London. Producing and Applying Electric Currents. 10d. (19 figs.) March 6.

CURRENT MEASUREMENT. — Two platinum electrodes pass through the botttom of a glass vessel which contains coloured water. A capillary tube, having its lower end bent upwards, dips below the surface of the water and rises vertically through a graduated glass tube. When a current of electricity is conducted into the vessel the liquid is decomposed and the gases force a portion of it up the capillary tube. The height to which the liquid rises in a given time forms a measure of the amount of electricity passing through the instrument.

*838.—T. Walker, London. Telegraph Cables, &c. 4d. March 11.

CONDUCTORS.—This invention relates in part to the use of insulated conductors in connection with " positive, negative, or neutral earth plates or batteries."

939.—W. Hooper, Mitcham, Surrey. Treating India-Rubber for Manufacturing Fabrics and Telegraphic Conductors. 4d. March 19.

CONDUCTORS.—A suitable composition consisting of india-rubber and oxide of zinc, in the proportion of 32 lbs. of the former to 16 lbs. of the latter, is applied to the metallic conductor through a die. When a strand of wires is insulated, the central wire is coated with india-rubber, or compound of india-rubber, into which the outer wires of the strand bed themselves. The india-rubber or compound must be kept at a temperature of 250° Fahrenheit during these processes.

1,253. — C. W. Siemens, London. (*Dr. W. Siemens, Berlin.*) Electrical Signalling Apparatus.) 5s. (63 figs.) April 17.

MAGNETO-ELECTRIC GENERATOR.—A cylindrical armature, with longitudinal grooves, containing a coil of insulated wire, is made to rotate between the poles of a series of permanent magnets, and the electric currents generated in the coils are collected by a vibrating spring, actuated by a lever. The above generator is for use in connection with telegraphic recording instruments.

ELECTRO - MAGNETS. — The inventor claims amongst other things the employment of flat sheet iron cores for electro-magnets, intended for the reception of quickly reversed currents.

1,336. — J. Rogers, London. Compound or Varnish. 4d. April 23.

INSULATING MATERIAL.—This invention, relates, in part, to the manufacture of a compound, consisting of a mixture of india-rubber or gutta-percha with certain residual products derived from cotton-seed oil, coal oil, &c., such compound being suitable for coating telegraphic wires.

1,750.—M. Gray, London. Manufacture of Insulated Electric Conductors. 8d. (7 figs.) May 27.

CONDUCTORS.—The successive lengths of cable are vulcanised to within a short distance of their ends, which ends are joined and the joint, together with the unvulcanised portions, is then cured. When vulcanised ends are to be joined the insulating material is removed for a short distance on either side of the joint, and replaced by uncured compound, which is cured in a suitable vessel.

2,012.—M. Gray, London ; L. Gibson, Silvertown, Essex. Coating Electric Conductors. 8d. (5 figs.) June 22.

CONDUCTORS.—The first part of this invention relates to improvements in nozzles used for feeding the wire through the dies of the covering machine. In order to secure uniformity in the thickness of the coating, the nozzles are each provided with an independent adjustment in a longitudinal direction, by which they can be advanced or withdrawn whilst the covering operation is proceeding.

ADHESIVE COMPOUND. — This consists of gum balata, resin, oil and pitch in suitable proportions, and is applied to the wires to facilitate the adhesion of the gutta-percha coating.

2,060. — F. H. Holmes, Gravesend, Kent. Electro-Magnetic and Magneto-Electric Machines. 1s. (7 figs.) June 26.

ELECTRO-MAGNETIC AND MAGNETO-ELECTRIC GENERATORS. Referring to the illustrations, Fig. 70

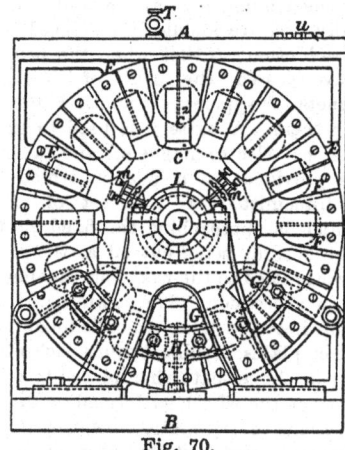

Fig. 70.

represents a side view and Fig. 71 a longitudinal section of the generator. The rotating magnets C^1, C^2, C^3, C^4, are constructed as follows :—When they are to be permanent magnets they consist of dished circular soft iron plates with radial steel arms welded on their peripheries : when electro-magnets, the plates and arms are of wrought or cast iron, and the space between their hollow faces is made to contain a coil of insulated wire D, wound on the axis parallel to the sides of the plates. In both cases the arms of each disc have an opposite polarity. The helices F have their iron cores and pole pieces split radially, and an insulating material inserted between the two portions, or the cores may consist of bundles of

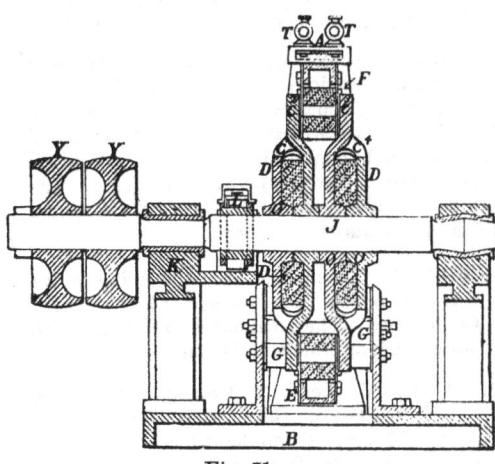

Fig. 71.

thick iron wires with their ends secured to iron plates. When great intensity is required the helices F may consist of two separate coils, an inner one of thick wire to excite the electro-magnets and an outer one of fine wire. When permanent magnets are used, fine wire helices alone are employed, and are composed of a series of volutes or spirals.

By means of suitable connections any number of pairs of the coils G can be used to excite the rotating electro-magnets, when it is desirable to use the whole of the current from the helices F. The commutator L has half of each tooth cut away in a direction parallel to the axis, and the space filled with an insulating material in order to interrupt the primary current when the greatest quantity of electricity is developed in the helices. The rubbers of the commutator are movable for the purpose of breaking connection between the coils G and the magnetic helices, so that the whole of the current from them may be discharged when required.

2,160.—T. J. Mayall, Roxburg, Massachusetts, U.S.A. **Electric Telegraph Cables.** 4d. July 8.

CONDUCTORS.—After the conductor is insulated, in the usual way, with india-rubber or gutta-percha, a wrapper of cloth or woven fabric "frictioned" with a solution of the insulating material is applied. When additional protection is required a further covering is applied, which consists, firstly, of pure rubber and, secondly, of a suitable vulcanised compound.

2,369.—S. M. Martin, Pinner; **S. A. Varley**, London. **Signalling in Railway Trains, &c.** 10d. (6 figs.) July 28.

CONDUCTORS.—A suitable number of strands, each consisting of one or more wires wound spirally round a hempen core, are arranged side by side, and bound together by a wrapping of hemp. They are then passed through a bath of melting marine glue or other compound.

*2,380. — J. R. Harper, London. **Strips of Zinc for Telegraphic Cables.** 4d. July 29.

CONDUCTORS. — This invention relates to an improved method of manufacturing strips or ribbons of zinc for the above purpose.

2,489.—F. Walton, Staines. **Treating Resins, &c.** 4d. August 8.

INSULATING MATERIAL.—100 lbs. Kowrie gum, 60 lbs. castor and 60 lbs. oxidised oil.

2,505.—M. Gray, London; **F. Hawkins**, Silver-town, Essex. **Telegraphic Cables.** 10d. (3 figs.) August 11.

INSULATING CONDUCTORS.—The usual method of coating a conductor with rubber compound, is to apply the same in strips, by means of grooved rollers, with cutting edges, which trim off the superfluous rubber. To avoid the waste consequent on this method, the inventor makes the edges of the grooves in the rollers blunt or rounded, and supplies the strips only of sufficient width to properly enclose the wire. A second part of the invention relates to preventing adhesion between covered wires when they are laid together.

*2,571.—A. Albini and **J. Vaglica**, London. **Electro-Magnets.** 4d.

ELECTRO-MAGNETS.—This invention consists in connecting the two coils parallel instead of in series.

*2,576. — D. G. Fitzgerald, London. **Constructing Electric Telegraphs, &c.** 4d. August 18.

CONDUCTORS.—Insulation is effected by means of good conductors of electricity, "so arranged as to generate an electromotive force which opposes the escape of the signalling current."

2,665.—N. J. Holmes, London. **Electric Telegraphs.** 6d. August 28.

MAGNETO-ELECTRIC GENERATOR.—Two soft iron armatures are fixed on an axle, and caused to oscillate before the ends of the soft iron cores of four coils, which are arranged one on either side of each pole of a permanent horseshoe magnet. The position of the armatures on the axle with reference to the coils is such, that one approaches whilst the other recedes, thus producing alternate currents of uniform intensity.

T

2,683. — C. F. Varley, Beckenham, Kent. Electric Telegraphs. 6d. (1 fig.) August 29.

CONDUCTORS.—The insulated conductor, is made to occupy the place of one of the helical strands in an ordinary cable, and should be previously served with tape to protect it against abrasion. The accompanying strands are hemp or other similar fibre. The invention also relates to a method of preserving vegetable fibres used in telegraph cables, by impregnating them with chloride of sodium or calcium, and then with a mixture of pitch with either of these materials.

*2,707.—J. H. Greener, London. Insulators. 4d. September 2.

INSULATORS.—These are made of enamelled iron, preferably in accordance with Patent No. 1,646 of 1863 (not included in these Abstracts).

*2,854.—A. M. Clark, London. (*Félix Gruet and Frédéric Gruet, Paris.*) Electro-Magnets. 6d. (4 figs.) September 16.

ELECTRO-MAGNETS.—This invention consists in combining a number of electro-magnets by joining their ends with a soft iron plate, there being an equal number of opposite poles connected to such plate.

*2,951.—E. Prevost, London. Electro-Magnets. 4d. September 25.

ELECTRO-MAGNETS.—The iron cores are heated and cooled in a bath of molten lead or fusible metal, the process being repeated several times, and they are then plunged in a hot bath of alcohol. Sal ammoniac is added when the metal, which has been previously polished, assumes a yellow colour. The whole is then allowed to cool together. Three-strand wire is preferred for winding these electro-magnets.

*3,051.—J. Aspinall, Harrow. Telegraphic and Other Ropes or Cables, &c. 4d. October 6.

CONDUCTORS.—This relates to the use of hair or animal fibre, saturated with gutta-percha, pitch, or resin, with which a silicious powder has been incorporated, for the purpose of repelling boring animals.

3,224.—E. O. W. Whitehouse, London. Protected Insulated Telegraph Wires. 4d. October 21.

CONDUCTORS.—Wires, which have been previously insulated, are protected with a coating, consisting entirely, or in part, of xyloidine in conjunction with volatile solvents, oils, resins, tar, &c.

*3,268.—W. Heasler, Greenwich, Kent. Coating Telegraphic Wire. 4d. October 26.

CONDUCTORS.—This invention relates to a method of applying to wires a coating of gutta-percha, india-rubber, or other insulating material by means of grooved wheels.

3,326.—A. M. Clark, London. (*J. H. Cazal, Paris.*) Sewing Machines. 10d. (4 figs.) October 31.

ELECTRO-MAGNETIC MOTOR.—The inventor claims the combination of an electro-magnetic motor with a sewing machine.

*3,329.—S. A. Varley, London. Generating Static Electricity. 4d. October 31.

ELECTRICAL GENERATOR.—Glass tubes, coated on their inner surfaces with a resinous cement, and mounted on bearings, are arranged in one plane and caused to revolve by suitable means. The ordinary rubbers are applied on one side, and collecting points on the other.

*3,490.—R. Green, London. Covering Telegraph Wires. 4d. November 17.

CONDUCTORS.—The conductors are enclosed in tubes of lead, or other suitable metal.

3,501.—C. W. Siemens, London. Fastening and Adjusting Telegraphic Line Wires. 8d. (13 figs.) November 18.

CONDUCTORS AND INSULATORS.—The insulator caps are provided with a cam or eccentric, carried by a pin, and situated between two studs or grooves, so that in one position of the cam the wire is gripped between its edge and the studs or grooves, but in another position the wire is released. The cam is secured in place either by making one portion of its edge circular, or by forming part of it in the shape of a nut, over which a plate fits, the plate having an arm abutting against a fixed projection.

*3,556.—W. A. Lyttle, Hammersmith. Electro-Telegraphic Apparatus. 4d. November 23.

CONDUCTORS. — Frederick Walton's patent oxidised oil solution is used, either pure or mixed with india-rubber or other suitable resinous solution, as an insulating coating for telegraphic wires.

MAGNETO - ELECTRIC GENERATOR. — Powerful electric currents, derived from a magneto-electric generator, are used to heat telegraphic wires, in order to free them from ice and snow in winter.

*3,661.—C. S. Rostaing, Paris. Electric Cables 6d. (8 figs.) December 2.

CONDUCTORS. — One or more conductors are wound helically round a central elastic cord of india-rubber, or other similar material. They are then covered with thin bands of insulating material wound in an opposite direction. Each succeeding coating is applied in the reverse direction to the last, the same principle being adopted when a metallic wire sheathing is added.

3,752.—T. Sturgeon, Manchester. Signalling in Railway Trains. 4d. December 9.

CONDUCTORS.—Conducting wires are connected by a metallic snap similar to an ordinary necklace snap, consisting of a socket and spring pin, with a projection which catches against a shoulder in the socket. In place of the socket a ring may be used having sharp edges so as to scrape a clean surface on the spring pin.

3,878.—W. F. Stanley, London. Machines for Exciting Frictional Electricity. 8d. (3 figs.) December 21.

ELECTRIC GENERATOR.— This relates to (1) a simple frame for a frictional plate electrical machine; (2) attaching the rubbers direct to the frame; (3) making the axle of wood; (4) making

the disc of sheet glass; (5) forming the collector of woven wire gauze.

3,938.—H. Clifford, Greenwich. Coating Submarine Cables. 4d. December 24.

CONDUCTORS.— The inventor claims the employment of powdered silica as a coating for cables to protect them from the attacks of submarine animals.

3,984.—D. Spill, London. Producing Compounds Containing Xyloidine. 4d. Dec. 31.

CONDUCTORS.—This invention relates to the production of a material for coating telegraph wire, consisting of a compound of xyloidine with camphor and oil.

1869.

222.—J. M. Merrick, jun., Massachusetts, U.S.A. Composition for Buttons, Insulators, Picture Frames, &c. 4d. January 25.

INSULATORS.—These are made of a composition of gum shellac and powdered silica. The latter should be in an impalpable state.

390.—F. Jenkin, Edinburgh. Apparatus for Producing Electric Light. 4d. February 8.

ELECTRIC LIGHT.—This invention is especially applicable to buoys and consists in the production of a rapid succession of sparks, due to a series of charges and discharges of a condenser. The condenser is placed on the buoy, and is charged from a battery ashore, the discharges being produced by the movement of a metal tongue actuated by the current, or by the motion of the buoy itself.

*391.—W. A. Lyttle, Hammersmith. Magneto-Electric Apparatus. 4d. February 9.

MAGNETS.—" Short circuiting," or intermittent suspension of the attractive force of permanent or electro-magnets, is effected by connecting their opposite poles with an armature, having coils at its two extremities, through which an intermittent current is sent. In winding electro-magnetic or magneto-electric coils, the wire is carried back to the commencing end on the completion of each helix, in order to make the direction round the axis and from end to end of the coil the same in each successive set of convolutions.

501.—D. G. Fitzgerald, London. Electric Telegraphs and Voltaic Batteries. 1s. (8 figs.) February 18.

CONDUCTORS.—This invention relates to " electrolytic insulation," or the insulation of conductors

by other conductors, so arranged as to generate an electromotive force which opposes the escape of the current.

INSULATORS.—In place of porcelain or other similar dielectric, " electrolytic insulators " or batteries are used.

531.—M. Gray, London. Covered Electrical Conductors. 4d. February 20.

CONDUCTORS.—A wire or strand, of sufficient strength to be used without a hempen or metallic envelope, is covered with pure india-rubber and india-rubber compound in the usual way. A tape or cotton strip is then lapped round before the compound is cured, over which a solution of rubber compound is applied; and, lastly, another strip, faced with the same solution (the prepared face inwards), is wound in an opposite direction to the first. The core is wound on a hollow drum, covered on the outside with a soft substance, such as cotton, and then cured. It is claimed that by these means the conductor is kept central.

797.—W. A. Lyttle, Hammersmith. Sustaining and Insulating Telegraph Wires. 4d. March 16.

INSULATORS. — This invention relates to the employment of enamelled iron for the stems of insulators, which may or may not have insulating caps. In the latter case a wrapping of paper, saturated with india-rubber or paraffin, is applied to the wire at the point of attachment. In the same way a number of wires may be carried by an enamelled iron frame.

917.—W. R. Lake, London. (*D. F. J. Lontin and E. L. C. D'Ivernois, Paris.*) Electro-Magnetic Machine. 1s. 2d. (5 figs.) March 25.

DYNAMO-ELECTRIC GENERATOR. — A number of

soft iron armatures, with flattened coils, and of horseshoe form, are disposed round a rotating shaft, with their poles arranged in a circle concentric with the axis of rotation. The field magnets, which are also flattened and of horseshoe form, have their poles placed close to the circumference of the circle described by the armature poles. By this method of construction it is claimed that a very moderate speed of rotation only is requisite.

Fig. 72 is a side view of a generator constructed according to this invention. e, e^1, are the armatures, b the shaft to which they are affixed, i, i, i, i, are the field magnets, and a the cast-iron frame.

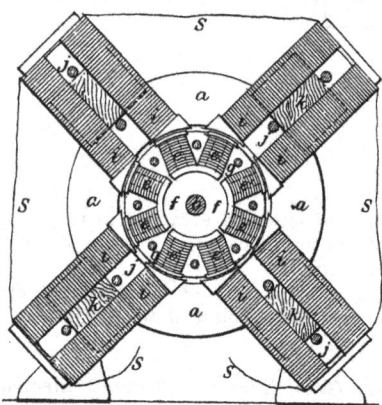

Fig. 72.

The field magnets are held in position and insulated from the frame a by the blocks of wood k. The shaft b carries two commutators, one of which serves to collect the current from part of the armature coils for the purpose of exciting the field magnets, and the other to connect the remaining armature coils with the terminals representing the outer circuit. The field magnets in the first instance are magnetised from any suitable source.

1,076. — J. Aspinall, Harrow. Telegraphic Cables, &c. 4d. April 9.

CONDUCTORS.—This invention relates partly to improvements on Patent No. 2,683 of 1868, and consists in applying to the cable, as it comes from the spinning machine, a serving of yarn, laid on in the opposite direction, and with a shorter lay than that of the cable. A further improvement consists in worming the cable with an elastic material. The inventor also claims the use of a compound of animal hair with vegetable fibre as a protection against sea-boring animals.

1,136.—J. H. Johnson, London. (*A. Cary, New York. M. G. Farmer, G. F. Milliken, and J. M. Batchelder, all of Boston, Massachusetts, U.S.A.*) **Manufacturing Compound Telegraph Wire.** 8d. (5 figs.) April 13.

CONDUCTORS.—This relates to machinery for sheathing wires with metallic ribbons, and afterwards closing the seam by drawing the compound conductor through a die. The central wire may be coated with tin before the sheathing process, in order to close the seam more effectually.

1,744. — F. H. Holmes, Mortlake, Surrey. Electro-Magnetic Machines. 10d. (10 figs.) June 5.

DYNAMO-ELECTRIC GENERATOR. — The inventor claims :—(1.) The construction of cores for the helices of split tubes, rivetted into soft iron end plates, the end plates also being split radially in order to prevent circular currents. (2.) The method of coupling up the helices in machines requiring not more than 8 horse-power to work them, which is done as follows :—In a machine of 20 helices, 10 alternate helices are put in series, and connected to the outer circuit. The other 10 helices are connected to a suitable switch-board, by means of which 6, 8, or the whole number of them may be made available for exciting the electro-magnets. (3.) The arrangement of the armature. On either side of a hollow brass boss—keyed to the shaft—is bolted a ring of horseshoe electro-magnets, the poles of which are equidistant in the same ring, and facing one another in opposite rings. The electro-magnets are further secured in position by a brass hoop which encircles them, and by two brass discs, fixed to the outside of each ring of electro-magnets, and screwed to their poles. (4.) Cooling the cores of the helices by the circulation of water during the action of the generator. (5.) A commutator having teeth of such a form that the roller, in passing over them makes a short circuit for about one-fourth of the time of passing one tooth. (6.) The mode of coupling up the helices of generators intended to work at a low speed, or to produce more than one light. As an example, if the helix ring be capable of holding 44 helices, 43 only are mounted on it, leaving on each side of a series, say, of 13 helices, a space equal to half that occupied by a single helix, so that when 30 poles of the electro-magnets coincide with the cores of 30 helices, 12 will be exactly midway between the 13 helices. The 30 helices are coupled up in two series of 15, each of which series will give a light. The remaining 13 are used for exciting the electro-magnets.

2,008.—A. Foucaut, Orleans, France. Telegraphic Cables. 6d. (2 figs.) July 2.

CONDUCTORS.—Metallic wires, insulated in the usual way, are covered with a fibrous material coated or impregnated with ceruse or carbonate of lead, and then enveloped in a conducting coating, preferably consisting of plumbago or metallic foil. When two or more cores are formed into a cable, they are wrapped round with a linen band saturated with a compound of ceruse with lithage oil, and afterwards served with tarred hemp and sheathed in the usual way. The inventor claims as an insulating material, in place of gutta-percha, a composition of ceruse or carbonate of lead finely powdered, and mixed with lithage oil and sawdust.

*2,160.—H. E. Newton, London. (*L. Bastet, New York, U.S.A.*) **Electro-Magnetic Engines.** 4d. July 17.

ELECTRO-MAGNETIC MOTOR. — Two circles of horseshoe electro-magnets are arranged on either side of a fly-wheel, which comprises a series of armatures carried on a hub and a rotating shaft. The number of armatures is greater than the number of electro-magnets, so that when the circuit is broken or closed from one magnet to the next, the distance through which the attracting force will have to act on the next armature will be equal only to the difference between the centres of magnets and armatures. A distributing wheel alternately breaks and closes the circuit. The electro-magnets are connected in pairs, three on each side of the fly-wheel, and these are again connected to the corresponding pairs on the opposite sides. In this way two sets of magnets on each side of the fly-wheel form a series, and the wires from each series are connected to their respective section of the distributing wheel. The surface of the distributing wheel is kept moist with water to prevent oxidation.

2,177.—A. M. Clark, London. (*P. A. Balestrini, Paris.*) **Electric Cables.** 8d. (8 figs.) July 19.

CONDUCTORS.—The feature of this invention is the formation of a cable consisting of a number of conductors, alternating with insulated cords, one of these cords forming a central core. Protecting coverings, of yarns, saturated with insulating or waterproof compounds, are afterwards applied.

2,235.—W. R. Lake, London. (*W. E. Simonds, Hartford, Connecticut, U.S.A.*) **Insulators.** 8d. (2 figs.) July 22.

INSULATORS.—On the iron supporting shank a cup is cast, or secured, the lip of which projects into, but does not touch, a groove formed in the bottom of the insulator proper. The lower part of the insulator, which may be of glass, hard rubber, or other suitable material, extends downwards so as to nearly touch the bottom of the cup. An insulating substance, preferably paraffin, while in a state of fusion, is poured into the cup. The invention it also applicable to inverted insulators.

*2,485.— J. Jones, London. Electromotive Power. 4d. August 19.

ELECTRO-MAGNETIC MOTOR.—This invention consists in equally distributing an uneven number of electro-magnets in a circle, and so arranging them that both ends of such magnets will be acted upon by the keepers, which are of a corresponding even number, and also equally distributed at each end of the magnets. The pole changes are effected preferably by a cam wheel and springs. The details are not very clear from the short description given.

2,525.—O. Varley and F. H. Varley, London. Transmitting and Recording Electric Signals. 1s. 10d. (26 figs.) August 25.

ELECTRO-MAGNETIC GENERATOR. — Fig. 73 illustrates this invention in a diagrammatic form. A and B are electro-magnets with flanges or pole-pieces C, D, E, P, attached to each end. An axle H, of gun-metal or other non-magnetic substance, placed parallel to and between the electro-magnets, carries on either end of it circular iron flanges I J, to which are screwed short iron bar electro-magnets K, L, M and N; the insulated wires of these magnets are attached to reversing commutators O P. The current generated in the coils of one of the armatures is made to flow through the wire G, and reinforce the magnetism already excited in the electro-magnets A B by the battery E F. The current generated in the

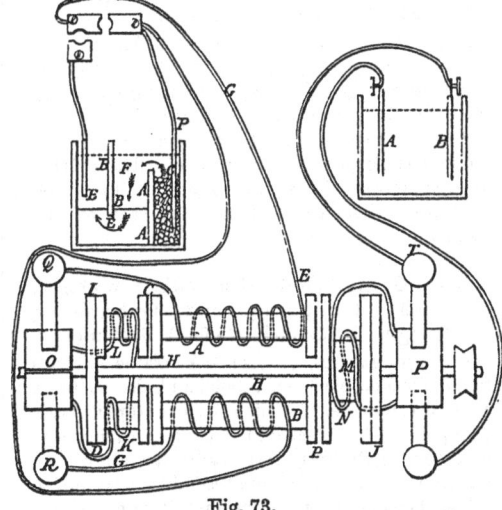

Fig. 73.

coils of the other armature is made to flow into the induction plates A B, which are inserted in the outer circuit. When the armatures are rotated a current flows into these plates until they are polarised, the resistance in circuit becoming augmented proportionally to the tension of the accumulated charge, thus reducing the current from the armatures, and consequently the magnetism of the field magnets, which would otherwise become so powerful as to retard the rotation of the armature. The induction plates therefore " act as valves shutting off the circuit when the machine is fully charged and relieving the labour of rotation."

SECONDARY BATTERY.—This invention comprises the use of the current due to the electric charge stored up in the "induction plates" above referred to. The secondary battery consists of "two or more pieces of carbon and palladium plates inserted in a cell filled with acid and water, or plates of metallic arsenic and carbon, or platinum substituted for the palladium or arsenic plates, or two lead or other metal plates, or two or more pieces of carbon. The currrent is made to generate hydrogen on the palladium, arsenic, or platinum plates, and oxygen on the carbon." Great efficiency is obtained by using palladium or arsenic on account of their capacity for storing hydrogen. The capacity of the carbon plates for oxygen is increased by submitting them to the action of fluoric acid, which dissolves the silica contained within them, and renders them porous.

CONDUCTORS.—Metallic wires are insulated by submitting them to the action of sulphuretted hydrogen, or by coating them with zinc, and afterwards oxidising it, or converting it into subchloride or carbonate. Oxides and sulphurets of metallic bases may be deposited on the wire by electrolysis with the same object.

2,603.—G. Henley, London. Telegraphs. 3s. 4d. (25 figs.) September 3.

MAGNETO-ELECTRIC GENERATOR.—This is the usual generator as applied to dial telegraph instruments; the only novelty consisting in having the lower flanges of the coils made of sheet iron instead of brass, in order to conserve the power of the magnet, over which they move. The driving band is pierced with equally pitched holes, which gear with studs screwed into the face of the pulley.

2,634.—G. Little, Rutherford Park, New Jersey, U.S.A. Composing and Transmitting Telegrams. 1s. 10d. (15 figs.) September 7.

ELECTRO-MAGNETIC MOTOR.—A frame carries two horseshoe electro-magnets, and also sustains a vertical spindle, on which the armature is fixed. The poles of the electro-magnets face each other, and are sufficiently far apart to allow the armature, which in section resembles a trefoil, to rotate between them. The circuit is closed for each magnet as the armature approaches, and broken, when the armature has reached its nearest point to the poles of the electro-magnet, by a commutator, which consists of three metal pins projecting from the bottom of the armature, and bearing successively against one of two springs in connection with the battery.

*2,643.—T. Walker, London. Electro Telegraphy. 4d. September 8.

CONDUCTORS.— Several conductors are formed into a cable, and the effects of induction are obviated by separating them by means of cocoa-nut fibre, or other suitable material. Cocoa-nut fibre is also used to cover single conductors.

2,712.—A. Collingridge, London. Stowing and Laying Submarine Telegraph Cables. 8d. (2 figs.) September 17.

CONDUCTORS.—Submarine cables are stowed on board ship by winding them on drums mounted so as to rotate on a horizontal axis. The drums are placed in a well open to the water at the bottom.

3,028.—J. M. A. Stroh, London. Electro-Magnetic Clocks, &c. 8d. (7 figs.) October 18.

ELECTRO-MOTOR.—A galvanometer needle within a flat coil is caused to rotate steadily in one direction by passage of currents, alternating in direction through the coil. Each electrical impulse carries the needle through one-half of a revolution.

In a second machine, a galvanometer needle is suspended over a coil wound on a soft iron ring. "By the action of the currents the ring is temporarily magnetised, changing its polarity at each inversion, and the needle is thereby propelled."

3,038.—C. E. Spagnoletti, London. Signalling. 10d. (5 figs.) October 18.

CONDUCTORS.—Each end of the conductor is bifurcated, and the two consist respectively of a spring clip and cone. The clip is formed of four springs which are made to embrace the cone when it is necessary to unite the conductors end to end.

3,070. — J. Buchanan, Gateshead-on-Tyne. Coiling Telegraph Cables or Ropes. 10d. (3 figs.) October 21.

CONDUCTORS.—This relates to a machine for coiling, uncoiling or paying out cable. The cable is led through a central tube to a radially traversing roller which also revolves, and thus lays down the cable in a succession of spirals.

3,102.—D. Spill, London. Preparing and Using Solvents of Xyloidine. 4d. October 26.

CONDUCTORS.—A compound suitable for covering and protecting wires, consists of xyloidine, castor oil, camphor, and a solvent in suitable proportions. Eight solvents are described in the specification.

3,147.—E. H. C. Monckton, London. Ocean Telegraphy. 4d. October 30.

CONDUCTORS.—These are constructed by coiling an insulated copper wire upon a previously insulated soft iron wire or rod. The wires may be insulated in the usual way or by dipping in pure cod-liver oil. Carbonaceous matter left after the distillation of tar, and sometimes combined with creosote, is also employed as the basis of an insulator.

3,178. — A. H. Brandon, Paris. (*Madame C. Errani, Paris.*) Motive-Power Engines. 8d. (4 figs.) November 2.

ELECTRIC GENERATOR.—As employed for igniting the charge in a gas engine; this generator consists of two discs of hardened india-rubber rotating close to each other but without actual contact. A rubber is pressed against the lateral surface of one disc, whilst induced electricity is obtained from the other.

3,236. — F. Jenkin, Edinburgh. Submarine Telegraph Cables. 4d. November 10.
CONDUCTORS —Deep sea cables are constructed with alternate strong and weak sections, in order to diminish their cost, but at the same time to provide for their recovery by grappling, in case of accidental injury.

3,274.—W. E. Gedge, London. (*J. Crougières, Ollionles, France.*) Composition for Preserving Metal, &c. 4d. November 13.
CONDUCTORS.—A composition for coating pipes, in which subterranean conductors are laid, consists of sulphur, coal tar, gutta-percha, red lead, white lead, pitch, resin, and spirits of turpentine in suitable proportions.

*3,363. — J. Burroughs, jun., Newark, New Jersey, U.S.A. Electro - Magnetic Machines. 4d. November 20.
ELECTRO-MAGNETIC MOTORS. — This invention consists in (1) the construction of electro-magnets "whose limbs are sectors or the approximates of segments of a circle;" (2) a method of binding the limbs of a magnet together with a tie bolt and washer; (3) compounding electro-magnets, having helices of plane surfaces in close contact with each other; (4) interposing layers of foil, or sheet metal, between the layers of the coil, in order to neutralise induction; (5) "the use of a shaft running through the revolving armature" of an electro-magnetic motor.

3,489.—F. C. Webb, London. Submarine Telegraphic Cables. 1s. 4d. (11 figs.) December 2.
CONDUCTORS.—To avoid damage to the insulated core by accidental piercing, and to guard against the attacks of the teredo, the cable is manufactured as follows:—The insulated conductor is first served with wet yarn, and then enclosed in a continuous metal sheath of soft steel in the form of a lapped tube, which is closed by a serving of tarred yarn. The invention includes the necessary machinery for conducting these operations.

3,587.—W. A. Marshall, London. Electric Telegraph Cables. 10d. (4 figs.) Dec. 11.
CONDUCTORS.—The conducting wire is wound with two servings of paraffined cotton cord, applied in opposite directions, and then drawn into a metallic tube, after which liquid paraffin wax is made to fill the space remaining between the wire and the inner surface of the tube. This last process is effected by exhausting the tube, and then admitting the paraffin wax in a liquid state. The inventor claims the apparatus for conducting this process.

3,637.—W. T. Henley, London. Protecting Telegraph Wires and Cables. 6d. (6 figs.) December 16.
CONDUCTORS.—For the protection of the shore ends of submarine cables, when they lie on an incline, they are enclosed in pipes made longitudinally in halves, and united, end to end, with ball and socket joints.

*3,696.—D. G. Fitzgerald, London. Electric Telegraphs. 4d. December 21.
CONDUCTORS.—This invention consists in the substitution for an "electrolytically" insulated cable, forming a voltaic combination of several elements, of a simple wire or wires, not coated with a dielectric.

3,778.—A. Matthiessen, London. Insulating Substance. 4d. December 31.
CONDUCTORS.—This invention relates to the use of a substance known as "ozokerite," as an insulating material for coating electric conducting wires. For this purpose it may be applied in its natural state or the products of its partial or complete distillation may be used, or the natural substance or these products may be mixed with gutta-percha, india-rubber, balata, or other known insulating substances, with or without fibrous material such as silk, wool, or mica, asbestos, and other similar substances.

*3,780.—H. E. Newton, London. (*L. Bastet, New York, U.S.A.*) Electro-Magnetic Engines. 4d. December 31.
ELECTRO-MAGNETIC MOTORS.—This specification is, in the main, indentical with No. 2,160 of 1869.

1870.

I7.—W. E. Newton, London. (*H. M. Paine, Newark, New Jersey, U.S.A.; and M. S. Frost, New York, U.S.A.*) Electro-Magnets. 1s. (14 figs.) January 3.
ELECTRO-MAGNETS.—The main feature of this invention is the construction and application of electro-magnets, the legs of which are made up of flat plates of sector-like form. For maintaining the space between the legs, and binding them together, there is an arrangment of a tie bolt and

washers, by which mode of construction, it is claimed, that economy and accuracy are attained. Figs 74 and 75 show respectively, a side view and longitudinal section of this form of electro-magnet.

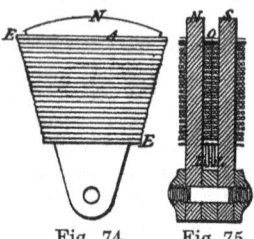

Fig. 74.　　　Fig. 75.

Another part of the invention relates to the employment of sheets of tin foil, or other ductile metal, arranged between adjacent coils as shown at O, Fig 75, for the purpose of "reversing by electrical induction the order of the currents," and thereby preventing the retarding effect, which is due to the mutual attraction of like currents, and detracts from the power of the electro-magnet.

ELECTRO-MAGNETIC MOTOR.—A series of electro-magnets, as above described, are attached to a wheel, with their poles so arranged as to lie in a circle around their centre of rotation, and another series of electro-magnets, with corresponding concave poles, are secured to a frame in which the wheel rotates. Fig. 76 is a side view of a

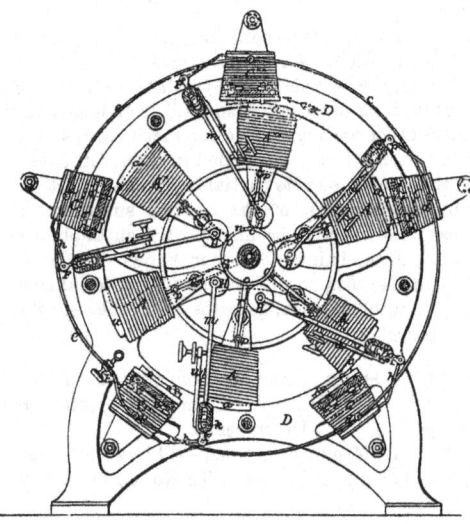

Fig. 76.

machine constructed according to this invention. The connections are arranged as follows.—All negative ends *d* of the magnet coils C are secured to a common conductor *e*, which is held by a pole connection E, attached to, but insulated from, the frame D ; the positive ends are in connection with adjustable circuit breakers F, having rollers

H resting upon a commutator, the metallic segments of which *o* are joined to the positive ends of the rotating magnet coils. The negative ends of these coils are secured to a guide plate, in electrical contact with the frame D, on which is placed the other terminal screw. The commutator is made adjustable with reference to the position of the rotating magnets, in order to avoid the retarding effect of the reflex current at the moment of breaking the circuit. This is done by breaking the circuit at some point intermediate to the coincidence of the poles, in which case the reflex current will aid in continuing the motion after the circuit is broken. The fixed magnets must be odd in number and the rotary magnets even, or *vice versâ*.

207. — A. M. Clark, London. (*G. H. Dervieu, Paris.*) Electro - Motor for Sewing Machines. 8d. (4 figs.) January 22.

COMMUTATOR.—This relates to improvements in the current distributor of that form of electro-magnetic motor, in which fixed electro-magnets act on a series of armatures, carried by a wheel. The invention consists in substituting, for the usual toothed wheel and spring blades, a roller constructed of some insulating material, on the periphery of which are inserted metal contact sections, against which two contact rollers are made to bear.

272.—R. Dick, Glasgow. Covering and Insulating Telegraph Wires. 1s. (24 figs.) January 31.

CONDUCTORS.—Gutta-percha solution is used as a cementing medium in applying insulating coatings of vulcanised india-rubber to conducting wires, and the invention is carried out, in one way, by preparing long narrow slips of vulcanised india-rubber, to which the gutta-percha is applied in the form of a hot solution, and by wrapping such prepared strips round the conducting wires by suitable means. The joint may be made either longitudinally or helically. By another method the wires are embedded in grooves formed in a continuous strip of vulcanised india-rubber, and cemented in place with gutta-percha solution.

This invention also includes the application of a first insulating coating to the conductor, by passing it first through heated shellac, and afterwards through a hopper containing dry, hot, hard-wood sawdust.

*274.—T. Walker, London. Insulating Telegraph Conductors. 4d. January 31.

CONDUCTORS. — Gutta-percha or india-rubber, dissolved in a volatile solvent, either pure or mixed with other substances, is employed as an insulating coating to electrical conductors.

*344.—H. Brooks, A. Brooks, T. Bestwick, and W. Bestwick, Manchester. Covering Telegraph Wires. 4d. February 7.

CONDUCTORS.—This relates to a method of covering wires with successive layers of tarred hemp, applied helically, each coating being wound in a contrary direction to the previous one.

*409.—J. Story, Paris, Kentucky, U.S.A. Submarine Telegraph Cables. 4d. February 11.

CONDUCTORS.—The central portion of the cable is a flat metal chain having on both sides small teeth or points. The chain is passed through hot pine tar and then tightly wrapped with a web, made of strong hemp cord, over which a coating of prepared gutta-percha is applied. The insulated conducting wires are bent in short curves, and placed on the flat sides of the chain, so that the curves fit into each other as they lie flat, and the wires are then cemented together with prepared gutta-percha. Over this are applied in succession a second hempen web, saturated with gutta-percha, a coating of sea sand, a preparation of pine tar and zinc paint, and another coating of sand. Over the sanded coating one or more thicknesses of prepared gutta-percha are cemented and the outside surface of the cable bound with small wire. The process is completed by the application of a varnish of pine pitch. Plates of iron, steel, or sheet metal may be substituted for the central chain.

627.—S. E. Phillips, Hackney Wick. Electric Cables. 4d. March 3.

CONDUCTORS.—The conducting wires are lapped with strips of gutta-percha, in one or more layers, and cemented and covered with shellac or thin tape of xylonite. They are finally covered with a waterproof material, preferably such as is described in Specifications Nos. 3,224 and 3,984 of 1869 (not included in these Abstracts).

787.—D. Spill, London. Compounds Containing Xyloidine. 4d. March 16.

CONDUCTORS.—This relates to improvements in the process, described in Specification No. 3,102 of 1869, for the preparation of compounds containing xyloidine, such compounds being suitable for covering conducting wires.

1,192.—G. Fenwick, Gateshead-on-Tyne. Ropes or Cables for Telegraphic or Other Purposes. 1s. 10d. (6 figs.) April 25.

CONDUCTORS.—To avoid putting a twist upon the protecting wires of cables, when covering them with a serving of hemp, or other fibrous material, the wire bobbins are placed in the centre of the drums carrying the yarns, and are so geared that the wire is held stationary with respect to the yarns during the serving process.

1,364.—W. Daniell and H. Lund, London. Electro-Magnetic Engine. 1s. (9 figs.) May 13.

ELECTRO-MAGNETIC MOTOR.—Two circular side frames are braced together by 12 stretchers, equally pitched around the circumference. These stretchers support 12 pairs of electro-magnets attached thereto by bolts, cast in the metal of their cores, and placed radially with respect to the circular side frames. Each pair of electro-magnets forms a horseshoe with N. and S. poles. Eight pairs of similar magnets, provided with curved and extended poles, as shown in Fig. 77, are attached by means of bridged connecting pieces to a boss affixed to a shaft running in bearings on the side frames. Fig. 78 illustrates

Fig. 78.

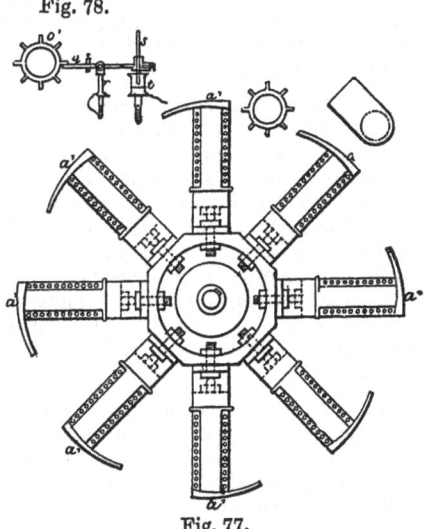

Fig. 77.

the method of breaking contact, and represents one of the distributing rings provided with teeth, which act on the lever *g*, raising the metallic point *s* out of the mercury cup *t*. The rotating shaft carries a boxwood cylinder on which are fixed three plain brass rings, two toothed rings, with eight teeth, and three toothed rings, with 12 teeth. The rings with 12 teeth break the current for the revolving magnets, and those with eight teeth effect the same for the fixed magnets. The exact working of this commutator is not clear from the description or drawing. Another part of the invention relates to the use of magnets of elliptical section arranged with the minor axis of the ellipse in the direction of the working of the magnets.

*1,445.—G. Zanni, London. Magneto-Electric Telegraph Apparatus. 4d. May 19.

MAGNETO-ELECTRIC GENERATOR.—The object of this invention is to obtain continuous currents in one direction. The generator consists of a cylindrical armature channelled longitudinally, so as to form a bobbin which is filled with covered wire, and rotates in a circular recess formed between the poles of the magnet. The bobbin may also be made of two iron discs, united by a flat strip of iron, and filled with wire, as in the former instance. This form of bobbin is caused to rotate either in front or at one side of the magnet poles, or two armatures may be used mounted on the same spindle, one on each side of the magnet.

1,555.— G. Stevens and J. Hendy, San Francisco, U.S.A. **Electro-Magnetic Motor for Sewing and other Machines.** 10d. (6 figs.) May 27.

ELECTRO-MAGNETS.—An insulated wire coil is surrounded by an iron cylinder, by which means, it is stated, the power of the coil is greatly augmented. A second coil, enclosed by another cylinder may be added.

1,574.—W. Beamish, London. **Iron Kerb for Telegraph Wires.** 4d. May 30.

CONDUCTORS.—For the protection of electrical conductors, when laid by the roadside, a cast-iron kerb is used, similar in form and size to the ordinary stone kerb. The kerb is formed in lengths united by tongues or bolts, and is provided with openings placed at convenient distances apart, which are covered with lids having internal hinges and scupper holes. Branches are inserted where desirable, and the top of the kerb is roughed or grooved.

1,576.—J. F. Wiles, London. **Electro-Magnetic Engines.** 10d. (9 figs.) May 30.

ELECTRO-MAGNETIC MOTOR.—A series of electro-magnets are arranged radially upon a disc, around which is an iron ring of circular or polygonal form, serving the purpose of a "keeper" to the said electro-magnets. The disc and keeper have only a small amount of difference in size, so that the poles of the electro-magnets are at no time far removed from the inner surface of the keeper. Each magnet is provided with a "double-action commutator," working upon two rings on the disc, in such a way that by reversing the connections the motor may also be reversed. Motion is transmitted from the revolving disc in any suitable manner. One form of this motor consists of a shallow iron cylinder supported upon pillars, with its axis vertical, and provided with a planed iron floor, through the centre of which passes a vertical spindle, having at its lower end a bevel wheel. Six pairs of magnets are fixed radially to a ring attached to a square on the vertical spindle, and furnished with an arrangement of slots so as to provide for its eccentricity of motion. The lower edge of the ring rests on balls or rollers, which run upon the planed iron floor of the cylinder, and thus reduce the friction. The commutator consists of two brass rings, faced on the inside with platinum, to each of which is attached a terminal screw. These rings are insulated from each other by ebonite, and also from the cylindrical keeper to which they are attached. One end of each magnet coil is connected to a platinum-faced brass disc fixed to the upper end of the spindle, but insulated therefrom; the other ends of the magnet coils are connected to the commutator rods, which are arranged one over each magnet, and are provided with helical springs to force their ends against the insulated brass rings already mentioned. The ends of these rods have a cross bar carrying at one extremity a platinum contact point, and at the other an ivory point. A third terminal is connected to an arm and roller, which bears against the platinum-faced disc on the spindle.

The action of the motor is as follows:—Each magnet as it comes in contact with the inner surface of the iron cylinder holds itself in contact therewith, whilst that previously attracted is set free by the ivory point on the commutator coming in contact with the brass rings, and thus releasing the platinum point through which the exciting current has been conveyed. In this way the magnets travel round within the cylinder and transmit their motion to the vertical spindle. The course of the current is through the arm and roller to the platinum-faced disc, thence to the magnet coils, from which it passes to the brass rings, by way of the commutator rods in turn, as they come into action. Other forms of the motor are illustrated; one with two magnet rings placed vertically upon the same horizontal shaft, and another with the magnets arranged as an octagon revolving within a nonagonal keeper.

1,584.—B. Hunt, London. (*L. Finger, Cambridge, Massachusetts, U.S.A.*) **Electro-Magnetic Apparatus for Moving Panoramas, &c.** 8d. (7 figs.) May 31.

ELECTRO-MAGNETIC MOTOR.—A series of helices with soft iron cores are arranged circularly, and parallel to each other, in suitable supports. The core ends project beyond the helices, and are of such dimensions that the space between any two adjacent ends is exactly equal to their own width. A shaft passes centrally through this arrangement, and carries two pairs of similar helices so placed that when the poles of one pair coincide with the poles of the fixed helices, those of the other pair are midway between the opposite ends of such helices. A commutator of the usual form distributes the current so as to excite the electro-magnet at the proper time to produce the desired rotation of the shaft.

1,626.—D. Spill, London. **Compounds for Insulating Telegraph Wires, &c.** 4d. June 4.

CONDUCTORS.—An insulating material is produced by the combination of one or more gums, resins, or gum resins with camphor, or with caoutchouc and camphor, or with balsams or fatty or volatile oils.

1,668.—Z. T. Gramme and E. L. C. D'Ivernois, Paris. Magneto-Electric Machines. 4d. June 9.

MAGNETO AND DYNAMO-ELECTRIC GENERATORS.—The invention consists in giving to the soft iron or other material, which is to be magnetised by the influence of one or more either permanent or electro-magnets, the "form of a solid or hollow ring cylinder, or other suitable endless shape, constructed either out of one piece of iron, or of a bundle of iron wires, and round the entire surface of which endless core is laid a series of coils of suitably insulated wire of copper or other good conductor of electricity." The wire of each coil is connected end to end, to the next succeeding one so as to constitute one large endless bobbin in which, when such bobbin is caused to revolve with a regular continuous motion, "continuous induction currents will be developed by the magnetic radiation of the poles of one or more either permanent or electro-magnets, to the influence of which the said endless bobbin is submitted, the continuity of the revolution of the bobbin procuring in this latter a continuous displacement or advancing of the magnetism," not requiring the use of pole changers or commutators. At each juncture of the coil ends is connected a rod of brass or other good conductor, which rods are insulated from each other, whilst two or more metal springs or other conducting rubbers, coming successively in contact with the free ends of each rod, serve to convey the induction currents in the required respective directions, and allow of applying, if required, part of such currents for charging the electro-magnets (in the case of a dynamo-electric generator) and of applying the currents to any useful purpose. Suitable driving gear is provided to cause either the endless bobbin or the magnets to revolve, and provision is made for changing the continuous into alternate currents. The coils may consist of metallic strips or ribbons in place of wire.

1,917.— E. L. C. D'Ivernois, Paris. Electric Light. 4d. July 7.

ELECTRIC LIGHT.—The object of this invention is to increase the intensity of an electric light without having to increase the power of the battery or generator. This is effected by sub-dividing the current, and, in place of one pair of carbon electrodes, using a number of such electrodes, so placed in relation to each other as to cause the several arcs to cross or intersect each other. Another method is to sub-divide the source of current by using either two or more smaller batteries or generators.

2,367. — M. Gray, London, and **F. Hawkins,** Silvertown, Essex. Marine Telegraph Cables. 8d. (3 figs.) August 29.

CONDUCTORS.—The protecting wires, forming the sheathing of submarine cables, are lapped round with strips of woven cloth saturated with a waterproof composition and applied to such wires before they are laid upon the cable. The wires should have at least two coatings, the inner one being by preference of cotton and the outer one of Hessian canvas, the strips being laid helically around the wires in opposite directions, and with sufficient side lap to ensure a good joint.

2,525.—J. B. Elkington, Birmingham. Wire for Electric Telegraph Conductors. 4d. September 20.

CONDUCTORS.—In order to obtain copper wire of high conductivity a plate of copper is deposited on a suitable mould by the electric current, and then cut into strips and drawn down into wire. Lengths of this wire are suspended in a depositing vat till they become about an inch thick, and the rods so obtained are again drawn into wire in the usual way. These rods are united by a slot and tongue formed in their ends, the joint being completed by immersion in the copper solution.

*3,069. — W. Thomson, Glasgow. Electric Telegraph Instruments and Clocks. 1s. 4d. November 23.

MAGNETS.—A steel core is surrounded with a coil, by means of which it can be powerfully magnetised in the beginning, and also have its magnetism restored, at any subsequent time, when it may have become weakened from various causes.

ELECTRO-MAGNETIC MOTOR. — A description of this motor will be found in the Abstract of Specification No. 252 of 1871.

(*Provisional Protection was not allowed for this invention.*)

3,245.—G. Zanni, London. Magnetic and Electric Telegraph Apparatus. 2s. (9 figs.) December 10.

MAGNETO-ELECTRIC GENERATOR.—This invention relates to the use of a magneto-electric generator driven by clockwork or other means, and applied to telegraph instruments. The drawings show the magnets of circular form, with a cylindrical armature between their poles.

*3,314.—J. Rogers, London. Insulating Telegraph Wires. 4d. December 10.

CONDUCTORS.—The wire is served with yarn and drawn through successive troughs containing alternately heated air and insulating material, after which it is passed through a trumpet-mouthed tube in order to compress it. The process is repeated as often as is desired. A number of such insulated wires can be laid together, and passed through a cauldron of insulating material previous to the application of a binding of yarns. A dusting of chalk or silicate is finally applied to prevent stickiness.

1871.

10.—C. A. McEvoy, London. Exploding Torpedoes. 1s. (6 figs.) January 3.

CONDUCTORS.—Joints in electrical conductors are made as follows:—Two or more metal discs screw together, and are provided with holes, through which the ends of the wires are run and turned over. For use under water the joint is enclosed in a wooden or ebonite case made in halves. When screwed together an india-rubber packing is compressed around the conductor, and effectually excludes water from the case.

131.— S. A. Varley, London. Telegraph Apparatus, &c. 6d. January 18.
(Provisional Protection not allowed.)

MAGNETO-ELECTRIC GENERATOR.—Two or more bundles of steel bar magnets are placed between two iron plates, called armature plates, and good contact is made between the plates and the polar faces of all the bars. In the centre of each armature plate is a boss. One of these bosses serves as a bearing for a shaft which carries a wheel with 13 teeth. Two iron bars wrapped with coils are attached at one end to the other boss, and approach the toothed wheel at either side. These are called the generator coils. "When the toothed wheel is revolved, the magnetic circuit is closed through the two bars alternately, and electric currents are developed in the convolutions of the insulated wire during the periods of magnetisation and demagnetisation. By an alternative arrangement, generator coils are connected to both armature plates, and the wheel works in a seat at an equal distance from both. When great power is required, two sets of magnets may be arranged with opposite poles facing each other. In such case one armature plate is divided into two plates, which form the armed poles of a compound magnet, in one of which the wheel axis is borne, while the generator coils are attached to the other. In some cases the magnets are formed of steel wire, and the iron ends attached by electro-deposition. For telegraphic relay purposes two teeth in the wheel are sufficient.

142.—W. R. Lake, London. (*P. S. Devlan, Jersey City, New Jersey, and J. P. Wendell and S. L. M. Tasker, both of Philadelphia.*) Electric Telegraph Cables. 8d. (6 figs.) January 18.

CONDUCTORS.—The wires are made of helical or "wave-like" form to ensure strength and elasticity, and are insulated with gutta-percha or other substance "mixed with any fibrous material." The outer portion of the cable is made of fibrous material saturated with some waterproof substance.

252.—W. Thomson, Glasgow. Electric Telegraph Instruments. 4s. 2d. (42 figs. January 31.

ELECTRO-MAGNETIC MOTOR. — This invention consists in improvements on Patent No. 2,147 of 1867. A series of soft iron bars arranged around a spindle, "like the staves of a barrel," but separated from one another by insulating material, are successively attracted by a horseshoe electro-magnet. A polygonal cam periodically raises the end of a straight spring and breaks connection between a platinum stud and contact piece, by which means the current is interrupted. A set screw raises or lowers this spring and thereby adjusts the length of the period of contact. The cam can be turned round on its axis for the purpose of adjusting the time when contact shall be made. The armature spindle has roller bearings to reduce friction. The speed of revolution is regulated by adding or subtracting resistance from the electro-magnet circuit, and sparking is obviated by using a permanent shunt connected with the ends of the electro-magnet coils.

GENERATOR.—This generator is combined with the above motor. The electro-phorous described in Specification No. 2,147 of 1867 is dispensed with, and the "replenisher" alone is used. The soft iron armatures of the motor together with a series of brass strips attached to them form the "carriers," and these are surrounded at a short distance by two insulated metal plates, one being connected to earth and the other serving as a receiver of the high potential charge. Springs are arranged to make contact with metal bars, attached to the carriers, and convey the charge of the carriers to the receiver. By the same springs and bars two opposite carriers are put in communication at a point when both are under cover of the two insulated plates, whereby each receives by induction an opposite charge. The negatively electrified carrier is then put to earth by the same means, and the other carrier conveys its charge to the receiver.

ELECTRO-MAGNETIC MOTOR. — An alternative arrangement to that already described consists of a straight electro-magnet attached at right angles to a spindle, rotating within a powerful cylindrical compound magnet. The current is alternately made and broken through the electro-magnet coils by a commutator and two sets of springs of the usual form. In some cases a chronometer is added to regulate the speed of rotation.

GENERATOR.—Attached to the spindle of the motor just described is an ebonite disc upon which is arranged a series of metal sectors partially enclosed in a metallic casing lined with thin vulcanite. This casing is made in two parts, one of which is in connection with the earth and the other forms a receiver as in the generator already described above. The necessary contacts are made by springs bearing against pins on the sectors, and a "dry pile" is employed to initiate the electrification of the inductor. A semi-insulating thread forms part of the conductor by which the charge is led away, its length being varied to oppose more or less resistance as required, and in this way an excess of charge is avoided.

537.—T. T. P. B. Warren, Mitcham, Surrey. Treating India-Rubber for Insulating Telegraph Wires. 10d. February 28.

CONDUCTORS.—The wires are first tinned, and then covered with adhesive coating, consisting of a compound of shellac, gutta-percha and Canada balsam, to which a small proportion of iodine may be added. India-rubber is applied in the usual manner, and "consolidated" by being coiled round drums, which are afterwards heated from the inside. The main feature of this invention consists in substituting for the usual vulcanising process a method of rendering the india-rubber durable, and this is effected by submitting the wire, coated and "consolidated" as described above, to the action of iodine, bromine, or chlorine dissolved in a vehicle which can exert no action on the india-rubber covering. Another part of the invention relates to the manufacture of a substance from certain vegetable oils, to be used as a "substitute for hemp" on electrical conductors. A compound is formed by adding to such substance asphalte, gum thus, shellac, colophane, plumbago, soap and French chalk, which is applied to conductors by suitable means, and then enveloped in a lapping of canvas or other fibrous material.

611.—R. S. Newall, Gateshead-on-Tyne. Manufacture of Ropes. 10d. (5 figs.) March 7.

CONDUCTORS.—This relates to machinery for the manufacture of cables. The yarn is wound into "cops" instead of upon bobbins, and the requisite number of cops are arranged in one or more concentric circles around a hollow axle, through which the insulated core is passed in the process of serving it.

***756.—T. G. Rylands**, Warrington. Stretching and Testing Wire. 4d. March 21.

CONDUCTORS.—Wire that is to be stretched or "killed" is passed around two pulleys, which have a differential motion with respect to each other, adjusted to the percentage of extension required.

776.—E. J. C. Welch, London. Generating Electricity for Discharging Fuses. 4d. March 22.

MAGNETO-ELECTRIC GENERATOR. — A metallic frame carries a permanent magnet, to the ends of which are attached soft iron cores having coils of insulated wire upon them. An armature of soft iron is arranged against the ends of the soft iron cores upon a spindle carried in bearings. On this spindle is fixed a collar, and on that part of the spindle which is on the side of the collar furthest from the armature, a metallic ram or weight capably of sliding freely thereon. Behind the weight, and between it and one of the bearings, is a spiral steel spring, just fitting, when fully open and at rest, between the weight and one of the bearings. The weight can be moved back by a rack, pinon and key, until it has fully compressed the springs, when the pinion drops out of gear and the weight is projected forward, like a shot from a spring gun, and forces the armature from the magnetic cores, thus inducing a powerful current in the coils. Provision is made for breaking the circuit between the coils and the leading wires when the armature is at a certain distance from the magnet.

1,150.—S. A. Varley, London. Electric Telegraph Apparatus, &c. 1s. 4d. (21 figs.) April 29.

MAGNETO-ELECTRIC GENERATOR.—This machine is substantially the same, and appears to be described in exactly the same words, as that included in Specification No. 131 of 1871. In the drawings are shown a number of cylindrical bar magnets retained between two plates, the whole having a general resemblance to the tubes and tube plates of a surface condenser. One tube plate has an upward extension, which carries two coiled cores, and the other serves as a bearing for a short vertical shaft, carrying a wheel with soft iron spokes or teeth, which pass before the polar faces of the coiled core, and in so doing complete the magnetic circuit.

1,208.—J. Wright, London. (*E. P. Minor, Schenectady, and B. F. Britton, New York, U.S.A.*) Waterproof Compounds. 4d. May 4.

CONDUCTORS. —An insulating material for coating conductors is composed of collodion, Venice turpentine, castor oil, shellac dissolved in alcohol, and glycerine in suitable proportions. The compound is also used as a non-conducting varnish for wooden insulators.

1,361—W. E. Newton, London. (*H. M. Paine, Newark, New Jersey, and M. S. Frost, New York, U.S.A.*) Electro-Magnetic Engines. 6d. (2 figs.) May 20.

ELECTRO-MAGNETIC MOTOR.—A wheel, consisting of a number of sector-shaped electro-magnets attached to a hub rolls around in contact with a series of iron sections or "armatures," supported on the inner surface of a cylindrical frame. The curvature of the electro-magnet poles corresponds with that of the armatures. A horizontal shaft is provided with a single crank the pin of which passes through a central hole in the hub of the electro-magnet wheel, and as this wheel rolls, by the successive action of the electro-magnets upon the surrounding armatures, it imparts an opposite rotary motion to the shaft. The commutator "peculiarly adapted" to the working of this motor is that described in Specification No. 17 of 1870.

1,364.—A. G. Day, Seymour, Connecticut, U.S.A. Insulating Compound. 4d. May 20.
CONDUCTORS.—An insulating material is formed by the combination of india-rubber with the artificial elastic substance described in Specification No. 1,010 of 1871, composed of sulphur and litharge in suitable proportions. The compound is applied to the wire through a die, and afterwards vulcanised by the application of heat.

*1,792.—H. A. Mallock, London. Electric Conductors. 4d. July 8.
CONDUCTORS. — A strand of any convenient number of iron wires encloses a single copper wire. The copper wire is tinned, and the iron wires coated with tin, galvanised, painted, varnished, or similarly protected.

1,842.—J. R. Croskey, London. (*A. McKinley, New York, U.S.A.*) Composition for Pavements, Roofing, &c. 4d. July 14.
CONDUCTORS.—A compound for coating electrical conductors consists of a combination, by means of heat, of bitumen, refuse gas lime or gypsum, burnt clay, and pulverised limestone.

2,057.—E. J. W. Parnacott, Leeds. Solidifying Oils and Manufacturing Floorcloth, &c. 10d. (5 figs.) August 4.
CONDUCTORS.—Vegetable oils are treated in a closed vessel with powdered and calcined sulphate of copper, or other "drying" substance, in order to effect their solidification. A greater amount of stiffness is attained by afterwards incorporating litharge with the mass by means of rollers. The compound is applied as an insulating covering for electrical conductors.

2,106.—W. R. Lake, London. (*A. G. De Wolfe, Seymour, Connecticut, U.S.A.*) Machine for Covering Wire and Manufacturing Tubes. 8d. (3 figs.) August 10.
CONDUCTORS.—This relates to a machine for coating electrical conductors with gutta-percha or other gummy substance. A screw works through a steam-jacketted cylinder and feeds the material in a liquid state, whilst the wire to be covered is passed through a die at the end of the cylinder, arranged at right angles to and intersecting its axis.

*2,172.—C. Wheatstone and J. M. A. Stroh, London. Telegraphs. 1s. 4d. (14 figs.) August 18.
MAGNETO-ELECTRIC GENERATOR.—This is adapted to ringing bells, and consists of a compound permanent magnet of circular form, to the poles of which the soft iron cores of two bobbins are fixed. A handle or crank, when depressed, partly rotates a spindle carrying a triple cam, thereby raising a hinged armature three several times from contact with the soft iron cores. By this means six reversed currents are generated within the bobbins and sent into the line wire. The cam is geared to the spindle by a click and ratchet wheel, so that as the handle returns to its first position the armature is not moved. A modification of this generator has a magnet of horseshoe form, and the armature is raised by a straight cam or rack attached to a horizontal "pull."

2,238.—W. R. Lake, London. (*M. H. Utley and A. Ross, Montreal, Canada.*) Electro-Magnetic Motor Engine. 10d. (11 figs.) August 25.
ELECTRO-MAGNETIC MOTOR.—A number of horseshoe electro-magnets are attached by their bent portions to the inner surfaces of two or more cylindrical metal frames, and arranged so that a series of lateral extensions, with which their poles are provided, lie in a circle. Within this circle, and concentric with it, is a brass wheel keyed to a shaft running, by preference, in conical bearings. A second series of electro-magnets is contained within the brass wheel, and has its pole-pieces dovetailed into, and turned flush with its outer surface. The number of magnets in each series is equal, and the poles of those within the wheel are as near as possible to those of the outer series, without actually touching them. When the wheel and frames are provided each with two circles of magnets, the magnets of one circle are placed intermediate with those of the other, so as to obviate the possibility of the motor stopping at a place where the attraction of the magnets would not be sufficient to start it again. There is a commutator for each series of opposed magnets, consisting of an ebonite disc, on which are mounted two metal plates, having extensions upon the rim of the disc, so arranged that the extensions of one plate lie between, but without touching, those of the other. The bobbins are connected in series, and the terminal wires are led one to each of the commutator plates just described. Springs bearing on these plates convey to them the current from the battery. The revolving magnets, as their poles are reversed, are successively attracted as they approach, and repelled as they leave, the fixed magnets, the polarity of the latter always remaining fixed. A switch is employed to reverse the current, and thus alter the direction of rotation. The other accessories of the motor are a condenser and governor. The former is constructed in the usual manner, and is used to receive the electricity induced in the bobbins at the moment of reversing the current. The governor consists of a revolving spindle, driven by the motor, and provided with

two weighted springs, which, under the centrifugal action due to excessive speed, move outwards and break the electrical circuit.

*2,256.—C. Davis and T. Struthers, London. Composition for Boot-Soling, Waterproofing, &c. 4d. August 28.

Conductors.—Cane or cocoa-nut fibre, bottoms of varnish, gutta-percha, and stone ochre are incorporated at a moderate heat, and the resulting compound employed, amongst other uses, to coat electrical conductors.

2,439. — W. R. Lake, London. (*E. Gassett, Boston, U.S.A.*) Electro-Magnetic Engine. 8d. (5 figs.) September 15.

Electro-Magnetic Motor. — The illustration shows an end view of the motor with the shaft

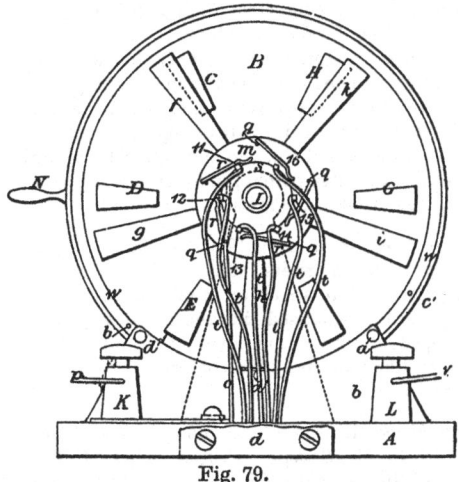

Fig. 79.

bearing removed. A base plate A supports a double circular frame B, between which are secured a series of six-bar electro-magnets, arranged radially, with their poles C, D, E, F, G, H, projecting through the frame. I is a horizontal shaft running in suitable bearings. To this shaft, on each side of the double frame, is secured a hub carrying five radial arms *f*, *g*, *h*, *i*, *k*, that revolve close to the electro-magnet poles. A commutator, also shown in the illustration, distributes the current in succession to each of the electro-magnets. The metal disc *m*, secured to the shaft by an insulating sleeve, is provided with a collar on which bears the spring *o* in connection with the battery terminal K. To the disc *m* the five flat springs *r* are attached, and these springs, as they revolve with the shaft, bear upon a fixed cylindrical block *s* of insulating material, having the metallic bars 11, 12, 13, 14, 15, 16, projecting above its surface. One end of each electro-magnet coil is connected to one of these bars, and the remaining ends are joined by a common wire to the second battery terminal L. In order to

regulate the speed of rotation, the relative position of the electro-magnets with respect to the radial arms, when the circuit is broken, may be varied by moving the frame B around the shaft by the handle N, the stop pins b^1, c^1, limiting such motion. By altering the position of these pins the direction of rotation of the shaft may be reversed. The inventor does not confine himself to the use of bar magnets, nor to the relative number of magnets and radial arms.

2,721.—G. Zanni, London. Magnetic Bells and Signals. 1s. (9 figs.) October 13.

Magneto-Electric Generator.—This invention relates to apparatus for ringing bells. The generator is contained in a small case for attachment to the wall, and consists of an ordinary cylindrical armature rotating in a recess formed between the poles of a permanent horseshoe or circular magnet. A crank of the usual kind gives rapid motion to the armature by means of a toothed wheel and pinion. The crank spindle drives the former by a click and ratchet wheel, so that as the crank returns to position the armature is thrown out of gear. A coiled spring effects the return. In another arrangement of the generator, a direct pull transmits motion to the armature by the intervention of a rack and toothed gear. A commutator and spring turn the currents in one direction.

2,800.—W. T. Henley, North Woolwich, and H. Horstman, London. Signalling. 8d. (2 figs.) October 20.

Magneto-Electric Generator.—For the automatic sending of signals on railways, the engine carries a soft iron armature or keeper which, in its progress, just clears the poles of a permanent magnet fixed between the rails. This magnet is wound with two coils of insulated wire, in which induced currents are generated when the keeper passes the poles. The arrangement may be varied by the permanent magnet being fixed on the locomotive, and the coils being wound on the keeper.

*3,324.—J. Petersen, Copenhagen. Protecting Submarine Cables, &c. 4d. December 8.

Conductors.—Shore ends of submarine cables are enclosed in an elastic covering formed of wires twisted together or interlaced. "To this covering a number of loose spiral (?) wires are attached, which, becoming embedded in the ground, serve as roots for holding the cable." To sink and protect a cable of small specific gravity it is covered with concrete blocks, made in halves and dovetailed together. These blocks are also provided with hooks and "root" wires. Another means of protection is the use of a sleeve formed by passing ropes or fibres through the wire covering mentioned above and coating the whole with bituminous substance.

*3,432.—F. Hurd, Wakefield, Yorks. Main Pipes for Air, Gas, &c. 4d. December 19.

CONDUCTORS.—Pipes for containing electrical conductors are made to form part of the footway or gutters, and are constructed with the upper surface cellular for the purpose of containing blocks of stone or asphalte. The pipes are united end to end by lugs and cotters, or by bolts and nuts.

1872.

86.—A. Robson, Lude, Blair Athole, N.B. Straining Wires. 8d. (6 figs.) January 11.

CONDUCTORS. — This relates to a method of straining and joining wires. The end of one wire is doubled and twisted back upon itself so as to form a shoulder against which the straining apparatus bears. This apparatus consists of a rectangular frame at one end of which is a winch with ratchet and pawl. The wires are passed through openings at each end and the required degree of tension put upon them by means of the winch, after which the frame is turned so as to lock the wires. The joint is then finished in the usual way. A gripping apparatus may be employed to retain the proper tension during the latter process.

*109.—W. A. Lyttle, Hammersmith. Telegraph Poles, &c. 4d. January 13.

INSULATORS.—The conductor is secured to the insulator by a U-shaped iron clamp having both legs screwed and provided with two nuts each. This clamp " is placed in the groove now provided round the top of the insulator with its legs projecting across and under the line wire." The conductor is gripped between the pairs of nuts. The back nut should have a transverse groove in its face to receive the wire, and the positions of the nuts should be such that when the wire is secured it is forced tightly against the insulator. The same clamp is made to hold a protecting cone of sheet iron.

121.—W. R. Lake, London. (*W. Radde, New York, U.S.A.*) Pipes and Pipe Joints for Gas, &c. 6d. (3 figs.) January 15.

CONDUCTORS.—Pipes for the protection of electrical conductors are made of an inner lining of vitreous material, an intermediate layer of hydraulic cement, bitumen or other suitable substance, and a protecting case of metal or wood. Joints are made with screwed sockets or elbows, having a lining of vitreous material, which becomes continuous with the lining of the pipe when the pipe ends are screwed home.

255.—A. M. Clarke, London. (*L. C. A. J. G. d'Arlincourt, Paris.*) Electro-Magnets. 2s. (35 figs.) January 25.

ELECTRO-MAGNETS.—The principle of this system of electro-magnets is the combination of two or more coils with a permanent magnet, or with an electro-magnet through which a permanent current passes, for the purpose of obtaining a rapid action and suppressing the residual magnetism. A number of arrangements are illustrated, chiefly applicable to telegraphs.

286.—G. W. R. Pigott, London. Covered Wire. 8d. (9 figs.) January 30.

CONDUCTORS. — This specification describes a machine for covering wires with yarns of cotton, silk or other fibrous substances, such covered wires being adapted presumably for electrical conductors. The invention also includes a process of protecting the wire from oxidation by passing it through a bath of sulphate of copper, or through an alloy of lead and tin, and subsequently through a solution of gutta-percha, india-rubber, or other insulating or adhesive matters.

288.—P. W. Seymour, Canning Town, Essex. Portable Magnets. 6d. (4 figs.) Jan. 30.

MAGNETS.—Thin strips or small chains of steel or iron are magnetised in the usual way, and enveloped in a protecting covering of india-rubber or other like material. The object is to obtain a flexible magnet.

375.—H. A. Bonneville, London. (*A. N. Allen and R. H. Dewey, Pittsfield, Mass., U.S.A.*) Lighting Gas by Electric Currents. 10d. (16 figs.) February 5.

MAGNETO-ELECTRIC GENERATOR.—A machine " of the most approved construction " is driven by a buchet wheel and current of air, to render incandescent a platinum coil over the burner.

FRICTIONAL ELECTRIC GENERATOR.—A glass disc revolves between rubbers.

*412.—T. Cockshott, Blackwall. Laying Submarine Telegraph Cables. 4d. February 8.

CONDUCTORS. — This invention consists in a method of connecting a cabled conductor to the chain cable of a ship, in order to prevent the two getting foul of each other. Another feature of the invention is the use of a cable containing two or more conductors, which are divided so as to form branches, and are led to as many separate stations.

473.—Sir C. Wheatstone and J. M. A. Stroh, London. Electro-Magnetic Telegraphs, &c. 1s. 6d. (15 figs.) February 15.

MAGNETO-ELECTRIC GENERATOR. — The generators

described in this specification are improvements on those forming part of Patent No. 2,172 of 1871. The object of the invention is to make the apparatus more compact. Attached to the poles of a compound circular permanent magnet are two bobbins with iron cores. A crank handle, when depressed, partly rotates a circular series of cams, which raise a hinged armature from contact with the bobbin cores. The cams work with the forward motion of the handle, through a ratchet and click, but on return of the handle the click runs on the ratchet and the armature is unmoved. The currents generated in the bobbins are conveyed to line by two springs which, together with an insulated metal sector upon the crank spindle, form a continuous conductor so long as the handle is being depressed, but when the pull is at rest the circuit is interrupted.

As in the former patent, above referred to, the generator is modified when required for use with a direct pull.

482.—E. T. Truman, London. Covering Wire. 6d. February 15.

CONDUCTORS.—Instead of applying gutta-percha and like dielectric to a conductor in successive layers, it is put on in a solid homogeneous coating. The tendency of the conducting wire to become eccentric to the insulating material is obviated by causing the drums on which it is wound to rotate on the axis of the wire, and at the same time to revolve in planes at right angles to the line of travel of the wire. An important feature in the invention is the employment of a delivery tube of glass or glazed earthenware, having a surface on which the gutta-percha will rotate when hot without adhering thereto. At the exit end of the glass tube is fitted a copper tube with open top, forming a trough, through which the wire passes. To prevent adhesion to its surface, the insulating material is lubricated with soapy water. The copper tube is refrigerated externally, and the process is completed when the conductor has arrived at its further end, the length of the tube being adjusted for this result.

783.—W. R. Lake, London. (*W. T. Batchelder, New York.*) **Electric Torch for Lighting Gas.** 8d. (3 figs.) March 14.

INDUCTIVE ELECTRIC GENERATOR. — This is an electroferous composed of a vulcanite and a leather covered disc.

829.—W. R. Lake, London. (*G. B. Field and E. W. Andrews, New York, U.S.A.*) **Printing Telegraphs.** 1s. (5 figs.) March 19.

ELECTRO - MAGNETIC MOTOR. — Four vertical electro-magnets are arranged at equal distances apart and around a spindle on which is fixed an armature, so adjusted as to rotate with the shaft and pass close to the poles of the electro-

magnets in a plane at right angles to their axes. A cam and springs form the commutator.

*832. — C. A. McEvoy, London. Connecting Electric-Cables and Conductors. 4d. March 19.

CONDUCTORS.—Insulated joints in electrical conductors are made by means of a hollow coupling, consisting of " two similar pieces with a socket at one end." The wire is first joined by soldering or by a separate metallic coupling. The joint is rendered watertight by the compression around the conductor of india-rubber washers.

*888.—W. Darlow, North Woolwich. Portable Magnets. 4d. March 23.

MAGNETS.—These are made in the form of " chains, wires, or parts," in order to obtain flexibility.

1,207.—C. Little, Rutherford Park, New Jersey, U.S.A. Electric Telegraph Apparatus. 4s. (32 figs.) April 22.

ELECTRO-MAGNETIC MOTOR.—A vertical rotating shaft carries three armatures in the form of cylinders, arranged parallel to the shaft and at equal distances apart. On opposite sides of the shaft are two horseshoe electro-magnets the poles of which are in a vertical line and very close to the armatures. An ordinary cylindrical commutator with two contact rollers turns the currents alternately into each of the electro-magnets, which, by their attraction on the cylindrical armatures, produce rotations of the shaft.

1,254.—J. H. Johnson, London. (*Z. T. Gramme, and E. L. C. d'Ivernois, Paris.*) **Magneto-Electric Machines.** 4d. April 26.

MAGNETO-ELECTRIC GENERATOR.—The invention consists in the application to machines, such as described in Specification 1,668 of 1870, " of one or more rubbers or conductors for collecting and conducting the currents, composed each of a number of wires united together in the form of a bundle, brush or sheaf " to conduct the current from and to the armature.

*1,279.—H. Highton, Putney. Electric Cables. 4d. April 29.

CONDUCTORS.—" This invention consists in the application to the conducting wires of cables which are covered with gutta-percha or caoutchouc, or similar materials, of a solution of paraffin, or ozokerit, or shellac for the purpose of filling up the minute pores and improving the insulation; the coated wire may be rubbed with a volatile solvent with the same object."

*1,320.— W. A. Lyttle, Hammersmith. Attaching Telegraph Wires to Insulators. 4d. May 2.

INSULATORS.—In the groove usually provided round the top of each insulator a U-shaped iron clamp is placed, having both of its legs screwed and furnished with nuts. An ebonite cross-piece, with a hole at each end is pushed over the legs and up against the insulator; the conducting wire is then lifted on to these legs, and firmly clipped between the cross-piece and its nuts. The same clip is made to support a protecting shield of metal. Modifications of this invention adapted to insulators of the inverted type are also described.

1,376.—D. G. Fitzgerald, Brixton, and B. C. Molloy, London. Treating Compound Substances by Electricity. 6d. May 6.

CARBON ELECTRODES.—These are formed by depositing carbon at a high temperature from ethylene or olefiant gas (bicarburetted hydrogen), or from other compounds rich in carbon, upon metallic or other surfaces. Plates of iron or unglazed earthenware, for instance, may be heated to a bright redness in a close casing, through which is passed a stream of coal gas, prepared at the lowest practicable temperature. The plates will become coated with a layer of carbon.

MAGNETO-ELECTRIC GENERATOR. — The use of such a machine for electro deposition is mentioned.

1,453.—T. A. Edison, Newark, N.J., U.S.A. Printing Telegraphs. 2s. 2d. (25 figs.) May 13.

ELECTRO-MAGNETIC MOTOR.—The instrument is driven by a small electro-magnetic motor.

1,473.—C. H. Siemens, London. Telegraph Poles and Fastenings. 1s. (22 figs.) May 15.

INSULATORS.—This is an improvement on Patent No. 3,501 of 1868. In place of the cam therein described the inventor uses a sector-shaped cam, similarly mounted, "presenting its periphery to the wire, which lies loosely on that part of the periphery which is nearest to the axis." With this arrangement a pull on the wire in either direction, brings the cam into action and causes it to grip the wire. Instead of a cam a double wedge-shaped key or cotter is used, or "a free roller" is placed "between the wire and a fixed concavely-curved surface." Another modification consists in the employment of a cam pivotted on an axis parallel to the wire, by which the wire is bent down between two fixed lugs.

Joints in electrical conducting wires are made by soldering their ends into a short metal tube, preferably perforated so as to admit the solder into the interior.

*1,628.—T. Slater, London. Apparatus for Obtaining Electric Light, &c. 4d. May 29.

ARC LAMP.—Each electrode is connected to a separate rack, which racks gear into the opposite insulated halves of a toothed wheel. An electro-magnet in the circuit of the electrodes, acts upon an armature in one arm of a spring gripping lever which loosely embraces the axis of the said wheel.

When the current is established through the electrodes the magnet attracts the arm of the lever, causing such lever to grip the axis and to impart a slight rotatory movement to the toothed wheel, thereby separating the electrodes more or less according to the arc required, the extent of this being regulated by an adjustable stop screw in the lever.

MAGNETO-ELECTRIC GENERATOR.—The machine comprises one or more pairs of rings of helices, each ring being composed of an annular core made up of a number of concentric rings of hoop iron lapped with insulated copper wire, so as to form a series of distinct helices, the terminal wire of each helix being connected with one of a corresponding number of studs disposed in a circle in a piece of insulating material which carries the aforesaid ring of helices. To this terminal wire is connected the wire of the next preceding helix, and so on throughout the series. On diametrically opposite sides of each pair of rings of helices, there is disposed a U-shaped permanent or electro-magnet wrapped with coils of insulated wire, and having its north pole opposite to the periphery of one ring of the pair, and its south pole opposite to the periphery of the other ring of the pair. As the studs revolve they come in contact successively with two springs, by which the currents from the helices are led around the magnets.

1,665.—W, Darlow, North Woolwich. Portable Magnets. 4d. June 1.

MAGNETS.—Flexible magnets are made by enclosing particles of iron or steel, mixed with an adhesive substance, in a suitable covering, which may consist of silk, felt, paper, or other like material.

1,753.—E. Gilbert, Edinburgh. Signalling in Railway Trains. 1s. 4d. (17 figs.) June 11

CONDUCTORS.—Detachable conductors for electric communications in railway trains, consist of a short length of insulated wire covered with a flexible envelope of plaited and painted hemp yarn. The joint is made with a plug and socket, through which a pin is passed to make it secure, or the plug may have a recess in which a stud is pressed by a blade spring on one side of the socket.

1,786.—J. G. Rolls, Durban, South Africa. Insulating Material. 4d. June 13.

CONDUCTORS.—This invention consists in the use of the gum of the euphorbia tree, as a substitute for gutta-percha and india-rubber, in insulating electrical conductors. It can be vulcanised if required.

1,847.—W. R. Lake, London. Copper-Covered and Copper-Cored Wire. 4d. June 19.

CONDUCTORS.—This relates to a process of combining copper, either internally or externally, with other metals to form a compound wire, which is "particularly suitable" for the transmission of electrical currents.

1,919.—C. W. Siemens, London. (*Dr. W. Siemens, Berlin.*) **Obtaining and Applying Magneto-Electric Currents.** 10d. (11 figs.) June 25.

MAGNETO-ELECTRIC GENERATOR. — Currents of considerable intensity are obtained by the motion of a portion of an electric circuit in a narrow space formed between the polar surfaces of a permanent or electro-magnet. A convenient arrangement for this purpose consists in causing one pole of the magnet to enter a cavity in the other pole, leaving a narrow annular space for the reception of a conducting coil. In a modification of this arrangement the coil may surround a ring connected by a strong stem with one pole of the magnet, while the other pole of the magnet forms a central stud within the ring, and also a ring-shaped extension surrounding the same. "The coil is in this case supported from a central axis, so arranged that a partly rotative motion may be imparted to it mechanically." The generator becomes a motor when alternating currents are directed through the coil.

1,961. — **R. Jobson,** Dudley. **Supports for Telegraph Insulators.** 8d. (3 figs.) June 29.

INSULATORS.—These insulators, which are fixed to their supports by means of a nut on the stem, have such stem provided with a thrust collar. By this invention the collar is formed in a better and more economical way.

2,085. — **C. A. McEvoy,** London. **Circuit Closers for Torpedoes.** 1s. (8 figs.) July 10.

CONDUCTORS.—This relates to joints in conductors. Similar joints have been described in Provisional Specification No. 832 of 1872. A hollow coupling is used which grips the outer projecting wires (in the case of cables), the conducting wires being joined within such coupling. For the latter purpose a metal tube is employed having two screw plugs, one at each end. The insulated wires are passed through the plugs and india-rubber washers are threaded upon them. The wires are inserted through metal discs and bent over; the plugs are then screwed home, so as to compress the india-rubber, and make a watertight joint. The main and outer coupling consists of two cup-shaped pieces of metal united by a box nut. Their ends, which take the sheathing wires, have recesses with hollow screwed plugs, through which such wires are passed and their ends turned outward, so as to be held by the plugs when screwed down. Several modifications of this joint are

illustrated and described, the coupling being of a simpler character when no sheathing wires exist.

2,260. — **C. Fairholme,** London. **Electric Signalling in Railway Trains.** 8d. (5 figs.) July 29.

CONDUCTORS.—The mechanical continuity of an electrical conductor is preserved, whilst it is divided into a number of insulated sections by making the joint between such sections contain a portion of insulating material. One form of this joint consists of a disc of glass, vulcanite or similar non-conducting substance, having a hole in its centre and a groove in its rim. One conductor is passed round the groove and the other through the central hole, their ends are then bent back and twisted upon themselves.

2,547.—H. Highton, M.A., Putney. **Submarine Cables, &c.** 4d. August 27.

CONDUCTORS.—A coating of boiled linseed oil, mixed with red-lead, is applied to the conductor over the insulating material, or in place of oil a vegetable tar may be used, or a mixture of both tar and oil. If desired, another coating of india-rubber or gutta-percha may afterwards be applied. The object of the invention is to fill any minute pores in the insulating material, and thus improve the insulation.

2,702.—G. A. A. Cunningham, and **T. P. C. Cunningham,** Liverpool. **Electro-Magnetic Motors.** 8d. (4 figs.) September 12.

ELECTRO-MAGNETIC MOTOR. — This relates to improvements in that type of motors in which the armatures, with their carrying arms and shaft, rotate under the action of electro-magnets, and consists (1) in constructing the electro-magnet coils from a compound wire, made up of two conductors laid parallel to each other, but separated and surrounded by an insulating material; (2) in disposing the armatures "on a series of radial arms whose terminations are not in the same line," and in placing the magnets in a line or lines, or *vice versâ*. A motor constructed in accordance with this invention is shown in the sheet of drawings. A frame supports three rows of electro-magnets, radially disposed with respect to the armatures. The armatures have seven arms, and are three in number, being set each slightly in advance of the other on a shaft. The same shaft carries a star or cam wheel, which, as it rotates, causes a series of springs to make and break contact with a corresponding set of insulated studs, from each of which a wire is led to one of the sets of electro-magnets. The frame carrying the insulated studs is movable around the shaft, so as to stop or reverse the action of the electro-magnets, and hence stop or reverse the motor. A "set of electro-magnets" consists of four soft iron cores, wound so as to form a double horseshoe magnet. The insulated studs are in connection with one battery pole, and the other ends of all the electro-magnet coils are joined to a common conductor attached to the other battery pole.

2,827.—C. J. A. Dick, Pittsburgh, U.S.A., and G. A. Dick, London. **Telegraphic Wires.** 4d. September 25.

CONDUCTORS. — An alloy of copper, tin and phosphorus, in suitable proportions, is drawn into a wire for use as an electrical conductor. Its advantages are, great tensile strength, and non-liability to corrosion. The alloy should be " free from oxides."

2,923.—C. H. Siemens, London. (*Partly a communication by Dr. W. Siemens, Berlin.*) **Telegraphic Instruments and Signalling Apparatus.** 3s. 4d. (21 figs.) October 3.

MAGNETO-ELECTRIC GENERATOR.—This invention includes the use of a magneto-electric generator, for signalling purposes on railways or for other purposes, consisting of an ordinary " Siemens cylindrical armature " rotating in a recess formed between the poles of a compound horseshoe permanent magnet.

2,973.—W. B. Chapin, Wickford, Rhode Island, U.S.A. **Operating Railway Brakes.** 1s. (10 figs.) October 9.

CONDUCTORS.—This relates to joints which may be readily unmade. Two forms are illustrated and described. One is an ordinary hook and eye, as used for joining the ends of lathe bands, and the other consists of a socket, formed by four converging wires, attached to the end of one conductor, into which the end of the second wire is inserted. A conical ferrule brings the wires tightly together.

*2,987.—M. Henry, London. (*H. J. Rogers, New York.*) **Sustaining and Insulating Telegraph Wires.** 4d. October 10.

INSULATORS.—A series of two or more insulators are used for each wire, so as to interpose more hindrance to leakage due to moisture.

3,167.—T. A. Redpath and W. A. Sherring, London. **Producing and Applying Magneto-Electric Power.** 8d. (3 figs.) October 25.

MAGNETO-ELECTRIC GENERATOR. — For working signals, actuating brakes, locking doors, and other purposes, a machine contained in one of the vehicles is driven from the axle.

*3,206.—W. A. Lyttle, Hammersmith. **Attaching Telegraph Wires to Insulators.** 4d. October 29.

INSULATORS.—The conducting wire is attached to the insulator with a U-shaped iron clip, similar to that described in Provisional Specification No.

1,320 of 1872. A wedge, wedges or roughened nuts are used to make the wire fast against the insulator. The shanks are screwed or roughened at the upper end, and provided with a collar at the lower end, which fits against the cross arm. Guards of wrought iron may be cast in the shank, and the U clip may carry a protecting hood.

3,221.—W. Darlow, North Woolwich. **Portable Magneto Apparatus.** 6d. (7 figs.) Oct. 31.

MAGNETS.—This relates to constructing compound magnets of thin sheet steel, enveloping them in a suitable protective covering, and employing as keepers to such magnets " a compound known as ' magnetine.' "

3,262.—G. Zanni, London. **Railway Signalling Apparatus.** 1s. 4d. (7 figs.) November 2.

MAGNETO-ELECTRIC GENERATOR.—The currents employed in the apparatus are produced by machines of this description.

3,299.—W. R. Lake, London. (*W.-W. Batchelder, New York, U.S.A.*) **Electric Torches.** 6d (2 figs.) November 7.

STATICAL ELECTRIC GENERATOR.—The same as that described in Specification No. 783 of 1872 similarly applied.

*3,462.—J. Thomas, Bangor, N. Wales. **Supporting and Protecting Telegraph Wires.** 4d. November 20.

INSULATORS.—Slate is used for the bearers and posts which support electrical conducting wires. The wire is passed through holes in the bearers, and insulators are thus dispensed with. For underground lines the wire is enclosed between pieces of slate grooved to receive it.

3,492.—C. E. Wetton, Cheltenham. **Portable Magneto Appliances.** 4d. November 22.

MAGNETS.—Flexible magnets are made in the usual way, but instead of being covered with india-rubber or like substance, are electro-plated with a non-oxidisable metal. A magnet thus prepared is proof against the " emanations ' sensible ' or ' insensible ' of the human body."

3,512.—C. Owen, London. **Signalling in Railway Trains.** 4d. November 23.

CONDUCTORS. — Electrical couplings for completing a circuit throughout a railway train, consists of " two copper wires covered with gutta-percha, and cemented between two layers webbing."

3,520. — J. C. Ramsden, Lightcliffe, Yorks. **Severing Double - Piled Fabrics.** 4d. November 23.

MAGNETO-ELECTRIC GENERATOR.—An improvement on Patent No. 5,010 of 1824, and consists in the use of a heated wire in place of the knives therein described. The wire may be heated by a current of electricity from a magneto-electric generator.

*3,666.—M. Evans, Glasgow. Signalling in Railway Trains. 4d. December 4.
MAGNETO-ELECTRIC GENERATOR.—For the purpose of signalling in railway trains, electric currents are generated by an apparatus consisting of " a series of permanent magnets, arranged on one or more axles of the engine or carriage wheels," " which magnets in rotating with the axles generate electricity."

3,692.—M. Henry, London. (*M. L. Ehrmann, Paris.*) Preserving Ships' Sheathing and Metal Plating. 4d. December 6.
MAGNETO-ELECTRIC GENERATORS.—This invention consists in the application of electric currents, produced by magneto-electric generators, for preserving the sheathing of ships. The same currents may be employed to produce the electric light at night. " Accumulators or secondary batteries may be used to get quantity or powerful currents."

3,736.—W. R. Lake, London. (*Z. G. Simmons, Kenosha, Wisconsin, U.S.A.*) Insulating Compound. 4d. December 9.
CONDUCTORS. — A compound suitable for insulating electrical conductors is made from one part coal-tar and two parts charcoal, sawdust, tan bark, or other organic fibrous body which is a poor conductor of electricity.

3,809.—S. W. Konn, London. (*A. M. Lodyguine.*) Electric Light. 8d. (9 figs.) December 14.
SEMI-INCANDESCENT LAMP.—" One or more conductors, which are preferably graphite, are enclosed in a species of lantern hermetically closed, and filled with nitrogen or other gas that does not support combustion, and are maintained

therein by any convenient method of support. To produce the electric current an ordinary magneto-electric machine is employed, and the

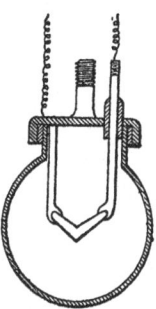

Fig. 80.

current is led into the lantern by means of a wire to one of the conductors (when the lantern contains more than one), and in passing through the conductor the electricity heats it to redness, and causes it to give out a brilliant light. The current is then led to a second graphite conductor," and thence to a third, and so on until all the conductors in a particular lamp have been illuminated, " when the current is led to the next lamp, and so on to the end of the series, which may consist of hundreds of conductors or lamps." The illustration shows one form of lamp in which the carbon is V-shaped, and is held in two metal clamps.

The conductors may also be straight and be held between clamps inserted at the top and bottom of the case. When there are two conductors in a lamp a claw attachment is provided to short-circuit the terminals, and so cut the conductors out of circuit. The lantern for the cylindrical stems is composed of a glass tube or cylinder with two ends.

NOTE.

By an oversight the accompanying illustration was omitted from the abridgment of the specification of Count Fontaine Moreau. It is referred to in that abridgment as Fig. 42, and described on page liii.

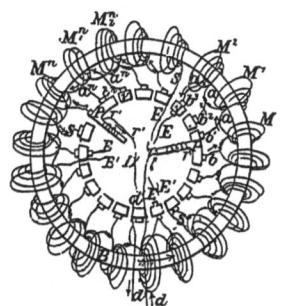

II.   Classified Index of English Patents, 1873-82.

SUBJECT MATTER INDEX

OF

PATENTS APPLIED FOR IN THE UNITED KINGDOM

AND

PATENTS GRANTED

RELATING TO

ELECTRIC LIGHTING.

January 1, 1873, to July 1, 1882.

LIST OF HEADINGS.

ARMATURES.
ARRANGING LIGHTS.
CANDLE LAMPS & CARBON HOLDERS.
CANDLES.
CONDUCTORS.
CURRENT METERS AND ELECTRO-
  METERS.
DIFFUSING LIGHT.
DYNAMOMETERS.
ELECTRODES (Carbons, &c.).
EXCITING DYNAMO-ELECTRIC GENE-
  RATORS.
FILAMENTS FOR INCANDESCENCE
  LAMPS.
GALVANOMETERS.
GENERATORS.
GLOBES.
INSULATING MATERIALS.
INSULATORS.
LAMPS (Arc).

LAMPS (Incandescence).
MAGNETS.
MISCELLANEOUS APPLICATION OF THE
  ELECTRIC LIGHT.
MOTORS.
PREVENTING REVERSAL OF POLARITY
  IN DYNAMO-ELECTRIC GENERATORS.
REFLECTORS.
REGULATING CURRENTS.
REGULATING GENERATORS & MOTORS.
RESISTANCES
RHEOSTATS.
SAFETY FUSES.
SECONDARY BATTERIES.
SECONDARY CURRENTS FOR LIGHTING.
SUB-DIVIDING AND DISTRIBUTING
  CURRENTS.
SWITCHES, COMMUTATORS & CIRCUIT
  CLOSERS.
TRANSMISSION OF POWER.

ARMATURES.

1878	4,988	Thompson and Thompson
1881	1,240	Brewer (Edison)
,,	5,551	Johnson (Griscom)

ARRANGING LIGHTS.

1880	3,637	Justice (Spalding)
1881	65	Justice (Spalding)

CANDLE LAMPS & CARBON HOLDERS.

1879	211	Rapieff
,,	4,718	Godfrey
1881	1,027	Berly
,,	4,533	Gibbs
,,	5,702	Swan

CANDLES.

1876	3,552	Applegarth (Jablochkoff)
1877	494	Jablochkoff
,,	1,996	Jablochkoff
1878	4,774	Pulvermacher
,,	4,960	Newton (Weston Dynamo-Electric Machine Co.)
,,	5,011	Cohné
1879	144	Morgan-Brown (McCartez and Seillieri)
,,	277	Cohné
,,	863	Abel (Jamin)
,,	1,122	Fox
,,	1,175	Imray (Société Generale d'Electricité)
,,	2,339	Clark (De Meritens & Co.)
,,	2,543	De Hamel
,,	4,796	Gatehouse
1880	3,404	Brewer (Desquiers)
1881	1,236	Berly

CONDUCTORS.

1873	441	Lake
,,	706	Newton
,,	1,970	Evans
,,	2,077	Rowett
,,	2,091	Bonneville (Radde)
,,	2,350	Clark (Achard)
,,	2,969	Moseley
,,	3,083	Wilkinson
,,	3,780	Hooper
,,	3,862	Gray
,,	3,863	Gray
,,	3,879	Stearns
,,	3,997	Hooper and Dunlop
,,	4,079	Rowett
,,	4,167	Madsen
,,	4,193	Rubery
1874	124	Truman
,,	265	Monckton
,,	447	Macintosh
,,	485	Harrop
,,	862	Imray (Waters)
,,	1,159	Bullivant
,,	1,766	Eustace

1874	1,855	Zanni
,,	3,172	Eustace
,,	3,771	Clark (Lartigue)
,,	3,800	Timmins
,,	4,475	Neave
1875	1,719	Ewen and James
,,	3,633	Lucas
,,	3,666	Lackersteen
,,	3,798	Phillips and Johnson
,,	4,115	Henley
,,	4,384	Smith
1876	557	Tyer
,,	1,944	Henley
,,	2,759	Rubery
,,	3,099	Potts
,,	3,533	Johnson and Phillips
,,	4,159	Hibell
,,	4,705	Newton (Menier)
1877	4,824	Hunt (Brooks)
1878	251	Siemens
,,	767	Conradi (Wiebe)
,,	1,195	Spalding
,,	1,196	Spalding
,,	1,937	Abbott
,,	3,603	Stanford (Phillips)
,,	3,622	Smith
,,	3,988	Fox
,,	4,043	Fox
,,	4,338	Harrison
,,	4,671	Scott
,,	4,686	Parod
,,	5,270	Lucas
,,	5,306	Edison
,,	5,321	Pierson
1879	307	Moseley
,,	399	Watson
,,	718	Jack and Greening
,,	2,016	Sanders and Danckwerth
,,	2,402	Edison
,,	2,629	Blondot and Bourdin
,,	2,692	Clark (Hermann)
,,	3,001	Moseley
,,	3,023	Johnson (Wheeler)
,,	3,728	Putnam
,,	3,778	Lake (Arbogast and McTighe)
,,	3,831	Jensen (Prall and Olrick)
,,	4,456	King (Linford)
,,	4,549	Bell
,,	4,572	Lake (Holmes and Greenfield)
,,	4,916	Newburn (Heins)
,,	5,056	Gray
1880	578	Edison
,,	695	Lake (Chinnock and Harrison)
,,	756	Reddie
,,	905	Loeffler
,,	925	Guest
,,	1,184	King
,,	1,295	Allen (Kimdson and Kane)
,,	1,407	Heaviside
,,	2,207	Wolff (Michalk)
,,	2,665	Lake (Lamb)

1880	3,310	Truman
,,	3,424	Barker (Jacques)
,,	3,880	Jensen (Edison)
,,	4,116	Lake (Maiche)
,	4,482	George
,;	4,674	Kendal
,,	4,755	Berly and Hulett
,,	5,083	Berthoud and Borel
,,	5,275	Fitzgerald
1881	264	Apps
,,	783	Perry and Ayrton
,,	785	Ayrton
,,	792	Jensen (Edison)
,,	879	Shippey
,,	1,235	Tambourin
,,	1,474	Newburn (Bourdin and Maltzoff)
,,	1,653	Johnson (Société La Force et la Lumière)
,,	1,809	Lake (Delany)
,,	1,873	Henley
,,	2,215	Allen
,,	2,217	Lake (Delany)
,	2,264	Barney
,,	2,437	Edmonds (Mowbray)
,,	2,532	Gourand (Delany and Johnson)
,,	2,542	Mackie
,,	2,573	Newton (Hussey and Dodd)
,,	2,592	Lake (Clark)
,,	2,807	Rangaro and Fleming
,,	3,177	Mayall
,,	3,254	Lake (Brooks)
,,	3,483	Brewer (Edison)
,,	3,790	Lake (Strohm)
,,	3,975	Smith
,,	4,058	Lake (Henck)
,,	4,093	Brewer (Delany and Johnson)
,,	4,296	Abel (Brasseur and Dejaer)
,,	4,409	Callender
,,	4,781	Woodward
,,	4,797	Gore
,,	4,885	Johnson and Phillips
,,	5,002	Vyle
,,	5,080	Crompton
,,	5,226	Brewtnall
,,	5,367	Lake (Maxim)
,,	5,599	Smith
,,	5,615	Culbertson and Brown
,,	5,661	Johnson (Labye and De Locht-Labye

CURRENT METERS & ELECTROMETERS.

1876	2,886	Dewar
1878	4,626	Fox
1879	2,402	Edison
,,	3,565	Elmore
1880	4,391	Jensen (Edison)
,,	5,004	Swan
1881	1,016	Brewer (Edison)
,,	4,454	Sprague
,,	4,472	Boys
	4,571	Brewer (Edison)

1881	4,576	Brewer (Edison)
,,	4,824	Carus-Wilson
,,	4,948	André
,,	5,499	Swan
,,	5,623	Carus-Wilson
,,	5,651	Fox

DIFFUSING LIGHT.

1878	4,031	Clark (Clémandot)
,,	4,079	Pulvermacher
,,	4,671	Scott
1879	259	Grieve
,,	350	Young and Freeman

DYNAMOMETERS.

| 1881 | 264 | Apps |
| ,, | 2,449 | Boys |

ELECTRODES (Carbons, &c.).

1876	4,597	Monckton
1877	2,982	Clark (Reynier)
,,	3,470	Harrison
,,	4,553	Gray
1878	471	Gray (Reynier)
,,	861	Scott
,,	3,011	Arnold and others
,,	3,470	Harrison
,,	4,553	Gray
,,	4,645	Sellon
,,	4,671	Scott
,,	4,774	Pulvermacher
,,	4,847	Cheesbrough (Sawyer and Man)
,,	5,060	Newton (De Mersanne)
,,	5,165	Clark (Regnard)
,,	5,307	Freeman
1879	65	Clark (Ducretet)
,,	178	De Meritens
,,	192	Remington
,,	211	Rapieff
,,	382	Imray
,,	523	Bousfield and Bousfield
,,	684	Wier
,,	788	Harding
,,	876	Spence
,,	925	Hedges
,,	927	Thompson
,,	1,207	Dillon
,,	1,287	Lake (Keith)
,,	1,622	Thompson
,,	1,971	Clark (Reynier)
,,	2,110	Siemens
,,	2,199	Siemens (Furstenhagen)
,,	2,267	Grout and Sennet
,,	2,321	Andrews
,,	2,340	Clark (Desnos)
,,	2,543	De Hamel
,,	3,272	Lake (Concornotti)
,,	3,355	Morgan-Brown (Poilier)
,,	3,875	Harrison
,,	4,400	Lake (Houston and Thompson)
1880	4,608	Heinrichs
,,	4,988	Hedges

1878	5,076	Johnson (Bertin and De Mersanne)
,,	5,078	Chamberlain (Hearington)
,,	5,123	Dering
,,	5,139	Brain
,,	5,306	Edison
1879	211	Rapieff
,,	332	Elphinstone and Vincent
,,	523	Bousfield and Bousfield
,,	565	Formby
,,	595	Courtenay
,,	644	Wiles
,,	749	Dubos
,,	805	Thompson (Gary)
,,	926	Boulton
,,	960	Simon (Schuckert)
,,	988	Birkhead
,,	1,207	Dillon
,,	1,387	Lake (Keith)
,,	1,949	Sellon and Edmunds
,,	2,321	Andrews
,,	2,402	Edison
,,	3,543	Courtenay
,,	3,565	Elmore
,,	4,400	Lake (Houston and Thompson)
,,	4,589	Heinrichs
,,	5,085	Wise (Bürgin)
,,	5,157	Joel
1880	33	Edison
,,	79	Werdermann
,,	315	Lake (Houston and Thompson)
,,	478	Morgan (Glouchoff)
,,	602	Edison
,,	842	Clark (Bouteilloux and Laing)
,,	849	Haddan (Brush)
,,	872	Fitzgerald
,,	885	Graddon
,,	886	Harrison
,,	1,136	Johnson (De Meritens)
,,	1,178	Perry
,,	1,259	Johnson (Griscom)
,,	1,385	Edison
,,	1,392	Lake (Maxim)
,,	1,580	Alexander (Zipernowsky)
,,	1,826	Gordon
,,	1,998	Lake (Seeley)
,,	2,272	Slater
,,	2,764	André
,,	2,895	Elphinstone and Vincent
,,	3,041	Gümpel
,,	3,324	Gümpel
,,	3,496	Harrison
,,	3,808	Willatt
,,	3,832	Elmore
,,	3,894	Jensen (Edison)
,,	3,928	Lake (Thomson)
,,	3,964	Jensen (Edison)
,,	3,971	Clark (Niaudet and Reynier)·
,,	4,005	Brewer (Cance)
,,	4,007	Zanni
,,	4,049	Reddie (Biloret and Mora)
,,	4,265	Lake (Hussey and Dodd)
1880	4,608	Heinrichs
,,	4,886	Hopkinson and Muirhead
,,	5,008	Wilde
,,	5,137	Henley
1881	48	Lake (Etéve)
,,	78	Gordon
,,	253	Gümpel
,,	497	Wilde
,,	803	Waller
,,	1,240	Brewer (Edison)
,,	1,447	Siemens (Siemens and Halske)
,,	1,787	Clark (Müller and Levett)
,,	1,961	Higgs
,,	2,013	Masson
,,	2,212	Barlow (De Meritens)
,,	2,375	Newton (Hussey and Dodd)
,,	2,416	Wolff (Jürgensen and Lorenz)
,,	2,482	Brewer (Edison)
,,	2,761	Groth (Lachansee)
,,	2,954	Jensen (Edison)
,,	3,231	Brewer (Edison)
,,	3,283	Pitt (Bear)
,,	3,394	Fox
,,	3,400	Johnson (Mignon and Rouart)
,,	3,441	Moffatt and Chichester
,,	3,456	Lake (Arcy)
,,	3,472	Harding and Hartmann
,,	3,804	Jensen (Edison)
,,	3,871	Harborow
,,	3,880	Lake (Dion)
,,	3,932	Jensen (Edison)
,,	4,019	Dering
,,	4,026	De Pass (La Société Anonyme des Cables Electriques)
,,	4,034	Jensen (Edison)
,,	4,057	Newton (Société Universelle d'Electricité Tommasi)
,,	4,107	Fahrig
,,	4,207	Barlow (De Meritens)
,,	4,271	Lake (Dobrochotoff, Maikoff and De Kabath)
,,	4,304	Aylesbury
,,	4,405	Clark (De Changy)
,,	4,508	Johnson (Parod)
,,	4,541	Kennedy
,,	4,552	Jensen (Edison)
,,	4,559	Newton
,,	4,582	Clark (De Changy)
,,	4,591	Haddan (Desaigne)
,,	4,592	Millar
,,	4,607	Joel
,,	4,820	Wise (Bürgin)
,,	4,825	Carus-Wilson
,,	4,854	Rogers
,,	5,104	Clark
,,	5,525	Akester
,,	5,536	Gordon
,,	5,551	Johnson (Griscom)
,,	5,566	Millar
,,	5,593	Powell (Lescuyer)
,,	5,600	Pitt (Starr)

Year	No.	Name
1878	4,558	Gray
,,	4,568	Mackenzie
,,	4,573	Zanni
,,	4,575	Tyler
,,	4,595	Heinrichs
,,	4,646	Sellon
,,	4,671	Scott
,,	4,690	Johnson (De Meritens)
,,	4,693	Reimenschneider and Christensen
,,	4,821	Sabine
,,	4,847	Cheesbrough (Sawyer and Man)
,,	4,849	Siemens (Von Alteneck)
,,	4,988	Thompson and Thompson
,,	5,011	Cohné
,,	5,044	Johnson (De Mersanne and Bertin)
,,	5,060	Newton (De Mersanne)
,,	5,105	Punshon
,,	5,110	Johnson (De Mersanne and Bertin)
,,	5,139	Brain
,,	5,152	Whyte
,,	5,165	Clark (Reynard)
,,	5,183	Lake (Fuller)
,,	5,197	Wilde
,,	5,281	Thompson and Earl
,,	5,307	Freeman
1879	27	Raworth
,,	65	Clark (Ducretet)
,,	81	Hedges
,,	83	Ladd (Jaspar)
,,	144	Morgan-Brown (McCarty and Seillieri)
,,	178	De Meritens
,,	192	Remington
,,	211	Rapieff
,,	245	Crompton and Willans
,,	299	Haddan (Molera and Cebrian)
,,	325	Paraire
,,	382	Imray
,,	416	Andrews
,,	427	Dubos
,,	454	Higgs
,,	523	Bousfield and Bousfield
,,	684	Wier
,,	740	Mori
,,	830	André
,,	863	Abel (Jamin)
,,	876	Spence
,,	947	Haddan (Brush)
,,	1,635	Mackenzie
,,	1,959	Hopkinson
,,	1,969	Longsdon (Krupp)
,,	2,000	Cougnet
,,	2,111	Lake (Puvilland and Raphaël)
,,	2,301	Werdermann
,,	2,322	Abel (Sedlaczek and Wikulill)
,,	2,652	Siemens (Alteneck)
,,	2,744	Whyte
,,	3,085	Mills (Million)
,,	3,509	Crompton
,,	3,697	Clark (Gérard-Lescuyer)
,,	3,750	Haddan (Brush)
1879	3,771	Brockie
,,	3,875	Harrison
,,	4,854	Hickley
,,	4,405	Newton (Tommasi)
,,	4,590	Harding
,,	4,718	Godfrey
,,	5,156	Pitt (Scribner)
,,	5,157	Joel
1880	75	Clark (Roguier)
,,	79	Werdermann
,,	231	Heinke and Lang
,,	350	De Puydt and Cougnet
,,	455	Blamires
,,	553	Wigner
,,	636	Clark (Pilleux)
,,	725	Imray (La Société Generale d'Electricité)
,,	842	Clark (Bouteilloux and Laing)
,,	1,397	Abel (Krizik and Piette)
,,	1,507	André
,,	1,552	Clark (Gérard-Lescuyer)
,,	1,553	Abel (Compagnie Generale d'Eclairage Electrique)
,,	1,649	Lake (Maxim)
,,	1,704	Haddan (Bureau)
,,	1,826	Gordon
,,	2,147	Wise (Mandon)
,,	2,236	Colhné
,,	2,252	André and Easton
,,	2,764	André
,,	2,980	Clark (Guest)
,,	3,509	Hopkinson
,,	4,191	Harding
,,	4,428	Johnson (Berjot)
,,	4,608	Heinrichs
,,	4,614	Siemens
,,	4,755	Berly and Hulett
,,	4,866	Lake (Maxim)
,,	4,914	Wise (Mandon)
,,	5,033	Johnson (De Meritens)
,,	5,137	Henley
1881	48	Lake (Etéve)
,,	153	Muirhead and Hopkinson
,,	218	Gordon
,,	253	Gümpel
,,	715	Tongue (Lacomine)
,,	774	Fyfe
,,	894	Sachs
,,	1,040	Common and Joel
,,	1,232	Upton
,,	1,236	Berly
,,	1,358	Harrison and Blagburn
,,	1,384	Lake (Holcombe)
,,	1,526	Andrews
,,	1,536	Auberville (Delaye)
,,	1,596	Reddie (Sedlaczek and Wikulill)
,,	1,670	Grimston
,,	1,685	Clark (Gérard-Lescuyer)
,,	1,922	Rogers
,,	1,942	Brockie
,,	1,968	Lake (Bouliguine)

1881	2,038	Haddan (Gülcher)
,,	2,198	Abel (Tschikoleff and Kleiber)
,,	2,344	Gadot
,,	2,369	Cohné
,,	2,402	Hawkes and Bowman
,,	2,495	Brewer (Edison)
,,	2,563	André
,,	2,572	Newton (Hussey and Dodd)
,,	2,788	Mills (Million)
,,	2,851	Lake (Wood)
,,	3,015	Lake (Greb)
,,	3,049	Haddan (Woolley)
,,	3,166	Morgan-Brown (Harding)
,,	3,214	Clark (Bouteilloux and Laing)
,,	3,305	Bright
,,	3,349	Reddie (Chertemps)
,,	3,362	Hopkinson
,,	3,441	Moffatt and Chichester
,,	3,455	Hubble (Partz)
,,	3,456	Lake (Arcy)
,,	3,473	Harling and Hartmann
,,	3,559	Harrison
,,	3,599	Lever
,,	3,635	Tubini
,,	3,668	Lake (Connolly)
,,	3,679	Pitt (Burrell)
,,	3,821	Fyfe and Main
,,	3,822	Tubini
,,	3,893	Lake (Hill)
,,	3,976	Jensen (Cance)
,,	4,011	Hunt (Brown)
,,	4,305	Haddan (Somzée)
,,	4,504	Brockie
,,	4,559	Newton
,,	4,617	Clark (Sheridan)
,,	4,775	Bonneville (Daft)
,,	4,777	Prentice
,,	4,820	Wise (Bürgin)
,,	4,855	Rogers
,,	4,948	André
,,	5,185	Brewer (Waterhouse)
,,	5,229	Lake (Williams)
,,	5,233	Lake (Williams)
,,	5,286	Sennett
,,	5,396	Varley
,,	5,477	Lake (De la Roche)
,,	5,490	Lake (Mondos)
,,	5,524	Kennedy
,,	5,660	Powell (Gérard-Lescuyer)
,,	5,738	Lorrain

LAMPS (Incandescence).

1875	2,410	Chauvin, Goizet and Aubry
1877	1,996	Jablochkoff
1878	3,985	Tyler
,,	3,988	Fox
,,	4,006	Jensen (Marcus)
,,	4,043	Fox
,,	4,180	Pulvermacher
,,	4,226	Edison
,,	4,283	Stokes

1878	4,388	Van Choate
,,	4,502	Brewer (Edison)
,,	4,626	Fox
,,	4,662	Sprague
,,	4,774	Pulvermacher
,,	5,306	Edison
1879	211	Rapieff
,,	454	Higgs
,,	1,122	Fox
,,	2,402	Edison
,,	3,587	Cheesbrough (Sawyer)
,,	4,576	Edison
,,	5,127	Edison
,,	5,157	Joel
,,	5,206	André
1880	18	Swan
,,	203	Clark
,,	250	Swan
,,	455	Blamires
,,	578	Edison
,,	925	Guest
,,	1,649	Lake (Maxim)
,,	1,840	Lake (Clingman)
,,	2,037	Clark (Drew)
,,	3,494	Fox
,,	3,765	Brewer (Edison)
,,	4,393	Lake (Maxim)
,,	4,495	Lake (Nichols)
,,	4,608	Heinrichs
,,	4,745	Gordon
,,	4,755	Berly and Hulett
,,	4,933	Swan
,,	4,988	Hedges
,,	5,014	Swan
,,	5,137	Henley
,,	5,275	Fitzgerald
1881	153	Muirhead and Hopkinson
,,	225	Fox
,,	539	Brewer (Edison)
,,	562	Jensen (Edison)
,,	689	Lake (Maxim)
,,	768	Brewer (Edison)
,,	792	Jensen (Edison)
,,	1,040	Common and Joel
,,	1,422	Crookes
,,	1,543	Fox
,,	1,653	Johnson (Société La Force et la Lumière)
,,	1,802	Jensen (Edison)
,,	1,918	Brewer (Edison)
,,	1,943	Brewer (Edison)
,,	2,079	Gimingham
,,	2,492	Jensen (Edison)
,,	2,612	Crookes
,,	2,833	André and Easton
,,	2,930	Ward
,,	3,122	Fox
,,	3,187	Lake (Nichols)
,,	3,189	Lake (Maxim)
,,	3,240	Gatehouse
,,	3,435	Wright

1881	3,437	Wright
,,	3,650	Pfannkuche
,,	3,711	Engel (Müller)
,,	3,799	Crookes
,,	3,890	Fitzgerald
,,	4,017	Hallett
,,	4,024	Morgan-Brown (Fox)
,,	4,174	Brewer (Edison)
,,	4,193	Gimingham
,,	4,202	Swan
,,	4,294	Schaeffer
,,	4,305	Haddan (Somzée)
,,	4,311	Johnson (Faure)
,,	4,383	Fox
,,	4,439	Jameson
,,	4,478	Harrison
,,	4,654	André
,,	4,659	Courtenay
,,	4,778	Wright
,,	4,939	St. George
,,	4,942	Pitt (Goulard and Gibbs)
,,	5,618	Graham
,,	5,632	Sellon

MAGNETS.

1873	402	Thorneloe
,,	618	Wilde
,,	706	Newton
,,	1,180	Johnson (Fontaine)
,,	3,461	Newton (Camacho)
1874	94	Newton (Stone)
,,	1,933	Van der Mez
,,	2,625	Slater
,,	3,509	Monckton
1875	1,800	Faulkner
,,	2,049	Nelson and Anderson (Paine and Paine)
,,	2,564	Hequet
1876	162	Harling
,,	433	Fahie (Fahie)
,,	731	Theiler and Theiler
,,	836	Jablochkoff
,,	3,537	Brewer (Lauritzen)
,,	4,597	Monckton
1877	851	Tommasi
,,	1,829	Werdermann
,,	3,981	Smith
,,	4,223	Zanni
,,	4,642	Werdermann (Boblique)
1878	162	Lake (McTighe)
,,	611	Bell
,,	1,927	Jensen (Cance)
,,	2,640	Leak and Edwards
,,	3,924	Van Choate
,,	4,903	Lonsdale
,,	4,961	Byshe
,,	5,189	Brain
,,	5,183	Lake (Fuller)
1879	312	Cranston
,,	394	Johnson (Gower and Roosevelt)
,,	1,476	Haddan (Rosebrugh)

1879	1,766	Hopkins
,,	2,617	Neale
,,	4,428	Brewer (Chambrier)
,,	4,555	Bell and Scarlett
,,	4,821	Elmore
1880	752	Cheesbrough
,,	872	Fitzgerald
,,	1,580	Alexander (Zipernowsky)
,,	1,585	Scarlett
,,	1,738	Russell
1881	200	Imray
,,	264	Apps
,,	2,375	Newton (Hussey and Dodd)
,,	4,448	Imray (Stearns)

MISCELLANEOUS APPLICATIONS OF THE ELECTRIC LIGHT.

1877	270	Varley (Signalling)
1881	879	Shippey (Signalling)
,,	4,812	Haddan (De Busscher) (For Locomotives)
,,	5,316	Laybourne (Lighting Carriages)

MOTORS.

1873	706	Newton
,,	1,000	D'Arros
,,	2,015	Newton
,,	2,618	Palmer
,,	3,078	Moore
1874	94	Newton (Stone)
,,	265	Monckton
,,	4,454	Bonneville (Chutaux)
1875	1,487	Courtenay
,,	2,043	Clark (Hussey)
,,	2,205	Clamond
,,	2,946	Faure
,,	3,243	Alexander (Bürgin)
,,	3,416	Johnson (Camacho)
,,	3,999	Browne (Kimball)
,,	4,118	Paine and Nelson (Paine)
1876	1,931	Clark (Bastet)
,,	2,034	Cole
,,	3,670	Faure
,,	4,269	Schaub
,,	4,595	Lake (Egger)
,,	4,597	Monckton
1877	732	Lovell
,,	1,829	Werdermann
,,	2,314	Thompson (Gary)
,,	3,750	Lake (Attwood)
,,	3,981	Smith
,,	4,256	Clark
,,	4,349	Wrigley
,,	4,435	Varley
1878	292	Andrews
,,	311	Newton (Hosmer)
,,	915	Spalding
,,	1,587	Pulvermacher
,,	1,927	Jensen (Cance)
,,	1,988	La Cour
,,	2,033	Clark (Ward and Ball)

1878	2,878	Harling and Bull
,,	2,930	Newton (Hosmer)
,,	3,134	Siemens (Siemens and Von Alteneck)
,,	3,338	Lüdeke and Thorman
,,	3,676	Siemens (Siemens and Von Alteneck)
,,	4,041	North
,,	4,094	Pulvermacher
,,	4,518	North
,,	4,611	Edwards and Normandy
,,	4,699	Melhado
,,	5,139	Brain
1879	211	Rapieff
,,	299	Haddan (Molera and Cebrian)
,,	513	Imray (Endrés)
,,	854	Port and Varley
,,	926	Boulton
,,	1,110	Finch
,,	1,445	Whitely
,,	2,481	Hopkinson
,,	4,534	Siemens
,,	4,589	Heinrichs
,,	4,653	Hopkinson
,,	4,696	Clark (Lamar)
,,	4,785	Smith (Chameroy)
,,	5,085	Wise (Bürgin)
1880	33	Edison
,,	583	Siemens
,,	602	Edison
,,	885	Graddon
,,	1,178	Perry
,,	1,244	Johnson (Griscom)
,,	1,259	Johnson (Griscom)
,,	1,339	Mewburn (Fouvielle)
,,	1,385	Edison
,,	2,764	André
,,	3,894	Jensen (Edison)
,,	3,928	Lake (Thomson)
,,	3,964	Jensen (Edison)
,,	4,009	Lorrain (Trouvé)
,,	4,825	Kesseler (Kuhlo)
1881	803	Waller
,,	1,636	Fox
,,	2,482	Brewer (Edison)
,,	2,954	Jensen (Edison)
,,	2,989	Hopkinson
,,	3,394	Fox
,,	3,857	Sueur
,,	4,034	Jensen (Edison)
,,	4,163	Thompson (Langley)
,,	4,592	Millar
,,	4,819	Wise
,,	4,825	Carus-Wilson

PREVENTING REVERSAL OF POLARITY IN DYNAMO-ELECTRIC GENERATORS.

1879	2,821	Zanni

REFLECTORS.

1874	3,509	Monckton

1878	446	Van der Weyde
,,	902	Neale
,,	142	Alder and Clarke
,,	4,317	Tillcard
,,	4,662	Sprague
,,	4,671	Scott
1879	389	Lake (Watson)
,,	3,425	Pitt (Mangin)

REGULATING CURRENTS.

1877	2,909	Edison
1878	3,988	Fox
,,	4,043	Fox
,,	4,047	Harding
,,	4,226	Edison
,,	4,283	Stokes
,,	4,304	Shea
,,	4,388	Van Choate
,,	4,626	Fox
,,	4,689	Welch
,,	5,053	Johnson (De Mersanne and Bertin)
,,	5,183	Lake (Fuller)
1879	27	Raworth
,,	1,692	Sellon and Edmonds
,,	1,791	Sellon and Edmonds
,,	2,019	Chrétien and Felix
,,	2,060	Blandy
,,	2,402	Edison
,,	3,793	Harrop
,,	3,875	Harrison
1880	33	Edison
,,	75	Clark (Roguier)
,,	315	Lake (Houston and Thomson)
,,	602	Edison
1881	1,835	Brush
,,	2,739	Newton (Gravier)
,,	3,032	Thomson
,,	3,655	Dunston and Pfannkuche
,,	5,665	Varley
,,	5,687	Carus-Wilson

REGULATING GENERATORS AND MOTORS.

1878	1,197	Spalding
1879	4,534	Siemens
1880	1,244	Johnson (Griscom)
1881	1,835	Haddan (Brush)
,,	2,618	Jameson
,,	3,409	Westinghouse
,,	4,496	Johnson (Société La Force et la Lumière)
,,	4,825	Carus-Wilson
,,	5,006	Wright and Ormiston

RESISTANCES.

1879	1,477	Haddan (Black and Rosebrugh)
1881	1,943	Brewer (Edison)
,,	2,263	Cuff
,,	2,572	Newton (Hussey and Dodd)
,,	5,743	Pfannkuche and Dunston

1879	3,479	Robinson		1881	4,057	Newton (Societe Universelle d'Electricite Tommasi)
,,	4,306	Wilkins		,,	4,478	Harrison
,,	4,653	Hopkinson		,,	4,792	Hubble (Gerard-Lescuyer)
,,	5,126	White		,,	5,096	Lake (Blake)
,,	5,337	Clark (Greenfield and Adie)				
1881	1,802	Jensen (Edison)				
,,	2,344	Gadot				

TRANSMISSION OF POWER.

1881	2,851	Lake (Wood)		1879	2,019	Chretien and Felix
,,	3,231	Brewer (Edison)		,,	4,653	Hopkinson
,,	3,369	Hedges		1881	4,128	Imray
,,	3,668	Lake (Conolly)		1879	200	Imray (Cabanellas)
,,	3,804	Jensen (Edison)		,,	253	Gumpel
,,	3,893	Lake (Hill)				

LIST OF

PATENTS APPLIED FOR IN THE UNITED KINGDOM

RELATING TO

ELECTRIC LIGHTING.

January 1 to June 30, 1882.

1882.

APPLICATIONS FOR LETTERS PATENT.

(Abbreviated Titles.)

14 Mackie. Apparatus used for electric lighting.

29 Fitzgerald. Secondary batteries.

49 Hopkinson. Measuring and recording quantity of electricity.

55 Perry. Distribution of electrical energy.

64 Groth (Gülcher). Magneto and dynamo-electric machines.

69 Liveing and Boys. Manufacture of incandescent lamps.

72 Kennedy. Secondary or reversible electric batteries.

84 Lake (Ball). Dynamo-electric machines.

85 Lake (Williams). Machines and apparatus for generating and utilising electricity for lighting, heating and other purposes.

95 Mackenzie. Electric lamps.

120 Liardet and Donnithorne. Apparatus for storing electrical energy.

129 Preece and James. Apparatus for electrically lighting railway trains.

130 Henley. Machinery and apparatus for obtaining, transmitting and applying electric currents for electric lighting, motive power and other useful purposes.

144 Haddan (Boettcher). Secondary galvanic batteries.

157 Hawkes. Apparatus for electric lighting.

169 Raison. Electro-motors and dynamo-electric machines.

185 Haddan (Morel). Electric accumulators.

195 Cradock. Apparatus employed in working telegraph cables,

224 Lake (Williams). Electric lighting apparatus.

231 Siemens (Jacob). Conductors.

232 Meyer. Permanent way for electric conductors.

234 Lake (Hussey and Dodd). Dynamo-electric machines.

245 Lake (Khotinsky). Apparatus for regulating electric currents.

252 Lake (La Société Universelle d'Electricité Tommasi). Electrical accumulators or secondary batteries.

289 Humphry's. Secondary batteries.

290 Haddan (Morel). Conductors.

305 Aronson. Electric lamps.

319 Sellon. Secondary batteries.

339 De Pass (Abakanowiez). Regulating electric lamps.

346 Crompton. Electric lamps and apparatus used in connection with electric lighting.

359 Aronson. Electric lamps.

361 Lake (Clark). Electrical conductors or cables.

377 Bright. Electric lamps.

386 Henley. Construction of cores for cables for sending electric currents.

392 Thompson (Union Electric Manufacturing Co.). Obtaining light by electricity, and regulating the electric current for the same.

497 Little. Electro-magnets and armatures for the same, and mechanism connected therewith for producing electro-magnetic motors and engines,

LIST OF PATENTS

GRANTED IN THE

UNITED STATES OF NORTH AMERICA

FOR INVENTIONS IN

1.—ARC ELECTRIC LIGHTS.
2.—INCANDESCENCE LIGHTS.
3.—ELECTRIC LIGHT, SYSTEM AND
  APPLIANCES.

4.—GENERATORS.
5.—METERS.
6.—ELECTRO-MAGNETS.
7.—REGULATORS.

I.—ARC ELECTRIC LIGHTS.

Including apparatus for producing and controlling light by the use of discontinuous conductors.

20,255	Collier, H. M., and Baker, H. N.	.. May 18, 1858
33,457	Way, J. T. ..	.. October 8, 1861
33,458	Way, J. T. ..	.. October 8, 1861
115,154	Berlioz, A. P.	.. May 23, 1871 .
123,923	Meynial, A. A.	.. February 20, 1872
147,827	Day, jun., M.	.. February 24, 1874
156,015	Day, jun., M.	.. October 20, 1874
184,553	Smith, P. E., Spruill, S. R., and Wood, W. R. ..	.. November 21, 1876
190,864	Jablochkoff, P.	.. May 15, 1877
191,177	Reynier, N. E.	.. May 22, 1877
194,500	Sawyer, W. E.	.. August 21, 1877
194,563	Sawyer, W. E.	.. August 28, 1877
196,425	Brush, C. F...	.. October 23, 1877
198,436	Wallace, W...	.. December 18, 1877
200,545	King, J. ..	.. February 19, 1878
203,411	Brush, C. F...	.. May 7, 1878
8,718	Brush, C. F. (re-issue)	May 20, 1879
203,844	Longworth, L. R.	.. May 21, 1878
205,962	Jenkins, P. O.	.. July 16, 1878
206,083	Bürgin, E. ..	.. July 16, 1878
9,548	Bürgin, E. (re-issue)	February 1, 1881
207,753	Lugo, O. ..	.. September 3, 1878
207,754	Lugo, O. ..	.. September 3, 1878
208,252	Maxim, H. S.	.. September 24, 1878
208,253	Maxim, H. S.	.. September 24, 1878
210,213	Rogers, J. H.	.. November 26, 1878
210,380	Weston, E. ..	.. November 26, 1878
212,183	Brush, C. F...	.. February 11, 1879
212,860	Tessie do Motay, C. M., and Stern, E...	March 4, 1879
214,242	Diehl, P. ..	.. April 15, 1879
214,514	Molera, E. J., and Cebrian, J. C.	.. April 22, 1879
214,515	Molera, E. J., and Cebrian, J. C.	.. April 22, 1879
214,516	Molera, E. J., and Cebrian, J. C.	.. April 22, 1879
215,733	Fuller, J. B...	.. May 27, 1879
215,910	Gilman, W. ..	.. May 27, 1879
216,760	Rogers, J. H.	.. June 24, 1879
217,744	Porter, J. C...	.. July 22, 1879
218,375	Fuller, J. B...	.. August 12, 1879
218,749	Jamin, J. C...	.. August 19, 1879
218,958	Gantt, N. B.	.. August 26, 1879
219,208	Brush, C. F.	.. September 2, 1879
219,209	Brush, C. F.	.. September 2, 1879
219,210	Brush, C. F.	.. September 2, 1879
219,211	Brush, C. F.	.. September 2, 1879
219,212	Brush, C. F.	.. September 2, 1879
219,213	Brush, C. F.	.. September 2, 1879
220,248	Manning, C. H.	.. October 7, 1879
220,287	Houston, E. J., and Thomson, E.	.. October 7, 1879
220,508	Thomson, E., and Houston, E. J.	.. October 14, 1879
220,728	Pendleton, D.	.. October 21, 1879
220,982	McCarty, W. F. C.	.. October 28, 1879
221,918	Holcombe, A. G.	.. November 25, 1879
222,503	Kipling, R. A.	.. December 9, 1879

223,495	Flanery, D. ..	.. January 13, 1880
223,646	Houston, E. J., and	
	Thomson E.	.. January 20, 1880
223,790	Winters, F., jun.	.. January 20, 1880
225,312	Weston, E.	.. March 9, 1880
227,025	Levison, W. G.	.. April 27, 1880
227,078	Van Depoele, C. J. ..	April 27, 1880
227,264	Keith, N. S. ..	.. May 4, 1880
227,478	Braunsdorf, J. E.	.. May 11, 1880
227,596	Tommasi, F.	.. May 11, 1880
229,246	Fuller, G. W., and	
	Mackintosh, E. D. ..	June 29, 1880
229,536	Kellogg, M. G.	.. July 6 1880
230,801	Moffat, R. R.	.. August 3, 1880
232,610	Jamin, J. C. ..	.. September 28, 1880
233,096	Holcombe, A. G.	.. October 12, 1880
233,236	Guest, J. H. ..	.. October 12, 1880
233,289	Siemens, C. W.	.. October 12, 1880
233,399	Brockie, J. ..	.. October 19, 1880
233,416	Jacobs, H. ..	.. October 19, 1880
233,589	Wood, J. J. ..	.. October 19, 1880
234,032	Harrison, C. W.	.. November 2, 1880

234,261	Finney, J. R.	.. November 9, 1880
234,456	Brush, C. F. ..	.. November 16, 1880
234,618	Sarcia, J. ..	.. November 16, 1880
234,770	Gordon, J. E. H.	.. November 23, 1880
234,835	Maxim, H. S.	.. November 23, 1880
235,203	Braunsdorf, J E.	.. December 7, 1880
235,258	Langley, J. W.	.. December 7, 1880
237,271	Haskins, C. D.	.. February 1, 1881
239,044	Heisler, C. ..	.. March 22, 1881
239,811	Levison, W. G.	.. April 5, 1881
240,210	Weston, E. ..	.. April 12, 1881
240,211	Weston, E. ..	.. April 12, 1881
240,280	Siemens, C. W.	.. April 19, 1881
240,781	Smithers, J. P.	.. April 26, 1881
240,795	Werdermann, R.	.. April 26, 1881
241,112	Woolley, L. G.	.. May 3, 1881
241,628	De Forest, D. W. ..	May 17, 1881
242,137	Keith, N. S...	.. May 31, 1881
242,747	Bureau, A. ..	.. June 14, 1881
243,196	Bernstein, A.	.. June 21, 1881
243,341	VonHefner-Alteneck,	
	F.	.. June 21, 1881

2.—INCANDESCENCE ELECTRIC LIGHTS.

Including apparatus for producing and controlling light by the use of continuous conductors.

20,706	Gardiner, S., jun.,	
	and Blossom, L. ..	June 29, 1858
166,877	Kosloff, S. A.	.. August 17, 1875
181,613	Woodward, H.	.. August 29, 1876
205,144	Sawyer, W. E., and	
	Man, A. ..	.. June 18, 1878
210,809	Sawyer, W. E., and	
	Man, A. ..	.. December 10, 1878
212,851	Jenkins, P. O.	.. March 4, 1879
213,643	Farmer, M. G.	.. March 25, 1879
214,636	Edison, T. A.	.. April 22, 1879
214,637	Edison, T. A.	.. April 22, 1879
217,792	Hall, A. W.	.. July 22, 1879
218,866	Edison, T. A.	.. August 26, 1879
219,628	Edison, T. A.	.. September 16, 1879
219,771	Sawyer, W. E.	.. September 16, 1879
223,524	Martin, A. J. ..	.. January 13, 1880
223,898	Edison, T. A.	.. January 27, 1880
224,329	Edison, T. A.	.. February 10, 1880
225,594	Guest, J. H. ..	.. March 16, 1880
227,118	Man, A. ..	.. May 4, 1880
227,227	Edison, T. A.	.. May 4, 1880
227,228	Edison, T. A.	.. May 4, 1880
227,229	Edison, T. A.	.. May 4, 1880
227,386	Sawyer, W. ..	.. May 11, 1880
227,387	Sawyer, W. ..	.. May 11, 1880
227,388	Sawyer, W. ..	.. May 11, 1880
227,389	Sawyer, W. ..	.. May 11, 1880
227,390	Sawyer, W. ..	.. May 11, 1880
228,122	Sawyer, W. ..	.. May 25, 1880

230,255	Edison, T. A.	.. July 20, 1880
230,310	Maxim, H. S.	.. July 20, 1880
230,953	Maxim, H. S.	.. August 10, 1880
230,954	Maxim, H, S.	.. August 10, 1880
232,357	Martin, A. J.	.. September 21, 1880
233,284	Sawyer, W. ..	.. October 12, 1880
233,346	Guest, J. H...	.. October 19, 1880
233,445	Swan, J. W.	.. October 19, 1880
234,345	Swan, J. W...	.. November 9, 1880
235,459	Sawyer, W. ..	.. December 14, 1880
236,883	Nichols, J. V.	.. January 18, 1881
237,198	Maxim, H. S.	.. February 1, 1881
237,732	Edison, T. A.	.. February 15, 1881
238,868	Edison, T. A.	.. March 15, 1881
239,149	Edison, T. A.	.. March 22, 1881
239,150	Edison, T. A.	.. March 22, 1881
239,153	Edison, T. A.	.. March 22, 1881
239,372	Edison, T. A., and	
	Batchelor, C.	.. March 29, 1881
239,373	Edison, T. A.	.. March 29, 1881
239,745	Edison, T. A.	.. April 5, 1881
241,366	Hussey, C. A.	.. May 10, 1881
241,430	Sawyer, W. E., and	
	Street, R. ..	.. May 10, 1881
242,051	Salathé, F., Brustlein,	
	J. E., and Sury, P.	May 24, 1881
242,896	Edison, T. A.	.. June 14, 1881
242,897	Edison, T. A.	.. June 14, 1881
242,930	Hussey, C. A.	.. June 14, 1881
242,984	Reynier, E. ..	.. June 14, 1881

3.—ELECTRIC LIGHTS: SYSTEMS AND APPLIANCES.

Including methods of distribution and regulation of electric currents for lighting, and apparatus connected therewith.

114,652	D'Ivernois, E. L. C.	May 9, 1871
194,111	Sawyer, W. E.	.. August 14, 1877
196,834	Sawyer, W. E.	.. November 6, 1877
205,303	Sawyer, W. E., and Man, A. ..	.. June 25, 1878
210,152	Sawyer, W. E., and Man, A.	.. November 19, 1878
210,317	Fuller, J. B. ..	.. November 26, 1878
210,543	Lambert, A. A.	.. December 3, 1878
211,262	Sawyer, W. E.	.. January 7, 1879
212,040	Molera, E. J., and Cebrian, J. C.	.. February 4, 1879
215,448	Frink, I. P. ..	.. May 20, 1879
218,055	Nitze, M. C. F.	.. July 29, 1879
218,167	Edison, T. A.	.. August 5, 1879
220,762	Huffman, S...	.. October 21, 1879
224,612	Sawyer, W...	.. February 17, 1880
227,226	Edison, T. A.	.. May 4, 1880
227,365	Knowles, E. R.	.. May 11, 1880
227,454	Stillman, T. B.	.. May 11, 1880
229,335	Sawyer, W. E., and Man, A. ..	.. June 29, 1880
229,476	Sawyer, W. E., and Man, A.	.. June 29, 1880
229,922	Siemens, C. W.	.. July 13, 1880
230,309	Maxim, H. S.	.. July 20, 1880
230,346	Sawyer, W. ..	.. July 20, 1880
230,360	Taber, F. A.	.. July 20, 1880
231,725	Lecoq, G. ..	.. August 31, 1880
233,831	Ball, C. M. ..	.. November 2, 1880
234,820	Sweanor, G...	.. November 23, 1880
235,460	Sawyer, W. E. and W.	December 14, 1880
235,913	Spalding, H. C.	.. December 28, 1880
235,914	Spalding, H. C.	.. December 28, 1880
236,460	Sawyer, W. E. and W.	January 11, 1881
236,478	Ball, C. M., and Guest, J. H. ..	.. January 11, 1881
237,361	Barr, F. ..	.. February 8, 1881
237,608	Sawyer, W. ..	.. February 8, 1881
239,147	Edison, T. A.	.. March 22, 1881
239,148	Edison, T. A.	.. March 22, 1881
239,151	Edison, T. A.	.. March 22, 1881
239,152	Edison, T. A.	.. March 22, 1881
239,311	Brush, C. F.	.. March 29, 1881
239,312	Brush, C. F.	.. March 29, 1881
242,483	Sisson, H. T.	.. June 7, 1881
242,498	Bardsley, J. ..	.. June 7, 1881
242,899	Edison, T. A.	.. June 14, 1881
242,900	Edison, T. A.	.. June 14, 1881
243,406	Sample, H. C., and Rabl, F. ..	.. June 28, 1881

4.—GENERATORS.

Including dynamic generators of electricity for producing currents by magneto-electric induction.

8,843	Sonnenburg, A., and Rechten, P.	.. March 30, 1852
10,175	Carpenter, C., jun..	November 1, 1853
11,415	Davis, A. ..	.. August 1, 1854
14,598	Carpenter, C., jun..	April 8, 1856
15,596	Shepard, E. C.	.. August 19, 1856
23,214	Burnap, W. H., and Bradshaw, J. A.	.. March 8, 1859
25,023	Marshall, M.	.. August 9, 1859
26,557	Beardslee, G. W.	.. December 27, 1859
26,558	Beardslee, G. W.	.. December 27, 1859
29,850	Baker, H. N.	.. September 4, 1860
58,960	Berlioz, A. P.	.. October 16, 1866
59,738	Wilde, H. ..	.. November 13, 1866
94,014	Lontin, D., and D'Ivernois, E. L. C.	August, 24, 1869
120,057	Gramme, Z. T., and D'Ivernois, E. L. C.	October 17, 1871
123,438	Allen, A. N...	.. February 6, 1872
124,216	Noble, B. G...	March 5, 1872
149,797	Siemens, E. W.	.. April 14, 1874
155,237	Hochhausen, W.	.. September 22, 1874
155,376	Heikel, O. ..	.. September 29, 1874
8,796	Heikel, O. (re-issue)	.. July 8, 1879
161,874	Farmer, M. G.	.. April 13, 1875
168,018	Heikel, O. ..	.. September 21, 1875
168,560	Drescher, L...	.. October 11, 1875
168,893	Fuller, J. B...	.. October 19, 1875
173,682	Smith, H. J...	.. February 15, 1876
175,361	Livingston, G. W.	.. March 28, 1876
179,184	Fuller, J. B...	.. June 27, 1876
180,082	Weston, E. ..	.. July 18, 1876
8,141	Weston, E. (re-issue)	March 26, 1878
181,342	Hochhausen, W.	.. August 22, 1876
181,553	Bell, A. G. ..	.. August 29, 1876
182,273	Gray, J. ..	.. September 19, 1876
184,377	Heikel, O. ..	.. November 14, 1876
184,966	Holcombe, A. G.	.. December 5, 1876
189,116	Lontin, D. F. J.	.. April 3, 1877
189,997	Brush, C. F...	.. April 24, 1877
9,410	Brush, C. F. (re-issue)	October 12, 1880
200,963	Anders, G. L.	.. March 5, 1878

5.—METERS.

Including devices for measuring the electric current.

6.—ELECTRO-MAGNETS.

Including special constructions and arrangements of electro-magnets and their armatures. (*See also in electric lights.*)

14,711	Coleman, A...	.. April 22, 1856
19,379	Parks, N. ..	.. February 16, 1858
29,761	Bradley, L. ..	.. August 28, 1860
29,862	Chester, C. T.	.. September 4, 1860
30,271	Vergnes, M...	.. October 2, 1860
32,478	Holcomb, A. G.	.. June 4, 1861
32,874	Jenness, J. T.	.. July 23, 1861
38,774	Van Choate, S. F. ..	June 2, 1863
40,133	Van Choate, S. F. ..	September 29, 1863
49,074	Bradley, L. ..	.. August 1, 1865
58,217	Clark, J. J., and Splitdorf, H.	.. September 25, 1866
58,439	Lyon, J. B., and Doll, A. ..	.. October 2, 1866
60,432	Shaffner, T. P.	.. December 11, 1866
66,001	Cabell, S. G.	.. June 25, 1867
71,863	Fairchild, J. M.	.. December 10, 1867
96,554	Davies, W. E.	.. November 9, 1869
102,856	Paine, H. M.	.. May 10, 1870
103,230	Paine, H. M.	.. May 17, 1870
103,231	Paine, H. M.	.. May 17, 1870
4,237	Paine, H. M. (re-issue) ..	.. January 24, 1871
103,440	Frost, M. S. ..	.. May 24, 1870
103,768	Paine, H. M.	.. May 31, 1870
105,653	D'Arlincourt, L. C. A. J. G. ..	.. July 26, 1870
106,418	Smith, W. W.	.. August 16, 1870
114,657	Edison, T. A.	.. May 9, 1871
114,658	Edison, T. A.	.. May 9, 1871
119,176	Prevost, E. ..	.. September 19, 1871
125,151	Tice, I. P. ..	.. April 2, 1872
130,795	Edison, T. A.	.. August 27, 1872
133,968	Davis, W. E ..	.. December 17, 1872
134,868	Edison, T. A.	.. January 14, 1873
135,690	Churchill, J. L.	.. February 11, 1873
146,444	Fontaine, H.	.. January 13, 1874
154,588	Davis, W. E.	.. September 1, 1874
160,120	Rice, M. A. ..	.. February 23, 1875
180,093	House, R. E.	.. July 25, 1876
200,929	Paine, E. L. ..	.. March 5, 1878
203,492	Paine, H. M...	.. May 7, 1878
204,141	Eaton, A. K.	.. May 28, 1878
226,485	Bunnell, J. H.	.. April 13, 1880
227,863	Wagner, P. ..	.. May 18, 1880
238,252	Russell, S. ..	.. March 1, 1881

7.—REGULATORS.

Including apparatus for the control of electric currents.

14,733	Phillips, N. M.	April 22, 1856
19,042	Phelps, G. M. (extended January 5, 1872)	January 5 1858
19,642	Lacassagne, J., and Theirs, R.	.. March 16, 1858
40,206	Wright, G. B.	.. October 6, 1863
49,842	Achard, F. F. A.	.. September 5, 1865
55,594	Averell, E. D.	.. June 12, 1866
60,663	Averell, E. D.	... January 1, 1867
76,965	Wood, E. A. ..	.. April 21, 1867
88,010	Chester, C. T., and S.	March 23, 1869
100,462	Sternberg, G. M.	.. March 1, 1870
4,326	Sternberg, G M. (re-issue) ..	.. April 4, 1871
105,272	Sternberg, G. M.	.. July 12, 1870
110,963	Fairbanks, H.	.. January 17, 1871
111,112	Edison, T. A.	.. January 24, 1871
119,541	Sternberg, G. M.	.. October 3, 1871
123,315	Wells, H. E., and Moran, D. H.	.. January 30, 1872
127,972	Graves, J. ..	.. June 18, 1872
129,085	Barjon, V. ..	.. July 16, 1872
138,993	Chapin, H. A.	.. May 20, 1873
140,287	Maertens, E.	.. June 24, 1873
142,688	Edison, T. A.	.. September 9, 1873
142,925	Lufbery, G. F	.. September 16, 1873
152,031	Baker, W. C...	.. June 16, 1874
152,444	Westland, C. S.	.. June 23, 1874
154,214	Bradford, J. M.	.. August 18, 1874
159,394	Crawford, W. A.	.. February 2, 1875
161,680	Guest, J. H...	.. April 6, 1875
166,471	Mac Connel, A.	.. August 10, 1875
168,548	Wolcott, C. C.	.. October 5, 1875
171,869	Sangster, J., and Grosvenor, W. S...	January 4, 1876
173,888	Boyden, E., and Pillsbury, G. H..	.. February 22, 1876
174,127	Field, S. D. ..	.. February 29, 1876
177,058	Boyden, E. ..	.. May 9, 1876
178,138	Gerrish, W. H.	.. May 30, 1876
182,859	Sangster, J., and Grosvenor, W. S...	October 3, 1876
182,977	Weston, E. ..	.. October 3, 1876
8,102	Weston, E. (re-issue)	February 26, 1878
183,569	Axford, H. W.	.. October 24, 1876
184,034	Chapin, H. A.	.. November 7, 1876
185,047	Rettig, J. E...	.. December 5, 1876
185,164	Bullough, J...	.. December 12, 1876

INDEX.

NAMES OF PATENTEES.

INDEX.

—◆—

SUBJECT MATTER.

Abstracts of the Specifications referred to will be found in the Appendix.

CONDUCTORS—	No.	Year
Poole	14,057	1852
Reid and Brett ...	14,166	1852
Bright	14,331	1852
Newton and Fuller ...	.17	1852
Dundonald	277	1852
Harrison	459	1852
Fowler	481	1852
Hunter	511	1852
Le Gras and Gilpin ...	566	1852
Henley	680	1852
Dundonald	740	1852
Mivand	750	1852
Physick	778	1852
Henley	185	1853
Brooman	231	1853
Price	303	1853
Allan	338	1853
Day	405	1853
Reid	1,023	1853
Smith	1,312	1853
Henley	1,779	1853
Allan	1,889	1853
Dering	1,909	1853
Wilkinson	2,307	1853
Callan	2,340	1853
Cadogan... ...	2,372	1853
Baudouin	2,710	1854
Guyard	83	1855
Johnson	191	1855
"	875	1855
Statham... ...	216	1856
Balestrini	2,124	1856
Harrison	2,483	1856
Mackintosh	2,707	1857
"	2,866	1857
Henry	2,868	1857
Henley	3,020	1857
Highton... ...	3,101	1857
Dering	18	1858
Chatterton	252	1858
Sievier	277	1858
Wilde	293	1858
Baudouin	333	1858
Hearder	444	1858
Rowett	782	1858
Chatterton	883	1858
Hancock... ...	1,268	1858
Newall	1,379	1858
Vasserot... ...	1,483	1858
Clark	1,491	1858
De Bergue	1,605	1858
Buckingham, Humfrey and Sykes ...	1,708	1858
De Bergue	1,740	1858
Crispin	1,742	1858
Molesworth	687	1859
Wilkes	217	1860
Siemens	519	1860
Newbold	912	1860
Reid	1,146	1860
Hooper	1,546	1860
Silas	3,103	1860
Moulton	260	1861
Parkes	353	1861
Morgan, Jay, Edwards and Tilston	748	1861
Searle	800	1861
Rowland... ...	1,113	1861
Morris, Weare and Monckton ...	2,661	1861
Dicey	3,170	1861
Siemens (Siemens) ...	59	1862
Chatterton and Smith	413	1862
Russell (Bensa and Anstruther) ...	665	1862
Fenton and Stubbs ...	2,238	1862
Brae	2,007	1863
Rowell	2,192	1863
Mulholland	2,386	1863
Wilde	3,006	1863
Henley	1,126	1864
Watson and Horwood	975	1865
Mulholland	2,025	1865
Piggott	2,213	1865
Macintosh	2,332	1865
Hubert	2,605	1865
Nicholl	480	1866
Bonneville	1,749	1866
Marshall	3,192	1866
Nicoll	694	1867
Gray	1,770	1867
Martin and Varley ...	315	1868
Walker	838	1868
Collingridge	2,712	1860

CONDUCTORS—	No.	Year
Jenkin	3,236	1869
Adhesive compound for, Chatterton	1,213	1859
Adhesive compound for, Daft	2,170	1859
Alloy for coating, Rowland...	1,113	1861
Aluminium cores, Baudouin	333	1858
Amalgamated, Duncan...	2,997	1858
Arranged circularly, Jaloureau	3,047	1860
Apparatus for straining, Poole	11,481	1846
Apparatus for straining, Robson	86	1872
Apparatus for sheathing on board ship, Smith	1,312	1853
Applying flexible varnish, Poole ...	14,057	1852
Applying several insulating coats simultaneously, Poole ...	14,057	1852
Attached to chains, Shepherd and Button	13,363	1850
Attaching air chambers to, Phillips ...	2,341	1865
Band of wires and rubber threads, Statham...	216	1856
Bent in short curves, Story	409	1870
Between bituminous coating, Dundonald...	277	1852
Between semi-circular troughs	405	1853
Bi-metallic, Dering ...	1,909	1853
Parkes	353	1861
Wilde	3,240	1862
Bituminous pipes for containing, Hargreaves	261	1862
Braided, Appold... ...	1,828	1858
Hall and Wells ...	2,411	1858
Johnson	2,626	1858
"	2,329	1866
Branch pipes for, Jaloureau	2,137	1858
Brass coated, Daft ...	2,170	1859
Bringing ends of, out of pipes, Cooke ...	7,614	1838
Buffers for coupling, Price	303	1853
Buoyant, Haye and Bloom...	2,467	1857
Searle	2,239	1858
Frost	1,152	1859
Carried on flat chain, Story	409	1870
Carrying overhead ...	Page 89	
Coated with thread and resin, Young and McNair	10,799	1845
Coated with thread and pitch, Young and McNair	10,799	1845
Coated with tar mapple	11,428	1846
Coated with indiarubber and sulphur, Poole	28	1852
Coated with rubber, yarn and wire, Henley ...	680	1852
Coated with several metals of high resistance, Sievier ...	277	1858
Coated with nickel, Sievier	277	1858
Coated with tin, Sievier	277	1858
Coated with Indian grass fibre, Rowett ...	782	1858
Coated with indiarubber and heated with sulphuric acid, Macintosh ...	1,090	1858
Coated with twine, De Bergue	1,605	1858
Coated with fibre and surrounded by hemp strands, Crispin ...	1,742	1858
Coated with drying oils, Walton	2,770	1860
Coated with cork dust, Chatterton and Smith	3,138	1860
Coated with indiarubber, Moulton ...	260	1861

CONDUCTORS—	No.	Year
Coated with leather, Branscombe	322	1861
Coated with spun glass, Hirsch	549	1861
Coated with indiarubber and guttapercha, Evans... ...	579	1861
Coated with pure indiarubber, Morgan, Jay, Edwards and Tilston	748	1861
Coated with tanned fibre, Duncan	1,329	1861
Coated with indiarubber, Roeber	1,406	1861
Coated with desiccated india-rubber, West ...	1,806	1861
Coated with Chatterton's compound, Siemens (Siemens) ...	59	1862
Coated with indiarubber and metallic oxides, West ...	194	1862
Coated with marine glue, Chatterton and Smith	413	1862
Coated with copper, Fenton and Stubbs ...	2,238	1862
Coated with asphalte, Fenton and Stubbs ...	2,238	1862
Coated with silk, Brooke	2,432	1862
Coated with graphite, Snider	2,461	1862
Coated with ballata, Hancock and Silver...	3,331	1862
Coated with indiarubber, Bonneville (Gérard)...	1,668	1863
Coated with collodion, Bonneville (Gérard)...	1,668	1863
Coating with shellac, Mulholland ...	2,386	1863
Coated with cotton, Hancock and Silver ...	3,092	1864
Coated with indiarubber, Mulholland...	2,025	1865
Coated with cork, Guy...	2,088	1865
Coated with canvas, Inkpen	2,326	1865
Coated with collodion, Macintosh	2,332	1865
Coated with asbestos, Bonneville (Nozan)...	2,530	1865
Coated with silica, Clifford	3,938	1868
Coated with foil, Foucaut	2,008	1869
Coating several wires at once, Allan ...	1,889	1853
Coatings marked in order to identify, Clark ...	1,491	1858
Coating with rubber cloth, Newton (Claes Vanderhest and Co.)...	2,674	1860
Coating copper with lead, Needham ...	637	1864
Coating, Abel ...	1,962	1865
Coating with indiarubber, Macintosh ...	2,332	1865
Coating with cork, Jones	2,209	1865
Coiling, Newall ...	1,091	1855
Coiled like warp threads, Nicholl	480	1866
Combination of iron or steel wires, Dering ...	18	1885
Combined with chain cables, Ker	329	1861
CONDUCTORS, Compound—		
Gilpin	779	1854
Physick	1,357	1854
Rogers	2,192	1855
Walker	1,797	1858
"	2,250	1858
Clark	2,891	1858
" (Rosencrantz) ...	2,552	1859
Varley	206	1860
Wilkes	217	1860
"	2,051	1860
Bonelli	2,457	1860
Parkes	353	1861
Wilde	3,240	1862
Boggett	2,416	1865
"	3,180	1865
"	3,299	1865
Varley	3,357	1865
Bonneville	1,749	1866
Lake	2,052	1866

a a